Energy Level Diagram of Electron Shells and Subshells of the Elements

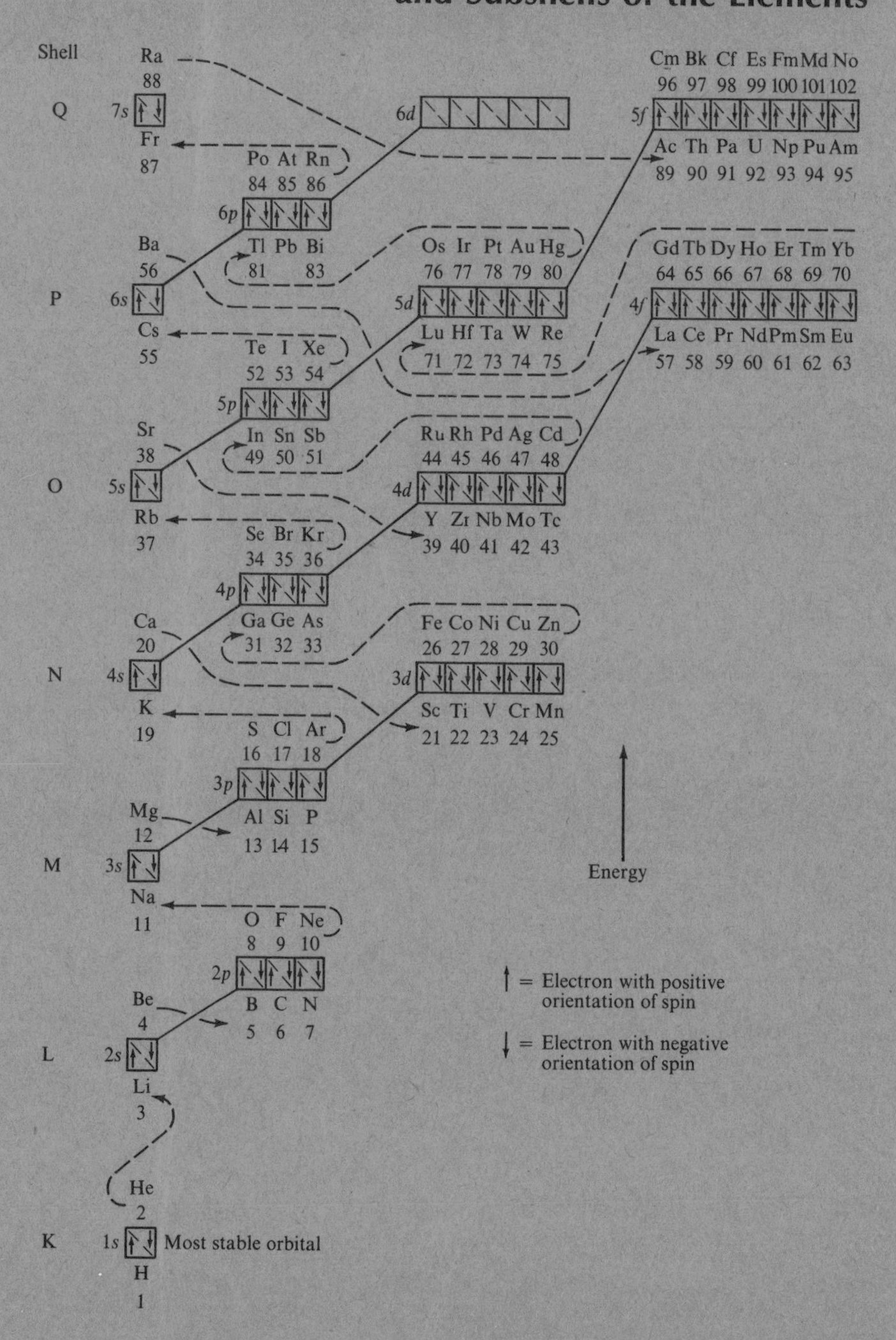

Siena College Library
Loudonville, N.Y.

CHEMISTRY

A SERIES OF BOOKS IN CHEMISTRY
Linus Pauling, Editor

CHEMISTRY

Linus Pauling
LINUS PAULING INSTITUTE OF SCIENCE AND MEDICINE

Peter Pauling
UNIVERSITY COLLEGE, LONDON

W. H. FREEMAN AND COMPANY
San Francisco

Library of Congress Cataloging in Publication Data

Pauling, Linus Carl, 1901–
Chemistry.

Bibliography: p.
Includes index.
1. Chemistry. I. Pauling, Peter, joint author.
II. Title.
QD31.2.P38 540 74-34071
ISBN 0-7167-0176-6

Printed in the United States of America

1 2 3 4 5 6 7 8 9

Contents

14 Biochemistry 442

15 Molecular Biology 495

16 Inorganic Complexes and Coordination Compounds 541

17 The Nature of Metals and Alloys 563

18 Lithium, Beryllium, Boron, and Silicon and Their Congeners 588

19 The Transition Metals and Their Compounds 622

Preface

The preface of the first edition of a book by one of us (*General Chemistry,* Linus Pauling, W. H. Freeman and Company, 1947) began with the following words: "Chemistry is a very large subject, which continues to grow, as new elements are discovered or made, new compounds are synthesized, and new principles are formulated. Nevertheless, despite its growth, the science can now be presented to the student more easily and effectively than ever before. In the past the course in general chemistry has necessarily tended to be a patchwork of descriptive chemistry and certain theoretical topics. The progress made in recent decades in the development of unifying theoretical concepts has been so great, however, that the presentation of general chemistry to the students of the present generation can be made in a more simple, straightforward, and logical way than formerly."

Chemistry has continued to grow. Since 1947 the most important advances in chemistry have involved a rapidly increasing understanding of the molecular basis of life. This field of knowledge is called molecular biology. In this book, *Chemistry,* we have striven to present to the student not only the basic principles of chemistry, coordinated with a considerable amount of inorganic and organic descriptive chemistry, but also an introduction to biochemistry and molecular biology. The book has been designed especially for students primarily interested in biology, medicine, human nutrition, and related fields. Some topics not directly pertinent to these fields, such as elementary particles and atomic nuclei, have been included for the benefit of the curious student. We hope that readers will find this book both useful and interesting.

9 January 1975

Linus Pauling
BIG SUR, CALIFORNIA
Peter Pauling
LONDON, ENGLAND

How to Look at Stereographic Drawings

Most people, with practice, can view the stereo drawings in this book without the aid of a special device.

The distance between the two parts of a stereo pair is about equal to the distance between the pupils of the eyes. The stereoscopic view is obtained by looking at the left drawing with the left eye and at the right drawing with the right eye. One way to achieve this view is to bring the book close to your face, touching your nose. You look at the drawings near your eyes (they are, of course, out of focus), and slowly move the book out to a position about 25 cm away, being careful to gaze steadily at the two halves of the drawing, so that the lines of sight of the two eyes remain parallel. Continue to gaze, moving the book by small amounts, until the two drawings fuse into a single stereo image. Success might require you to gaze a minute or two, and perhaps begin over several times. With practice you can achieve the fusion of the images in a few seconds. A vertically held piece of cardboard, shielding half of the field of vision for each eye, may be used.

CHEMISTRY

1

Chemistry and Matter

The rapid progress true Science now makes occasions my regretting sometimes that I was born so soon. It is impossible to imagine the heights to which may be carried, in a thousand years, the power of man over matter. O that moral Science were in as fair a way of improvement, that men would cease to be wolves to one another, and that human beings would at length learn what they now improperly call humanity.—

BENJAMIN FRANKLIN,
in a letter to the chemist
Joseph Priestley, 8 February 1780.

Why study chemistry? An important reason is indicated in the foregoing statement by Benjamin Franklin—it is through chemistry and her sister sciences that the power of man, of mind, over matter is obtained. Nearly two hundred years ago Franklin said that science was making rapid progress. We know that the rate of progress of science has become continually greater, until now the nature of the world in which we live has been greatly changed, through scientific and technical progress, from that of Franklin's time.

Science plays such an important part in the modern world that no one can now feel that he understands the world in which he lives unless he has an understanding of science.

The science of chemistry deals with *substances*. At this point in the study of chemistry we shall not define the word substance in its scientific sense, but shall assume that you have a general idea of what the word means. Common examples of substances are water, sugar, salt, copper, iron, oxygen—you can think of many others.

A century and a half ago it was discovered by an English chemist, Sir Humphry Davy (1778–1829), that common salt can be separated, by passing electricity through it, into a soft, silvery metal, to which he gave the name sodium, and a greenish-yellow gas, which had been discovered some time earlier, named chlorine. Chlorine is a corrosive gas, which attacks many metals, and irritates the mucous membranes of the nose and throat if it is inhaled. That the substance salt is composed of a metal (sodium) and a corrosive gas (chlorine) with properties quite different from its own properties is one of the many surprising facts about the nature of substances that chemists have discovered.

A sodium wire will burn in chlorine, producing salt. The process of combination of sodium and chlorine to form salt is called a *chemical reaction.* Ordinary fire also involves a chemical reaction, the combination of the fuel with oxygen in the air to form the products of combustion. For example, gasoline contains compounds of carbon and hydrogen, and when a mixture of gasoline and air burns rapidly in the cylinders of an automobile a chemical reaction takes place, in which the gasoline and the oxygen of the air react to form carbon dioxide and water vapor (plus a small amount of carbon monoxide), and at the same time to release the energy that moves the automobile. Carbon dioxide and carbon monoxide are compounds of carbon and oxygen, and water is a compound of hydrogen and oxygen.

Chemists study substances in order to learn as much as they can about their properties (their characteristic qualities) and about the reactions that change them into other substances. Knowledge obtained in this way has been found to be extremely valuable. It not only satisfies man's curiosity about himself and about the world in which he lives, but it also can be applied to make the world a better place to live in, to make people happier, by raising their standards of living, ameliorating the suffering due to ill health, and enlarging the sphere of their activities.

Let us consider some of the ways in which a knowledge of chemistry has helped man in the past and may help him in the future.

It was discovered centuries ago that preparations could be made from certain plants, such as poppies and coca, which, when taken by a human being, serve to deaden pain (are analgesics). From these plants chemists isolated pure substances, morphine and cocaine, which have the pain-deadening property. These substances have, however, an undesirable property, that of inducing a craving for them that sometimes leads to drug addiction. Chemists then investigated morphine and cocaine, to learn their chemical structure, and then made in the laboratory a great number of other substances, somewhat similar in structure, and tested these substances for their powers of deadening pain and of producing addiction. In this way some drugs that are far more valuable than the natural ones have been discovered; substances with 10,000 times the potency of morphine have been made.

A related story is that of the discovery of general anesthetics. In 1800 Humphry Davy, as a young man just beginning his scientific career, tested many gases on himself by inhaling them. (He was lucky that he did not kill himself, because one of the gases he inhaled is very poisonous.) He discovered that one gas produced a state of hysteria when inhaled, and that people under the influence of this gas, which was given the name laughing gas, seemed not to suffer pain when they fell down or bumped into an object. He suggested its use in surgery in the following words: "As nitrous oxide, in its extensive operation, seems capable of destroying physical pain, it may probably be used with advantage in surgical operations." His suggestion, however, remained unheeded for nearly half a century. Then in 1844 nitrous oxide was used for the extraction of a tooth by Dr. Horace Wells in Hartford, Conn., and two years later the first surgical operation under diethyl ether anesthesia was carried out (in Massachusetts General Hospital, Boston). Ether, chloroform, and nitrous oxide were soon brought into general use. The discovery of anesthesia was a great discovery, not only because it relieves pain, but also because it permits delicate surgical operations to be carried out that would be impossible if the patients remained conscious.

The rubber industry may be mentioned as an example of a chemical industry. This industry began when it was discovered that raw rubber, a sticky material made from the sap of the rubber tree, could be converted into vulcanized rubber, which has superior properties (greatly increased strength, freedom from stickiness), by mixing it with sulfur and heating it. During recent years artificial materials similar to rubber (called synthetic rubber) have been made, which are in many ways better than natural rubber. The synthetic rubbers are made from petroleum or natural gas.

The steel industry is another great chemical industry. Steel, which consists mainly of the metal iron, is our most important structural material. It is made from iron ore by a complex chemical process. In the United States the production of steel is carried on at the rate of about 2000 lbs per person per year.

Chemistry plays such an important part in the life of twentieth-century man that this age may properly be called the chemical age.

1-1. The Study of Chemistry

Chemistry has two main aspects: **descriptive chemistry,** *the discovery and tabulation of chemical facts;* and **theoretical chemistry,** *the formulation of theories that, upon verification, unify these facts and combine them into a system.**

It is not possible to obtain a sound knowledge of chemistry simply by learning theoretical chemistry. Even if a student were to learn all the chemical theory that is known he would not have a knowledge of the science, because a major part of chemistry (many of the special properties of individual substances) has not yet been well incorporated into chemical theory. It is accordingly necessary for the student to learn a number of the facts of descriptive chemistry simply by memorizing them. The number of these facts that might be memorized is enormous, and increases rapidly year by year, as new discoveries are made. In this book a selection from the more important facts is presented. *You should learn some of these facts by studying them, and by frequently referring to them and renewing your knowledge of them. You should also learn as much about chemistry as possible from your own experience in the laboratory and from your observations of chemical substances and chemical reactions in everyday life.*

A special effort has been made in this book to present the subject of chemistry in a logical and simple manner, and to correlate descriptive chemistry with the theories of chemistry. It is therefore necessary that the theoretical sections of the book be carefully studied and thoroughly understood. Read each chapter with care. Examine the arguments to be sure that you understand them.

1-2. Matter

The universe is composed of **matter** *and* **radiant energy.**

The chemist is primarily interested in matter, but he must also study radiant energy—light, x-rays, radio waves—in its interaction with sub-

*The broad field of chemistry may also be divided in other ways. An important division of chemistry is that into the branches *organic chemistry* and *inorganic chemistry*. Organic chemistry is the chemistry of the compounds of carbon, especially those that occur in plants and animals. Inorganic chemistry is the chemistry of the compounds of elements other than carbon. Each of these branches of chemistry is in part descriptive and in part theoretical. Many other branches of chemistry, which in general are parts of organic chemistry and inorganic chemistry, have also been given names; for example, analytical chemistry, physical chemistry, biochemistry, nuclear chemistry, industrial chemistry. Their nature is indicated by their names.

stances. For example, he may be interested in the color of substances, which is produced by their absorption of light.

Matter consists of all of the materials around us—gases, liquids, solids. This statement is really not a definition. The dictionary states that matter is "that of which a physical object is composed; material." Then it defines material and physical object as matter, so that we are back where we started. The best course that we can follow is to say that no one really knows how to define matter, but that we agree to start out by using the word. Often in science it is necessary to begin with some undefined words.

Mass and Weight

All matter has *mass*. Chemists are interested in the masses of materials, because they want to know how much material they need to use to prepare a certain amount of a product.

The **mass** *of an object is the quantity that measures its resistance to change in its state of rest or motion.*

The mass of an object also determines its *weight*. The weight of an object is only a measure of the *force* with which the object is attracted by the earth. This force depends upon the mass of the object, the mass of the earth, and the position of the object on the earth's surface, especially the distance of the object from the center of the earth. Since the earth is slightly flattened at its poles, the distance of its surface at the North Pole or South Pole from its center is less than that at the equator. In consequence the weight of an object as measured by a spring balance, which measures the force, is greater at the North Pole or South Pole than at the equator. For example, if your weight, measured by a spring balance, is 150.0 lbs at the equator, it would be 150.8 lbs at the North Pole, measured on the same spring balance—nearly a pound more. Your mass, however, is the same.

The mass of an object remains the same at the North Pole as at the equator, and it can easily be determined, at any place on the earth's surface, by comparison with a standard set of masses. For small objects a *chemical balance* is used. Since the weights of two bodies of equal mass are the same at any place on the earth's surface, these bodies will balance one another when placed on the two pans of a balance with arms of equal length.

It is common practice to refer to the masses of objects as their weights. It might be thought that confusion would arise from the practice of using the word weight to refer both to the mass of an object and to the force with which the object is attracted by the earth. In general it does not, but if there is danger of confusion you should use the word mass.

The standard masses (standard weights) in the metric system are calibrated (checked) by comparison with the standard kilogram in Paris (Appendix I).* The IS unit of mass is the *kilogram*. The abbreviation for gram is g, and for kilogram kg (1 kg = 1000 g).

1-3. Kinds of Matter

As we look about us we see material objects, such as a stone wall or a table. The chemist is primarily interested not in the objects themselves, but in the kinds of matter of which they are composed. He is interested in wood as a material (a kind of matter), whether it is used for making a table or a chair. He is interested in granite, whether it is in a stone wall or in some other object. Indeed, his interest is primarily in those properties (characteristic qualities) of a material that are independent of the objects containing it.

The word **material** *is used in referring to any kind of matter, whether homogeneous or heterogeneous.*

A **homogeneous** *material is a material with the same properties throughout.*

A **heterogeneous** *material consists of parts with different properties.*

Wood, with soft and hard rings alternating, is obviously a heterogeneous material, as is also granite, in which grains of three different species of matter (the minerals† quartz, mica, and feldspar) can be seen.

Heterogeneous materials are mixtures of two or more homogeneous materials. For example, each of the three minerals quartz, mica, and feldspar that constitute the rock granite is a homogeneous material.

Phases

A material system (that is, a limited part of the universe) may be described in terms of the *phases* constituting it. A phase is a homogeneous part of a system, separated from other parts by physical boundaries. For example, if a flask is partially full of water in which ice is floating, the system comprising the contents of the flask consists of three phases: ice

---

*There are many systems of weights and measures, which are ordinarily used in different countries. In order to avoid confusion, all scientists use the *metric system* or an expanded and improved version of the metric system called the *International System* (IS, or sometimes SI, for *Système International*), which is described in Appendix I, in their scientific work. In general, we shall use the IS system in this book, but an occasional exercise or example may be given in the American system.

†A *mineral* is any homogeneous material occurring naturally as a product of inorganic processes (that is, not produced by a living organism).

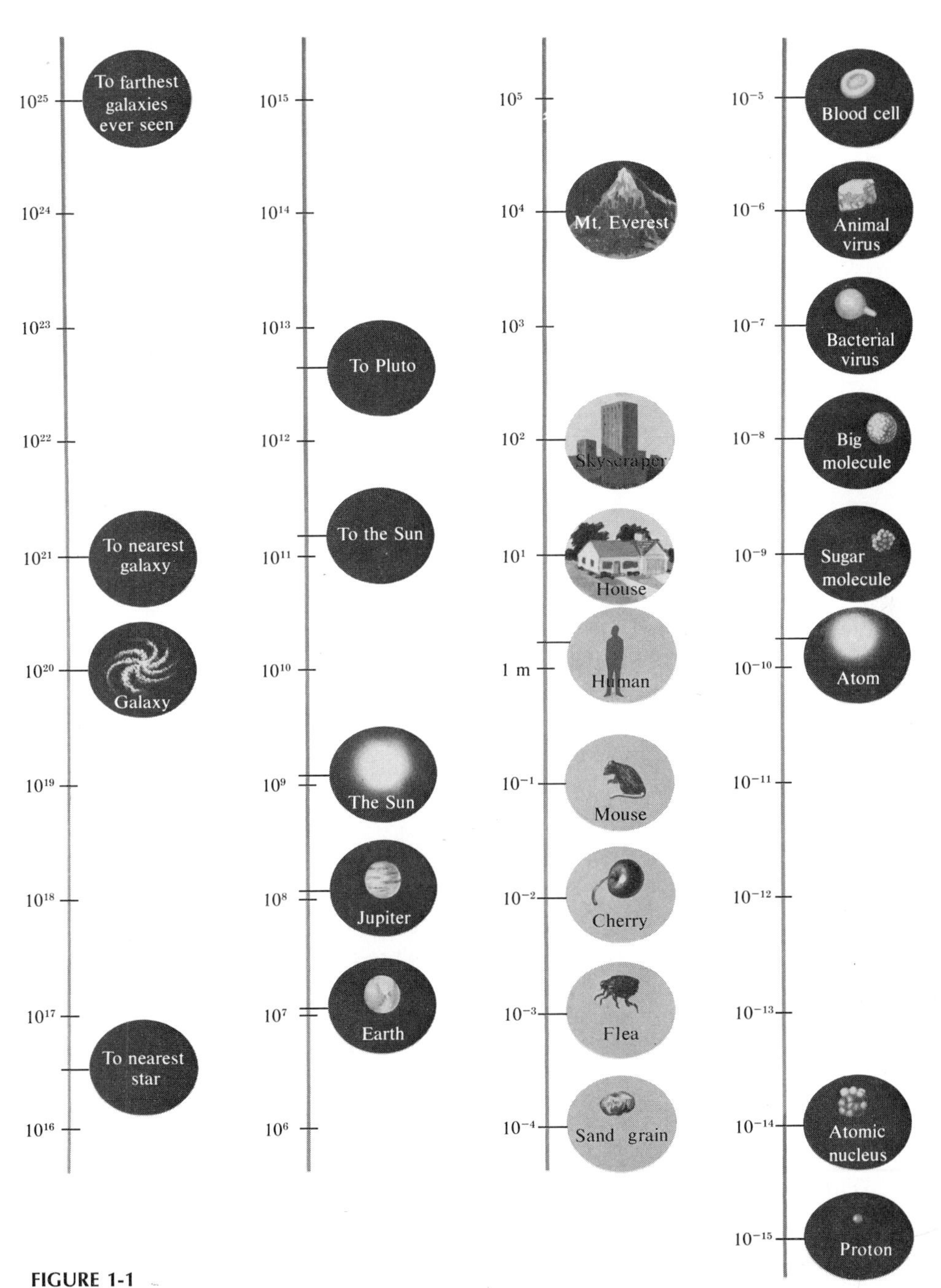

FIGURE 1-1
A diagram showing dimensions of objects, from 10^{-15} m (the diameter of a proton) to 10^{25} m (the radius of the known universe).

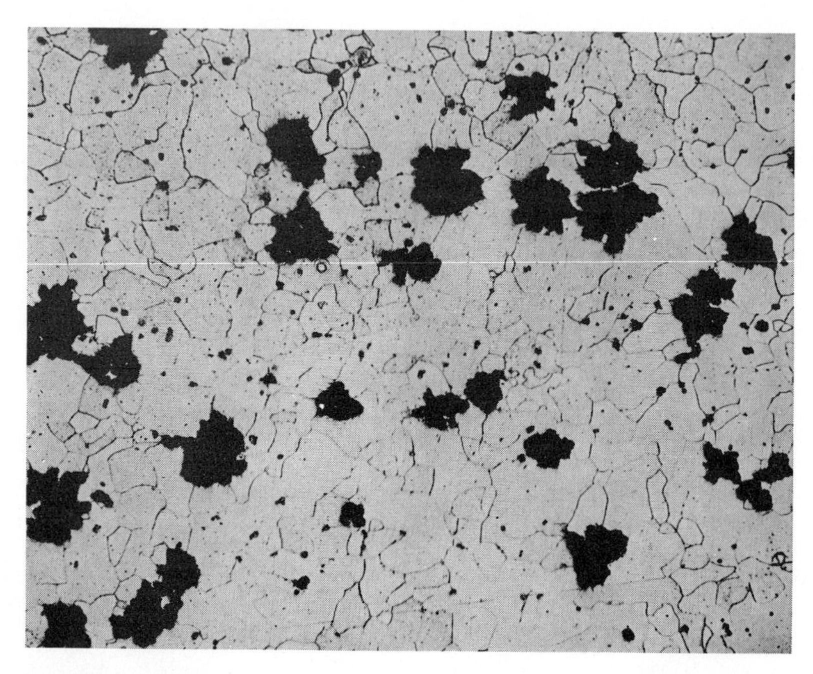

FIGURE 1-2
A photomicrograph (linear magnification 100 times—that is, magnified by a linear factor of 100) of a polished and etched surface of a specimen of malleable cast iron, showing small grains of iron and roughly spherical particles of graphite (carbon). The grains of iron look somewhat different from one another because of different illumination. (From Malleable Founders' Society.)

(a solid phase); water (a liquid phase); and air (a gaseous phase). A piece of malleable cast iron can be seen with a microscope to be a mixture of small grains of iron and particles of graphite (a form of carbon); it hence consists of two phases, iron and graphite (Figure 1-2).

A phase in a system comprises all of the parts that have the same properties and composition. Thus, if there were several pieces of ice in the system discussed above, they would constitute not several phases, but only one phase, the ice phase.

Kinds of Definition

Definitions may be either precise or imprecise. The mathematician may define precisely the words that he uses; in his further discussion he then adheres rigorously to the defined meaning of each word. But the words that are used in describing nature, which is itself complex, may not be capable of precise definition. In giving a definition for such a word the effort is made to describe the accepted usage.

For example, sometimes it is difficult to decide whether a material is homogeneous or heterogeneous. A specimen of granite, in which grains of three different species of matter can be seen, is obviously a mixture. An emulsion of fat in water (a suspension of small droplets of fat in the water, as in milk) is also a mixture. The heterogeneity of a piece of granite is obvious to the eye. The heterogeneity of milk can be seen if a drop of milk is examined under a microscope.

Substances and Solutions

Let us now define the words *substance* and *solution.*

A **substance** *is a homogeneous material with definite chemical composition.*

A **solution** *is a homogeneous material that does not have a definite composition.**

Pure salt, pure sugar, pure iron, pure copper, pure sulfur, pure water, pure oxygen, and pure hydrogen are representative substances. Quartz is also a substance.

On the other hand, a solution of sugar in water is not a substance according to this definition: it is, to be sure, homogeneous, but it does not satisfy the second part of the above definition, inasmuch as its composition is not definite, but is widely variable, being determined by the amount of sugar that happens to have been dissolved in a given amount of water. Gasoline is also not a pure substance; it is a solution of several substances.

In practice all substances are impure to some degree, and for many chemical purposes the degree of purity and the methods of purification and analysis must be specified. Furthermore, there are many materials classified as substances in which the chemical composition may vary over narrow limits. An example is the iron sulfide that is made by heating iron and sulfur together. This homogeneous material when made in different ways ranges in composition from 35% to 39% sulfur.

Elements and Compounds

Substances are classified as *elementary substances* or *compounds.*

A substance that can be decomposed into two or more substances is a **compound.**

A substance that cannot be decomposed is an **elementary substance** (an **element**).†

---

*The word solution is commonly used for liquid solutions. Chemists also refer to gaseous solutions (mixtures of two or more pure gases) and to solid solutions (such as a gold-copper alloy).

†The discovery of radioactivity made it necessary to change these definitions slightly (see Chapter 20).

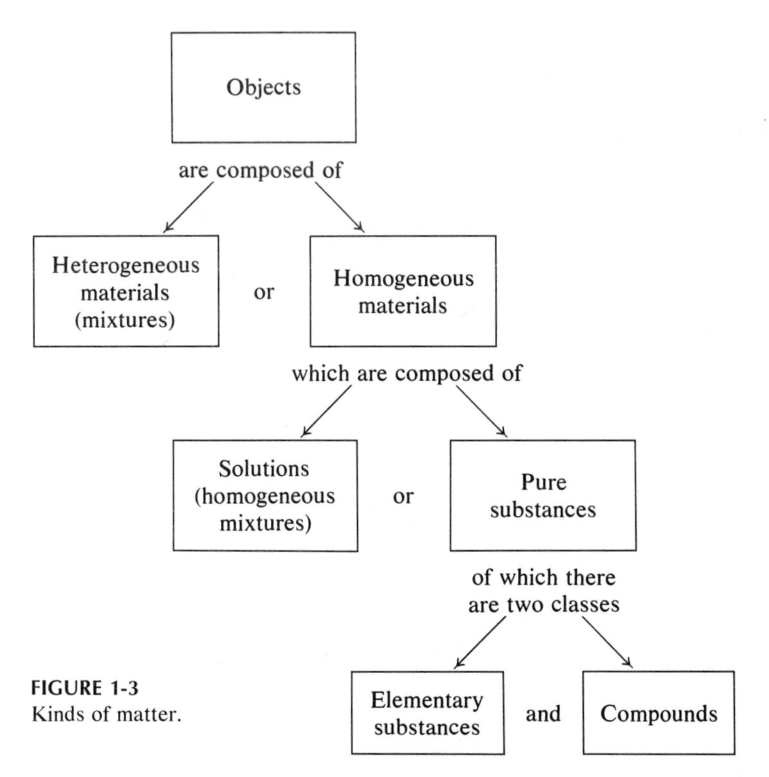

FIGURE 1-3
Kinds of matter.

Salt can be decomposed by an electric current into two substances, sodium and chlorine. Hence salt is a compound.

Water can be decomposed by an electric current into two substances, hydrogen and oxygen. Hence water is a compound.

Mercuric oxide can be decomposed by heat, to form mercury and oxygen. Hence mercuric oxide is a compound.

No one has ever succeeded in decomposing sodium, chlorine, hydrogen, oxygen, or mercury into other substances.* Hence these five substances are accepted as elementary substances (elements).

At the present time (1975) 106 elements are known. Several hundred thousand compounds of these elements have been found in nature or made in the laboratory.

The process of decomposing a compound into two or more simpler substances is sometimes called *analysis*. The reverse process, of forming a substance by combining two or more substances, is called *synthesis*.

The composition of a compound can be determined by analysis. For example, a *qualitative analysis* of salt might be carried out by decompos-

*In this discussion the word substance is considered not to include electrons and atomic nuclei. Atoms of sodium and other elements can be decomposed into fundamental particles (Chapters 3 and 20).

ing it with an electric current and identifying the products as sodium and chlorine; the chemist could then say that the salt is a compound of the two elements sodium and chlorine. To carry out a *quantitative analysis* he would have to weigh the substances; he could then report the composition as 39.4% sodium, 60.6% chlorine.

Our classification of matter is summarized in Figure 1-3. You may find it worthwhile to examine this chart carefully. Can you define all of the words? Can you give two or three examples of each of the six kinds of materials that might constitute an object? Can you think of one or two materials that are hard to classify?

Example 1-1. Classify the following materials as homogeneous or heterogeneous, substance or solution, compound or element:
(*a*) ice
(*b*) diamond
(*c*) maple syrup
(*d*) pure sugar
(*e*) whole blood

Solution.
(*a*) Ice is homogeneous, a substance, and a compound of the elements oxygen and hydrogen.
(*b*) The mineral diamond is a homogeneous crystalline substance, a form of the element carbon.
(*c*) Maple syrup is a homogeneous solution of sugar and other substances in water.
(*d*) Pure sugar is homogeneous, a substance, and a compound of the elements carbon, hydrogen, and oxygen.
(*e*) Whole blood is heterogeneous, containing red and white blood cells, other cells, and many substances in solution.

1-4. The Physical Properties of Substances

The study of the properties of substances constitutes an important part of chemistry, because their properties determine the uses to which they can be put.

The **properties** *of substances are their characteristic qualities.*

The **physical properties** *are those properties of a substance that can be observed without changing the substance into other substances.*

Let us again use sodium chloride, common salt, as an example of a substance. We have all seen this substance in what appear to be different forms—table salt, in fine grains; salt in the form of crystals a quarter of an inch in diameter, for use with ice for freezing ice cream; and natural crystals of rock salt an inch or more across. Despite their obvious difference, all of these samples of salt have the same fundamental properties.

In each case the crystals, small or large, are naturally bounded by square or rectangular *crystal faces* of different sizes, but with each face always at right angles to each adjacent face. The *cleavage* of the different crystals of salt is the same: when crushed, the crystals always break (cleave) along planes parallel to the original faces, producing smaller crystals similar to the larger ones. The different samples, dissolved in water, have the same salty *taste*. Their *solubility* is the same: at room temperature 36 g of salt can be dissolved in 100 g of water. The *density* of the salt is the same, 2.16 g cm^{-3}. The density of a substance is the mass (weight) of a unit volume (1 cubic centimeter) of the substance.*

There are other properties besides density and solubility that can be measured precisely and expressed in numbers. Such another property is the *melting point,* the temperature at which a solid substance melts to form a liquid. On the other hand, there are also interesting physical properties of a substance that are not so simple in nature. One such property is the *malleability* of a substance—the ease with which a substance can be hammered out into thin sheets. A related property is the *ductility*—the ease with which the substance can be drawn into a wire. *Hardness* is a similar property: we say that one substance is less hard than the second substance when it is scratched by the second substance. The *color* of a substance is an important physical property.

It is customary to say that under the same external conditions all specimens of a particular substance have the same physical properties (density, hardness, color, melting point, crystalline form, etc.). Sometimes, however, the word substance is used in referring to a material without regard to its state. For example, ice, liquid water, and water vapor may be referred to as the same substance. Moreover, a specimen containing crystals of rock salt and crystals of table salt may be called a mixture, even though the specimen may consist entirely of one substance, sodium chloride. This lack of definiteness in usage seems to cause no confusion in practice.

1-5. The Chemical Properties of Substances

The **chemical properties** *of a substance are those properties that relate to its participation in chemical reactions.*

Chemical reactions *are the processes that convert substances into other substances.*

---

*The unit of density in the SI system is kg m^{-3}. We shall, however, often use the customary unit g cm^{-3}.

Thus sodium chloride has the property of changing into a soft metal, sodium, and a greenish-yellow gas, chlorine, when it is decomposed by passage of an electric current through it. It also has the property, when it is dissolved in water, of producing a white precipitate when a solution of silver nitrate is added to it; and it has many other chemical properties.

Iron has the property of combining readily with the oxygen in moist air, to form iron rust; whereas an alloy* of iron with chromium and nickel (stainless steel) is found to resist this process of rusting. It is evident from this example that the chemical properties of materials are important in engineering.

Many chemical reactions take place in the kitchen. When biscuits are made with use of sour milk and baking soda there is a chemical reaction between the baking soda and a substance in the sour milk, lactic acid, to produce the gas carbon dioxide, which leavens the dough by forming small bubbles in it. And, of course, a great many chemical reactions take place in the human body. Foods that we eat are digested in the stomach and intestines. Oxygen in the inhaled air combines with a substance, hemoglobin, in the red cells of the blood, and then is released in the tissues, where it takes part in many different reactions. Many biochemists and physiologists are engaged in the study of the chemical reactions that take place in the human body.

Most substances have the power to enter into many chemical reactions. The study of these reactions constitutes a large part of the study of chemistry. Chemistry may be defined as *the science of substances—their structure, their properties, and the reactions that change them into other substances.*

Example 1-2. Which of the following processes would you class as chemical reactions?
(*a*) The boiling of water.
(*b*) The burning of paper.
(*c*) The preparation of sugar syrup by adding sugar to hot water.
(*d*) The formation of rust on iron.
(*e*) The manufacture of salt by evaporation of sea water.

Solution. The burning of paper and the formation of rust on iron are chemical reactions. The boiling of water, the preparation of sugar syrup, and the manufacture of salt by evaporation of sea water are changes of state that are not classed as chemical reactions.

*An *alloy* is a metallic material containing two or more elements. It may be either homogeneous or heterogeneous (a mixture of grains of two or more kinds). If homogeneous, it may be either a pure compound or a solid solution, or even a liquid solution—many alloys of mercury and other metals are liquid.

1-6. Energy and Temperature

The concept of *energy* is as difficult to define as that of matter. Energy is involved in doing work, or in heating an object. A boulder at the top of a mountain has *potential energy*. As it rolls down the mountainside, its potential energy is changed into the *kinetic energy* of its motion. If it were to fall into a lake, and be slowed down by the friction of its motion through water, part of its kinetic energy would be changed by friction into *heat,* which then would raise the temperature of the boulder and of the water. In addition, part of its kinetic energy would be transferred to the water, and would evidence itself in waves radiating from the point of impact.

Another important kind of energy is *radiant energy*. Visible light, infrared radiation, ultraviolet radiation, x-rays, and radio waves are radiant energy. They are all closely similar in nature (see Sections 3-10, 3-12).

When a mixture of gasoline vapor and air is burned, energy is liberated—energy that can do the work of propelling an automobile, and that in addition causes an increase in temperature of the engine and the exhaust gases. This energy is said to have been stored up in the gasoline and air as *chemical energy*.

The Law of Conservation of Energy

It has been found that *whenever energy of one form disappears an equivalent amount of energy of other forms is produced.* This principle is called the *law of conservation of energy.**

All chemical reactions are accompanied by either the liberation of energy or the absorption of energy. Usually this energy is in the form of heat. If some substances when mixed together in a flask undergo a chemical reaction with liberation of heat, the contents of the flask become warmer. If, on the other hand, they undergo a chemical reaction with absorption of heat, the contents of the flask become colder. These facts can be described by saying that every substance has a certain *heat content,*† and that in general the heat contents of the products of a reaction differ from the heat contents of the reactants. In accordance with the law of conservation of energy, the *heat of the reaction* is the difference in heat contents of the products and the reactants, both at standard temperature. For example, a mixture of gasoline and oxygen has a greater total heat content than the products of their reaction, which are carbon

*This law is a special case of the more general *law of conservation of mass-energy,* which will be discussed in Sections 20-19 and 20-22.

†The heat content of a substance is called its enthalpy; see Section 6-12 and Appendix VI.

dioxide and water, at the same temperature. In consequence, some heat is liberated during the reaction, raising the temperature of the products and of other materials in contact with them.

Under some conditions chemical energy is liberated during a chemical reaction in forms other than heat. For example, the chemical energy stored up in an explosive may do work, in breaking a stone cliff into fragments. The chemical energy in the substances of an electric battery is converted into electric energy during the operation of the battery. Some of the chemical energy in a fuel may be converted into radiant energy as the fuel burns.

Temperature

If two objects are placed in contact with one another, heat may flow from one object to the other one. *Temperature* is the quality that determines the direction in which heat flows—it flows from the object at higher temperature to the object at lower temperature.

Temperatures are ordinarily measured by means of a thermometer such as the ordinary mercury thermometer, consisting of a quantity of mercury in a glass tube. The temperature scale used by scientists is the

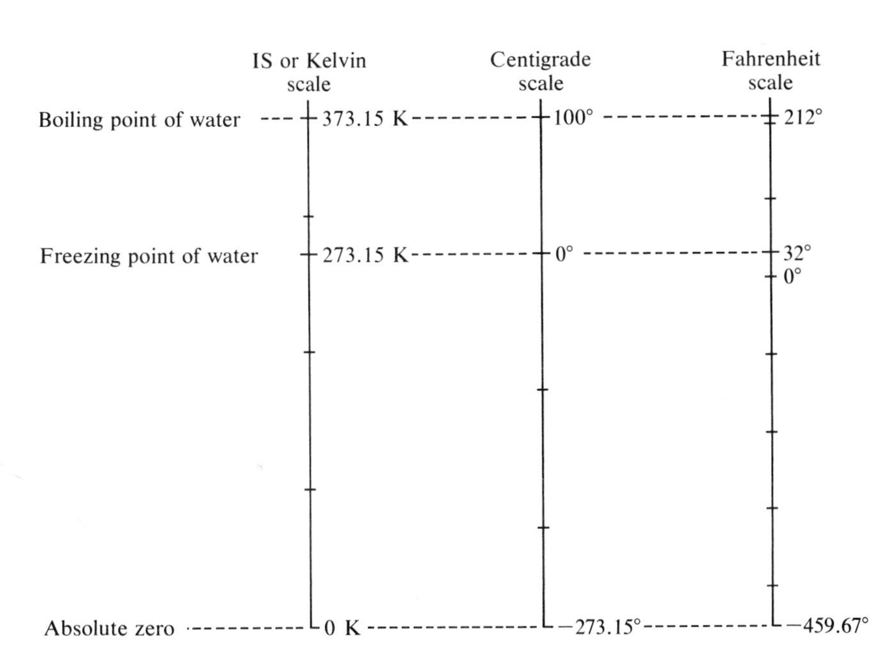

FIGURE 1-4
Comparison of Kelvin, Centigrade, and Fahrenheit scales of temperature.

centigrade scale or *Celsius scale;* it was introduced by Anders Celsius, a Swedish professor of astronomy, in 1742. On this scale the temperature of freezing water is 0°C and the temperature of boiling water is 100°C.

On the *Fahrenheit scale,* used in everyday life in English-speaking countries, the freezing point of water is 32°F and the boiling point of water is 212°F. On this scale the freezing point and the boiling point differ by 180°, rather than the 100° of the centigrade scale.*

The relation between the centigrade scale and the Fahrenheit scale is indicated in Figure 1-4. To convert temperatures from one scale to another, you need only remember that the Fahrenheit degree is $\frac{100}{180}$ or $\frac{5}{9}$ of the centigrade degree, and that 0°C is the same temperature as 32°F. You may find it convenient to use the method described in Exercise 1-14.

Example 1-3. A school room may be kept at 68°F. What is this temperature on the centigrade scale?

Solution. 68°F is 36°F (that is, 68° − 32°) above the freezing point of water. This number of Fahrenheit degrees is equal to $\frac{5}{9} \times 36 = 20$°C. Since the freezing point of water is 0°C, the temperature of the room is 20°C.

The Kelvin Temperature Scale

About 200 years ago scientists noticed that a sample of gas that is cooled decreases in volume in a regular way, and they saw that if the volume were to continue to decrease in the same way it would become zero at about −273°C. The concept was developed that this temperature, −273°C (more accurately, −273.15°C), is the minimum temperature, the *absolute zero.* A new temperature scale was then devised by Lord Kelvin, a great British physicist (1824–1907). The *Kelvin scale* is defined in such a way as to permit the laws of thermodynamics to be expressed in simple form.

The IS temperature scale is the Kelvin scale with a new definition of the degree. The absolute zero is taken to be 0 K and the triple point of water is taken to be 273.16 K. (The triple point of water is the temperature at which pure liquid water, ice, and water vapor are in equilibrium.) With this definition of the degree, the boiling point of water at one atmosphere pressure is 373.15 K and the freezing point of water saturated with

---

*The Fahrenheit scale was devised by Gabriel Daniel Fahrenheit (1686–1736), a natural philosopher who was born in Danzig and settled in Holland. He invented the mercury thermometer in 1714; before then alcohol had been used as the liquid in thermometers. As the zero point on his scale he took the temperature produced by mixing equal quantities of snow and ammonium chloride. His choice of 212° for the boiling point of water was made in order that the temperature of his body should be 100°F. The normal temperature of the human body is 98.6°F; perhaps Fahrenheit had a slight fever while he was calibrating his thermometer.

air at one atmosphere pressure is 273.15 K. Hence the IS Kelvin temperature is 273.15 K greater than the centigrade temperature. Note that the degree sign is not used in giving the Kelvin temperature.

Heat

The IS unit of energy and heat is the *joule*. In chemistry the *calorie* has been used extensively as the unit of energy, and most thermochemical reference books use the calorie or kilocalorie (1000 cal). The calorie is now defined as 4.184 J:

$$1 \text{ cal} = 4.184 \text{ J}$$

$$1 \text{ kcal} = 1000 \text{ cal} = 4.184 \text{ kJ}$$

The calorie, to within ordinary requirements of accuracy, is the amount of heat (energy) required to raise the temperature of 1 g of liquid water by 1°C.

Example 1-4. Into a flask containing 100 g of water at 18.0°C, with a small amount of hydrochloric acid dissolved in it, there was poured 100 g of water, also at 18.0°C, containing a small amount of sodium hydroxide. The temperature of the mixed solution increased to 24.5°C. Neglecting the effect of the substances dissolved in the water and the loss of heat to the flask, calculate how much heat (how many joules) was produced by the reaction of the acid and the sodium hydroxide.

Solution. A total of 200 g water is raised in temperature by $24.5 - 18.0 = 6.5$°C. To do this requires $200 \times 6.5 = 1300$ calories. 1300 calories $= 1300 \times 4.184 = 5440$ joules $= 5.44$ kJ. Hence 5.44 kJ of heat energy was produced by the reaction.

1-7. Pressure

In chemical work it is often necessary to know not only the temperature at which an experiment is carried out, but also the *pressure*. For example, the large-scale industrial preparation of ammonia is carried out at high pressure, because the chemical reaction does not proceed satisfactorily at ordinary pressure.

Pressure is force per unit area. Pressure may be measured in newtons*

*The newton (N) is the IS unit of force. It is defined as the force necessary to accelerate a mass of 1 kilogram by 1 m s^{-2}. 1 N $= 10^5$ dynes.

per square meter, or in pounds per square inch, or in other units. The atmosphere of the earth exerts a pressure on all objects at the surface of the earth. The pressure of the atmosphere is 101.325 kN m^{-2}.

Another unit of pressure that is often used is the *atmosphere* (abbreviation atm). The pressure 1 atm is the average pressure at the surface of the earth (at sea level) that is due to the weight of the air.

The pressure of the atmosphere can be measured by means of a *barometer.* A simple barometer is made by filling a long glass tube, which is closed at one end, with mercury, being careful that no air remains entrapped, and then inverting the open end of the tube under the surface of some mercury in a cup. If the tube is longer than 760 mm, the surface of the mercury at the upper end of the tube drops until the height of the mercury column, measured from the level of mercury in the cup, is just enough to balance the atmospheric pressure. This occurs when the weight of the column of mercury per unit area is equal to the pressure of the atmosphere.

Pressure is often reported as the height of the column of mercury required to balance it. For example, the pressure 1 atm is equal to 760 mm of mercury (abbreviated as mm Hg).

The units used to measure pressure are summarized in the following equation:

$$1 \text{ atm} = 760 \text{ mm Hg} = 14.7 \text{ pounds per square inch} = 101.325 \text{ kN m}^{-2}.$$

Example 1-5. Pressure can also be reported in grams per square centimeter. The density of mercury is 13.595 g cm^{-3}. What is 1 atm pressure in g cm^{-2}? (Remember that 1 atm = 76 cm Hg.)

Solution. The weight over 1 cm^2 at the bottom of a column of mercury 76 cm high is 13.595 g $cm^{-3} \times 76$ cm^3 = 1033 g. Hence the pressure is 1033 g cm^{-2}.

1-8. Solids, Liquids, and Gases

Materials may exist as solids, liquids, or gases. A specimen of solid, such as a piece of ice, has a definite volume and also has rigidity. It retains its shape even when acted on by an outside force, provided that the force is not great enough to break the specimen. A liquid, such as a portion of water in a cup, has a definite volume, but adjusts its shape to the shape of the bottom part of its container. A gas, such as steam (water vapor) in the cylinder of a steam engine, has neither definite shape nor definite volume—it changes its shape and also its volume with change in the shape and volume of the container.

Ice, water, and water vapor represent the same chemical substance, water substance, in three different states. Ice is the *solid state* (*crystalline state*), water the *liquid state*, and water vapor the *gaseous state*.

Scientists usually distinguish between *crystalline solids* and *noncrystalline solids*.

A **crystal** *is a homogeneous material* (either a pure substance or a solution) *that, as a result of its regular internal structure, has spontaneously assumed the shape of a figure bounded by plane faces.**

For example, when a solution of salt evaporates, small cubes of solid salt form. These cubes, which are bounded by plane square faces, are crystals.

Most solid substances are crystalline in nature. Sometimes the individual crystals, with plane faces and sharp edges and corners, are visible to the naked eye; sometimes they can be seen only under a microscope.

Some solids, such as charcoal, do not show any crystalline character even when examined with a microscope of high power; these solids are called *amorphous solids* (the word amorphous means without shape).

Certain other materials, of which sealing wax and glass are examples, are called *supercooled liquids*. When a stick of sealing wax, which is hard and brittle at room temperature, is gradually warmed, it begins to soften and finally becomes a mobile liquid. As it is being cooled it shows a gradual change from a mobile liquid to a viscous liquid, and then to a solid. Even at room temperature it might be described as a liquid which is so viscous that it flows only extremely slowly.

EXERCISES

1-1. What is the difference between mass and weight?

1-2. A sphere of plutonium metal (one of the five crystalline modifications) 7.4 cm in diameter weighs 3.92 kg. What is the density of this form of plutonium?

1-3. Classify the following materials as homogeneous or heterogeneous:

pure gold	air	glass	granite
milk	ice	sugar	quartz
wood	gasoline	coffee	snow

1-4. According to the definition of mineral (first footnote, Section 1-3), is the ice in a glacier to be classified as a mineral?

*Another definition of a crystal in terms of the atomic theory will be given later.

1-5. Classify the following homogeneous materials as substances or solutions:

rainwater	ocean water	oxygen	honey
air	salt	mercury	vodka

1-6. What is the evidence proving that water is a compound, and not an element? What is the chemical evidence indicating that oxygen is an element, and not a compound? Why is the word "proving" used in the first of the preceding sentences, and "indicating" in the second?

1-7. The name holosiderite is given to metallic meteorites, which are alloys of iron and nickel. Many homogeneous holosiderites have been analyzed. They have been found to contain various amounts of nickel, between 6% and 10%. Is holosiderite an iron-nickel compound, or is it a solid solution?

1-8. When the substance calcite is heated it forms lime and carbon dioxide. Is calcite an elementary substance or a compound? Can you say from the foregoing information whether lime is an elementary substance or a compound?

1-9. When diamond is heated in a vacuum (no other material present) it is converted completely into graphite. Does this prove that diamond is a compound?

1-10. Classify the following statements as referring to physical properties or chemical properties:
(*a*) Silver tarnishes.
(*b*) Lead has a high density.
(*c*) Metals are ductile.
(*d*) Gasoline burns in air.

1-11. A kilogram of gold (2.2 lbs) occupies the volume 51.5 cm^3. What is the density of gold? (Answer: 19.4 g cm^{-3}.)

1-12. Archimedes (born about 287 B.C.) is said to have discovered a way of checking the possible adulteration of gold with copper in a crown made for King Hiero of Syracuse, Sicily. His method was to compare the volume of water displaced by the crown with the volumes displaced by equal weights of pure gold and pure copper. Let us suppose that the crown weighed 1000 g and displaced the volume 71.5 cm^3.
(*a*) What is the density of the crown?
(*b*) To what percentage of gold in the gold-copper alloy does this density correspond? Assume that there is no change in volume when the two metals are mixed (melted together) to form the alloy. The density of gold is 19.40 g cm^{-3} and the density of copper is 8.99 g cm^{-3}.

1-13. Mercury freezes at about −40°C. What is this temperature on the Fahrenheit scale?

1-14. A simple way to convert from the Fahrenheit to the centigrade scale or from the centigrade to the Fahrenheit scale is to add 40°, multiply by 5/9 or 9/5, respectively, and then subtract 40°. Verify this statement by use of the definitions of the two scales.

1-15. How much heat is needed to raise the temperature of 100 g of water from 10°C to 50°C? (Answer: 16.7 kJ.)

1-16. The melting point of pure iron is 1535°C. What is this temperature on the Fahrenheit scale?

1-17. One liter of boiling water is poured into a vessel containing three liters of water at 20°C. What is the temperature of the water after stirring? (Ignore the heat loss to the container.)

1-18. The density of water is about 1 g cm^{-3}. At what depth would a diver have to descend under the surface of a lake in order that the pressure acting on him would be 3 atm, rather than the 1 atm that is due to the weight of the air? What is this depth in feet?

2

The Atomic and Molecular Structure of Matter

The properties of any kind of matter are most easily and clearly learned and understood when they are correlated with its structure, in terms of the molecules, atoms, and still smaller particles that compose it. This subject, the atomic structure of matter, will be taken up in this chapter.

2-1. Hypotheses, Theories, and Laws

When it is first found that an idea explains or correlates a number of facts, the idea is called a *hypothesis*. A hypothesis may be subjected to further tests and to experimental checking of deductions that may be made from it. If it continues to agree with the results of experiment the hypothesis is dignified by the name of *theory* or *law*.

A theory, such as the atomic theory, usually involves some idea about the nature of some part of the universe, whereas a law may represent a summarizing statement about observed experimental facts. For example, there is a law of the constancy of the angles between the faces of crystals. This law states that whenever the angles between corresponding faces of various crystals of a pure substance are measured they are found to have the same value. The law simply expresses the fact that the angles between corresponding faces on a crystal of a pure substance are found to have the same value whether the crystal is a small one or a large one; it does not in any way explain this fact. An explanation of the fact is given by the atomic theory of crystals, the theory that in crystals the atoms are arranged in a regular order (as described later in this chapter).

It may be mentioned that chemists and other scientists use the word theory in two somewhat different senses. The first meaning of the word is that described above, namely, a hypothesis that has been verified. The second use of the word theory is to represent a systematic body of knowledge, compounded of facts, laws, theories in the limited sense described above, deductive arguments, etc. Thus by the atomic theory we mean not only the idea that substances are composed of atoms, but also all the facts about substances that can be explained and interpreted in terms of atoms and the arguments that have been developed to explain the properties of substances in terms of their atomic structure.

2-2. The Atomic Theory

The most important of all chemical theories is the atomic theory. In 1805 the English chemist and physicist John Dalton (1766–1844), of Manchester, England, stated the hypothesis that *all substances consist of small particles of matter, of several different kinds, corresponding to the different elements.* He called these particles atoms, from the Greek word *atomos,* meaning indivisible. This hypothesis gave a simple explanation or picture of previously observed but unsatisfactorily explained relations among the weights of substances taking part in chemical reactions with one another. As it was verified by further work in chemistry and physics, Dalton's atomic hypothesis became the atomic theory. The existence of atoms is now accepted as a fact.

The rapid progress of our science during the current century is well illustrated by the increase in our knowledge about atoms. In a popular textbook of chemistry written in the early years of the twentieth century atoms were defined as the "imaginary units of which bodies are aggregates." The article on "Atom" in the 11th edition of the *Encyclopaedia Britannica,* published in 1910, ends with the words "The atomic theory has been of priceless value to chemists, but it has more than once happened in the history of science that a hypothesis, after having been useful in the discovery and the coordination of knowledge, has been abandoned and replaced by one more in harmony with later discoveries. Some distinguished chemists have thought that this fate may be awaiting the atomic theory. . . . But modern discoveries in radioactivity are in favor of the existence of the atom, although they lead to the belief that the atom is not so eternal and unchangeable a thing as Dalton and his predecessors had imagined." Now, only two-thirds of a century later, we have precise knowledge of the structure and properties of atoms and molecules. Atoms and molecules can no longer be considered "imaginary."

Dalton's Arguments in Support of the Atomic Theory

The concept of atoms is very old. The Greek philosopher Democritus (about 460–370 B.C.), who had adopted some of his ideas from earlier philosophers, stated that the universe is composed of void (vacuum) and atoms. The atoms were considered to be everlasting and indivisible—absolutely small, so small that their size could not be diminished. He considered the atoms of different substances, such as water and iron, to be fundamentally the same, but to differ in some superficial way; atoms of water, being smooth and round, could roll over one another, whereas atoms of iron, being rough and jagged, would cling together to form a solid body.

The atomic theory of Democritus was pure speculation and was much too general to be useful. Dalton's atomic theory, however, was a hypothesis that explained many facts in a simple and reasonable way.

In 1785 the French chemist Antoine Laurent Lavoisier (1743–1794) showed clearly that there is no change in mass during a chemical reaction—the mass of the products is equal to the mass of the reacting substances.

In 1799 another general law, the **law of constant proportions,** was enunciated by the French chemist Joseph Louis Proust (1754–1826). The law of constant proportions states that *different samples of a substance contain its elementary constituents* (*elements*) *in the same proportions.* For example, it was found by analysis that the two elements hydrogen and oxygen are present in any sample of water in the proportion by weight 1:8. One gram of hydrogen and 8 g of oxygen combine to form 9 g of water.

Dalton stated the hypothesis that elements consist of atoms, all of the atoms of one element being identical, and that compounds result from the combination of a certain number of atoms of one element with a certain number of atoms of another element (or, in general, from the combination of atoms of two or more elements, each in definite number). In this way he could give a simple explanation of the law of conservation of mass, and also of the law of constant proportions.

A **molecule** *is a group of atoms bonded to one another.* If a molecule of water is formed by the combination of two atoms of hydrogen with one atom of oxygen, the mass of the molecule should be the sum of the masses of two atoms of hydrogen and an atom of oxygen, in accordance with the law of conservation of mass. The definite composition of a compound is then explained by the definite ratio of atoms of different elements in the molecules of the compound.

Dalton also formulated another law, the **law of simple multiple proportions.*** This law states that *when two elements combine to form more than one compound, the weights of one element that combine with the same weight of the other are in the ratios of small integers.* It is found by experiment that, whereas water consists of hydrogen and oxygen in the weight ratio 1:8, hydrogen peroxide consists of hydrogen and oxygen in the ratio 1:16. The weights of oxygen combined with the same weight of hydrogen, 1 g, in water and hydrogen peroxide are 8 g and 16 g; that is, they are in the ratio of the small integers 1 and 2. This ratio can be explained by assuming that twice as many atoms of oxygen combine with an atom of hydrogen in hydrogen peroxide as in water. This situation is illustrated in Figure 2-1, which shows the symbols used by Dalton to represent the atoms of some elements and the molecules of compounds.

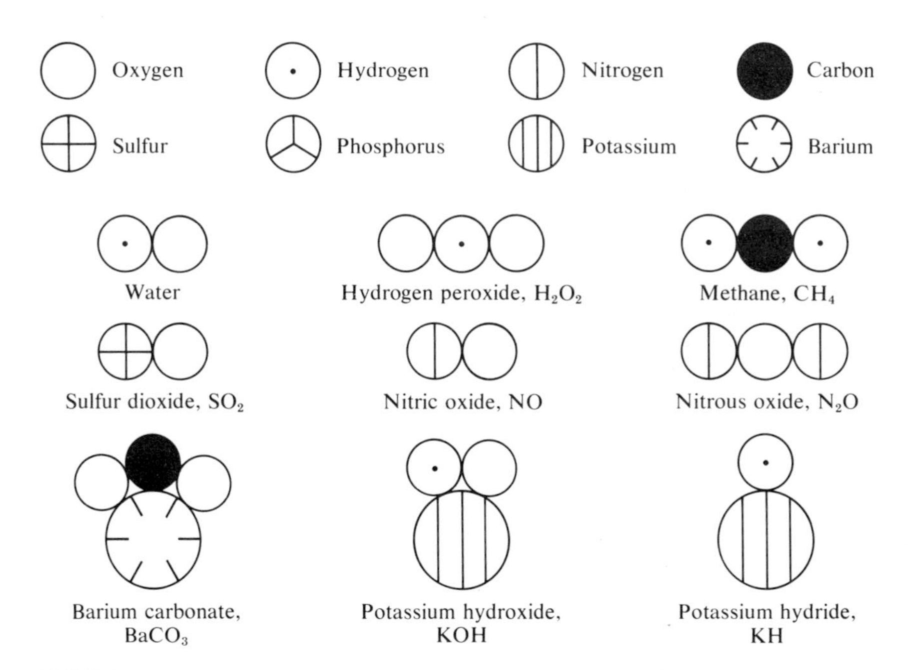

FIGURE 2-1
Atomic symbols and molecular formulas used by John Dalton, about 1803. Some of Dalton's formulas were wrong.

*The discovery of the law of simple multiple proportions was the first great success of Dalton's atomic theory. This law was not induced from experimental results, but was derived from the theory, and then tested by experiments.

Dalton had no way of determining the correct formulas of compounds, and he arbitrarily chose formulas to be as simple as possible: for example, he assumed that the molecule of water consisted of one atom of hydrogen and one atom of oxygen, as shown in the figure, whereas in fact it consists of two atoms of hydrogen and one of oxygen.

2-3. Modern Methods of Studying Atoms and Molecules

During the second half of the nineteenth century chemists began to discuss the properties of substances in terms of assumed structures of the molecules—that is, of definite arrangements of the atoms relative to one another. Precise information about the atomic structure of molecules and crystals of many substances was finally obtained during the recent period, beginning about 1913. The physicists have developed many powerful methods of investigating the structure of matter. One of these methods is the interpretation of the *spectra* of substances (see Figure 19-6). A flame containing water vapor, for example, emits light that is characteristic of the water molecule; this is called the spectrum of water vapor. Measurements of the lines in the water spectrum have been made and interpreted, and it has been found that the two hydrogen atoms in the molecule are about 97 pm from the oxygen atom.* Moreover, it has been shown that the two hydrogen atoms are not on opposite sides of the oxygen atom, but that the molecule is bent, the angle formed by the three atoms being 105°. The distances between atoms and the angles formed by the atoms in many simple molecules have been determined by spectroscopic methods.

Also, the structures of many substances have been determined by the methods of diffraction of electrons and diffraction of x-rays. In the following pages we shall describe many atomic structures that have been determined by these methods. The x-ray diffraction method of determining the structure of crystals is discussed in Appendix IV.

2-4. The Arrangement of Atoms in a Crystal

Most solid substances are crystalline in nature. Sometimes the particles of a sample of solid substance are themselves single crystals, such as the cubic crystals of sodium chloride in table salt. Sometimes these single crystals are very large; occasionally crystals of minerals several meters in diameter are found in nature.

*The distance between a hydrogen atom and the oxygen atom in the water molecule is given as 97 pm. The picometer is the modern unit of length for molecules; it is 10^{-12} meter. In older books and journals the Ångström (symbol Å) is used. The Ångström has the value 1×10^{-10} m; that is, 1 Å = 100 pm.

In our discussion we shall use *copper* as an example. Crystals of copper as large as a centimeter on edge are found in deposits of copper ore. An ordinary piece of the metal copper does not consist of a single crystal of copper, but of an aggregate of crystals. The crystal grains of a specimen of a metal can be made clearly visible by polishing the surface of the metal, and then etching the metal lightly with an acid. Often the grains are small, and can be seen only with the aid of a microscope (Figure 2-2), but sometimes they are large, and can be easily seen with the naked eye, as in some brass doorknobs.

FIGURE 2-2
A polished and etched surface of a piece of cold-drawn copper bar, showing the small crystal grains that compose the ordinary metal. Magnification 200 × (200-fold linearly). The small round spots are gas bubbles.

It has been found by experiment (Appendix IV) that *every crystal consists of atoms arranged in a three-dimensional pattern that repeats itself regularly*. In a crystal of copper all of the atoms are alike, and they are arranged in the way shown in Figures 2-3 and 2-4. This is a way in which spheres of uniform size may be packed together to occupy the smallest volume. This structure, called the *cubic closest-packed structure,* was assigned to the copper crystal by W. L. Bragg in 1913.

You must remember while looking at Figures 2-3 and 2-4 that the atoms are shown greatly enlarged relative to the crystal. Even if the crystal were a small one, with edges only about 0.1 mm long, there would still be about 400,000 atoms in a row along each edge.

It is the **regularity of arrangement** *of the atoms in a crystal that gives to the crystal its characteristic properties, in particular the property of growing in the form of polyhedra.* (A polyhedron is a solid figure bounded by plane faces.) The faces of crystals are defined by surface layers of atoms, as shown in Figures 2-3 and 2-4. These faces lie at angles to one another that have definite characteristic values, the same for all specimens of the same substance. The sizes of the faces may vary from specimen to specimen, but the angles between them are always constant. The

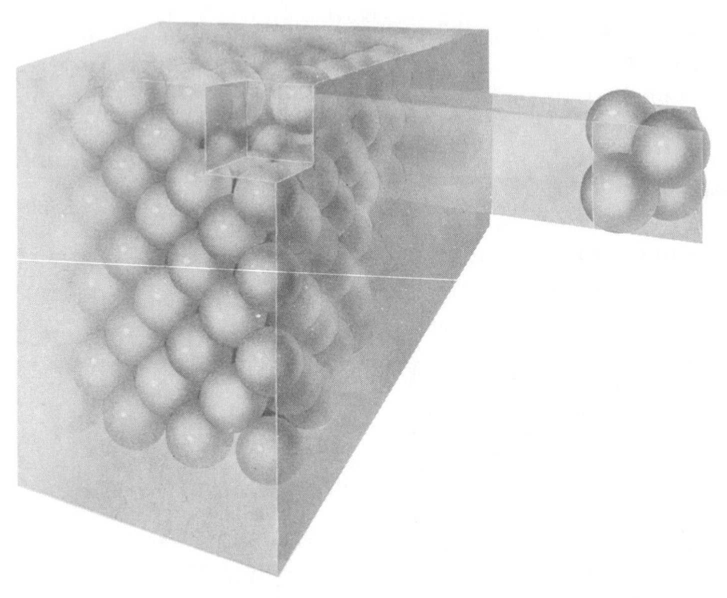

FIGURE 2-3
The arrangement of atoms in a crystal of copper. The small cube, containing four copper atoms, is the unit of structure; by repeating it the entire crystal is obtained.

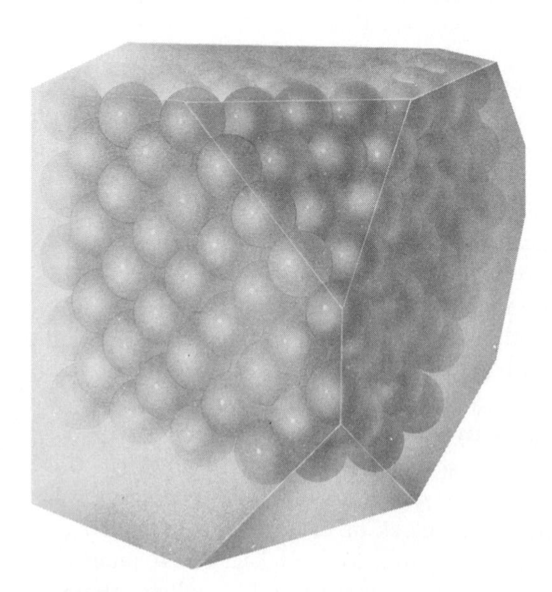

FIGURE 2-4
Another atomic view of a copper crystal, showing small octahedral faces and large cube faces.

principal surface layers shown in Figures 2-3 and 2-4 for copper correspond to the faces of a cube (*cubic faces* or *cube faces*); these faces are always at right angles with one another. The smaller surface layer, obtained by cutting off a corner of a cube, is called an *octahedral face*. Native copper, found in deposits of copper ore, often is in the form of crystals with cubic and octahedral faces.

Atoms are not hard spheres, but are soft, so that by increased force they may be pushed more closely together (be compressed). This compression occurs, for example, when a copper crystal becomes somewhat smaller in volume under increased pressure. The sizes that are assigned to atoms correspond to the distances between the center of one atom and the center of a neighboring atom of the same kind in a crystal under ordinary circumstances. The distance from a copper atom to each of its twelve nearest neighbors in a copper crystal at room temperature and atmospheric pressure is 255 pm; this is called the *diameter* of the copper atom in metallic copper. The radius of the copper atom is half this value.

2-5. The Description of a Crystal Structure

Chemists often make use of the observed shapes of crystals to help in the identification of substances. The description of the shapes of crystals is part of the subject of the science of *crystallography*. The method of studying the structure of crystals by the diffraction of x-rays, which was discovered by the German physicist Max von Laue (1879–1960) in 1912 and developed by the British physicists W. H. Bragg (1862–1942) and W. L. Bragg (1890–1971), has become especially valuable in recent decades. Much of the information about molecular structure that is given in this book has been obtained by the x-ray diffraction technique.

The basis of the description of the structure of a crystal is the *unit of structure*. For cubic crystals the unit of structure is a small cube, which, when repeated parallel to itself in such a way as to fill space, reproduces the entire crystal.

The way in which this is done can be seen from a two-dimensional example. In Figure 2-5 there is shown a portion of a square lattice. The unit of structure of this square lattice is a square; when this square is repeated parallel to itself in such a way as to fill the plane, we obtain a sort of two-dimensional crystal. In this case there are present a lattice of atoms of one sort, represented by small spheres at the intersections of the lattice lines, and a lattice of atoms of another sort, represented by larger spheres at the centers of the unit squares. We might describe the structure by the use of coordinates x and y, giving the positions of the atoms relative to an origin at the corner of the unit of structure, with x and y taken as fractions of the edges of the unit of structure, as indicated in the figure. The atom represented by the small sphere would then have the coordinates $x = 0$, $y = 0$, and the atom at the center of the square would have the coordinates $x = \frac{1}{2}$, $y = \frac{1}{2}$.

Similarly, for a cubic crystal the unit of structure can be taken as a cube, which when reproduced in parallel orientation would fill space to produce a cubic lattice, as shown in Figure 2-6. The unit of structure could be described, for a cubic crystal, by giving the value of the edge of the unit, a, and the values of the coordinates x, y, and z for each atom, as fractions of the edge of the unit. Thus, for the cubic closest-packed structure, represented by metallic copper, the unit of structure is cubic, with edge $a = \sqrt{2} \times 255$ pm, and with four atoms per unit, with coordinates $x = 0$, $y = 0$, $z = 0$; $x = 0$, $y = \frac{1}{2}$, $z = \frac{1}{2}$; $x = \frac{1}{2}$, $y = 0$, $z = \frac{1}{2}$; and $x = \frac{1}{2}$, $y = \frac{1}{2}$, $z = 0$, as shown in Figure 2-7. Often these coordinates are written without giving the symbols x, y, and z; it is then said that there are four copper atoms in the unit, at 0, 0, 0; 0, $\frac{1}{2}$, $\frac{1}{2}$; $\frac{1}{2}$, 0, $\frac{1}{2}$; $\frac{1}{2}$, $\frac{1}{2}$, 0. These are called the *coordinates* of the atoms in the unit cube.

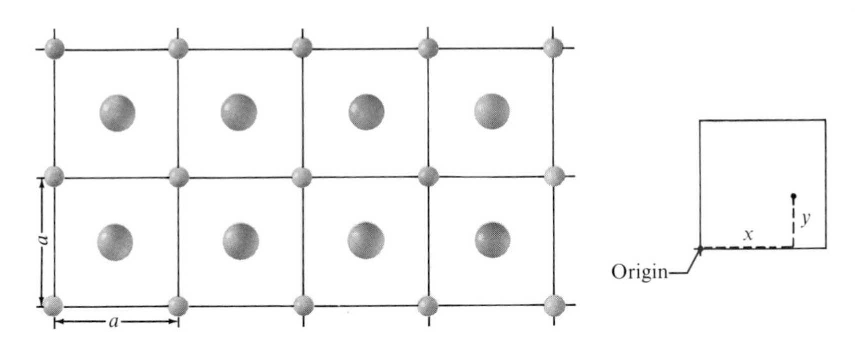

FIGURE 2-5
Arrangement of atoms in a plane. The unit of structure is a square. Small atoms have the coordinates 0, 0 and large atoms the coordinates $\frac{1}{2}$, $\frac{1}{2}$.

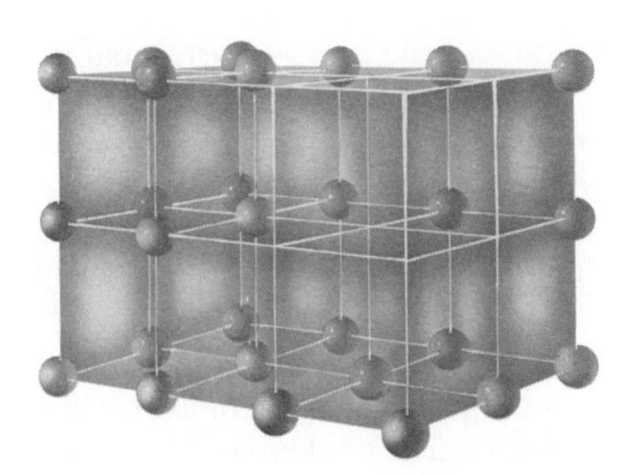

FIGURE 2-6
The simple cubic arrangement of atoms. The unit of structure is a cube, with one atom per unit, its coordinates being 0, 0, 0.

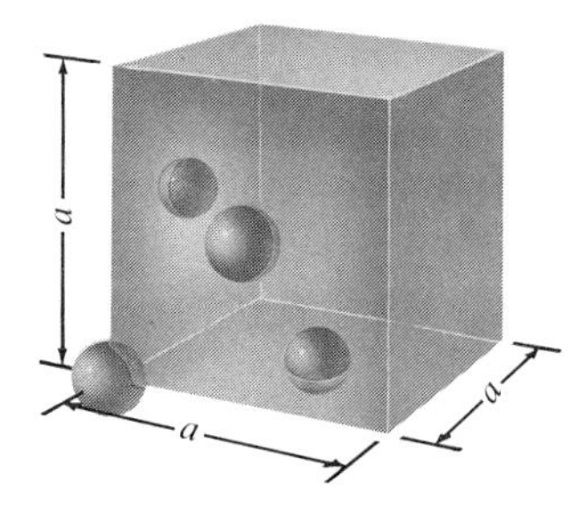

FIGURE 2-7
The cubic unit of structure for the face-centered cubic arrangement, corresponding to cubic closest packing of spheres. There are four atoms in the unit, with coordinates 0, 0, 0; 0, $\frac{1}{2}$, $\frac{1}{2}$; $\frac{1}{2}$, 0, $\frac{1}{2}$; $\frac{1}{2}$, $\frac{1}{2}$, 0.

Note that in the unit cube shown in Figure 2-7 an atom is represented at only one of the eight corners. Of course, when this unit cube is surrounded by other unit cubes, atoms are placed at the seven other corners, these atoms being formally associated with the adjacent unit cubes.

The unit of structure of a crystal other than a cubic crystal is a parallelepiped. In the case of the most general sort of crystal, a triclinic crystal (see the paragraph after Example 2-2 below), the parallelepiped is a general one. It can be described by giving the values of a, b, and c, the lengths of the three edges, and also the values of α, β, and γ, the angles between pairs of edges.

Example 2-1. The metal iron is cubic, with $a = 286$ pm, and with two iron atoms in the unit cube, at 0, 0, 0 and $\frac{1}{2}$, $\frac{1}{2}$, $\frac{1}{2}$. How many nearest neighbors does each iron atom have, and how far away are they?

Solution. We draw a cubic unit of structure, with edge 286 pm, as shown in Figure 2-8, and we indicate in it the positions 0, 0, 0 and $\frac{1}{2}$, $\frac{1}{2}$, $\frac{1}{2}$. When cubes of this sort are reproduced parallel to one another, we see that we obtain the structure shown in Figure 2-9; this is called the *body-centered arrangement.* It is seen that the atom at $\frac{1}{2}$, $\frac{1}{2}$, $\frac{1}{2}$ is surrounded by eight

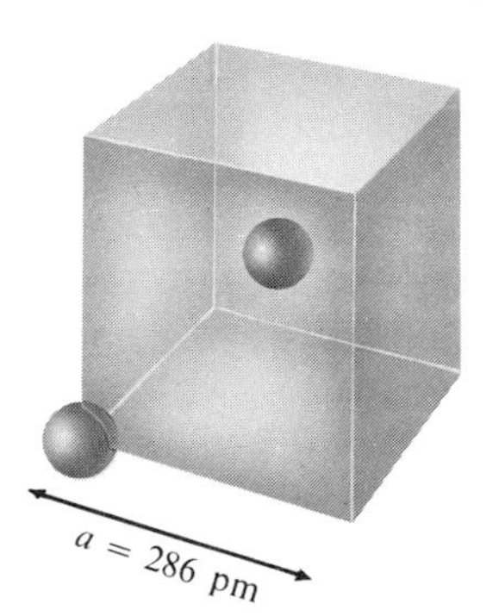

FIGURE 2-8
The unit of structure corresponding to the cubic body-centered arrangement. There are two atoms in the unit, with coordinates 0, 0, 0 and $\frac{1}{2}$, $\frac{1}{2}$, $\frac{1}{2}$.

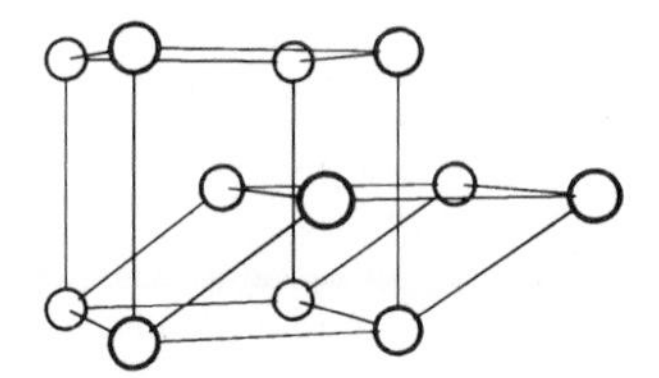
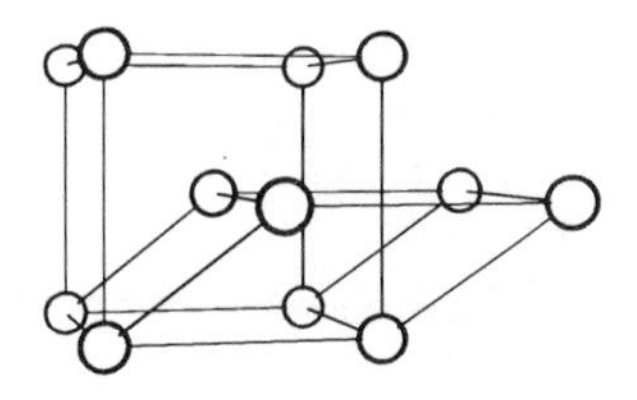

FIGURE 2-9
Stereoscopic view of the body-centered cubic structure, found for iron and some other metals. [The stereo drawings in this book can be viewed by looking at the right drawing with the right eye and the left drawing with the left eye from a distance of a few inches. It may help to hold a stiff piece of paper upright between the drawings. A moment or two may be required before the viewer learns to integrate the two images.]

atoms, the atom at 0, 0, 0 and seven similar atoms. Also, the atom at 0, 0, 0 is surrounded by eight atoms. In each case the surrounding atoms are at the corners of a cube. This situation is described by saying that each atom in the body-centered arrangement has *ligancy* 8 (or *coordination number* 8).

To calculate the interatomic distance, we note that, by the theorem of Pythagoras, the square of the distance is equal to $(a/2)^2 + (a/2)^2 + (a/2)^2$, and hence the distance itself is equal to $\sqrt{3}a/2$. Thus the distance between each iron atom and its neighbors is found to be $1.732 \times 286/2 = 248$ pm. The metallic radius of iron is hence 124 pm.

Example 2-2. The English mathematician and astronomer Thomas Harriot (1560–1621), who was tutor to Sir Walter Raleigh and who traveled to Virginia in 1585, was interested in the atomic theory of substances. He believed that the hypothesis that substances consist of atoms was plausible, and capable of explaining some of the properties of matter. His writings contain the following propositions:

"9. The more solid bodies have Atoms touching on all Sydes.
"10. Homogeneall bodies consist of Atoms of like figure, and quantitie.
"11. The waight may increase by interposition of lesse Atoms in the vacuities betwine the greater.
"12. By the differences of regular touches (in bodies more solid), we find that the lightest are such, where euery Atom is touched with six others about it, and greatest (if not intermingled) where twelve others do touch euery Atom."

Assuming that the atoms can be represented as hard spheres in contact with one another, what difference in density would there be between the two structures described in the above proposition 12?

Solution. The structure where every atom is in contact with six others about it that Harriot had in mind is probably the simple cubic arrangement, shown in Figure 2-6. In this arrangement of atoms the unit of structure is a cube, containing one atom, which can be assigned the coordinates 0, 0, 0. Each atom is then in contact with six other atoms, which are at the distance d from it. The volume of the unit cube is accordingly d^3. If the mass of the atom is M, the density for this arrangement is M/d^3.

The more dense structure referred to by Harriot, where twelve atoms are in contact with each atom, is the cubic closest-packed arrangement described in the preceding section. (Harriot had apparently discovered that there is no way of packing equal hard spheres in space that gives a greater density than is given by this arrangement.) The cubic unit of structure for this arrangement contains four atoms. Its edge, a, is equal to $2^{1/2}d$, and its volume to $2^{3/2}d^3$. The mass contained in the unit cube is $4M$, and the density is accordingly $4M/2^{3/2}d^3$, or $2^{1/2}M/d^3$.

We thus have found that the dense structure described by Harriot has density $\sqrt{2} = 1.414$ times that of the less dense structure; it is accordingly 41.4% denser than the less dense structure.

The Six Crystal Systems

Every crystal can be classified in one of the six crystal systems, called cubic (or isometric), hexagonal, tetragonal, orthorhombic, monoclinic, and triclinic. Some characteristics of crystal systems are shown in Figure 2-10.

The symmetry of the crystals representing these different systems is such that their units of structure may be chosen in special ways, except for triclinic crystals. These special ways involve restrictions on the relative values of the three edges and values of the three angles as follows:

Cubic crystals: three equal edges, with length a, at right angles to one another.
Tetragonal crystals: two equal edges, with length a, and a third edge, with different length c, all at right angles to one another.
Trigonal and hexagonal crystals: two equal edges, with length a, at 120° to one another, and a third edge, with length c, at right angles to the other two.
Orthorhombic crystals: three edges, with unequal lengths a, b, and c, all at right angles to one another.
Monoclinic crystals: two edges, a and c, at the angle β with one another, and a third, b, at right angles to a and c.
Triclinic crystals: three edges, a, b, and c, with angles α, β, and γ between them.

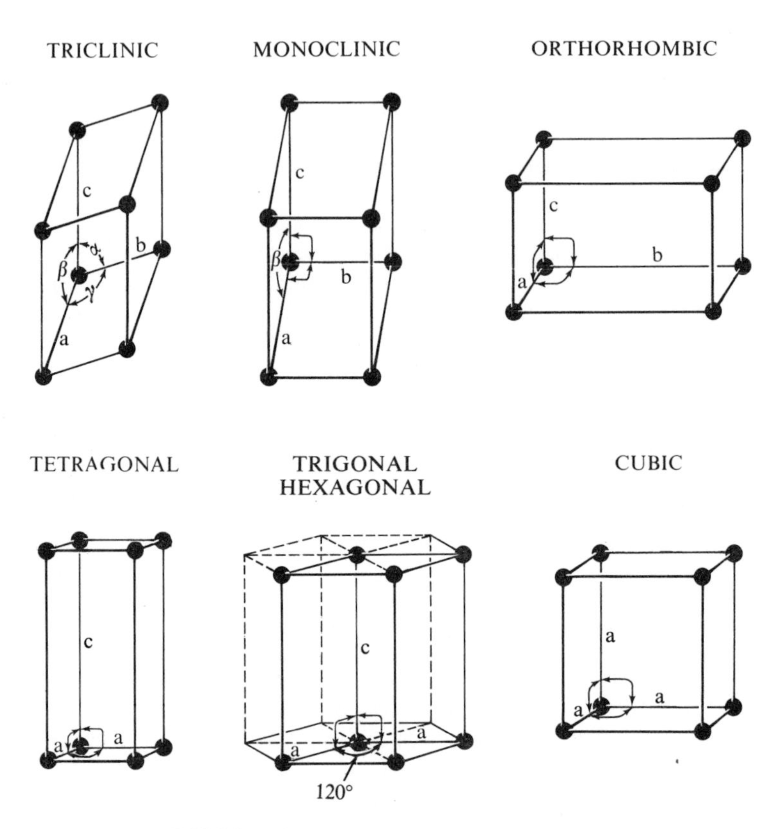

FIGURE 2-10
Representative unit cells of the crystal systems.

2-6. The Molecular Structure of Matter

Molecular Crystals

The crystal of copper, which we have discussed as an example of a kind of matter, is built up of *atoms* arranged in a regular pattern. We shall now discuss crystals that contain *discrete groups of atoms* (distinct groups), which are called *molecules*. These crystals are called *molecular crystals*.

An example of a molecular crystal is shown in the upper left part of Figure 2-11, which is a drawing representing the structure of a crystal of the blackish-gray solid substance *iodine*. It is seen that the iodine atoms are grouped together in pairs, to form molecules containing two atoms each. Iodine is used as an example in this section and the following ones because its molecules are simple (containing only two atoms), and because it has been thoroughly studied by scientists.

The distance between the two atoms of iodine in the same molecule of this molecular crystal is less than the distances between atoms in different molecules. The two iodine atoms in each molecule are only 270 pm apart, whereas the smallest distance between iodine atoms in different molecules is 354 pm.

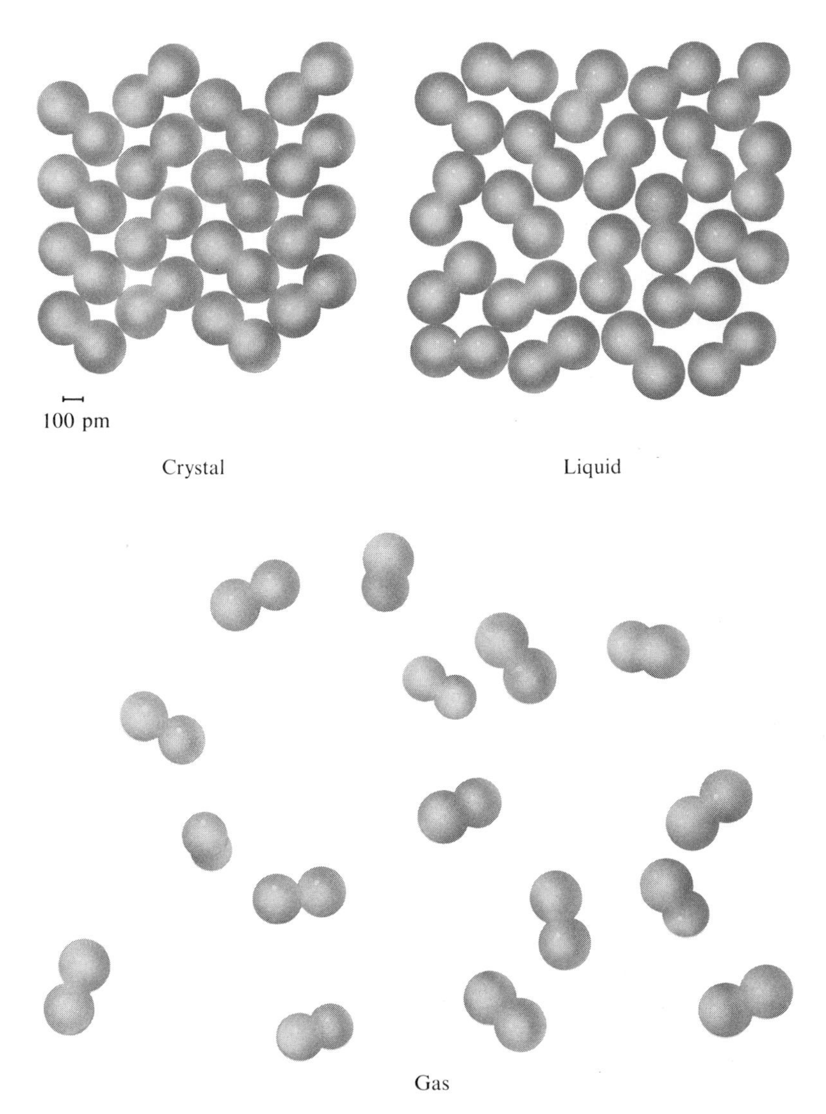

FIGURE 2-11
Crystal, liquid, and gaseous iodine, showing diatomic molecules I_2.

The forces acting between atoms within a molecule are very strong, and those acting between molecules are weak. As a result of this, it is hard to cause the molecule to change its shape, whereas it is comparatively easy to roll the molecules around relative to one another. For example, under pressure a crystal of iodine decreases in size: the molecules can be pushed together until the distances between iodine atoms in different molecules have decreased by several percent; but the molecules themselves retain their original size, with no appreciable change in interatomic distance within the molecule. When a crystal of iodine at low temperature is heated it expands, so that each of the molecules occupies a larger space in the crystal; but the distance between the two iodine atoms in one molecule stays very close to the normal 270 pm.

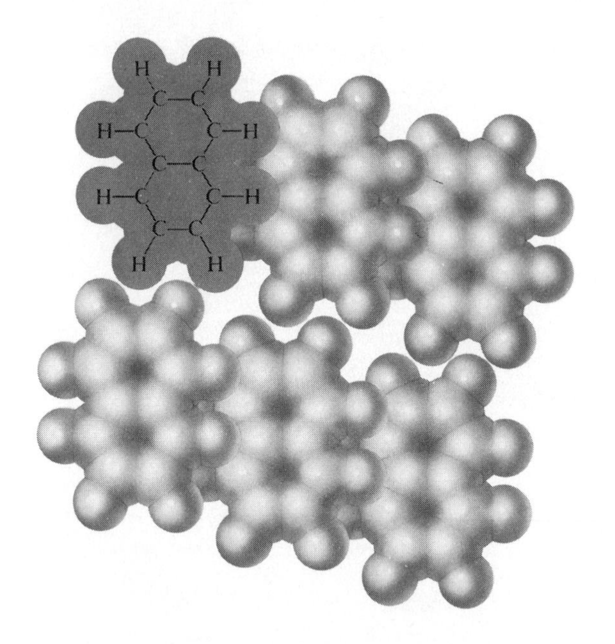

FIGURE 2-12
A portion of a crystal of naphthalene, showing molecules $C_{10}H_8$.

The molecules of different chemical substances contain varying numbers of atoms, bonded tightly together. An example of a more complicated molecule is shown in Figure 2-12, which represents a portion of a crystal of *naphthalene*. The molecule of naphthalene contains ten carbon atoms, arranged in two hexagonal rings that have one edge in common, and eight hydrogen atoms. Naphthalene is a rather volatile substance, with a characteristic odor. In the form of moth balls, it is used as a moth repellent. The properties of naphthalene are determined by the structure of its molecules.

Photographs of Molecules Made with the Electron Microscope

It has recently become possible to see and to photograph molecules. They are too small to be seen with a microscope using ordinary visible light, which cannot permit objects much smaller in diameter than the wavelength of light, about 500 nm,* to be seen. An instrument, the electron microscope, has been developed that permits objects a thousand times smaller in diameter to be seen. It uses beams of electrons in place of beams of light. Its linear magnifying power is about 1,000,000, as compared with about 1000 for the light microscope. It is accordingly possible to see objects as small as 500 pm in diameter with the electron microscope.

Two photographs made with the electron microscope are reproduced here, as Figures 2-13 and 2-14. They show molecules of viruses that cause disease in tomato plants. Each bushy stunt virus molecule is about 23 nm in diameter. It consists of about 750,000 atoms. The necrosis virus molecules are somewhat smaller, about 19.5 nm in diameter. In each photograph the individual molecules can be clearly seen.

FIGURE 2-13
Electron micrograph of crystals of necrosis virus protein, showing individual molecules in ordered arrangement. Linear magnification 65,000. [From R. W. G. Wyckoff.]

*The nanometer (nm) is 10^{-9} meter (m).

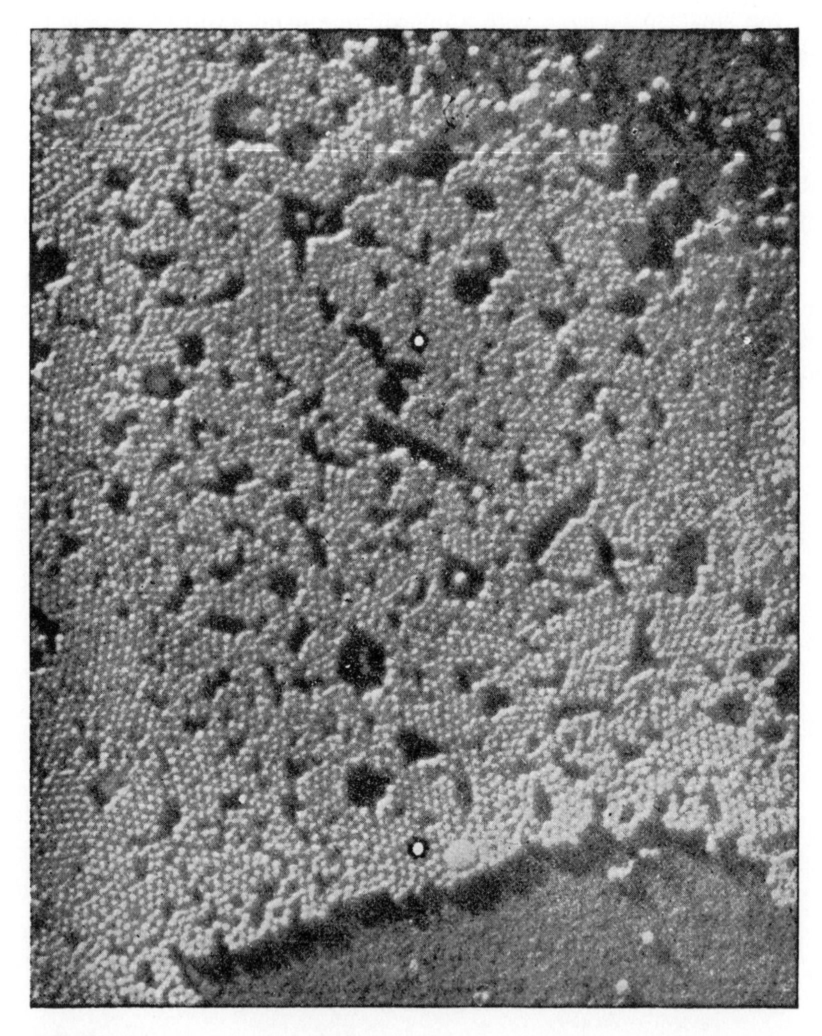

FIGURE 2-14
Electron micrograph of a single layer of tomato bushy stunt virus molecules. The photograph was made to show added contrast by depositing a very thin layer of gold on the specimen at a small angle, giving the impression of shadows cast by the molecules. Linear magnification 55,000. [From Price, Williams, and Wyckoff, *Arch. Biochem.* **7**, 175 (1946).]

2-7. Evaporation of Crystals. The Nature of a Gas

At a very low temperature the molecules in a crystal of iodine lie rather quietly in their places in the crystal (Figure 2-11). As the temperature increases the molecules become more and more agitated; each one bounds back and forth more and more vigorously in the little space left for it by its neighbors, and each one strikes its neighbors more and more strongly as it rebounds from them.

A molecule on the surface of the crystal is held to the crystal by the forces of attraction that its neighboring molecules exert on it. Attractive forces of this kind, which are operative between all molecules when they are close together, are called *van der Waals attractive forces;* this name is used because it was the Dutch physicist J. D. van der Waals (1837–1923) who first gave a thorough discussion of intermolecular forces in relation to the nature of gases and liquids.

These attractive forces are quite weak, much weaker than the forces between the atoms in one molecule. Hence occasionally a certain molecule may become so agitated as to break loose from its neighbors, and to fly off into the surrounding space. If the crystal is in a vessel, there will soon be present in the space within the vessel through this process of evaporation a large number of these free molecules, each moving in a straight-line path, and occasionally colliding with another molecule or with the walls of the vessel to change the direction of its motion. These free molecules constitute *iodine vapor* or *iodine gas* (Figure 2-11). The gas molecules are very much like the molecules in the crystal, their interatomic distance being practically the same; but the distances between molecules are much larger in a gas than in a crystal.

It may seem surprising that molecules on the surface of a crystal should evaporate directly into a gas, instead of going first through the stage of being in a liquid layer; but in fact the process of slow evaporation of a crystalline substance is not uncommon. Solid pieces of camphor or of naphthalene (as used in moth balls, for example) left out in the air slowly decrease in size, because of the evaporation of molecules from the surface of the solid. Snow may disappear from the ground without melting, by evaporation of the ice crystals at a temperature below that of their melting point. Evaporation is accelerated if a wind is blowing, to take the water vapor away from the immediate neighborhood of the snow crystals, and to prevent the vapor from condensing again on the crystals.

The characteristic feature of a gas is that *its molecules are not held together, but are moving about freely, in a volume rather large compared with the volume of the molecules themselves.* The attractive forces between the molecules still operate whenever two molecules come close together, but usually these forces are negligibly small because the molecules are far apart.

Because of the freedom of motion of its molecules a specimen of gas does not have either definite shape or definite size. *A gas shapes itself to its container.*

Gases at ordinary pressure are very dilute—the molecules themselves constitute only about one one-thousandth of the total volume of the gas, the rest being empty space. Thus 1 g of solid iodine has a volume of about 0.2 cm^3 (its density* is 4.93 g cm^{-3}), whereas 1 g of iodine gas at 1 atm pressure and at the temperature 184°C (its boiling point) has a volume of 148 cm^3, over 700 times greater. The volume of all of the molecules in a gas is accordingly very small compared with the volume of the gas itself at ordinary pressure. On the other hand, the diameter of a gas molecule is not extremely small compared with the distance between molecules; in a gas at room temperature and 1 atm pressure the average distance from a molecule to its nearest neighbors is about ten times its molecular diameter, as indicated in the drawing of gaseous iodine, Figure 2-11.†

The Vapor Pressure of a Crystal

A crystal of iodine in an evacuated vessel will gradually change into iodine gas by the evaporation of molecules from its surface. Occasionally one of these free gas molecules will again strike the surface of the crystal, and it may stick to the surface, held by the van der Waals attraction of the other crystal molecules. This is called *condensation* of the gas molecules.

The rate at which molecules evaporate from a crystal surface is proportional to the area of the surface, but is essentially independent of the pressure of the surrounding gas, whereas the rate at which gas molecules strike the crystal surface is proportional to the area of the surface and also proportional to the concentration of molecules in the gas (the number of gas molecules in unit volume).

If some iodine crystals are put into a flask, which is then stoppered and allowed to stand at room temperature, it will soon be seen that the gas in the flask has become violet in color, showing that a quantity of iodine has evaporated. After a while it will be evident that the process of evaporation has apparently ceased, because the intensity of coloration of the gas will no longer increase, but will remain constant. This steady state is reached when the concentration of gas molecules becomes so great that the rate at which gas molecules strike the crystal surface and stay there is just

*It was mentioned in Section 1-4 that the density of a substance is the mass (weight) of a unit volume of the substance; in the metric system, grams per cubic centimeter or kilograms per cubic meter.

†You will remember that a cube 1 inch on edge has a diameter one tenth as great as that of a cube 10 inches on edge, an area one one-hundredth as great, and a volume one one-thousandth as great.

equal to the rate at which molecules leave the crystal surface. *The corresponding gas pressure is called the* **vapor pressure** *of the crystal.*

A steady state of such a sort is an example of *equilibrium*. It must be recognized that equilibrium does not represent a situation in which nothing is happening, but rather a situation in which opposing reactions are taking place at the same rate, so as to result in no overall change. This is indicated in Figure 2-15.

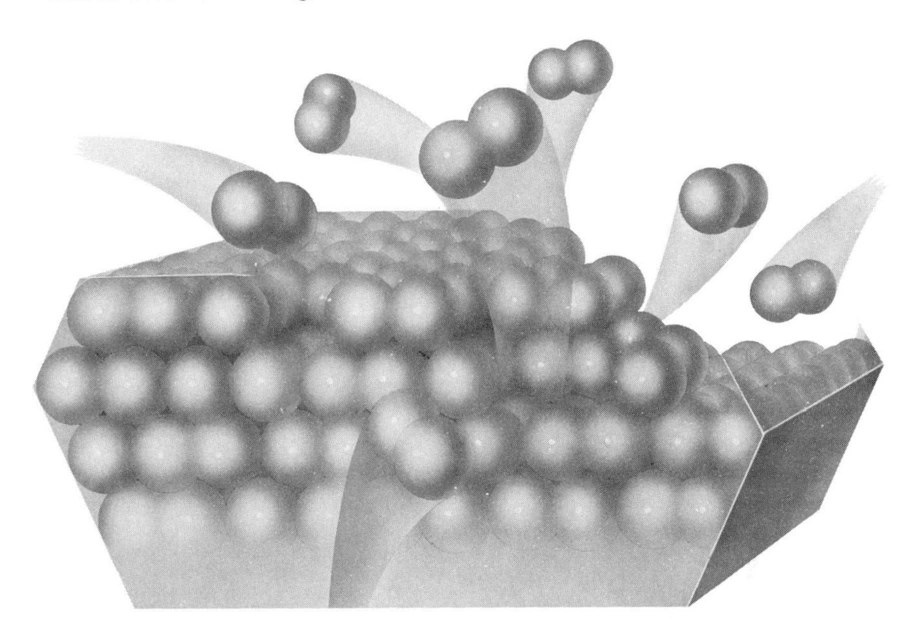

FIGURE 2-15
Equilibrium between molecules evaporating from an iodine crystal and gas molecules depositing on the crystal.

2-8. The Nature of a Liquid

When iodine crystals are heated to 114°C they melt, forming liquid iodine. The temperature at which the crystals and the liquid are in equilibrium—that is, at which there is no tendency for the crystals to melt or for the liquid to freeze—is called the *melting point* of the crystals, and the *freezing point* of the liquid. This temperature is 114°C for iodine.

Liquid iodine differs from the solid (crystals) mainly in its *fluidity*. It is like the gas in being able to adjust itself to the shape of its container. However, like the solid, and unlike the gas, it has a definite volume, 1 g occupying about 0.2 cm^3.

From the molecular viewpoint the process of melting can be described in the following way. As a crystal is heated its molecules become increasingly agitated, and move about more and more vigorously; but this thermal agitation does not carry any one molecule a significant distance away from the position fixed for it by the arrangement of its neighbors in the crystal. At the melting point the agitation finally becomes so great as to cause the molecules to slip by one another and to change somewhat their location relative to one another. They continue to stay close together, but do not continue to retain a regular fixed arrangement; instead, the grouping of molecules around a given molecule changes continually, sometimes being much like the close packing of the crystal, in which each iodine molecule has twelve near neighbors, and sometimes considerably different, the molecule having only ten or nine or eight near neighbors, as shown in Figure 2-11. Thus in a liquid, as in a crystal, the molecules are piled rather closely together; but whereas a crystal is characterized by regularity of atomic or molecular arrangement, a liquid is characterized by randomness of structure. The randomness of structure usually causes the density of a liquid to be somewhat less than that of the corresponding crystal; that is, the volume occupied by the liquid is usually somewhat greater than that occupied by the crystal.

The Vapor Pressure and Boiling Point of a Liquid

A liquid, like a crystal, is, at any temperature, in equilibrium with its own vapor when the vapor molecules are present in a certain concentration. The pressure corresponding to this concentration of gas molecules is called the *vapor pressure of the liquid* at the given temperature.

The vapor pressure of every liquid increases with increasing temperature. *The temperature at which the vapor pressure reaches a standard value (usually 1 atm) is called the* **boiling point** *of the liquid.* At this temperature it is possible for bubbles of the vapor to appear in the liquid and to escape to the surface.

The vapor pressure of liquid iodine reaches 1 atm at 184°C. Hence 184°C is the boiling point of iodine.

Other substances undergo similar changes when they are heated. When copper melts, at 1083°C, it forms liquid copper, in which the arrangement of the copper atoms shows the same sort of randomness as that of the molecules of liquid iodine. Under 1 atm pressure copper boils at 2310°C to form copper gas; the gas molecules are single copper atoms.

Note that it is customary to refer to the particles that move about in a gas as molecules even though each one may be only a single atom, as in the case of copper.

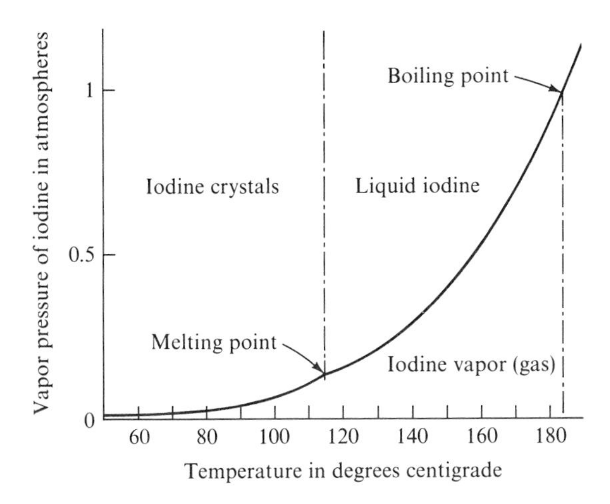

FIGURE 2-16
A graph showing the vapor-pressure curve of iodine crystal and the vapor-pressure curve of liquid iodine. The melting point of the crystal is the temperature at which the crystal and the liquid have the same vapor pressure, and the boiling point of the liquid (at 1 atm pressure) is the temperature at which the vapor pressure of the liquid equals 1 atm.

The Dependence of Vapor Pressure on Temperature

It has been found by experiment that the vapor pressure of crystals and liquids increases as the temperature is raised. Curves showing the vapor pressure of iodine crystals and liquid iodine are shown in Figure 2-16.

2-9. The Meaning of Temperature

In the preceding discussion the assumption has been made that molecules move more rapidly and violently at any given temperature than at a lower one. This assumption is correct—the temperature of a system is a measure of the vigor of motion of all the atoms and molecules in the system.

With increase in temperature there occurs increase in violence of molecular motion of all kinds. Gas molecules rotate more rapidly, and the atoms within a molecule oscillate more rapidly relative to one another. The atoms and molecules in liquids and solids carry out more vigorous vibrational motions. This vigorous motion at high temperatures may result in chemical reaction, especially decomposition of substances. Thus when iodine gas is heated to about 1200°C at 1 atm pressure about one-half of the molecules dissociate (split) into separate iodine atoms (Figure 2-17).

You can get a better understanding of many of the phenomena of chemistry by remembering that the absolute temperature is a measure of the vigor of the motion of atoms and molecules.

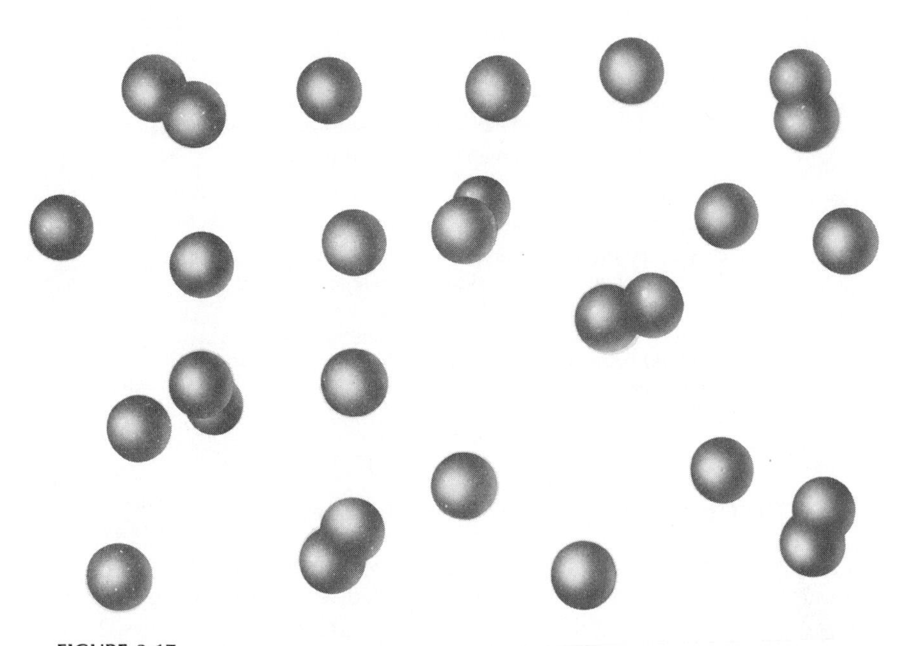

FIGURE 2-17
Iodine vapor at elevated temperature; this vapor contains both diatomic molecules (I_2) and monatomic molecules (I) of iodine.

EXERCISES

2-1. Classify the following statements as hypotheses, theories, laws, or facts:
(*a*) The interior of the moon consists of granite and similar silicate rocks.
(*b*) With a few exceptions, substances increase in volume on melting.
(*c*) The core of the earth is composed of a metallic form of hydrogen, which has not yet been prepared in the laboratory.
(*d*) Hydrogen, oxygen, nitrogen, and neon are all gases under ordinary conditions.
(*e*) All crystals are composed of atoms or molecules arranged in a regular way.
(*f*) Ordinary matter is composed of atoms.

2-2. Of what value are the law of constant proportions and the law of multiple proportions in arguing for the acceptance of the atomic theory?

2-3. Give an example of a solid material that is crystalline and of one that is not crystalline. Though the study of crystals for their own sake is not now as important as it was toward the end of the nineteenth century, why do you think there are now thousands of crystallographers analyzing the structures of crystals?

2-4. The crystal sodium chloride has a cubic unit of structure, with $a = 562.8$ pm. There are four sodium atoms in the unit of structure, with coordinates 0, 0, 0; 0, $\frac{1}{2}$, $\frac{1}{2}$; $\frac{1}{2}$, 0, $\frac{1}{2}$; $\frac{1}{2}$, $\frac{1}{2}$, 0; and also four chlorine atoms, with coordinates $\frac{1}{2}$, $\frac{1}{2}$, $\frac{1}{2}$; $\frac{1}{2}$, 0, 0; 0, $\frac{1}{2}$, 0; 0, 0, $\frac{1}{2}$. Make a drawing showing the cubic unit of structure and the positions of the atoms. How many nearest neighbors does each atom have, what is the interatomic distance for these neighbors, and what polyhedron is formed by them? (This arrangement of atoms, called the sodium chloride arrangement, is a common one for salts.)

2-5. The mineral fluorite, CaF_2, has a cubic unit of structure, with $a = 545$ pm. The unit cube contains four calcium atoms, with coordinates $\frac{1}{4}$, $\frac{1}{4}$, $\frac{1}{4}$; $\frac{1}{4}$, $\frac{3}{4}$, $\frac{3}{4}$; $\frac{3}{4}$, $\frac{1}{4}$, $\frac{3}{4}$; and $\frac{3}{4}$, $\frac{3}{4}$, $\frac{1}{4}$; and eight fluorine atoms, with coordinates 0, 0, 0; 0, $\frac{1}{2}$, $\frac{1}{2}$; $\frac{1}{2}$, 0, $\frac{1}{2}$; $\frac{1}{2}$, $\frac{1}{2}$, 0; $\frac{1}{2}$, 0, 0; 0, $\frac{1}{2}$, 0; 0, 0, $\frac{1}{2}$; and $\frac{1}{2}$, $\frac{1}{2}$, $\frac{1}{2}$. Make a drawing showing the positions of the atoms. How many nearest neighbors does each calcium atom have? each fluorine atom? What is the Ca-F distance? (Answer: 8, 4, 236 pm.)

2-6. The crystal cesium chloride, CsCl, is cubic. The cubic unit of structure has $a = 411$ pm. The atomic positions are Cs at 0, 0, 0 and Cl at $\frac{1}{2}$, $\frac{1}{2}$, $\frac{1}{2}$.
(*a*) Make a drawing showing the atomic positions.
(*b*) What is the smallest Cs-Cl distance? (Answer: 356 pm.)
(*c*) How many chlorine atoms are at this distance from each cesium atom?
(*d*) How many cesium atoms are at this distance from each chlorine atom?
(*e*) What is the smallest distance between chlorine atoms? How many chlorine atoms are at this distance?

2-7. The metal tin containing a small amount of impurity is reported to form hexagonal crystals with $a = b = 320$ pm and $c = 298$ pm, and with one atom of tin per unit cell. The coordinates of this atom can be taken to be $x = 0$, $y = 0$, $z = 0$.

(*a*) Make a drawing showing the axes *a* and *b*, at 120° to one another, in the plane $z = 0$. Outline the rhomb that is the base of one unit cell and locate the tin atom assigned to this cell.

(*b*) Outline the eight rhombs touching the first one. How many near neighbors in this plane does each tin atom have?

(*c*) At what distances are these neighbors?

(*d*) How many near neighbors does a tin atom have in the direction of the *c* axis (positive and negative)? At what distance? (Answer: (*b*) Six; (*c*) All at 320 pm; (*d*) Two, both at 298 pm.)

2-8. Define vapor pressure of a crystal, and also vapor pressure of a liquid. Can you think of an argument showing that these two vapor pressures of a substance must be equal at the melting point?

2-9. The vapor pressure of solid carbon dioxide at its melting point, −56.5°C, is 5 atm. How do you explain the fact that solid carbon dioxide when used for packing ice cream does not melt to form a liquid carbon dioxide? If you wanted to make some liquid carbon dioxide, what would you have to do?

2-10. Carbon dioxide (dry ice) consists of CO_2 molecules. These molecules are linear, with the carbon atom in the center. Make three drawings, representing your concepts of carbon dioxide gas, carbon dioxide liquid, and carbon dioxide crystal.

2-11. What is the effect of increase in pressure on the boiling point of a liquid? Estimate the boiling point of liquid iodine at a pressure of $\frac{1}{2}$ atm (see Figure 2-16).

3

The Electron and the Nuclei of Atoms

In the preceding chapter we have discussed the atomic theory and have seen that some of the properties of substances can be explained by this theory. The two substances copper and iodine, which were used as the principal examples in the discussion, have different properties because their atoms are different.

Chemists of the nineteenth century asked whether it might be possible to understand the differences between atoms of different elements, such as copper and iodine, but they were not able to answer the question. During the period 1897 to 1911, however, it was discovered that atoms themselves are composed of still smaller particles. The discovery of the components of atoms and the investigation of the structure of atoms—the ways in which atoms of different kinds are built of the smaller particles—constitute one of the most interesting stories in the history of science. Moreover, knowledge about the structure of atoms has during recent years permitted the facts of chemistry to be systematized in a striking way, making the subject easier to understand and to remember. The student of chemistry can be helped greatly in mastering his subject by first obtaining a good understanding of atomic structure.

The particles that constitute atoms are *electrons* and *atomic nuclei*. Electrons and atomic nuclei carry electric charges, and these electric charges are in large part responsible for the properties of the particles and for the structure of atoms. We shall accordingly begin this chapter with a discussion of the nature of electricity.

3-1. The Nature of Electricity

The ancient Greeks knew that when a piece of amber is rubbed with wool or fur it achieves the power of attracting light objects, such as feathers or bits of straw. This phenomenon was studied by William Gilbert (1540–1603), Queen Elizabeth I's physician, who invented the adjective *electric* to describe the force of attraction, after the Greek word *elektron*, meaning amber. Gilbert and many other scientists, including Benjamin Franklin, investigated electric phenomena, and during the nineteenth century many discoveries about the nature of electricity and of magnetism (which is closely related to electricity) were made.

It was found that if a rod of sealing wax, which behaves in the same way as amber, is rubbed with a woolen cloth, and a rod of glass is rubbed with a silken cloth, an electric spark will pass between the sealing-wax rod and the glass rod when they are brought near one another. Moreover, it was found that a force of attraction operates between them. If the sealing-wax rod that has been electrically charged by rubbing with a woolen cloth is suspended from a thread, as shown in Figure 3-1, and the charged glass rod is brought near one end of it, this end will turn toward

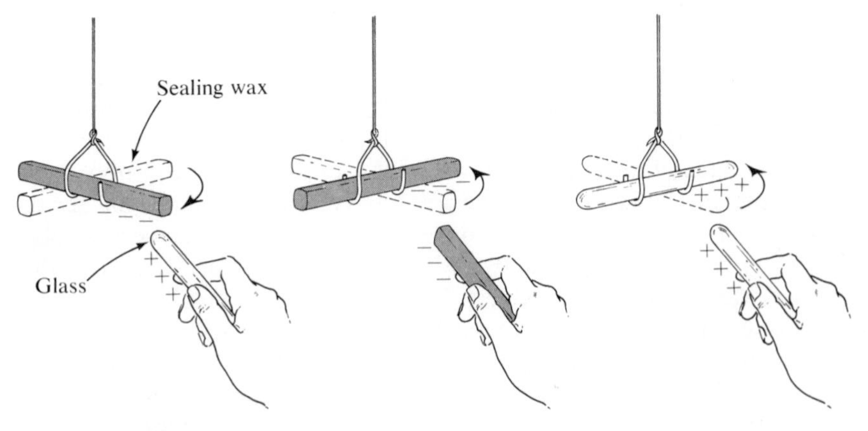

FIGURE 3-1
Experiments showing the attraction of unlike charges of electricity and the repulsion of like charges.

the glass rod. An electrified sealing-wax rod is repelled, however, by a similar sealing-wax rod, and also an electrified glass rod is repelled by a similar glass rod (Figure 3-1).

Through the experimental study of such phenomena the ideas were developed that there are two kinds of electricity, which were called resinous electricity (that which is picked up by the sealing-wax rod) and vitreous electricity (that which is picked up by the glass rod), and that the two kinds of electricity attract one another, whereas each kind repels itself. Franklin simplified this picture of electricity somewhat, by assuming that only one kind of electricity can flow from an object to another object. He assumed that when a glass rod is rubbed with a silken cloth this electric "fluid" is transferred from the cloth to the glass rod, and he described the glass rod as *positively charged,* meaning that it had an excess of the electric fluid. He described the cloth as having a deficiency of the electric fluid, and being *negatively charged.* He pointed out that he did not really know whether the electric fluid had been transferred from the silken cloth to the glass rod or from the glass rod to the silken cloth, and that accordingly the decision to describe vitreous electricity as positive (involving an excess of electric fluid) was an arbitrary one. We now know, in fact, that when the glass rod is rubbed with a silken cloth negatively charged particles, the electrons, are transferred from the glass rod to the silken cloth, and that Franklin thus made the wrong decision in his assumption.

3-2. The Electron

The idea that there are electric particles in substances was proposed, as a hypothesis, by G. Johnstone Stoney, an English scientist. Stoney knew that substances can be decomposed by an electric current—for example, water can be decomposed into hydrogen and oxygen in this way. He also knew that Michael Faraday had found that a definite amount of electricity is needed to liberate a certain amount of an element from one of its compounds. (The experiment carried out by Faraday will be discussed in Chapter 11.) In 1874, after thinking about these facts, Stoney stated that they indicate that *electricity exists in discrete units,* and that these units are associated with atoms. In 1891 he suggested the name *electron* for his postulated unit of electricity. The discovery of the electron by experiment was made in 1897 by Sir Joseph John Thomson (1856–1940), in Cambridge University, England.*

*The experiments that led to the discovery of the electron are described in Section 3-7.

The Properties of the Electron

The electron is a particle with a negative electric charge of magnitude -0.1602×10^{-18} coulombs.

The mass of the electron is 0.9108×10^{-30} kg, which is 1/1837 of the mass of the hydrogen atom.

The electron is very small. The radius of the electron has not been determined exactly, but it is known to be far less than 1×10^{-15} m.

In 1925 it was discovered that the electron spins about an axis and that it has a magnetic moment. The spin and magnetic moment are discussed in Section 5-5.

3-3. The Flow of Electricity in a Metal

Knowledge of the existence of electrons permits us to discuss some of the properties of electricity in a simple way.

In a metal or similar conductor of electricity there are electrons that have considerable freedom of motion and that move along between the atoms of the metal when an electric potential difference is applied. A direct current of electricity passing along a copper wire is a *flow of electrons* along the wire.

Let us call to mind the analogy between the flow of electricity along a wire and the flow of water in a pipe. *Quantity* of water is measured in liters or cubic feet; quantity of electricity is usually measured in *coulombs* (ampere seconds). *Rate of flow,* or *current,* of water, the quantity passing a given point of the pipe in unit time, is measured in liters per second, or cubic feet per second; current of electricity is measured in *amperes* (coulombs per second). The rate of flow of water in a pipe depends on the *difference in the pressures* at the two ends of the pipe, with atmospheres or pounds per square inch as units. The current of electricity in a wire depends on the *electric potential difference* or *voltage drop* between its ends, which is usually measured in *volts*. The definitions of the unit of quantity of electricity (the coulomb) and the unit of electric potential (the volt) have been made by international agreement.

An electric generator is essentially an electron pump, which pumps electrons out of one wire and into another. A generator of direct current pumps electrons continually in the same direction and one of alternating current reverses its pumping direction regularly, thus building up electron pressure first in one direction and then in the other. A 60-cycle generator reverses its pumping direction 120 times per second.

Example 3-1. An ordinary electric light bulb is operated under conditions such that one ampere is passing through the filament. How many electrons pass through the filament each second? (Remember that the charge of the electron is -0.1602×10^{-18} C).

Solution. One ampere is one coulomb per second. The number of electrons passing per second is one coulomb divided by the charge of the electron in coulombs:

$$\frac{1 \text{ C s}^{-1}}{0.1602 \times 10^{-18} \text{ C}} = 6.242 \times 10^{18} \text{ electrons per second}$$

3-4. The Nuclei of Atoms

In 1911 the British physicist Ernest Rutherford carried out some experiments* that showed that every atom contains, in addition to one or more electrons, another particle, called the *nucleus* of the atom. Every nucleus has a positive electric charge. It is very small, being only about 10^{-14} m in diameter, and it is very heavy—the lightest nucleus is 1836 times as heavy as an electron.

There are many different kinds of nuclei; those of the atoms of one element are different from those of every other element. The nucleus of the hydrogen atom has the same electric charge as the electron, but with opposite sign, positive instead of negative. The nuclei of other atoms have positive charges that are multiples of this fundamental charge.

3-5. The Proton and the Neutron

The *proton* is the simplest atomic nucleus. It is the nucleus of the most abundant kind of hydrogen atom, which is the lightest of all atoms.

The proton has an electric charge 0.1602×10^{-18} C. This charge is exactly the same as that of the electron, except that it is positive, whereas the charge of the electron is negative.

The mass of the proton is 1.672×10^{-27} kg. This is 1836 times the mass of the electron.

The *neutron* was discovered by the English physicist James Chadwick in 1932. The mass of the neutron is 1.675×10^{-27} kg, which is 1839 times the mass of the electron. The neutron has no electric charge.

It is customary for chemists to use an *atomic mass unit* or *dalton* (d), which is approximately the mass of the proton. Both the proton and the neutron have masses that are approximately one atomic mass unit.

---

*These experiments are described in later sections of this chapter.

3-6. The Structure of Atomic Nuclei

Several hundred different kinds of atomic nuclei are known to exist. Together with the electrons that surround them, they make up the atoms of the different chemical elements.

Although the detailed structures of nuclei are not completely known, physicists agree in accepting the idea that they can all be described as being built up of protons and neutrons.

Let us first discuss, as an example, the *deuteron.* This is the nucleus of the *heavy hydrogen atom,* or *deuterium atom.* The deuteron has the same electric charge as the proton, but has about twice the mass of the proton. It is thought that the deuteron is made of one proton and one neutron, as indicated in Figure 3-2.

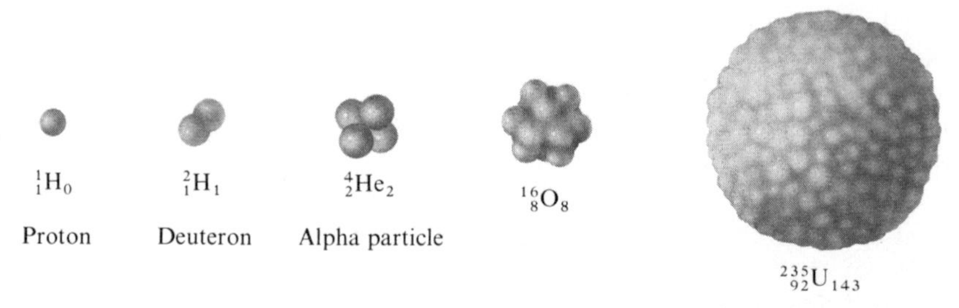

FIGURE 3-2
Hypothetical structures of some atomic nuclei. We do not yet know just how these nuclei are constructed out of elementary particles, but it is known that nuclei are 3 to 15 fm in diameter, and are, accordingly, very small even compared with atoms.

The nucleus of the helium atom, which is also called the *alpha particle* or *helion,* has electric charge twice as great as that of the proton, and mass about four times as great as that of the proton. It is thought that the alpha particle is composed of two protons and two neutrons.

In Figure 3-2 there is also shown a drawing representing the nucleus of an oxygen atom, composed of eight protons and eight neutrons. The electric charge of this nucleus is eight times the electric charge of the proton. This electric charge would accordingly be neutralized by the negative charges of eight electrons. The mass of this oxygen nucleus is about 16 d.

There is also shown in the figure a hypothetical drawing of the nucleus of a uranium atom. This nucleus is composed of 92 protons and 143 neutrons. The electric charge of this nucleus is 92 times that of the proton; it would be neutralized by the negative charges of 92 electrons. The mass of this nucleus is about 235 times the mass of the proton.

In thinking about atoms and atomic nuclei, you must remember that the drawings of atomic nuclei in Figure 3-2 correspond to a magnification ten thousand times greater than the drawings of atoms and molecules that are shown elsewhere in this book. The nuclei are very small, even compared with atoms.

3-7. The Experiments That Led to the Discovery of the Electron

Many interesting experiments involving electricity were carried out by physicists during the nineteenth century. These experiments ultimately led to the discovery of the electron. In order to understand them it is necessary to know something about the way in which the motion of an electrically charged particle is affected by other electric charges or by a magnet.

The Interaction of an Electric Charge with Other Electric Charges and with Magnets

An electric charge is said to be surrounded by an *electric field,* which exercises a force, either of attraction or of repulsion, on any other electric charge in its neighborhood. The strength of an electric field can be measured by determining the force that operates on a unit of electric charge.

In experimental work use is often made of an apparatus like that shown in Figure 3-3, in which two large parallel plates of metal are held a small, constant distance from one another. By use of a battery or generator of

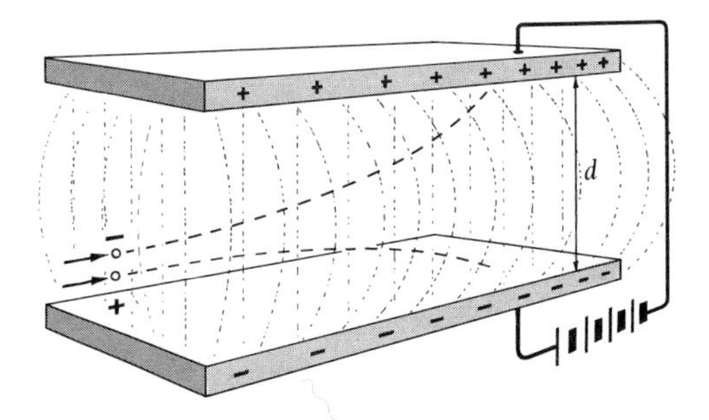

FIGURE 3-3
The motion of an electrically charged particle in the uniform electric field between charged plates.

electricity, one of these parallel plates is charged positively (that is, some electrons are taken away from it), and the other is charged negatively.

A wire or plate that has an excess of positive charge is called an *anode*. A wire or plate that has an excess of negative electric charge is called a *cathode*. In Figure 3-3 the upper plate is the anode and the lower plate is the cathode.

A particle with negative electric charge placed between the plates would be attracted toward the upper plate and repelled from the lower plate. It would accordingly move in the direction of the upper plate. Similarly, a particle with positive electric charge placed between the plates would move toward the lower plate.

The force exerted on a positive charge by the electric field between the plates has the same effect as the force exerted on a mass by the gravitational field of the earth. Accordingly, a positively charged particle shot into the region between the plates, as indicated in Figure 3-3, would fall to the bottom plate along the path indicated by the dashed line, in the same way that a rock thrown horizontally would fall toward the surface of the earth.

You know that a piece of iron or steel can be magnetized, to form a *magnet*, and that the magnet has the power of attracting other pieces of iron. A magnet also has the power of exerting a force on any electrically charged particle that moves by it. A magnet can therefore also be used to study particles.

No force is exerted on a stationary electric charge by a constant magnetic field. A force is exerted on an electric charge moving in a magnetic field. This is illustrated in Figure 3-4. The poles of the magnet are marked *N* (north-seeking pole) and *S* (south-seeking pole); the lines of force are indicated as going from the north-seeking pole to the south-seeking pole.

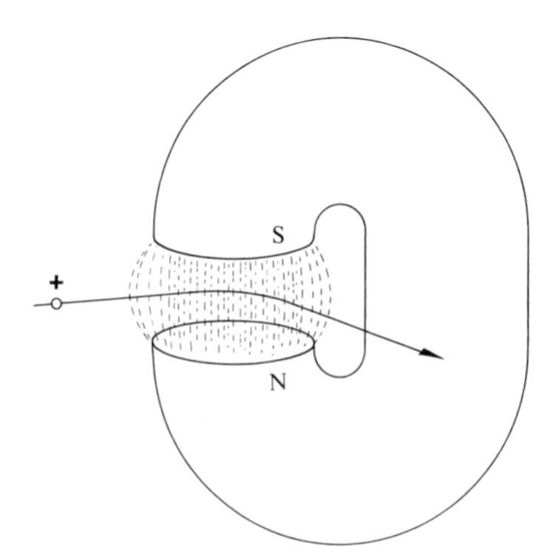

FIGURE 3-4
The path of a moving electric charge in a magnetic field.

A positively charged particle is shown as moving through the field from the left to the right. The nature of electricity and magnetism is such that a force operates that is proportional to the strength of the magnetic field, the quantity of electric charge on the particle, and the speed of the particle; the direction of this force is at right angles to the plane formed by the direction of motion of the moving particle and the direction of the lines of force of the magnetic field, its sense being out of the plane of the paper. This causes the moving charged particle to be deflected to the front, as indicated in the drawing.

The Explanation of Electricity and Magnetism

You may well ask how scientists explain the fact that an electron is repelled by another electron, that in general two electrically charged bodies repel or attract one another. How do scientists explain the still more extraordinary fact that an electric charge that is moving through a magnetic field is pushed to one side by interaction of its charge with the field? The answer to these questions is that there is no explanation. These properties of electric charges and of electric and magnetic fields are simply a part of the world in which we live.

The discovery that a lateral force is exerted on an electric charge moving through a magnetic field has led to very important practical applications. The ordinary electric generator (dynamo) produces electricity because of this fact. In an electric generator a wire is moved rapidly through a magnetic field, in a direction perpendicular to the lengthwise direction of the wire. Moving the wire (including the electrons in it) through the magnetic field causes the electrons to be set in motion, relative to the atoms in the wire, in the direction toward one end of the wire—namely, the end indicated by the rule of Figure 3-4. In this way a flow of electrons (current of electricity) is produced in the wire. Practically all the electric power that is used in the world is produced by this method, the energy for operating the electric generators (for moving the wire through the magnetic field) being provided by the fall of water in the earth's gravitational field, by the combustion of coal or oil to drive steam engines, or by nuclear reactions. A small amount of electric power is produced directly from chemical energy, as will be discussed in Chapter 11.

The Discovery of the Electron

During the nineteenth century many physicists carried out experiments on the conduction of electricity through gases. For example, if a glass tube about 50 cm long is fitted with electrodes and a potential of about 10,000 volts is applied between the electrodes, no electricity is at first conducted between the electrodes. If, however, some of the air in the tube is pumped out, electricity begins to be conducted through the tube. While the electricity is being conducted through the tube light is emitted by the gas in the tube. You are familiar with this phenomenon, because you have seen many neon lamps in street signs. These neon lamps contain the gas neon, or some other gas, which is caused to emit light when electricity is conducted through the gas.

As the pressure of gas in the tube is further decreased, a dark space appears in the neighborhood of the cathode, and alternate light and dark regions are observed in the rest of the tube. At still lower pressure the dark space increases in size until it fills the whole tube. At this pressure no light is given out by the gas that is still present in very small quantity within the tube, but the glass of the tube itself glows with a faint greenish light.

It was discovered that the greenish light coming from the glass is due to the bombardment of the glass by rays liberated at the cathode. These rays, called *cathode rays,* travel in straight lines from the cathode to the glass. This is shown by the experiment illustrated in Figure 3-5: an object placed within the tube, such as the cross shown in this figure, casts a shadow on the glass—the glass fluoresces everywhere except in the region of this shadow.

It was shown by the French scientist Jean Perrin (1870–1942) in 1895 that these cathode rays consist of particles with a negative electric charge, rather than a positive charge. His experiment is illustrated in Figure 3-6. He introduced a shield with a slit in the tube, so as to form a beam of cathode rays. He also placed a fluorescent screen* in the tube, so that the path of the beam could be followed by the trace of the fluorescence. When a magnet was placed near the tube the beam was observed to be deflected in the direction corresponding to the presence of a negative charge on the particles.

J. J. Thomson then carried out some experiments that permitted him to make some quantitative statements about the particles that constitute the cathode rays. He used the apparatus shown in Figure 3-7, in which a beam of cathode rays can be affected either by a magnet that is brought up beside the tube, or by an electric field, produced by applying an electric potential to the two metal plates in the tube, or by both the magnet and the electric field. The effect on the beam of cathode rays was observed by use of a fluorescent screen. The results of his experiment convinced Thomson that the cathode-ray particles constitute a form of matter different from ordinary forms of matter. The particles were indicated by

*A fluorescent screen is a sheet of paper or glass coated with a substance that sends out a flash of light when it is struck by a fast-moving electron or other energetic particle.

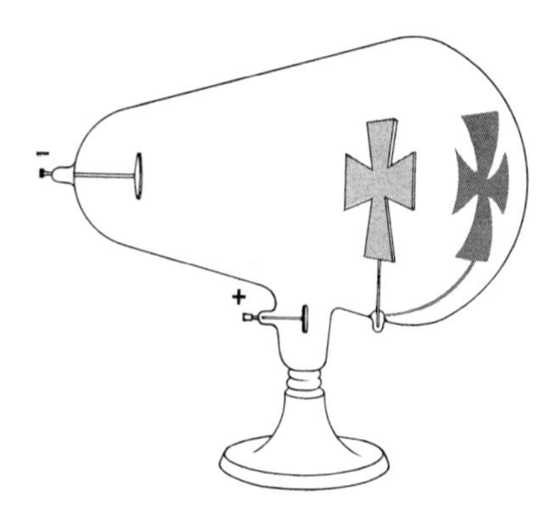

FIGURE 3-5
Experiment showing that cathode rays, starting from the cathode at the left, move through the Crookes tube in straight lines.

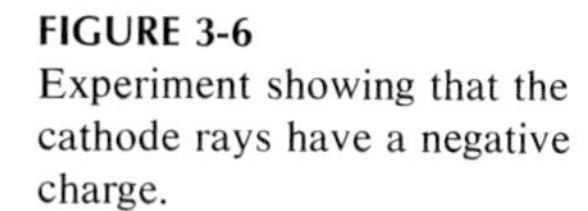
FIGURE 3-6
Experiment showing that the cathode rays have a negative charge.

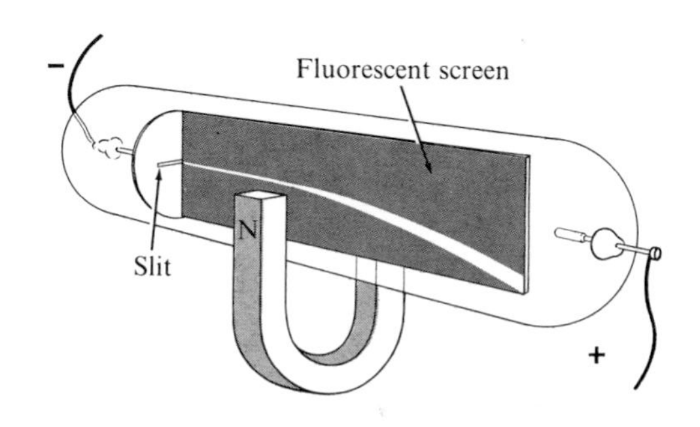

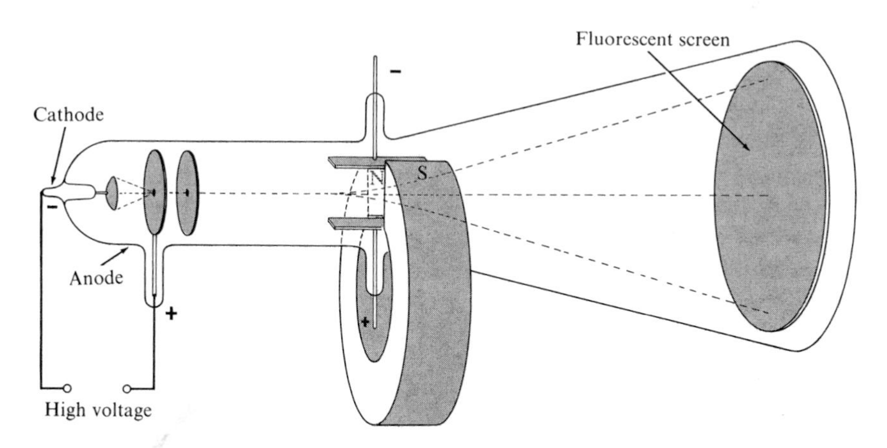

FIGURE 3-7
The apparatus used by J. J. Thomson to determine the ratio of electric charge to mass of the cathode rays, through the simultaneous deflection of the rays by an electric field and a magnetic field.

Thomson's experiments to be much lighter than atoms. Later and more accurate experiments showed that the mass of the cathode-ray particle is only 1/1837 times the mass of the hydrogen atom.

Although other investigators had carried out important experiments on cathode rays, the quantitative experiments by Thomson provided the first convincing evidence that these rays consist of particles (electrons) much lighter than atoms, and Thomson is hence given the credit for discovering the electron.

The Determination of the Charge of the Electron

After the discovery of the electron by Thomson, many investigators worked on the problem of determining accurately the charge of the electron. The American physicist R. A. Millikan (1868–1953), who began his experiments in 1906, was the most successful of the earlier experimenters. By means of his oil-drop experiment he determined in 1909 the value of the charge of the electron to within 1%.

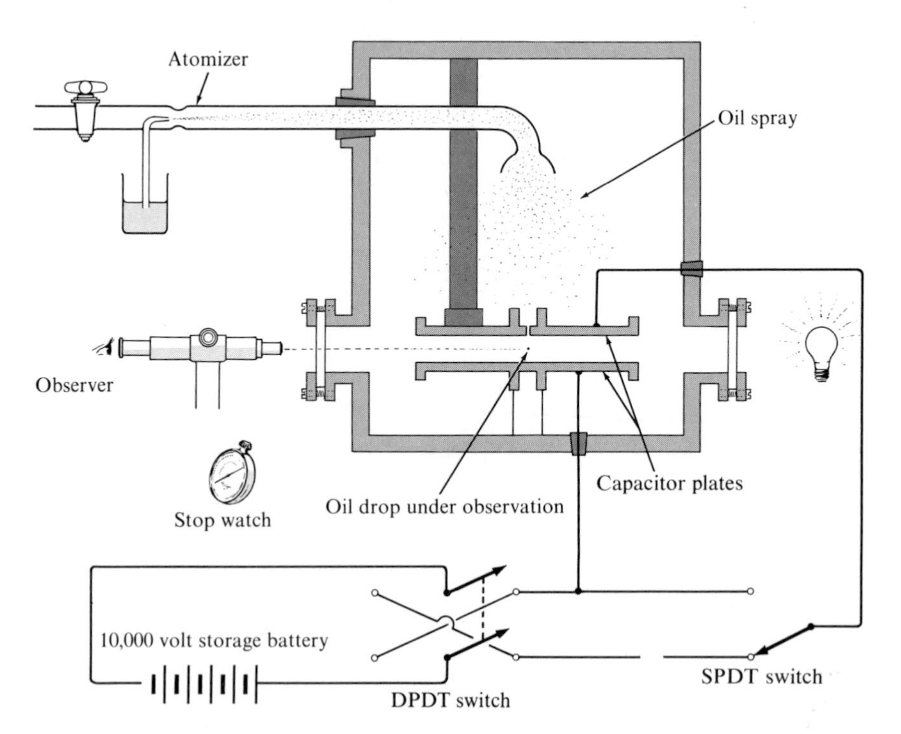

FIGURE 3-8
A diagram of the apparatus used by R. A. Millikan in determining the charge of the electron by the oil-drop method.

The apparatus that he used is illustrated in Figure 3-8. Small drops of oil are formed by a sprayer, and some of them attach themselves to electrons that have been separated from molecules by action of a beam of x-rays. The experimenter watches one of these small oil drops through a microscope. He first measures the rate at which it falls in the earth's gravitational field. The small drops fall at a rate determined by their size, and measurement of the rate of fall of a drop permits the investigator to calculate the size.

When the electric field is turned on, by charging the plates above and below the region where the oil drops are moving, some of the drops, which carry no electric charge, continue to fall as before. Other drops, carrying electric charges, change their speed, and may rise, being pulled up by the attraction of the electric charge for the oppositely charged upper plate in the apparatus. The changed speed of a drop that has been watched falling is then observed. From these measurements, the magnitude of the electric charge on the drop can be calculated.

Since Millikan carried out his work a number of other methods have been developed for determining the charge of the electron, and its value is now known to about 0.001%.

3-8. The Discovery of X-rays and Radioactivity

Several great scientific discoveries were made in a period of a few years, beginning in 1895. These discoveries made great changes in chemistry as well as in physics. X-rays were discovered in 1895, radioactivity was discovered in 1896, the new radioactive elements polonium and radium were isolated in the same year, and the electron was discovered in 1897.

Wilhelm Konrad Röntgen (1845–1923), Professor of Physics in the University of Würzburg, Germany, reported in 1895 that he had discovered a new kind of rays, which he called x-rays. These rays are produced when electricity is passed through an evacuated tube. The rays radiate from the place where the electrons strike the glass. They have the power of passing through matter that is opaque to ordinary light, and of exposing a photographic plate. Within a few weeks after the announcement of this great discovery, x-rays were being used by physicians for the investigation of patients with broken bones and other disorders.

Soon after the discovery of x-rays the French physicist Henri Becquerel (1852–1908) investigated some minerals containing uranium. He found that these minerals emit rays that, like x-rays, can pass through black paper and other opaque materials and expose a photographic plate. He also found that the radiation produced by the uranium minerals

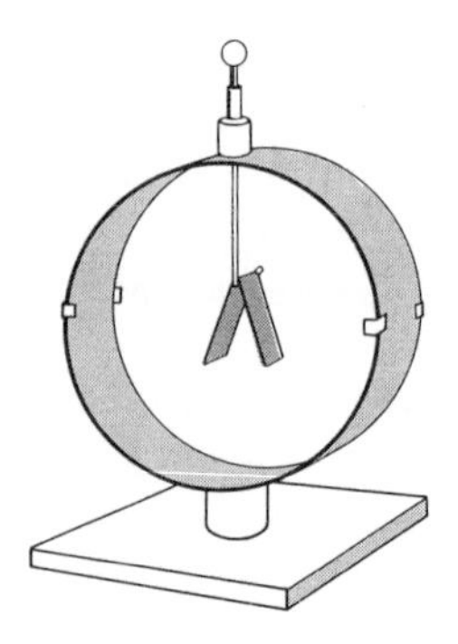

FIGURE 3-9
A simple electroscope. When an electric charge is present on the gold foil and its support, the two leaves of the foil separate, because of the repulsion of like electric charges.

could, like x-rays, discharge an electroscope (Figure 3-9), by making the air conductive.

Marie Sklodowska Curie (1867–1934) then began a systematic investigation of "Becquerel radiation," using the electroscope as a test. She investigated many substances, to see if they were similar to uranium in producing rays. She found that natural pitchblende, an ore of uranium, is several times more active than purified uranium oxide.

With her husband, Professor Pierre Curie (1859–1906), she began to separate pitchblende into fractions and to determine their activity in discharging the electroscope. She isolated a fraction that was 400 times more active than uranium. This fraction consisted largely of bismuth sulfide. Since pure bismuth sulfide is not radioactive, she assumed that a new, strongly radioactive element, similar in chemical properties to bismuth, was present as a contaminant. This element, which she named *polonium,* was the first element discovered through its properties of radioactivity. In the same year, 1896, the Curies isolated another new radioactive element, which they named *radium.*

In 1899 Ernest Rutherford reported that the radiation from uranium is of at least two distinct types, which he called alpha radiation and beta radiation. A French investigator, P. Villard, soon reported that a third kind of radiation, gamma radiation, is also emitted.

Alpha, Beta, and Gamma Rays

The experiments showing the presence of three kinds of rays emitted by natural radioactive materials are illustrated by Figure 3-10. The rays, formed into a beam by passing along a narrow hole in a lead block, traverse a strong magnetic field. They are affected in three different ways, showing that the three kinds of rays have different electric charges. Alpha rays carry a positive electric charge. Beta rays carry a negative electric charge, and are deflected by a magnet in the opposite direction to the alpha rays. Gamma rays do not carry an electric charge, and are not deflected by the magnet.

Rutherford found that the alpha rays, after they are slowed down, produce the gas helium. Further studies made by him showed definitely that the *alpha rays are the positively charged parts of helium atoms,* moving at high speeds. The *beta rays are electrons,* also moving at high speeds—they are similar in nature to the cathode rays produced in an electric discharge tube. *Gamma rays are a form of radiant energy, similar to visible light.* They are identical with x-rays produced in an x-ray tube operated at very high voltage.

The identification of the positively charged alpha particles with helium atoms was made by Rutherford by an experiment in which he allowed alpha particles to be shot through a thin metal foil into a chamber, and later was able to show that helium was present in the chamber. He could, moreover, correlate the amount of helium in the chamber with the number of alpha particles that had passed through the foil.

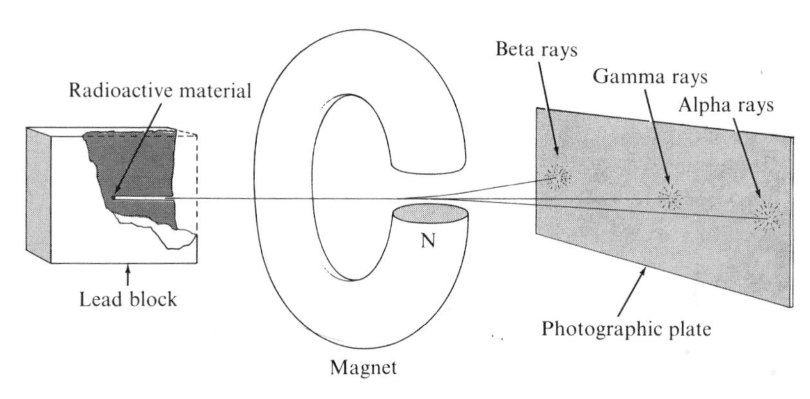

FIGURE 3-10
The deflection of alpha rays and beta rays by a magnetic field.

3-9. The Discovery of the Nuclei of Atoms

In 1911 Rutherford pointed out that an experiment showed that most of the mass of atoms is concentrated in particles that are very small in size compared with the atoms themselves.

The experiment consisted in bombarding a film of some substance, a piece of metal foil, with a stream of fast-moving alpha particles, and observing the direction in which the alpha particles rebound from the atoms.

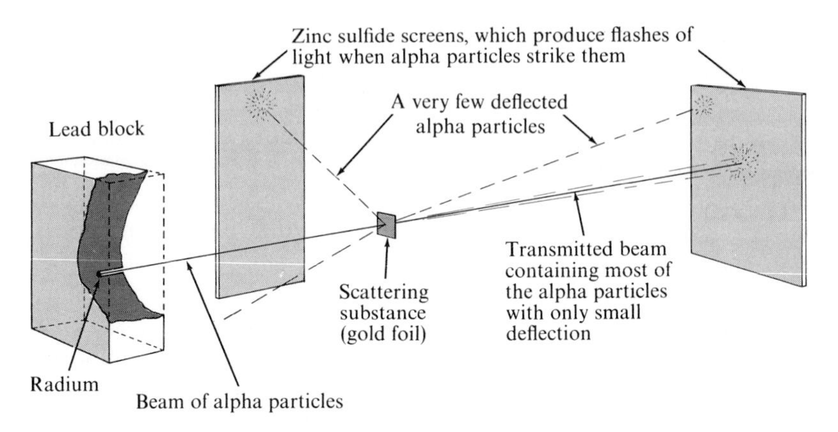

FIGURE 3-11
A diagram representing the experiment that showed that atoms contain very small, heavy atomic nuclei.

The nature of the experiment is indicated by the drawing in Figure 3-11. A piece of radium emits alpha particles in all directions. A narrow hole in a lead block defines a beam of the alpha particles. This beam of alpha particles then passes through the metal foil, and the directions in which the alpha particles continue to move are observed. The direction in which an alpha particle moves can be detected by use of a screen coated with zinc sulfide. When an alpha particle strikes the screen a flash of light is sent out.

If the atoms bombarded with alpha particles were solid throughout their volume we should expect all of the alpha particles in the beam to be deflected to some extent. Actually, however, Rutherford observed that most of the alpha particles passed through the metal foil without appreciable deflection: in one experiment, in which the alpha particles were sent through a gold foil 400 nm thick, so that they penetrated about 1000 layers of atoms, only about one alpha particle in 100,000 was deflected. This one usually showed a great deflection, often through more than 90°, as indicated in the figure. When foil twice as thick was taken, it was found that about twice as many alpha particles showed deflection through large angles, with most of them still passing straight through.

These experimental results can be understood if the assumption is made that *most of the mass of the atom is concentrated into a very small particle,* which Rutherford called the atomic nucleus. If the alpha particle were also very small, then the chance of collision of these two very small particles as the alpha particle passed through the atom would be small. Most of the alpha particles could pass through the foil without striking any atomic nucleus, and these alpha particles would not then be deflected.

Since about one particle in 100,000 is deflected on passing through a foil consisting of 1000 atom layers, only about one particle in 100,000,000 would be deflected by a single layer of atoms. Rutherford concluded from this that the heavy nucleus has a cross-sectional area only 0.00000001 as great as the cross-sectional area of the atom, and hence that the diameter of the nucleus is only 1/10,000 as great as the diameter of the atom (the square root of 0.00000001 is 1/10,000).

The picture of the atom that has been developed from this experiment and similar experiments is indeed an extraordinary one. If we could magnify a piece of gold leaf by the linear factor 1,000,000,000—a billionfold—we would see it as an immense pile of atoms about two feet in diameter, each atom thus being about as big as a bushel basket. Practically the entire mass of each atom would, however, be concentrated in a single particle, the nucleus, about 0.001 inch in diameter, like an extremely small grain of sand. This nucleus would be surrounded by electrons, equally small, and moving very rapidly about. Rutherford's experiment would correspond to shooting through a pile of these bushel-basket atoms a stream of minute grains of sand, each of which would continue in a straight line unless it happened to collide with one of the minute grains of sand representing the nuclei of the atoms. It is obvious that the chance of such a collision would be very small. (The alpha particles are not deflected by the electrons in the atoms, because they are very much heavier than the electrons.)

Because of the new knowledge about the nature of atoms that it led to, Rutherford's experiment must be considered one of the most important experiments that any man has ever made.

3-10. The Quantum Theory of Light. The Photon

Light (electromagnetic radiation) plays such an important part in many of the phenomena of chemistry that it is necessary to discuss its nature and some of its properties.

Waves and Their Interference

During the nineteenth century it was recognized that light can be produced by moving an electric charge back and forth in an oscillatory manner. The motion of the electric charge produces an oscillatory change in the electric field surrounding the charge, and this change is transmitted through space with the velocity of light, 3.00×10^8 m s^{-1}.

The nature of the wave motion is represented by the sine curve shown in Figure 3-12. This curve might represent, for example, the instantaneous contour of waves on the surface of the ocean. The distance between one crest and an adjacent crest is called the *wavelength,* usually

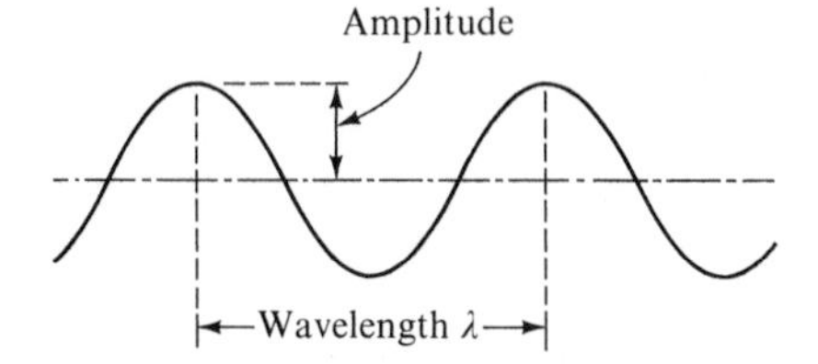

FIGURE 3-12
Diagram representing wave motion.

represented by the symbol λ (Greek letter lambda). The *amplitude* of the wave is the height of the crest, which is also the depth of the trough, with reference to the average level. If the waves are moving with the velocity c m s^{-1}, the frequency of the waves, represented by the symbol ν (Greek letter nu), is equal to c/λ; that is, it is the number of waves that pass by a fixed point in unit time (1 s). The dimensions of the wavelength are those of length. The dimensions of frequency, number of waves per second, are [time^{-1}]. We see that the product of wavelength and frequency has the dimensions [length] [time^{-1}]—that is, the dimensions of velocity. The equation connecting wavelength λ, frequency ν, and velocity c is

$$\lambda\nu = c \tag{3-1}$$

For a light wave the sine curve shown in Figure 3-12 is considered to represent the magnitude of the electric field in space. The electric field of a light wave is perpendicular to the direction of motion of the beam of light.

The phenomenon of *interference of waves* is used to determine the wavelength of light waves and x-rays. This phenomenon can be illustrated by Figures 3-13 and 3-14. In Figure 3-13 there is shown a set of water waves approaching a jetty in which there is a small opening. The

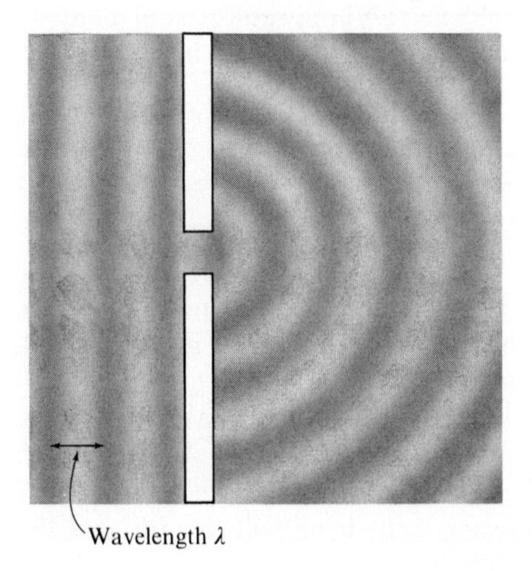

FIGURE 3-13
Diagram representing waves on the surface of water, from the left, striking a pier; waves propagated through the opening in the pier then spread out in circles.

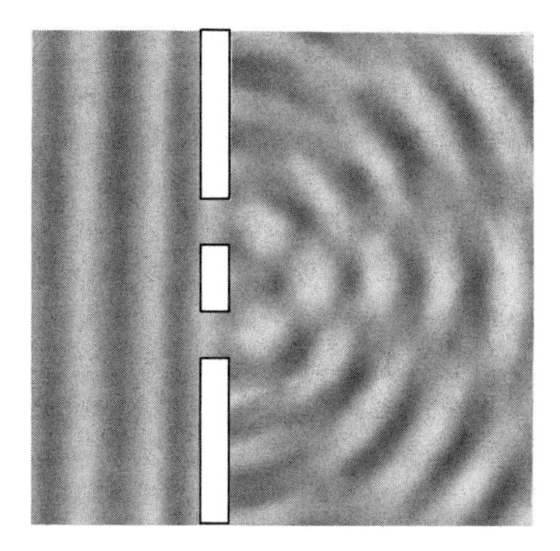

FIGURE 3-14
The interference and reinforcement of two sets of circular waves, from two openings.

waves that strike the jetty dissipate their energy among the rocks of the jetty, but the part of the waves that strikes the opening causes a disturbance on the other side of the jetty. This disturbance is in the form of a set of circular waves that spread out from the opening of the jetty. The wavelength of these circular waves is the same as the wavelength of the incident water waves. When light or x-rays strike atoms, part of the energy of the incident light is scattered by the atoms. Each atom scatters a set of circular waves. If two atoms that are excited by the same incident waves scatter light, as illustrated in Figure 3-14, there are certain directions in which the circular waves (spherical waves for atoms in three-dimensional space) from the two scattering centers reinforce one another, producing waves with twice the amplitude of either set, and other directions in which the trough of one set of waves coincides with the crest of the other set, and interference occurs. The directions of reinforcement and interference for two sets of circular waves are shown in the figure.

The wavelengths in the radio region are found to be 1 meter or more, in the microwave region around 1 cm, in the infrared region about 800 nm to 1 mm, in the visible region 380 to 800 nm, in the ultraviolet region about 10 nm to 380 nm, in the x-ray region about 10 pm to 10 nm, and in the gamma-ray region less than about 10 pm. In the infrared, visible, and ultraviolet regions the wavelengths have been determined by the use of a prism or a ruled grating (Figure 3-15), and the wavelengths of x-rays and gamma rays have been determined by diffraction from a crystal grating. The whole spectrum of light waves (electromagnetic waves) is shown in Figure 19-6, and the sequence of colors in the visible region is also shown, in the diagram next to the top one in this figure.

When gases are heated or are excited by the passage of an electric spark, the atoms and molecules in the gases emit light of definite wavelengths. The light that is emitted by an atom or molecule under these conditions is said to constitute its *emission spectrum*. The emission spectra of the alkali metals, mercury, and neon are shown in Figure 19-6. The emission spectra of elements, especially of the metals, can be used for identifying them, and *spectroscopic chemical analysis* is an important technique of analytical chemistry. The *spectroscope* is an instrument, using a ruled grating or a prism, for analyzing light into its constituent wavelengths and determining their values. A simple spectroscope is shown in Figure 3-15. An instrument of this sort was used by the German chemist Robert Wilhelm Bunsen (1811–1899) to discover rubidium and cesium, in 1860. The instrument had been invented by the physicist Kirchhoff just the year before, and cesium was the first element to be discovered with its use.

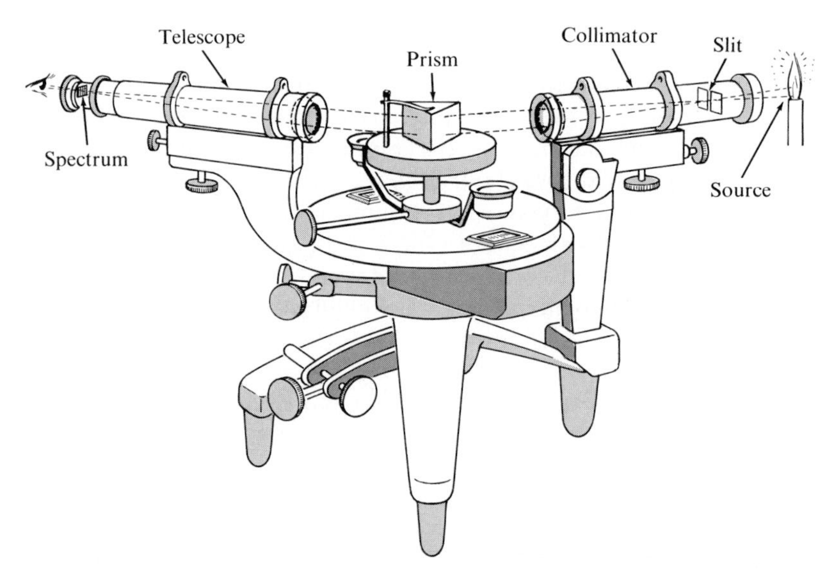

FIGURE 3-15
A simple spectroscope. The light from the source is refracted into a spectrum by use of a glass prism; it could instead be diffracted into a spectrum by use of a ruled grating, in place of the prism.

Planck's Constant and the Photon

During the last years of the nineteenth century it was found that the light that emerges through a hole from the hollow center of a hot body does not show characteristic emission lines, but has a smooth distribution of intensity with wavelength, characteristic of the temperature but independent of the nature of the hot body. The theoretical physicists who were interested in the problem of the emission of light by hot bodies, during the years before 1900, found that they were unable to account for the observed intensity distribution on the basis of the emission and absorption of light by vibrating molecules in the hot body. The German physicist Max Planck (1858–1947) then discovered that a satisfactory theory could be formulated if the assumption were made that the hot body cannot emit or absorb light of a given wavelength in arbitrarily small amount, but must emit or absorb a certain quantum of energy of light of that wavelength. Although Planck's theory did not require that the light itself be considered as consisting of bundles of energy—*light quanta* or *photons*—it was soon pointed out by Einstein (in 1905) that other evidence supports this concept.

The amount of light energy of wavelength λ absorbed or emitted by a solid body in a single act was found by Planck to be proportional to the frequency ν (equal to c/λ):

$$E = h\nu \tag{3-2}$$

In this equation E is the amount of energy of light with frequency ν emitted or absorbed in a single act, and h is the constant of proportionality. *This constant h is a very important constant; it is one of the fundamental constants of nature, and the basis of the whole quantum theory.* It is called **Planck's constant.** Its value is

$$\mathbf{h = 0.66252 \times 10^{-33}\ J\ s} \tag{3-3}$$

(The units of h, J s, have the dimensions of energy times time, as is required by Equation 3-2.)

We see that light of short wavelength consists of large bundles of energy and light of long wavelength of small bundles of energy. Some of the experiments in which these bundles of energy express their magnitudes will be discussed in the following section.

The Photoelectric Effect

In 1887 the German physicist Heinrich Hertz (1857–1894), who discovered radiowaves, observed that a spark passes between two metal electrodes at a lower voltage when ultraviolet light is shining on the electrodes than when they are not illuminated. It was then discovered by J. J. Thomson in 1898 that negative electric charges are emitted by a metal surface on which ultraviolet light impinges. A simple experiment to show this effect is represented in Figure 3-16. An electroscope is negatively charged, and ultraviolet light is allowed to fall on the zinc plate

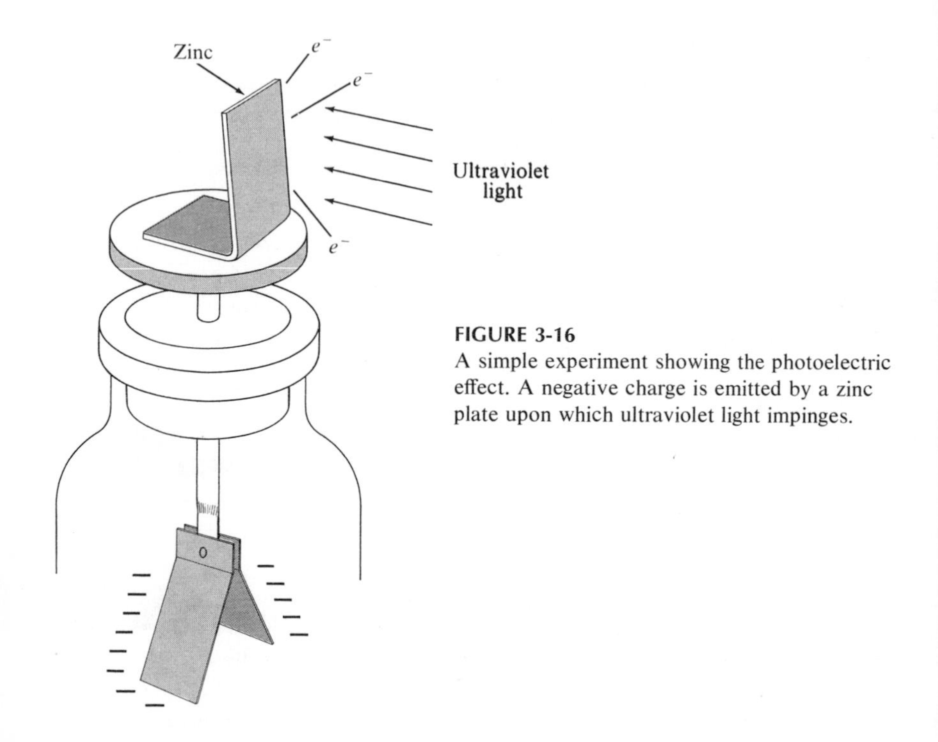

FIGURE 3-16
A simple experiment showing the photoelectric effect. A negative charge is emitted by a zinc plate upon which ultraviolet light impinges.

in contact with it. The leaves of the electroscope fall, showing that the negative electric charge is being removed under the action of the ultraviolet light. If the electroscope has a large positive charge the leaves do not fall, showing that positive charges are not emitted under similar conditions. An uncharged electroscope becomes charged when the metal plate is illuminated with ultraviolet light, and the charge that remains on the leaves of the electroscope is a positive charge, showing that negative charges have left the metal.

J. J. Thomson was able to show that the negative electric charge that leaves the zinc plate under the influence of ultraviolet light consists of electrons. The emission of electrons by action of ultraviolet light or x-rays is called the *photoelectric effect.* The electrons that are given off by the metal plate are called *photoelectrons;* they are not different in character from other electrons.

A great deal was learned through study of the photoelectric effect. It was soon found that visible light falling on a zinc plate does not cause the emission of photoelectrons, whereas ultraviolet light with a wavelength shorter than about 350 nm does cause their emission. The maximum wavelength that is effective is called the *photoelectric threshold.*

Substances differ in their photoelectric thresholds: the alkali metals are especially good photoelectric emitters, and their thresholds lie in the visible region; that for sodium is about 650 nm, so that visible light is effective with this metal except at the red end of the spectrum.

It was discovered that the photoelectrons are emitted with extra kinetic energy, depending upon the wavelength of the light. An apparatus somewhat like the photoelectric cell shown in Figure 3-17 can be used for this purpose. In this apparatus the photoelectrons that are emitted when the metal is illuminated are collected by a collecting electrode, and the number of them that strike the electrode can be found by measuring the current that flows along the wire to the electrode. A potential difference can be applied between the electrode and the emitting metal. If the collecting electrode is given a slight negative potential, which requires work to be done on the electrons to transfer them from the emitting metal to the collecting electrode, the flow of photoelectrons to the collecting electrode is stopped if the incident light has a wavelength close to the threshold, but it continues if the incident light has a wavelength much shorter than the threshold wavelength. By increasing the negative charge on the collecting electrode the potential difference can be made great enough to stop the flow of photoelectrons to the electrode.

These observations were explained by Einstein in 1905, by his theory of the photoelectric effect. He assumed that the light that impinges on the metal plate consists of photons with energy $h\nu$, and that when the light is absorbed by the metal all of the energy of one photon is converted into energy of a photoelectron. However, the electron must have a certain amount of energy to escape from the metal. This may be represented by the symbol E_i (the energy of ionizing the metal). The remaining energy is kinetic energy of the photoelectron. The *Einstein photoelectric equation* is

$$h\nu = E_i + \tfrac{1}{2}mv^2 \tag{3-4}$$

where m is the mass of the electron and v is its velocity. This famous

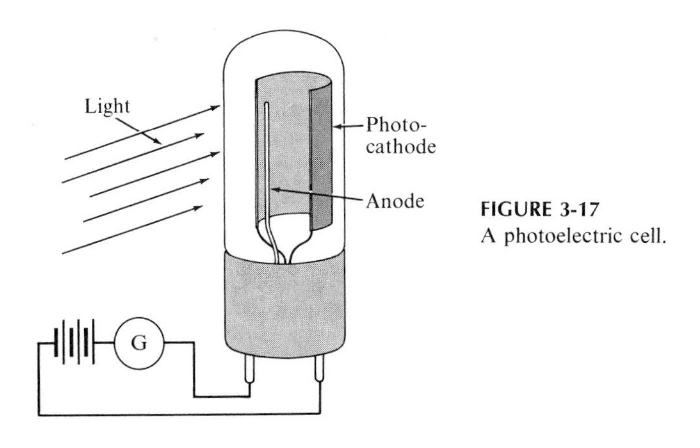

FIGURE 3-17
A photoelectric cell.

equation states that the energy of the light quantum, $h\nu$, is equal to the energy required to remove the electron from the metal, E_i, plus the kinetic energy imparted to the electron, $\frac{1}{2}mv^2$. The success of this equation in explaining the observations of the photoelectric effect was largely responsible for the acceptance of the idea of light quanta.

It is difficult to measure the velocity of the electrons directly. Instead, the energy quantity $\frac{1}{2}mv^2$ is measured by measuring the potential difference, V, which is necessary to keep the photoelectrons from striking the collecting electrode; the product of the potential difference V and the charge of the electron, e, is the amount of work done against the electrostatic field, and when V has just the value required to prevent the electrons from reaching the collecting plate the following relation holds:

$$eV = \tfrac{1}{2}mv^2$$

Introducing this in the preceding equation, we obtain

$$eV = h\nu - E_i$$

or

$$V = \frac{h\nu}{e} - \frac{E_i}{e} \tag{3-5}$$

The Photoelectric Cell

The photoelectric cell is used in talking motion pictures, television, automatic door-openers, and many other practical applications. The cell may be made by depositing a thin layer of an alkali metal on the inner surface of a small vacuum tube, as shown in Figure 3-17. The collecting electrode is positively charged, so that the photoelectrons are attracted to it. Illumination of the metal surface by any radiation with wavelength shorter than the photoelectric threshold causes the emission of photoelectrons and a consequent flow of electric current through the circuit. The current may be registered on an ammeter. It is found that the magnitude of the current is proportional to the intensity of the light.

Example 3-2. How much energy is there in one photon with wavelength 650 nm?

Solution. The amount of energy in a photon is $h\nu$, where h is Planck's constant and ν is the frequency of the light. The frequency of light of wavelength λ is c/λ; hence

$$\nu = \frac{3 \times 10^8}{650 \times 10^{-9}} = 4.62 \times 10^{14} \text{ cycles/sec or hertz}$$

Thus we obtain

$$\text{energy of photon} = h\nu = 0.6625 \times 10^{-33}\ \text{J s} \times 4.62 \times 10^{14}\ \text{Hz}$$
$$= 0.306 \times 10^{-18}\ \text{J}$$

Example 3-3. What retarding voltage would be required to stop the flow of photoelectrons produced by light of wavelength 650 nm from a sodium metal surface?

Solution. The photoelectric threshold of sodium metal is 650 nm. Accordingly, the photoelectrons that are produced have no kinetic energy; the amount of energy in the photon is just enough to remove the electron from the metal. Hence an extremely small retarding potential would stop the flow of photoelectrons under these conditions.

Example 3-4. What retarding potential would be required to stop the flow of photoelectrons in a photoelectric cell with sodium metal illuminated with light of wavelength 325 nm?

Solution. If, using the method of the solution of Example 3-2, we calculate the energy of a light quantum with wavelength 325 nm, we obtain the value 0.612×10^{-18} J. This result can be obtained, in fact, with little calculation by noting that this wavelength is just half that of the photoelectric threshold, 650 nm; hence the frequency ν is twice as great and the energy $h\nu$ is also twice as great as in Example 3-2.

Of this total amount of energy in the light quantum, the amount 0.306×10^{-18} J is used to remove the electron from the metal. The remaining amount, 0.306×10^{-18} J, is kinetic energy of the photoelectron. The retarding potential that would slow the electron down to zero speed is such that its product with the charge of the electron is equal to this amount of energy:

$$eV = 0.306 \times 10^{-18}\ \text{J}$$

$$V = \frac{0.306 \times 10^{-18}\ \text{J}}{0.1602 \times 10^{-18}\ \text{C}} = 1.91\ \text{volt}$$

Hence the retarding potential necessary to prevent the flow of photoelectrons in the sodium photoelectric cell illuminated with wavelength 325 nm is 1.91 volts.

The Production of X-rays

In an x-ray tube electrons from the cathode are speeded up to high velocity by a potential difference V. Their kinetic energy then becomes equal to the energy quantity eV. When such a fast-moving electron strikes the anode it is quickly brought down to low velocity, perhaps to velocity zero. If it is brought to velocity zero, the whole of its energy eV is converted into x-radiation (light), with energy $h\nu$ and corresponding frequency ν. The frequency of this radiation can hence be calculated from the photoelectric equation, $eV = h\nu$. (The ionization energy of the metal, E_i, can be neglected in this case, because it is a small energy quantity in comparison with the others.) If the electron is not slowed down completely the frequency of the x-ray quantum that is emitted will be somewhat smaller than the limiting value.

This process of creation of photons from the energy of fast-moving electrons is called the *inverse photoelectric effect.*

Example 3-5. An x-ray tube is operated at 50,000 volts. What is the short-wavelength limit of the x-rays that are produced?

Solution. The energy of an electron that strikes the anode in the x-ray tube is eV. The value of e is 0.1602×10^{-18} C and that of V is 50,000 volts. The value of eV is accordingly 8.010×10^{-15} J. This is equal to $h\nu$; hence for ν we have

$$\nu = \frac{8.010 \times 10^{-15}}{0.6625 \times 10^{-33}} = 12.091 \times 10^{18} \text{ Hz}$$

The wavelength λ is obtained by dividing the velocity of light by this quantity:

$$\lambda = \frac{c}{\nu} = \frac{3 \times 10^{8}}{12.091 \times 10^{18} \text{ Hz}} = 24.8 \times 10^{-12} \text{ m} = 24.8 \text{ pm}$$

Hence the short-wavelength limit of an x-ray tube operated at 50,000 volts is calculated to be 24.8 pm.

It is interesting to note that the preceding calculation can be simplified by combining all the steps into a single equation:

$$\text{short-wavelength limit (in nm)} = \frac{1240}{\text{accelerating potential (in volts)}} \qquad (3\text{-}6)$$

This equation states that a quantum of light with wavelength 1240 nm (in the near infrared) has the same energy as an electron that has been accelerated by a potential difference of 1 volt. This energy quantity is sometimes called 1 *electron volt,* with abbreviation 1 eV.

The relations among the commonly used units of energy are summarized in Appendices I and II.

Example 3-6. A beam of light with wavelength 650 nm and carrying the energy 1×10^{-2} J per second falls on a photoelectric cell, and is completely used in the production of photoelectrons. (This is about the energy of the light from the sun and sky on a bright day that strikes an area of 1 cm^2.) What is the magnitude of the photoelectric current that then flows in the circuit of which the photoelectric cell is a part?

Solution. The energy of one quantum of light with wavelength 650 nm is 0.306×10^{-18} J. Hence there are 1×10^{-2} J/$0.306 \times 10^{-18} = 3.27 \times 10^{16}$ photons in the amount of light carrying 1×10^{-2} J of radiant energy, and this number of photons impinges on the metal of the photoelectric cell every second. The same number of photoelectrons would be produced. Multiplying by the charge of the electron, 0.1602×10^{-18} C, we obtain 5.22×10^{-3} as the number of coulombs transferred per second. One ampere is a flow of electricity at the rate of 1 coulomb per second; hence the current that is produced by the beam of light is 5.22×10^{-3} A—that is, 5.22 milliamperes. (See also Exercise 3-21.)

3-11. The Wave Character of the Electron

Until 1924 the observed properties of the electron were considered to justify describing it as a small electrically charged particle, similar except for size to a ball bearing carrying an electric charge. In that year the wave character of the electron was discovered by the French physicist Louis de Broglie (born 1892). While making a theoretical study of the quantum theory, to serve as his thesis for the doctor's degree from the University of Paris, he found that a striking analogy between the properties of electrons and the properties of photons could be recognized if a moving electron were to be assigned a wavelength. This wavelength is now called the de Broglie wavelength of the electron.

The equation for the wavelength of the electron is

$$\lambda = \frac{h}{mv} \tag{3-7}$$

In this equation λ is the wavelength of the electron, h is Planck's constant, m is the mass of the electron, and v is the velocity of the electron. It is seen that according to this equation a stationary electron has infinite wavelength, and the wavelength decreases with increase in the velocity of the electron.

Example 3-7. What is the wavelength of an electron with 13.6 eV of kinetic energy?

Solution. We must solve for the momentum, mv, of the electron. The kinetic energy of the electron is

$$\begin{aligned}\tfrac{1}{2}mv^2 &= 13.6 \text{ eV} \times 0.1602 \times 10^{-18} \text{ C} \\ &= 2.18 \times 10^{-18} \text{ J}\end{aligned}$$

which leads to

$$mv^2 = 4.36 \times 10^{-18}$$

Multiplying both sides of this equation by m, we obtain

$$m^2v^2 = 4.28 \times 10^{-18} \times 0.9108 \times 10^{-30}$$
$$= 3.96 \times 10^{-48}$$

By taking the square root of each side of this equation we obtain

$$mv = 1.99 \times 10^{-24} \text{ kg m s}^{-1}$$

By use of the de Broglie equation we can now obtain the value for the wavelength:

$$\lambda = \frac{h}{mv} = \frac{0.66252 \times 10^{-33} \text{ J } s}{1.99 \times 10^{-24} \text{ kg m s}^{-1}}$$
$$= 3.33 \times 10^{-10} \text{ m} = 333 \text{ pm.}$$

Accordingly, we have found that the de Broglie wavelength of an electron that has been accelerated by a potential difference of 13.6 V is 333 pm.

We can now calculate easily the wavelength of an electron with 100 times as much kinetic energy—that is, an electron that has been accelerated by a potential difference of 1360 V. Since the energy is proportional to the square of the velocity, such an electron has a velocity ten times that of a 13.6-eV electron, and, according to the de Broglie equation, its wavelength is 1/10 as great. Thus the wavelength of a 1360-eV electron is 33.3 pm.

The Direct Experimental Verification of the Wavelength of the Electron

The wave character of moving electrons was established beyond question by the work of the American physicist C. J. Davisson (1881–1958) and the English physicist G. P. Thomson (born 1892). These investigators found that electrons that are scattered by crystals produce a diffraction pattern, similar to that produced by x-rays scattered by crystals, and, moreover, that the diffraction pattern corresponds to the wavelength given by the de Broglie equation.

The penetrating power of electrons through matter is far less than that of x-rays with the same wavelength. It is accordingly necessary to reflect the beam of electrons from the surface of a crystal (as was done by Davisson and his collaborators, using a single crystal of nickel), or to shoot a stream of high-speed electrons through a very thin crystal or layer of crystalline powder (as was done by Thomson).

The structure of crystals can be investigated by the electron-diffraction method as well as by the x-ray-diffraction method. The electron-diffraction method has been especially useful in studying the structure of very thin films on the surface of a crystal. For example, it has been shown that when argon is adsorbed on a clean face of nickel crystal the argon atoms occupy only one-quarter of the positions formed by triangles of nickel atoms (in the octahedral face of the cubic closest-packed crystal; Figure 2-7). The structure of very thin films of metal oxide that are formed on the surface of metals, and that protect them against further corrosion, has been studied by this method.

The electron-diffraction method is also very useful for determining the structure of gas molecules. The way in which the diffraction pattern is formed is illustrated by Figure 3-14, which corresponds to the scattering of waves by a diatomic molecule. The molecules in a gas have different orientations, and the diffraction pattern is accordingly somewhat blurred. It consists of a series of rings. Knowledge of the wavelength of the electrons and measurement of the diameters of these rings permit calculations of the interatomic distances in the molecules. Since the discovery of the electron-diffraction method the structures of several hundred molecules have been determined in this way.

The Wavelength of the Neutron. Neutron Diffraction

Not only electrons, but also protons and neutrons (Section 3-5) and other particles have wave character. Their wavelength is given by the de Broglie equation with use of the appropriate value for the mass. The relative scattering powers of atoms of different kinds in a crystal for neutrons are different from those for x-rays. In consequence, the study of the neutron diffraction patterns of crystals gives information supplementing that given by the x-ray diffraction patterns. Neutron diffraction has been found to be especially valuable for locating hydrogen atoms in a crystal containing heavier atoms and for the study of magnetic substances.

3-12. What Is Light? What Is an Electron?

During recent years many people have asked the following questions: Does light *really* consist of waves, or of particles? Is the electron *really* a particle, or is it a wave?

These questions cannot be answered by one of the two alternatives. Light is the name that we have given to describe a part of nature. The name refers to all of the properties that light has, to all of the phenomena that are observed in a system containing light. Some of the properties of light resemble those of waves, and can be described in terms of a wavelength. Other properties of light resemble those of particles, and can be described in terms of a light quantum, having a certain amount of energy, $h\nu$, and a certain mass $h\nu/c^2$. A beam of light is neither a sequence of waves nor a stream of particles; it is both.

In the same way, an electron is neither a particle nor a wave, in the ordinary sense. In many ways the behavior of electrons is similar to that expected of small particles, with mass m and electric charge $-e$. But electrons differ from ordinary particles, such as ball bearings, in that they also behave as though they had wave character, with wavelength given by the de Broglie equation. The electron, like the photon, has to be described as having the character both of a particle and of a wave.

After the first period of adjustment to these new ideas about the nature of light and of electrons, scientists became accustomed to them, and found that they could usually predict when, in a certain experiment, the behavior of a beam of light would be determined mainly by its wavelength, and when it would be determined by the energy and mass of the photon; that is, they would know when it was convenient to consider light as consisting of waves, and when to consider it as consisting of particles, the photons. Similarly, they learned when it was convenient to consider an electron as a particle, and when as a wave. In some experiments the wave character and the particle character both contribute significantly, and it is then necessary to carry out a careful theoretical treatment, using the equations of quantum mechanics, in order to predict how the light or the electron will behave.

You may ask two other questions: Do electrons exist? What do they look like?

The answer to the first question is that electrons do exist: "electron" is the name that scientists have used in discussing certain phenomena, such as the beam in the electric discharge tube studied by J. J. Thomson, the carrier of the unit electric charge on the oil drops in Millikan's apparatus, the part that is added to the neutral fluorine atom to convert it into a fluoride ion. The second question—what does the electron look like?—cannot be answered. No one knows how to look at an electron—it is too small to be seen by scattering ordinary visible light from it, and unless somebody discovers some better way of studying nature than is now known, this question will remain unanswered. However, it is possible to say some things about what the proton and the neutron look like. By the study of the scattering of rapidly moving electrons by protons and neutrons information has been obtained about the distribution in space of the electric charge in these particles. The results of these studies are described in Section 20-4.

3-13. The Uncertainty Principle

The *uncertainty principle,* an important relation that is a consequence of quantum mechanics, was discovered by the German physicist Werner Heisenberg (born 1901) in 1927. Heisenberg showed that as a result of the wave-particle duality of matter it is impossible to carry out simultaneously a precise determination of the position of a particle and of its velocity. He also showed that it is impossible to determine exactly the energy of a system at an instant of time.

The uncertainty principle in quantum mechanics is closely related to an uncertainty equation between frequency and time for any sort of waves. Let us consider, for example, a train of ocean waves passing a buoy that is anchored at a fixed point. An observer on the buoy could measure the frequency (number of waves passing the buoy per unit time) at time t_0 by counting the number of crests and troughs passing the buoy between the times $t_0 - \Delta t$ and $t_0 + \Delta t$, dividing by 2 to obtain the number of waves in time $2\Delta t$, and by $2\Delta t$ to obtain the frequency (ν), which is defined as the number of waves in unit time:

$$\nu = \frac{\text{number of crests plus number of troughs}}{2 \times 2\Delta t}$$

This measurement is an average for a period of time $2\Delta t$ in the neighborhood of t_0; we may describe it as being for the time t_0 with uncertainty Δt. There is also an uncertainty in the frequency. The crest A (Figure 3-18) might or might not be counted, and similarly the trough Z. Hence there is an uncertainty of about 2 in the number of crests plus the number of troughs, and hence of about $1/2\Delta t$ in the frequency:

$$\Delta\nu = \frac{2}{2 \times 2\Delta t} = \frac{1}{2\Delta t}$$

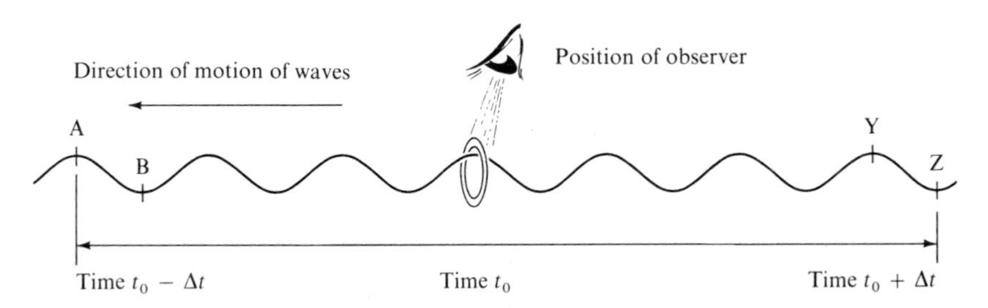

FIGURE 3-18
A diagram illustrating the uncertainty in determining the frequency by counting the number of waves passing a point during a period of time.

We may rewrite this equation as $\Delta\nu \times \Delta t = \frac{1}{2}$. A more detailed discussion based on an error-function definition of $\Delta\nu$ and Δt leads to the *uncertainty equation for frequency and time* in its customary form:

$$\Delta\nu \times \Delta t = \frac{1}{2\pi} \tag{3-8}$$

By use of quantum theory this equation can be at once converted into the uncertainty equation for energy and time for photons. The energy of a photon with frequency ν is $h\nu$. The uncertainty in frequency $\Delta\nu$ when multiplied by h is the uncertainty in energy ΔE:

$$\Delta E = h\Delta\nu$$

By substituting this relation in Equation 3-8 we obtain the energy-time uncertainty equation:

$$\Delta E \times \Delta t = \frac{h}{2\pi} \tag{3-9}$$

It has been found by analysis of many experiments on the basis of quantum theory that this relation holds for any system. Only by making the measurement of the energy of any system over a long period of time can the error in the measured energy of the system be made small.

Example 3-8. The yellow D lines, with wavelengths 589.0 and 589.6 nm, are emitted by sodium gas when it is excited by an electric discharge. The rate at which the D lines are emitted has been determined by measuring the intensity of the lines as absorption lines in sodium gas. It is such as to correspond to a mean life of the excited states of 1.6×10^{-6} s. (Most excited states of atoms have about this mean life.) Does this value of the mean life lead to a broadening of the spectral lines?

Solution. We use the uncertainty principle here on the argument that the train of waves corresponding to the photon emitted by an atom is emitted during a period of time approximately equal to the mean life of the excited state of the atom. The frequency corresponding to wavelength $\lambda = 0.589 \times 10^{-6}$ m is $\nu = c/\lambda = 3 \times 10^8/0.589 \times 10^{-6} = 5.09 \times 10^{14}$ s^{-1}. The uncertainty in frequency corresponding to the uncertainty in time $\Delta t = 1.6 \times 10^{-8}$ s is given by Equation 3-8 as $\Delta\nu = 1/(2\pi\Delta t) = 1/(2\pi \times 1.6 \times 10^{-8}) = 1.00 \times 10^7$ s^{-1}. Hence the relative uncertainty in frequency is $\Delta\nu/\nu = 1.00 \times 10^7/5.09 \times 10^{14} = 1.96 \times 10^{-8}$. This is also the value of $\Delta\lambda/\lambda$, the relative uncertainty in wavelength. Hence we obtain $\Delta\lambda = 1.96 \times 10^{-8} \times 0.589 \times 10^{-6} = 1.15 \times 10^{-14}$ m.

The uncertainty in frequency of these lines thus gives a line width that is only about 1/50,000,000 of the wavelength. This line broadening is usually masked by other effects. However, some excited states of atoms, with energy greater than the ionization energy, have mean life only 10^{-12} s, because of very rapid decomposition into an electron and a positive ion, and the spectral lines with one of these states as upper state are about 0.1 nm broad.

The uncertainty principle between position and momentum of a particle states that the product of the uncertainty in one of the coordinates x, y, z describing the position, such as Δx, and the uncertainty in the corresponding component of the momentum, $\Delta(mv_x)$, is equal to or greater than $h/2\pi$:

$$\Delta x \times \Delta(mv_x) \geqq \frac{h}{2\pi} \tag{3-10}$$

EXERCISES

3-1. What important simplification in the understanding of the nature of electricity did Benjamin Franklin propose? What assumption did he make that is now known to be incorrect?

3-2. Who named the electron and when? Who experimentally discovered it and when? Where?

3-3. How many electrons are there in 1 kilogram of electrons? How many electrons are in 1 coulomb?

3-4. When did Lord Rutherford provide evidence that atoms have very small, very heavy nuclei? What does "very heavy" mean in this context?

3-5. How many protons are there in a kilogram of protons? How many neutrons are there in a kilogram of neutrons? How do these numbers compare with the number of electrons in a kilogram of electrons? What is the electric charge (in C) of the proton and of the electron?

3-6. What are the components of the hydrogen nucleus, the deuteron, and the helion?

3-7. What is an electric field? A magnetic field?

3-8. State the law of attraction or repulsion between two electrically charged particles. How is the force between the two charged particles changed
(*a*) if the distance between them is doubled?
(*b*) if the charge of one particle is doubled?
(*c*) if the charges of both particles are doubled?

3-9. Describe the motion of an electron moving transversely between two parallel plates carrying opposite electric charges. Also describe the motion of an electron moving transversely between the poles of a magnet.

3-10. Describe the experiments of Jean Perrin and J. J. Thomson that led to the discovery of the electron.

3-11. Describe Millikan's oil-drop experiment. Why is a knowledge of the value of the viscosity of air needed in order for the value of the charge of the electron to be calculated from the measurements?

3-12. What unexpected observations led to the discovery of x-rays and radioactivity? Who discovered them and when?

3-13. What are alpha, beta, and gamma rays? What is the difference between gamma rays and x-rays?

3-14. Describe the Rutherford experiment, and explain why the observations indicate that most of the mass of an atom is concentrated in a very small particle, the nucleus.

3-15. What is the relationship between wavelength, wave velocity, and frequency? What units are used for these quantities?

3-16. Draw a linear diagram of the spectrum of electromagnetic waves, indicating wavelength, frequency, and common names for the various regions.

3-17. What is a photon and what is its energy in terms of wavelength? Give units.

3-18. What is the energy of a quantum of blue light of wavelength 450 nm? Of a quantum of red light of wavelength 700 nm?

3-19. What is the minimum wavelength of the x-rays emitted by a million-volt x-ray tube?

3-20. Calculate the velocity with which the electrons would move in the apparatus used by J. J. Thomson, operated at an accelerating voltage of 6000 V. (You may neglect the relativistic change in mass.)

3-21. In Example 3-6 of this chapter it is calculated that the light that falls on 1 cm^2 surface on a bright day could produce a photocurrent of 5.22 mA. *(a)* To what voltage does the wavelength 650 nm (an average for sunlight) correspond? *(b)* Assuming an average of 6 hours of sunlight and 10% efficiency of the solar electricity plant, calculate the average power that a plant using 100 km^2 of desert area would produce. (Answers: *(a)* 1.91 V, *(b)* 25 MW.)

4

Elements and Compounds. Atomic and Molecular Masses

One of the most important parts of chemical theory is the division of substances into the two classes *elementary substances* and *compounds*. This division was achieved in 1787 by the French chemist Antoine Laurent Lavoisier (1743–1794), on the basis of the quantitative studies that he had made during the preceding fifteen years of the masses of the substances (reactants and products) involved in chemical reactions. Lavoisier defined a compound as a substance that can be decomposed into two or more other substances, and an elementary substance (or element) as a substance that can not be decomposed. In his *Traité Elémentaire de Chimie* [Elementary Treatise on Chemistry], published in 1789, he listed 33 elements, including 10 that had not yet been isolated as elementary substances, but were known as oxides, the compound nature of which was correctly surmised by Lavoisier. Since the discovery of the electron and the atomic nucleus the definitions of elementary substances and compounds have been revised in the ways presented in the following paragraphs.

4-1. The Chemical Elements

A kind of matter consisting of atoms that all have nuclei with the same electric charge is called an *element*.

For example, all of the atoms that contain nuclei with the charge $+e$, each nucleus having one electron attached to it to neutralize its charge, comprise the element hydrogen, and all of the atoms that contain nuclei with the charge $+92e$ comprise the element uranium.

An *elementary substance* is a substance that is composed of atoms of one element only. An elementary substance is commonly called an element.

A *compound* is a substance that is composed of atoms of two or more different elements. These atoms of two or more different elements must be present in a definite numerical ratio, since compounds are defined as having a definite composition.

Atomic Number

The electric charge of the nucleus of an atom, in units equal to the charge of the proton, is called the atomic number of the atom. It is usually given the symbol Z, the electric charge of a nucleus with atomic number Z being Z times e, with the charge of the proton equal to e, and the charge of the electron equal to $-e$. Thus the simplest atom, that of hydrogen, has atomic number 1; it consists of a nucleus with electric charge e, and an electron with electric charge $-e$.

The Assignment of Atomic Numbers to the Elements

Soon after the discovery of the electron as a constituent of matter it was recognized that elements might be assigned atomic numbers, representing the number of electrons in an atom of each element, but the way of doing this correctly was not known until 1913. In that year H. G. J. Moseley (1887–1915), a young English physicist working in the University of Manchester, found that the atomic number of any element could be determined by the study of the x-rays emitted by an x-ray tube containing the element. By a few months of experimental work he was able to assign their correct atomic numbers to many elements.

It was found that the x-rays produced by an x-ray tube contain lines of definite wavelengths, characteristic of the material in the target of the x-ray tube. Moseley measured the wavelengths produced by a number of different elements, and found that they change in a regular way. The wavelengths of the two principal x-ray lines of the elements from aluminum to zinc (omitting the gas argon) are shown in Figure 4-1.

The regularity in the wavelengths can be shown more strikingly by plotting the square root of the reciprocals of the wavelengths of the two x-ray lines for the various elements arranged in the proper sequence, which is the sequence of the atomic numbers of the elements. In a graph of this sort, called a Moseley diagram, the points for a given x-ray line lie on a straight line. The Moseley diagram for the elements for aluminum to zinc is shown in Figure 4-2. It was easy for Moseley to assign the cor-

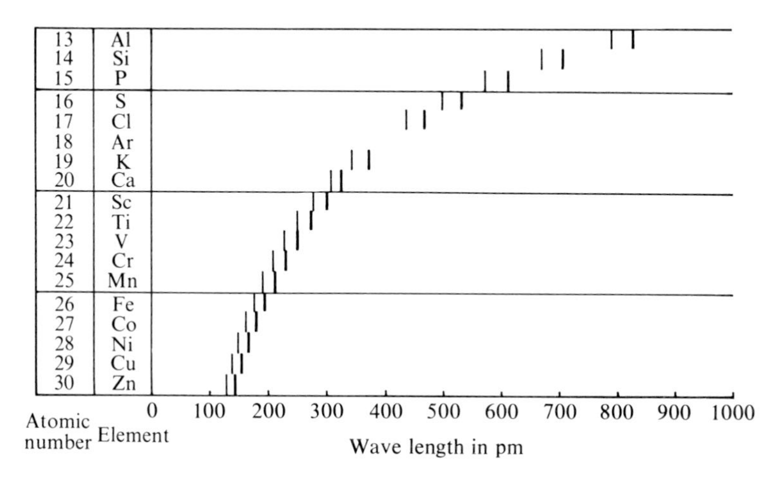

Figure 4-1
Diagram showing the regular change of wavelength of x-ray emission lines for a series of elements.

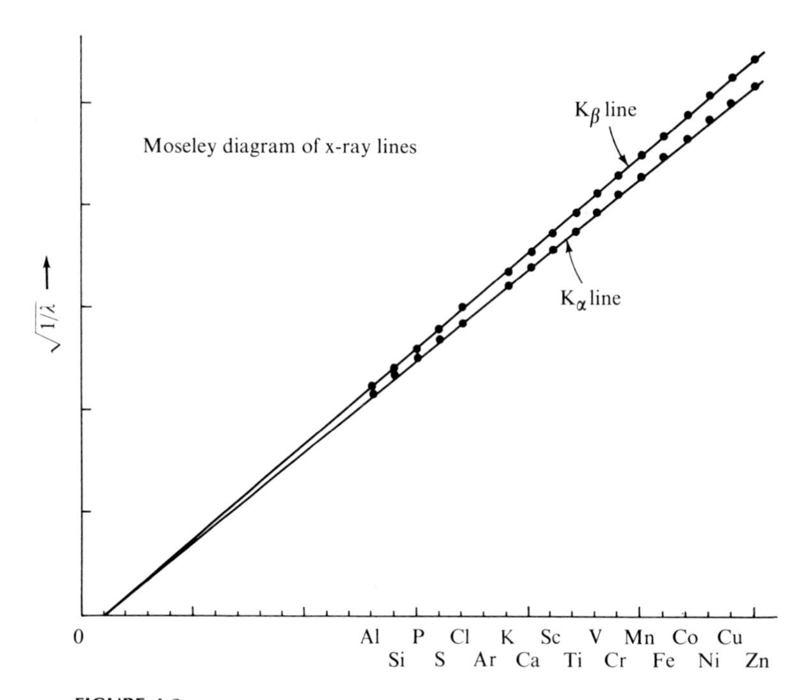

FIGURE 4-2
A graph of the reciprocal of the square root of the wavelengths of x-ray lines, for the K_α line and the K_β line, of elements, plotted against the order of the elements in the periodic table. This graph, called the Moseley diagram, was used by Moseley in determining the atomic numbers of the elements.

rect atomic numbers to the elements with use of a diagram of this sort (see the discussion of the Bohr theory in Chapter 5).

Isotopes

In a partially evacuated tube through which an electric discharge is passing the collision of high-velocity electrons with atoms or molecules may knock one or more electrons from an atom or molecule, leaving it with a positive electric charge. It is called a *cation.* (Note that a cation is attracted toward the cathode, the negatively charged terminal; an *anion,* which is a negatively charged atom or molecule, is attracted toward the anode, the positive terminal. These names were introduced by Michael Faraday in 1834, in his discussion of the conduction of electricity by aqueous solutions of salts.) The cations, attracted toward a perforated cathode, pass through the perforation and emerge on the other side as a beam of positively charged particles. In 1912 J. J. Thomson, using an apparatus with electric and magnetic fields (somewhat similar to that shown in Figure 3-7), found the cations of neon ($Z = 10$) to be of two kinds, one with mass about 20 times the proton mass and the other with mass about 22 times the proton mass. They are called isotopes (from the Greek *isos,* the same, and *topos,* place—in the table of elements), and are said to differ in *mass number,* which is given the symbol A. The element neon as found in nature (the atmosphere) contains 89.97% of the isotope with $A = 20$, 9.73% with $A = 22$, and 0.30% of a third isotope, with $A = 21$, not noticed by Thomson.

All known elements have two or more isotopes. In some cases (such as aluminum) only one isotope occurs naturally, the others being unstable. The maximum number of stable isotopes of any element is 10, possessed by tin.

The chemical properties of all the isotopes of an element are essentially the same. These properties are determined in the main by the atomic number of the nucleus, and not by its mass.

The word *nucleide* (sometimes written *nuclide*) is used for the kind of matter involving nuclei with given values of Z and A. Nucleides of the same element are isotopes.

The Names and Symbols of the Elements

The names of the elements are given in Table 4-1. The chemical symbols of the elements, used as abbreviations for their names, are also given in the table. These symbols are usually the initial letters of the names, plus another letter when necessary. In some cases the initial letters of Latin names are used: Na for sodium (natrium), K for potassium (kalium), Fe for iron (ferrum), Cu for copper (cuprum), Ag for silver (argentum), Au for gold (aurum), Hg for mercury (hydrargyrum), Sn for tin (stannum), Sb for antimony (stibium), and Pb for lead (plumbum). The system of chemical symbols was proposed by the Swedish chemist Jöns Jakob Berzelius (1779–1848) in 1811.

The elements are also shown in a special arrangement, the *periodic table,* at the front of the book and in Table 5-1.

A symbol is used to represent an atom of an element, as well as the element itself. The symbol I represents the element iodine, and also may be used to mean the elementary substance. However I_2 is the customary formula for the elementary substance, because it is known that elementary iodine consists of molecules containing two atoms in the solid and liquid states as well as in the gaseous state (except at very high temperature). In formulas showing composition or molecular structure the numerical subscript of the symbol of an element gives the number of atoms of the element in the molecule.

4-2. Chemical Reactions

The formula of a compound should give as much as possible of the information known about its composition and structure. The formula of benzene is written C_6H_6, not CH; it is known that the benzene molecule contains six carbon atoms and six hydrogen atoms. The formula of crystalline copper sulfate pentahydrate is written $CuSO_4(H_2O)_5$ or $CuSO_4 \cdot 5H_2O$, to show that it contains the sulfate group SO_4 and that five molecules of water are easily removed; the name also reflects these facts.

The equations for chemical reactions can be correctly written if the nature of the products is known. For example, in the reaction of a rocket propellant made of carbon and potassium perchlorate, $KClO_4$, the products may be potassium chloride, KCl, and either carbon monoxide or carbon dioxide, or a mixture of the two. It would probably be wise to write two equations, corresponding to two reactions:

$$KClO_4 + 4C \rightarrow KCl + 4CO$$

$$KClO_4 + 2C \rightarrow KCl + 2CO_2$$

For each of these equations the same number of atoms of each element is shown on the right side as on the left side: the equations are balanced. Writing a balanced equation for a reaction is often the first step in solving a chemical problem.

TABLE 4-1
International Atomic Masses

Element	Symbol	Atomic Number	Atomic Mass	Element	Symbol	Atomic Number	Atomic Mass
Actinium	Ac	89	[227]*	Holmium	Ho	67	164.930
Aluminum	Al	13	26.9815	Hydrogen	H	1	1.00797†
Americium	Am	95	[243]	Indium	In	49	114.82
Antimony	Sb	51	121.75	Iodine	I	53	126.9044
Argon	Ar	18	39.948	Iridium	Ir	77	192.2
Arsenic	As	33	74.9216	Iron	Fe	26	55.847‡
Astatine	At	85	[210]	Khurchatovium	Kh	104	[260]
Barium	Ba	56	137.34	Krypton	Kr	36	83.80
Berkelium	Bk	97	[247]	Lanthanum	La	57	138.91
Beryllium	Be	4	9.0122	Lawrencium	Lw	103	[257]
Bismuth	Bi	83	208.980	Lead	Pb	82	207.19
Boron	B	5	10.811†	Lithium	Li	3	6.939
Bromine	Br	35	79.909‡	Lutetium	Lu	71	174.97
Cadmium	Cd	48	112.40	Magnesium	Mg	12	24.312
Calcium	Ca	20	40.08	Manganese	Mn	25	54.9380
Californium	Cf	98	[249]	Mendelevium	Md	101	[256]
Carbon	C	6	12.01115†	Mercury	Hg	80	200.59
Cerium	Ce	58	140.12	Molybdenum	Mo	42	95.94
Cesium	Cs	55	132.905	Neodymium	Nd	60	144.24
Chlorine	Cl	17	35.453‡	Neon	Ne	10	20.183
Chromium	Cr	24	51.996‡	Neptunium	Np	93	[237]
Cobalt	Co	27	58.9332	Nickel	Ni	28	58.71
Copper	Cu	29	63.54	Niobium	Nb	41	92.906
Curium	Cm	96	[247]	Nitrogen	N	7	14.0067
Dysprosium	Dy	66	162.50	Nobelium	No	102	[256]
Einsteinium	Es	99	[254]	Osmium	Os	76	190.2
Erbium	Er	68	167.26	Oxygen	O	8	15.9994†
Europium	Eu	63	151.96	Palladium	Pd	46	106.4
Fermium	Fm	100	[253]	Phosphorus	P	15	30.9738
Fluorine	F	9	18.9984	Platinum	Pt	78	195.09
Francium	Fr	87	[223]	Plutonium	Pu	94	[242]
Gadolinium	Gd	64	157.25	Polonium	Po	84	[210]
Gallium	Ga	31	69.72	Potassium	K	19	39.102
Germanium	Ge	32	72.59	Praseodymium	Pr	59	140.907
Gold	Au	79	196.967	Promethium	Pm	61	[147]
Hafnium	Hf	72	178.49	Protactinium	Pa	91	[231]
Helium	He	2	4.0026	Radium	Ra	88	[226]

Element	Symbol	Atomic Number	Atomic Mass	Element	Symbol	Atomic Number	Atomic Mass
Radon	Rn	86	[222]	Tellurium	Te	52	127.60
Rhenium	Re	75	186.2	Terbium	Tb	65	158.924
Rhodium	Rh	45	102.905	Thallium	Tl	81	204.37
Rubidium	Rb	37	85.47	Thorium	Th	90	232.038
Ruthenium	Ru	44	101.07	Thulium	Tm	69	168.934
Samarium	Sm	62	150.35	Tin	Sn	50	118.69
Scandium	Sc	21	44.956	Titanium	Ti	22	47.90
Selenium	Se	34	78.96	Tungsten	W	74	183.85
Silicon	Si	14	28.086†	Uranium	U	92	238.03
Silver	Ag	47	107.870‡	Vanadium	V	23	50.942
Sodium	Na	11	22.9898	Xenon	Xe	54	131.30
Strontium	Sr	38	87.62	Ytterbium	Yb	70	173.04
Sulfur	S	16	32.064†	Yttrium	Y	39	88.905
Tantalum	Ta	73	180.948	Zinc	Zn	30	65.37
Technetium	Tc	43	[97]	Zirconium	Zr	40	91.22

*A value given in brackets is the mass number of the most stable known isotope.

†The atomic mass varies because of natural variations in the isotopic composition of the element. The observed ranges are boron, ±0.003; carbon, ±0.00005; hydrogen, ±0.00001; oxygen, ±0.0001; silicon, ±0.001; sulfur, ±0.003.

‡The atomic mass is believed to have an experimental uncertainty of the following magnitude: bromine, ±0.002; chlorine, ±0.001; chromium, ±0.001; iron, ±0.003; silver, ±0.003. For other elements the last digit given is believed to be reliable to ±5.

4-3. Nucleidic and Atomic Masses

The accepted unit of mass for nucleides is the *dalton,* symbol d. The dalton is defined as exactly $\frac{1}{12}$ of the mass of the neutral atom ${}^{12}_{6}C_6$. The nucleidic mass of ^{12}C is exactly 12.00000 d. The dalton is approximately 1.66033×10^{-27} kg.

In a quantitative discussion of an ordinary chemical reaction the relative amounts of various substances that react or are formed can be calculated from the numbers of atoms of different elements and the masses of the atoms. If an element is present as a mixture of isotopes, it is the average atomic mass that is needed. It is this average atomic mass for an element that is called the *chemical atomic mass.* The unit for atomic mass is the dalton.

The name atomic mass is here used in place of atomic weight. There might, however, be some advantage to retaining the old name atomic weight in referring to the natural mixtures of isotopes that are involved in most chemical reactions, and it seems likely that this usage will continue for some time.

The History of the Atomic Mass Scale

John Dalton chose the value 1 for hydrogen as the base of his scale of atomic masses. The Swedish chemist J. J. Berzelius used 100 for oxygen, and the Belgian chemist J. S. Stas (1813–1891), who carried out many quantitative analyses of compounds, proposed 16 for oxygen (the natural mixture of isotopes), and this base was used for many years. For several decades nucleidic masses were expressed on a scale (called the physical scale) based on $\frac{1}{16}$th the mass of the neutral atom ^{16}O; the chemical atomic-weight unit was then 1.000272 times the physical atomic-mass unit. This period of confusion was brought to an end in 1961 by the acceptance of $\frac{1}{12}$th the mass of ^{12}C as the unit for both atomic masses and nucleidic masses.

4-4. Avogadro's Number. The Mole

Avogadro's number (N) is defined as the number of carbon-12 atoms in exactly 12 g of carbon 12. Its value is approximately 0.60229×10^{24}. It was named for the Italian physicist Amedeo Avogadro, whose work is discussed in Section 4-10.

A *mole* of a substance is defined as Avogadro's number of molecules of the substance. Thus a mole of water, H_2O, is the quantity of water containing N H_2O molecules. The molecular mass of water (the sum of the atomic masses of the atoms in the molecule: $2 \times 1.00797 + 1 \times 15.9994$, from Table 4-1) is 18.0153. From the definitions of atomic masses and the mole we see that a mole of water is 18.0153 g.

A mole of iodine atoms is 126.9044 g of iodine, and a mole of iodine molecules (I_2) is 253.8088 g of iodine. Sometimes, to prevent confusion, a mole of atoms of an element is called a *gram-atom*. The mass of a mole of a compound for which a formula is written is sometimes taken to be the gram-formula-mass (gfm), the number of grams equal to the sum of the atomic masses corresponding to the formula, whether or not the formula is a correct molecular formula for the substance. Thus the molar mass of liquid acetic acid, for which the formula $CH_3COOH(l)$ is written, is taken as 60.05, even though it is likely that the liquid contains some *dimers* (double molecules, $(CH_3COOH)_2$), as does the vapor.

Often, as shown above, the state of aggregation of a substance is represented by appended letters: Cu(c) refers to crystalline copper (c standing for crystal; sometimes, for solid, s is used), Cu(l) to liquid copper, and Cu(g) to gaseous copper. A substance in solution is sometimes represented by its formula followed by the name of the solvent in parentheses; (aq) is used for aqueous solutions.

4-5. Examples of Mass-relation Calculations

The way to work a mass-relation problem is to think about the problem, in terms of atoms and molecules, and then to decide how to carry out the calculations. You should not memorize any rule about these problems—such rules are apt to confuse you and to cause you to make mistakes.

The way to work these problems is best indicated by some examples.

In general, chemical problems may be solved by using a slide rule for the numerical work. This gives about three reliable figures in the answer, which is often all that is justified by the accuracy of the data. Sometimes the data are more reliable, and logarithms or long-hand calculations might be used to obtain the answer with the accuracy required. Unless the problem requires unusual accuracy, you may round values of atomic mass off to the first decimal point.

Example 4-1. What is the percentage of lead in galena, PbS? Calculate to 0.1%.

Solution. The formula mass of PbS is obtained by adding the atomic masses of lead and sulfur, which we obtain from the table inside the back cover:

$$\begin{aligned} \text{Mass of one lead atom (1 Pb)} &= 207.2 \text{ d} \\ \text{Mass of one sulfur atom (1 S)} &= 32.1 \text{ d} \\ \text{Mass of 1 PbS} &= 239.3 \text{ d} \end{aligned}$$

Hence 239.3 d of PbS contains 207.2 d of lead. We see that 100.0 g of PbS would contain

$$\frac{207.2 \text{ d of lead}}{239.3 \text{ d of galena}} \times 100.0 \text{ g of galena} = 86.6 \text{ g of lead}$$

Hence the percentage of lead in galena is 86.6%.

Example 4-2. A propellant for rockets can be made by mixing powdered potassium perchlorate, $KClO_4$, and powdered carbon (carbon black), C, with a little adhesive to bind the powdered materials together. What mass of carbon should be mixed with 1000 g of potassium perchlorate in order that the products of the reaction be KCl and CO_2?

Solution. Taking the equation for the reaction as

$$KClO_4 + 2C \longrightarrow KCl + 2CO_2$$

we first calculate the formula mass of $KClO_4$:

$$\begin{aligned} \text{Mass of 1K} &= 39.1 \\ \text{Mass of 1Cl} &= 35.5 \\ \text{Mass of 4O} = 4 \times 16.0 &= 64.0 \\ \text{Mass of } KClO_4 &= 138.6 \end{aligned}$$

The atomic mass of carbon is 12.0; the mass 2C is 24.0. Hence the mass of carbon required is 24.0/138.6 times the mass of potassium perchlorate:

$$\frac{24.0(\text{C})}{138.6(\text{KClO}_4)} \times 1000 \text{ g } (\text{KClO}_4) = 173 \text{ g } (\text{C})$$

Hence about 173 g of carbon* is required for 1000 g of potassium perchlorate.

Example 4-3. In a determination of the atomic mass of iron, 7.59712 g of carefully purified ferric oxide, Fe_2O_3, was reduced by heating in a stream of hydrogen, and found to yield 5.31364 g of metallic iron. Given the atomic mass of oxygen, to what value of the atomic mass of iron does this result lead?

Solution. The difference between the mass of ferric oxide and the mass of iron, 2.28348 g, is the mass of the oxygen contained in the sample of ferric oxide. From the formula we see that two thirds of this quantity is the mass of the number of oxygen atoms equal to the number of iron atoms in the sample, 5.31364 g. The atomic mass of oxygen is 15.9994 (Table 4-1). Hence the atomic mass of iron is

$$\frac{5.31364}{2/3 \times 2.28348} \times 15.9994 = 55.8457$$

This value for the atomic mass of iron, 55.8457, is from an investigation made by G. P. Baxter and C. R. Hoover fifty years ago. It agrees closely with the value in Table 4-1: 55.847.

---

*Note the convention in chemistry that "173 g of carbon" is singular in number; it means a quantity of carbon weighing 173 g, rather than 173 separate grams of carbon.

Example 4-4. An oxide of arsenic contains 65.2% arsenic. What is its simplest formula?

Solution. In 100 g of this oxide of arsenic there are contained, according to the reported analysis, 65.2 g of arsenic and 34.8 g of oxygen. If we divide 65.2 g by the gram-atomic mass of arsenic, 74.9 g, we obtain 0.870 as the number of gram-atoms of arsenic. Similarly, by dividing 34.8 g by the gram-atomic mass of oxygen, 16 g, we obtain 2.17 as the number of gram-atoms of oxygen in 100 g of this oxide of arsenic. Hence the numbers of atoms of arsenic and oxygen in the compound are in the ratio 0.870 to 2.17. If we set this ratio, 0.870/2.17, on the slide rule, we see that it is very close to 2/5, being 2/4.99. Hence the simplest formula is As_2O_5.

We say that this is the simplest formula in order not to rule out the possibility that the substance contains more complex molecules, such as As_4O_{10}, in which case it would be proper to indicate in the formula the larger numbers of atoms per molecule.

4-6. Determination of Atomic Masses by the Chemical Method

It is hard to overestimate the importance of the table of atomic masses. Almost every activity of a chemist involves the use of atomic masses in some way. For nearly two centuries successive generations of chemists have carried out experiments in the effort to provide more and more accurate values of the atomic masses, in order that chemical calculations could be carried out with greater accuracy.

Until recently almost all atomic-mass determinations were made by the chemical method. This method consists in determining the amount of the element that will combine with one gram-atom of oxygen or of another element with known atomic mass. One example, which was important in the development of the theory of radioactivity, follows.

Example 4-5. A 1.0000-g sample of lead sulfide, PbS, prepared from an ore of uranium (Katangan curite) was found to give 0.8654 g of metallic lead on reduction. Another 1.0000-g sample of lead sulfide prepared from an ore of thorium (Norwegian thorite) was found to give 0.8664 g of metallic lead on reduction. Assuming the atomic mass of sulfur to be 32.064, calculate the two values of the atomic mass of lead.

Solution. For the curite lead sulfide we see that the Pb/S ratio of atomic masses is 0.8654/0.1346, and that the atomic mass of lead is accordingly $0.8654/0.1346 \times 32.064 = 206.15$. For thorite lead sulfide a similar calculation gives $0.8664/0.1336 \times 32.064 = 207.94$.

Results of this sort were obtained in 1914 by the American chemist Theodore William Richards (1868–1928). They confirmed the deductions about the existence of isotopes that had been made in 1913 by the English chemist Frederick Soddy (1877–1956) from the study of radioactive transformations of uranium and thorium.

4-7. Determination of Atomic Masses by Use of the Mass Spectrograph

The apparatus used by J. J. Thomson in discovering two isotopes of neon, described in Section 4-1, was a simple form of mass spectrograph. Modern mass spectrographs have been used in attacking many physical and chemical problems, including that of determining nucleidic masses and the abundance ratios of isotopes.

The principle of the modern mass spectrograph can be illustrated by the simple apparatus shown in Figure 4-3.

At the left is a chamber in which positive ions are formed by an electric discharge, and then accelerated toward the right by an electric potential. The ions passing through the first pinhole have different velocities; in the second part of the apparatus a beam of ions with approximately a certain velocity is selected, and allowed to pass through the second pinhole, the ions with other velocities being stopped. (We shall not attempt to describe the construction of the velocity selector.) The ions passing through the second pinhole then move on between two metal plates, one of which has a positive electric charge and the other a negative charge. The ions accordingly undergo an acceleration toward the negative plate, and are deflected from the straight path A that they would pursue if the plates were not charged.

The force acting on an ion between these plates is proportional to $+ze$,

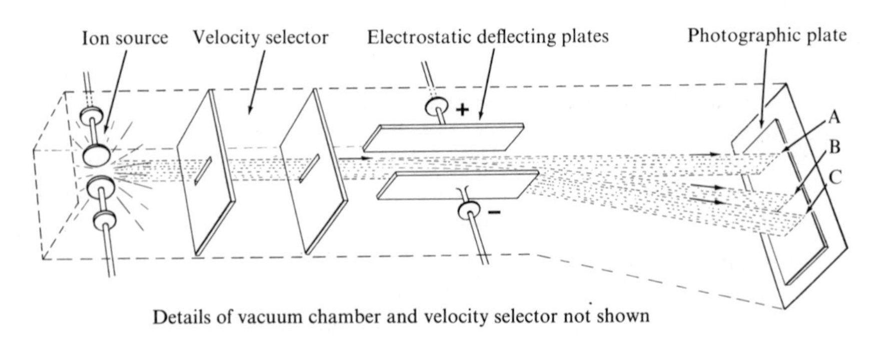

FIGURE 4-3
Diagram of a simple mass spectrograph.

its electric charge (z being the number of missing electrons), and its inertia is proportional to its mass M. The amount of deflection is hence determined by ze/M, the ratio of the charge of the ion to its mass.

Of two ions with the same charge, the lighter one will be deflected in this apparatus by the greater amount. The beam C might accordingly represent the ion C^+, with charge $+e$ and mass 12 d (the atomic mass of carbon), and the beam B the heavier ion O^+, with the same charge but with mass 16.

Of two ions with the same mass, the one with the greater charge will be deflected by the greater amount. Beams B and C might represent O^{++} and O^{+++}, respectively.

By measuring the deflection of the beams, relative values of ze/M for different ions can be determined. Since e is constant, relative values of ze/M for different ions are also inverse relative values of M/z: therefore this method permits the direct experimental determination of the relative masses of atoms, and hence of their atomic masses.

The value of the integer z—the degree of ionization of the ions—can usually be fixed from knowledge of the substances present in the discharge tube; thus neon gives ions with $M/z = 20$ and 22 ($z = 1$), 10 and 11 ($z = 2$), and so on.

Instead of the mass spectrograph described above, others of different design, using both an electric field and a magnetic field, are usually used. These instruments are designed so that they focus the beam of ions with a given value of M/z into a sharp line on a slit in front of an ion-beam detector which drives a fast pen recorder through an amplifier. Such a mass spectrometer scans a large range of values of M/z by varying the electric or magnetic field in a few seconds. These instruments are of great value in chemical analysis as the ion source breaks the compound to be analyzed into many fragments of varying sizes whose masses can be determined.

Modern types of mass spectrographs have an accuracy of about one part in 200,000 and a resolving power of 20,000 (that is, they are able to separate ion beams with values of M/z differing by only one part in 20,000). The great accuracy of modern mass spectrographs makes the mass-spectrographic method of determining atomic weights more useful and important at present than the chemical method.

Mass-spectrographic comparisons with ^{12}C or ^{16}O are made in the following way. An ion source that produces ions both of carbon or oxygen and of the element to be investigated is used; the lines of carbon or oxygen and of the element in such states of ionization that their M/z values are nearly the same are then obtained; thus for ^{32}S, ^{33}S, and ^{34}S the lines for the doubly ionized atoms would lie near the line for singly ionized oxygen. Accurate relative measurements of these lines can then be made.

Examples of the Determination of Atomic Masses with the Mass Spectrograph

For a single element, with only one isotope, the value of the atomic mass of that isotope is the atomic mass of the element. Thus for gold, which consists entirely of one stable isotope, ^{197}Au, the mass-spectrographic mass is reported to be 196.967. This value has been accepted by the International Committee on Atomic Weights.

For elements with two or more stable isotopes the atomic mass can be calculated from the masses of the isotopes and values of their relative amounts, as shown in the following example.

Example 4-6. Silver has two stable isotopes. Their masses as determined by use of the mass spectrograph were found to be 106.902 and 108.900, with relative amounts (numbers of nuclei) 51.35% and 48.65%, respectively. What is the atomic mass of silver, from this investigation?

Solution. The average atomic mass is $0.5135 \times 106.902 + 0.4865 \times 108.900$. We see that this can be rewritten as $106.902 + 0.4865(108.900 - 106.902) = 106.902 + 0.4865 \times 1.998 = 106.902 + 0.972 = 107.874$. The value of the atomic mass of silver is accordingly calculated to be 107.874.

4-8. Determination of Nucleidic Masses by Nuclear Reactions

A tremendous amount of information about nucleidic masses has been obtained by studying the energy released in nuclear reactions. A discussion of nuclear reactions is given in Chapter 20. The following example illustrates the general method.

Example 4-7. The fraction 0.0118% of the element potassium in its natural occurrence is the radioactive isotope ${}^{40}_{19}\mathrm{K}_{21}$. It undergoes beta decay:

$$ {}^{40}_{19}\mathrm{K}_{21} \longrightarrow {}^{40}_{20}\mathrm{Ca}^{+}_{20} + e^{-} $$

The kinetic energy of the emitted electron (the beta ray) has been measured by use of a magnetic beta-ray spectrometer (measurement of the curvature of the path of the beta ray in a magnetic field) and found to be 1.32 MeV. The mass of ^{40}Ca, which is stable and constitutes 97% of natural calcium, is 39.96259 d, as determined with the mass spectrograph. What is the mass of ^{40}K?

Solution. The energy of the reaction $\mathrm{Ca}^{+} + e^{-} \longrightarrow \mathrm{Ca}$ (6eV) is negligible. Hence the energy given out in the reaction $^{40}\mathrm{K} \longrightarrow {}^{40}\mathrm{Ca}$ is 1.32

MeV, which, with $e = 0.1602 \times 10^{-18}$ C, equals 0.211×10^{-12} J. We use the equation $E = mc^2$ to convert this energy to mass. Division by $c^2 = (2.9979 \times 10^8)^2$ gives 2.35×10^{-30} kg as the amount of mass converted into kinetic energy. The dalton is 1.660×10^{-27} kg. Hence the decrease in mass from ^{40}K to ^{40}Ca is $2.35 \times 10^{-30}/1.660 \times 10^{-27} = 0.00142$ d, and the mass of ^{40}K is $39.96259 + 0.00142 = 39.96401$ d.

4-9. The Discovery of the Correct Atomic Masses. Isomorphism

In the early years of the atomic theory there was no secure knowledge of the true relative masses of different elements. Dalton assigned atomic masses in such a way as to lead to simple formulas for compounds. Many chemists continued to write HO for water until 1858. In that year a principle discovered much earlier (1811) by Avogadro was applied so effectively by Stanislao Cannizzaro as to convince most chemists that the atomic masses obtained by its use could be accepted as correct. This principle will be discussed in Section 4-10.

Some other methods for assigning correct atomic masses were also developed during the first half of the nineteenth century. One of them involves the phenomenon of isomorphism.

In 1819 the German chemist Eilhard Mitscherlich (1794–1863) discovered isomorphism, the existence of different crystalline substances with essentially the same crystal form, and suggested his rule of isomorphism, which states that isomorphous crystals have similar chemical formulas.

As an example of isomorphism we may consider the minerals rhodochrosite, $MnCO_3$, and calcite, $CaCO_3$. Crystals of these two substances resemble one another closely, as shown in Figure 4-4. They both belong to the hexagonal crystal system (Section 2-5), and they both have

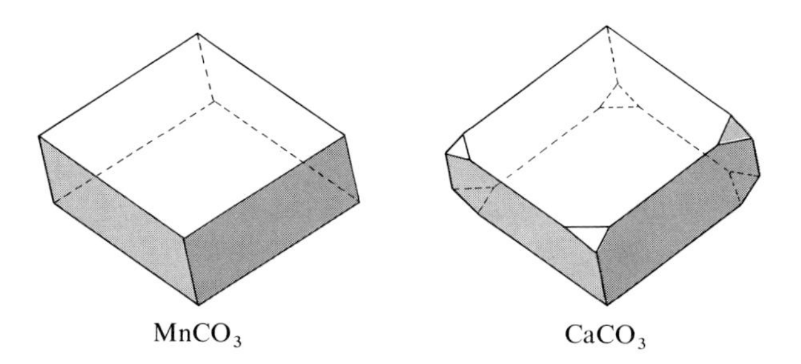

FIGURE 4-4
Isomorphous crystals of rhodochrosite and calcite (hexagonal system).

excellent rhombohedral cleavage. The larger of the two angles of a rhombic face of the cleavage rhombohedron has the value 102°50′ for rhodochrosite and 101°55′ for calcite. These facts justified the description of the two crystals as isomorphous more than a century ago. When x-ray diffraction was discovered it was possible to verify that the crystals have the same structure.

An illustration of the use of the rule of isomorphism is given by the work of the English chemist Henry E. Roscoe in determining the correct atomic weight of vanadium. Berzelius had attributed the atomic mass 68.5 to vanadium in 1831. In 1867 Roscoe noticed that the corresponding formula for the hexagonal mineral vanadinite was not analogous to the formulas of other hexagonal minerals apparently isomorphous with it:

		Axial ratio
Apatite	$Ca_5(PO_4)_3F$	$c/a = 1.363$
Hydroxyapatite	$Ca_5(PO_4)_3OH$	1.355
Pyromorphite	$Pb_5(PO_4)_3Cl$	1.362
Mimetite	$Pb_5(AsO_4)_3Cl$	1.377
Vanadinite	$Pb_5(VO_3)_3Cl$ (wrong)	1.404

The formula for vanadinite analogous to the other formulas is $Pb_5(VO_4)_3Cl$. On reinvestigating the compounds of vanadium Roscoe found that this latter formula is indeed the correct one, and that Berzelius had accepted VO, vanadium monoxide, as the elementary substance. The atomic mass of vanadium now accepted is 50.942.

There are occasional exceptions to the rule of isomorphism. An example is provided by hydroxyapatite, which in the form of its crystals is classed as isomorphous with apatite, but which has an extra atom in its formula. The explanation of this deviation from the rule is that the hydrogen atom is smaller than other atoms, and in the hydroxyapatite crystal the hydrogen atoms occupy interstices among the larger atoms that are vacant in the apatite crystal.

4-10. Avogadro's Law

A major part of chemical (and physical) theory was developed in connection with the experimental investigation of the properties of gases. An example is the evaluation of the correct atomic masses of elements by use of Avogadro's law.

In 1805 Gay-Lussac began a series of experiments to find the volume percentage of oxygen in air. The experiments were carried out by mixing a certain volume of hydrogen with air and exploding the mixture, and then testing the remaining gas to see whether oxygen or hydrogen had been present in excess. He was surprised to find a very simple relation: 1000 ml of oxygen required just 2000 ml of hydrogen, to form water. Continuing the study of the volumes of gases that react with one another, he found that 1000 ml of hydrogen chloride combined exactly with 1000 ml of ammonia, and that 1000 ml of carbon monoxide combined with 500 ml of oxygen to form 1000 ml of carbon dioxide. On the basis of these observations he formulated the law of combining volumes: the volumes of gases that react with one another or are produced in a chemical reaction are in the ratios of small integers.

Such a simple empirical law as this called for a simple theoretical interpretation, and in 1811 Amedeo Avogadro (1776–1856), professor of physics in the University of Turin, proposed a hypothesis to explain the law. Avogadro's hypothesis was that *equal numbers of molecules are contained in equal volumes of all dilute gases under the same conditions.* This hypothesis has been thoroughly verified to within the accuracy of approximation of real gases to ideal behavior, and it is now called a law—*Avogadro's law.**

During the last century Avogadro's law provided the most satisfactory and the only reliable way of determining which multiples of the equivalent masses of the elements should be accepted as their atomic masses; the arguments involved are discussed in the following sections. But the value of this law remained unrecognized by chemists from 1811 to 1858. In this year Stanislao Cannizzaro (1826–1910), an Italian chemist working in Geneva, showed how to apply the law systematically, and immediately the uncertainty about the correct atomic masses of the elements and the correct formulas of compounds disappeared. Before 1858 many chemists used the formula HO for water and accepted 8 as the atomic weight of oxygen; since that year H_2O has been accepted as the formula for water by everyone.†

Avogadro's Law and the Law of Combining Volumes

Avogadro's law requires that the volumes of gaseous reactants and products (under the same conditions) be approximately in the ratios of small integers; the numbers of molecules of reactants and products in a chemical

*Dalton had considered and rejected the hypothesis that equal volumes of gases contain equal numbers of atoms; the idea that elementary substances might exist as polyatomic molecules (H_2, O_2) did not occur to him.

†The failure of chemists to accept Avogadro's law during the period from 1811 to 1858 seems to have been due to a feeling that molecules were too "theoretical" to deserve serious consideration.

reaction are in integral ratios, and the same ratios represent the relative gas volumes. Some simple diagrams illustrating this for several reactions are given in Figure 4-5. Each cube in these diagrams represents the volume occupied by four gas molecules.

Standard Conditions

It is customary to refer the volumes of gases to conditions of temperature 0°C and pressure 1 atm. This temperature and pressure are called *standard conditions*. A sample of gas is said to be reduced to standard conditions when its volume is calculated at this temperature and pressure.

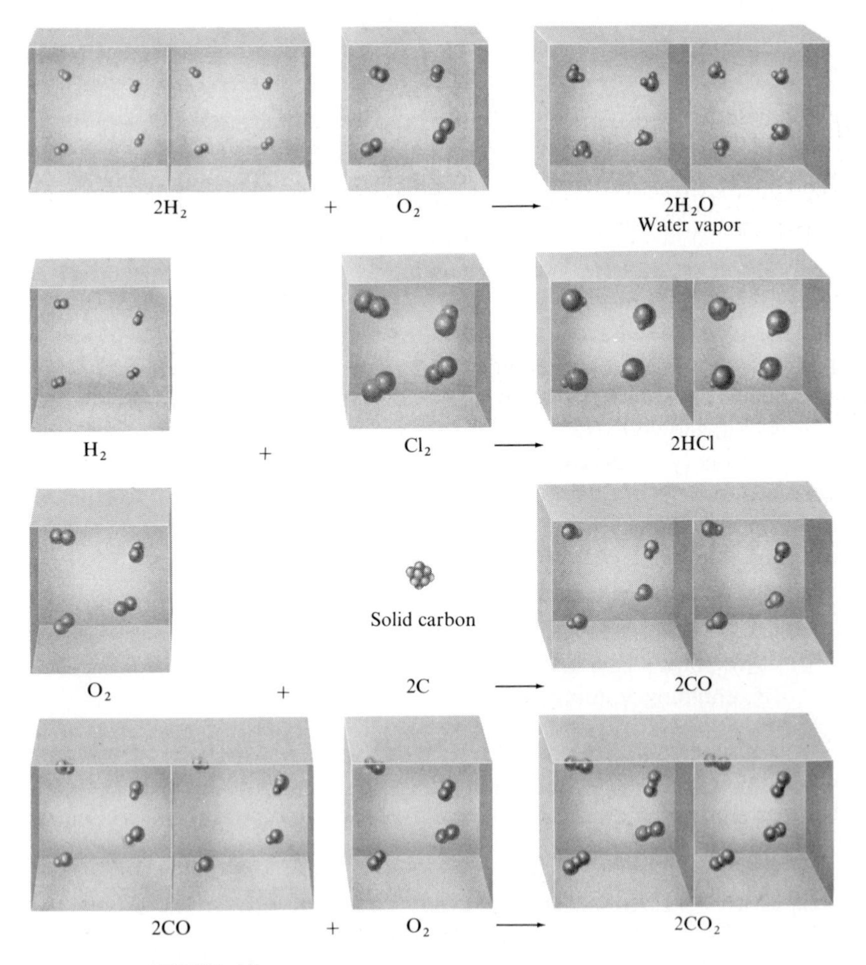

FIGURE 4-5
The relative volumes of gases involved in chemical reactions.

The Use of Avogadro's Law in the Determination of the Correct Atomic Weights of Elements

The way in which Avogadro's law was applied by Cannizzaro in 1858 for the selection of the correct approximate atomic masses of elements was essentially the following. Let us accept as the molecular mass of a substance the mass in grams of 22.4 liters of the gaseous substance reduced to standard conditions. (Any other volume could be used—this would correspond to the selection of a different base for the atomic mass scale.) Then it is probable that, of a large number of compounds of a particular element, at least one compound will have only one atom of the element per molecule; the mass of the element in the standard gas volume of this compound is its atomic mass.

For gaseous compounds of hydrogen the mass per standard volume and the mass of the contained hydrogen per standard volume are given in the table below. In these and all other compounds of hydrogen the minimum mass of hydrogen in the standard gas volume is found to be 1 g, and the mass is always an integral multiple of the minimum mass; hence 1 can be accepted as the atomic mass of hydrogen. The elementary substance hydrogen then is seen to consist of diatomic molecules H_2, and water is seen to have the formula H_2O_x, with x still to be determined.

	Mass of gas (in grams)	Mass of contained hydrogen (in grams)
Hydrogen (H_2)	2	2
Methane (CH_4)	16	4
Ethane (C_2H_6)	30	6
Water (H_2O)	18	2
Hydrogen sulfide (H_2S)	34	2
Hydrogen cyanide (HCN)	27	1
Hydrogen chloride (HCl)	36	1
Ammonia (NH_3)	17	3
Pyridine (C_5H_5N)	79	5

For oxygen compounds the following similar table of experimental data can be set up. From the comparison of oxygen and water in this table it can be concluded rigorously that the oxygen molecule contains two atoms or a multiple of two atoms; we see that the standard volume of oxygen contains twice as much oxygen (32 g) as is contained by the standard volume of water vapor (16 g of oxygen). The data for the other compounds provide no evidence that the atomic mass of oxygen is less than 16; hence this value may be adopted. Water thus is given the formula H_2O.

	Mass of gas (in grams)	Mass of contained oxygen (in grams)
Oxygen (O_2)	32	32
Water (H_2O)	18	16
Carbon monoxide (CO)	28	16
Carbon dioxide (CO_2)	44	32
Nitrous oxide (N_2O)	44	16
Nitric oxide (NO)	30	16
Sulfur dioxide (SO_2)	64	32
Sulfur trioxide (SO_3)	80	48

Note that this application of Avogadro's law provided rigorously only a maximum value of the atomic mass of an element. The possibility was not eliminated that the true atomic mass was a submultiple of this value.

Example 4-8. What is the density of acetylene at 0°C and 1 atmosphere pressure?

Solution. Acetylene, C_2H_2, has a molecular mass of 26.04. At 0°C and 1 atmosphere 26.04 g acetylene occupies 22.4 liters. The density is therefore 26.04 g/22.4 l = 1.16 g l^{-1}.

Example 4-9. A certain compound of carbon and hydrogen contains 81.8% carbon. This compound is a gas with density 1.96 g per liter at standard conditions. What is its formula?

Solution. The molecular mass of the compound is $1.96 \times 22.4 = 43.9$ d. The number of atoms of carbon in the compound is $(0.818 \times 43.9)/12 = 3$ and the number of atoms of hydrogen is $[(1 - 0.818) \times 43.9]/1 = 8$. The formula is thus C_3H_8.

4.11. The Perfect-gas Equation

It is interesting that it was not until the early years of the seventeenth century that the word "gas" was used. This word was invented by a Belgian physician, J. B. van Helmont (1577–1644), to fill the need caused by the new idea that different kinds of "airs" exist. Van Helmont discovered that a gas (the gas that we now call carbon dioxide) is formed when limestone is treated with acid, and that this gas differs from air in that when respired it does not support life and that it is heavier than air. He also found that the same gas is produced by fermentation, and that it is present in the Grotto del Cane, a cave in Italy in which dogs were observed to become unconscious (carbon dioxide escaping from fissures in the floor displaces the air in the lower part of the cave).

During the seventeenth and eighteenth centuries other gases were discovered, including hydrogen, oxygen, and nitrogen, and many of their properties were investigated. It was not until nearly the end of the eighteenth century, however, that these three gases were recognized as elements. When Lavoisier recognized that oxygen is an element, and that combustion is the process of combining with oxygen, the foundation of modern chemistry was laid.

Gases differ remarkably from liquids and solids in that the volume of a sample of gas depends in a striking way on the temperature of the gas and the applied pressure. The volume of a sample of liquid water, say 1 kg of water, remains essentially the same when the temperature and pressure are changed somewhat. Increasing the pressure from 1 atm to 2 atm causes the volume of a sample of liquid water to decrease by less than 0.01% and increasing the temperature from 0°C to 100°C causes the volume to increase by only 2%. On the other hand, the volume of a sample of air is cut in half when the pressure is increased from 1 atm to 2 atm, and it increases by 36.6% when the temperature is changed from 0°C to 100°C.

We can understand why these interesting phenomena attracted the attention of scientists during the early years of development of modern chemistry through the application of quantitative experimental methods of investigation of nature, and why many physicists and chemists during the past century have devoted themselves to the problem of developing a sound theory to explain the behavior of gases. A part of this theory is presented in Appendix III.

In addition to the desire to understand this part of the physical world, there is another reason, a practical one, for studying the gas laws. This reason is concerned with the measurement of gases. The most convenient way to determine the amount of material in a sample of a solid is to weigh it on a balance. This can also be done conveniently for liquids; or we may measure the volume of a sample of a liquid, and, if we want to know its weight, multiply the volume by its density, as found by a previous experiment. The method of weighing is, however, not conveniently used for gases, because their densities are very small; volume measurements can be made much more accurately and easily by the use of containers of known volume. It is partly for this reason that study of the pressure-volume-temperature properties of gases is a part of chemistry.

It has been found by experiment that *at low density all ordinary gases behave in nearly the same way*. The nature of this behavior is described by the *perfect-gas laws* (often referred to briefly as the *gas laws*).

The Dependence of Gas Volume on Pressure

Experiments on the volume of a gas as a function of the pressure have shown that, for nearly all gases, the volume of a sample of gas at constant temperature is inversely proportional to the pressure; that is, the product of pressure and volume under these conditions is constant:

$$PV = \text{constant (temperature constant, moles of gas constant)}$$

This equation expresses Boyle's law. The law was inferred from experimental data by the English natural scientist Robert Boyle (1627–1691) in 1662.

Whereas all of the common gases, such as oxygen, hydrogen, nitrogen, carbon monoxide, carbon dioxide, and the rest, behave in the way described by Boyle's law, there are some gases that do not. One of these is the gas nitrogen dioxide, NO_2, the molecules of which can combine to form double molecules, of dinitrogen tetroxide, N_2O_4. A sample of this gas under ordinary conditions contains some molecules NO_2 and some molecules N_2O_4. When the pressure on the sample of the gas is changed the number of molecules of each of these kinds changes, causing the volume to depend on the pressure in a complicated way, rather than in the simple way described by Boyle's law. This effect will be discussed in Chapter 10.

The Dependence of Gas Volume on Temperature

In 1787 the French physicist Jacques Alexandre Charles (1746–1823) reported that different gases expand by the same fractional amount for the same rise in temperature. Dalton in England continued these studies in 1801, and in 1802 Joseph Louis Gay-Lussac (1778–1850) extended the work, and determined the amount of expansion per degree centigrade. He found that all gases expand by $\frac{1}{273}$ of their volume at 0°C for each degree centigrade that they are heated above this temperature. Thus a sample of gas with volume 273 ml at 0°C has the volume 274 ml at 1°C and the same pressure, 275 ml at 2°C, 373 ml at 100°C, and so on.

We now state the law of the dependence of the volume of a gas on temperature, the *law of Charles and Gay-Lussac*, in the following way:

if the pressure and the number of moles of a sample of gas remain constant, the volume of the sample of gas is proportional to the absolute temperature:

$$V = \text{constant} \times T \text{ (pressure constant, number of moles constant)}$$

The dependence of volume on the absolute temperature is a direct proportionality, whereas the volume is inversely proportional to the pressure. The nature of these two relations is illustrated in Figure 4-6.

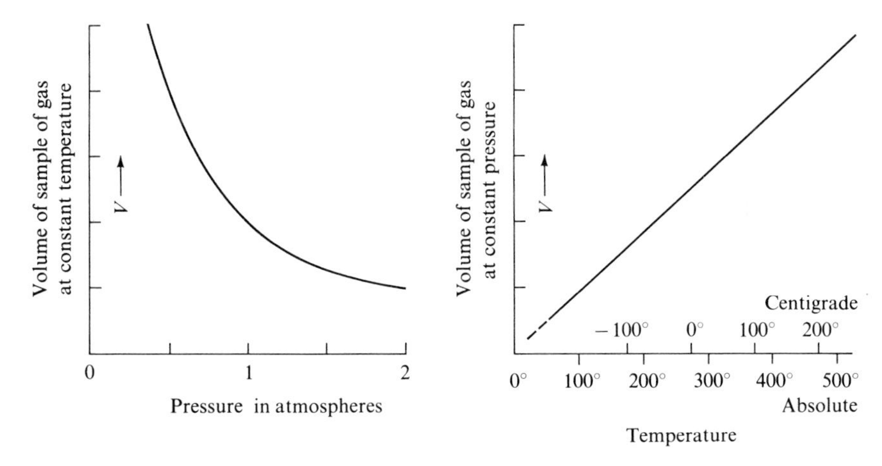

FIGURE 4-6
Curves showing, at the left, the dependence of the volume of a sample of gas at constant temperature and containing a constant number of molecules on the pressure, and, at the right, the dependence of the volume of a sample of gas at constant pressure and containing a constant number of molecules on the temperature.

The Complete Perfect-gas Equation

Boyle's law, the law of Charles and Gay-Lussac, and Avogadro's law can be combined into a single equation,

$$PV = nRT$$

In this equation P is the pressure acting on a given sample of gas, V is the volume occupied by the sample of gas, n is the number of moles of gas in the sample, R is a quantity called the *gas constant*, and T is the absolute temperature.

The gas constant R has a numerical value depending on the units in which it is measured (that is, the units used for P, V, n, and T). If P is measured in atmospheres, V in liters, n in moles, and T in degrees Kelvin, the value of R is 0.0820 liter atmospheres per degree mole (more accurately 0.08206 l atm K^{-1} $mole^{-1}$); R also has the value 8.3146 J K^{-1} $mole^{-1}$. The value of R/N (N is Avogadro's number) is called *Boltzmann's constant, k;* its value is 13.805×10^{-24} J K^{-1}.

The volume of one mole of a perfect gas at standard conditions is 22.415 liters (22.415×10^{-3} m^3).

Some of the ways in which the perfect-gas equation can be used in the solution of chemical problems are illustrated in the following examples.

Example 4-10. A sample of gas is found by measurement to have the volume 1000 ml at the pressure 730 mm of mercury. What would be its volume at normal atmospheric pressure, 760 mm of mercury?

Solution. Let P_1 be the initial pressure, 730 mm Hg, and V_1 be the volume, 1000 ml. Let P_2 be the changed pressure, 760 mm Hg, and V_2 be the changed volume, which we wish to determine. From Boyle's law we know that the product PV remains constant; hence we write

$$P_1V_1 = P_2V_2$$

or

$$730 \text{ mm Hg} \times 1000 \text{ ml} = 760 \text{ mm Hg} \times V_2$$

Solving for V_2, we obtain

$$V_2 = \frac{730 \text{ mm Hg}}{760 \text{ mm Hg}} \times 1000 \text{ ml} = 960 \text{ ml}$$

There is another way of solving the problem that involves more thinking, and that may help to prevent errors. We know that Boyle's law is of such a form that the volume changes by a factor equal to the ratio of the two pressures. We can hence obtain the final volume by multiplying the initial volume by the ratio $\frac{730}{760}$. (We know that the ratio $\frac{760}{730}$ is not the correct one to multiply by, because *increase* in pressure always causes a *decrease* in volume, and the factor must accordingly be *less* than 1.) Thus we obtain as the desired volume

$$\frac{730}{760} \times 1000 \text{ ml} = 960 \text{ ml}$$

Example 4-11. To what temperature would a sample of gas, held at constant pressure, have to be heated in order to have double the volume that it has at 0°C?

Solution. At constant pressure, the volume of a gas is proportional to the absolute temperature. 0°C = 273 K. Therefore the volume will be twice as great at 2×273 K = 546 K, or 546 K − 273 K = 273°C.

Example 4-12. How much helium should be put into a balloon of 10,000-liter capacity that is meant to rise without loss of gas to a height where the pressure is 200 mm of mercury and the temperature is −10°C, if it starts at sea level (1 atm pressure) at a temperature of 25°C?

Solution. The perfect-gas equation shows that for a given sample of a gas (n is constant) PV/T = a constant.
Therefore

$$\frac{10{,}000 \times 200}{263} = \frac{V \times 760}{298}$$

$$V = \frac{298}{263} \times \frac{200}{760} \times 10{,}000 = 3{,}080 \text{ liters}$$

Example 4-13. Calculate the volume occupied at 20°C and 1 atm pressure by the gas evolved from 100 cm^3 of solid carbon dioxide, which has density 1.53 g cm^{-3}.

Solution. One hundred cm^3 of solid CO_2 weighs 153 g. The molecular weight of CO_2 = 44; therefore n, the number of moles of CO_2 in 100 cm^3 of the solid = 153/44 = 3.477.

$$PV = nRT \text{ or } V = nRT/P$$

$$= (3.477 \text{ mole} \times 0.08206 \text{ l atm K}^{-1} \text{ mole}^{-1} \times 293 \text{ K})/1 \text{ atm}$$

$$= 8.36 \text{ liters}$$

EXERCISES

4-1. What is an element? Is it possible to prove rigorously by chemical methods that a substance is an element? Is it possible to prove rigorously by chemical methods that a substance is a compound?

4-2. Describe two chemical experiments that would prove that water is not an element. Can you devise a chemical proof that oxygen is an element?

4-3. Gasoline burns to form water and carbon dioxide. Does this prove rigorously that gasoline is not an element?

4-4. Use the Moseley diagram (Figure 4-2) to make a prediction of the approximate value of the wavelength of the K_α x-ray emission line of argon. Can you surmise why it was not measured by Moseley?

4-5. Define isotope, nucleide, mass number, N, A, Z, the dalton.

4-6. What are the atomic number and approximate atomic mass of the element each of whose nuclei contains 81 protons and 122 neutrons? Give the complete symbol for this nucleide, including chemical symbol, atomic number, mass number, and neutron number.

4-7. How many protons and how many neutrons are there in the nucleus of the isotope of cobalt with mass number 60? Of the isotope of nickel with mass number 60? Of the isotope of plutonium with mass number 238?

4-8. An atom of ^{90}Sr emits a beta ray. What are the atomic number and mass number of the resulting nucleus? What element is it? This nucleus also emits a beta ray. What nucleus does it produce?

4-9. Argon, potassium, and calcium all have nucleides with mass number 40. How many protons and how many neutrons constitute each of the three nuclei?

4-10. What are the relative advantages of basing the chemists' atomic mass scale on H = 1.000, $^{16}O = 16.00000$, or $^{12}C = 12.00000$?

4-11. What would be the implications of defining Avogadro's number as 1.00000×10^{24}? What units would have to be changed?

4-12. By counting the flashes of light produced by alpha particles when they strike a screen coated with zinc sulfide, Sir William Ramsay and Professor Frederick Soddy found that 1 g of radium gives off 13.8×10^{10} alpha particles (nuclei of helium atoms) per second. They also measured the amount of helium gas produced in this way, finding 0.158 cm^3 (at 0°C and 1 atm) per year per gram of radium. At this temperature and pressure 1 l of helium weighs 0.179 g. Avogadro's number of helium atoms weighs 4.003 g (the atomic mass of helium is 4.003). From these data calculate an approximate value of Avogadro's number.

4-13. What is the weight percentage composition of the elementary constituents of ethanol, C_2H_5OH?

4-14. Ethanol burns according to the reaction $C_2H_5OH + 3O_2 \rightarrow 2\,CO_2 + 3H_2O$. What weight of water is produced by burning 1 kg of ethanol?

4-15. How much coal (assumed to be pure carbon) is needed to reduce one ton of Fe_2O_3 to iron? How much iron is produced?

4-16. Exactly what is meant by the statement that the atomic mass of samarium is 150.35?

4-17. Thallium occurs in nature as ^{203}Tl and ^{205}Tl. From the atomic mass of natural thallium, 204.39, calculate the nucleide composition of thallium. The atomic mass of ^{203}Tl is 202.97 and that of ^{205}Tl is 204.97.

4-18. The density of NaCl(c) is 2.165 g cm^{-3}. Calculate the molar volume, and, with use of Avogadro's number, the volume of the unit cube, containing four sodium atoms and four chlorine atoms, and the value of a, the edge of the unit cube. (This is the method used by the Braggs in their original determination of the wavelengths of x-rays.)

4-19. What methods have been used to determine atomic masses?

4-20. What is Avogadro's number? What is Avogadro's law?

4-21. The number of people on the earth is about 4×10^9. What volume of gas at standard conditions contains this number of molecules?

4-22. Disulfur decafluoride, S_2F_{10}, has standard boiling point 29°C. What is the density of the gas at this temperature (1 atm pressure)? What is the ratio of its density to that of hydrogen?

4-23. What is the volume in cubic feet at standard conditions of one ounce-molecular-weight of a gas?* (Answer: 22.4.)

4-24. What is the weight in ounces of 22.4 cu ft of carbon dioxide at standard conditions? (Answer: 44.)

4-25. The density of hydrogen cyanide at standard conditions is 1.29 g $liter^{-1}$. Calculate the apparent molecular mass of hydrogen cyanide vapor.

---

*It is interesting in this connection that the master craftsmen of Lübeck defined the ounce as one one-thousandth of the weight of one cubic foot of ice-cold water.

5

The Chemical Elements, the Periodic Law, and the Electronic Structure of Atoms

The 106 known elements include some with which everyone is familiar and many that are rare. At room temperature some of the elementary substances are gases, some are liquids, and some are solids.* They show great variety in their chemical properties and in the nature of the compounds that they form. In consequence, the study of chemistry is not simple or easy; to obtain a reasonably broad knowledge of general chemistry it is necessary to learn many facts.

The facts of chemistry cannot be completely coordinated by a unifying theory. Nevertheless, the development of chemical theories has now proceeded far enough to be of great aid to the student, who can simplify his task of learning about the properties and reactions of substances by correlating this information with theories, such as the theory of atomic structure, which has been discussed in the preceding chapters, and the periodic law, which we shall now consider.

---

*The elements that are gases at standard conditions (0°C and 1 atm) are hydrogen, helium, nitrogen, oxygen, fluorine, neon, chlorine, argon, krypton, xenon, and radon. The only elements that are liquids at standard conditions are bromine and mercury.

5-1. The Periodic Law

Let us first recall from Chapters 3 and 4 that atoms are built of particles of three kinds: protons, neutrons, and electrons. The nucleus of each atom is made of protons and neutrons. The number of protons (the atomic number) determines the electric charge of the nucleus, and the total number of protons and neutrons (the mass number) determines its mass. In a neutral atom the number of electrons surrounding the nucleus is equal to the atomic number.

The periodic law states that *the properties of the chemical elements are not arbitrary, but depend upon the structure of the atom and vary with the atomic number in a systematic way.* The important point is that this dependence involves a crude periodicity that shows itself in the recurrence of characteristic properties.

For example, the elements with atomic numbers 2, 10, 18, 36, 54, and 86 are all chemically inert gases. Similarly, the elements with atomic numbers one greater—namely, 3, 11, 19, 37, 55, and 87—are all light metals that are very reactive chemically. These six metals, lithium (3), sodium (11), potassium (19), rubidium (37), cesium (55), and francium (87), all react with chlorine to form colorless compounds that crystallize in cubes and show a cubic cleavage. The chemical formulas of these salts are similar: LiCl, NaCl, KCl, RbCl, CsCl, and FrCl. The composition and properties of other compounds of these six metals are correspondingly similar, and different from those of other elements.

The comparison of the observed chemical and physical properties of elements and their compounds with the atomic numbers of the elements accordingly indicates that, after the first two elements, hydrogen and helium, which constitute the **very short period** (the word period is used for a sequence of elements), there are the **first short period** of eight elements (from helium, atomic number 2, to neon, 10), the **second short period** of eight elements (to argon, 18), the **first long period** of eighteen elements (to krypton, 36), the **second long period** of eighteen elements (to xenon, 54), and then the **very long period** of 32 elements (to radon, 86). In case that enough new elements of very large atomic number are made in the future it may well be found that there is another very long period of 32 elements ending in another inert gas, with atomic number 118.

5-2. The Periodic Table

The periodic recurrence of properties of the elements with increasing atomic number may be effectively emphasized by arranging the elements in a table, called the *periodic table* or *periodic system* of the elements.

TABLE 5-1
The Periodic System of the Elements

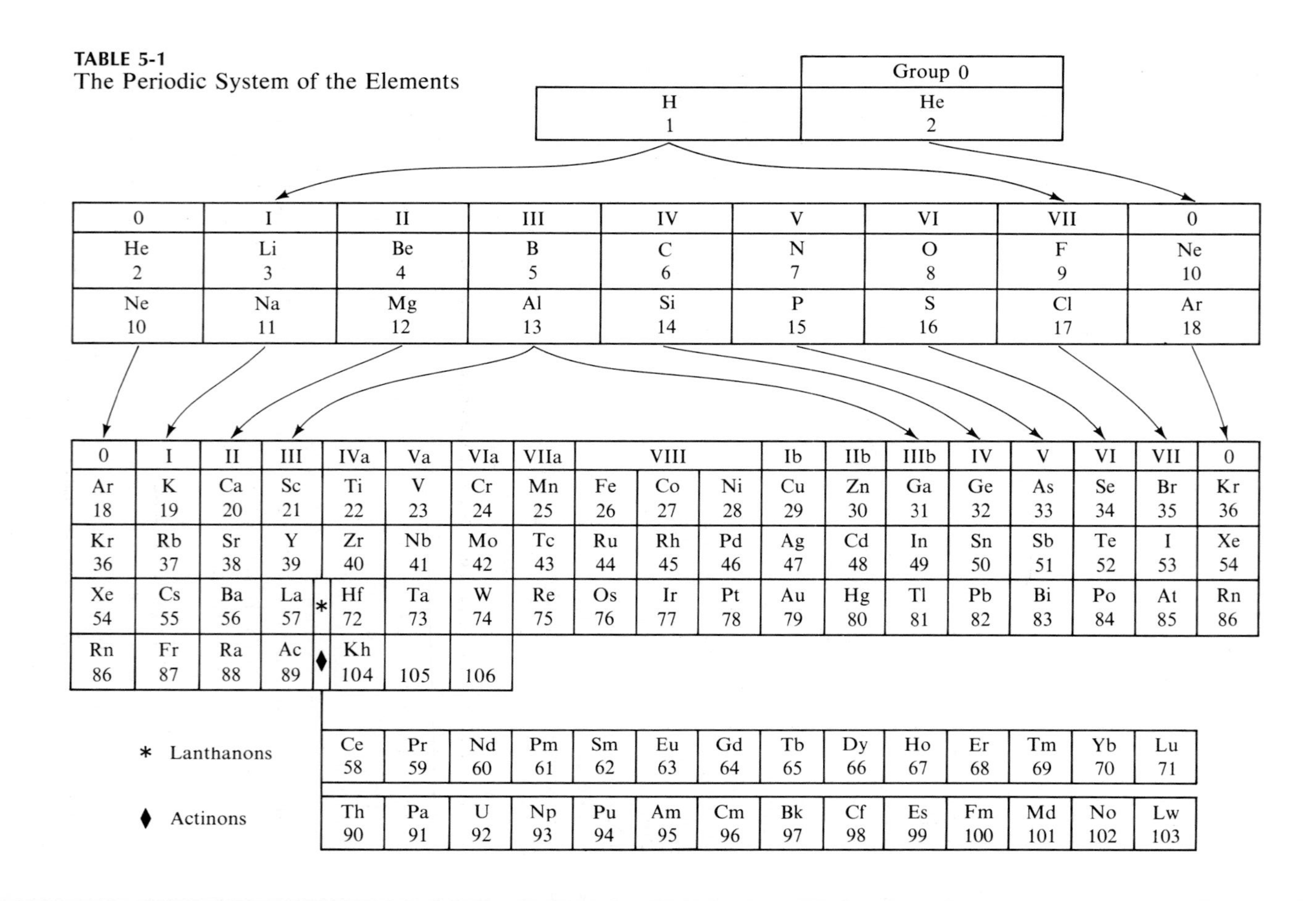

	Group 0
H 1	He 2

0	I	II	III	IV	V	VI	VII	0
He 2	Li 3	Be 4	B 5	C 6	N 7	O 8	F 9	Ne 10
Ne 10	Na 11	Mg 12	Al 13	Si 14	P 15	S 16	Cl 17	Ar 18

0	I	II	III	IVa	Va	VIa	VIIa	VIII			Ib	IIb	IIIb	IV	V	VI	VII	0
Ar 18	K 19	Ca 20	Sc 21	Ti 22	V 23	Cr 24	Mn 25	Fe 26	Co 27	Ni 28	Cu 29	Zn 30	Ga 31	Ge 32	As 33	Se 34	Br 35	Kr 36
Kr 36	Rb 37	Sr 38	Y 39	Zr 40	Nb 41	Mo 42	Tc 43	Ru 44	Rh 45	Pd 46	Ag 47	Cd 48	In 49	Sn 50	Sb 51	Te 52	I 53	Xe 54
Xe 54	Cs 55	Ba 56	La 57 *	Hf 72	Ta 73	W 74	Re 75	Os 76	Ir 77	Pt 78	Au 79	Hg 80	Tl 81	Pb 82	Bi 83	Po 84	At 85	Rn 86
Rn 86	Fr 87	Ra 88	Ac 89 ♦	Kh 104	105	106												

* Lanthanons	Ce 58	Pr 59	Nd 60	Pm 61	Sm 62	Eu 63	Gd 64	Tb 65	Dy 66	Ho 67	Er 68	Tm 69	Yb 70	Lu 71
♦ Actinons	Th 90	Pa 91	U 92	Np 93	Pu 94	Am 95	Cm 96	Bk 97	Cf 98	Es 99	Fm 100	Md 101	No 102	Lw 103

Several alternative forms of the periodic table have been proposed and used. We shall base the discussion of the elements and their properties in this book on the table shown as Table 5-1 (it is also reproduced inside the front cover of the book).

The Development of the Periodic Table

The differentiation of chemical substances into two groups, elements and compounds, was achieved at the end of the eighteenth century. A long time was required for the recognition of the fact that the elements can be classified in the way now described by the periodic law.

The most important step in the development of the periodic table was taken in 1869, when the Russian chemist Dmitri I. Mendelyeev (1834–1907) made a thorough study of the relation between the atomic masses of the elements and their physical and chemical properties. Mendelyeev proposed a periodic table containing seventeen columns, resembling Table 5-1 with the end columns (labeled 0) missing (these elements had not yet been discovered at that time). In 1871 Mendelyeev and the German chemist Lothar Meyer (1830–1895), who was working independently, proposed another table, with eight columns, obtained by splitting each of the long periods into a period of seven elements, an eighth group containing the three central elements (such as iron, cobalt, nickel), and a second period of seven elements. The first and second periods of seven were later distinguished by use of the letters "a" and "b" attached to the group symbols, which were the Roman numerals. This nomenclature of the periods (Ia, IIa, IIIa, IVa, Va, VIa, VIIa, VIII, Ib, IIb, IIIb, IVb, Vb, VIb, VIIb) appears, slightly revised, in the present periodic table.

The periodic table in the second form proposed by Mendelyeev (the "short-period" form) remained popular for many years, but has now been largely replaced by the "long-period" form, used in this book, which is in better agreement with the new knowledge about the electronic structure of atoms.

The periodic law was accepted immediately after its proposal by Mendelyeev because of his success in making predictions with its use which were afterward verified by experiment. In 1871 Mendelyeev found that by changing seventeen elements from the positions indicated by the atomic masses that had then been assigned to them into new positions, their properties could be better correlated with the properties of the other elements. He pointed out that this change indicated the existence of small errors in the previously accepted atomic masses of several of the elements, and large errors for several others, to the compounds of which incorrect formulas had been assigned. Further experimental work verified Mendelyeev's revisions.

A very striking application of the periodic law was made by Mendelyeev. He was able to predict the existence of six elements that had not yet been discovered, corresponding to vacant places in his table. He named these elements eka-boron, eka-aluminum, eka-silicon, eka-manganese, dvi-manganese, and eka-tantalum (Sanskrit: *eka,* first; *dvi,* second).

Three of these elements were soon discovered (they were named scandium, gallium, and germanium by their discoverers), and it was found that their properties and the properties of their compounds are very close to those predicted by Mendelyeev for eka-boron, eka-aluminum, and eka-silicon, respectively. Since then the elements technetium, rhenium, and protactinium have been discovered or made artificially, and have been found to have properties similar to those predicted for eka-manganese, dvi-manganese, and eka-tantalum. A comparison of the properties predicted by Mendelyeev for eka-silicon and those determined experimentally for germanium is given below.

MENDELYEEV'S PREDICTIONS FOR EKA-SILICON *(1871)*	OBSERVED PROPERTIES OF GERMANIUM *(discovered in 1886)*
Atomic mass about 72.	Atomic mass 72.59.
Es will be obtained from EsO_2 or K_2EsF_6 by reaction with sodium.	Ge is obtained by reaction of K_2GeF_6 and sodium.
Es will be a dark gray metal, with high melting point and density 5.5 g cm^{-3}.	Ge is gray, with melting point 958°C and density 5.36 g cm^{-3}.
Es will be slightly attacked by acids, such as hydrochloric acid, HCl, and will resist alkalies, such as sodium hydroxide, NaOH.	Ge is not dissolved by HCl or NaOH, but is dissolved by concentrated nitric acid, HNO_3.
On heating Es, it will form the oxide EsO_2, with high melting point and density 4.7 g cm^{-3}.	Ge reacts with oxygen to give GeO_2, m.p. 1100°C, density 4.70 g cm^{-3}.
A hydrated EsO_2 soluble in acid and easily reprecipitated is expected.	$Ge(OH)_4$ dissolves in dilute acid and is reprecipitated on dilution or addition of base.
The sulfide, EsS_2, will be insoluble in water but soluble in ammonium sulfide.	GeS_2 is insoluble in water and dilute acids, but readily soluble in ammonium sulfide.
$EsCl_4$ will be a volatile liquid, with boiling point a little under 100°C and density 1.9 g cm^{-3}.	$GeCl_4$ is a volatile liquid, with b.p. 83°C and density 1.88 g cm^{-3}.

Example 5-1. In the article on Chemistry in the Ninth Edition of The Encyclopaedia Britannica (published in 1878) the author, H. S. Armstrong, states that Mendelyeev had recently proposed that uranium be assigned the atomic mass 240 in place of the old value 120 that had been assigned to it by Berzelius, but that he himself preferred 180. Mendelyeev was

right. The correct formula of pitchblende, an important ore or uranium, is U_3O_8. What formula was written for pitchblende by *(a)* Berzelius, *(b)* Armstrong?

Solution. With the atomic mass 120 assigned to uranium by Berzelius the atoms of uranium would have only half the mass assigned to them by Mendelyeev, and accordingly Mendelyeev's formula U_3O_8 would become U_6O_8 or, in simplest form, U_3O_4. Similarly, Armstrong's value 180 (three quarters of 240) leads to the formula UO_2.

5-3. Description of the Periodic Table

The horizontal rows of the periodic table consist of a very short period (containing hydrogen and helium, atomic numbers 1 and 2), two short periods of 8 elements each, two long periods of 18 elements each, a very long period of 32 elements, and an incomplete period.

The properties of elements change in a systematic way through a period: this is indicated in Figure 5-1, which shows the density of the elements, in the crystalline state, as a function of the atomic number. It is seen that there are five pronounced minima (low points) in the density curve. They occur for the elements sodium (11), potassium (19), rubidium (37), cesium (55), and francium (87). These five elements together

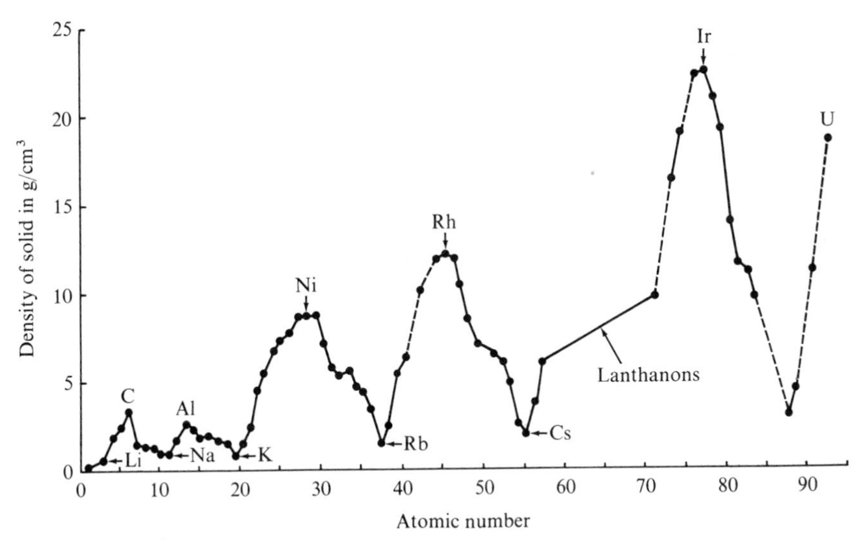

FIGURE 5-1
The density of the elements in the solid state, in g cm^{-3}. The symbols of the elements at high and low points of the jagged curve are shown.

with lithium constitute a group of elements that are strikingly similar in their properties (Section 18-3).

The **vertical columns** *of the periodic table,* with connections between the short and long periods as shown, *are the* **groups** *of chemical elements.* Elements in the same group may be called *congeners;* these elements have closely related physical and chemical properties.

The groups I, II, and III are considered to include the elements in corresponding places at the left side of all the periods in Table 5-1, and IV, V, VI, and VII the elements at the right side. The central elements of the long periods, called the *transition elements,* have properties differing from those of the elements of the short periods; these elements are discussed separately, as groups IVa, Va, VIa, VIIa, VIII (which, for historical reasons, includes three elements in each long period), Ib, IIb, and IIIb.

The very long period is compressed into the table by removing fourteen elements, the *rare-earth metals* or *lanthanons* (elements resembling lanthanum, $Z = 57$), from $Z = 58$ to $Z = 71$, and representing them separately below. The elements from $Z = 90$ to $Z = 103$, called the *actinons* (elements resembling actinium, $Z = 89$), are listed below the lanthanons.

The elements on the left side and in the center of the periodic table are **metals.** These elementary substances have the characteristic properties called *metallic properties*—high electric and thermal conductivity, metallic luster, the capability of being hammered or rolled into sheets (malleability) and of being drawn into wire (ductility). The elements on the right side of the periodic table are **nonmetals,** the elementary substances not having metallic properties.

The metallic properties are most pronounced for elements in the lower left corner of the periodic table, and the nonmetallic properties are most pronounced for elements in the upper right corner (omitting the noble gases). The transition from metals to nonmetals is marked by the *elements with intermediate properties,* which occupy a diagonal region extending from a point near the upper center to the lower right corner. These elements, which are called **metalloids,** include boron, silicon, germanium, arsenic, antimony, tellurium, and polonium.

The groups of elements may be described briefly in the following way:

Group 0, the argonons: The elements of this group, helium, neon, argon, krypton, xenon, and radon, are nearly completely unreactive chemically; they form only a few chemical compounds. A discussion of the argonons is given in the following sections of this chapter.

Group I, the alkali metals: The alkali metals—lithium, sodium, potassium, rubidium, cesium, and francium—are light metals which are very reactive chemically. Many of their compounds have important uses in industry and in life. The alkali metals and their compounds are discussed in Chapter 18. The word alkali is derived from an arabic word meaning ashes (compounds of these metals were obtained from wood ashes).

Group II, the alkaline-earth metals: These metals—beryllium, magnesium, calcium, strontium, barium, and radium—and their compounds are discussed in Chapter 18.

Group III, the boron or aluminum group: Boron is a metalloid, whereas aluminum and its other congeners are metals. The properties of boron and its congeners are discussed in Chapter 18.

Group IV, carbon and silicon: The chemistry of carbon is described in Chapters 7, 8, and 9 and in greater detail in Chapters 13 to 15. The chemistry of silicon and the other elements of this group is described in Chapter 18.

Group V, the nitrogen or phosphorus group: Nitrogen and phosphorus are nonmetals, their congeners arsenic and antimony are metalloids, and bismuth is usually classed as a metal. The chemistry of nitrogen, phosphorus, and the other elements of the group is described in Chapters 7 and 8.

Group VI, the oxygen group: Oxygen and its congeners sulfur and selenium are nonmetals, whereas tellurium and polonium are classed as metalloids. The chemistry of oxygen and its congeners is discussed in Chapters 7 and 8.

Group VII, the halogen group: The halogens—fluorine, chlorine, bromine, iodine, and astatine—are the class of the most strongly nonmetallic elements. They are very reactive chemically, and form many compounds. Their chemistry is discussed in Chapters 7 and 8. The word halogen is from the Greek words *hals*, salt, and *genes*, producing.

The discussion in the immediately following chapters will be largely restricted to these elements, whose chemistry can be systematized by comparison of their electronic structures with those of the argonons. The remaining elements are called the transition elements. They are discussed in Chapter 19.

5-4. The Argonons

The first element in the periodic table, hydrogen, is a reactive substance that forms a great many compounds. The chemistry of hydrogen is discussed in Chapter 7. Helium, the second element (atomic number 2), is much different; it is a gas with the very striking chemical property that *it forms no chemical compounds whatever,* but exists only in the free state. Its atoms will not even combine with one another to form polyatomic molecules, but remain as separate atoms in the gas, which is hence described as containing monatomic molecules. Because of its property of remaining aloof from other elements it is called a "noble" gas.

This lack of chemical reactivity is the result of an extraordinary stability of the electronic structure of the helium atom. This stability is characteristic of the presence of two electrons close to an atomic nucleus.

The other elements of the zero group—neon, argon, krypton, xenon, and radon—are also chemically inert. The small tendency of these inert elements to form chemical compounds is similarly due to the great stability of their electronic structures. These extremely stable electronic structures are formed by 2, 10, 18, 36, 54, and 86 electrons about a nucleus.

These six gases are called the argonons (or sometimes the noble gases or inert gases). Their names, except radon, are from Greek roots: *helios,* sun; *neos*, new; *argos*, inert; *kryptos*, hidden; *xenos*, stranger. Radon is named after radium, from which it is formed by radioactive decomposition. The properties of the argonons are given in Table 5-2. Note the regular dependence of melting point and boiling point on atomic number.

Helium

Helium is present in very small quantities in the atmosphere. Its presence in the sun is shown by the occurrence of its spectral lines in sunlight. These lines were observed in 1868, long before the element was discovered on earth, and the lines were ascribed to a new element, which was named helium* by Sir Norman Lockyer (1836–1920).

Helium occurs as a gas entrapped in some uranium minerals, from which it can be liberated by heating. It is also present in natural gas from some wells, especially in Texas and Canada; this is the principal source of the element.

Helium is used for filling balloons and for mixing with oxygen (in place of the nitrogen of the air) for breathing by divers, in order to avoid the "bends," caused by gas bubbles formed by release of the nitrogen of the atmosphere that had dissolved in the blood under increased pressure, and to avoid the narcotic action (anesthetizing action) of nitrogen under pressure. Its chief use is as a liquid to provide extreme cold for scientific purposes and electrical superconductivity.

---

*The ending "ium," which is otherwise used only for metallic elements, is due to Lockyer's incorrect surmise that the new element was a metal. "Helion" would be a better name, as its ending is consistent with those of the names of the other argonons.

TABLE 5-2
Properties of the Argonons

	Symbol	Atomic Number	Atomic Mass	Melting Point	Boiling Point
Helium	He	2	4.003	−272.2°C*	−268.9°C
Neon	Ne	10	20.183	−248.67°	−245.9°
Argon	Ar	18	39.944	−189.2°	−185.7°
Krypton	Kr	36	83.80	−157°	−152.9°
Xenon	Xe	54	131.30	−112°	−107.1°
Radon	Rn	86	222	−77°	−61.8°

*At 26 atm pressure. At smaller pressures helium remains liquid at still lower temperatures.

Neon

The second argonon, neon, occurs in the atmosphere to the extent of 0.002%. It is obtained, along with the other noble gases (except helium), by the distillation of liquid air (air that has been liquefied by cooling).

When an electric current is passed through a tube containing neon gas at low pressure, the atoms of neon are caused to emit light with their characteristic spectral lines. This produces a brilliant red light, used in advertising signs (neon signs). Other colors for signs are obtained by the use of helium, argon, and mercury, sometimes in mixtures with neon or with one another.

Argon

Argon composes about 1% of the atmosphere. It is used in incandescent light bulbs to permit the filament to be heated to a higher temperature, and thus to produce a whiter light than would be practical in a vacuum. The argon decreases the rate at which the metallic filament evaporates, by keeping vaporized metal atoms from diffusing away from the filament and permitting them to reattach themselves to it. Argon is also extensively used in industry to provide a chemically inert atmosphere, especially in welding and in making pure metals and alloys.

Krypton, Xenon and Radon

Krypton and xenon, which occur in very small quantities in the air, have not found any significant use. Xenon is a good anesthetic agent, but it is too expensive for general use (it has been used in two major operations on human beings).

In 1962 and 1963 several compounds of xenon were synthesized. The first one to be reported (by the English chemist Neil Bartlett) was xenon hexafluoroplatinate, $XePtF_6$, a yellow crystalline substance. Later (1963) he reported the synthesis of the corresponding rhodium compound, $XeRhF_6$. Scientists in the Argonne National Laboratory, and later other investigators, prepared several xenon fluorides, including XeF_2, XeF_4, and XeF_6. Each compound is a colored solid substance that reacts vigorously with water. The products of the reaction of XeF_4 and XeF_6 with water are unstable (explosive) solid substances with composition $Xe(OH)_4$, XeO_3, and H_4XeO_6 (perxenic acid). Some compounds of radon and krypton (RnF_4, KrF_4) have also been synthesized.

Radon, which is produced steadily by radium, is radioactive and is used in the treatment of cancer. It has been found that the rays given off by radioactive substances are often effective in controlling this disease. A convenient way of administering this radiation is to pump the radon that has been produced by a sample of radium into a small gold tube, which is then placed in proximity to the tissues to be treated.

The Discovery of the Argonons

The story of the discovery of argon provides an interesting illustration of the importance of attention to minor discrepancies in the results of scientific investigations.

For over a hundred years it was thought that atmospheric air consisted, aside from small variable amounts of water vapor and carbon dioxide, solely of oxygen (21% by volume) and nitrogen (79%). In 1785 the English scientist Henry Cavendish (1731–1810) investigated the composition of the atmosphere. He mixed oxygen with air and then passed an electric spark through the mixture, to form a compound of nitrogen and oxygen, which was dissolved in a solution in contact with the gas (Figure 5-2). The sparking was continued until there was no further decrease in volume, and the oxygen was then removed from the residual gas by treatment with another solution. He found that after this treatment only a small bubble of air remained unabsorbed, not more than $\frac{1}{120}$ of the original air. Although Cavendish did not commit himself on the point, it seems to have been assumed by chemists that if the sparking had been continued for a longer time there would have been no residue, and Cavendish's experiment was accordingly interpreted as showing that only oxygen and nitrogen were present in the atmosphere.

Then in 1894, more than 100 years later, Lord Rayleigh began an investigation involving the careful determination of the densities of the gases hydrogen, oxygen, and nitrogen. To prepare nitrogen he mixed dried air with an excess of ammonia, NH_3, and passed the mixture over red-hot

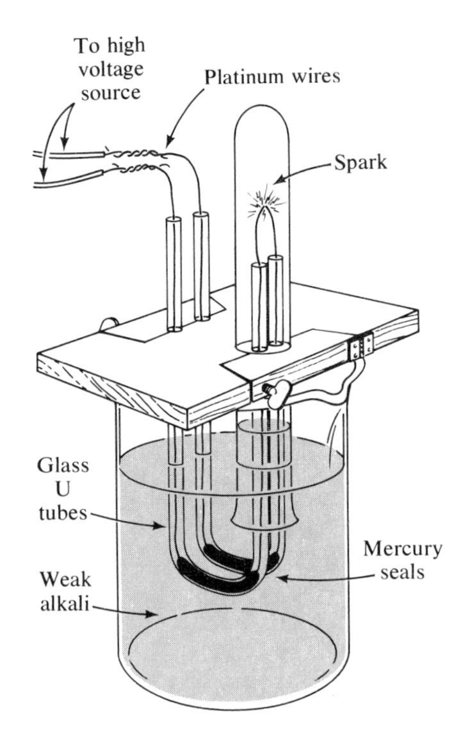

FIGURE 5-2
The Cavendish apparatus, as used in the investigation of the composition of air.

copper. Under these conditions the oxygen reacts with ammonia, according to the equation

$$4NH_3 + 3O_2 \rightarrow 6H_2O + 2N_2$$

The excess ammonia is then removed by bubbling the gas through sulfuric acid. The remaining gas, after drying, should have been pure nitrogen, derived in part from the ammonia and in part from air. The density of this gas was determined. Another sample of nitrogen was made simply by passing air over red-hot copper, which removed the oxygen by combining with it to form copper oxide:

$$O_2 + 2Cu \rightarrow 2CuO$$

When the density of this gas was determined it was found to be about 0.1% greater than that from the sample of ammonia and air. In order to investigate this discrepancy, a third sample of nitrogen was made by the reaction of ammonia and pure oxygen. It was found that this sample of nitrogen had a density 0.5% less than that of the second sample.

Further investigations showed that nitrogen prepared entirely from air had a density 0.5% greater than nitrogen prepared from ammonia or in any other chemical way. Nitrogen obtained from air was found to have density 1.2572 g liter^{-1} at 0°C and 1 atm, whereas nitrogen made by chemical methods has density 1.2505 g liter^{-1}. Rayleigh and Ramsay then repeated Cavendish's experiment, and showed by spectroscopic analysis that the residual gas was indeed not nitrogen but a new element. They then searched for the other argonons and discovered them.

5-5. The Electronic Structure of Atoms

The argonons are strange elements. They are different from all other elements—they form very few compounds, whereas every other element forms many compounds.

This peculiarity of the argonons is explained by the **electronic structure** of their atoms—the way in which the electrons move about the atomic nuclei. This is the subject that we shall now consider, beginning with the electronic structure of the simplest element, hydrogen.

The knowledge about the structure of atoms that is presented in the following paragraphs has been obtained largely by physicists from the study of spectral lines. The understanding of atomic structure was obtained during the years between 1913 and 1925. It was in 1913 that Niels Bohr (1885–1962), the great Danish physicist, developed his simple theory of the hydrogen atom (Section 5-7), which during the following twelve years was expanded and refined into our present theory of atomic structure.

The detailed mathematical theory of quantum mechanics—the modern mathematical theory of the properties of electrons and other small particles—is not suited to study by the beginning student. However, the picture of the electronic structure of atoms that is provided by this theory is easy to understand and to learn. Knowledge of this electronic structure is important to the student of chemistry.

The Electronic Structure of the Hydrogen Atom

The smallest and lightest nucleus is the proton. The proton carries one unit of positive charge, and with one electron, which carries one unit of negative charge, it forms a hydrogen atom.

Soon after the development of the concept of the nuclear atom some idea was gained as to the way in which a proton and an electron are combined to form a hydrogen atom. Because of the attraction of the oppositely charged electron and proton, the electron might be expected to revolve in an orbit about the much heavier proton in a way similar to that in which the earth revolves about the sun. Bohr suggested that the orbit of the electron in the normal hydrogen atom should be circular, with radius 53 pm (see Section 5-7). The electron was calculated to be going around in this orbit with the constant speed 2.18×10^6 m s^{-1}, which is a little less than 1% of the speed of light.

As a result of studies made by many physicists, this picture is now known to be nearly but not quite right. The electron does not move in a definite orbit, but rather in a somewhat random way, so that it is sometimes very close to the nucleus and sometimes rather far away. Moreover, it moves mainly toward the nucleus or away from it, and it travels in all directions about the nucleus instead of staying in one plane. Although it does not stay just 53 pm from the nucleus, this is its most probable distance. By moving around rapidly it effectively occupies all the space within about 100 pm from the nucleus, and so gives the hydrogen atom an effective radius of about 100 pm. It is because of this motion of electrons that atoms, which are made of particles only about 10^{-3} pm in diameter, act as solid objects several hundred pm in diameter. The speed of the electron in the hydrogen atom is not constant; but its root-mean-square average is the Bohr value 2.18×10^6 m s^{-1}.

Thus we can describe the free hydrogen atom as having a heavy nucleus at the center of a sphere defined by the space filled by the fast-moving electron in its motion about the nucleus. This sphere is about 200 pm in diameter.

Because of the nature of the equations of quantum mechanics that describe the electron in the normal hydrogen atom, it has been decided that it is not right to say that the electron moves about the nucleus in an orbit. Instead, the electron is said to occupy an *orbital*. The orbital that is occupied by the electron in the normal state (most stable state) of the hydrogen atom is called the 1s orbital. The number 1 is the value of the *principal quantum number n*.

There is only one orbital for $n = 1$. There are other possible orbitals for the hydrogen atom, corresponding to $n = 2$, $n = 3$, etc. A hydrogen atom in which the electron occupies one of these other orbitals is unstable; it is said to be in an *excited state*. A large amount of energy is required to change the hydrogen atom from its normal state to the first excited state ($n = 2$), three-quarters as much as to remove the electron completely. The diameter of the atom for this excited state is about 800 pm, four times as great as for the normal state. The orbitals with $n = 2, 3, 4$, etc. are occupied by electrons in heavier atoms.

The Spin of the Electron

It was discovered in 1925 by two Dutch physicists, G. E. Uhlenbeck and S. A. Goudsmit, that *the electron has a spin*—it rotates about an axis in a way that can be compared with the rotation of the earth about an axis through its north pole and south pole. The amount of the spin (angular momentum) is the same for all electrons, but the orientation of the axis can change. With respect to a specified direction, such as the direction of the earth's magnetic field, a free electron can orient itself in either one of only two ways: it lines up either parallel to the field, or antiparallel (with the opposite orientation).

In general the motion of electricity produces a magnetic field. The spin of the electron is no exception—the electron produces a magnetic field corresponding to the magnetic moment that would be expected for the rotation of negative electricity. The spinning electron can be described as a small magnet that can orient itself in a magnetic field so that its component along the field direction is either $+\mu_B$ or $-\mu_B$, where μ_B is the Bohr magneton, 0.927×10^{-23} J T^{-1}.* The spin of an electron in a magnetic field can be made to change from positive to negative orientation by absorption of microwave radiation of suitable frequency. This is the basis of the technique of *electron spin resonance spectroscopy,* which during the years since 1945 has provided much information about electronic structure.

The Pauli Exclusion Principle

Two electrons can occupy the same orbital only if their spins are opposed; that is, oriented in opposite directions.

This sentence is a statement of the *Pauli exclusion principle.* W. Pauli (Austrian physicist, 1900–1959) was the first man to notice that an electron excludes another electron with the same spin orientation from the orbital it occupies. Only two electrons can occupy one orbital, and they must have opposite spins. Two electrons with opposed spins occupying the same orbital are called an *electron pair.*

The Electronic Structure of the Argonons

The distributions of electrons in atoms of the noble gases have been determined by physicists by experimental and theoretical methods that are too complex to be discussed here. The results obtained are shown in Figure 5-3. It is seen that *for the atoms neon, argon, krypton, and xenon the electrons are arranged about the atomic nuclei in two or more concentric shells.*

The **helium atom** contains two electrons, each of which carries out motion about the helium nucleus similar to that of the one electron in the hydrogen atom. These two electrons occupy the same orbital, the 1*s* orbital, and in accordance with the Pauli exclusion principle their spins are opposed.

*One joule tesla^{-1} equals 10^3 erg gauss^{-1}.

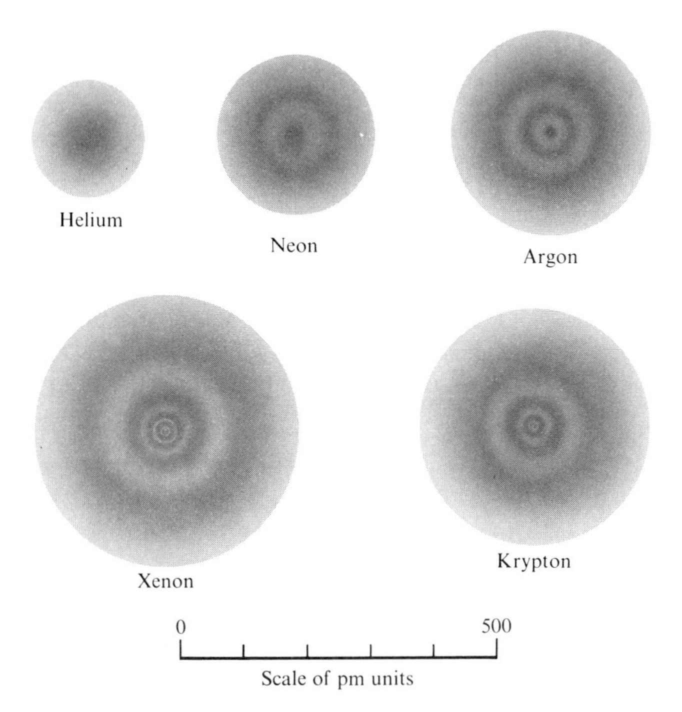

FIGURE 5-3
Drawing of electron distributions in argonon atoms, showing successive electron shells.

The symbol $1s^2$ is used to express the electron configuration of the normal helium atom. The superscript 2 means that two electrons occupy the $1s$ orbital. The electrons can be described as forming a ball of negative electricity near the nucleus. Its diameter is only about half that of the hydrogen atom, because of the doubled value of the nuclear charge.

These two electrons are said to constitute a **completed helium shell** (also called a **completed *K* shell**).

All of the atoms heavier than hydrogen have a completed helium shell, consisting of two $1s$ electrons ($1s^2$) close to the nucleus. The diameter of the helium shell is inversely proportional to the atomic number; for radon ($Z = 86$) it is only about 2 pm.

The **neon atom** has two shells. First, it has a helium shell of two electrons, with diameter about 20 pm, as shown in Figure 5-3, and around this shell an **outer shell of eight electrons,** called the **neon shell** or ***L* shell.** The diameter of this outer shell is about 200 pm.

These two shells, reduced in size, appear in argon, krypton, and xenon (Figure 5-3), together with additional shells. The nature of these shells is discussed in the following paragraphs.

Shells and Subshells of Electrons

Around 1920, while they were developing the theory of atomic spectra (line spectra and x-ray spectra of the elements), physicists discovered that the successive shells, after the helium shell, contain orbitals of more than one kind.

The K shell consists of only one orbital, the $1s$ orbital, described in the preceding section. The L shell consists of four orbitals and two subshells. The $2s$ subshell consists of only one orbital, the $2s$ orbital. The $2p$ subshell consists of three $2p$ orbitals. An electron in a $2s$ orbital is somewhat more stable and somewhat closer to the nucleus than an electron in one of the $2p$ orbitals, as is indicated in the energy diagram, Figure 5-4. The three $2p$ orbitals have the same energy.

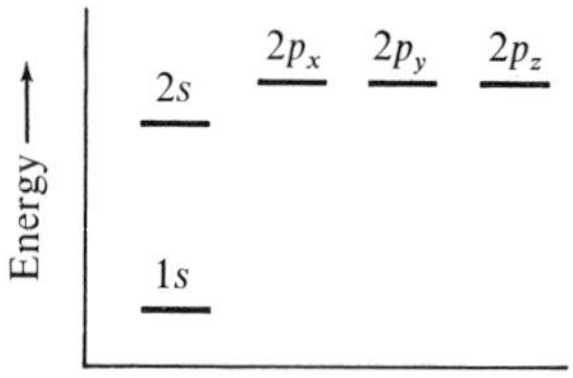

Energy →
$2s$ $2p_x$ $2p_y$ $2p_z$
$1s$

FIGURE 5-4
Diagram showing the relative stability of the $1s$, $2s$, $2p_x$, $2p_y$, and $2p_z$ orbitals. The vertical coordinate also measures the relative average distances of the electrons from the nucleus.

The $2s$ orbital, like the $1s$ orbital, corresponds to an electron distribution that is spherically symmetrical. The electron distribution for a $2p$ orbital is not spherically symmetrical, but is concentrated about an axis, as shown in Figure 5-5. The characteristic axes of the three $2p$ orbitals in an atom are at right angles to one another, and they can be taken as the x axis, the y axis, and the z axis, respectively, as indicated in Figure 5-5. The three $2p$ orbitals can be given the symbols $2p_x$, $2p_y$, and $2p_z$.

In accordance with the Pauli exclusion principle, each of these orbitals can be occupied by two electrons, which must have their spins opposed. Hence the completed $2s$ subshell contains two electrons (one electron pair) and the complete $2p$ subshell contains six electrons (three electron pairs, one for each of the three $2p$ orbitals). The completed L shell accordingly contains eight electrons (four electron pairs).

The symbol for the completed $2s$ subshell is $2s^2$ and that for the completed $2p$ subshell is $2p_x^2 2p_y^2 2p_z^2$, which is usually simplified to $2p^6$. The symbol for the completed L shell is $2s^2 2p_x^2 2p_y^2 2p_z^2$, usually written as $2s^2 2p^6$.

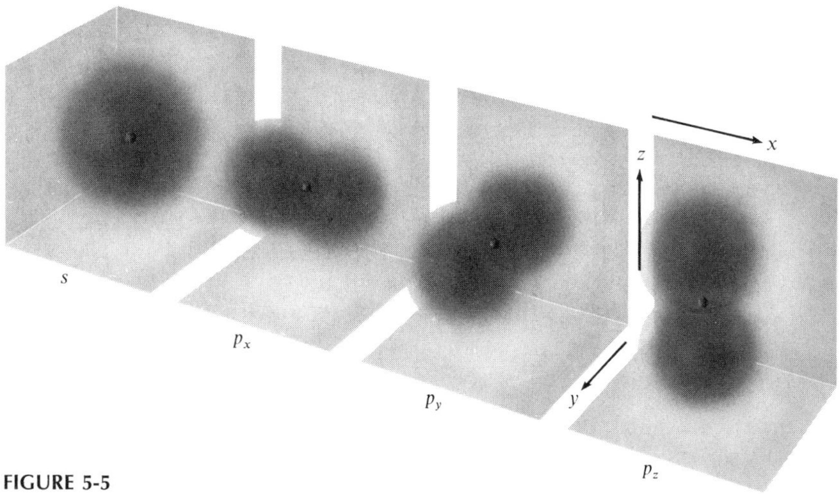

FIGURE 5-5
Representation of the relative magnitudes of the s orbital and the three p orbitals in dependence on angle.

The M shell, with principal quantum number $n = 3$, consists of nine orbitals and three subshells. In addition to the $3s$ subshell (one orbital) and the $3p$ subshell (three orbitals), it contains a $3d$ subshell, with five $3d$ orbitals.

The next shell, the N shell, contains these subshells plus an additional one, the $4f$ subshell, which has seven orbitals.

The letters K, L, M, N, O, P that are used for electron shells correspond to the successive values of the principal quantum number n: K to $n = 1$, L to $n = 2$, M to $n = 3$, and so on. It is as a result of historical accident that this sequence of letters begins with K rather than A. Also, the letters s, p, d, f have a curious origin: they are the initial letters of the adjectives *sharp, principal, diffuse,* and *fundamental,* which happened to be used by the spectroscopists to describe the spectra that they had observed. These letters accordingly are not abbreviations of words that describe the orbitals or subshells in a significant way.

The successive electron shells (K, L, M, etc.) described above are those used by physicists. Chemists have found it convenient to use a different classification, corresponding to a different way of grouping the subshells together. In Figure 5-6 the approximate sequence of energy values for all orbitals that are occupied by electrons in atoms in their normal states are shown. It is seen that the energy values for the physicists' shells overlap. For example, the energy of a $3d$ electron (in the M shell) is about the same as that of a $4s$ electron (in the N shell). Because of their interest in sets of electrons with the same energy, chemists have found it convenient to assign the five $3d$ orbitals to $4s$ and $4p$, rather than to $3s$ and

TABLE 5-3
Shells and Subshells of Electrons

Chemists' Name for Shell	Symbol for Electron Configuration of Completed Shell, Showing Subshells
Helium shell	$1s^2$
Neon shell	$2s^2 2p^6$
Argon shell	$3s^2 3p^6$
Krypton shell	$3d^{10} 4s^2 4p^6$
Xenon shell	$4d^{10} 5s^2 5p^6$
Radon shell	$4f^{14} 5d^{10} 6s^2 6p^6$
Eka-radon shell	$5f^{14} 6d^{10} 7s^2 7p^6$

$3p$, and to make similar assignments for $4d$, $5d$, $6d$, $4f$, and $5f$, as indicated in Figure 5-6.

The successive shells are named for the argonons in which they first are completed. The number of electrons in each shell is equal to the number of elements in the period of the periodic table that is completed by the corresponding noble gas: 2 for the helium shell, 8 each for the neon shell and the argon shell, 18 each for the krypton shell and the xenon shell, and 32 each for the radon shell and the eka-radon shell. The symbols for the completed shells are given in Table 5-3.

Electron Shells and the Periods of the Periodic Table

The successive electron shells listed in Table 5-3 involve the following numbers of electrons: 2, 8, 8, 18, 18, 32, 32.

These numbers are equal to the numbers of elements in the successive periods of the periodic system (the last period incomplete).

We see that each of the two short periods, with eight elements, corresponds to the filling of two subshells, an s subshell (one orbital) and a p subshell (three orbitals).

The next two periods, the long periods, correspond to the filling not only of the next sets of these orbitals (giving $4s^2 4p^6$ and $5s^2 5p^6$), but also of sets of d orbitals ($3d^{10}$ and $4d^{10}$). It is the introduction of the ten d electrons that increases these periods to 18 elements each.

The very long period, which ends in radon, involves not only $5d^{10}$ and $6s^2 6p^6$, but also $4f^{14}$.

The Octet

In every one of the argonons except helium the outermost electrons, with the maximum value of the principal quantum number, are a set of eight (four pairs) with the symbol ns^2np^6. This set of eight electrons is called an *octet*.

It is found that most of the properties of elements close to the argonons in the periodic table can be discussed in a simple and satisfactory way in terms of the octet and the four corresponding orbitals ns, np_x, np_y, and np_z. (For other elements, discussed for the most part in Chapter 19, the d orbitals must also be taken into consideration.)

The Electronic Structure of the Elements of the First Short Period

Each of the elements from lithium to fluorine has an inner shell, $1s^2$. Lithium has in addition a $2s$ electron. Accordingly its electron configuration is $1s^22s$.

The electronic structure of an atom may be represented by an *electron-dot symbol*, in which the electrons of the outer shell (or outer octet) are represented by dots and the nucleus and inner electrons by the chemical symbol of the element. The electron-dot symbol of lithium, Li·, shows only the outermost electron, which is called the valence electron.

The next element, beryllium, has two valence electrons, both of which occupy the $2s$ orbital if the atom is in its normal state. The normal beryllium atom has the electron-dot symbol Be: and the electron configuration $1s^22s^2$. The two dots together represent two electrons with opposed spins in one orbital.

The electron-dot symbols for the normal states of the eight elements of the first short period are

$$\text{Li}\cdot \quad \text{Be}: \quad :\text{B}\cdot \quad :\dot{\text{C}}\cdot \quad :\dot{\underset{\cdot}{\text{N}}}\cdot \quad :\dot{\underset{\cdot\cdot}{\text{O}}}\cdot \quad :\ddot{\underset{\cdot\cdot}{\text{F}}}\cdot \quad :\ddot{\underset{\cdot\cdot}{\text{Ne}}}:$$

Note that two or three $2p$ electrons occupy different orbitals, and remain unpaired. The normal state of carbon corresponds to the configuration $1s^22s^22p_x2p_y$, and not to $1s^22s^22p_x{}^2$.

The electron configurations for the normal states of the elements lithium to neon are given in Table 5-4.

The electron configurations for the congeners of these elements are the same except for the increased value of the quantum period n. For example, sulfur, the congener of oxygen in the next period, has the configuration $3s^23p^4$ for its valence electrons.

The use of electronic structures in correlating the properties of substances will be illustrated in the following chapters.

Example 5-2. What is the general electronic configuration of the octet of valence electrons of an argonon? Why is an octet especially stable?

TABLE 5-4
Electron Configurations, Li to Ne

Atom	Number of Electrons per Orbital*				Electron Configuration*
	$2s$	$2p_x$	$2p_y$	$2p_z$	
Li	1				$2s$
Be	2				$2s^2$
B	2	1			$2s^2 2p$
C	2	1	1		$2s^2 2p^2$
N	2	1	1	1	$2s^2 2p^3$
O	2	2	1	1	$2s^2 2p^4$
F	2	2	2	1	$2s^2 2p^5$
Ne	2	2	2	2	$2s^2 2p^6$

*The inner electrons are not shown; for all of these atoms they are a $1s^2$ pair.

Solution. The electronic configuration of the octet of electrons of an argonon is ns^2np^6. By use of quantum mechanics there is an explanation of the great stability of the octet; in terms of elementary chemistry there is no explanation, it is a fact of nature.

Example 5-3. In an electric arc between carbon electrodes some of the carbon atoms are raised to an excited state to which the spectroscopists have assigned the electron configuration $1s^2 2s 2p_x 2p_y 2p_z$ (also written $2s2p^3$). What electron-dot symbol would you write for this state of the carbon atom?

Solution. The electron-dot symbol for the normal state of the carbon atom is $:\dot{\mathrm{C}}\cdot$ implying that two electrons have spins opposed in the $2s^2$ subshell. In the excited state with electron configuration $1s^2 2s 2p_x 2p_y 2p_z$ with no spin-paired electrons among the valence electrons a reasonable electron-dot formula is

$$\cdot\underset{\cdot}{\dot{\mathrm{C}}}\cdot$$

5-6. An Energy-level Diagram

A diagram representing the distribution of all electrons in all atoms is given in Figure 5-6.

Each orbital is represented by a square. The most stable orbital (its electrons being held most tightly by the nucleus) is the $1s$ orbital, at the bottom of the diagram. Energy is required to lift an electron from a stable orbital to a less stable one, above it in the diagram.

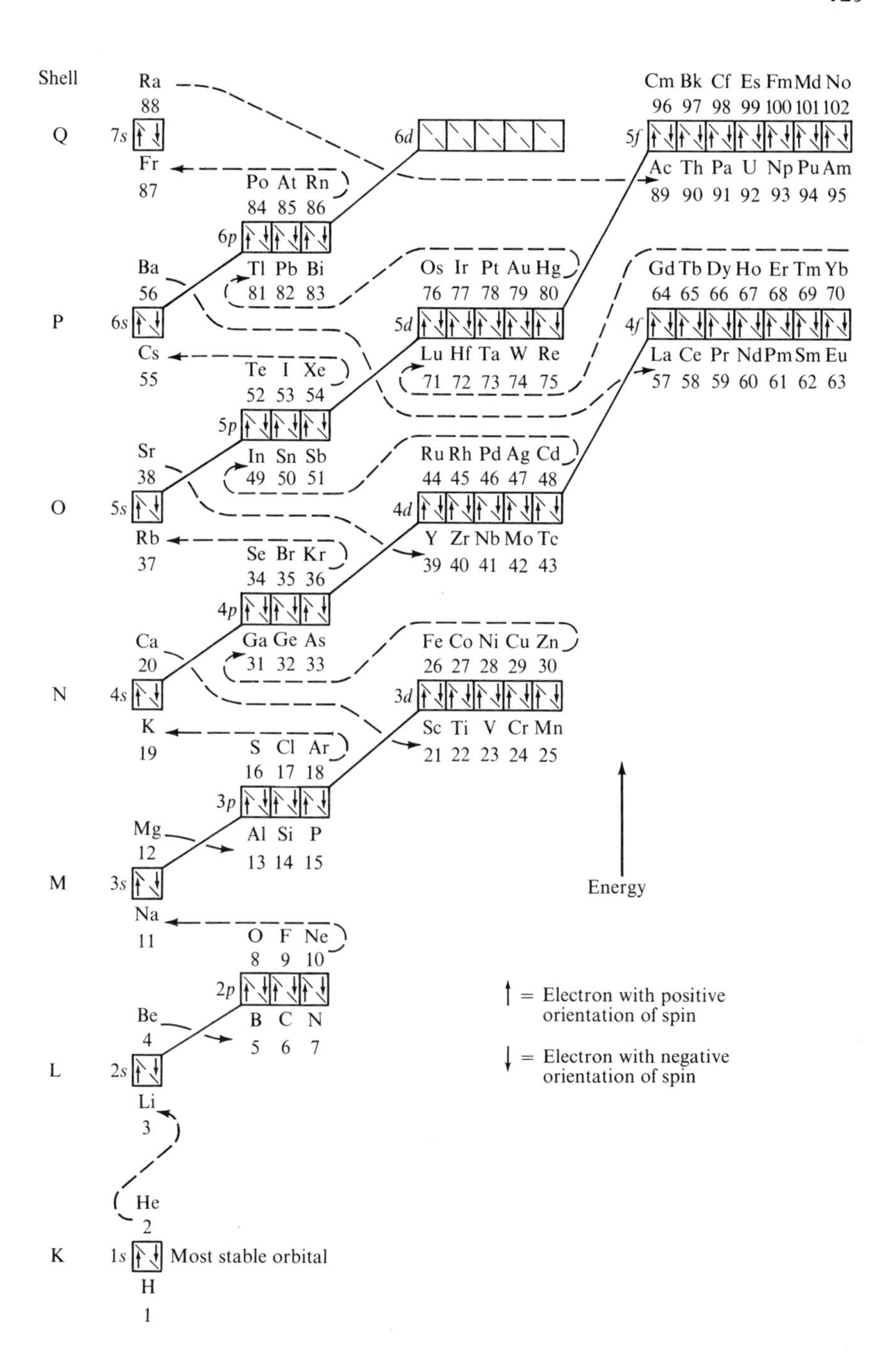

FIGURE 5-6
Energy-level diagram of electron shells and subshells of the elements.

The electrons are shown being introduced in sequence; the first and second in the $1s$ orbital, the next two in the $2s$ orbital, the next six in the $2p$ orbitals, and so on. The sequence is indicated by arrows. The symbol and atomic number of each element are shown adjacent to the outermost electron (least tightly held electron) in the neutral atom.

The electron configuration is indicated by the sequence along this path, up to the symbol of the element. Thus the electron configuration of nitrogen is $1s^2 2s^2 2p^3$ and that of scandium is $1s^2 2s^2 2p^6 3s^2 3p^6 4s^2 3d$.

For the heavier atoms two or more electron configurations may have nearly the same energy, and there is some arbitrariness in the diagram shown in Figure 5-6. The configuration shown for each element is either that of the most stable state of the free atom (in a gas) or of a state close to the most stable state.

Example 5-4. What is the electronic configuration of iron in its normal state (ground state)?

Solution. From the order of subshells in Figure 5-6 we see that the electronic configuration of iron is $1s^2 2s^2 2p^6 3s^2 3p^6 4s^2 3d^6$.

Example 5-5. How many electron pairs are there in iron? Which orbitals do they occupy? How many unpaired electrons are there? Which orbitals do they occupy?

Solution. By use of the results of Example 5-4 or by reference to Figure 5-6 and counting boxes of electron pairs there are 11 electron pairs in iron, occupying orbitals $1s^2 2s^2 2p^6 3s^2 3p^6 4s^2 3d^2$. Then there are four unpaired electrons occupying the other four $3d$ orbitals.

5-7. The Bohr Theory of the Structure of the Hydrogen Atom

Most of our knowledge of the electronic structure of atoms has been obtained by the study of the light given out by atoms when they are excited by high temperature or by an electric arc or spark. The light that is emitted by atoms consists of lines of certain frequencies; it is described as the *line spectrum* of the atom (see Figure 19-6).

The careful study of line spectra began about 1880. Early investigators made some progress in the interpretation of spectra, in recognizing regularities in the frequencies of the lines: the frequencies of the spectral lines of the hydrogen atom, for example, show an especially simple relationship with one another, which will be discussed below. The regularity is evident in the reproduction of a part of the spectrum of hydrogen in Figure 5-7. It was not until 1913, however, that the interpretation of

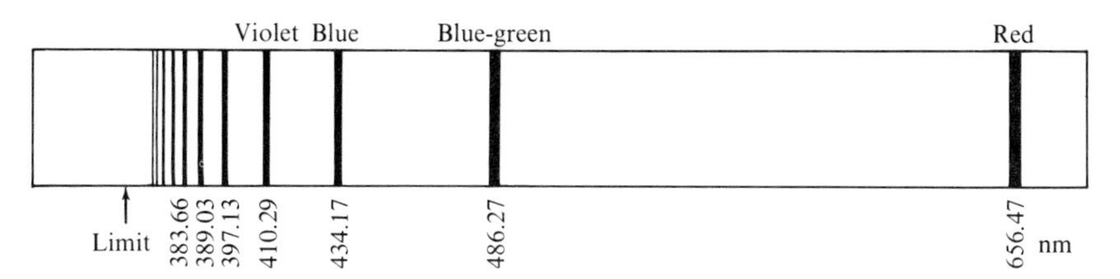

FIGURE 5-7
The Balmer series of spectral lines of atomic hydrogen. The line at the right, with the longest wavelength, is H_α. It corresponds to the transition from the state with $n = 3$ to the state with $n = 2$.

the spectrum of hydrogen in terms of the electronic structure of the hydrogen atom was achieved. In this year Niels Bohr successfully applied the quantum theory to this problem, and laid the basis for the extraordinary advance in our understanding of the nature of matter that has been made during the past sixty years.

The Quantum Theory of the Hydrogen Atom

The hydrogen atom consists of an electron and a proton. The interaction of their electric charges, $-e$ and $+e$, respectively, corresponds to inverse-square attraction, in the same way that the gravitational interaction of the earth and the sun corresponds to inverse-square attraction. If Newton's laws of motion were applicable to the hydrogen atom we should accordingly expect that the electron, whose mass is small compared with that of the nucleus, would revolve about the nucleus in an elliptical orbit, in the same way that the earth revolves about the sun. The simplest orbit for the electron about the nucleus would be a circle, and Newton's laws of motion would permit the circle to be of any size, as determined by the energy of the system.

After the discovery of the electron and the proton, this model was considered by physicists interested in atomic structure, and it became evident that the older theories of the motion of particles (Newton's laws of motion) and of electricity and magnetism could not apply to the atom. If the electron were revolving around the nucleus it should, according to electromagnetic theory, produce light, with the frequency of the light equal to the frequency of revolution of the electron in the atom. This emission of light by the moving electron is similar to the emission of radiowaves by the electrons that move back and forth in the antenna of a radio station. However, with the continued emission of energy by the atom, in the form of light, the electron would move in a spiral approaching more and more closely to the nucleus, and the frequency of its motion about the nucleus would become greater and greater. Accordingly, the older (classical) theories of motion and of electromagnetism would require that hydrogen atoms produce a spectrum of light of all wavelengths (a *continuous spectrum*). This is contrary to observation: the spectrum of hydrogen, produced in a discharge tube containing hydrogen atoms (formed by dissociation of hydrogen molecules), consists of lines, as shown in Figure 5-7. Moreover, it is known that the volume occupied by a hydrogen atom in a solid or liquid substance corresponds to a diameter of about 200 pm, whereas the older theory of the hydrogen atom contained no mechanism for preventing the electron from approaching more and more closely to the nucleus, and the atom from becoming far smaller than 200 pm in diameter.

A hint as to the way to solve this difficulty had been given to Bohr by Planck's quantum theory of emission of light by a hot body, and by Einstein's theory of the photoelectric effect and the light quantum. Both Planck and Einstein assumed that light of frequency ν is not emitted or absorbed by matter in arbitrarily small amounts, but only in quanta of energy $h\nu$. If a hydrogen atom in which the electron is revolving about the nucleus in a large circular orbit emits a quantum of energy $h\nu$, the electron must then be in a much different (smaller) circular orbit, corresponding to an energy value of the atom $h\nu$ less than its initial energy. Bohr accordingly assumed that *the hydrogen atom can exist only in certain states,* which are called the **stationary states** of the atom. He assumed that one of these states, the *ground state* or *normal state,* represents the minimum energy possible for the atom; it is accordingly the most stable state of the atom. The other states, with an excess of energy relative to the ground state, are called the *excited states* of the atom. He further assumed, in agreement with the earlier work of Planck and Einstein, that when an atom changes from a state of energy E'' to a state with energy E' the difference in energy $E'' - E'$ is equal to the energy of the light quantum that is emitted. This equation,

$$h\nu = E'' - E' \tag{5-1}$$

is called the **Bohr frequency rule;** it gives the value of the frequency of the light that is emitted when an atom changes from an excited state with energy E'' to a lower state with energy E'.

The same equation also applies to the absorption of light by atoms. The frequency of the light absorbed in the transition from a lower state to an upper state is equal to the difference in energy of the upper state and the lower state divided by Planck's constant. The equation also applies to the emission and absorption of light by molecules and more complex systems.

Bohr also discovered a method of calculating the energy of the stationary states of the hydrogen atom, with use of Planck's constant. He found that the correct values of the energies of the stationary states were obtained if he assumed that the orbits of the electrons are circular and that the angular momentum of the electron has for the normal state the value $h/2\pi$, for the first excited state the value $2h/2\pi$, for the next excited state the value $3h/2\pi$, and so on.

In general, the angular momentum of the electron in the circular orbit about the nucleus (the *Bohr orbit*) was represented by Bohr as having the value

$$\text{Angular momentum} = \frac{nh}{2\pi}, \qquad \text{with } n = 1, 2, 3, \ldots \tag{5-2}$$

the number n introduced in this way in the Bohr theory is called the *principal quantum number* of the Bohr orbit.

The radius of the Bohr orbit is found to be equal to n^2a_0, in which

$$a_0 = h^2/4\pi^2me^2 = 53 \text{ pm}$$

In this equation m is the mass of the electron and e is the electronic charge. Thus the radius of the Bohr orbit for the normal state of the hydrogen atom is 53 pm, that for the first excited state is four times as great, that for the next excited state nine times as great, and so on, as illustrated in Figure 5-8.

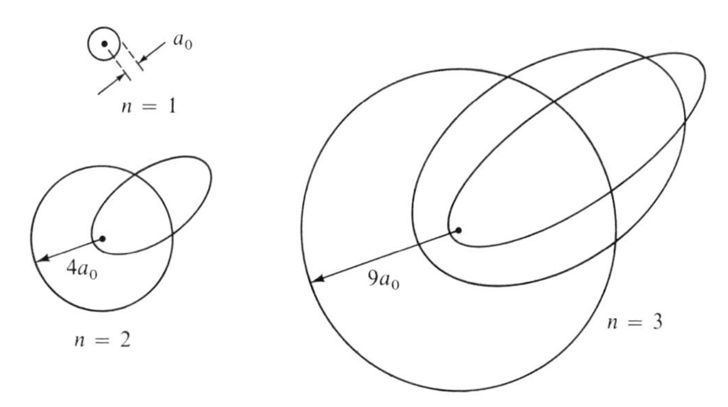

FIGURE 5-8
Bohr orbits for an electron in the hydrogen atom. These circular and elliptical orbits were involved in the Bohr theory. They do not provide a correct description of the motion of the electron in the hydrogen atom. According to the theory of quantum mechanics, which seems to be essentially correct, the electron moves about the nucleus in the hydrogen atom in roughly the way described by Bohr, but the motion in the normal state ($n = 1$) is not in a circle, but is radial (in and out, toward the nucleus and away from the nucleus). The most probable distance of the electron from the nucleus, according to quantum mechanics, is the same as the radius of the Bohr orbit.

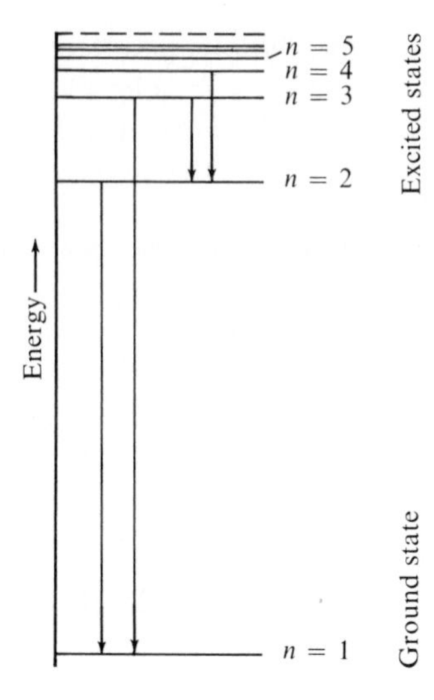

FIGURE 5-9
Energy-level diagram for the hydrogen atom.

The energy-level diagram for the hydrogen atom is shown in Figure 5-9. The energy of the atom in the nth stationary state is given in the Bohr theory by the equation

$$E_n = -\frac{2\pi^2 me^4}{n^2h^2} \tag{5-3}$$

Making use of the Bohr frequency rule, we obtain from this expression the following equation for the wavelength of the light emitted or absorbed on transition between the n' stationary state and the n'' stationary state:

$$\frac{1}{\lambda} = \frac{2\pi^2 me^4}{ch^3}\left(\frac{1}{n'^2} - \frac{1}{n''^2}\right) \tag{5-4}$$

On introducing the numerical values of the mass of the electron, the charge of the electron, the velocity of light, and Planck's constant, we obtain

$$\frac{1}{\lambda} = 109{,}678 \left(\frac{1}{n'^2} - \frac{1}{n''^2}\right) \text{cm}^{-1*} \tag{5-5}$$

This equation, with n'' and n' given various integral values, accounts for the line spectrum of hydrogen.

*The cm^{-1} (reciprocal centimeter or wave number) is a unit of frequency and energy generally used by spectroscopists.

Example 5-6. It is found that a tube containing hydrogen atoms in their normal state does not absorb any light in the visible region, but only in the far ultraviolet. The absorption line of longest wavelength has $\lambda = 121.6$ nm. What is the energy of the excited state of the hydrogen atom that is produced from the normal state by the absorption of a quantum of this light?

Solution. The frequency of the absorbed light is $\nu = c/\lambda = (2.998 \times 10^8 \text{ m s}^{-1}/(1.216 \times 10^{-7}\text{m}) = 2.467 \times 10^{15}$ Hz. The energy of a light quantum is $h\nu$. This is just the energy of the excited state relative to the normal state of the hydrogen atom. Accordingly, the answer to our problem is

$$h\nu = 0.6624 \times 10^{-33} \text{ J s} \times 2.467 \times 10^{15} \text{ Hz}$$

$$= 1.634 \times 10^{-18} \text{ J}$$

This can be converted into electron volts in the usual way; the answer is 10.20 eV. This result is also obtained simply by applying Equation 3-6 of Chapter 3:

$$1240/121.6 \text{ nm} = 10.20 \text{ eV}.$$

Example 5-7. From Figure 5-7, what is the energy of the transition of the hydrogen atom from the state $n = 2$ to the state $n = 3$?

Solution. Figure 5-7 shows that the wavelength of the H_α line in the Balmer series of spectral lines of hydrogen corresponding to the transition from $n = 3$ to $n = 2$ (or from $n = 2$ to $n = 3$, which is absorption rather than emission) is 656.47 nm. By use of Equation 3-6 the energy is found to be 1240/656.47 nm = 1.889 eV.

5-8. Excitation and Ionization Energies

Interesting verification of Bohr's idea about stationary states of atoms and molecules was provided by some electron-impact experiments carried out by James Franck (1882–1964) and Gustav Hertz (1887–1963) during the years 1914 to 1920. They were able to show that when a fast-moving electron collides with an atom or molecule it bounces off with only small loss of kinetic energy, unless it has a high enough speed to be able to raise the atom or molecule from its normal electronic state to an excited electronic state, or even to ionize the atom or molecule, by knocking an electron out of it.

The apparatus that they used is indicated diagrammatically in Figure 5-10. Electrons are boiled out of the hot filament, and are accelerated toward the grid by the accelerating potential difference V_1. Many of these electrons pass through the perforations in the grid, and strike the collecting plate, which is held at a negative voltage relative to the grid. The

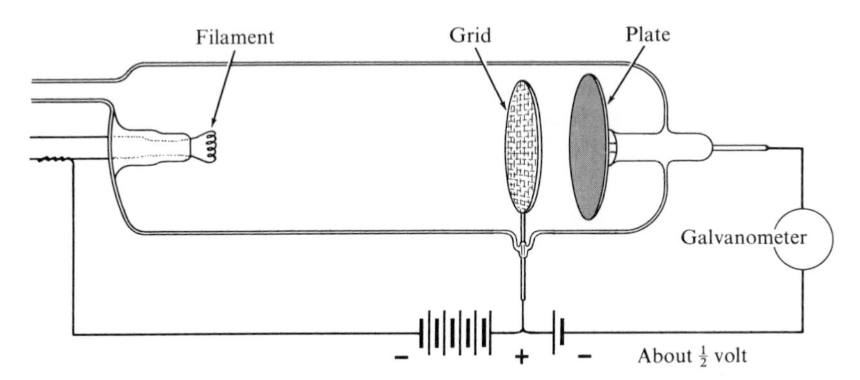

FIGURE 5-10
Apparatus for electron-impact experiments of the sort carried out by Franck and Hertz.

electrons are able to move against the electrostatic field between the grid and the collecting plate because of the kinetic energy that they have gained while being accelerated from the filament to the grid. Even if there are some atoms or gas molecules in the space between the filament and the grid, the electrons may bounce off from them without much loss in energy.

If, however, the accelerating voltage V_1 is great enough that the kinetic energy picked up by the electron exceeds the excitation energy of the atom or molecule with which the electron may collide, then the electron on impact with the atom or molecule may raise it from the normal state to the first excited state. The colliding electron will retain only the kinetic energy equal to its original kinetic energy minus the excitation energy of the atom or molecule with which it has collided. It may then not have enough residual kinetic energy to travel to the plate, against the opposing field, and the current registered by the galvanometer to the plate may show a decrease, as the accelerating voltage is increased.

With hydrogen atoms in the tube, for example, no change in the plate current would be registered on the galvanometer until the accelerating voltage reaches 10.2 V. At this accelerating voltage the electrons obtain from the field between the filament and the grid just enough energy to raise a normal hydrogen atom to the first excited state—that is, to change the quantum number from $n = 1$ to $n = 2$. There then occurs a decrease in the plate current. The voltage 10.2 V is called a *critical voltage* or *critical potential* for atomic hydrogen. Other critical potentials occur, corresponding to the other excited states, and a large one occurs at 13.60 V. This critical voltage, 13.60 V, corresponds to the energy, 13.60 eV, required to remove an electron completely from the hydrogen atom; that is, it corresponds to the energy required to convert a normal hydrogen atom into a proton plus an electron far removed from it. The voltage 13.60 V is called the *ionization potential* of the hydrogen atom, and the amount of energy 13.60 eV is called the *ionization energy* of the hydrogen atom.

The discussion of ionization energies of atoms will be continued in Section 6-9.

5-9. Magnetic Moments of Atoms and Molecules

It was discovered by Faraday that most substances when placed in a magnetic field develop a magnetic moment opposed to the field. Such a substance is said to be diamagnetic. Substances that develop a moment parallel to the field are called paramagnetic substances (or ferromagnetic substances, if the amount of magnetic polarization is large in weak fields and approaches a constant value as the field strength increases).

A sample of a diamagnetic substance placed in an inhomogeneous magnetic field is acted on by a force that tends to push it into the region where the field is weaker. A sample of paramagnetic substance is pulled into the stronger field.

The explanation of diamagnetism is that the application of a magnetic field induces a motion of the electrons in an atom or monatomic ion in a direction such as to give rise to a magnetic dipole with orientation opposed to the field. All substances show this diamagnetic effect.

The magnetic susceptibility of a diamagnetic or paramagnetic substance is defined by the equation

$$M = \chi H$$

in which M is the induced moment in unit amount (unit volume, unit mass, or one mole) of substance, χ is the magnetic susceptibility (χ_v, χ_m, or χ_{molar}, respectively), and H is the strength of the magnetic field.

It was shown by Pierre Curie in 1895 that paramagnetic susceptibility is strongly dependent on temperature, and for many substances is inversely proportional to the absolute temperature. The equation

$$\chi_{\mathrm{molar}} = \frac{C_{\mathrm{molar}}}{T} + D$$

is called Curie's law, and the constant C_{molar} is called the molar Curie constant; D, which is usually negative, represents the contribution of diamagnetism.

The first term in this equation can be evaluated by use of the Boltzmann principle, with the assumption that the substance contains permanent magnetic dipole moments that can orient themselves in the magnetic field. This theoretical treatment was carried out by the French scientist Paul Langevin in 1905. He derived the equation

$$C_{\text{molar}} = \frac{N\mu^2}{3k}$$

in which μ is the value of the magnetic dipole moment per atom or molecule.

By application of this equation to the observed magnetic susceptibilities of paramagnetic substances, measured over a range of temperature, the values of magnetic moments can be determined. From these values the number of unpaired electrons in the molecules of substances can be determined.

EXERCISES

5-1. State the periodic law.

5-2. Who proposed the periodic law and the periodic table and when?

5-3. How many groups and how many periods are there in the periodic table?

5-4. Without looking at the periodic table, but by remembering the number of elements in each row (2, 8, 8, 18, 18, 32), deduce what elements have atomic numbers 9, 10, 11, 17, 19, 35, 37, 54.

5-5. Sketch a plan of the periodic table, and fill in from memory the symbols and atomic numbers of the first eighteen elements and the remaining alkali metals, halogens, and argonons.

5-6. By extrapolation with use of the data given in Table 5-2, predict approximate values of the atomic weight, melting point, and boiling point of element 118. What would you expect its chemical properties to be?

5-7. Where was helium first detected? What is the principal source of this element at present?

5-8. What predictions would you make about the formula, color, solubility, taste, and melting point of the compound that would be formed by reaction of chlorine and element 119?

5-9. What are the most important metallic properties? In what part of the periodic table are the elements with metallic properties?

5-10. Classify the following elements as metals, metalloids, or nonmetals: potassium, arsenic, aluminum, xenon, bromine, silicon, phosphorus.

5-11. Neon, argon, krypton, and xenon crystallize in cubic closest packing (Section 2-4), with $a = 452$, 543, 559, and 618 pm, respectively. To what values of the density do these values of the edge of the unit cube correspond?

5-12. What are the values of the effective atomic radius of neon, argon, krypton, and xenon given by the information in the preceding exercise?

5-13. What is an electron orbital? How do the various orbitals differ?

5-14. What is electron spin and what is Pauli's exclusion principle?

5-15. What letters are used as alternative description of the principal quantum numbers?

5-16. How many subshells are there in a shell with principal quantum number n? What letters are used to describe subshells?

5-17. *(a)* Without referring to the text, draw electron-dot symbols of normal states of atoms of the elements from sodium to argon, showing 1 to 8 electrons of the outer shell.
(b) What electrons in these atoms are not represented by dots in these symbols?

5-18. The alkali metals, group I of the periodic table, are Li, Na, K, Rb, Cs, and Fr. Their atomic numbers are 3, 11, 19, 37, 55, and 87, respectively. How do they differ in electronic structure from the argonons that precede them in the periodic table?

5-19. Write electron-dot symbols for the alkali metals.

5-20. *(a)* What is the electron configuration of fluorine? (Refer to Figure 5-6, and show all nine electrons. Remember that there are three $2p$ orbitals in the subshell.)
(b) How many electron pairs are there in the atom? Which orbitals do they occupy?
(c) How many unpaired electrons are there? Which orbital does it occupy?

5-21. *(a)* What are the electron configurations of beryllium and boron, as shown in Figure 5-6?
(b) What electron-dot symbols do they correspond to?
(c) What are the customary chemical electron-dot symbols for these atoms? (They show a larger number of unpaired electrons.)

5-22. Write the electron configuration for the element with $Z = 103$, showing all 103 electrons. To what orbital do you assign the last electron? Why?

5-23. The Bohr orbit for the normal hydrogen atom has radius 53.0 pm. What is the radius for the first excited orbit, with $n = 2$? For the orbit with $n = 3$?

5-24. In Section 5-8 it is pointed out that the first change in plate current in the Franck-Hertz experiment with atomic hydrogen in the apparatus would occur when the voltage reached 10.2 V. To what change in principal quantum number of the electron in the hydrogen atom does this excitation correspond? At what voltage would the next excitation occur? (Answer: $n = 1$ to $n = 2$; 12.09 V.)

5-25. How can one experimentally determine the number of unpaired electrons in an atom?

6

Covalence and Electronic Structure

For over a century chemists have systematized the formulas of compounds by assigning certain combining powers, *valences*, to the elements. The valence of an element was formerly described as the number of valence bonds formed by an atom of the element with other atoms.

The effort to obtain a clear understanding of the nature of valence and of chemical combination in general has led in recent years to the dissociation of the concept of valence into several new concepts—especially *covalence* and *ionic valence*. We shall discuss these concepts and the related question of the nature of the chemical bond in the present chapter.

6-1. The Nature of Covalence

The atoms of most molecules are held tightly together by a very important sort of bond, the *shared-electron-pair bond* or *covalent bond*. This bond is so important, so nearly universally present in substances that Professor Gilbert Newton Lewis of the University of California (1875–1946), who discovered its electronic structure, called it *the* chemical bond.

It is the covalent bond that is represented by a dash in the valence-bond formulas, such as Br—Br and

```
     Cl
     |
Cl—C—Cl
     |
     Cl
```

that have been written by chemists for nearly a hundred years.

Modern chemistry has been greatly simplified through the development of the theory of the covalent bond. It is now easier to understand and to remember chemical facts—by connecting them with our knowledge of the nature of the chemical bond and the electronic structure of molecules—than was possible fifty years ago. It is accordingly wise for the student of chemistry to study this chapter carefully, and to get a clear picture of the covalent bond.

6-2. Covalent Molecules

The Hydrogen Molecule

The simplest example of a covalent molecule is the hydrogen molecule, H_2. For this molecule the electronic structure H:H is written, indicating that the two electrons are shared between the two hydrogen atoms, forming the bond between them. This structure corresponds to the valence-bond structure H—H.

By the study of its spectrum and by calculations made on the basis of the theory of quantum mechanics, the hydrogen molecule has been shown to have the structure represented in Figure 6-1. The two nuclei are firmly

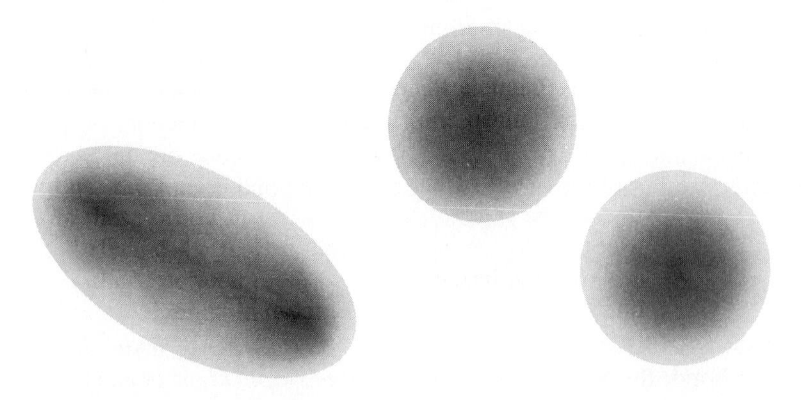

FIGURE 6-1
The electron distribution in a hydrogen molecule and in two hydrogen atoms. The two nuclei in the molecule are 74 pm apart.

held at a distance of about 74 pm apart—they oscillate relative to each other with an amplitude of a few picometers at room temperature, and with a somewhat larger amplitude at higher temperatures. The two electrons move very rapidly about in the region of the two nuclei, their time-average distribution being indicated by the shading in the figure. It can be seen that the motion of the two electrons is largely concentrated into the small region between the two nuclei. (The nuclei are in the positions where the electron density is greatest.) *The two electrons held jointly by the two nuclei constitute the chemical bond between the two hydrogen atoms in the hydrogen molecule.*

There is a very strong tendency for atoms of the stronger metals and the nonmetals to achieve the electron number of an argonon by losing or gaining one or more electrons. It was pointed out by Lewis that the same tendency is operating in the formation of molecules containing covalent bonds, and that the electrons in a covalent bond are to be counted for each of the bonded atoms.

Thus the hydrogen atom, with one electron, can achieve the helium structure by taking up another electron, to form the hydride anion, $H:^-$, as in the salt lithium hydride, Li^+H^-. But the hydrogen atom can also achieve the helium structure by sharing its electron with the electron of another hydrogen atom, to form a shared-electron-pair bond. Each of the two atoms thus contributes one electron to the shared electron pair. The shared electron pair is to be counted first for one hydrogen atom, and then for the other; if this is done, it is seen that in the hydrogen molecule each of the atoms has the helium structure:

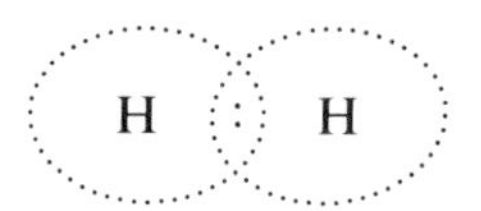

The Covalent Bond in Other Molecules

The covalent bond in other molecules is closely similar to that in the hydrogen molecule. For each covalent bond a pair of electrons is needed; also, two orbitals are needed, one of each atom.

The covalent bond consists of a pair of electrons shared between two atoms, and occupying two stable orbitals, one of each atom.

For example, reference to the energy-level diagram (Figure 5-6 or inside the front cover) shows that the carbon atom has four stable orbitals in its *L* shell, and four electrons that may be used in bond formation. Hence it may combine with four hydrogen atoms, each of which has one stable orbital (the 1*s* orbital) and one electron, forming four covalent bonds:

```
   H                          H
   ..                         |
H:C:H    equivalent to    H—C—H
   ..                         |
   H                          H
```

In this molecule each atom has achieved an argononic structure; the shared electron pairs are to be counted for each of the atoms sharing them. The carbon atom, with four shared pairs in the L shell and one unshared pair in the K shell, has achieved the neon structure, and each hydrogen atom has achieved the helium structure.

It has been found that the atoms of the principal groups of the periodic table (that is, all atoms except the transition elements) usually have an argononic structure in their stable compounds.

Stable molecules and complex ions usually have structures such that each atom has the electronic structure of an argonon, the shared electrons of each covalent bond being counted for each of the two atoms connected by the covalent bond.

The argonon atoms, except for helium, have eight electrons in the outermost shell, occupying four orbitals (one *s* orbital and three *p* orbitals). These eight electrons are called the *octet.* When an atom achieves an argononic structure, either by transferring electrons to or from other atoms or by sharing electron pairs with other atoms, it is said to *complete the octet.*

6-3. The Structure of Covalent Compounds

The electronic structure of molecules of covalent compounds involving the principal groups of the periodic table can usually be written by counting the number of valence electrons in the molecule and then distributing the valence electrons as unshared electron pairs and shared electron pairs in such a way that each atom achieves an argononic structure.

It is often necessary to have some experimental information about the way in which the atoms are bonded together. This is true especially of organic compounds. Thus there are two compounds with the composition C_2H_6O, ethyl alcohol and dimethyl ether. The chemical properties of these two substances show that one of them, ethyl alcohol, contains one hydrogen atom attached to an oxygen atom, whereas dimethyl ether does not contain such a hydroxyl group. The structures of these two isomeric molecules are shown in Figure 6-2.

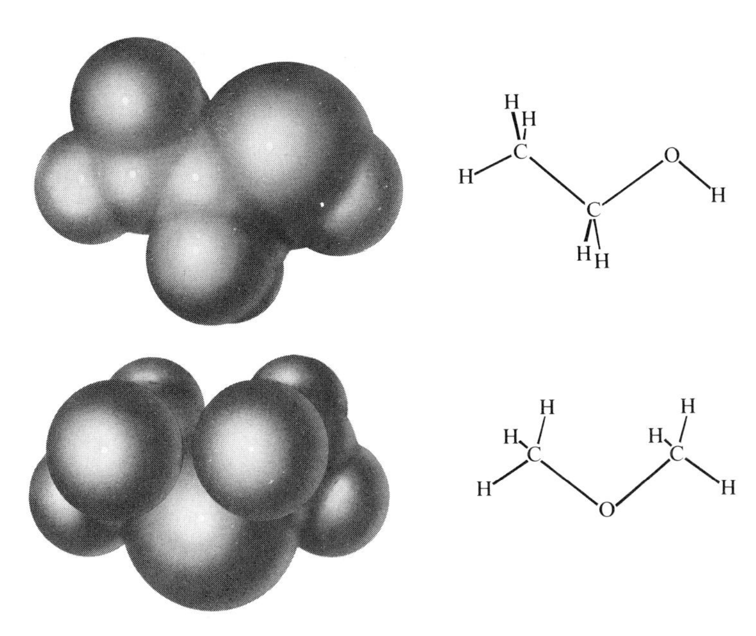

FIGURE 6-2
The structures of the isomeric molecules ethyl alcohol, CH_3CH_2OH, and dimethyl ether, $(CH_3)_2O$.

Compounds of Hydrogen with Nonmetals

Let us consider first the structure expected for a compound between hydrogen and fluorine, the lightest element of the seventh group. Hydrogen has a single orbital and one electron. Accordingly, it could achieve the helium configuration by forming a single covalent bond with another element. Fluorine has seven electrons in its outer shell, the *L* shell. These seven electrons occupy the four orbitals of the *L* shell. They accordingly constitute three electron pairs in three of the orbitals and a single electron in the fourth orbital. Hence fluorine also can achieve an argononic configuration by forming a single covalent bond with use of its odd electron. We are thus led to the following structure for the hydrogen fluoride molecule:

$$\mathrm{H}:\ddot{\underset{\cdot\cdot}{\mathrm{F}}}:$$

In this molecule, the hydrogen fluoride molecule, there is a single covalent bond that holds the hydrogen atom and the fluorine atom firmly together.

It is often convenient to represent this electronic structure by using a dash as a symbol for the covalent bond instead of the dots representing the shared electron pair. Sometimes, especially when the electronic structure of the molecule is under discussion, the unshared pairs in the outer shell of each atom are represented, but often they are omitted:

$$\mathrm{H{-}\ddot{\underset{..}{F}}:} \qquad \text{or} \qquad \mathrm{H{-}F}$$

The other halogens form similar compounds:

$\mathrm{H{-}\ddot{\underset{..}{Cl}}:}$	$\mathrm{H{-}\ddot{\underset{..}{Br}}:}$	$\mathrm{H{-}\ddot{\underset{..}{I}}:}$
Hydrogen chloride	Hydrogen bromide	Hydrogen iodide

These substances are strong acids: when they are dissolved in water the proton leaves the molecule, and attaches itself to a water molecule to form a hydronium ion, H_3O^+, the halogen being left as a halide ion,

$$\mathrm{:\ddot{\underset{..}{Cl}}:^-} \qquad \mathrm{:\ddot{\underset{..}{Br}}:^-} \qquad \text{or} \qquad \mathrm{:\ddot{\underset{..}{I}}:^-}$$

Hydrogen fluoride is a weak acid.

Elements of the sixth group (oxygen, sulfur, selenium, tellurium) can achieve an argononic structure by forming two covalent bonds. Oxygen has six electrons in its outer shell. These can be distributed among the four orbitals by putting two unshared pairs in two of the orbitals and an odd electron in each of the other two orbitals. These two odd electrons can be used in forming covalent bonds with two hydrogen atoms, to give a water molecule, with the following electronic structure:

$$\begin{array}{c}\mathrm{H}\\ \mathrm{:\ddot{\underset{..}{O}}:H}\end{array} \qquad \text{or} \qquad \begin{array}{l}\;\;\mathrm{H}\\ \;\;|\\ \mathrm{:\underset{..}{O}{-}H}\end{array}$$

If a proton is removed, a hydroxide ion, OH^-, is formed:

$$\left[\mathrm{:\ddot{\underset{..}{O}}{-}H}\right]^-$$

If a proton is added to a water molecule (attaching itself to one of the unshared electron pairs) a hydronium ion, OH_3^+, is formed:

$$\left[\begin{array}{l}\;\;\mathrm{H}\\ \;\;|\\ \mathrm{:O{-}H}\\ \;\;|\\ \;\;\mathrm{H}\end{array}\right]^+$$

All three of the hydrogen atoms in the hydronium ion are held to the oxygen atom by the same kind of bond, a covalent bond.

In hydrogen peroxide, H_2O_2, each oxygen atom achieves the neon configuration by forming one covalent bond with the other oxygen atom and one covalent bond with a hydrogen atom:

```
  H   H
  |   |
 :O — O:
  ..  ..
```

Hydrogen sulfide, hydrogen selenide, and hydrogen telluride have the same electronic structure as water:

```
  H          H           H
  |          |           |
 :S—H       :Se—H       :Te—H
  ..         ..          ..
```

Nitrogen and the other fifth-group elements, with five outer electrons, can achieve the argononic configuration by forming three covalent bonds. The structures of ammonia, phosphine, arsine, and stibine are the following:

```
  H          H          H           H
  |          |          |           |
 :N—H       :P—H       :As—H       :Sb—H
  |          |          |           |
  H          H          H           H
```

The ammonia molecule can attach a proton to itself, to form an ammonium ion, NH_4^+, in which all four hydrogen atoms are held to the nitrogen atom by covalent bonds:

```
 [    H    ]+
 |    |    |
 | H—N—H  |
 |    |    |
 [    H    ]
```

In the ammonium ion all four of the *L* orbitals are used in forming covalent bonds. The formation of the ammonium ion from ammonia is similar to the formation of the hydronium ion from water.

The Electronic Structure of Some Other Compounds

Electronic structures of other molecules containing covalent bonds may be readily written, by keeping in mind the importance of completing the octets of atoms of nonmetallic elements. The structures of some compounds of nonmetallic elements with one another are shown below:

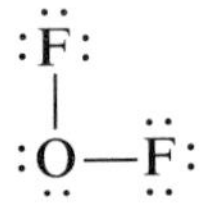

```
 :F:
  |
 :O—F:
```

Oxygen difluoride

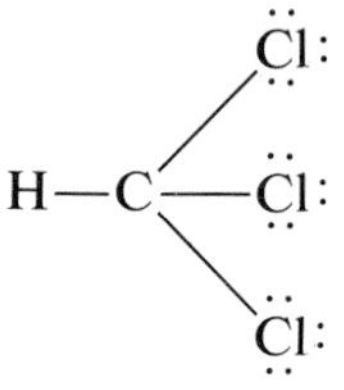

```
        Cl:
       /
  H—C——Cl:
       \
        Cl:
```

Chloroform

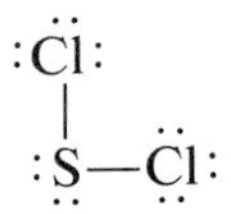

```
 :Cl:
  |
 :S—Cl:
```

Sulfur dichloride

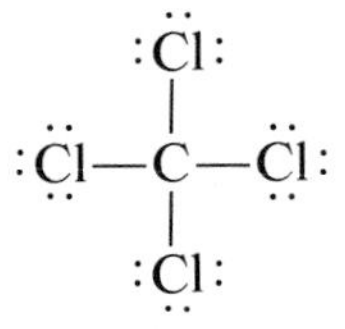

```
       :Cl:
        |
  :Cl—C—Cl:
        |
       :Cl:
```

Carbon tetrachloride

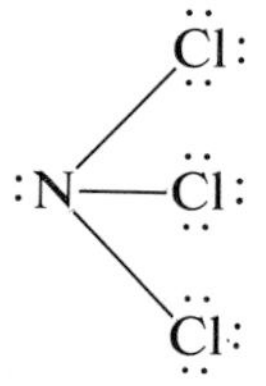

```
       Cl:
      /
  :N——Cl:
      \
       Cl:
```

Nitrogen trichloride*

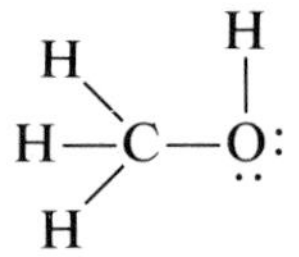

```
  H      H
   \     |
  H—C—O:
   /
  H
```

Methyl alcohol

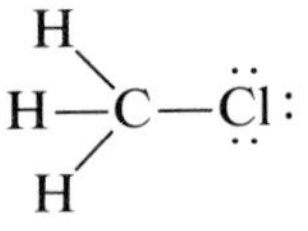

```
  H
   \
  H—C—Cl:
   /
  H
```

Methyl chloride

```
  H       H
   \     /
  H—C—C—H
   /     \
  H       H
```

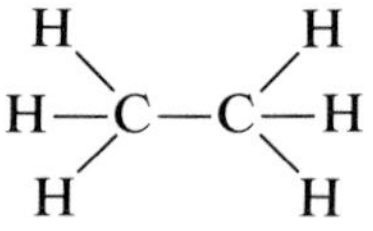

Ethane

Example 6-1. Silicon and hydrogen form the compound silane, SiH_4. *(a)* What is its electronic structure? *(b)* What orbitals of the silicon atom are used in forming the four covalent bonds?

---

*Note that sometimes an effort is made in drawing the structure of a molecule to indicate the spatial configuration; the structure shown here for nitrogen trichloride is supposed to indicate that the molecule is pyramidal, with the chlorine atoms approximately at three corners of a tetrahedron about the nitrogen atom. The spatial configuration of molecules is discussed in the following section.

Solution. (*a*) Silicon is a congener of carbon and silane is analogous to methane, CH_4. Four electrons come from the silicon atom and one electron from each hydrogen atom, leading to the valence electronic structure

$$\begin{array}{c} \mathrm{H} \\ \mathrm{H}:\underset{\cdot\cdot}{\overset{\cdot\cdot}{\mathrm{Si}}}:\mathrm{H} \\ \mathrm{H} \end{array}$$

(*b*) Silicon is in the second short period. Therefore the four valence electrons are in the *M* shell, with principal quantum number 3.

Example 6-2. Write the electronic structure of ethyl chloride, C_2H_5Cl. What argononic structure does each atom achieve?

Solution. Hydrogen has one electron in the valence shell, carbon four, and chlorine seven. Each atom can achieve an argononic structure with the electronic structure

$$\begin{array}{c} \mathrm{H}\ \ \mathrm{H}\qquad \\ \mathrm{H}:\underset{\cdot\cdot}{\overset{\cdot\cdot}{\mathrm{C}}}:\underset{\cdot\cdot}{\overset{\cdot\cdot}{\mathrm{C}}}:\underset{\cdot\cdot}{\overset{\cdot\cdot}{\mathrm{Cl}}}: \\ \mathrm{H}\ \ \mathrm{H}\qquad \end{array}$$

Hydrogen is in the very short period, carbon in the first short period, and chlorine in the second short period. The argononic structures are therefore those of helium, neon, and argon, respectively.

6-4. The Direction of Valence Bonds in Space

In 1874 it was discovered that the four bonds formed by a carbon atom are directed in space toward the four corners of a tetrahedron. This discovery was made through the effort to explain the observed effects of some substances on polarized light, as described in the following paragraphs.

Optical Activity

When a beam of ordinary light is passed through a crystal of calcite, it is split into two beams. Each of these beams is a beam of plane-polarized light; the vibrating electric field of the light lies in one plane for one of the beams and in the plane at right angles to it for the other beam.

A prism made of two pieces of calcite cut in a certain way and cemented together has the property of permitting only one beam to pass through; the other is reflected to the side and absorbed in the darkened side of the prism. Such a prism (called a Nicol prism) can be used to form a beam of plane-polarized light, and also to determine the orientation of the plane of polarization. In the instrument called the polarimeter, shown in Figure 6-3, the first prism defines the beam. If there is nothing between the two prisms to rotate the plane of polarization, the beam will pass through the second prism if it has the same orientation as the first, but will be absorbed if it is oriented at right angles to the first.

In 1811 the French physicist Dominique François Jean Arago (1786–1853) discovered that a quartz crystal has the power of rotating the plane of polarization of the beam of polarized light passing through it. Some quartz crystals were found to rotate the plane of polarization to the right (so that the second prism has to be turned clockwise, viewed in the direction opposite to the path of the light, in order that the beam pass through it), and others to the left. These are called dextrorotatory crystals and levorotatory crystals, respectively.

The two kinds of quartz crystals also differ in their face development, as shown in Figure 6-4. They are mirror images of one another—one can be described as a right-handed crystal and one as a left-handed crystal.

Right-handed and Left-handed Molecules

The French physicist Jean Baptiste Biot (1774–1862) then found that some liquids are optically active (that is, have the power of rotating the plane of polarization). For example, turpentine was found to be levorotatory, and an aqueous solution of sucrose (cane sugar, $C_{12}H_{22}O_{11}$) was found to be dextrorotatory. The substances that were found to be optically active in solution were all organic compounds, produced by plants or animals.

A puzzling observation was then made. It was found that two kinds of tartaric acid were deposited from wine lees. These two kinds of tartaric acid are closely similar in their properties, but they show the astonishing difference that one is dextrorotatory, and the other is completely without rotatory power. How could there be two molecules with the same composition but with such greatly different power of interacting with polarized light?

The answer to this puzzle was found in 1844 by the great French chemist Louis Pasteur (1822–1895). He added sodium hydroxide and ammonium hydroxide to the solution of the optically inactive tartaric acid and allowed the solution to evaporate, so that crystals of sodium ammonium tartrate, $NaNH_4C_4H_4O_6$, were formed. On examining the crystals he first noticed that they appeared to be identical with the crystals similarly made from the optically active tartaric acid. Then, as he continued to scrutinize them carefully, he suddenly recognized that only half of them were truly identical; the others were their mirror images

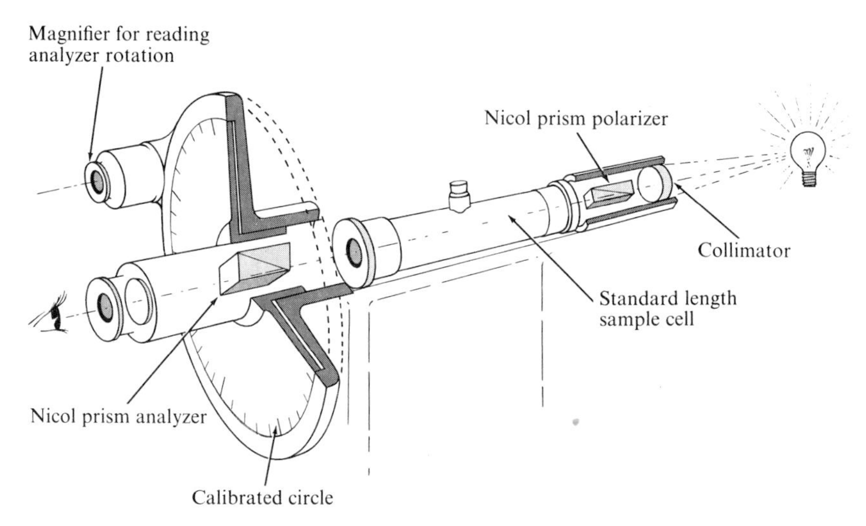

FIGURE 6-3
The polarimeter, an instrument used to determine the rotation of the plane of polarization of a beam of plane-polarized light by an optically active substance.

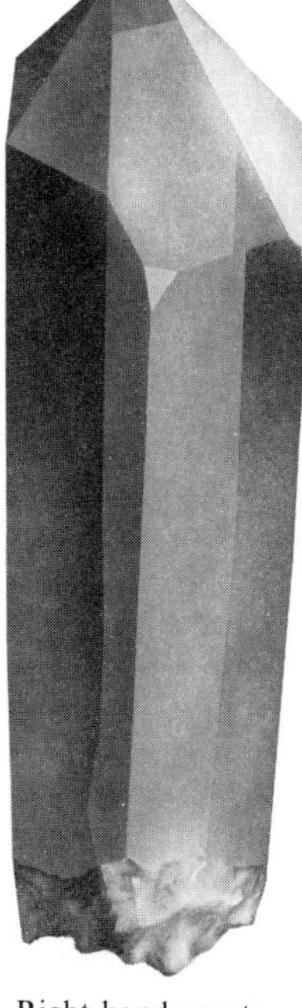
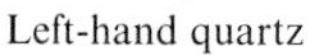

FIGURE 6-4
Right-handed and left-handed quartz crystals.

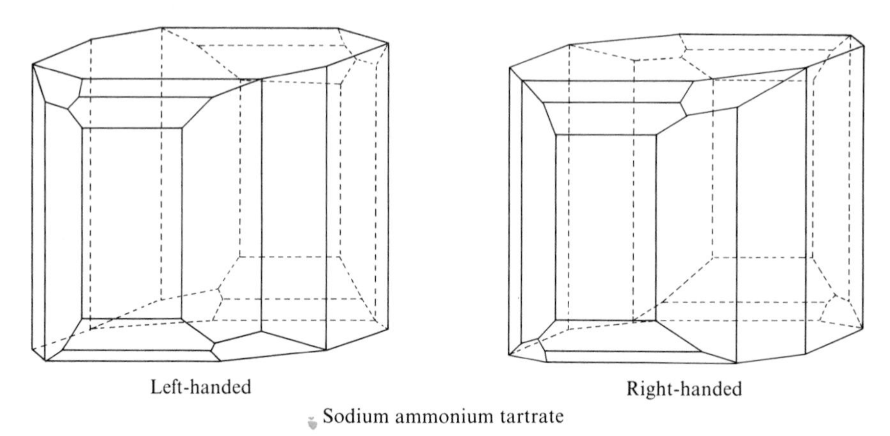

FIGURE 6-5
Right-handed and left-handed crystals of sodium ammonium tartrate.

(Figure 6-5). He separated the two kinds of crystals by hand and dissolved them in water. One of the solutions was dextrorotatory and the other levorotatory, with the same rotatory power, except for sign.

It was accordingly evident that the atoms in the tartaric acid molecule arrange themselves in a structure that does not have a plane of symmetry or a center of symmetry; hence there is a right-handed arrangement of atoms and there also is a left-handed arrangement, the mirror image of the first.

In 1844 the possibility of discovering these arrangements (the three-dimensional structures of the molecules) was so small that neither Pasteur nor any other chemist attacked the problem. But within fifteen years the correct atomic weights were accepted and the correct formulas were assigned, the concept of the chemical bond was developed, and the quadrivalence of carbon was established. The term "chemical structure" was used for the first time in 1861 by the Russian chemist Alexander M. Butlerov (1828–1886), who wrote that it is essential to find the way in which each atom is linked to other atoms in the molecules of substances.

A few more years passed. Many students of chemistry learned about the quadrivalence of carbon, about structural formulas, and about right-handed and left-handed molecules. Then two of them, the young Dutch chemist Jacobus Hendricus van't Hoff (1852–1911) and the young French chemist Jules Achille le Bel (1847–1930), saw in 1874 that *no* structure in which the atoms lie in a single plane can lead to optical activity. A planar molecule is its own mirror image, because the plane is itself a plane of symmetry for the molecule. For example, the substance fluorochlorobromomethane, CHFClBr, can be resolved into a dextrorotatory variety and a levorotatory variety. Hence it cannot be correctly represented by the planar formula

$$
\begin{array}{c}
\mathrm{F} \\
| \\
\mathrm{Cl-C-H} \\
| \\
\mathrm{Br}
\end{array}
$$

The four bonds formed by the carbon atom in this molecule cannot lie in one plane, but instead must extend toward the corners of a tetrahedron. There must be two kinds of fluorochlorobromomethane molecules, identical except for handedness, each the exact mirror image of the other (Figure 6-6).

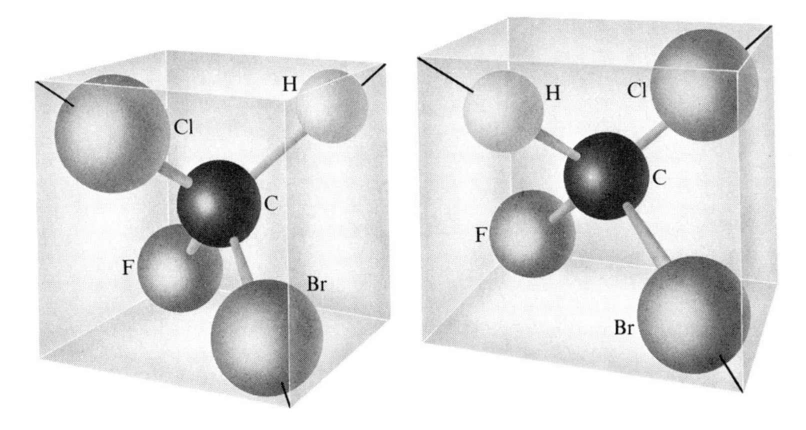

FIGURE 6-6
Right-handed and left-handed molecules of fluorochlorobromomethane. The molecule to the left is S and that to the right is R.

This was the birth of the tetrahedral carbon atom and of stereochemistry (the chemistry of three-dimensional space, structural chemistry). It led to the rapid development of chemical structure theory and of chemistry as a whole.

A pair of right-handed and left-handed molecules is called an *enantiomeric pair,* and the two substances they compose are called *enantiomers* (from the Greek *enantios,* opposite, and *meros,* part). The symbols D and L are used as prefixes to distinguish the two substances that constitute an enantiomeric pair.

A more precise description of the absolute configuration of the asymmetric carbon atom is the R and S convention. The basic rules are simple.

Each substituent (possibly polyatomic) is assigned a number from 1 to 4, corresponding to the atomic number of its atom nearest the central carbon atom (if necessary the next nearest, and so on), with value 1 for the greatest atomic number. When the asymmetric carbon atom is viewed from the direction opposite the substituent with value 4, if the groups with values $1 \rightarrow 2 \rightarrow 3$ are in the clockwise direction the carbon atom is R and if in the counterclockwise direction it is S. In Figure 6-6 the bromine atom has weight 1, chlorine 2, fluorine 3, and hydrogen 4. The drawing of fluorochlorobromomethane to the left in Figure 6-6 is S and that to the right is R.

A discussion of right-handed and left-handed molecules in living organisms is given in Section 14-3.

Example 6-3. Fluorochloromethane, CH_2FCl, does not comprise an enantiomeric pair of substances; it is a single optically inactive substance. Explain this fact by making a drawing of the molecule and indicating its plane of symmetry.

Solution. The carbon atom is tetrahedral. A drawing is

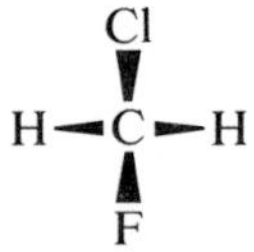

```
    Cl
    ▼
H◄─C─►H
    ▲
    F
```

The plane of symmetry is that defined by the F, C, and Cl atoms.

Example 6-4. Three substances correspond to the following formula:

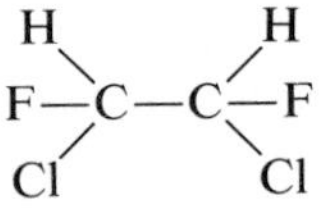

```
 H           H
   \       /
F—C—C—F
   /       \
Cl           Cl
```

Two of them constitute an enantiomeric pair and the third is optically inactive. Make a drawing of the three-dimensional structure of the molecule of the inactive substance.

Solution. The two enantiomeric molecules are the RR and SS forms of the formula given. The optically inactive substance is composed of RS or SR molecules, which are identical because of a mirror plane of symmetry bisecting the C—C bond.

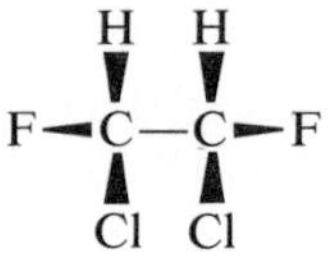

```
    H   H
    ▼   ▼
F◄─C—C─►F
    ▲   ▲
   Cl  Cl
```

6-5. Tetrahedral Bond Orbitals

In a molecule such as methane (Figure 6-7) or carbon tetrachloride, in which the four bonds are equivalent, the bond angles have the value 109°28′. In an asymmetric molecule such as CHFClBr the angles differ somewhat from this value, but only by a few degrees. It has been found by experiment (x-ray diffraction, electron diffraction, microwave spectroscopy) that these angles usually lie between 106° and 113°, with the average value for the six bond angles close to 109°28′.

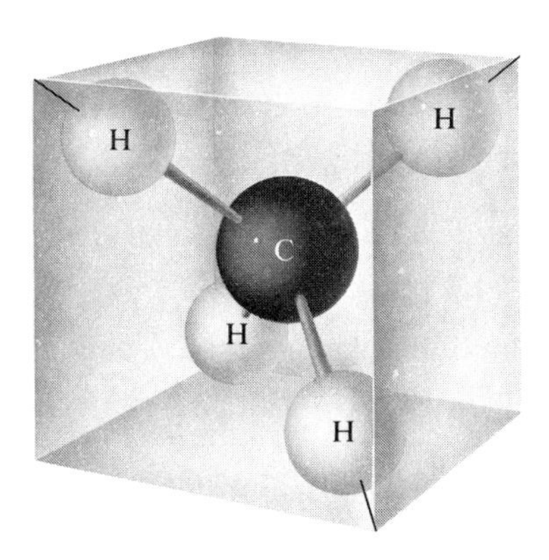

FIGURE 6-7
The methane molecule, CH_4.

Each of the four bonds formed by the carbon atom involves one of the four orbitals of the *L* shell (Section 6-2). These orbitals are given in Chapter 5 as the 2*s* orbital and the three 2*p* orbitals. We might hence ask whether or not the bonds to the four hydrogen atoms are all alike. Would not the 2*s* electron form a bond of one kind, and the three 2*p* electrons form bonds of a different kind?

Chemists have made many experiments to answer this question, and have concluded that the four bonds of the carbon atom are identical. A theory of the tetrahedral carbon atom was developed in 1931. According to this theory, the *theory of hybrid bond orbitals,* the 2*s* orbital and the three 2*p* orbitals of the carbon atom are hybridized (combined) to form four *tetrahedral bond orbitals* (or sp^3 orbitals). They are exactly equivalent to one another, and are directed toward the corners of a regular tetrahedron, as shown in Figure 6-8. Moreover, the nature of *s* and *p* orbitals and their hybrids is such that of all possible hybrid orbitals of *s* and *p* the tetrahedral orbitals are the best suited for forming strong bonds. Accordingly the tetrahedral arrangement of the bonds is the stable one.

Some molecules are known in which the bond angles are required by the molecular structure to differ greatly from the tetrahedral value. These

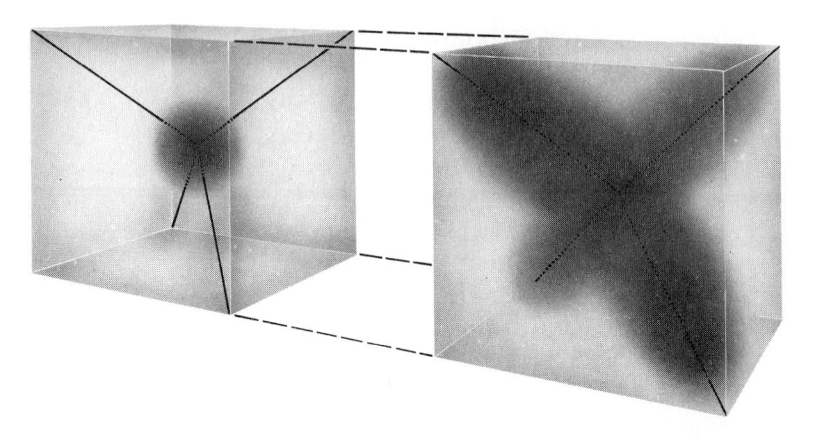

FIGURE 6-8
Diagram illustrating (left) the $1s$ orbital in the K shell of the carbon atom, and (right) the four tetrahedral orbitals of the L shell.

molecules can be described as containing bent bonds, and as being strained. The substance cyclopropane, C_3H_6, is an example. The cyclopropane molecule contains a ring of three carbon atoms, as shown in Figure 6-9. Each bond is bent through nearly 50°. The molecules are less stable by about 100 kJ mole^{-1} (33 kJ mole^{-1} per bond) than corresponding unstrained molecules, such as those of cyclohexane, C_6H_{12} (Sec. 7-2).

The Carbon-carbon Double Bond

Sometimes two valence bonds of an atom are used in the formation of a double bond with another atom. There is a double bond between two carbon atoms in the molecules of ethylene (ethene), C_2H_4:

```
H       H
 \     /
  C=C
 /     \
H       H
```

Ethylene

Such a double bond between two atoms may be represented by two tetrahedra sharing two corners: that is, sharing an edge, as shown in Figure 6-10. The amount of bending of the two bonds constituting the double bond is indicated in Figure 6-11.

It is interesting to note that the four other bonds that the two carbon atoms in ethylene can form lie in the same plane, at right angles to the plane containing the two bent bonds.

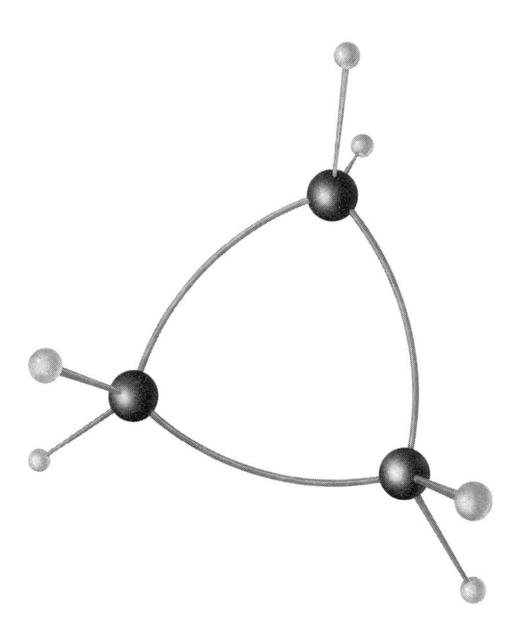

FIGURE 6-9
The molecule of cyclopropane, C_3H_6, showing the bent carbon-carbon bonds.

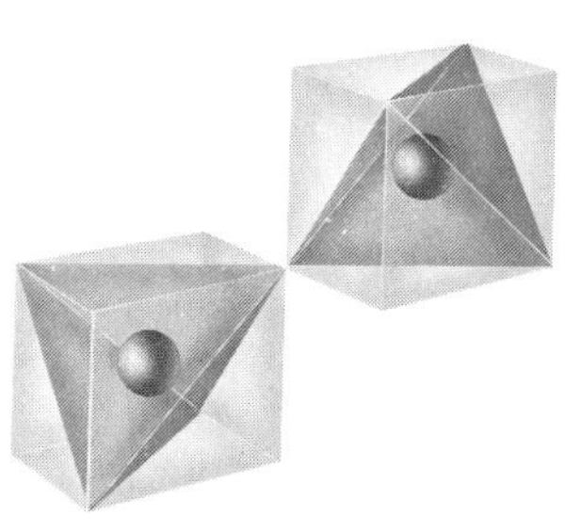

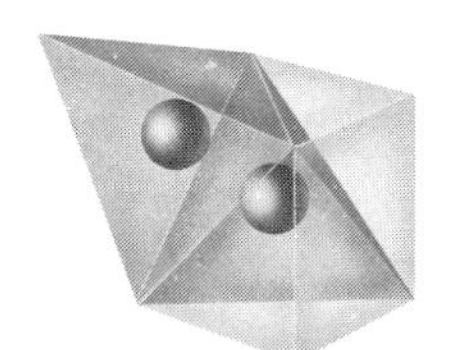

FIGURE 6-10
Tetrahedral atoms forming single, double, and triple bonds.

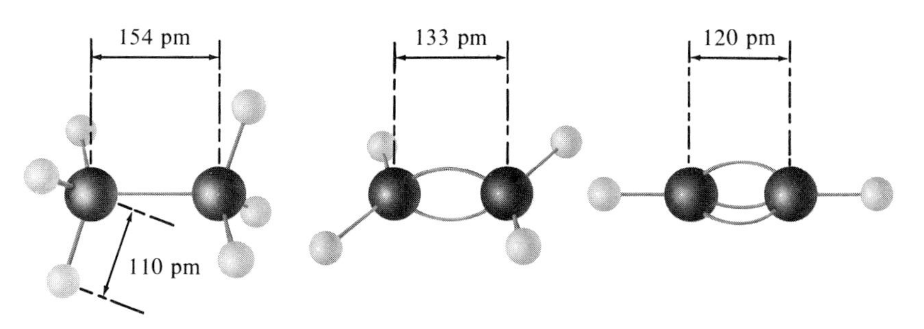

FIGURE 6-11
Valence-bond models of ethane, C_2H_6, ethylene, C_2H_4, and acetylene, C_2H_2.

The Carbon-carbon Triple Bond

In acetylene (ethyne), C_2H_2, there is a triple bond between the two carbon atoms:

$$H—C≡C—H$$

Acetylene

The triple bonds between two atoms may be represented by two tetrahedra sharing a face (Figures 6-10, 6-11, and 6-12). This causes the acetylene molecule to be linear.

Bond Lengths

Spectroscopic studies have led to the determination of the carbon-carbon bond length (distance between the nuclei of the two carbon atoms), in ethane, ethylene, and acetylene. The values are 154 pm for the single bond in ethane (and also in other molecules containing the C—C bond), 133 pm for the double bond, and 120 pm for the triple bond.

It is interesting that these values for C=C and C≡C are within 2 pm of the values that correspond to bent bonds with the normal single-bond length 154 pm and at tetrahedral angles, as indicated in Figure 6-11. This agreement supports the bent-bond description of the double bond and the triple bond.

Multiple Bonds and the Periodic Table

It is found that multiple bonds are often formed by elements in the first row of the periodic table (boron, carbon, nitrogen, oxygen), and less often by the heavier elements. For example, the nitrogen molecule, N_2, has the structure :N≡N:, whereas the phosphorus molecule, P_4, contains six bent single bonds.

Example 6-5. The diacetylene molecule has the formula C_4H_2. What is its structural formula? What is the spatial arrangement of the six atoms in the molecule relative to one another?

Solution. Only one structural formula is consistent with the valence requirements of carbon and hydrogen and the formula C_4H_2. It is H—C≡C—C≡C—H. The atoms of the molecule must be linear, that is, all lie on a single line.

Example 6-6. Cyclopropane has the formula C_3H_6. One other substance (called propylene) has the same formula. What is its structural formula? How many of the six hydrogen atoms lie in the plane of the three carbon atoms?

FIGURE 6-12
Drawing showing the relation of the tetrahedron and the octahedron to the cube. These polyhedra are important in molecular structure.

Solution. The structural formula of propylene must be

```
                H
                |  H
                | /
     H          C—H
      \        /
       C == C
      /       \
     H         H
```

The double bond requires the three hydrogen atoms connected to the double bonded carbon atoms and the third carbon atom to be coplanar with the two carbon atoms. One hydrogen connected to the third carbon atom may lie in this plane but need not. If it does, the other two hydrogen atoms cannot lie in the plane.

Example 6-7. Three isomers of dichloroethylene, $C_2H_2Cl_2$, exist. Can you assign structural formulas to them?

Solution.

```
H       Cl        H       H        H       Cl
 \     /           \     /          \     /
  C=C               C=C              C=C
 /     \           /     \          /     \
H       Cl        Cl      Cl       Cl      H
    1                 2                3
```

6-6. The Description of Molecules by Sigma and Pi Orbitals

In the preceding section we have discussed the double bond in ethylene and the triple bond in acetylene as involving two or three bent bonds. These bent bonds are described as involving tetrahedral orbitals of each of the two carbon atoms. Each bent bond is similar to the carbon-carbon single bond in ethane, $H_3C—CH_3$, except that it is bent. There is another description of these molecules that has become popular in recent years. This description involves use of hybrid orbitals different from the tetrahedral sp^3 hybrid orbitals described above and shown in Figure 6-8.

An orbital pointing toward an adjacent orbital is called a σ orbital (sigma orbital), and the bond involving the σ orbitals of two atoms is called a σ bond (sigma bond; the Greek letter σ is analogous to s). Thus in methane the four tetrahedral orbitals of the carbon atom are σ orbitals, and the four C—H bonds are σ bonds.

Let us assume that in ethylene, $H_2C{=}CH_2$, each carbon atom forms three hybrid bond orbitals from its $2s$, $2p_x$, and $2p_y$ orbitals (Figure 5-5). These three sp^2 hybrid orbitals have their maximum concentration in three directions in the xy plane, indicated by the heavy bond lines in Figure 6-13, which represent the σ bonds in the ethylene molecule.

Each carbon atom has one additional electron and one additional orbital, the p_z orbital. The p_z orbital extends above and below the xy plane, as shown in Figures 5-5 and 6-13. The two p_z electrons are assumed to interact to give another covalent bond between the carbon atoms, involving the overlapping of the upper parts of the two p_z orbitals and also the overlapping of the lower parts. This bond is represented by the two lighter lines in Figure 6-13. Each of these lighter lines may be said to represent a half bond; together they represent a shared-electron-pair bond formed by the two p_z electrons. The bond is called a π bond (pi bond), the two p_z orbitals are called π orbitals, and the two bonding electrons are called π electrons.

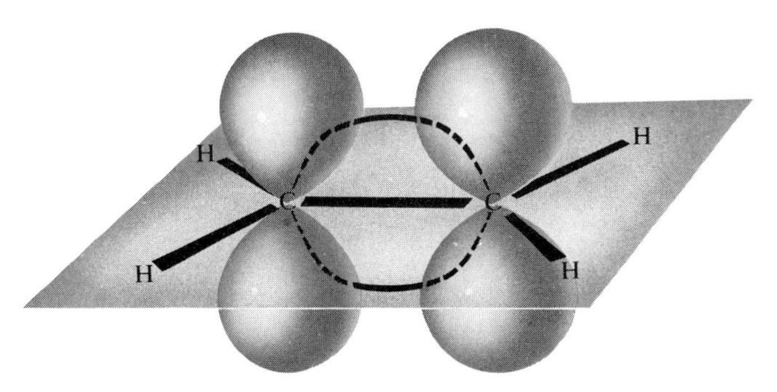

FIGURE 6-13
σ bonds and π bonds in ethylene.

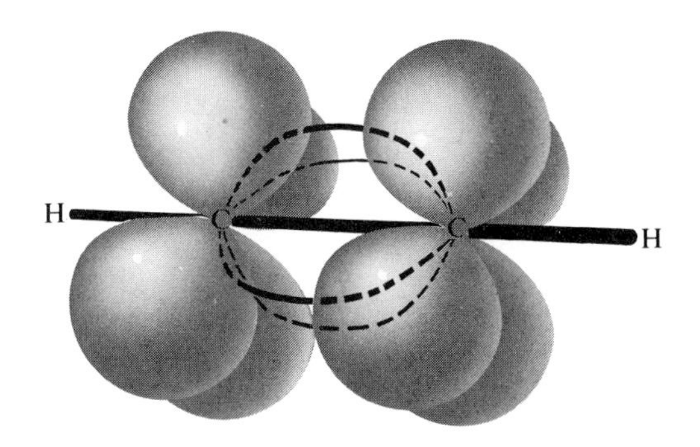

FIGURE 6-14
σ, π-bond structure of acetylene.

The analogous structure for acetylene, HC≡CH, is shown in Figure 6-14. It involves a σ bond from each carbon atom to the attached hydrogen atom and another σ bond and two π bonds between the two carbon atoms. The two σ orbitals of each carbon atom are sp_x hybrids, pointing in opposite directions. The two π bonds involve the p_y and p_z orbitals of the carbon atoms.

Bent Bonds and σ,π Bonds Compared

The two descriptions of ethylene and acetylene, the bent-bond description and the σ,π-bond description, represent in fact essentially the same structure. Quantum mechanical calculations carried out for a bent-bond wave function and a σ,π-bond wave function give the same results. It is essentially a matter of personal choice that determines the use of one description or the other.

For example, each description requires that the ethylene molecule be planar. In the bent-bond description the plane perpendicular to the plane of the molecule is the plane of the two bent bonds, and in the σ,π-bond description it is the plane of the two π orbitals. In each case rotation of the two ends of the molecule out of the plane decreases the overlap of the atomic orbitals and weakens the bonding.

As described in the preceding section, the bent-bond picture provides an explanation of the observed decrease in carbon-carbon distance from 154 pm in ethane to 133 pm in ethylene and 120 pm in acetylene. The σ,π-bond picture does not, because it is not evident that there is a relation between bond lengths for sp^3 σ bonds, sp^2 σ bonds, and π bonds. With a little experience, however, it becomes possible to discuss the properties of substances in terms of their structure in a satisfactory way on the basis of either of the two descriptions.

6-7. Bond Orbitals with Large *p* Character

In a molecule such as ammonia, NH_3, with structural formula

```
      H
     /
  :N—H
     \
      H
```

the bond orbitals of the nitrogen atom are not tetrahedral orbitals, but instead have mainly the character of the three $2p$ orbitals. Quantum mechanical calculations and nuclear magnetic resonance experiments (which measure the interaction energy of the nuclear spin magnetic moment with the valence electrons) agree in allocating the unshared electron pair to a hybrid orbital that is largely $2s$ in character (about 79%). The three bond orbitals have about 93% $2p$ character and 7% $2s$ character.

It was pointed out in Chapter 5 that a $2s$ electron is more stable than a $2p$ electron. The difference in energy is about 750 kJ mole^{-1}. The nitrogen atom, $:\dot{N}\cdot$,is 750 kJ mole^{-1} more stable if the pair of electrons is in the $2s$ orbital ($2s^2 2p_x 2p_y 2p_z$) than if it is in one of the $2p$ orbitals ($2s 2p_x^2 2p_y 2p_z$). Hence it tends to retain the $2s$ pair in forming compounds, and to use the $2p_x$, $2p_y$, and $2p_z$ orbitals for the bonding electrons.

The three $2p$ orbitals are represented in Figure 5-5. The $2p_x$ orbital extends in two opposite directions along the x axis, and can be used in forming a bond along the x axis, the $2p_y$ orbital in forming a bond along the y axis, and the $2p_z$ orbital in forming a bond along the z axis. Hence *the bonds formed by p orbitals are approximately at 90° to one another.* With some *s* character to the bond orbitals the bond angles increase, reaching 109°28′ for tetrahedral orbitals, which have 25% *s* character.

The experimental values of bond angles for atoms with unshared electron pairs usually lie between 90° and 109°. For example, the spectroscopically determined value for NH_3 is 107°, for H_2O 104.5°, for PH_3 93°, for H_2S 92°, and for H_2Se 91°. In NH_3, H_2O, and SCl_2, in which the Cl—S—Cl angle is 102°, part of the increase in angle above 90° is because of crowding between the two atoms attached to the central atom (see Section 6-15).

Example 6-8. The ordinary form of sulfur (orthorhombic sulfur, yellow crystals) contains octatomic molecules, S_8. Discuss the structure of the S_8 molecule.

Solution. The sulfur atom, with four stable outer orbitals and six outer electrons, can form single covalent bonds with two other atoms. These bonds may hold the molecule together either into a ring, such as an S_8 ring, or into a very long chain, with the two end atoms having an abnormal structure:

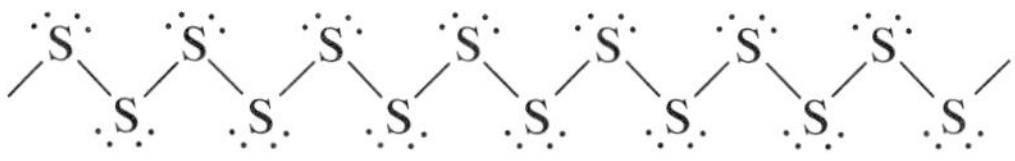

```
  :S:   :S:   :S:   :S:   :S:   :S:   :S:
 /   \ /   \ /   \ /   \ /   \ /   \ /   \ /
      :S:   :S:   :S:   :S:   :S:   :S:   :S:
```

The bond orbitals have a large amount of *p* character, and hence the bond angle has the expected value 90° or a little larger. The S_8 ring must accordingly be puckered, as shown in Figure 6-15 (a planar octagon has angles 135°, and is ruled out). The x-ray study of the crystal has shown the S—S—S angle to have the value 102°, with bond length 206 pm.

When sulfur is melted it is converted into a straw-colored liquid, which also consists of the staggered ring S_8. However, when molten sulfur is heated to a temperature considerably above its melting point it becomes deep red in color and extremely viscous, so that it will not pour out of the test tube when it is inverted. This change in properties is the result of the formation of very large molecules containing hundreds of atoms in a long chain—the S_8 rings break open, and then combine together in a "high polymer."* The deep red color is caused by the abnormal atoms at the end of the chains, which are forming only one bond instead of the two bonds that a sulfur atom is expected to form. The great viscosity of the liquid is due to the interference with molecular motion caused by entanglement of the long chains of atoms with one another.

---

*A polymer is a molecule made by combination of two or more identical smaller molecules. A high polymer is made by the combination of a great many smaller molecules. The words dimer, trimer, tetramer, etc. are used for molecules obtained by combining two, three, four, etc. identical smaller molecules.

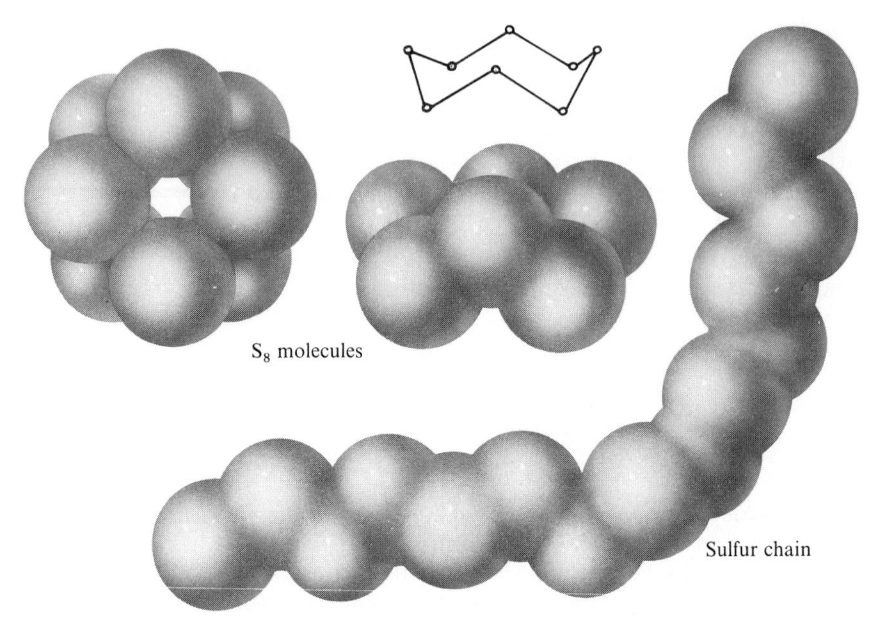

FIGURE 6-15
The S_8 ring, and a long chain of sulfur atoms.

Example 6-9. Discuss the structures of N_2 and P_4.

Solution. The nitrogen atom, lacking three electrons of a completed octet, may complete the octet by forming three covalent bonds. It does this in elementary nitrogen by forming a triple bond in the molecule N_2. Three electron pairs are shared by the two nitrogen atoms:

$$:N:::N: \quad \text{or} \quad :N{\equiv}N:$$

This bond is extremely strong, and the N_2 molecule is a very stable molecule.

Phosphorus gas at very high temperatures consists of P_2 molecules, with a similar structure, $:P{\equiv}P:$. At lower temperatures, however, phosphorus forms a molecule containing four atoms, P_4. This molecule has the structure shown in Figure 6-16. The four phosphorus atoms are arranged at the corners of a regular tetrahedron. Each phosphorus atom forms covalent bonds with the three other phosphorus atoms. This P_4 molecule exists in phosphorus vapor, in solutions of phosphorus in carbon disulfide and other nonpolar solvents, and in solid white phosphorus. In other forms of elementary phosphorus (red phosphorus, black phosphorus) the atoms are bonded together into larger aggregates, with P—P—P angles about 102°, as expected for *p* bonds.

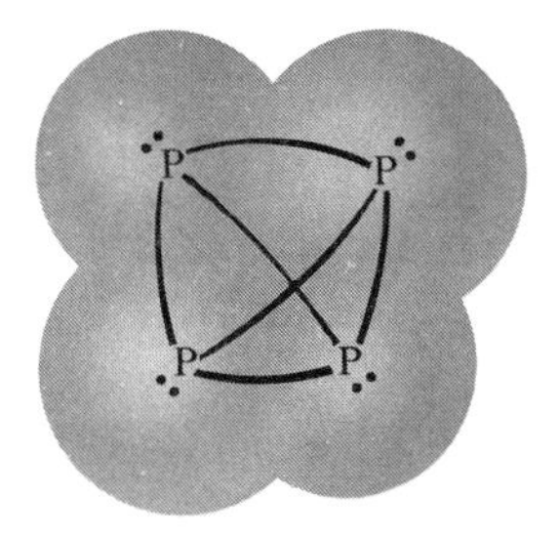

FIGURE 6-16
The P_4 molecule.

6-8. Resonance

For many molecules a simple valence-bond structure can be assigned that accounts satisfactorily for the properties of the substance (Examples 6-8 and 6-9). For other molecules, such as ozone, O_3, however, no single structure is satisfactory. It has been found that for these molecules a satisfactory description can be presented with use of two or more valence-bond structures. This concept in structural chemistry is called the *theory of resonance.*

Ozone is a blue gas which has a characteristic odor (its name is from the Greek *ozein,* to smell). It is a stronger oxidizing agent than ordinary oxygen. It is formed in air by the passage of electric sparks or an electric arc. Ozone is largely the cause of the odor that is observed around electrical machinery. Most people can detect this odor at a concentration of one part in 100,000,000 in air. Ozone is a normal constituent of the upper atmosphere, where it is formed by action of cosmic rays. It oxidizes organic substances in the atmosphere, and by reacting with hydrocarbons (gasoline vapor) produces the irritating substances in the smog of Los Angeles and other regions.

The density of ozone gas is 50 percent greater than that of ordinary oxygen, O_2, showing that the ozone molecule is triatomic (Figure 6-17).

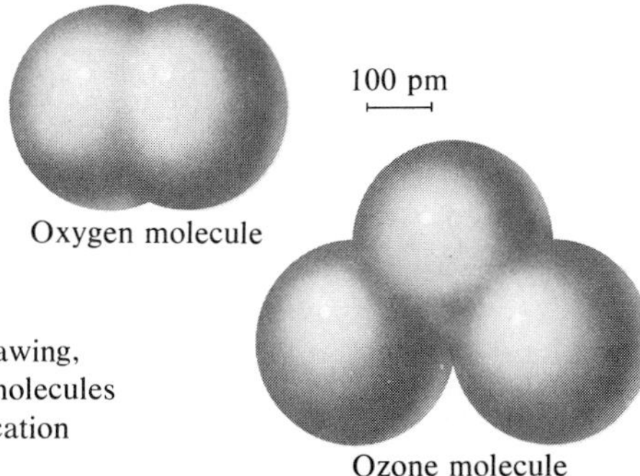

FIGURE 6-17
Molecules of oxygen and ozone. This drawing, like most of the drawings of atoms and molecules in this book, is made with linear magnification about 60,000,000.

A rather reasonable structure for O_3 is the equilateral triangle,

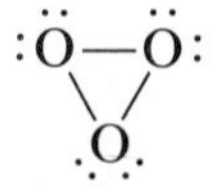

with each atom forming two single bonds. Spectroscopic studies, however, have shown that there is an O—O—O bond angle of 117°, which is expected between a single bond and double bond formed by orbitals with large *p* character. We accordingly assign the structure

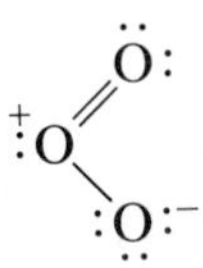

in which each atom has the neon structure (four outer electron pairs, either shared or unshared).

The spectroscopic structure determination has, however, shown the two oxygen-oxygen bonds to have the same length, 128 pm. Hence the above electronic structure does not, by itself, represent the molecule in a satisfactory way.

The resonance description of ozone involves two structures (equivalent, with single and double bonds interchanged), written together in braces:

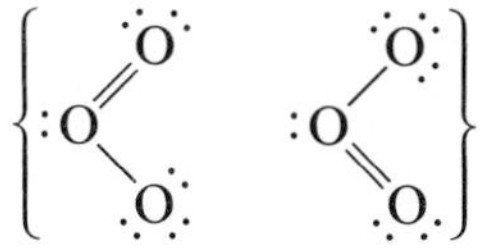

The molecule has a structure represented by the superposition of the two structures shown; that is, each bond is a *hybrid* of a single covalent bond and a double covalent bond.

In these structures one of the end atoms of the molecule resembles a fluorine atom in that it completes its octet by sharing only one electron pair. It may be considered to be the negative ion, $:\ddot{O}\cdot^-$, which forms one covalent bond. The central oxygen atom of the ozone molecule resembles a nitrogen atom, and may be considered to be the positive ion, $:\dot{O}\cdot^+$, which forms three covalent bonds (one double bond and one single bond).

The theory of resonance is especially important for benzene and other organic compounds. We shall discuss it further in later chapters.

Example 6-10. The carbonate ion, CO_3^{--}, is a stable anion found in many inorganic salts. The carbon atom is bonded to the three oxygen atoms and it is known that the three oxygen atoms are equivalent. What is the valence bond structure of the ion? Discuss the nature of the carbon-oxygen bonds.

Solution. The carbon atom contributes four electrons, each oxygen contributes six electrons, and two electrons are contributed by the double negative charge of the ion. We can draw a valence bond structure

```
      :O:⁻
      /
:O=C
      \
      :O:⁻
```

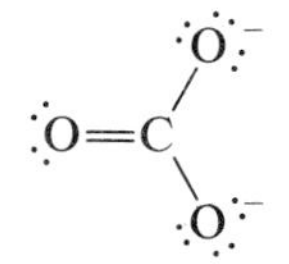

which uses all the available valence electrons and gives each atom an argononic structure. The three carbon-oxygen bonds are not equivalent, however; one is a double bond and the other two are single bonds. Therefore, we must describe the carbonate ion as a resonance hybrid:

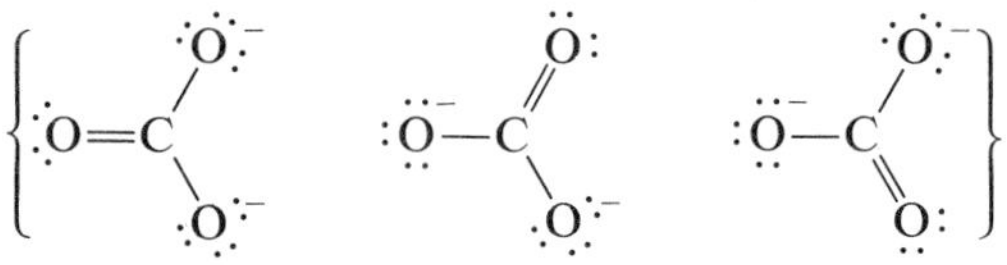

```
 ⎧       :O:⁻             O:               :O:⁻  ⎫
 ⎪       /               //                /     ⎪
 ⎨ :O=C          ⁻:O—C           ⁻:O—C          ⎬
 ⎪       \               \                \\     ⎪
 ⎩       :O:⁻             :O:⁻             O:    ⎭
```

in which each carbon-oxygen bond has bond order $1\frac{1}{3}$. The double bond in the resonating system requires the molecule to be planar, as ethylene is planar. The angle between a single bond and a double bond as in ethylene is $(360 - 109.5)/2 = 125°$. The O—C—O angle in the carbonate ion is thus expected to be between 109.5° and 125°, and the equivalence of the three oxygen atoms requires it to be 120°.

6-9. Ionic Valence

The British scientist Henry Cavendish (1731–1810) reported that the electric conductivity of water is greatly increased by dissolving salt in it. In 1884 the young Swedish scientist Svante Arrhenius (1859–1927) published his doctor's dissertation, which included measurements of the electric conductivity of salt solutions and his ideas as to their interpretation. These ideas were rather vague, but he later made them more precise and then published a detailed paper on ionic dissociation in 1887. Arrhenius assumed that in a solution of sodium chloride in water there are present sodium ions, Na^+, and chloride ions, Cl^-. When electrodes are put into such a solution the sodium ions are attracted toward the cathode and move in that direction, and the chloride ions are attracted toward the anode and move in the direction of the anode. The motion of these ions through the solution, in opposite directions, provides the mechanism of conduction of the current of electricity by the solution.

The presence of hydrated ions such as $Na^+(aq)$, $Mg^{++}(aq)$, $Al^{+++}(aq)$, $S^{--}(aq)$, and $Cl^-(aq)$, as well as complex ions such as $SO_4^{--}(aq)$, in aqueous solutions has been verified by the study of the properties of the solutions. Many of these ions have an electric charge such as to give the atom the electron number of the nearest argonon. The number of electrons removed from or added to the atom is called its ionic valence: +1 for Na^+, for example, and −1 for Cl^-.

The alkali metals (in group I of the periodic table) are unipositive because their atoms contain one more electron than an argonon atom, and this electron is easily removed, to produce the corresponding cation, Li^+, Na^+, K^+, Rb^+, and Cs^+. The ease with which the outermost electron can be removed from an atom of an alkali metal is shown by the values of the first ionization energy given in Table 6-1 (in kJ $mole^{-1}$) and in Figure 6-18.

The halogens (in group VII of the periodic table) are uninegative because each of their atoms contains one less electron than an argonon atom, and readily gains an electron, producing the corresponding anion, F^-, Cl^-, Br^-, and I^-. The energy that is liberated when an extra electron is attached to an atom to form an anion is called the *electron affinity* of the

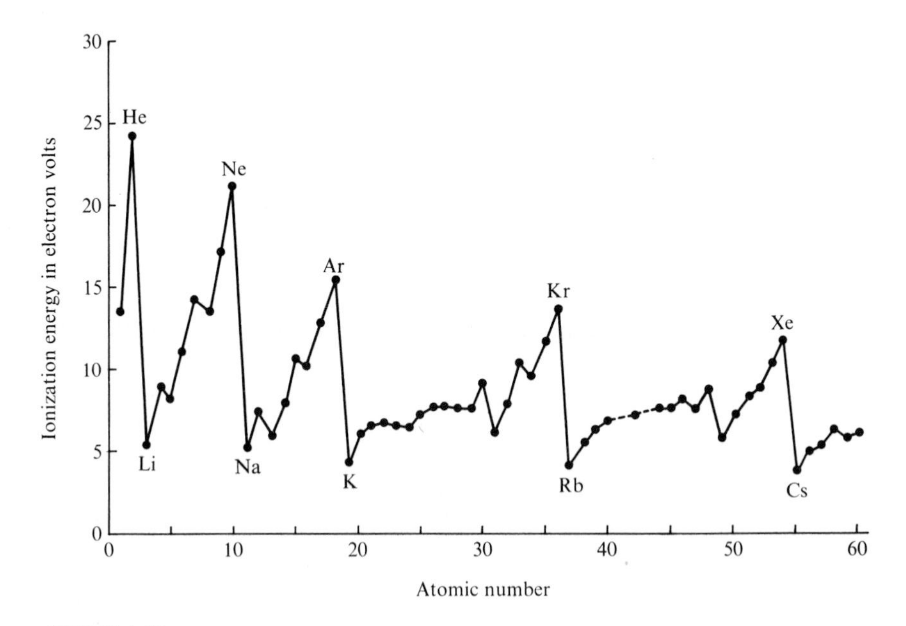

FIGURE 6-18
The ionization energy, in electron-volts, of the first electron of atoms from hydrogen, atomic number 1, to neodymium, atomic number 60. Symbols of the elements with very high and low ionization energy are shown in the figure.

TABLE 6-1
Ionization Energy and Electron Affinity of Univalent Elements

Element	First Ionization Energy	Element	Electron Affinity
H	1312 kJ mole^{-1}	H	71 kJ mole^{-1}
Li	520	F	333
Na	496	Cl	350
K	419	Br	330
Rb	403	I	300
Cs	376		

atom. Values of electron affinities of the halogens, given in Table 6-1, are larger than those of other atoms.*

The atoms of group II of the periodic table, by losing two electrons, can also produce ions with argononic structures: these ions are Be^{++}, Mg^{++}, Ca^{++}, Sr^{++}, and Ba^{++}. The alkaline-earth elements are hence bipositive in valence. The elements of group III are tripositive, those of group VI are binegative, and so on.

The formulas of binary salts of these elements can thus be written from knowledge of the positions of the elements in the periodic table:

$$Na^+F^- \quad Na^+Br^- \quad K^+I^- \quad Ca^{++}(F^-)_2 \quad Ba^{++}(Cl^-)_2$$

$$Al^{+++}(Cl^-)_3 \quad (Na^+)_2O^{--} \quad Ca^{++}O^{--} \quad (Al^{+++})_2(O^{--})_3$$

Ionic compounds are formed between the strong metals in groups I and II and the strong nonmetals in the upper right corner of the periodic table. In addition, ionic compounds are formed containing the cations of the strong metals and the anions of acids, especially of the oxygen acids.

It will be pointed out later in this chapter that the description of compounds as aggregates of ions is an approximation. The electronic structure of molecules and crystals usually described as ionic involves only a partial transfer of electrons from the metal atoms to the nonmetal atoms. Nevertheless, the discussion of ionic valence in relation to the argononic electron configurations, as given above, is an important and useful part of chemical theory.

In 1913 the structure of NaCl(c) was determined by x-ray diffraction, and it was found that there are no discrete Na—Cl molecules in the

*It is surprising that the electron affinity of fluorine is less than that of chlorine. The first electron affinity of oxygen is 140 kJ mole^{-1}, and that of OH is 175 kJ mole^{-1}.

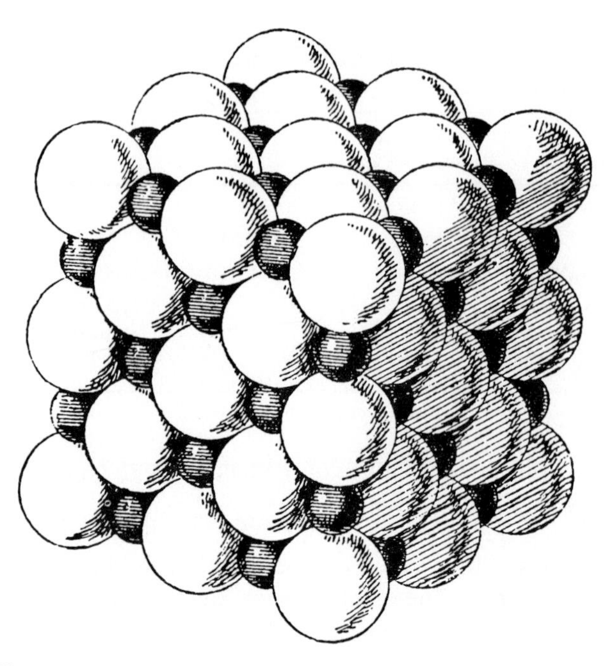

FIGURE 6-19
The sodium chloride structure. The cubic unit cell contains 4 Na at 0, 0, 0; 0, $\frac{1}{2}$, $\frac{1}{2}$; $\frac{1}{2}$, 0, $\frac{1}{2}$; and $\frac{1}{2}$, $\frac{1}{2}$, 0, and 4 Cl at $\frac{1}{2}$, $\frac{1}{2}$, $\frac{1}{2}$; $\frac{1}{2}$, 0, 0; 0, $\frac{1}{2}$, 0; and 0, 0, $\frac{1}{2}$. The structure is based on a face-centered cubic lattice. This figure is from an early paper by William Barlow.

crystal (Figure 6-19). Instead, each sodium atom is equidistant from six neighboring chlorine atoms, and each chlorine atom is similarly surrounded by six sodium atoms. It was at once recognized that the crystal can be described as an aggregate of sodium cations and chloride anions, and that each ion is bonded to each of its six neighbors by an electrostatic or ionic bond with bond-number (or strength) $\frac{1}{6}$. The alkali hydrides (LiH to CsH) and most of the alkali halides crystallize with the NaCl structure.

6-10. Ionic Radii

The electron distributions in alkali ions and halide ions are shown in Figure 6-20. It is seen that these ions are closely similar to the corresponding argonons, which are shown, drawn to the same scale, in Figure 5-3. With increase in nuclear charge from $+9e$ for fluoride ion to $+11e$ for sodium ion the electron shells are drawn closer to the nucleus, so that the sodium ion is about 30% smaller than the fluoride ion. The neon atom is intermediate in size.

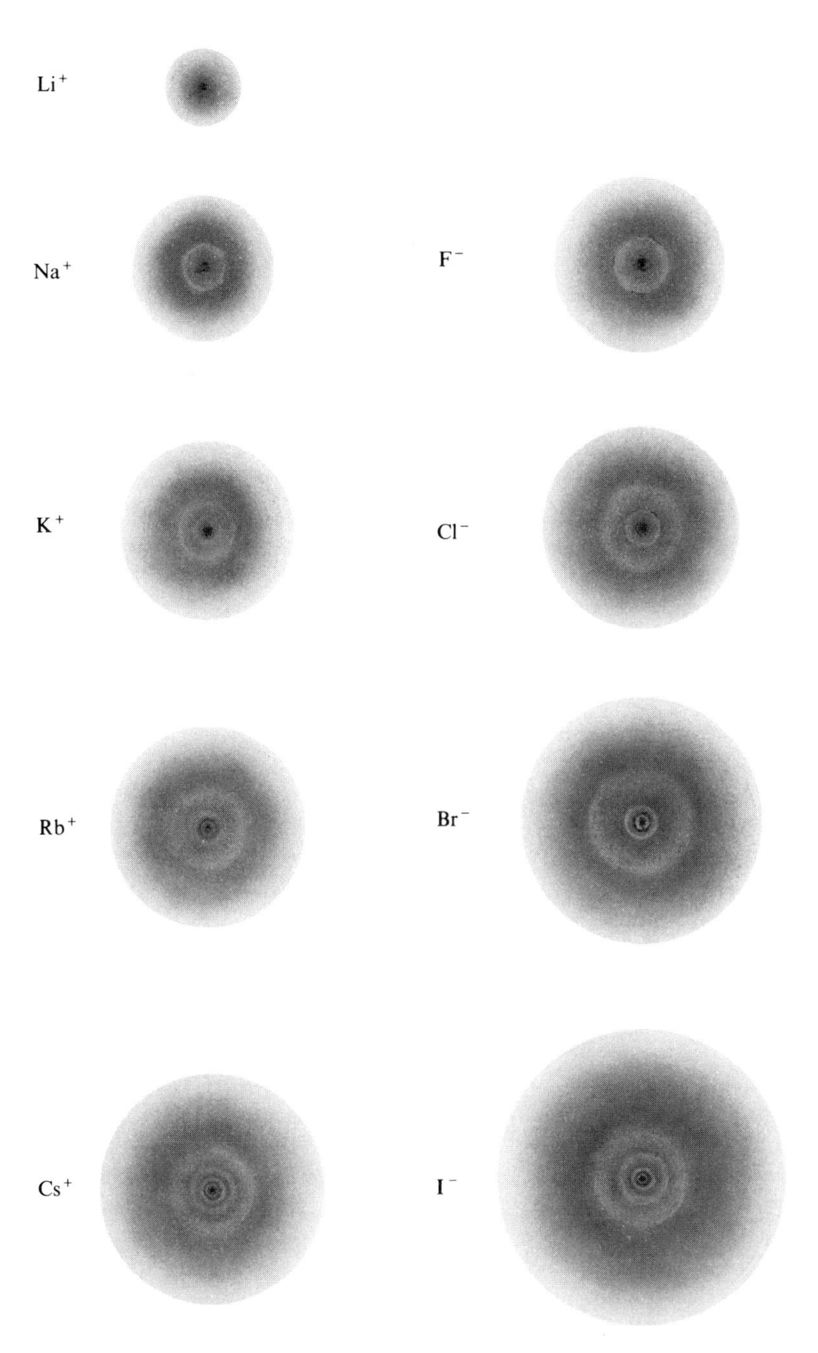

FIGURE 6-20
The electron distribution in alkali ions and halide ions.

Atoms and ions do not have a sharply defined outer surface. Instead, the electron distribution function usually reaches a maximum for the outer shell and then decreases asymptotically toward zero with increasing distance from the nucleus. It is possible to define a set of crystal radii for ions such that the radii of two ions with similar electronic structures are proportional to the relative extensions in space of the electron distribution functions for the two ions, and that the sum of two radii is equal to the contact distance of the two ions in the crystal. Figure 6-21 shows the relative sizes of various ions with argonon structures, chosen in this way. Some values of ionic radii are given in Table 6-2.

These radii give the observed cation-anion distance in crystals in which cation and anion have the structure of the same argonon, such as Na^+F^- (both ions with the neon structure) and K^+Cl^- (both with the argon structure). The observed distances, $Na^+—F^- = 231$ pm and $K^+—Cl^- = 314$ pm, are equal to the sums of the corresponding radii. In other crystals, in which the anions are almost in contact, the observed distance is larger than the radius sum.

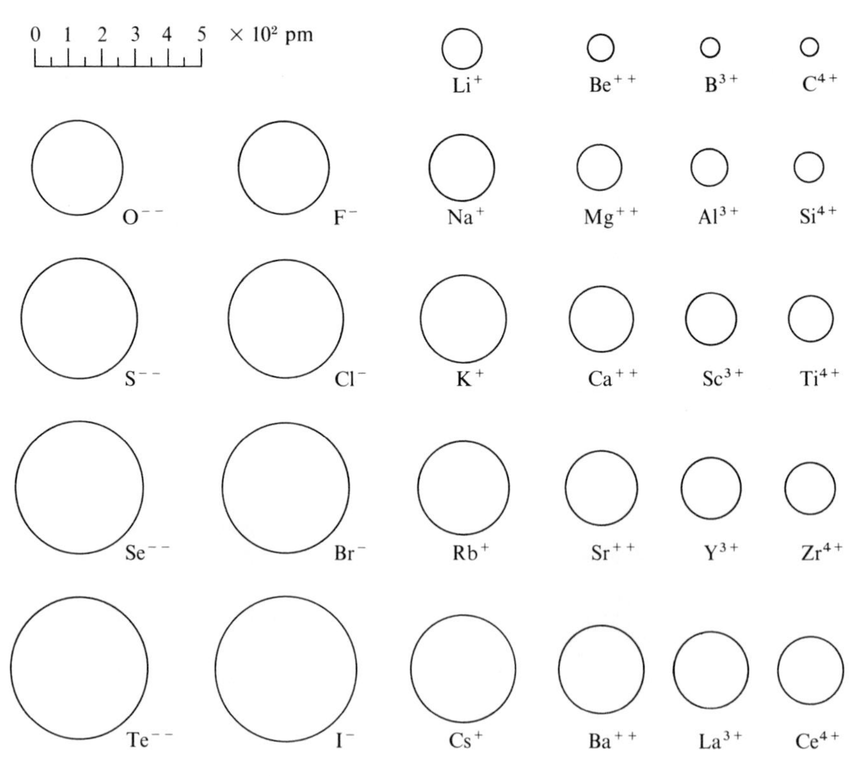

FIGURE 6-21
A drawing representing the ionic radii of ions.

TABLE 6-2
Crystal Radii of Some Ions

Ion	Radius	Ion	Radius	Ion	Radius	Ion	Radius	Ion	Radius	Ion	Radius
				Li^+	60 pm	Be^{++}	31 pm	B^{3+}	20 pm	C^{4+}	15 pm
O^{--}	140 pm	F^-	136 pm	Na^+	95	Mg^{++}	65	Al^{3+}	50	Si^{4+}	41
S^{--}	184	Cl^-	181	K^+	133	Ca^{++}	99	Sc^{3+}	81	Ge^{4+}	53
Se^{--}	198	Br^-	195	Rb^+	148	Sr^{++}	113	Y^{3+}	93	Sn^{4+}	71
Te^{--}	221	I^-	216	Cs^+	169	Ba^{++}	135	La^{3+}	115	Pb^{4+}	84

6-11. The Partial Ionic Character of Covalent Bonds

Often a decision must be made as to whether a molecule is to be considered as containing an ionic bond or a covalent bond. There is no question about a salt of a strong metal and a strong nonmetal; an ionic structure is to be written for it. Thus for lithium chloride we write

$$Li^+Cl^- \quad \text{or} \quad Li^+:\ddot{\underset{..}{Cl}}:^-$$

Similarly there is no doubt about nitrogen trichloride, NCl_3, an oily molecular substance composed of two nonmetals. Its molecules have the covalent structure

$$:N\begin{matrix} \diagup :\ddot{\underset{..}{Cl}}: \\ —:\ddot{\underset{..}{Cl}}: \\ \diagdown :\ddot{\underset{..}{Cl}}: \end{matrix}$$

Between LiCl and NCl_3 there are the three compounds $BeCl_2$, BCl_3, and CCl_4. Where does the change from an ionic structure to a covalent structure occur?

The answer to this question is provided by the theory of resonance. *The transition from an ionic bond to a normal covalent bond in a series of compounds such as those mentioned in the preceding sentence does not occur sharply, but gradually.*

Often only the covalent structure is shown, and the chemist bears in mind that the covalent bonds have a certain amount of ionic character. These bonds are called *covalent bonds with partial ionic character.*

For example, the hydrogen chloride molecule may be assigned the resonating structure

$$\left\{ H^{+} : \ddot{\underset{..}{Cl}} :^{-} \qquad H - \ddot{\underset{..}{Cl}} : \right\}$$

This is usually represented by the simple structural formula

$$H - \ddot{\underset{..}{Cl}} : \qquad \text{or} \qquad H - Cl$$

In practice it is customary to indicate bonds between the highly electropositive metals and the nonmetals as ionic bonds, and bonds between nonmetals and nonmetals or metalloids as covalent bonds, which are understood to have a certain amount of partial ionic character.

6-12. The Electronegativity Scale of the Elements

It has been found possible to assign to the elements numbers representing their power of attraction for the electrons in a covalent bond, by means of which the amount of partial ionic character of the bond may be estimated. This power of attraction for the electrons in a covalent bond is called the *electronegativity* of the element. In Figure 6-22 the elements other than the transition elements and the rare-earth metals are shown on an *electronegativity scale*. The electronegativity values are also given in Table 6-3. The symbol x is used for electronegativity.

The scale extends from cesium, with electronegativity 0.7, to fluorine, with electronegativity 4.0. Fluorine is by far the most electronegative element, with oxygen in second place, and nitrogen and chlorine in third place. Hydrogen and the metalloids are in the center of the scale, with electronegativity values close to 2. The metals have values about 1.7 or less.

The electronegativity scale as drawn in Figure 6-22 is seen to be similar in a general way to the periodic table, but deformed by pushing the top to the right and the bottom to the left. In describing the periodic table we have said that the strongest metals are in the lower left corner and the strongest nonmetals in the upper right corner of the table; because of this deformation, the electronegativity scale shows the metallic or nonmetallic character of an element simply as a function of the value of the horizontal coordinate, the electronegativity.

A rough relation between the electronegativity difference $x_A - x_B$ (or $x_B - x_A$) and the amount of partial ionic character of the bond between atoms A and B is known from the values of the electric dipole moment

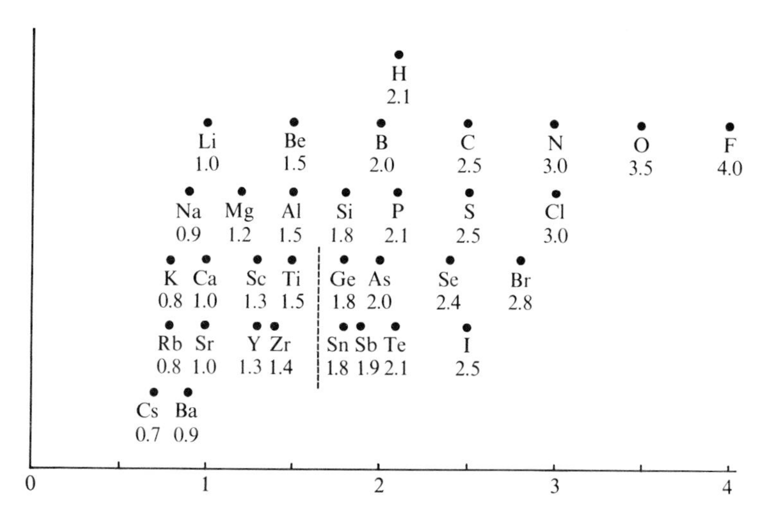

FIGURE 6-22
The electronegativity scale. The dashed line indicates approximate values for the transition metals.

TABLE 6-3
Values of the Electronegativity of Elements

						H										
						2.1										
Li	Be	B											C	N	O	F
1.0	1.5	2.0											2.5	3.0	3.5	4.0
Na	Mg	Al											Si	P	S	Cl
0.9	1.2	1.5											1.8	2.1	2.5	3.0
K	Ca	Sc	Ti	V	Cr	Mn	Fe	Co	Ni	Cu	Zn	Ga	Ge	As	Se	Br
0.8	1.0	1.3	1.5	1.6	1.6	1.5	1.8	1.9	1.9	1.9	1.6	1.6	1.8	2.0	2.4	2.8
Rb	Sr	Y	Zr	Nb	Mo	Tc	Ru	Rh	Pd	Ag	Cd	In	Sn	Sb	Te	I
0.8	1.0	1.2	1.4	1.6	1.8	1.9	2.2	2.2	2.2	1.9	1.7	1.7	1.8	1.9	2.1	2.5
Cs	Ba	La–Lu	Hf	Ta	W	Re	Os	Ir	Pt	Au	Hg	Tl	Pb	Bi	Po	At
0.7	0.9	1.0–1.2	1.3	1.5	1.7	1.9	2.2	2.2	2.2	2.4	1.9	1.8	1.9	1.9	2.0	2.2
Fr	Ra	Ac	Th	Pa	U	Np–No										
0.7	0.9	1.1	1.3	1.4	1.4	1.4–1.3										

(distribution of electric charge) in diatomic molecules, calculated from the measured values of the dielectric constant of substances. This relation is given in Table 6-4 and Figure 6-23.

The farther away two elements are from one another in the electronegativity scale (horizontally in Figure 6-22), the greater is the amount of ionic character of a bond between them. When the separation on the scale is 1.7 the bond has about 50% ionic character. If the separation is greater than this, it would seem appropriate to write an ionic structure for the substance, and if it is less to write a covalent structure. However, it is not necessary to adhere rigorously to this rule.

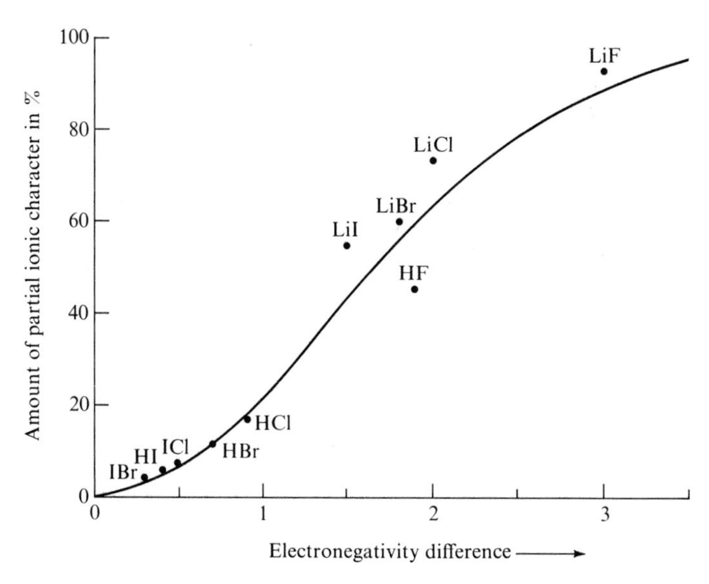

FIGURE 6-23
A diagram showing the relation between the ionic character of a bond and the electronegativity difference of the two atoms that are bonded together.

TABLE 6-4
Relation between Electronegativity Difference and Amount of Partial Ionic Character of Bonds

$x_A - x_B$	Amount of Ionic Character	$x_A - x_B$	Amount of Ionic Character
0.2	1%	1.8	55%
.4	4	2.0	63
.6	9	2.2	70
.8	15	2.4	76
1.0	22	2.6	82
1.2	30	2.8	86
1.4	39	3.0	89
1.6	47	3.2	92

Heat of Reaction and Electronegativity Difference

Some chemical reactions take place with the evolution of heat, and some with the absorption of heat. The reactions that take place with the evolution of heat are called *exothermic reactions*, and those that take place with the absorption of heat are called *endothermic reactions*. Of course, any reaction that is exothermic when it takes place in one direction is endothermic when it takes place in the reverse direction.

One method of determining the heat of reaction, with use of a calorimeter, is described in Appendix VI.

It has been found that for many reactions, especially those involving reactant molecules and product molecules with single bonds between the atoms, the heat of reaction can be correlated with the electronegativity values of the atoms.

Let us discuss hydrogen iodide, HI, as an example. Atoms of hydrogen and iodine, although quite different, are approximately equal in electronegativity. In the molecule H—I the two atoms exert about the same attraction on the shared electron pair that constitutes the covalent bond between them. This bond is accordingly much like the covalent bonds in the elementary molecules H—H and I—I. It is hence not surprising that the energy of the H—I bond is very nearly the average of the energies of the H—H bond and the I—I bond. The heat of formation of HI from the gas molecules H_2 and I_2 is only 6kJ $mole^{-1}$:

$$\tfrac{1}{2}H_2(g) + \tfrac{1}{2}I_2(g) \longrightarrow HI(g) + 6\text{kJ mole}^{-1}$$

The other hydrogen halide molecules have larger amounts of electronegativity difference (0.7 for HBr, 0.9 for HCl, 1.9 for HF), and of partial ionic character, and their heats of formation also increase greatly in the same order:

$$\tfrac{1}{2}H_2(g) + \tfrac{1}{2}Br_2(g) \longrightarrow HBr(g) + 51\text{ kJ mole}^{-1}$$

$$\tfrac{1}{2}H_2(g) + \tfrac{1}{2}Cl_2(g) \longrightarrow HCl(g) + 92\text{ kJ mole}^{-1}$$

$$\tfrac{1}{2}H_2(g) + \tfrac{1}{2}F_2(g) \longrightarrow HF(g) + 269\text{ kJ mole}^{-1}$$

This means that the bonds in these hydrogen halide molecules are stronger than the average of the bonds in the molecules of the elementary substances, and that the increased strength (bond energy) is determined by the electronegativity difference of the two atoms. *The greater the separation of two elements on the electronegativity scale, the greater is the strength of the bond between them.* The extra stability is the resonance energy between the normal covalent structure and the ionic structure.

Values of bond energies are given in Appendix V, along with a detailed discussion of heats of reaction in relation to electronegativity differences. The equation that gives approximate values for the heat of formation of many substances from the elements in their standard states is

$$Q_f = \text{heat of formation (in kJ mole}^{-1})$$

$$= 100\Sigma(x_A - x_B)^2 - 6.5\Sigma(x_A - x_B)^4 - 235n_N - 106n_O \qquad (6\text{-}1)$$

Here the summation is to be taken over all the bonds represented by the formula of the compound. The equation applies only to substances in which the bonds are single bonds. Nitrogen (N_2) and oxygen (O_2) contain multiple bonds, and their molecules are more stable, by 470 kJ mole^{-1} and 212 kJ mole^{-1}, respectively, than they would be if the molecules contained single bonds. The last two terms in Equation 6-1 give the correction for this extra stability: n_N is the number of nitrogen atoms and n_O the number of oxygen atoms in the substance.

It was suggested by the American chemist Robert S. Mulliken that the electronegativity of an element is proportional to the average of the first ionization energy and the electron affinity of its atoms (Table 6-1). Values of the first ionization energy of F, Cl, Br, and I are 1687, 1257, 1149, and 1013 kJ mole^{-1}, respectively. The factor with which the sum of the first ionization energy and the electron affinity, in kJ mole^{-1}, is multiplied is 0.00192. This factor leads to values within 0.1 unit of the values given in Table 6-3 for the halogens.

Example 6-11. Which of the two following substances is predicted to form from the elements by a strongly exothermic reaction: PI_3, PF_3?

Solution. Phosphorus and iodine differ by only 0.3 in electronegativity; hence the formation of PI_3 is predicted to be only slightly exothermic. The extra P—I bond energy due to partial ionic character, $100(x_A - x_B)^2$ kJ mole^{-1}, is $100 \times 0.3^2 = 9$ kJ mole^{-1} per P—I bond; hence we predict $P + \frac{3}{2}I_2 \longrightarrow PI_3 + 27$ kJ mole^{-1}.

For formation of PF_3 a strongly exothermic reaction is expected, because of the large electronegativity difference, 1.9, which leads to $3 \times (100 \times 1.9^2 - 6.5 \times 1.9^4) = 829$ kJ mole^{-1}: $P(c) + \frac{3}{2}F_2(g) \longrightarrow PF_3(g) + 829$ kJ mole^{-1}. Hence PF_3 would be formed by a strongly exothermic reaction.

Example 6-12. A mixture of aluminum powder and iron(III) oxide, Fe_2O_3, can be ignited; the reaction

$$2Al(c) + Fe_2O_3(c) \longrightarrow 2Fe(c) + Al_2O_3(c)$$

then takes place, producing much heat (the product is molten iron).

Would you predict that magnesium could be similarly made from MgO by use of aluminum?

Solution: We may consider the two reactions

$$4Al(c) + 3O_2(g) \longrightarrow 2Al_2O_3(c) + Q_1$$

and

$$6Mg(c) + 3O_2(g) \longrightarrow 6MgO(c) + Q_2$$

Here Q_1 is the heat evolved in the first reaction and Q_2 is the heat evolved in the second reaction. In each reaction, as written, twelve metal-oxygen single bonds are formed. The electronegativity of Al is 1.5 and that of Mg is 1.2. Hence the electronegativity difference of Al and O, 2.0, is less than that of Mg and O, 2.3, and accordingly Q_1 is less than Q_2. By subtracting the second equation from the first we obtain the result

$$4Al(c) + 6MgO(c) \longrightarrow 6Mg(c) + 2Al_2O_3(c) + Q_1 - Q_2$$

Since Q_1 is less than Q_2, this reaction is endothermic, and probably would not take place. Accordingly magnesium probably could not be made by igniting a mixture of aluminum and magnesium oxide.

Heat Content. Enthalpy

It has been found by experiment that it is possible to assign to every substance at standard conditions a numerical value of a certain quantity, represented by the symbol H, such that the heat absorbed during a chemical reaction carried out at constant temperature and constant pressure can be found by subtracting the sum of the values of H for the reactants from the sum for the products of the reaction. The names used for the quantity H are *heat content* and *enthalpy*. These names are equivalent to one another.

The symbol ΔH is used to express the change in enthalpy (or heat content) of a system accompanying a change in state, such as a chemical reaction. Thus a positive value of ΔH means that heat is absorbed from the surroundings during the reaction. The symbol ΔH°_{298} is used for the change in enthalpy accompanying a change (reaction) at 298.16 K (25°C) and 1 atm pressure. For example, we may write the equations for the combustion of carbon as follows:

$$\text{C (graphite)} + O_2(g) \longrightarrow CO_2(g) \qquad \Delta H^\circ_{298} = -94{,}052 \text{ cal/mole}$$

$$\text{C (graphite)} + \tfrac{1}{2}O_2(g) \longrightarrow CO(g) \qquad \Delta H^\circ_{298} = -26{,}416 \text{ cal/mole}$$

(The temperature 25°C has been accepted by chemists as the customary one for determination of thermochemical quantities.)

The name *enthalpy of reaction* is used for the quantity ΔH°. (ΔH°_{298} is the enthalpy of reaction at 298.16 K.) We see that the enthalpy of reaction is negative for exothermic reactions and positive for endothermic reactions.

In a preceding paragraph we have described the heat of a reaction as the amount of heat evolved or absorbed when the reaction takes place at constant temperature and pressure. Two mutually contradictory definitions of heat of reactions are used at the present time in textbooks and reference books. For over a century it has been customary to define the heat of a reaction (heat of combustion, heat of formation, heat of solution) as the heat evolved in the process; that is, as $-\Delta H°$. On the other hand, heats of fusion and vaporization have been defined as the heat absorbed during fusion or vaporization. During the last few years many chemists have adopted the definition of heat of reaction as the heat absorbed in the process. This usage is to be found, for example, in the valuable reference book *Selected Values of Chemical Thermodynamic Properties,* Circular of the U.S. Bureau of Standards No. 500 (1952), in which values of heats of formation of compounds from elements in their standard states and some other properties of substances are given.

In order to avoid the confusion that might result from the present lack of agreement about the sign in the definition of heat of reaction, we shall in this textbook make use instead of the term enthalpy of reaction, defined as $\Delta H°$.

6-13. The Electroneutrality Principle

A useful principle in writing electronic structures for substances is the *electroneutrality principle.* This principle is that *stable molecules and crystals have electronic structures such that the electric charge of each atom is close to zero.* "Close to zero" means between -1 and $+1$.

That this principle is a reasonable one may be seen by the consideration of values of the ionization energy and electron affinity of atoms. The electron affinity of atoms of nonmetals is about 350 kJ mole^{-1} for the first electron added, to convert the atom $:\ddot{\underset{\cdot\cdot}{\mathrm{F}}}\cdot$ into the anion $:\ddot{\underset{\cdot\cdot}{\mathrm{F}}}:^-$, or the atom $:\dot{\underset{\cdot\cdot}{\mathrm{O}}}\cdot$ into the anion $:\ddot{\underset{\cdot\cdot}{\mathrm{O}}}\cdot^-$ (Section 6-9). But there is in general no significant affinity for a second electron to convert $:\ddot{\underset{\cdot\cdot}{\mathrm{O}}}\cdot^-$ into $:\ddot{\underset{\cdot\cdot}{\mathrm{O}}}:^{--}$, even when it would complete an octet. The repulsion of the two negative charges decreases the attraction for the second electron to nearly zero. Similarly, the values of the first ionization energy of metal atoms lie between about 400 and 800 kJ mole^{-1}, but the second ionization energy is 1500 kJ mole^{-1} or more. It is accordingly unlikely that an atom in a stable molecule would have either a double negative charge or a double positive charge.

The use of the electroneutrality principle in assigning electronic structures to molecules and crystals is discussed in the following examples and in the following section and later chapters.

Example 6-13. Should the hydrogen cyanide molecule be assigned the structure HCN or the structure HNC?

Solution. The electronic structure $H{-}C{\equiv}N{:}$ makes the atoms nearly neutral. The partial ionic character of the bonds (4% for H—C and 7% for each C—N) leads to the charge +0.04 on H, +0.17 on C, and −0.21 on N. These charges are small, and are compatible with the electroneutrality principle. For HNC the structure $H{-}N{\equiv}C{:}$, which completes the octet about N and C, assigns four valence electrons to N and five to C, and hence corresponds to N^+ and C^-. The partial ionic character of the bonds then leads to the charges +0.04 on H, +0.75 on N, and −0.79 on C. These charges on N and C are much larger than for the structure $H{-}C{\equiv}N{:}$, and correspond to instability. Hence HCN is the preferable structure.

Example 6-14. What is the electronic structure of the anesthetic gas nitrous oxide, N_2O?

Solution. A ring structure is not likely, because of the strain of the bent bonds. The linear structure $\overset{--}{:\ddot{\underset{\cdot\cdot}{N}}}{-}\overset{+}{N}{\equiv}\overset{+}{O}{:}$ completes the octet for each atom, but we reject it because of the double negative charge on the end nitrogen atom. The two other structures that complete the octet for each atom are ${:}N{\equiv}\overset{+}{N}{-}\overset{-}{\underset{\cdot\cdot}{\ddot{O}}}{:}$ and $\overset{-}{:\ddot{N}}{=}\overset{+}{N}{=}\ddot{O}{:}$, each of which has formal charges on two atoms as shown. These two structures look equally good, and we conclude that the molecule can best be described as the resonance hybrid with the two structures contributing about equally.

6-14. Transargononic Structures

Sometimes an atom forms so many covalent bonds as to surround itself with more than four electron pairs; it assumes a transargononic structure. An example is phosphorus pentachloride, PCl_5; in the molecule of this substance the phosphorus atom is surrounded by five chlorine atoms, with each of which it forms a covalent bond (with some ionic character):

$$\begin{array}{ccc} & :\ddot{\underset{\cdot\cdot}{Cl}}: & \\ :\ddot{\underset{\cdot\cdot}{Cl}} \diagdown & | & \\ & P & {-}\,:\ddot{\underset{\cdot\cdot}{Cl}}: \\ :\ddot{\underset{\cdot\cdot}{Cl}} \diagup & | & \\ & :\ddot{\underset{\cdot\cdot}{Cl}}: & \end{array}$$

The phosphorus atom in this compound seems to be using five of the nine orbitals of the *M* shell, rather than only the four most stable orbitals, which are occupied by electrons in the argon configuration. It seems likely that of the nine or more orbitals in the *M* shell, the *N* shell, and the *O* shell four are especially stable but one or more others may occasionally be utilized.

The difference in electronegativity of chlorine and phosphorus is 0.9, which corresponds to 18% of partial ionic character. Accordingly, an alternative description of the PCl_5 molecule is that the phosphorus atom forms four covalent bonds, using only the four orbitals of the outer shell, and one ionic bond to Cl^-, and that the four covalent bonds resonate among the five positions, so that each chlorine atom is held by a bond with 80% covalent and 20% ionic character.

The oxygen acids, such as H_2SO_4, may be assigned similar transargononic structures:

```
        ..
      H—O:
        |    ..
   :O═S—O:
    ..  ‖    |
       :O    H
        ..
```

The stability of transargononic structures will be discussed in Sections 7-8 and 8-1.

6-15. The Sizes of Atoms and Molecules. Covalent Radii and van der Waals Radii

Interatomic distances (bond lengths) in molecules and crystals can be determined by the methods of spectroscopy (including microwave spectroscopy), x-ray diffraction, electron diffraction, neutron diffraction, and nuclear magnetic resonance. The description of these methods is beyond the scope of this book. During the past forty years the bond lengths have been determined for many hundreds of substances, and their values have been found to be useful in the discussion of the electronic structures of molecules and crystals.

It has been found that usually the bond length for the single bond A—B is, to within about 3 pm, equal to the average of the bond lengths A—A and B—B. For example, the average of C—C (154 pm, Section 6-5) and Cl—Cl (198 pm) is $\frac{1}{2} \times (154 + 198) = 176$ pm. The value of C—Cl found by the investigation of CCl_4 by the electron diffraction method is 176 pm; hence the *single-bond covalent radii* 77 pm for C and

TABLE 6-5
Single-bond Covalent Radii

C	77 pm	N	70 pm	O	66 pm	F	64 pm
Si	117	P	110	S	104	Cl	99
Ge	122	As	121	Se	117	Br	114
Sn	140	Sb	141	Te	137	I	133

99 pm for Cl can be added together in three ways to give the three observed bond lengths C—C, Cl—Cl, and C—Cl.

Values of the single-bond covalent radii of nonmetallic elements are given in Table 6-5. The value for hydrogen is 30 pm for all bonds other than H—H (the H—H bond length, 74 pm, corresponds to a larger radius for hydrogen than the value used for other bonds).

It was mentioned in Section 6-5 that the C═C and C≡C bond lengths are 21 pm and 34 pm, respectively, less than the C—C bond length. Approximately the same shortening is found for other double and triple bonds. For example, for C—N the bond length 147 pm is given by the sum of the radii for carbon and nitrogen (Table 6-5), and the value $147 - 34 = 113$ pm would be expected for C≡N. The value observed in H—C≡N is 115 pm, in reasonably good agreement with the result of the calculation.

Bond lengths for hybrid structures have intermediate values.

Example 6-15. The bond lengths 113 pm for nitrogen-nitrogen and 119 pm for nitrogen-oxygen are observed in nitrous oxide, N_2O. What do these values indicate about the structure of the molecule?

Solution. Expected bond lengths (Table 6-5), with use of −21 pm for a double bond and −34 pm for a triple bond, are 119 pm for N═N, 106 pm for N≡N, 136 pm for N—O, and 115 pm for N═O. We see that the observed values indicate that the nitrogen-nitrogen bond is intermediate between a double bond and a triple bond and that the nitrogen-oxygen bond is intermediate between a single bond and a double bond. This comparison accordingly supports the conclusion reached in Example 6-14 that the structure is a resonance hybrid of $:N\equiv\overset{+}{N}-\underset{\cdot\cdot}{\overset{\cdot\cdot}{O}}:^{-}$ and $^{-}:\overset{\cdot\cdot}{N}=\overset{+}{N}=\overset{\cdot\cdot}{O}:$

Van der Waals Radii

The Dutch physicist J. D. van der Waals (1837–1923) found that in order to explain some of the properties of gases it was necessary to assume that molecules have a well-defined size, so that two molecules begin to undergo strong repulsion when, as they approach, they reach a certain distance from one another. For example, the deviations of the argonons from ideal behavior and other properties such as viscosity lead to the assignment of effective radii between 100 pm and 200 pm to their molecules. These radii are called the *van der Waals radii* of the atoms.

It has been found that the effective sizes of molecules packed together in liquids and crystals can be described by assigning similar van der Waals radii to each atom in the molecule. Values of these radii are given in Table 6-6.

The values are seen to be about 80 pm larger than the corresponding single-bond covalent radii. This difference is illustrated in Figure 6-24, which represents two chlorine molecules in van der Waals contact (packed together in a crystal or colliding in the liquid or gas). Each chlorine atom is surrounded by four outer electron pairs. One pair is shared with the other chlorine atom in the same Cl_2 molecule. The point midway between the two nuclei, 99 pm from each nucleus, represents the

TABLE 6-6
Van der Waals Radii of Atoms

H	110 pm	N	150 pm	O	140 pm	F	135 pm
		P	190	S	185	Cl	180
		As	200	Se	200	Br	195
		Sb	220	Te	220	I	215
		Half-thickness of aromatic molecule, such as benzene or naphthalene					170

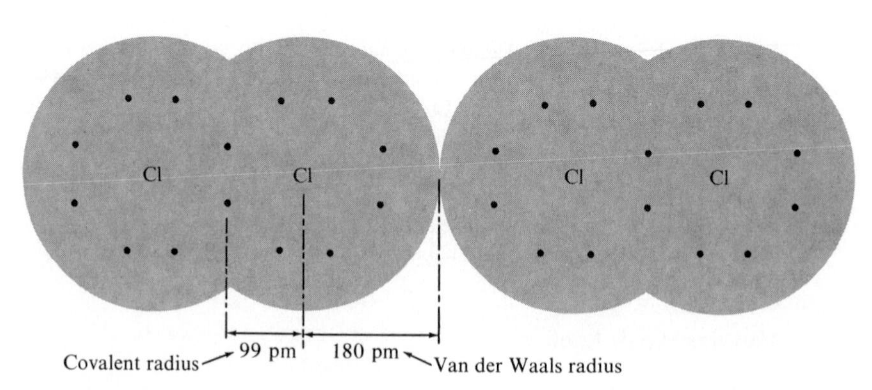

FIGURE 6-24
Two chlorine molecules in van der Waals contact, illustrating the difference between van der Waals radius and covalent radius.

average position of the shared pair. The three unshared pairs about each nucleus are also about the same distance (equal to the covalent radius) from the nucleus. When two nonbonded chlorine atoms are in contact there are two unshared pairs in the region between the nuclei; the van der Waals radius defines the region that includes the major part of the electron distribution function for the unshared pairs.

In making drawings of atoms or molecules, the van der Waals radii may be used in indicating the volume within which the electrons are largely contained. For ions, the ionic radii (crystal radii) discussed in Section 6-10 may be used. The van der Waals radius of an atom and the ionic radius of its negative ion are essentially the same. For example, the van der Waals radius of chlorine is 180 pm, and the ionic radius of the chloride ion is 181 pm.

The covalent radii have a different meaning and a different use. The sum of the single-bond covalent radii for two atoms is equal to the distance between the atoms when they are connected by a single covalent bond. The single-bond covalent radius of an atom may be considered to be the distance from the nucleus to the average position of the shared electron pair, whereas the van der Waals radius extends to the outer part of the region occupied by the electrons of the atom, as indicated in Figure 6-24. The effective radius of an atom in a direction that makes only a small angle with the direction of a covalent bond formed by the atom is smaller than the van der Waals radius in directions away from the bond. For example, in the carbon tetrachloride molecule the chlorine atoms are only 290 pm apart, and yet the properties of the substance indicate that there is no great strain, even though this distance is much less than the van der Waals diameter 360 pm.

6-16. Oxidation Numbers of Atoms

The nomenclature of inorganic chemistry is based upon the assignment of numbers (positive or negative) to the atoms of the elements. These numbers, called *oxidation numbers,* are defined in the following way.

The oxidation number of an atom is a number that represents the electric charge that the atom would have if the electrons in a compound were assigned to the atoms in a certain conventional way.

The assignment of electrons is somewhat arbitrary, but the conventional procedure, described below, is useful because it permits a simple statement to be made about the valences of the elements in a compound without considering its electronic structure in detail and because it can be made the basis of a simple method of balancing equations for oxidation-reduction reactions.

An oxidation number may be assigned to each atom in a substance by the application of simple rules. These rules, though simple, are not completely unambiguous. Although their application is usually a straightforward procedure, it sometimes requires considerable chemical insight and knowledge of molecular structure. The rules are given in the following statements.

1. The oxidation number of a monatomic ion in an ionic substance is equal to its electric charge.
2. The oxidation number of an atom in an elementary substance is zero.
3. In a covalent compound of known structure, the oxidation number of each atom is the charge remaining on the atom when each shared electron pair is assigned completely to the more electronegative of the two atoms sharing it. An electron pair shared by two atoms of the same element is usually split between them.
4. The oxidation number of an element in a compound of uncertain structure may be calculated from a reasonable assignment of oxidation numbers to the other elements in the compound.

The application of the first three rules is illustrated by the following examples; the number by the symbol of each atom is the oxidation number of that atom:

$Na^{+1}Cl^{-1}$	$Mg^{+2}(Cl^{-1})_2$	$(B^{+3})_2(O^{-2})_3$
$H_2{}^0$	$O_2{}^0$	C^0 (diamond or graphite)
H^{+1} (hydrogen cation)	$(O^{-2}H^{+1})^-$ (hydroxide ion)	
$N^{-3}(H^{+1})_3$	$Cl^{+1}F^{-1}$	$C^{+4}(O^{-2})_2$
$C^{+2}O^{-2}$	$C^{-4}(H^{+1})_4$	$K^{+1}Mn^{+7}(O^{-2})_4$

Fluorine, the most electronegative element, has the oxidation number -1 in all of its compounds with other elements.

Oxygen is second only to fluorine in electronegativity, and in its compounds it usually has oxidation number -2; examples are $Ca^{+2}O^{-2}$, $(Fe^{+3})_2(O^{-2})_3$, $C^{+4}(O^{-2})_2$. Oxygen fluoride, OF_2, is an exception; in this compound, in which oxygen is combined with the only element that is more electronegative than it is, oxygen has the oxidation number $+2$. Oxygen has oxidation number -1 in hydrogen peroxide, H_2O_2, and other peroxides.

Hydrogen when bonded to a nonmetal has oxidation number +1, as in $(H^{+1})_2O^{-2}$, $(H^{+1})_2S^{-2}$, $N^{-3}(H^{+1})_3$, $(P^{-2})_2(H^{+1})_4$. In compounds with metals, such as $Li^{+1}H^{-1}$ and $Ca^{+2}(H^{-1})_2$, its oxidation number is −1, corresponding to the electronic structure $H{:}^{-1}$ for a negative hydrogen ion with completed *K* shell (helium structure).

Oxidation Number and Chemical Nomenclature

The principal classification of the compounds of an element is made on the basis of its oxidation state. In our discussions of the compounds formed by the various elements or groups of elements in the following chapters of this book we begin by a statement of the oxidation states represented by the compounds. The compounds are grouped together in classes, representing those with the principal element in the same oxidation state. For example, in the discussion of the compounds of iron they are divided into two classes, representing the compounds of iron in oxidation state +2 and those in oxidation state +3, respectively.

The nomenclature of the compounds of the metals is also based upon their oxidation states. At the present time there are two principal nomenclatures in use. We may illustrate the two systems of nomenclature by taking the compounds $FeCl_2$ and $FeCl_3$ as examples. In the older system a compound of a metal in the lower of two important oxidation states is named by use of the name of the metal (usually the Latin name) with the suffix *ous*. Thus the salts of iron in oxidation state +2 are ferrous salts; $FeCl_2$ is called ferrous chloride. The compounds of a metal in the higher oxidation state are named with use of the suffix *ic*. The salts of iron in oxidation state +3 are called ferric salts; $FeCl_3$ is ferric chloride.

Note that the suffixes *ous* and *ic* do not tell what the oxidation states are. For copper compounds, such as CuCl and $CuCl_2$, the compounds in which copper has oxidation number +1 are called cuprous compounds, and those in which it has oxidation number +2 are called cupric compounds.

A new system of nomenclature for inorganic compounds was drawn up by a committee of the International Union of Chemistry in 1940. According to this system the value of the oxidation number of a metal is represented by a Roman numeral given in parentheses following the name (usually the English name rather than the Latin name) of the metal. Thus $FeCl_2$ is given the name iron(II) chloride, and $FeCl_3$ is given the name iron(III) chloride. These names are read simply by stating the numeral after the name of the metal: thus iron(II) chloride is read as "iron two chloride."

It may be noted that it is not necessary to give the oxidation number of a metal in naming a compound if the metal forms only one principal series of compounds. The compound $BaCl_2$ may be called barium chloride rather than barium(II) chloride, because barium forms no stable compounds other than those in which it has oxidation number +2. Also, if one oxidation state is represented by many compounds, and another by only a few, the oxidation state does not need to be indicated for the compounds of the important series. Thus the compounds of copper with oxidation number +2 are far more important than those of copper with the oxidation number +1, and for this reason $CuCl_2$ may be called simply copper chloride, whereas CuCl would be called copper(I) chloride.

We shall in general make use of the new system of nomenclature in the following chapters of our book, except that, for convenience, we shall use the old nomenclature for the following common metals:

Iron: +2, ferrous; +3, ferric

Copper: +1, cuprous; +2, cupric (or copper)

Mercury: +1, mercurous; +2, mercuric

Tin: +2, stannous; +4, stannic

Compounds of metalloids and nonmetals are usually given names in which the numbers of atoms of different kinds are indicated by prefixes. The compounds PCl_3 and PCl_5, for example, are called phosphorus trichloride and phosphorus pentachloride, respectively. We shall use the name dinitrogen trioxide for N_2O_3, and similar names for N_2O_4, N_2O_5, and other such compounds, although the prefix di is often omitted in general usage.

Oxidation and Reduction

An increase in the oxidation number of an element is called *oxidation* of the element. A decrease (or reduction) in the oxidation number of an element is called *reduction* of the element. For example the rusting of metallic iron to form Fe_2O_3 involves oxidation of iron because it changes the oxidation number from 0 to +3. The reverse process, making metallic iron by the reaction of Fe_2O_3 with carbon, involves reduction of iron because the oxidation number changes from +3 to 0. Oxidation is the removal of electrons and reduction is the addition of electrons.

EXERCISES

6-1. What is the covalent bond? Describe with reference to the molecules H_2 and CH_4.

6-2. What is the usual electronic structure of a covalently bound atom?

6-3. How are the electrons of a covalent bond divided between the two bound atoms in determining the electronic structures of each atom?

6-4. Write the electronic structure of a compound of hydrogen and an element from each of the fifth, sixth, and seventh groups.

6-5. The substance pentamethylenetetrazole, $C_6H_{10}N_4$, which is used in the treatment of mental illness, contains a ring of six carbon atoms and one nitrogen atom, fused with a five-membered ring (that is, with two atoms in common). The large ring contains a sequence of five CH_2 groups. Draw a structural formula for the substance on the basis of this information, with each atom completing the helium or neon shell. How many double bonds are there in the molecule?

6-6. Write electronic structures for the molecules NH_3 and BF_3. These molecules combine to form the addition compound H_3NBF_3. What is the electronic structure of this compound? What similarity is there in the electronic rearrangements in the following chemical reactions?

$$NH_3 + H^+ \rightarrow NH_4^+$$

$$NH_3 + BF_3 \rightarrow H_3NBF_3$$

6-7. Spiropentane, C_5H_8, contains two three-membered rings with a common corner. Make a drawing showing the arrangement of bonds in the molecule. How many isomers would you expect for dimethylspiropentane, $C_5H_6(CH_3)_2$? How many of them would be members of enantiomeric pairs?

6-8. White phosphorus and sulfur combine, when heated, to form the compound P_4S_3, which is used in making safety matches. The molecule has been shown by x-ray studies to have a threefold axis of symmetry, and the low value (near zero) of the heat of formation indicates that the atoms have their normal covalences. Assign an electronic structure to the molecule.

6-9. What is the Pasteur method for separating the enantiomers of a substance?

6-10. Assuming four-coordination, is there any configuration of the carbon atom other than tetrahedral consistent with the observation that enantiomers of organic compounds exist?

6-11. Acetylcholine,

$$\begin{array}{c} \quad CH_3 \qquad\qquad\qquad\qquad\quad O \\ \quad | \qquad\qquad\qquad\qquad\qquad \| \\ CH_3—N^+—CH_2—CH_2—O—C—CH_3 \\ \quad | \\ \quad CH_3 \end{array}$$

is an important transmitter substance in the nervous system. Do enantiomers of this cation exist?

6-12. What atomic orbitals are used to form tetrahedral hybrid bond orbitals? What are the angles between these hybrid orbitals?

6-13. Three different substances are known with the molecular formula C_3H_4. Draw their structural formulas and describe their spatial arrangements.

6-14. Allene has the structural formula $H_2C{=}C{=}CH_2$. Do all of the atoms lie in the same plane? Would you predict that 1,3-difluoro-allene, $HFC{=}C{=}CHF$, exists as an enantiomeric pair of substances or as a single optically inactive substance?

6-15. What is the σ-π orbital description of diacetylene, propylene, and allene?

6-16. What is the angle between pure p bonds? Toward what value does this angle tend as the s orbital mixes with the p orbitals to form hybrid bond orbitals?

6-17. Define and describe chemical bond resonance.

6-18. Write the resonating electronic structures for the nitrate ion, NO_3^-; the nitrite ion NO_2^-; carbon dioxide, CO_2; carbon disulfide, CS_2; nitrous oxide, N_2O. What shapes do these molecules have and what are the expected values of the bond angles? Note that there are three valence-bond structures possible for CO_2 that give each atom an argononic structure.

6-19. The crystal structure of NaCl is cubic (isometric) with one type of ion in a face-centered lattice (cubic closest packing, Sections 2-4 and 2-5) with origin at 0, 0, 0 in the cubic unit cell, and the other ion on a face-centered lattice with origin at 0, 0, $\frac{1}{2}$ (or $\frac{1}{2}$, 0, 0 or 0, $\frac{1}{2}$, 0), as shown in Figure 6-19. What is the expected cubic unit cell edge of crystals of (a) KCl, (b) CsBr, (c) NaI, from the values of the ionic radii? What is the distance between the anion and cation in each of these crystals?

6-20. Is there a sharp division between compounds with covalent bonds and those with ionic bonds?

6-21. What does the electronegativity of an element mean?

6-22. What is the percentage of ionic character in the bonds of methane, ethane, ammonia, and ethyl alcohol?

6-23. By reference to the electronegativity scale, arrange the following binary compounds in rough order of their stability, placing those you think would be especially stable at the top of the list, and the most unstable at the bottom of the list:

Phosphine, PH_3	Cesium fluoride, CsF
Aluminum oxide, Al_2O_3	Sodium iodide, NaI
Hydrogen iodide, HI	Nitrogen trichloride, NCl_3
Lithium fluoride, LiF	Selenium diiodide, SeI_2

6-24. Methyl cyanide and methyl isocyanide have the same composition. Their standard enthalpies of formation are −88 and −150 kJ mole^{-1}, respectively. Which of the two is $H_3C—C{\equiv}N$:? (Make use of information in Appendix V.)

6-25. By use of the partial ionic character of the bonds, calculate the charge on the atoms of NH_3 and BF_3. Show how the electroneutrality principle explains some of the stability of the addition product H_3NBF_3.

6-26. Assign electronic structures to the following gas molecules *(a)* from the extreme ionic point of view; *(b)* from the extreme covalent point of view; *(c)* from use of electronegativity values to determine partial ionic character of bonds. To what extent is there agreement with the electroneutrality principle? LiI, $MgCl_2$, $Pb(CH_3)_4$, H_3CF, HCN, H_2CO, P_2H_4.

6-27. Assign resonance structures, with each atom argononic, to the following molecules and ions: NO_3^- (nitrate ion); NO_2^+ (nitronium ion); H_3BO_3 (boric acid); O_3; H_3CNO_2 (nitromethane).

6-28. Briefly discuss the following facts in relation to the electronegativity values of the elements: *(a)* in 1787 Lavoisier did not include sodium, potassium, calcium, and aluminum in his list of known elementary substances; *(b)* fluorine combines with all other elements except the lighter argonons; *(c)* ethyl chloride, C_2H_5Cl, can be made by adding HCl(g) to ethanol, C_2H_5OH; *(d)* the most stable transargononic compounds are fluorides; *(e)* many metals can be made by reaction of their chlorides with sodium; *(f)* NCl_3 is explosive, whereas NF_3 is stable; *(g)* the polymer $(CF_2)_n$ is highly resistant to attack by corrosive chemicals; *(h)* the boron hydrides are used as fuel for rocket propulsion.

6-29. What is the total length of the acetylene molecule? Of the diacetylene molecule?

6-30. What is the diameter of the carbon tetrachloride molecule? Of the tetramethylammonium ion?

6-31. Assign oxidation numbers to each element in Al_2O_3, HCl, S_8, $CuSO_4{\cdot}5H_2O$, C_2H_5OH, Cu_2O, K_2MnO_4.

6-32. What oxidation states do the following names refer to: ferrous, cuprous, mercurous, stannous, stannic, mercuric, cupric, ferric?

6-33. Potassium burns in oxygen to form the orange-yellow paramagnetic crystalline substance potassium superoxide, KO_2. The paramagnetism results from the spin of the unpaired electron of the superoxide ion, O_2^-, which has bond length 128 pm. Discuss the electronic structure of the superoxide ion in relation to this bond length. (The bond length in hydrogen peroxide is 146 pm and that in the normal state of the oxygen molecule is 121 pm.)

6-34. The standard enthalpy of formation of $CO_2(g)$ is -394 kJ mole^{-1}, and that of $CS_2(g)$ is $+115$ kJ mole^{-1}. Calculate the enthalpies of formation of the molecules from the monatomic elements (Table V-3). By comparison with the C=O and C=S bond-energy values (Table V-2), calculate the resonance energy for each of these molecules. (Answer: 157, 204 kJ mole^{-1}.)

6-35. Make a similar calculation of the resonance energy for carbon oxysulfide, OCS(g), standard enthalpy of formation -137 kJ mole^{-1}.

6-36. Assuming $x = 2.1$ for H, evaluate x for C, N, P, As, and S from values of the standard enthalpy of formation given in tables in Appendix VI, by using $100(x - x_H)^2$ kJ mole^{-1} as the increase in bond energy resulting from partial ionic character.

6-37. From the standard enthalpies of formation of $C_2H_6(g)$ and $C_2H_5OH(g)$ given in Table VI-6 and of O(g) given in Table V-3 evaluate $\Delta H°$ for the reaction

$$C_2H_6(g) + O(g) \rightarrow C_2H_5OH(g)$$

From this value and the C—H and O—H bond energies in Table V-1 calculate a value for the C—O bond energy. (Answer: 353 kJ mole^{-1}.)

6-38. Use the method of the preceding exercise to obtain $\Delta H°$ for the reaction

$$C_2H_6(g) + O(g) \rightarrow (CH_3)_2O(g)$$

and another value for the C—O bond energy. Can you suggest a structural explanation for the difference in the two values of this bond energy? (Answer: 347 kJ mole^{-1}.)

6-39. Using $100(x_A - x_B)^2 - C(s_A - x_B)^4$ for the extra bond energy from partial ionic character and the values of x in Table 6-3, calculate the coefficient C by comparing the O—H bond energy with the average of O—O and H—H (Table V-1). Repeat the calculation for C—O, C—F, and H—F. (Answer: 5.9, 6.5, 6.9, 7.3; av. 6.7 kJ mole^{-1}.)

6-40. *(a)* Write electronic structural formulas for H_2O, H_2O_2, H_2S, H_2S_2, H_2S_3, H_2S_4, H_2S_5.
(b) What is the structural explanation of the instability of hydrogen peroxide and the stability of the corresponding compound dihydrogen disulfide, as expressed by the enthalpies of decomposition:

$$H_2O_2(l) \longrightarrow H_2O(l) + \tfrac{1}{2}\,O_2(g) + 98 \text{ kJ mole}^{-1}$$

$$H_2S_2(l) \longrightarrow H_2S(g) + \tfrac{1}{8}\,S_8(c) - 3 \text{ kJ mole}^{-1}$$

Note that this explanation applies also to the nonexistence of H_2O_3, H_2O_4, H_2O_5; the analogous sulfur compounds are stable.

6-41. From the standard enthalpies of formation of $CH_4(g)$, $CH_3Cl(g)$, $CH_2Cl_2(g)$, $CHCl_3(g)$, $C(g)$, in Appendix VI, $Cl(g)$ in Appendix V, and the value 415 kJ mole^{-1} for the C—H bond energy (Appendix V), calculate the four successive C—Cl bond-energy values. (Answers: 325, 324, 330, 325 kJ mole^{-1}.)

6-42. In Example 6-5 in the text it is stated that the linear (diacetylene) formula is the only one consistent with the valence requirements of carbon and hydrogen and the formula C_4H_2.
(a) What properties of the bonds formed by the carbon atom would you use to eliminate the structure

```
H         H
 \       /
  C === C
  |     |
  C ≡≡≡ C
```

from serious consideration?
(b) There are three other possible structures that can be similarly eliminated. What are they?

6-43. Assign transargononic structures to the following molecules and ions: PCl_3F_2; $TeCl_4$; SF_6; S_2F_{10}; $HClO_4$; $Te(OH)_6$. What outer orbitals are used for bonds and unshared pairs by the transargononic atoms?

7

The Nonmetallic Elements and Some of Their Compounds

In the first section of this chapter some of the properties of the elements hydrogen, carbon, nitrogen, phosphorus, arsenic, antimony, bismuth, oxygen, sulfur, selenium, tellurium, fluorine, chlorine, bromine, and iodine are described. The following sections are devoted to some of their compounds with one another, especially the single-bonded normal-valence compounds. Compounds of nonmetals with oxygen are discussed in the following chapter.

7-1. The Elementary Substances

Hydrogen

Hydrogen is a very widely distributed element. It is found in most of the substances that constitute living matter, and in many inorganic substances. There are more compounds of hydrogen known than of any other element, carbon being a close second.

Free hydrogen, H_2, is a colorless, odorless, and tasteless gas. It is the lightest of all gases, its density being about one-fourteenth that of air. Its melting point (−259°C or 14 K) and boiling point (−252.7°C) are very low, only those of helium being lower. Liquid hydrogen, with density 0.070 g cm^{-3}, is, as might be expected, the lightest of ail liquids. Crystalline hydrogen, with density 0.088 g cm^{-3}, is also the lightest of all crystalline substances. Hydrogen is very slightly soluble in water; 1 liter of water at 0°C dissolves only 21.5 ml of hydrogen gas under 1 atm pressure. The solubility decreases with increasing temperature, and increases with increase in the pressure of the gas.

The electronic structure of the hydrogen molecule has been discussed in Section 6-2.

In the laboratory hydrogen may be easily made by the reaction of an acid such as sulfuric acid, H_2SO_4, with a metal such as zinc. The equation for the reaction is

$$H_2SO_4(aq) + Zn(c) \longrightarrow ZnSO_4(aq) + H_2(g)$$

Hydrogen can also be prepared by the reaction of some metals with water or steam. Sodium and the other alkali metals react very vigorously with water, so vigorously as to generate enough heat to ignite the liberated hydrogen. An alloy of lead and sodium, which reacts less vigorously, is sometimes used for the preparation of hydrogen.

Much of the hydrogen that is used in industry is produced by the reaction of iron with steam. The steam from a boiler is passed over iron filings heated to a temperature of about 600°C. The reaction that occurs is

$$3Fe(c) + 4H_2O(g) \longrightarrow Fe_3O_4(c) + 4H_2(g)$$

After a mass of iron has been used in this way for some time, it is largely converted into iron oxide, Fe_3O_4. The iron can then be regenerated by passing carbon monoxide, CO, over the heated oxide:

$$Fe_3O_4(c) + 4CO(g) \longrightarrow 3Fe(c) + 4CO_2(g)$$

There is, of course, nothing special about sodium and iron (except their low cost and availability) that causes them, rather than other metals, to be used for the preparation of hydrogen. Other metals with electronegativity about the same as that of sodium ($x = 0.9$) react with water as vigorously as sodium, and metals with electronegativity about the same as that of iron ($x = 1.8$) react with steam in about the same way as iron. Although other structural features, which we shall discuss later, are involved in chemical reactivity, bond energy, which is determined by electronegativity differences, is the most important one.

Elementary Carbon

Carbon occurs in nature in its elementary state in two allotropic forms: *diamond,* one of the hardest substances known,* which often forms beautiful transparent and highly refractive crystals, used as gems; and *graphite,* a soft, black crystalline substance, used as a lubricant and in the "lead" of lead pencils. *Bort* and *black diamond (carbonado)* are imperfectly crystalline forms of diamond, which do not show the cleavage characteristic of diamond crystals. Their density is slightly less than that of crystalline diamond, and they are tougher and somewhat harder. They are used in diamond drills and saws and other grinding and cutting devices. Other uses of diamonds also depend upon their great hardness. For example, diamonds with a tapering hole drilled through them are used for drawing wires. Charcoal, coke, and carbon black (lampblack) are microcrystalline or amorphous forms of carbon. The density of diamond is 3.51 g cm^{-3} and that of graphite is 2.26 g cm^{-3}.

The great hardness of diamond is explained by the structure of the diamond crystal, as determined by the x-ray diffraction method. In the diamond crystal (Figure 7-1) each carbon atom is surrounded by four other carbon atoms, which lie at the corners of a regular tetrahedron about it. A structural formula can be written for a small part of a diamond crystal:

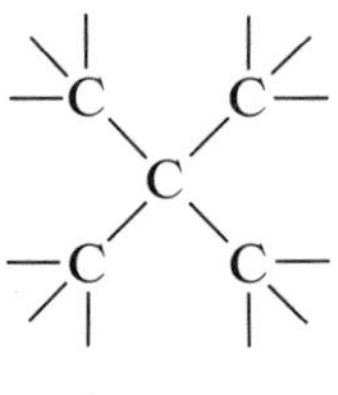

```
 \|     |/
 —C     C—
    \  /
     C
    /  \
 —C     C—
 /|     |\
```

Valence bonds connect each carbon atom with four others. Each of these four is bonded to three others (plus the original one), and so on throughout the crystal. The entire crystal is a giant molecule, held together by covalent bonds. To break the crystal, many of these bonds must be broken; this requires a large amount of energy, and hence the substance is very hard. The bond length in diamond has the single-bond value, 154 pm.

The commercial manufacture of diamonds was begun in the period around 1950, after techniques for obtaining very high pressures (over 50,000 atm) at high temperatures (2000°C) had been developed. The crystallization of the artificial diamonds is favored by the addition of a small amount of a metal such as nickel. It is significant that the length of the edge of the unit cube of the nickel crystal, containing four nickel atoms

*Some of the boron carbides are harder than diamond, as are also bort and black diamond.

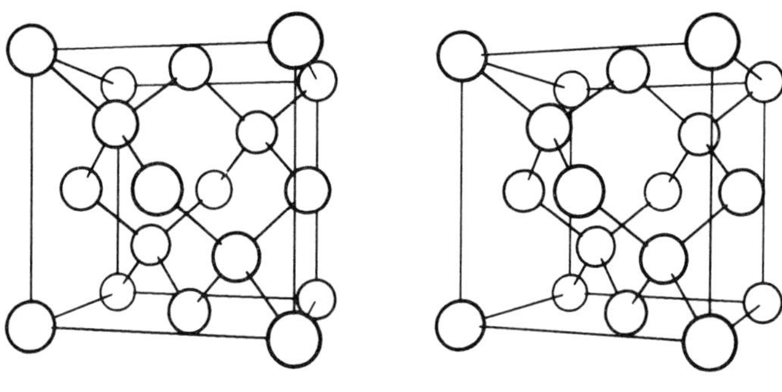

FIGURE 7-1
The structure of diamond (stereo).

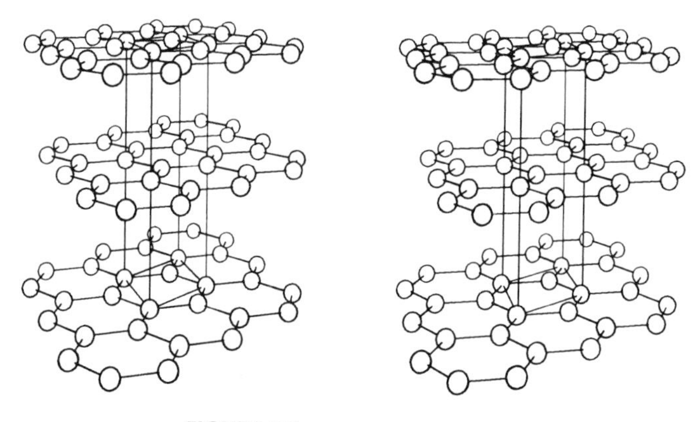

FIGURE 7-2
The structure of graphite (stereo).

in cubic closest packing, is 352 pm, nearly equal to that, 356 pm, of the unit cube of the diamond crystal, which contains eight carbon atoms in the arrangement shown in Figure 7-1. Artificial diamonds contain some nickel atoms replacing pairs of carbon atoms.

The structure of graphite is shown in Figure 7-2. It is a layer structure. Each atom forms two single bonds and one double bond with its three nearest neighbors. The bonds resonate among the positions in each layer in such a way as to give each bond two-thirds single-bond character and one-third double-bond character. The interatomic distances in the layer are 142 pm, which is intermediate between the single-bond value, 154 pm, and the double-bond value, 133 pm. The distance between layers is 340 pm, over twice the bond length in a layer. The crystal of graphite can be described as built of giant flat molecules, loosely held together in a pile. The layers can be easily separated; hence graphite is a soft substance, which is even used as a lubricant. The lubricating property is determined to some extent by the presence of water; the mechanism of this dependence is not well understood.

Nitrogen and Its Congeners

Elementary nitrogen occurs in nature in the atmosphere, of which it constitutes 78% by volume. It is a colorless, odorless, and tasteless gas, composed of diatomic molecules, N_2. At 0°C and 1 atm pressure a liter of nitrogen weighs 1.2506 g. The gas condenses to a colorless liquid at −195.8°C, and to a white solid at −209.96°C. Nitrogen is slightly soluble in water, 1 liter of which dissolves 23.5 ml of the gas at 0°C and 1 atm. Some of the properties of nitrogen and other group-V elements are given in Table 7-1.

The nitrogen atom, lacking three electrons of a completed octet, may achieve the octet by forming three covalent bonds. It does this in elementary nitrogen by forming a triple bond in the molecule N_2. Three electron pairs are shared by the two nitrogen atoms:

$$:N:::N: \qquad \text{or} \qquad :N{\equiv}N:$$

This bond is extremely strong, and the N_2 molecule is very stable.

Phosphorus was discovered in 1669 by a German alchemist, Henning Brand, in the course of his search for the Philosopher's Stone. Brand heated the residue left on evaporation of urine, and collected the distilled phosphorus in a receiver. The name given the element (from Greek *phosphoros,* giving light) refers to its property of glowing in the dark.

TABLE 7-1
Properties of the Elements of Group V

	Atomic Number	Atomic Mass	Melting Point	Boiling Point	Density of Solid	Color	Covalent Radius	Van der Waals Radius
N	7	14.0067	−209.8°C	−195.8°C	1.026 g cm^{-3}	White	70 pm	150 pm
P	15	30.9738	44.1°	280°	1.81	White	110	190
As	33	74.9216	814°*	715°†	5.73	Gray	121	200
Sb	51	121.75	630°	1380°	6.68	Silvery white	141	220
Bi	83	208.980	271°	1470°	9.80	Reddish white	151	230

*At 36 atm
†It sublimes.

Elementary phosphorus is made by heating calcium phosphate with silica and carbon in an electric furnace. The silica forms calcium silicate, displacing tetraphosphorus decoxide, P_4O_{10}, which is then reduced by the carbon. The phosphorus leaves the furnace as vapor, and is condensed under water to *white phosphorus.*

Phosphorus vapor is tetratomic: in the P_4 molecule each atom has one unshared electron pair and forms a single bond with each of its three neighbors (Figure 6-16).

At high temperatures the vapor is dissociated slightly, forming some diatomic molecules, P_2, with the structure $:P{\equiv}P:$, analogous to that of the nitrogen molecule.

Phosphorus vapor condenses to liquid white phosphorus, which freezes to solid white phosphorus, a soft, waxy, colorless material, soluble in carbon disulfide, benzene, and other nonpolar solvents. Both solid and liquid white phosphorus contain the same P_4 molecules as the vapor.

White phosphorus is metastable, and it slowly changes to a stable form, *red phosphorus,* in the presence of light or on heating. White phosphorus usually has a yellow color because of partial conversion to the red form. The reaction takes several hours even at 250°C; it can be accelerated by the addition of a small amount of iodine, which serves as a catalyst.* Red phosphorus is far more stable than the white form – it does not catch fire in air at temperatures below 240°C, whereas white phosphorus ignites at about 40°C, and oxidizes slowly at room temperature, giving off a white light ("phosphorescence"). Red phosphorus is not poisonous, but white phosphorus is very poisonous, the lethal dose being about 0.15 g; it causes necrosis of the bones, especially those of the jaw. White phosphorus burns are painful and slow to heal. Red phosphorus cannot be converted into white phosphorus except by vaporizing it. It is not appreciably soluble in any solvent. When heated to 500° or 600°C red phosphorus slowly melts (if under pressure) or vaporizes, forming P_4 vapor.

Several other allotropic forms of the element are known. One of these, *black phosphorus,* is formed from white phosphorus under high pressure. It is still less reactive than red phosphorus, and is the stable form of the element (standard enthalpy, relative to white phosphorus, -43 kJ mole^{-1}; that of red phosphorus is -18 kJ mole^{-1}).

The explanation of the properties of red and black phosphorus lies in their structure. These substances are high polymers, consisting of giant molecules extending throughout the crystal. In order for such a crystal to melt or to dissolve in a solvent, a chemical reaction must take place. This

*A substance with the property of accelerating a chemical reaction without itself undergoing significant change is called a *catalyst,* and is said to *catalyze* the reaction; see Section 10-5.

chemical reaction is the rupture of some P—P bonds and formation of new ones. Such processes are very slow. The structure of red phosphorus is not known in detail; that of black phosphorus involves puckered layers, somewhat similar to those of arsenic shown in Figure 7-3, but with a different sort of puckering.

Elementary arsenic exists in several forms. Ordinary *gray arsenic* is a semimetallic substance, steel-gray in color, with density 5.73 and melting point (under pressure) 814°C. It sublimes rapidly at about 450°C, forming gas molecules As_4, similar in structure to P_4. An unstable yellow crystalline allotropic form containing As_4 molecules, and soluble in carbon disulfide also exists. The gray form has the layer structure shown in Figure 7-3, in which each atom forms three covalent bonds with neighboring atoms in the same layer. Elementary antimony and bismuth crystallize with the same layer structure.

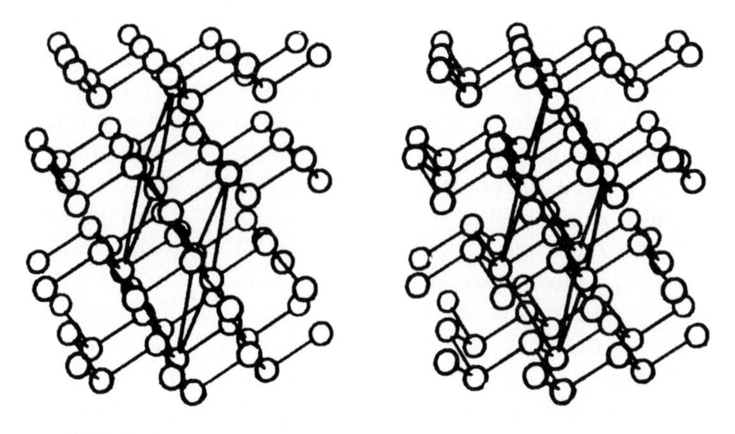

FIGURE 7-3
A stereo view of a portion of an arsenic crystal. Each atom is attached by single bonds to three other atoms in a puckered layer.

Example 7-1. It is stated above that the black phosphorus is more stable than white phosphorus by 43 kJ mole^{-1}; that is, the conversion of white phosphorus to black phosphorus is exothermic:

$$\tfrac{1}{4}P_4\ \text{(c,white)} \longrightarrow P\text{(c,black)} + 43\ \text{kJ mole}^{-1}$$

Assuming that the enthalpy difference is the strain energy of the bent bonds in P_4 (Figure 6-16), evaluate the strain energy per bent bond.

Solution. The strain energy of the P_4 molecule is $4 \times 43 = 172$ kJ mole^{-1}. There are six bent bonds in the molecule. Hence the strain energy per bent bond is $172/6 = 29$ kJ mole^{-1}.

Example 7-2. On the basis of the statement that red phosphorus is more stable than white phosphorus by 18 kJ mole^{-1}, suggest a possible structure for red phosphorus.

Solution. We see that the strain energy in red phosphorus per P_4 is $4(43 - 18) = 100$ kJ mole^{-1}. The strain energy is about 58% of the value for the tetrahedral P_4 molecule in white phosphorus. This molecule has twelve 60° bond angles. If one bond is broken, the P groups can form chains:

```
        P           P
···—P<  |  >P—P<  |  >P—···
        P           P
```

In these chains each P_4 group has only six 60° bond angles; the other six can be close to the normal value 102° found in black phosphorus. Hence we conclude that red phosphorus probably has this structure.

Oxygen and Its Congeners

Oxygen is the most abundant element in the earth's mantle. It constitutes by weight 89% of water, 23% of air (21% by volume), and nearly 50% of the common minerals (silicates). The element consists of diatomic molecules. with the structure described below. It is a colorless, odorless gas, which is slightly soluble in water: 1 liter of water at 0° dissolves 48.9 ml of oxygen gas at 1 atm pressure. Its density at 0°C and 1 atm is 1.429 g liter^{-1}. Oxygen condenses to a pale blue liquid as its boiling point, −183.0°C, and on further cooling freezes, at −218.4°C, to a pale blue crystalline solid.

Oxygen may be easily prepared in the laboratory by heating potassium chlorate, $KClO_3$:

$$2KClO_3 \rightarrow 2KCl + 3O_2(g)$$

The reaction proceeds readily at a temperature just above the melting point of potassium chlorate if a small amount of manganese dioxide, MnO_2, is mixed with it. Although the manganese dioxide accelerates the rate of evolution of oxygen from the potassium chlorate, it itself is not changed.

Oxygen is made commercially mainly by the distillation of liquid air. Nitrogen is more volatile than oxygen, and tends to evaporate first from liquid air. Nearly pure oxygen can be obtained by properly controlling the conditions of the evaporation. The oxygen is stored and shipped in steel cylinders, at pressures of 100 atm or more. Oxygen is also made commercially, together with hydrogen, by the electrolysis of water.

Some of the properties of oxygen and its congeners are listed in Table 7-2.

TABLE 7-2
Properties of Oxygen, Sulfur, Selenium, and Tellurium

	Atomic Number	Atomic Mass	Melting Point	Boiling Point	Density	Co-valent Radius	Ionic Radius, X^{--}
Oxygen (gas)	8	15.9994	−218.4°C	−183.0°C	1.429 g $liter^{-1}$	66 pm	140 pm
Sulfur (orthorhombic)	16	32.064	119.25° 112.8°*	444.6°	2.07 g cm^{-3}	104	184
Selenium (gray)	34	78.96	217°	685°	4.79	117	198
Tellurium (gray)	52	127.60	450°	1087°	6.25	137	221

*For monoclinic sulfur and (rapidly heated) orthorhombic sulfur, respectively.

We might expect the O_2 molecule to contain a double bond:

$$:\ddot{O}::\ddot{O}: \quad \text{or} \quad :\ddot{O}=\ddot{O}:$$

Instead, only one shared pair is formed, leaving two unshared electrons:

$$:\underset{\cdot}{\ddot{O}}-\underset{\cdot\cdot}{\dot{O}}:$$

These two unshared electrons are responsible for the paramagnetism of oxygen. It has been found by study of the oxygen spectrum that the force of attraction between the oxygen atoms is much greater than that expected for a single covalent bond. This shows that the unpaired electrons are really involved in the formation of bonds of a special sort. The oxygen molecule may be said to contain a single covalent bond plus two *three-electron bonds,* and its structure may be written as

$$:O\overset{\cdots}{\underset{\cdots}{—}}O:$$

Elementary sulfur exists in several allotropic forms. Ordinary sulfur is a yellow solid substance, which forms crystals with orthorhombic symmetry; it is called *orthorhombic sulfur* or, usually, *rhombic sulfur.* It is insoluble in water, but soluble in carbon disulfide (CS_2), carbon tetrachloride, and similar nonpolar solvents, giving solutions from which well-formed crystals of sulfur can be obtained.

At 112.8°C orthorhombic sulfur melts to form a straw-colored liquid. This liquid crystallizes in a monoclinic crystalline form, called *monoclinic sulfur.* The sulfur molecules in both orthorhombic sulfur and monoclinic sulfur, as well as in the straw-colored liquid, are S_8 molecules, with a staggered-ring configuration (Figure 6-15). The formation of this large molecule (and of the similar molecules Se_8 and Te_8) is the result of the tendency of the sixth-group elements to form two single covalent bonds, instead of one double bond. Diatomic molecules S_2 are formed by heating sulfur vapor (S_6 and S_8 at lower temperatures) to a high temperature, but these molecules are less stable than the large molecules containing single bonds. This fact is not isolated, but is an example of the generalization that stable double bonds and triple bonds are formed readily by the light elements carbon, nitrogen, and oxygen, but only rarely by the heavier elements (Section 6-5). Carbon disulfide, CS_2, and other compounds containing a carbon-sulfur double bond are the main exceptions to this rule.

Monoclinic sulfur is the stable form above 95.5°C, which is the *equilibrium temperature (transition temperature* or *transition point)* between it and the orthorhombic form. Monoclinic sulfur melts at 119.25°C.

The elementary substances selenium and tellurium differ from sulfur in their physical properties in ways expected from their relative positions in the periodic table. Their melting points, boiling points, and densities are higher, as shown in Table 7-2. The stable forms of selenium and tellurium (gray) involve a hexagonal packing of infinitely long chains, each chain having a three-fold screw axis of symmetry. The red allotropic forms of selenium consist of Se_8 molecules.

The increase in metallic character with increase in atomic number is striking. Sulfur is a nonconductor of electricity, as is the red allotropic form of selenium. The gray form of selenium has a small but measurable electronic conductivity, and tellurium is a semiconductor, with conductivity a fraction of 1% of that of metals. An interesting property of the gray form of selenium is that its electric conductivity is greatly increased during exposure to visible light. This property is used in "selenium cells" for the measurement of light intensity, and is the basis of the xerographic method of copying documents.

The Halogens

A halogen atom such as fluorine can achieve an argononic structure by forming a single covalent bond with another halogen atom:

$$:\ddot{\underset{..}{F}}-\ddot{\underset{..}{F}}: \qquad :\ddot{\underset{..}{Cl}}-\ddot{\underset{..}{Cl}}: \qquad :\ddot{\underset{..}{Br}}-\ddot{\underset{..}{Br}}: \qquad :\ddot{\underset{..}{I}}-\ddot{\underset{..}{I}}:$$

This single covalent bond holds the atoms together into diatomic molecules, which are present in the elementary halogens in all states of aggregation—crystal, liquid, and gas. Some physical properties of the halogens are given in Table 7-3.

TABLE 7-3
Properties of the Halogens

	Atomic Number	Atomic Mass	Color and Form	Melting Point	Boiling Point	Ionic Radius*	Co-valent Radius	Heat of Disso-ciation
F_2	9	18.9984	Pale yellow gas	−223°C	−187°C	136 pm	64 pm	153kJ $mole^{-1}$
Cl_2	17	35.453	Greenish yellow gas	−101.6°	−34.6°	181	99	243
Br_2	35	79.909	Reddish brown liquid	−7.3°	58.7°	195	114	193
I_2	53	126.9044	Grayish black lustrous solid	113.5°	184°	216	127	151

*For negatively charged ion with ligancy 6, such as Cl^- in the NaCl crystal.

Fluorine, the lightest of the halogens, is the most reactive of all the elements, and it forms compounds with all the elements except the lighter argonons. This great reactivity may be attributed to the large value of its electronegativity. Substances such as wood and rubber burst into flame when held in a stream of fluorine, and even asbestos (a silicate of magnesium and aluminum) reacts vigorously with it and becomes incandescent. Platinum is attacked only slowly by fluorine. Copper and steel can be used as containers for the gas; they are attacked by it, but become coated with a thin layer of copper fluoride or iron fluoride, which then protects them against further attack.

Because its electronegativity is greater than that of any other element, we cannot expect that fluorine could be prepared by reaction of any other element with a fluoride. It can, however, be made by electrolysis of fluorides, since the oxidizing power (electron affinity) of an electrode can be increased without limit by increasing the applied voltage. (We shall discuss this matter in Chapter 11.) It was by the electrolysis of a solution of KF in liquid HF that fluorine was first obtained, by the French chemist Henri Moissan (1852–1907), in 1886.

Chlorine (from Greek *chloros,* greenish yellow), the most common of the halogens, is a greenish-yellow gas, with a sharp irritating odor. It was first made by the Swedish chemist K. W. Scheele in 1774, by the action of manganese dioxide on hydrochloric acid. It is now manufactured on a large scale by the electrolysis of a strong solution of sodium chloride.

The element bromine (from Greek *bromos,* stench) occurs in the form of compounds in small quantities in seawater and in natural salt deposits. It is an easily volatile, dark reddish-brown liquid with a strong, disagreeable odor and an irritating effect on the eyes and throat. It produces painful sores when spilled on the skin. The free element can be made by treating a bromide with an oxidizing agent, such as chlorine.

The element iodine (from Greek *iodes*, violet) occurs as iodide ion, I^-, in very small quantities in seawater, and, as sodium iodate, $NaIO_3$, in deposits of Chile saltpeter. It is made commercially from sodium iodate obtained from saltpeter, from kelp, which concentrates it from the seawater, and from oil-well brines.

The free element is an almost black crystalline solid with a slightly metallic luster. On gentle warming it gives a beautiful blue-violet vapor. Its solutions in chloroform, carbon tetrachloride, and carbon disulfide are also blue-violet in color, indicating that the molecules I_2 in these solutions closely resemble the gas molecules. The solutions of iodine in water containing potassium iodide and in alcohol (tincture of iodine) are brown; this change in color suggests that the iodine molecules have undergone chemical reaction in these solutions. The brown compound KI_3, potassium triiodide, is present in the first solution, and a compound with alcohol in the second.

7-2. Hydrides of Nonmetals. Hydrocarbons

The elements carbon, nitrogen, oxygen, and fluorine and their congeners form simple hydrides with composition corresponding to their normal covalences (CH_4, NH_3, H_2O, HF). Some properties of these hydrides are given in Table 7-4.

The tetrahydrides (CH_4 to SnH_4) have a regular tetrahedral structure, corresponding to the use of sp^3 tetrahedral bond orbitals by the central atom (bond angles 109.5°, Section 6-5). The other hydrides have smaller bond angles, approaching 90°, the value for p bond orbitals (Section 6-7).

The stability of these hydrides is determined mainly by the difference in electronegativity of hydrogen and the other element, as has been discussed in Section 6-12 and Appendix V. In Figure 7-4 values are shown of the heat of formation per bond (per hydrogen atom) of the hydrides in the gaseous state from the elements in their standard states, compared with the calculated function $100(x - 2.1)^2 - 6.5(x - 2.1)^4$ kJ mole^{-1} (here 2.1 is the electronegativity of hydrogen). The values for H_2S, H_2Se, and HI would be brought closer to the curve by correction for the van der Waals attraction in the crystalline (standard) state of sulfur, selenium, and iodine. The low values for NH_3 and H_2O, shown by the open circles, are corrected (as indicated by the arrow) to points near the curve by adding the multiple-bond correction for nitrogen and oxygen discussed in Section 6-12.

TABLE 7-4
Some Properties of Hydrides of Nonmetallic Elements

Formula	Melting Point	Boiling Point	Density of Liquid	Standard Enthalpy (g)	Bond Length	Bond Angle
CH_4	−183°C	−161°C	0.54 g cm^{-3}	−75 kJ mole^{-1}	109 pm	109.5°
SiH_4	−185°	−112°	0.68	−62	148	109.5°
GeH_4	−165°	−90°	1.52		153	109.5°
SnH_4	−150°	−52°			170	109.5°
NH_3	−78°	−33°	0.82	−46	101	107.3°
PH_3	−133°	−85°	0.75	9	142	93.1°
AsH_3	−114°	−55°		171	152	91.8°
SbH_3	−88°	−17°	2.26		171	91.3°
BiH_3		22°				
H_2O	0°	100°	1.00	−242	96	104.5°
H_2S	−86°	−61°		−20	133	92.2°
H_2Se	−64°	−42°	2.12	86	146	91.0°
H_2Te	−49°	−2°	2.57	154	167	90°
HF	−92°	19°	0.99	−269	92	
HCl	−112°	−84°	1.19	−92	127	
HBr	−89°	−67°	1.78	−36	141	
HI	−51°	−35°	2.85	26	161	

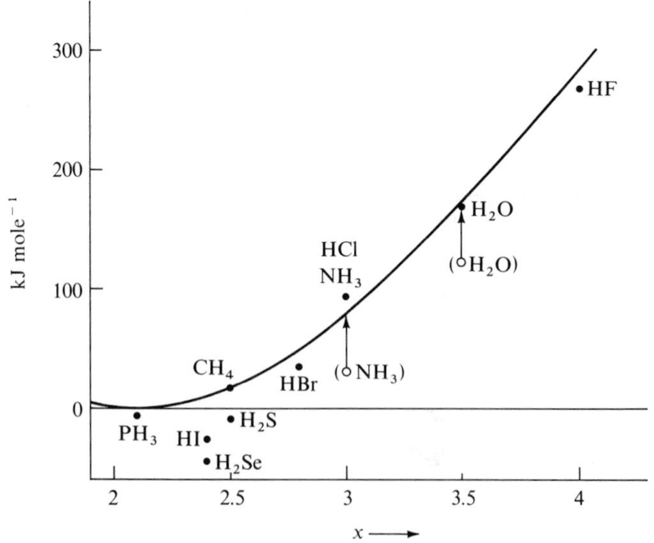

FIGURE 7-4
Standard heat of formation per bond of hydrides (g) of nonmetallic elements, compared with values calculated by the electronegativity equation.

This correction is especially important for nitrogen. The nitrogen molecule, $:N{\equiv}N:$, is 469 kJ mole^{-1} more stable than a hypothetical single-bonded form of the element, and in consequence the heat of formation of $NH_3(g)$ from $\frac{1}{2}N_2(g)$ and $\frac{3}{2}H_2(g)$ is only 46 kJ mole^{-1}, rather than $46 + 235 = 281$ kJ mole^{-1}.

Methane and Other Alkanes

The *hydrocarbons* are compounds of hydrogen and carbon alone. *Methane*, CH_4, is the first member of a series of hydrocarbons called the *methane series* or *paraffin series*. Some of these compounds are listed in Table 7-5. They are called *alkanes*.

Natural gas, from oil wells or gas wells, is usually about 85% methane. The gas that rises from the bottom of a marsh is methane (plus some carbon dioxide and nitrogen), formed by the anaerobic (air-free) fermentation of vegetable matter.

Methane is used as a fuel. It is also used in large quantities for the manufacture of carbon black, by combustion with a limited supply of air:

$$CH_4 + O_2 \longrightarrow 2H_2O + C$$

The methane burns to form water, and the carbon is deposited as very finely divided carbon, which finds extensive use as a filler in rubber for automobile tires.

TABLE 7-5
Some Physical Properties of Normal Alkanes

Substance	Formula	Melting Point	Boiling Point	Density of Liquid
Methane	CH_4	−183°C	−161°C	0.54 g cm^{-3}
Ethane	C_2H_6	−172	−88	.55
Propane	C_3H_8	−190	−45	.58
Butane	C_4H_{10}	−135	−1	.60
Pentane	C_5H_{12}	−130	36	.63
Hexane	C_6H_{14}	−95	69	.66
Heptane	C_7H_{16}	−91	98	.68
Octane	C_8H_{18}	−57	126	.70
Nonane	C_9H_{20}	−54	151	.72
Decane	$C_{10}H_{22}$	−30	174	.73
Pentadecane	$C_{15}H_{32}$	10	271	.77
Eicosane	$C_{20}H_{42}$	38		.78
Triacontane	$C_{30}H_{62}$	70		.79

The name paraffin means "having little affinity." The compounds of this series are not very reactive chemically. They occur in petroleum. *Ethane* has the structure

```
 H       H
  \     /
H—C—C—H
  /     \
 H       H
```

It is a gas (Table 7-5) which occurs in large amounts in natural gas from some wells. *Propane,* the third member of the series, has the structure

```
  H  H    H  H
   \ |    | /
 H—C      C—H
      \  /
       C
      / \
     H   H
```

It is an easily liquefied gas, used as a fuel.

In the structural formula for propane there is a chain of three carbon atoms bonded together. The next larger alkane, *butane,* C_4H_{10}, can be obtained by replacing a hydrogen atom at one end of the chain by a methyl group,

```
      H
     /
—C—H
     \
      H
```

Its formula is obtained by adding CH_2 to that of propane. These hydrocarbons, with longer and longer chains of carbon atoms, are called the normal alkanes (*n*-alkanes).

The lighter members of the paraffin series are gases, the intermediate members are liquids, and the heavier members are solid substances. The common name petroleum ether refers to the pentane-hexane-heptane mixture, used as a solvent and in dry cleaning. Gasoline is the heptane-to-nonane mixture (C_7H_{16} to C_9H_{20}), and kerosene the decane-to-hexadecane mixture ($C_{10}H_{22}$ to $C_{16}H_{34}$). Heavy fuel oil is a mixture of paraffins containing twenty or more carbon atoms per molecule. The lubricating oils, petroleum jelly ("Vaseline"), and solid paraffin are mixtures of still larger paraffin molecules.

The phenomenon of isomerism is shown first in the paraffin series by butane, C_4H_{10}. Isomerism (Section 6-3) is the existence of two or more compound substances having the same composition but different properties. The difference in properties is usually the result of difference in the way that the atoms are bonded together. There are two isomers of butane, called normal butane (*n*-butane) and isobutane. These substances have the structures shown in Figure 7-5; normal butane has a "straight chain" (actually the carbon chain is a zigzag chain, because of the tetrahedral nature of the carbon atom), and the isobutane molecule has a branched chain. In general, the properties of these isomers are rather similar; for example, their melting points are −135°C and −145°C, respectively. The branched-chain hydrocarbons are more stable than their straight-chain isomers [standard enthalpy −126 kJ mole^{-1} for *n*-$C_4H_{10}(g)$, −135 for *iso*-$C_4H_{10}(g)$; −146 for *n*-$C_5H_{12}(g)$, −154 for *iso*-$C_5H_{12}(g)$, −166 for *neo*-$C_5H_{12}(g)$; neopentane is tetramethylmethane, $C(CH_3)_4$]. The greater stability of the branched chains than the straight chains may be attributed

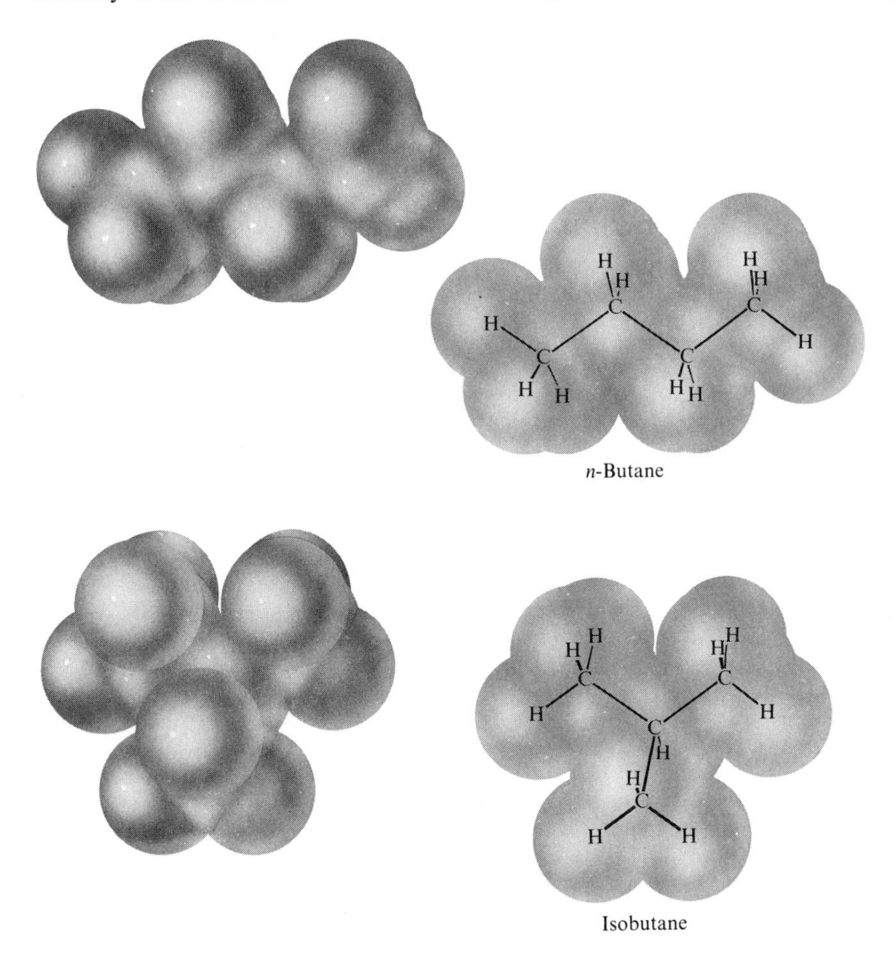

FIGURE 7-5
The structure of the isomers normal butane and isobutane.

to their more compact structure, which leads to greater van der Waals stabilization (attraction of pairs of nonbonded atoms).

The normal (straight-chain) hydrocarbons "knock" badly when burned in a high-compression gasoline engine, whereas the highly branched hydrocarbons, which burn more slowly, do not knock. The "octane number" (the antiknock rating) of a gasoline is measured by comparing it with varying mixtures of *n*-heptane and a highly branched octane, 2,2,4-trimethylpentane, with structural formula

$$
\begin{array}{ccccccc}
H_3C & & CH_3 & & CH_3 \\
 & \diagdown & \diagup & H & \diagup \\
 & & C & & C \\
 & \diagup & \diagdown & \diagup & \diagdown \\
H_3C & & C & & CH_3 \\
 & & H_2 & &
\end{array}
$$

The octane number is the percentage of this octane in the mixture that has the same knocking properties as the gasoline being tested.

The substance *tetraethyl lead,* $Pb(C_2H_5)_4$, is widely used in gasoline as an antiknock agent. Gasoline containing it is called *ethyl gasoline.* Its use is now being abandoned because of the toxicity of the lead introduced into the atmosphere. Compounds of lead that are ingested or inhaled damage the brain and peripheral nervous system.

Restricted Rotation about Single Bonds

Until about thirty-five years ago chemists assumed that the two ends of a molecule such as ethane, $H_3C—CH_3$, could rotate freely about the single bond connecting them. This assumption of free rotation about single bonds was made because all efforts to obtain isomers of substances such as 1,2-dichloroethane, $H_2ClC—CH_2Cl$, had failed. Then in 1937 the American chemists J. D. Kemp and K. S. Pitzer showed that the experimental value of the entropy of ethane requires that there be an energy barrier with height about 12.5 kJ mole^{-1}, restricting rotation of one methyl group in ethane relative to the other.

Many experimental values of the height of the potential barrier have been obtained through use of the techniques of microwave spectroscopy (study of absorption spectra of gas molecules in the wavelength region about 1 cm). The values 13.8 kJ mole^{-1} for $H_3C—CH_2F$ and 13.3 kJ mole^{-1} for $H_3C—CHF_2$ are nearly the same as for ethane; those for $H_3C—CH_2Cl$ and $H_3C—CH_2Br$ are somewhat higher, both 14.9 kJ mole^{-1}. In every case the stable configuration is the *staggered conformation* (bonds on opposite sides of the C—C axis, as illustrated in Figure

7-5). The unstable configuration, obtained by rotating a methyl group 60° about the C—C bond, is called the *eclipsed conformation.*

Spectroscopic studies have shown that 1,2-dichloroethane in the gas phase or in the solution is a mixture of three isomers. All three have staggered conformations; viewed along the C—C axis, they have the following aspects:

```
      Cl              H               H
  H   |   H       H   |   H       H   |   H
    \ | /           \ | /           \ | /
      C               C               C
    / | \           / | \           / | \
  H   |   H      Cl   |   H       H   |   Cl
      Cl              Cl              Cl
```

(The second and third constitute an enantiomeric pair, Section 6-4.) The energy of the barrier that restricts rotation is so low that the isomers are converted into one another too rapidly to permit their separation in the laboratory.

The cause of restricted rotation is still somewhat uncertain. The small dependence on the size of the atoms attached to the carbon atoms indicates that it is not mainly steric hindrance (hindrance resulting from contact of the atoms). The most likely explanation is that the barrier results from the repulsion of the H—C bond electrons of one methyl group and those of the other – that is, repulsion of the outer bonds. This explanation is supported by the values of the barrier height for $H_3C—NH_2$ and $H_3C—OH$, 7.9 and 4.5 kJ mole^{-1}, respectively.

Cyclic Hydrocarbons

It was mentioned in Section 6-5 that some hydrocarbon molecules involve rings of carbon atoms. The simplest cyclic hydrocarbon is *cyclopropane* (also called trimethylene), C_3H_6, with the structure shown in Figure 6-9. It is a colorless gas, with melting point −126.6°C, boiling point −34.4°C, and standard enthalpy of formation 20.4 kJ mole^{-1}. It is a good anesthetic agent, but it is dangerous: its mixtures with air may explode if ignited by an electrostatic spark.

Cyclohexane, C_6H_{12}, is a colorless liquid (boiling point 81°C, melting point 6.5°C) that is obtained in the distillation of petroleum and is used as a solvent. Its structure is that of a puckered hexagonal ring, with unstrained angles (tetrahedral, 109.5°), normal bond lengths (C—C = 154 pm, C—H = 110 pm), and the stable (staggered) orientation around each carbon-carbon bond. Its standard enthalpy, for $C_6H_{12}(g)$, is −126 kJ mole^{-1}. This value may be taken as the normal value for an unstrained

$(CH_2)_n$ molecule; it can be written as $-21n$ kJ mole^{-1}. For a hypothetical unstrained cyclopropane molecule we thus expect the standard enthalpy -63 kJ mole^{-1}. The difference between this value and the observed standard enthalpy of cyclopropane, 38 kJ mole^{-1}, is 101 kJ mole^{-1}, the strain energy of the bent bonds in cyclopropane.

The strain energy of cyclopropane makes it more reactive than cyclohexane. Cyclopropane reacts with hydrogen at 80°C, in the presence of a catalyst (finely divided platinum):

$$C_3H_6(g) + H_2(g) \xrightarrow[\text{Pt, 80°C}]{} C_3H_8(g)$$

The symbols Pt and 80°C are written below the arrow in this equation to show that platinum and a temperature of 80°C are needed to cause the reaction to take place.

The ring molecule cyclobutane, C_4H_8, reacts similarly, to form butane, at a somewhat higher temperature:

$$C_4H_8(g) + H_2(g) \xrightarrow[\text{Pt, 120°C}]{} C_4H_{10}(g)$$

The larger rings, such as cyclohexane, are not opened by hydrogenation below 200°C.

Other 3-ring molecules, ethylene imine,

```
H2C——CH2
   \  /
    N
    H
```

and ethylene oxide,

```
H2C——CH2
   \  /
    O
```

have about the same strain as cyclopropane, and have high chemical reactivity. The chemical reactivity of the epoxy group,

```
  |   |
—C——C—
   \ /
    O
```

permits the cross-linking of large molecules to take place during the setting of epoxy glues.

Hydrocarbons Containing Double Bonds and Triple Bonds

The substance *ethylene*, C_2H_4, consists of molecules

```
H       H
 \     /
  C=C
 /     \
H       H
```

in which there is a double bond between the two carbon atoms. This double bond confers upon the molecule the property of much greater chemical reactivity than is possessed by the alkanes. For example, whereas chlorine, bromine, and iodine do not readily attack the alkane hydrocarbons, they easily react with ethylene; a mixture of chlorine and ethylene reacts readily at room temperature in the dark, and with explosive violence in light, to form the substance *dichloroethane,* $C_2H_4Cl_2$:

$$C_2H_4 + Cl_2 \rightarrow C_2H_4Cl_2$$

or

$$\begin{array}{c}H \quad\quad H \\ \diagdown \quad \diagup \\ C{=}C \\ \diagup \quad \diagdown \\ H \quad\quad H\end{array} + Cl{-}Cl \rightarrow \begin{array}{c}H \quad\quad\quad Cl \\ \diagdown \quad\quad \diagup \\ H{-}C{-}C{-}H \\ \diagup \quad\quad \diagdown \\ Cl \quad\quad\quad H\end{array}$$

In the course of this reaction the double bond between the two carbon atoms has become a single bond, and the single bond between the two chlorine atoms has been broken. Two new bonds—single bonds between a chlorine atom and a carbon atom—have been formed. We may use the bond-energy values of Tables V-1 and V-2 to estimate the heat of the reaction:

REACTANT BOND ENERGIES		PRODUCT BOND ENERGIES	
C=C	615	C—C	344
Cl—Cl	243	2C—Cl	656
	858 kJ mole^{-1}		1000 kJ mole^{-1}

We see that the bonds in the product molecules are more stable than those in the reactant molecules by 142 kJ mole^{-1}. Hence this reaction is exothermic, with a moderately large amount of heat evolved on reaction, 142 kJ mole^{-1}.

A reaction of this sort is called an *addition reaction. An addition reaction is a reaction in which a molecule adds to a molecule containing a double bond, converting the double bond into a single bond.*

Because of this property of readily combining with other substances such as the halogens, ethylene and related hydrocarbons are said to be *unsaturated.* Ethylene is the first member of a homologous series of hydrocarbons, called the *alkenes.*

Ethylene is a colorless gas (b.p. −104°C) with a sweetish odor. It can be made in the laboratory by heating ethyl alcohol, C_2H_5OH, with concentrated sulfuric acid, preferably in the presence of a catalyst (such as silica) to increase the rate of the reaction. Concentrated sulfuric acid is a strong dehydrating agent, which removes water from the alcohol molecule:

$$C_2H_5OH \xrightarrow[H_2SO_4]{} C_2H_4 + H_2O$$

Ethylene is made commercially by passing alcohol vapor over a catalyst (aluminum oxide) at about 400°C. The reaction is endothermic; a small amount of heat is absorbed when it takes place:

$$C_2H_5OH \rightarrow C_2H_4 + H_2O - 47 \text{ kJ mole}^{-1}$$

Endothermic chemical reactions are in general favored by heating the reactants.

Ethylene has the interesting property of causing green fruit to ripen, and it is used commercially for this purpose. It is also used as an anesthetic.

Cis and Trans Isomers

The ethylene molecule is planar (Section 6-5), with greatly restricted rotation around the double bond. In consequence, a disubstituted ethylene such as 1,2-dichloroethylene, CHCl=CHCl, exists as two isomers, called *cis*-1,2-dichloroethylene and *trans*-1,2-dichloroethylene:

```
H         H        H         Cl
 \       /          \       /
  C=====C            C=====C
 /       \          /       \
Cl        Cl       Cl        H
     Cis               Trans
```

These two substances have different properties: the *cis* isomer has m.p. −80.5°C, b.p. 59.8°C, density of liquid 1.291 g cm^{-3}; and the *trans* isomer has m.p. −50°C, b.p. 48.5°C, density of liquid 1.265 g cm^{-3}.

The potential energy barrier restricting rotation about the double bond has been found by experiment to be about 200 kJ mole^{-1}.*

Acetylene, H—C≡C—H, is the first member of a homologous series of hydrocarbons containing triple bonds. Aside from acetylene, these substances (called *alkynes*) have not found wide use, except for the manufacture of other chemicals.

---

*The height of the barrier can be estimated from the spectroscopic value of the frequency of the torsional vibration of the molecule (twisting vibration of one end relative to the other end), and from the activation energy (Chapter 10) of the *cis-trans* isomerization reaction.

Acetylene is a colorless gas (b.p. −84°C), with a characteristic garlic-like odor. It is liable to explode when compressed in a pure state, and is usually kept in solution under pressure in acetone. It is used as a fuel, in the oxyacetylene torch and the acetylene lamp, and is also used as the starting material for making other chemicals.

Acetylene is most easily made from *calcium carbide* (calcium acetylide, CaC_2). Calcium carbide is made by heating lime (calcium oxide, CaO) and coke in an electric furnace:

$$CaO + 3C \longrightarrow CaC_2 + CO(g)$$

Calcium carbide is a gray solid that reacts vigorously with water to produce calcium hydroxide and acetylene:

$$CaC_2 + 2H_2O \longrightarrow Ca(OH)_2 + C_2H_2(g)$$

The existence of calcium carbide and other carbides with similar formulas and properties shows that acetylene is an acid, with two replaceable hydrogen atoms. It is an extremely weak acid, however, and its solution in water does not taste acidic.

Acetylene and other substances containing a carbon-carbon triple bond are very reactive. They readily undergo addition reactions with chlorine and other reagents, and they are classed as unsaturated substances.

7-3. Aromatic Hydrocarbons. Benzene

An important hydrocarbon is *benzene,* which has the formula C_6H_6. It is a volatile liquid (m.p. 5.5°C, b.p. 80.1°C, density 0.88 g cm^{-3}). Benzene and other hydrocarbons similar to it in structure are called the *aromatic hydrocarbons.** Their derivatives are called aromatic substances—many of them have a characteristic aroma (agreeable odor). Benzene itself was discovered in 1825 by Faraday, who found it in the illuminating gas made by heating oils and fats.

For many years there was discussion about the structure of the benzene molecule. The German chemist August Kekulé (1829–1896) in 1862 proposed that the six carbon atoms form a regular hexagon in space, the six hydrogen atoms being bonded to the carbon atoms, and forming a larger hexagon. Kekulé suggested that, in order for a carbon atom to show its normal quadrivalence, the ring contains three single bonds and

*The non-cyclic hydrocarbons discussed in the preceding section are called *aliphatic hydrocarbons*.

three double bonds in alternate positions, as shown below. A structure of this sort is called a Kekulé structure.

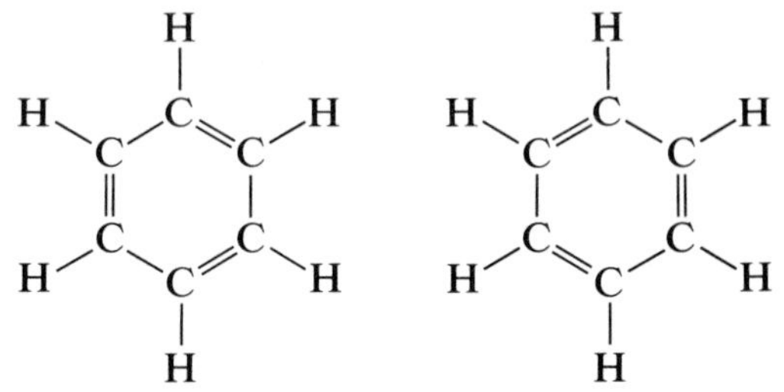

Other hydrocarbons, derivatives of benzene, can be obtained by replacing the hydrogen atoms by methyl groups or similar groups. Coal tar and petroleum contain substances of this sort, such as *toluene,* C_7H_8, and the three *xylenes,* C_8H_{10}. These formulas are usually written $C_6H_5CH_3$ and $C_6H_4(CH_3)_2$, to indicate the structural formulas:

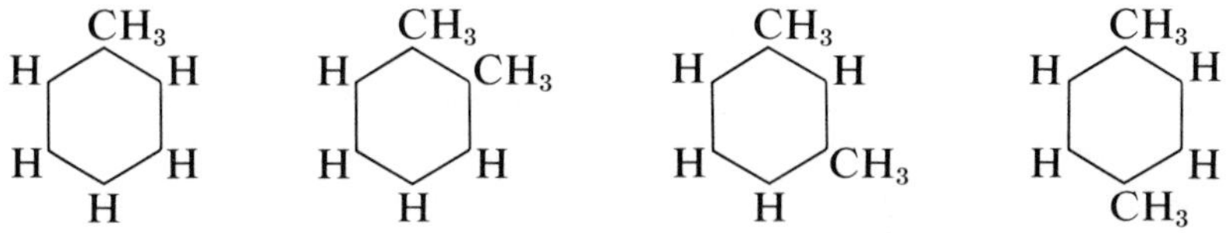

Toluene

Ortho-xylene (*o*-xylene)

Meta-xylene (*m*-xylene)

Para-xylene (*p*-xylene)

In these formulas the benzene ring of six carbon atoms is shown simply as a hexagon. This convention is used by organic chemists, who often also do not show the hydrogen atoms, but only other groups attached to the ring.

It is to be noted that we can draw two Kekulé structures for benzene and its derivatives. For example, for ortho-xylene the two Kekulé structures are

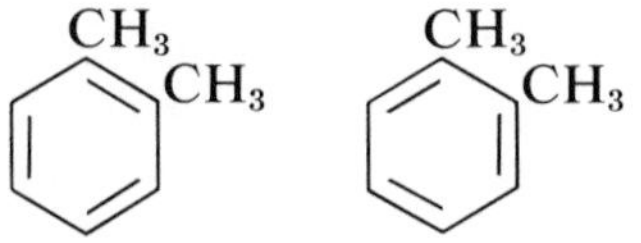

In the first structure there is a double bond between the carbon atoms to which methyl groups are attached, and in the second there is a single bond in this position. The organic chemists of a century ago found it impossible, however, to separate two isomers corresponding to these formulas. To explain this impossibility of separation Kekulé suggested that the molecule does not retain one Kekulé structure, but rather slips easily from one to the other. In the modern theory of molecular structure the ortho-xylene molecule is described as a hybrid of these two structures, with each bond between two carbon atoms in the ring intermediate in character between a single bond and a double bond. Even though this resonance structure is accepted for benzene and related compounds, it is often convenient simply to draw one of the Kekulé structures, or just a hexagon, to represent a benzene molecule.

The structure of the benzene molecule was determined by the electron diffraction method in 1929 and the following years. It is a planar hexagon with carbon-carbon bond length 140 pm (C—H bond length 106 pm). This value for a bond with 50% double-bond character is reasonable in comparison with the values 154 pm for C—C, 133 pm for C=C, and 142 pm for $33\frac{1}{3}\%$ double-bond character (graphite). The planar configuration is required by the properties of the double bond (Section 6-5).

Benzene and its derivatives are extremely important substances. They are used in the manufacture of drugs, explosives, photographic developers, plastics, synthetic dyes, and many other substances. For example, the substance *trinitrotoluene*, $C_6H_2(CH_3)(NO_2)_3$, is an important explosive (TNT). The structure of this substance is

CH$_3$
O$_2$N NO$_2$
H H
NO$_2$

In addition to benzene and its derivatives, there exist many other aromatic hydrocarbons, containing two or more rings of carbon atoms. *Naphthalene*, $C_{10}H_8$, is a solid substance with a characteristic odor; it is used as a constituent of moth balls and in the manufacture of dyes and other organic compounds. *Anthracene*, $C_{14}H_{10}$, and *phenanthrene*, $C_{14}H_{10}$, are isomeric substances containing three rings fused together. These substances are also used in making dyes, and derivatives of them are important biological substances (cholesterol, sex hormones; see Chapter 14). For naphthalene, anthracene, and phenanthrene we may write the following structural formulas:

Naphthalene Anthracene Phenanthrene

These molecules also have hybrid structures: the structural formulas shown above do not represent the molecules completely, but are analogous to one Kekulé structure for benzene.

Resonance Energy of Benzene

The heat evolved when a molecule of hydrogen is added to a double bond is about 120 kJ mole^{-1}. For cyclohexene, for example, the value determined by experiment is 119.6 kJ mole^{-1}:

$$\begin{array}{ccc} & CH & \\ H_2C & & CH \\ | & & | \\ H_2C & & CH_2 \\ & CH_2 & \end{array} + H_2 \rightarrow \begin{array}{ccc} & CH_2 & \\ H_2C & & CH_2 \\ | & & | \\ H_2C & & CH_2 \\ & CH_2 & \end{array} + 119.6\ \text{kJ mole}^{-1}$$

If the benzene molecule had one Kekulé structure, we might well expect that the heat of hydrogenation of its three double bonds would be approximately three times the heat of hydrogenation of the one double bond in cyclohexene, $3 \times 119.6 = 358.8$ kJ mole^{-1}:

$$\begin{array}{ccc} & CH & \\ HC & & CH \\ \| & & | \\ HC & & CH \\ & CH & \end{array} + 3H_2 \rightarrow \begin{array}{ccc} & CH_2 & \\ H_2C & & CH_2 \\ | & & | \\ H_2C & & CH_2 \\ & CH_2 & \end{array} + 358.8\ \text{kJ mole}^{-1}\ \text{(incorrect)}$$

The experimental value of the heat of hydrogenation is, however, 150 kJ mole^{-1} smaller:

$$C_6H_6(g) + 3H_2(g) \longrightarrow C_6H_{12}(g) + 208.4\ \text{kJ mole}^{-1}$$

We conclude that *the benzene molecule is 150 kJ mole^{-1} more stable than it would be if it were represented by a single Kekulé structure, with each of the three double bonds similar to the double bond in cyclohexene.* This extra stabilizing energy of 150 kJ mole^{-1} is called the *resonance energy* of benzene. It is attributed to the fact that the benzene molecule is not satisfactorily represented by a single Kekulé structure, but instead can be reasonably well described as a hybrid of the two Kekulé structures.*

*It would, of course, be surprising if the benzene molecule in its normal state were actually *less* stable than the hypothetical molecule with a single Kekulé structure; we would then ask why the molecule was prevented from assuming this more stable structure. The theory of resonance is based upon a theorem in quantum mechanics that the normal state of an atom or molecule is the most stable of all possible states.

The resonance energy of benzene makes the substance far less reactive chemically than alkenes or other unsaturated substances. For example, the reaction of addition of one hydrogen molecule to benzene to form cyclohexadiene,

```
       CH
     //  \
   HC     CH2
   |      |
   HC     CH2
     \\  /
       CH
```

is endothermic, not exothermic. The properties of benzene and other aromatic substances reflect the stability conferred upon them by the resonance energy.

7-4. Hydrazine, Hydrogen Peroxide, and Related Hydrides

Many other nonmetal hydrides are known in which there are two or more nonmetal atoms connected to one another by single bonds. For example, a solution of sodium sulfide, Na_2S, dissolves sulfur, S_8, to form a series of polysulfides, Na_2S_2, Na_2S_3, Na_2S_4, . . . , and the hydrogen compounds H_2S_2, H_2S_3, H_2S_4, . . . are liberated on addition of hydrochloric acid. We would expect, in agreement with observation, that a reaction such as $8H_2S + S_8 \rightarrow 8HS{-}SH$ would have only a very small value of ΔH, because the product molecules contain the same bonds (16H—S, 8S—S) as the reactant molecules. The various polysulfides are accordingly about as stable (in relation to the elementary nonmetal) as are the alkanes.

For nitrogen and oxygen, however, the corresponding reactions

$$H_2O(g) + \tfrac{1}{2}O_2(g) \rightarrow H_2O_2(g)$$

$$\tfrac{4}{3}NH_3(g) + \tfrac{1}{3}N_2(g) \rightarrow N_2H_4(g)$$

are strongly endothermic. The reverse reactions are exothermic:

$$H_2O_2(g) \rightarrow H_2O(g) + \tfrac{1}{2}O_2(g) + 109 \text{ kJ mole}^{-1}$$

$$N_2H_4(g) \rightarrow \tfrac{4}{3}NH_3(g) + \tfrac{1}{3}N_2(g) + 155 \text{ kJ mole}^{-1}$$

Both hydrogen peroxide and hydrazine may be described as energy-rich molecules, because of the presence of the relatively unstable O—O and N—N bonds. Both substances are used as rocket propellants. Higher analogues, such as H_2O_3 and N_3H_5, have not been made.

When barium oxide, BaO, is heated to a dull red heat in a stream of air it adds oxygen to form a higher oxide, BaO_2, barium peroxide:

$$2BaO + O_2 \rightarrow 2BaO_2$$

Hydrogen peroxide, H_2O_2, is made by treating barium peroxide with sulfuric acid or phosphoric acid, and distilling:*

$$BaO_2 + H_2SO_4 \rightarrow BaSO_4 + H_2O_2$$

Pure hydrogen peroxide is a colorless liquid, about as mobile as water, and with density 1.47 g cm^{-3}, melting point $-0.4°C$, and boiling point 151°C. It is a very strong oxidizing agent, which spontaneously oxidizes organic substances. Its uses are in the main determined by its oxidizing power.

Commercial hydrogen peroxide is an aqueous solution, sometimes containing a small amount of a stabilizer, such as phosphate ion, to decrease its rate of decomposition to water and oxygen by the reaction

$$2H_2O_2 \rightarrow 2H_2O + O_2$$

Drug-store hydrogen peroxide is a 3% solution (containing 3 g H_2O_2 per 100 g), for medical use as an antiseptic, or a 6% solution, for bleaching hair. A 30% solution and an 85% solution are used in chemical industries.

Oxygen in hydrogen peroxide is assigned the oxidation number -1. In the above reaction one oxygen atom of each molecule is oxidized to oxidation number 0 and the other is reduced to oxidation number -2. Such a reaction is called an *auto-oxidation-reduction reaction.* Hydrogen peroxide can act as an oxidizing agent and also as a reducing agent. It is its oxidizing power that permits it to be used for bleaching hair and is responsible for its effectiveness as a mild antiseptic. Oil paintings that have been discolored by the formation of lead sulfide, PbS, which is black, from the lead hydroxycarbonate (called white lead) in the paint may be bleached by washing with hydrogen peroxide:

$$PbS(c) + 4H_2O_2(aq) \rightarrow PbSO_4(c) + 4H_2O(l)$$

Its action as a reducing agent is shown, for example, by its decolorization of permanganate ion in acidic solution:

$$2MnO_4^- + 5H_2O_2 + 6H^+ \rightarrow 2Mn^{++} + 5O_2(g) + 8H_2O$$

*A method involving organic compounds is used in industry.

Hydrogen Sulfide and the Sulfides

Hydrogen sulfide, H_2S, is analogous to water. Its electronic structure is

```
    H
    |
  :S—H
   ··
```

It is far more volatile (m.p. −85.5°C, b.p. −60.3°C) than water. It is appreciably soluble in cold water (2.6 liters of gas dissolves in 1 liter of water at 20°C), forming a slightly acidic solution. The solution is slowly oxidized by atmospheric oxygen, giving a milky precipitate of sulfur.

Hydrogen sulfide has a powerful odor, resembling that of rotten eggs. It is very poisonous, and care must be taken not to breathe the gas while using it in the analytical chemistry laboratory.

Hydrogen sulfide is readily prepared by action of hydrochloric acid on ferrous sulfide:

$$2HCl(aq) + FeS(c) \rightarrow FeCl_2(aq) + H_2S(g)$$

The *sulfides* of the alkali and alkaline-earth metals are colorless substances easily soluble in water. The sulfides of most other metals are insoluble or only very slightly soluble in water, and their precipitation under varying conditions is an important part of the usual scheme of qualitative analysis for the metallic ions. Many metallic sulfides occur in nature; important sulfide ores include FeS, Cu_2S, CuS, ZnS, Ag_2S, HgS, and PbS.

7-5. Ammonia and Its Compounds

Ammonia, NH_3, is an easily condensable gas (b.p. −33.4°C; m.p. −77.7°C). readily soluble in water. The gas is colorless and has a pungent odor, often detected around stables and manure piles, where ammonia is produced by decomposition of organic matter. The solution of ammonia in water, called ammonium hydroxide solution (or sometimes *aqua ammonia*), contains the molecular species NH_3, NH_4OH (ammonium hydroxide), NH_4^+, and OH^-. Ammonium hydroxide is a weak base, and is only slightly ionized to ammonium ion, NH_4^+, and hydroxide ion OH^-:

$$NH_3 + H_2O \rightleftarrows NH_4OH \rightleftarrows NH_4^+ + OH^-$$

The ammonium ion has the configuration of a regular tetrahedron. The

NH_4^+ ion can be described as having four electrons in four tetrahedral sp^3 orbitals. In the ammonium hydroxide molecule the ammonium ion and the hydroxide ion are held together by a hydrogen bond (Section 9-6).

The Preparation of Ammonia

Ammonia is easily made in the laboratory by heating an ammonium salt, such as ammonium chloride, NH_4Cl, with a strong alkali, such as sodium hydroxide or calcium hydroxide:

$$2NH_4Cl + Ca(OH)_2 \longrightarrow CaCl_2 + 2H_2O + 2NH_3(g)$$

The gas may also be made by warming concentrated ammonium hydroxide.

The principal commercial method of production of ammonia is the *Haber process,* the direct combination of nitrogen and hydrogen under high pressure (several hundred atmospheres) in the presence of a catalyst (usually iron, containing molybdenum or other substances to increase the catalytic activity). The gases used must be specially purified, to prevent "poisoning" the catalyst. The yield of ammonia at equilibrium (Chapter 10) is less at a high temperature than at a lower temperature. The gases react very slowly at low temperatures, however, and the reaction became practical as a commercial process only when a catalyst was found which speeded up the rate satisfactorily at 500°C. Even at this relatively low temperature the equilibrium is unfavorable if the gas mixture is under low pressure, less than 0.1% of the mixture at 1 atm being converted to ammonia. Increase in the total pressure favors the formation of ammonia; at 500 atm pressure the equilibrium mixture is over one-third ammonia.

Smaller amounts of ammonia are obtained as a by-product in the manufacture of coke and illuminating gas by the distillation of coal, and are made by the cyanamide process. In the *cyanamide process* a mixture of lime and coke is heated in an electric furnace, forming calcium acetylide (calcium carbide), CaC_2:

$$CaO + 3C \longrightarrow CO + CaC_2$$

Nitrogen, obtained by fractionation of liquid air, is passed over the hot calcium acetylide, forming calcium cyanamide, $CaCN_2$:

$$CaC_2 + N_2 \longrightarrow CaCN_2 + C$$

Calcium cyanamide may be used directly as a fertilizer, or may be converted into ammonia by treatment with steam under pressure:

$$CaCN_2 + 3H_2O \rightarrow CaCO_3 + 2NH_3$$

Ammonium Salts

The ammonium salts are similar to the potassium salts and rubidium salts in crystal form, molar volume, color, and other properties. This similarity is due to the close approximation in size of the ammonium ion (radius 148 pm) to these alkali ions (radius of K^+, 133 pm, and of Rb^+, 148 pm). The ammonium salts are all soluble in water, and are completely ionized in aqueous solution.

Ammonium chloride, NH_4Cl, is a white salt, with a bitter, salty taste. It is used in dry batteries and as a flux in soldering and welding. Ammonium sulfate, $(NH_4)_2SO_4$, is an important fertilizer; and ammonium nitrate, NH_4NO_3, mixed with other substances, is used as an explosive and as a fertilizer.

Liquid Ammonia as a Solvent

Liquid ammonia (b.p. −33.4°C) has a high dielectric constant, and is a good solvent for salts, forming ionic solutions. It also has the unusual power of dissolving the alkali metals and alkaline-earth metals without chemical reaction, to form blue solutions that have an extraordinarily high electric conductivity and a metallic luster. These metallic solutions slowly decompose, with evolution of hydrogen, forming amides, such as sodium amide, $NaNH_2$:

$$2Na + 2NH_3 \rightarrow 2Na^+ + 2NH_2^- + H_2$$

The amides are ionized in the solution into sodium ion and the amide ion,

$$\left[:N\begin{matrix} \diagup H \\ \diagdown H \end{matrix} \right]^-$$

which is analogous to the hydroxide ion in aqueous systems. The ammonium ion in liquid ammonia is analogous to the hydronium ion in aqueous systems (see Chapters 9 and 12).

7-6. Other Normal-valence Compounds of the Nonmetals

The halogens form covalent compounds with most of the nonmetallic elements (including each other) and the metalloids. These compounds are usually molecular substances, with the relatively low melting points and boiling points characteristic of substances with small forces of intermolecular attraction.

An example of a compound involving a covalent bond between a halogen and a nonmetal is chloroform, $CHCl_3$. In this molecule the carbon atom is attached by single covalent bonds to one hydrogen atom and three chlorine atoms. Chloroform is a colorless liquid, with a characteristic sweetish odor, b.p. 61°C, density 1.498 g ml^{-1}. Chloroform is only slightly soluble in water, but it dissolves readily in alcohol, ether, and carbon tetrachloride.

The halides of carbon and its congeners are tetrahedral (sp^3 bond orbitals). Those of nitrogen and oxygen and their congeners have bond angles near 100°, corresponding to p bond orbitals with a small amount of s character.

The melting points, boiling points, bond length, and bond angles of some chlorides are given in Table 7-6. Some values of the standard enthalpy are given in Appendix VI. The stability of the substances is determined by the electronegativity difference of the bonded atoms, with correction for compounds of nitrogen and oxygen.

Many of these substances react readily with water, to form a hydride of one element and a hydroxide of the other:

$$ClF + H_2O \rightarrow HClO + HF$$

$$PCl_3 + 3H_2O \rightarrow P(OH)_3 + 3HCl$$

In general, in a reaction of this sort, called *hydrolysis,* the more electronegative element combines with hydrogen, and the less electronegative element combines with the hydroxide group. This rule is seen to be followed in the above examples.

Resonance in Fluorinated Hydrocarbons

An interesting difference in properties is observed between methyl fluoride and methylene difluoride, trifluoromethane, and tetrafluoromethane. It is found that successive steps in chlorination of methane are accompanied by nearly the same enthalpy change (see Exercise 6-37), to within ±4 kJ $mole^{-1}$. This constancy for reactions in which C—H is

TABLE 7-6
Properties of Some Chlorides of Nonmetals

	CCl_4	NCl_3	Cl_2O	ClF
m.p.	−23°	−40°	−20°	−154°C
b.p.	77°	70°	4°	−100°
Bond length	177 pm	173 pm	169 pm	163 pm
Bond angle	109.5°	110°	110°	
	$SiCl_4$	PCl_3	SCl_2	Cl_2
m.p.	−70°	−112°	−78°	−102°
b.p.	60°	74°	59°	−34°
Bond length	201 pm	204 pm	200 pm	199 pm
Bond angle	109.5°	100.0°	102°	
	$GeCl_4$	$AsCl_3$		BrCl
m.p.	−50°	−18°		
b.p.	83°	130°		
Bond length	209 pm	216 pm		214 pm
Bond angle	109.5°	99°		
	$SnCl_4$	$SbCl_3$	$TeCl_2$	ICl
m.p.	−33°	73°	209°	27°
b.p.	114°	223°	327°	97°
Bond length	232 pm	238 pm	234 pm	230 pm
Bond angle	109.5°	99°	99°	

converted to C—Cl is expected from the bond-energy principle. For the fluoromethanes, however, the values differ greatly in the sequence of reactions:

	$CH_4(g)$	$CH_3F(g)$	$CH_2F_2(g)$	$CHF_3(g)$	$CF_4(g)$
$\Delta H°_f(198°K)$	−75	−234	−449	−691	−923 kJ mole^{-1}
Difference	−159	−215	−242	−232	

The enthalpy value for CH_3F leads to the value 443 kJ mole^{-1} for the C—F bond energy. This value is given in Table V-1. The other fluoromethanes are stabilized by resonance with structures other than the normal valence-bond structure.

The observed C—F bond length in methyl fluoride and ethyl fluoride is 138.5 pm, which is close to the value expected for a single bond. Significant shortening is found, however, for the other molecules, to 135.8 pm for CH_2F_2, 133.4 pm for CHF_3, and 132.0 pm for CF_4. These values suggest an increasing amount of double-bond character, such as would result from resonance among the following structures for methylene fluoride:

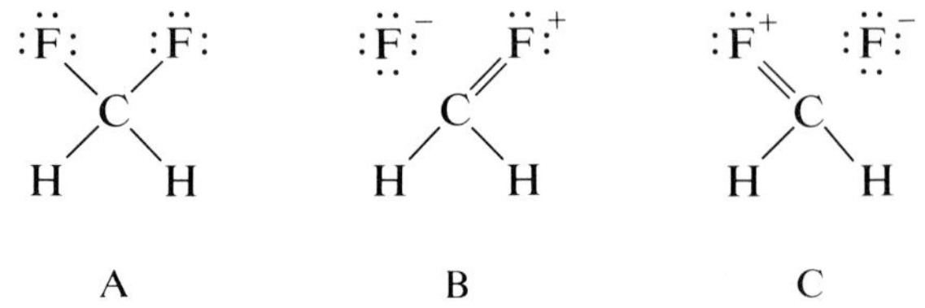

```
 :F:   :F:        :F:⁻   F:⁺        :F⁺    :F:⁻
   \   /              \\ /          \\     
    C                 C               C
   / \               / \             / \
  H   H             H   H           H   H

    A                 B               C
```

The electronegativity difference of carbon and fluorine corresponds to 43% partial ionic character for the C—F bond, freeing a carbon orbital for double-bond formation with the other fluorine atom. With use of 443 kJ $mole^{-1}$ for the C—F bond energy for the normal-valence structures (A for CH_2F_2) we obtain the following values for the resonance energy with structures such as B and C: 56 kJ $mole^{-1}$ for CH_2F_2, 139 kJ $mole^{-1}$ for CHF_3, and 212 kJ $mole^{-1}$ for CF_4.

A significant amount of resonance energy is also found for CH_2ClF (25 kJ $mole^{-1}$), $CHClF_2$ (69), $CClF_3$ (132), $CHCl_2F$ (36), CCl_2F_2 (69), and CCl_3F (21). In these molecules some decrease in bond length is observed for C—Cl, but not so great as for C—F.

The decreased chemical reactivity resulting from this resonance energy has great practical importance. The chloromethanes, such as carbon tetrachloride, are toxic, because of the ease with which they hydrolyze. The fluorochloromethanes do not hydrolyze in this way, and can be used safely in the home and in industrial plants. There was rapid development of the home refrigerator industry after the discovery of these substances, especially the freons CCl_3F (b.p. 23.8°C) and CCl_2F_2 (b.p. −30°C). They are used also as the vehicle for aerosols.

The bond lengths in $SiCl_4$ and other halides of the heavier elements are usually less than the sum of the single-bond radii; for $SiCl_4$ the difference is 16 pm. It is likely that this difference is to be interpreted as showing that the bonds have some double-bond character, resulting from resonance of the sort discussed above.

Substitution Reactions

Methane and other paraffins react with chlorine and bromine when exposed to sunlight or when heated to a high temperature. When a mixture of methane and chlorine is passed through a tube containing a catalyst (aluminum chloride, $AlCl_3$, mixed with clay) heated to about 300°C, the following reactions occur:*

$$CH_4 + Cl_2 \rightarrow CH_3Cl + HCl$$

$$CH_3Cl + Cl_2 \rightarrow CH_2Cl_2 + HCl$$

$$CH_2Cl_2 + Cl_2 \rightarrow CHCl_3 + HCl$$

$$CHCl_3 + Cl_2 \rightarrow CCl_4 + HCl$$

In each of these reactions a chlorine molecule, with structural formula Cl—Cl, is split into two chlorine atoms; one chlorine atom takes the place of a hydrogen atom bonded to carbon, and the other combines with the displaced hydrogen atom to form a molecule of hydrogen chloride, H—Cl. By use of values of bond energies from Table V-1, we calculate for the heat of each of these reactions the value $328 + 432 - 243 - 415 = 102$ kJ $mole^{-1}$. The reactions are not as strongly exothermic as the reaction of addition of chlorine to a double bond (142 kJ $mole^{-1}$).

Chemical reactions such as these four are called *substitution reactions. A substitution reaction* is the replacement of one atom or group of atoms in a molecule by another atom or group of atoms. The four chloromethanes are substitution products of methane. Substitution reactions and addition reactions (Section 7-2) are extensively used in practical organic chemistry.

Some physical properties of the chloromethanes are given in Table 7-7; their enthalpies of formation from the elements are given in Table VI-6. All four are colorless, with characteristic odors and with low boiling points, increasing with the number of chlorine atoms in the molecule. The chloromethanes do not ionize in water.

TABLE 7-7
Some Physical Properties of the Chloromethanes

Substance	Formula	Melting Point	Boiling Point	Density of Liquid
Methyl chloride	CH_3Cl	−98°C	−24°C	0.92 g ml^{-1}
Dichloromethane	CH_2Cl_2	−97	40	1.34
Chloroform	$CHCl_3$	−64	61	1.50
Carbon tetrachloride	CCl_4	−23	77	1.60

*The relative amounts of the four products may be varied somewhat by changing the ratio of methane and chlorine in the gas mixture used.

Chloroform and carbon tetrachloride are used as solvents; carbon tetrachloride is an important dry-cleaning agent. Chloroform was formerly used as a general anesthetic.

Care must be taken in the use of carbon tetrachloride that no large amount of its vapor is inhaled, because it damages the liver.

7-7. DDT and Other Chlorinated Aromatic Compounds

Many compounds of chlorine with benzene and related substances have found use in medicine, public health, and agriculture. An outstanding example is DDT, which is used as an insecticide and pediculicide. The amount distributed over the earth in thirty years after its introduction in 1940 is about 3 billion kilograms. Its use is now restricted because of its toxicity to humans, birds, and other animals.

DDT, 1,1,1-trichloro-2,2-bis(*p*-chlorobenzene)ethane, has the structural formula

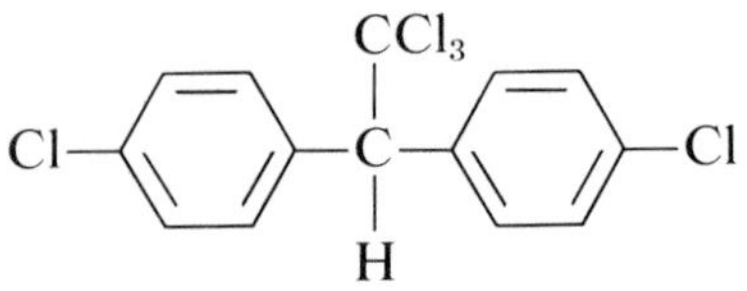

CCl_3

Cl—C_6H_4—C(H)—C_6H_4—Cl

It is synthesized by the reaction of chloral hydrate, $Cl_3CCH(OH)_2$, with chlorobenzene, C_6H_5Cl, in the presence of concentrated sulfuric acid, which favors the reaction through its affinity for water (Section 8-2).

Another example is hexachlorophene, 2,2-methylene-bis(3,4,6-trichlorophenol), with formula

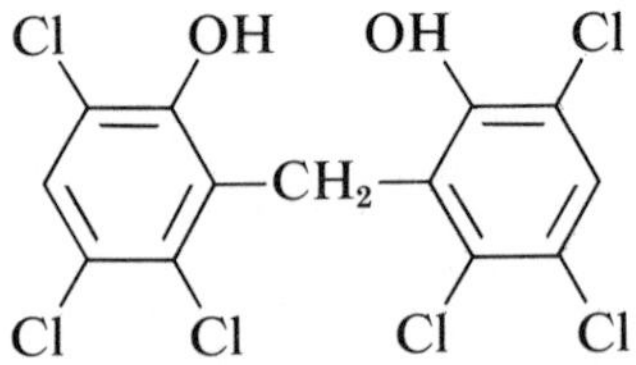

Cl OH OH Cl

—CH_2—

Cl Cl Cl Cl

It is made by reaction of 2 moles of 2,4,5-trichlorophenol with 1 mole of formaldehyde, $H_2C{=}O$, in concentrated sulfuric acid. For about thirty years it has been widely used as a disinfectant, especially in germicidal soaps. Its use is now restricted because of its toxicity.

7-8. Some Transargononic Single-bonded Compounds

When chlorine is passed over phosphorus it reacts with it to form phosphorus trichloride:

$$\tfrac{1}{4}P_4(c) + \tfrac{3}{2}Cl_2(g) \longrightarrow PCl_3(g) + 279 \text{ kJ mole}^{-1}$$

Another product, phosphorus pentachloride, is also formed:

$$PCl_3(g) + Cl_2(g) \longrightarrow PCl_5(g) + 92 \text{ kJ mole}^{-1}$$

In the PCl_5 molecule the phosphorus atom has a transargononic structure, with five shared electron pairs in the outer shell. It forms five covalent bonds, with the bond orbitals formed by hybridization of a 3*d* orbital with the 3*s* orbital and the three 3*p* orbitals. The valence-bond structure for the molecule is

```
      Cl
      |  Cl
      | /
  Cl—P
      | \
      |  Cl
      Cl
```

The molecule has the trigonal bipyramidal structure, with three equally-spaced chlorine atoms around the equator and one at each pole (Figure 7-6).

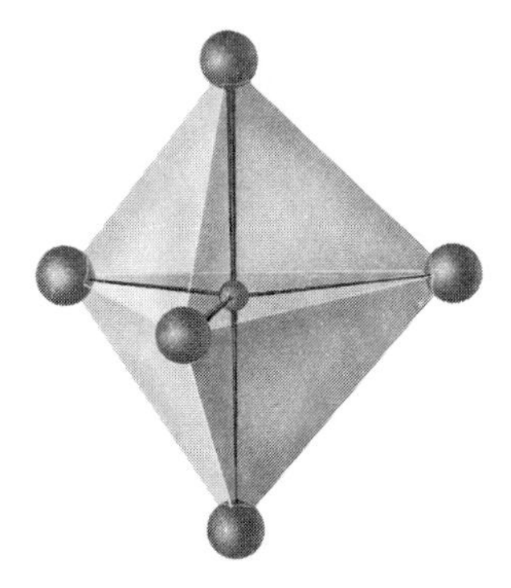

FIGURE 7-6
The structure of the molecule PCl_5, showing the arrangement of the five chlorine atoms at the corners of a trigonal bipyramid about the phosphorus atom.

The P—Cl bond energy in PCl_3 is 317 kJ mole^{-1}:

$$P(g) + 3Cl(g) \longrightarrow PCl_3(g) + 3 \times 317 \text{ kJ mole}^{-1}$$

The effective P—Cl bond energy for each of the two added bonds in PCl_5 is 165 kJ mole^{-1}:

$$PCl_3(g) + 2Cl(g) \longrightarrow PCl_5(g) + 2 \times 165 \text{ kJ mole}^{-1}$$

A transargononic P—Cl bond is 152 kJ mole^{-1} less stable than an argononic P—Cl bond. This smaller stability results from the smaller stability of a 3*d* electron in phosphorus than a 3*p* electron, partially neutralized by the greater bond-forming power of *spd* hybrid bond orbitals (greater overlap) than of *p* bond orbitals. In addition, a significant contribution to the normal state of PCl_5 is made by ionic structures such as

Cl⁻ Cl
Cl—P⁺
| Cl
Cl

The importance of ionic structures in stabilizing transargononic compounds is shown by the great stability of transargononic fluorides. The bond energy for each of the two added fluorine atoms in PF_5 is 425 kJ mole^{-1}:

$$PF_3(g) + 2F(g) \longrightarrow PF_5(g) + 2 \times 425 \text{ kJ mole}^{-1}$$

This value is only 61 kJ mole^{-1} less than the normal (argononic) P—F bond energy, 486 kJ mole^{-1} (in PF_3), a far smaller difference than for P—Cl (152 kJ mole^{-1}). It is not surprising that the fluorides predominate among the transargononic molecules and ions, as listed in Table 7-8.

Most of the known single-bonded transargononic molecules and ions are listed in Table 7-8. Those with small molecular weight, especially the fluorides, are gases at room temperature (SF_6 has boiling point −62°C; PCl_5(c) sublimes at 160°C and melts, under pressure, at 168°C).

SF_6 and other molecules and ions with ligancy six and no unshared electron pairs in the outer shell of the central atom have the six bonds directed towards the corners of a regular octahedron. In BrF_5 the five bonds are directed toward five corners of a regular octahedron, the sixth corner presumably being occupied by an unshared pair. The unshared pair, in an orbital that is largely *s* in character, occupies more space than a bonding pair; the angle between the bond to the polar fluorine atom and

TABLE 7-8
Formulas of Single-bonded Transargononic Molecules and Ions

SiF_6^{--}					
GeF_6^{--}					
SnF_6^{--}	$SnCl_6^{--}$	$SnBr_6^{--}$	SnI_6^{--}	$Sn(OH)_6^{--}$	
PF_5	PCl_5	PF_6^-	PCl_6^-		
PF_3Cl_2	PBr_3Cl_2	PCl_3F_2	PCl_3Br_2		
AsF_6^-					
SbF_6^-	$SbCl_5$	$SbCl_6^-$	$SbBr_6^-$	$Sb(OH)_6^-$	
SF_4	SF_6	S_2F_{10}	SCl_4		
SeF_4	SeF_6	$SeCl_4$	$SeBr_4$	$SeCl_6^{--}$	$SeBr_6^{--}$
TeF_4	TeF_6	$TeCl_4$	$TeBr_4$	$TeCl_6^{--}$	$Te(OH)_6$
ClF_3	ClF_5				
BrF_3	BrF_5	BrF_4^-			
IF_5	IF_7	ICl_3	ICl_2^-	ICl_4^-	

that to any of the four equatorial fluorine atoms is less than 90° (about 86°). In BrF_4^- the four bonds lie in a plane, with the two unshared pairs occupying the other two octahedral directions.

An interesting and not fully understood fact is that phosphorus and antimony form more stable transargononic compounds than their intermediate congener, arsenic.

The stability of transargononic compounds increases with increase in electronegativity difference of the ligated atom and the central atom, as shown by the following comparison:

$$ClF(g) + 2F_2(g) \longrightarrow ClF_5(g) + 188 \text{ kJ mole}^{-1}$$

$$BrF(g) + 2F_2(g) \longrightarrow BrF_5(g) + 370 \text{ kJ mole}^{-1}$$

$$IF(g) + 2F_2(g) \longrightarrow IF_5(g) + 727 \text{ kJ mole}^{-1}$$

The elements of the first period (C, N, O) do not form stable transargononic compounds. This general difference in their chemical properties from their heavier congeners is clearly explained by the nonexistence of 2*d* orbitals.

Transargononic compounds involving multiple bonds are discussed in the following chapter.

EXERCISES

7-1. Relate the valence, number of neighboring bonded atoms, and bond type of elementary hydrogen, carbon, nitrogen, oxygen, and fluorine and their congeners.

7-2. Plot the melting point and boiling point of elementary hydrogen, nitrogen, oxygen, and fluorine and their congeners as a function of their atomic numbers. Explain any discontinuities. Do you expect the melting and boiling points of elementary carbon to be high or low? Why?

7-3. How can hydrogen and oxygen be prepared easily in the laboratory? What are the reactions?

7-4. How are elementary hydrogen, nitrogen, phosphorus, oxygen, and chlorine prepared commercially?

7-5. Why is red phosphorus more stable than white phosphorus and less stable than black phosphorus?

7-6. How is the correction to the heat of formation per bond for NH_3 and H_2O in Figure 7-4 for multiple bonding in N_2 and O_2 applied?

7-7. What is the general formula of an alkane? How do alkenes and alkynes differ from alkanes?

7-8. What are the names and formulae of the first ten members of the paraffin series of hydrocarbons? Which are gases at room temperature and 1 atm pressure?

7-9. Why should highly branched hydrocarbons burn more slowly than straight chain hydrocarbons?

7-10. What is the height of the energy barrier restricting rotation about the carbon-carbon single bond in ethane? In monofluoroethane? What is the likely explanation of this barrier?

7-11. What are the relative stabilities and chemical formulas of cyclopropane, cyclobutane, and cyclohexane?

7-12. Why does the addition reaction of chlorine with ethylene proceed?

7-13. Compare the energy barrier of conversion of *cis*-1,2-dichloroethylene to *trans*-1,2-dichloroethylene to the barrier of rotation about the carbon-carbon single bond of ethane.

7-14. At one time it was thought all organic compounds were derived from living organisms. Discuss the implications of the formation of acetylene from the reaction of calcium carbide and water on theories of the origin of life. You need not include the implications of the fact that the organism that drops the calcium carbide in the water is living.

7-15. What characterizes an aromatic hydrocarbon with respect to other hydrocarbons?

7-16. Assuming that the resonance energy is proportional to $(n - 1)$ where n is the number of possible Kekulé structures, calculate the resonance energies of naphthalene, anthracene, and phenanthrene from the value for benzene. The observed values are 314, 440, and 459 kJ $mole^{-1}$, respectively.

7-17. *(a)* What combination of molecules of benzene and ethylene must be taken to represent the formula of naphthalene, $C_{10}H_8$, with the proper numbers of carbon-carbon single bonds and double bonds? *(b)* The heat of combustion of benzene (liquid) is 3273 kJ $mole^{-1}$, and that of ethylene (gas) is 1387 kJ $mole^{-1}$. Calculate a value for naphthalene with the assumption that the resonance energy of naphthalene is twice that of benzene. The experimental value for naphthalene (crystal) is 5157 kJ $mole^{-1}$. Note that agreement with the calculated value provides support for the assumption.

7-18. Hydrazine and hydrogen peroxide may be described as "energy-rich molecules." What does this mean?

7-19. What is an auto-oxidation-reduction reaction? What are the oxidation numbers of the atoms of the reactant and products in the auto-oxidation-reduction reaction of hydrazine?

7-20. Are the sulfides of the alkali, alkaline-earth, and other metals soluble or insoluble in water?

7-21. What are the molecular or ionic species present in a solution of ammonia in water. Make a realistic drawing of each molecular species.

7-22. Compare the sizes of the ammonium ion with those of the alkali ions.

7-23. What is hydrolysis? What compounds would the hydrolysis of $TeCl_2$ produce?

7-24. By extrapolation, what would you expect the melting point, boiling point, bond length, and bond angle of $PbCl_4$, $BiCl_3$, $PoCl_2$ and AtCl to be?

7-25. What are the formulas of methyl fluoride, methylene difluoride, trifluoromethane and tetrafluoromethane?

7-26. Write all the resonating structures for trifluoromethane.

7-27. How does the electronegativity of fluorine make the freons useful?

7-28. How are substitution reactions different from addition reactions? Give examples. Which type of reaction produces more heat?

7-29. What orbitals are used in the formation of the hybrid valence bonds of transargononic compounds?

7-30. What effects stabilize transargononic compounds?

7-31. Why are there many fluorides among transargononic compounds?

7-32. Why do elements of the first short period not form the central atom of transargononic compounds?

8

Oxygen Compounds of Nonmetallic Elements

Of the 2468 inorganic compounds considered important enough to be listed in a table of physical properties in a reference book, over 49% (1220 compounds) involve nonmetallic elements bonded to oxygen atoms. Most of the oxygen-containing compounds of nonmetallic elements have transargononic structures, and it is the variety of these structures that makes a greater contribution than any other structural feature to the richness of inorganic chemistry.

In Section 8-1 of this chapter we shall examine the oxycompounds of chlorine, as an example. The following sections present a survey of the oxycompounds of other nonmetallic elements.

8-1. Oxycompounds of the Halogens

Chlorine forms a normal-valence oxide, Cl_2O $\left(\begin{array}{l} :\ddot{\mathrm{Cl}}: \\ \;\;| \\ :\ddot{\mathrm{O}}-\ddot{\mathrm{Cl}}: \end{array}\right)$, a normal-valence oxygen acid, $HClO$ $\left(\begin{array}{l} \;\mathrm{H} \\ \;\;| \\ :\ddot{\mathrm{O}}-\ddot{\mathrm{Cl}}: \end{array}\right)$, and a rather large number of transargononic oxycompounds.

Dichlorine monoxide, Cl_2O, is a yellow gas obtained by passing chlorine over mercuric oxide:

$$2Cl_2(g) + HgO(c) \longrightarrow HgCl_2(c) + Cl_2O(g)$$

The gas condenses to a liquid at about 4°C. It is the anhydride of hypochlorous acid: that is, it reacts with water to give hypochlorous acid:

$$Cl_2O(g) + H_2O(l) \longrightarrow 2HClO(aq)$$

The standard enthalpy of $Cl_2O(g)$ is 24 kJ $mole^{-1}$.

An interesting transargononic chlorine oxide is *dichlorine heptoxide,* Cl_2O_7, a colorless liquid with melting point −91°C and boiling point 82°C, which can be made by mixing tetraphosphorus decoxide, P_4O_{10}, and perchloric acid, $HClO_4$. We might write an argononic structural formula for Cl_2O_7:

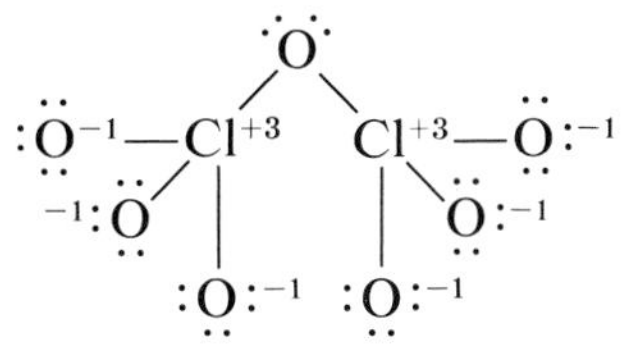

```
                 ..
                 :O:
       ..       /   \        ..
     :O:⁻¹—Cl⁺³      Cl⁺³—O:⁻¹
       ..   ..  /|    |\ ..  ..
       ⁻¹:O:  |    |  :O:⁻¹
           ..     |    |     ..
            :O:⁻¹ :O:⁻¹
             ..    ..
```

This formula is, however, unsatisfactory in that it places a large electric charge, +3, on each chlorine atom. Moreover, the electron-diffraction determination of the structure of the molecule has given the value 142 pm for the length of the six outer Cl—O bonds. This value is 28 pm less than the single-bond value 170 pm found in Cl_2O and the two central bonds in Cl_2O_7, and supports the assignment to the molecule of the transargononic structure

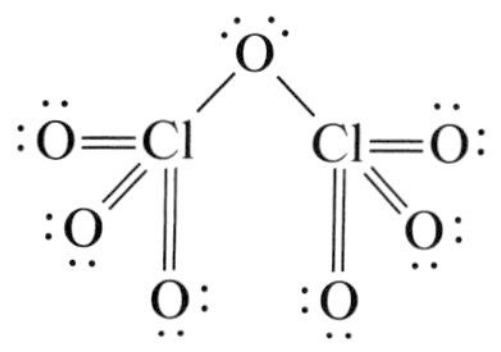

```
                ..
                :O:
      ..       /   \       ..
     :O=Cl         Cl=O:
       //|         |\\
     :O  ||        ||  O:
      ..  ||        ||  ..
          O:      :O
          ..        ..
```

For this structure each chlorine atom has covalence 7, corresponding to its group in the periodic table. In the formation of seven covalent bonds the chlorine atom may make use of three 3*d* orbitals, together with its 3*s* and 3*p* orbitals.

The standard enthalpy of $Cl_2O_7(g)$ is 265 kJ mole^{-1}. The liquid is not a very sensitive explosive, but it explodes on percussion or ignition.

From the enthalpy values of the substances and of O(g) (Appendix VI) we write the equation

$$Cl_2O(g) + 6O(g) \longrightarrow Cl_2O_7(g) + 6 \times 217 \text{ kJ mole}^{-1}$$

The bond energy for a transargononic Cl=O double bond is accordingly 217 kJ mole^{-1}.

Other experimental values of heats of reaction show that the value of the transargononic Cl=O bond energy is nearly the same in other molecules as in Cl_2O_7. An example is the oxidation of ClF to ClO_3F:

$$ClF(g) + 3O(g) \longrightarrow ClO_3F(g) + 3 \times 239 \text{ kJ mole}^{-1}$$

The increase in value over Cl_2O_7 may be attributed to the greater ionic character of the Cl—F than of the Cl—O single bond, thus releasing an added part of an *sp* orbital.

Oxidation Numbers of the Halogens

The halogens other than fluorine form stable compounds corresponding to nearly all values of the oxidation number from −1 to +7, as shown in the accompanying chart.

+7		$HClO_4$, Cl_2O_7		H_5IO_6
+6		Cl_2O_6		
+5		$HClO_3$	$HBrO_3$	HIO_3, I_2O_5
+4		ClO_2	BrO_2	IO_2
+3		$HClO_2$		
+2				
+1	HOF	HClO, Cl_2O	HBrO, Br_2O	HIO
0	F_2	Cl_2	Br_2	I_2
−1	HF, F^-	HCl, Cl^-	HBr, Br^-	HI, I^-

(Highly reactive molecules known only in the dilute gas phase or trapped in a crystal or supercooled liquid, such as OF and ClO, are not listed.)

The Oxygen Acids of Chlorine

From the preceding discussion it is not surprising that the transargononic oxygen acids $HClO_2$, $HClO_3$, and $HClO_4$ exist, as well as the normal-valence oxygen acid HClO (correctly written HOCl).

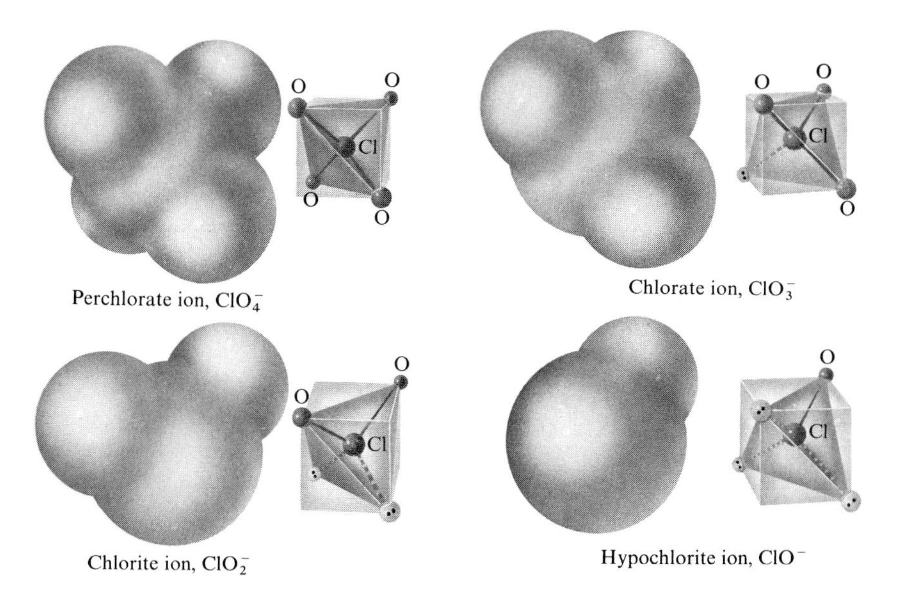

FIGURE 8-1
The structure of ions of the four oxygen acids of chlorine.

These oxygen acids and their anions have the following names:

$HClO_4$, perchloric acid	ClO_4^-, perchlorate ion
$HClO_3$, chloric acid	ClO_3^-, chlorate ion
$HClO_2$, chlorous acid	ClO_2^-, chlorite ion
$HClO$, hypochlorous acid	ClO^-, hypochlorite ion

The structures of the four anions are shown in Figure 8-1.

The electronic structures shown below, which are in agreement with the electroneutrality principle but involve making use of the 3*d* orbitals for the chlorine atom (except for hypochlorous acid), may be assigned to the four acids:

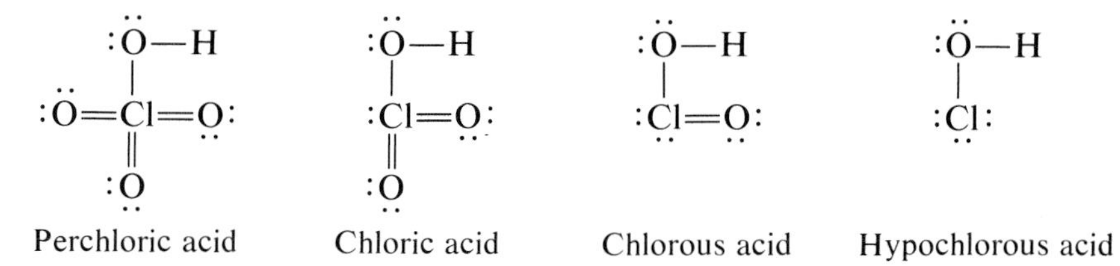

Perchloric acid   Chloric acid   Chlorous acid   Hypochlorous acid

In the following sections these acids, their salts, and the oxides of chlorine are discussed in the order of increasing oxidation number of the halogen.

Hypochlorous Acid and the Hypochlorites

Hypochlorous acid, $HClO$, and most of its salts are known only in aqueous solution; they decompose when the solution is concentrated. A mixture of chloride ion and hypochlorite ion is formed when chlorine is bubbled through a solution of sodium hydroxide:

$$Cl_2 + 2OH^- \longrightarrow Cl^- + ClO^- + H_2O$$

A solution of *sodium hypochlorite,* $NaClO$, made in this way or by electrolysis of sodium chloride solution is a popular household sterilizing and bleaching agent. The hypochlorite ion is an active oxidizing agent, and its oxidizing power is the basis of its sterilizing and bleaching action.

Chlorous Acid and the Chlorites

When chlorine dioxide, ClO_2 (discussed below), is passed into a solution of sodium hydroxide or other alkali a chlorite ion and a chlorate ion are formed:

$$2ClO_2 + 2OH^- \longrightarrow ClO_2^- + ClO_3^- + H_2O$$

This is an auto-oxidation-reduction reaction, the chlorine with oxidation number +4 in chlorine dioxide being reduced and oxidized simultaneously to oxidation numbers +3 and +5. Pure sodium chlorite, $NaClO_2$, can be made by passing chlorine dioxide into a solution of sodium peroxide:

$$2ClO_2 + Na_2O_2 \longrightarrow 2Na^+ + 2ClO_2^- + O_2$$

In this reaction the peroxide oxygen serves as a reducing agent, decreasing the oxidation number of chlorine from +4 to +3.

Sodium chlorite is an active bleaching agent, used in the manufacture of textile fabrics.

Chloric Acid and Its Salts

Chloric acid, $HClO_3$, is an unstable acid that, like its salts, is a strong oxidizing agent. The most important salt of chloric acid is *potassium chlorate,* $KClO_3$, which is made by passing an excess of chlorine through a hot solution of potassium hydroxide or by heating a solution containing hypochlorite ion and potassium ion:

$$3ClO^- \longrightarrow ClO_3^- + 2Cl^-$$

The potassium chlorate can be separated from the potassium chloride formed in this reaction by crystallization, its solubility at low temperatures being much less than that of the chloride (3 g and 28 g, respectively, per 100 g of water at 0°C).

Potassium chlorate is a white crystalline substance, which is used as the oxidizing agent in matches and fireworks and in the manufacture of dyes.

A solution of the similar salt *sodium chlorate,* $NaClO_3$, is used as a weed killer. Potassium chlorate would be as good as sodium chlorate for this purpose; however, sodium salts are cheaper than potassium salts, and for this reason they are often used when only the anion is important. Sometimes the sodium salts have unsatisfactory properties, such as deliquescence (attraction of water from the air to form a solution), which make the potassium salts preferable for some uses, even though more expensive.

All the chlorates form sensitive explosive mixtures when mixed with reducing agents; *great care must be taken in handling them.* The use of sodium chlorate as a weed killer is attended with danger, because combustible material such as wood or clothing that has become saturated with the chlorate solution will ignite by friction after it has dried. Also, *it is very dangerous to grind a chlorate with sulfur, charcoal, or other reducing agent.*

Perchloric Acid and the Perchlorates

Potassium perchlorate, $KClO_4$, is made by heating potassium chlorate just to its melting point:

$$4KClO_3 \longrightarrow 3KClO_4 + KCl$$

At this temperature very little decomposition with evolution of oxygen occurs in the absence of a catalyst. Potassium perchlorate may also be made by long-continued electrolysis of a solution of potassium chloride, potassium hypochlorite, or potassium chlorate.

Potassium perchlorate and other perchlorates are oxidizing agents, somewhat less vigorous and less dangerous than the chlorates. Potassium perchlorate is used in explosives, such as the propellent powder of the bazooka and other rockets. This powder is a mixture of potassium perchlorate and carbon together with a binder; the equation for the principal reaction accompanying its burning is

$$KClO_4 + 4C \longrightarrow KCl + 4CO$$

Anhydrous magnesium perchlorate, $Mg(ClO_4)_2$, and barium perchlorate,

$Ba(ClO_4)_2$, are used as drying agents *(desiccants)*. These salts have a very strong attraction for water. Nearly all the perchlorates are highly soluble in water; potassium perchlorate is exceptional for its low solubility, 0.75 g in 100 g of water at 0°C. Sodium perchlorate, $NaClO_4$, made by the electrolytic method, is used as a weed killer; it is safer than sodium chlorate. In general the mixtures of perchlorates with oxidizable materials are less dangerous than the corresponding mixtures of chlorates.

Perchloric acid, $HClO_4 \cdot H_2O$, is a colorless liquid made by distilling, under reduced pressure, a solution of a perchlorate to which sulfuric acid has been added. The perchloric acid distills as the monohydrate, and it cools to form crystals of the monohydrate. These crystals are isomorphous with ammonium perchlorate, NH_4ClO_4, and the substance is presumably hydronium perchlorate, $(H_3O)^+(ClO_4)^-$.

Chlorine Oxides

The oxides ClO, ClO_2, ClO_3 (or Cl_2O_6), Cl_2O_7, and ClO_4 (possibly Cl_2O_8) are known, in addition to the normal-valence oxide Cl_2O, which has been discussed above.

Chlorine monoxide, ClO, has been characterized by analysis of its band spectrum. Its bond length, 155 pm, is intermediate between that for a single bond, 169 pm, and that for a double bond, about 142 pm (as in Cl_2O_7). Its bond energy, 269 kJ mole^{-1}, is 59 kJ mole^{-1} greater than that for the Cl—O single bond. This additional bond energy is ascribed to the formation of a three-electron bond, in addition to a single bond, and the electronic structure is written as $:\dot{\text{Cl}}\overset{\cdots}{—}\text{O}:$, corresponding to resonance between the two structures $:\ddot{\underset{\cdot\cdot}{\text{Cl}}}—\dot{\underset{\cdot\cdot}{\text{O}}}:$ and $:\dot{\underset{\cdot\cdot}{\text{Cl}}}—\ddot{\underset{\cdot\cdot}{\text{O}}}:$.

Chlorine dioxide, ClO_2, is the only known compound of quadripositive chlorine. It is a reddish-yellow gas, which is very explosive, decomposing readily to chlorine and oxygen. The violence of this decomposition makes it very dangerous to add sulfuric acid or any other strong acid to a chlorate or to any dry mixture containing a chlorate.

Chlorine dioxide can be made by carefully adding sulfuric acid to potassium chlorate, $KClO_3$. It would be expected that this mixture would react to produce chloric acid, $HClO_3$, and then, because of the dehydrating power of sulfuric acid, to produce the anhydride of chloric acid, Cl_2O_5:

$$KClO_3 + H_2SO_4 \rightarrow KHSO_4 + HClO_3$$

$$2HClO_3 \rightarrow H_2O + Cl_2O_5$$

Dichlorine pentoxide, Cl_2O_5, however, is very unstable—its existence has never been verified. If it is formed at all, it decomposes at once to give chlorine dioxide and oxygen:

$$2Cl_2O_5 \longrightarrow 4ClO_2 + O_2$$

The molecule has a triangular structure, with O—Cl—O angle 118° and bond lengths 149 pm. We assign it the structure $\begin{array}{l} :\ddot{O} \\ \ \ \| \\ \underset{\cdot\cdot}{Cl}\overset{\cdots}{-}\underset{\cdot\cdot}{O}: \end{array}$, with interchange of the two kinds of bonds (resonance).

From the value 269 kJ mole^{-1} for the $Cl\overset{\cdots}{-}O$ bond energy of ClO and the value 216 kJ mole^{-1} for the transargononic Cl=O in Cl_2O_7 we would expect about -485 kJ mole^{-1} for the enthalpy of $ClO_2(g)$ relative to Cl(g) and 2O(g); the experimental value is -497 kJ mole^{-1}.

The Oxygen Acids and Oxides of Bromine

Bromine forms only two stable oxygen acids—hypobromous acid and bromic acid—and their salts:

HBrO, hypobromous acid	KBrO, potassium hypobromite
$HBrO_3$, bromic acid	$KBrO_3$, potassium bromate

Their preparation and properties are similar to those of the corresponding compounds of chlorine. They are somewhat weaker oxidizing agents than their chlorine analogues.

The bromite ion, BrO_2^-, has been reported to exist in solution. During many years no effort to prepare perbromic acid or any perbromate had succeeded; the preparation of perbromic acid has recently been reported.

Three very unstable oxides of bromine, Br_2O, BrO_2, and Br_3O_8, have been described. The structure of Br_3O_8 is not known.

None of the oxygen compounds of bromine has found important practical use.

The Oxygen Acids and Oxides of Iodine

Iodine reacts with hydroxide ion in cold alkaline solution to form the *hypoiodite ion,* IO^-, and iodide ion:

$$I_2 + 2OH^- \longrightarrow IO^- + I^- + H_2O$$

The solution when warmed reacts further to form iodate ion, IO_3^-:

$$3IO^- \longrightarrow IO_3^- + 2I^-$$

The salts of hypoiodous acid and iodic acid may be made in these ways. *Iodic acid* itself, HIO_3, is usually made by oxidizing iodine with concentrated nitric acid:

$$I_2 + 10HNO_3 \longrightarrow 2HIO_3 + 10NO_2 + 4H_2O$$

Iodic acid is a white solid, which is only very slightly soluble in concentrated nitric acid; it accordingly separates out during the course of the reaction. Its principal salts, potassium iodate, KIO_3, and sodium iodate, $NaIO_3$, are white crystalline solids.

Periodic acid has the normal formula H_5IO_6, with an octahedral arrangement of the oxygen atoms around the iodine atom, as shown in Figure 8-2. This difference in composition from its analogue perchloric acid, $HClO_4$, results from the large size of the iodine atom, which permits this atom to coordinate six oxygen atoms about itself, instead of four. The ligancy of iodine in periodic acid is hence 6.

There exists a series of periodates corresponding to the formula H_5IO_6 for periodic acid, and also a series corresponding to HIO_4. Salts of the first series are dipotassium trihydrogen periodate, $K_2H_3IO_6$, silver periodate, Ag_5IO_6, and so on. Sodium periodate, $NaIO_4$, a salt of the second series, occurs in small amounts in crude Chile saltpeter.

The two forms of periodic acid, H_5IO_6 and HIO_4 (the latter being unstable, but forming stable salts), represent the same oxidation state of

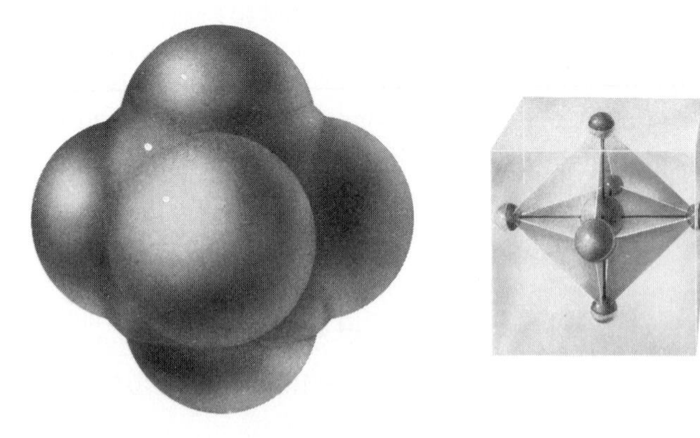

FIGURE 8-2
The periodate ion, IO_6^{5-}.

iodine, +7. The equilibrium between the two forms is a hydration reaction:

$$HIO_4 + 2H_2O \rightleftharpoons H_5IO_6$$

Iodine pentoxide, I_2O_5, is obtained as a white powder by gently heating either iodic acid or periodic acid:

$$2HIO_3 \longrightarrow I_2O_5 + H_2O$$

$$2H_5IO_6 \longrightarrow I_2O_5 + 5H_2O + O_2$$

The anhydride of periodic acid, I_2O_7, seems not to be stable; its preparation has never been reported.

The lower oxide of iodine, IO_2, can be made by treating an iodate with concentrated sulfuric acid and then adding water. This oxide is a paramagnetic yellow solid.

The Oxidizing Strength of the Oxygen Compounds of the Halogens

Elementary fluorine, F_2, is able to oxidize the halogen ions of its congeners to the free halogens, by reactions such as

$$F_2 + 2Cl^- \longrightarrow 2F^- + Cl_2$$

Fluorine is more electronegative than the other elements, and it accordingly is able to take electrons away from the anions of these elements. Similarly, chlorine is able to oxidize both bromide ion and iodide ion, and bromine is able to oxidize iodide ion:

$$Cl_2 + 2Br^- \longrightarrow 2Cl^- + Br_2$$

$$Cl_2 + 2I^- \longrightarrow 2Cl^- + I_2$$

$$Br_2 + 2I^- \longrightarrow 2Br^- + I_2$$

The order of strength as an oxidizing agent for the elementary halogens is accordingly $F_2 > Cl_2 > Br_2 > I_2$.

At first sight there seems to be an anomaly in the reactions involving the free halogens and their oxygen compounds. Thus, although chlorine is able to liberate iodine from iodide ion, iodine is able to liberate chlorine from chlorate ion, according to the reaction

$$I_2 + 2ClO_3^- \longrightarrow 2IO_3^- + Cl_2$$

In this reaction, however, it is to be noted that elementary iodine is acting as a reducing agent, rather than as an oxidizing agent. During the course of the reaction the oxidation number of iodine is increased, from 0 to +5, and that of chlorine is decreased, from +5 to 0. The direction in which the reaction takes place predominantly is accordingly that which would be predicted by the electronegativity scale; iodine, the heavier halogen and less electronegative element, tends to have a high positive oxidation number, and chlorine tends to have a low oxidation number. (Remember that in this case, as in nearly all chemical reactions, we may be dealing with chemical equilibrium. The foregoing statement is to be interpreted as meaning that at equilibrium there are present in the system larger amounts of iodate ion and free chlorine than of chlorate ion and free iodine.)

Similarly, hypochlorite ion, ClO^-, can oxidize bromine to hypobromite ion, and hypobromite ion can oxidize iodine to hypoiodite ion. These regularities fail, however, for the higher oxidation states of bromine: $HBrO_2$, $HBrO_3$, and $HBrO_4$ are very much less stable than their chlorine and iodine analogs. No satisfactory explanation of this property of bromine has been advanced. Selenium and arsenic in their high oxidation states show somewhat similar deviations in properties from their lighter and heavier congeners.

8-2. Oxycompounds of Sulfur, Selenium, and Tellurium

The transargononic oxycompounds of sulfur are more stable than those of chlorine, and those of phosphorus are still more stable. Perchloric acid and the perchlorates are strong oxidizing agents, whereas sulfuric acid and the sulfates are weak oxidizing agents, and phosphoric acid and the phosphates are still weaker. This difference in properties corresponds to the electronegativity values $x = 3$ for Cl, 2.5 for S, 2.1 for P, with Δx (relative to oxygen) = 0.5 for Cl, 1.0 for S, 1.4 for P. The following representative heats of reaction reflect the increasing values of Δx:

$$HCl(g) + 2O_2(g) \longrightarrow HClO_4(l) + 8 \text{ kJ mole}^{-1}$$

$$H_2S(g) + 2O_2(g) \longrightarrow H_2SO_4(l) + 790 \text{ kJ mole}^{-1}$$

$$H_3P(g) + 2O_2(g) \longrightarrow H_3PO_4(l) + 1250 \text{ kJ mole}^{-1}$$

The stable compounds of sulfur, selenium, and tellurium correspond to several values of oxidation number, from –2 to +6, as shown in the chart:

+6	SO_3, H_2SO_4, SF_6	H_2SeO_4, SeF_6	TeO_3, $Te(OH)_6$, TeF_6
+4	SO_2, H_2SO_3	SeO_2, H_2SeO_3	TeO_2
+2			
0	S_8, S_2	Se	Te
−2	H_2S, S^{--}	H_2Se	H_2Te

Oxides of Sulfur

The normal-valence oxide of sulfur, SO, is much less stable than the transargononic oxides SO_2 and SO_3. The heats of formation have the following values:

$$\tfrac{1}{8}S_8(c) + \tfrac{1}{2}O_2(g) \longrightarrow SO(g) - 7 \text{ kJ mole}^{-1}$$

$$\tfrac{1}{8}S_8(c) + O_2(g) \longrightarrow SO_2(g) + 297 \text{ kJ mole}^{-1}$$

$$\tfrac{1}{8}S_8(c) + \tfrac{3}{2}O_2(g) \longrightarrow SO_3(g) + 396 \text{ kJ mole}^{-1}$$

From the first two equations we see that the decomposition of sulfur monoxide to sulfur dioxide and sulfur is strongly exothermic:

$$2SO(g) \longrightarrow \tfrac{1}{8}S_8(c) + SO_2(g) + 311 \text{ kJ mole}^{-1}$$

It is accordingly not surprising that sulfur monoxide is not known as a stable substance, but only as a highly reactive molecule in a very dilute gas or frozen in a matrix. It has the structure $:S\overset{\cdots}{\underset{\cdots}{-}}O:$, with two electrons with parallel spins, thus resembling the molecules O_2 and S_2.

Sulfur dioxide, SO_2, is formed by burning sulfur or a sulfide, such as pyrite (FeS_2):

$$S + O_2 \longrightarrow SO_2$$

$$4FeS_2 + 11O_2 \longrightarrow 2Fe_2O_3 + 8SO_2$$

It is colorless, and has a characteristic choking odor. Its melting point is –75°C and its boiling point –10°C.

Sulfur dioxide is conveniently made in the laboratory by adding a strong acid to solid sodium hydrogen sulfite:

$$H_2SO_4 + NaHSO_3 \longrightarrow NaHSO_4 + H_2O + SO_2$$

It may be purified and dried by bubbling it through concentrated sulfuric acid.

The electronic structure of sulfur dioxide is

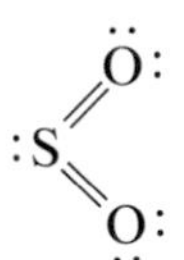

In this structure use is made of one 3*d* orbital, as well as the 3*s* orbital and the three 3*p* orbitals. The observed sulfur-oxygen bond length, 143 pm, is a little less than the value 149 pm expected for a double bond. The angle O—S—O has the value 119.5°.

Sulfur dioxide is used in great quantities in the manufacture of sulfuric acid, sulfurous acid, and sulfites. It destroys fungi and bacteria, and is used as a preservative in the preparation of dried prunes, apricots, and other fruits. A solution of calcium hydrogen sulfite, $Ca(HSO_3)_2$, made by reaction of sulfur dioxide and calcium hydroxide, is used in the manufacture of paper pulp from wood. The solution dissolves lignin, a substance that cements the cellulose fibers together, and liberates these fibers, which are then processed into paper.

Sulfur trioxide, SO_3, is formed in very small quantities when sulfur is burned in air. It is usually made by oxidation of sulfur dioxide by air, in the presence of a catalyst. The reaction of its formation from the elements is exothermic, but less exothermic, per oxygen atom, than that of sulfur dioxide. The equilibrium

$$SO_2(g) + \tfrac{1}{2}O_2(g) \rightleftarrows SO_3(g)$$

is such that at low temperatures a satisfactory yield of SO_3 can be obtained; the reaction proceeds nearly to completion. The rate of the reaction, however, is so small at low temperatures as to make the direct combination of the substances unsuitable as a commercial process. At higher temperatures, at which the rate is satisfactory, the yield is low because of the unfavorable equilibrium.

The solution to this problem was the discovery of certain catalysts (platinum, vanadium pentoxide) that speed up the reaction without affecting the equilibrium. The catalyzed reaction proceeds not in the gaseous mixture, but on the surface of the catalyst, as the gas molecules strike it. In practice, sulfur dioxide, made by burning sulfur or pyrite, is mixed with air and passed over the catalyst at a temperature of 400° to 450°C. About 99% of the sulfur dioxide is converted into sulfur trioxide under these conditions. It is used mainly in the manufacture of sulfuric acid.

Sulfur trioxide is a corrosive gas, which combines vigorously with water to form sulfuric acid:

$$SO_3(g) + H_2O(l) \longrightarrow H_2SO_4(l) + 130 \text{ kJ mole}^{-1}$$

It also dissolves readily in sulfuric acid, to form *oleum* or *fuming sulfuric acid,* which consists mainly of disulfuric acid, $H_2S_2O_7$ (also called pyrosulfuric acid):

$$SO_3 + H_2SO_4 \rightleftarrows H_2S_2O_7$$

Sulfur trioxide condenses at 44.5°C to a colorless liquid, which freezes at 16.8°C to transparent crystals. The substance is polymorphous, these crystals being the unstable form (the α-form). The stable form consists of silky asbestos-like crystals, which are produced when the α-crystals or the liquid stands for some time, especially in the presence of a trace of moisture (Fig. 8-3). There exist also one or more other forms of this substance, which are hard to investigate because the changes from one form to another are very slow. The asbestos-like crystals slowly evaporate to SO_3 vapor at temperatures above 50°C.

The sulfur trioxide molecule, in the gas phase, the liquid, and the α-crystals, has the electronic structure

```
           ..
           O:
          //
  :O=S
   ..     \\
           O:
           ..
```

The molecule is planar, and the bonds have the same length, 143 pm, as in the sulfur dioxide molecule.

The properties of sulfur trioxide may be in large part explained as resulting from the instability of the sulfur-oxygen double bond, relative to two single bonds. Thus by reaction with water one double bond of sulfur trioxide can be replaced by two single bonds, in sulfuric acid:

```
     ..      ..
  H—O:    :O—H
      \   /
        S
      //  \\
   :O      O:
    ..     ..
```

The increased stability of the product is reflected in the large amount of heat evolved in the reaction.

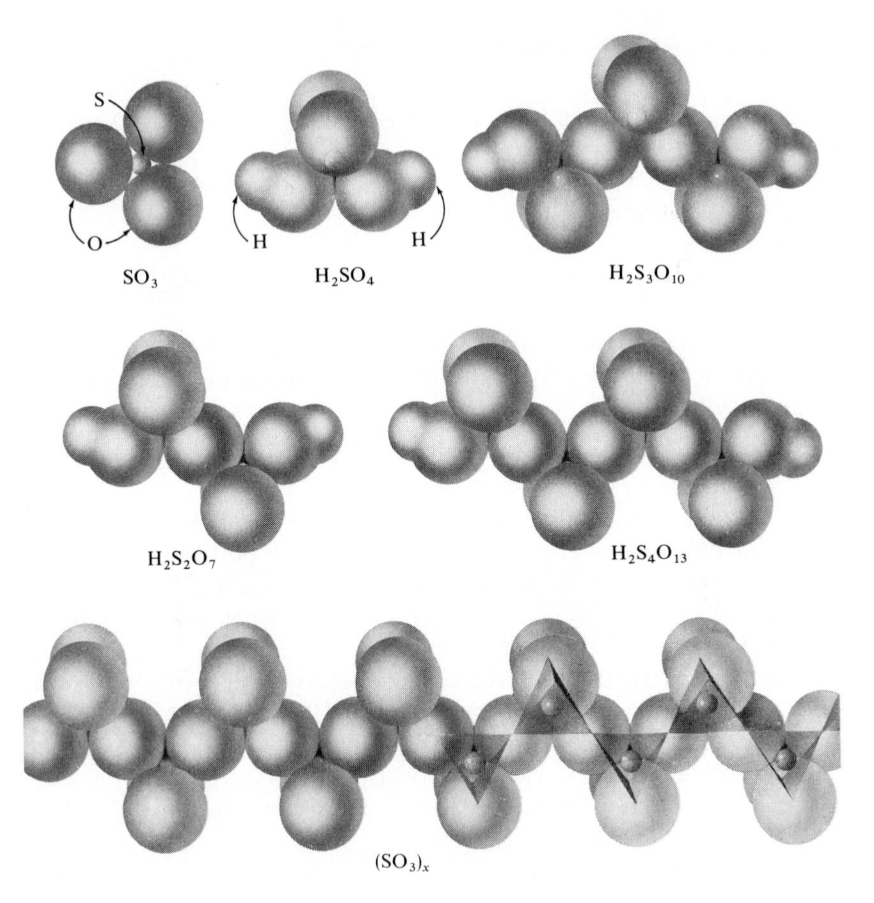

FIGURE 8-3
Sulfur trioxide and some oxygen acids of sulfur.

Sulfurous Acid

A solution of sulfurous acid, H_2SO_3, is obtained by dissolving sulfur dioxide in water. Both sulfurous acid and its salts, the sulfites, are active reducing agents. They form sulfuric acid, H_2SO_4, and sulfates on oxidation by oxygen, the halogens, hydrogen peroxide, and similar oxidizing agents.

The structure of sulfurous acid is

$$\begin{array}{lll} & & :\ddot{O}-H \\ & \diagup & \\ :S & = & \ddot{O}: \\ & \diagdown & \\ & & :\ddot{O}-H \end{array}$$

Sulfuric Acid and the Sulfates

Sulfuric acid, H_2SO_4, is one of the most important of all chemicals, finding use throughout the chemical industry and related industries. It is a heavy, oily liquid (density 1.838 g cm^{-3}), which fumes slightly in air, as the result of the liberation of traces of sulfur trioxide, which then combine with water vapor to form droplets of sulfuric acid. When heated, pure sulfuric acid yields a vapor rich in sulfur trioxide, and then boils, at 338°C, with the constant composition 98% H_2SO_4, 2% water. This is the ordinary "concentrated sulfuric acid" of commerce.

Concentrated sulfuric acid is very corrosive. It has a strong affinity for water, and a large amount of heat is liberated when it is mixed with water, as the result of the formation of hydronium ion:

$$H_2SO_4 + 2H_2O \rightleftarrows 2H_3O^+ + SO_4^{--}$$

In diluting it, the concentrated acid should be poured into water in a thin stream, with stirring; *water should never be poured into the acid*, because it is apt to spatter and throw drops of acid out of the container. The diluted acid occupies a smaller volume than its constituents, the effect being a maximum at $H_2SO_4 + 2H_2O$ $[(H_3O)_2{}^+(SO_4)^{--}]$.

The Chemical Properties and Uses of Sulfuric Acid

The uses of sulfuric acid are determined by its chemical properties, as an acid, a dehydrating agent, and an oxidizing agent.

Sulfuric acid has a high boiling point, 330°C, which permits it to be used with salts of more volatile acids in the preparation of these acids. Nitric acid, for example, can be made by heating sodium nitrate with sulfuric acid:

$$NaNO_3 + H_2SO_4 \rightarrow NaHSO_4 + HNO_3$$

The nitric acid distills off at 86°C. Sulfuric acid is also used for the manufacture of soluble phosphate fertilizers, of ammonium sulfate for use as a fertilizer, and of other sulfates, and in the manufacture of many chemicals and drugs. Steel is usually cleaned of iron rust (is "pickled") by immersion in a bath of sulfuric acid before it is coated with zinc, tin, or enamel. Sulfuric acid is used as the electrolyte in ordinary lead sulfate electric storage cells.

Sulfuric acid has such a strong affinity for water as to make it an effective dehydrating agent. Gases that do not react with the substance may be dried by being bubbled through sulfuric acid. The dehydrating power of the concentrated acid is great enough to cause it to remove hydrogen and oxygen as water from organic compounds, such as sugar:

$$\underset{\text{Sugar (sucrose)}}{C_{12}H_{22}O_{11}} \xrightarrow{H_2SO_4} 12C + 11H_2O$$

Many explosives, such as glyceryl trinitrate (nitroglycerine), are made by reaction of organic substances with nitric acid, producing the explosive substance and water:

$$\underset{\text{Glycerine}}{C_3H_5(OH)_3} + 3HNO_3 \xrightarrow{H_2SO_4} \underset{\text{Glyceryl trinitrate}}{C_3H_5(NO_3)_3} + 3H_2O$$

These reversible reactions are made to proceed to the right by mixing the nitric acid with sulfuric acid, which by its dehydrating action favors the products. Two other examples are given in Section 7-7.

Hot concentrated sulfuric acid is an effective oxidizing agent, the product of its reduction being sulfur dioxide. It will dissolve copper, and will even oxidize carbon:

$$Cu + 2H_2SO_4 \rightarrow CuSO_4 + 2H_2O + SO_2$$

$$C + 2H_2SO_4 \rightarrow CO_2 + 2H_2O + 2SO_2$$

The solution of copper by hot concentrated sulfuric acid illustrates a general reaction—*the solution of an unreactive metal in an acid under the influence of an oxidizing agent.* The reactive metals are oxidized to their cations by hydrogen ion, which is itself reduced to elementary hydrogen; for example,

$$Zn + 2H^+ \rightarrow Zn^{++} + H_2(g)$$

Copper does not undergo this reaction. It can be oxidized to cupric ion, however, by a stronger oxidizing agent, such as chlorine or nitric acid or, as illustrated above, hot concentrated sulfuric acid.

Sulfates

Sulfuric acid combines with bases to form normal sulfates, such as K_2SO_4, potassium sulfate, and hydrogen sulfates or acid sulfates, such as $KHSO_4$, potassium hydrogen sulfate.

The less soluble sulfates occur as minerals: these include $CaSO_4 \cdot 2H_2O$ (gypsum), $SrSO_4$, $BaSO_4$ (barite), and $PbSO_4$. Barium sulfate is the least soluble of the sulfates, and its formation as a white precipitate is used as a test for sulfate ion.

Common soluble sulfates include $Na_2SO_4 \cdot 10H_2O$, $(NH_4)_2SO_4$, $MgSO_4 \cdot 7H_2O$ (Epsom salt), $CuSO_4 \cdot 5H_2O$ (blue vitriol), $FeSO_4 \cdot 7H_2O$, $(NH_4)_2Fe(SO_4)_2 \cdot 6H_2O$ (a well-crystallized, easily purified salt used in analytical chemistry in making standard solutions of ferrous ion), $ZnSO_4 \cdot 7H_2O$, $KAl(SO_4)_2 \cdot 12H_2O$ (alum), $NH_4Al(SO_4)_2 \cdot 12H_2O$ (ammonium alum), and $KCr(SO_4)_2 \cdot 12H_2O$ (chrome alum).

The Thio or Sulfo Acids

Sodium thiosulfate, $Na_2S_2O_3 \cdot 5H_2O$ (incorrectly called "hypo," from an old name, sodium hyposulfite), is a substance used in photography. It is made by boiling a solution of sodium sulfite with free sulfur:

$$\underset{\text{Sulfite ion}}{SO_3^{--}} + S \rightarrow \underset{\text{Thiosulfate ion}}{S_2O_3^{--}}$$

Thiosulfuric acid, $H_2S_2O_3$, is unstable, and sulfur dioxide and sulfur are formed when a thiosulfate is treated with acid.

The structure of the thiosulfate ion, $S_2O_3^{--}$, is interesting in that the two sulfur atoms are not equivalent. This ion is a sulfate ion, SO_4^{--}, in which one of the oxygen atoms has been replaced by a sulfur atom (Fig. 8-4). The central sulfur atom may be assigned oxidation number +6, and the attached sulfur atom oxidation number −2.

Thiosulfate ion is easily oxidized, especially by iodine, to tetrathionate ion, $S_4O_6^{--}$:

$$2S_2O_3^{--} \rightarrow S_4O_6^{--} + 2e^-$$

or

$$2S_2O_3^{--} + I_2 \rightarrow S_4O_6^{--} + 2I^-$$

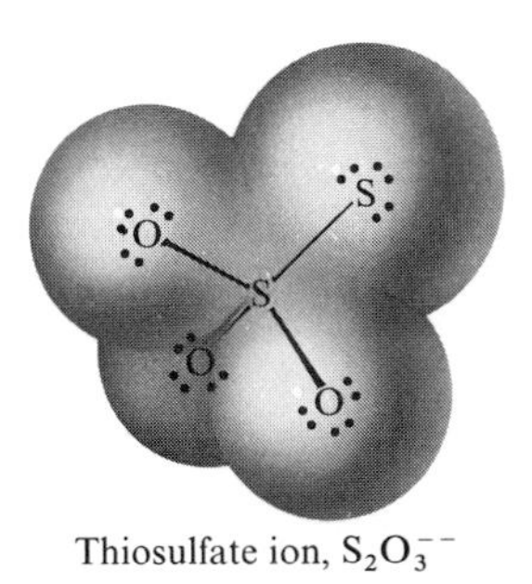

Thiosulfate ion, $S_2O_3^{--}$

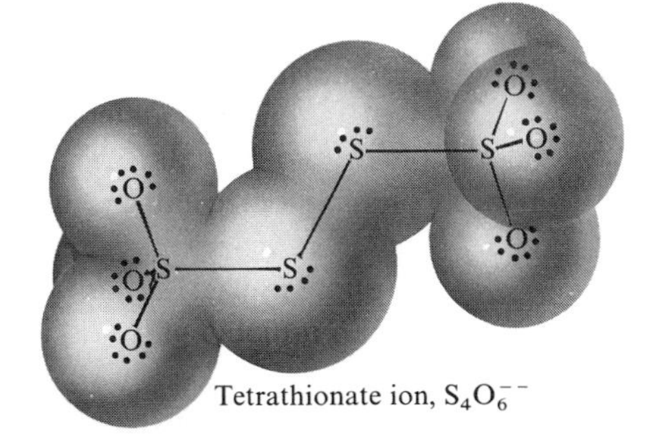

Tetrathionate ion, $S_4O_6^{--}$

FIGURE 8-4
The thiosulfate ion and the tetrathionate ion.

This reaction, between thiosulfate ion and iodine, is very useful in the quantitative analysis of oxidizing and reducing agents.

Selenium and Tellurium

The transargononic compounds of selenium closely resemble those of sulfur. The selenates, salts of selenic acid, H_2SeO_4, are much like the sulfates. Telluric acid, however, has the formula $Te(OH)_6$, in which the large central atom has increased its ligancy from 4 to 6, as the iodine atom does in H_5IO_6.

8-3. Oxycompounds of Phosphorus, Arsenic, Antimony, and Bismuth

Phosphorus and its heavier congeners form stable compounds corresponding to various values of the oxidation number between −3 and +5, as shown in the following chart:

	P	As	Sb	Bi
+5	P_4O_{10}, H_3PO_4, PCl_5	As_2O_5, H_3AsO_4	Sb_2O_5, $HSb(OH)_6$, $SbCl_5$	Bi_2O_5
+4				
+3	P_4O_6, H_2HPO_3, PCl_3	As_4O_6, H_3AsO_3, $AsCl_3$	Sb_4O_6, H_3SbO_3, $SbCl_3$, Sb^{+++}	Bi_4O_6, $BiCl_3$, Bi^{+++}
+2				
+1	HH_2PO_2			
0	P_4	As	Sb	Bi
−1				
−2	P_2H_4			
−3	PH_3, PH_4^+	AsH_3	SbH_3	BiH_3

In addition, the properties of a number of highly reactive simple molecules, including PH, PH_2, PO, PS, and PN, have been determined by spectroscopic studies.

Oxides of Phosphorus

Tetraphosphorus hexoxide, P_4O_6, is made by burning phosphorus in a limited supply of oxygen or air. It has melting point 22.5°C and boiling point 173.1°C. Its molecular structure, shown in Figure 8-5, is that of a normal-valence compound. Its standard enthalpy, −2145 kJ mole^{-1} for

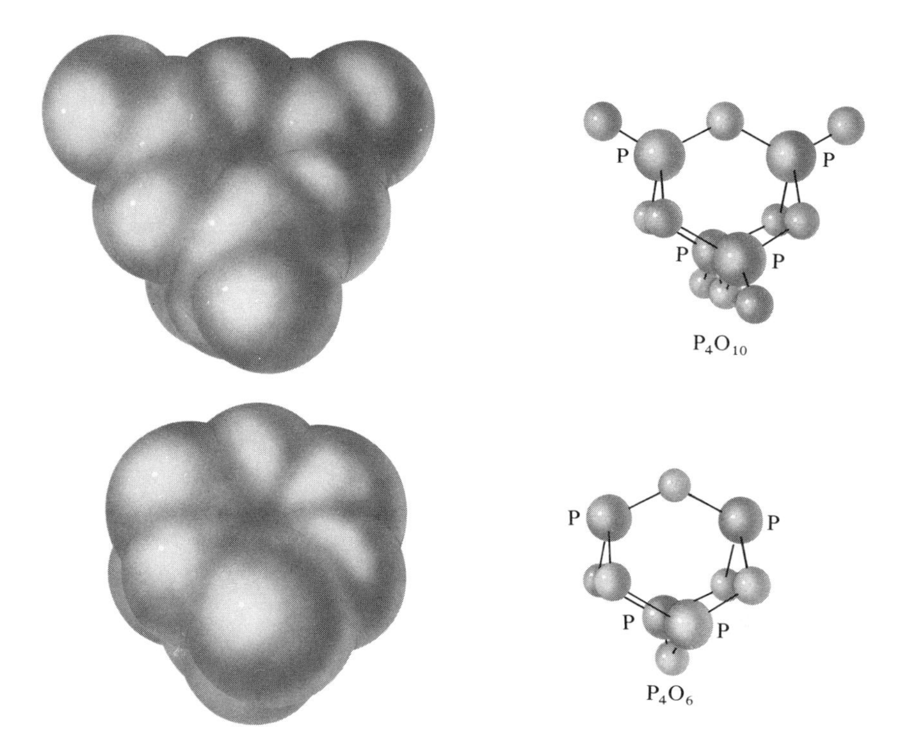

FIGURE 8-5
Molecules of the oxides of phosphorus.

$P_4O_6(g)$, leads to 415 kJ mole^{-1} for the P—O bond energy. The observed bond length is 166 pm, with bond angles 99° (O—P—O) and 128° (P—O—P).

Tetraphosphorus decoxide, P_4O_{10}, is formed when phosphorus is burned with a free supply of air. It reacts with water with great violence, to form phosphoric acid, and it is used in the laboratory as a drying agent for gases; it is the most efficient drying agent known. Its standard enthalpy, −2834 kJ mole^{-1} for $P_4O_{10}(g)$, corresponds to the value 584 kJ mole^{-1} for the energy of the transargononic P═O double bond:

$$P_4O_6(g) + 4O(g) \longrightarrow P_4O_{10}(g) + 4 \times 584 \text{ kJ mole}^{-1}$$

The molecule has the structure shown in Figure 8-5, with bond lengths P—O = 160 pm, P═O = 140 pm.

Phosphoric Acid

The most important acid of phosphorus is phosphoric acid, H_3PO_4 (also called *orthophosphoric acid*). Pure phosphoric acid is a deliquescent crystalline substance with melting point 42°C. It is made by dissolving tetraphosphorus decoxide in water. Commercial phosphoric acid (86% H_3PO_4) is a viscous liquid.

Phosphoric acid is a weak acid. It is a stable substance, without effective oxidizing power.

Orthophosphoric acid forms three series of salts, with one, two, and three of its hydrogen atoms replaced by metal atoms. The salts are usually made by mixing phosphoric acid and the metal hydroxide or carbonate, in proper proportion. Sodium dihydrogen phosphate, NaH_2PO_4, is slightly acidic in reaction. It is used (mixed with sodium hydrogen carbonate) in baking powder, and also for treating boiler water to prevent formation of scale. Disodium hydrogen phosphate, Na_2HPO_4, is slightly basic in reaction. Trisodium phosphate, Na_3PO_4, is strongly basic. It is used as a detergent (for cleaning woodwork and other surfaces) and for treating boiler water.

Phosphates are valuable fertilizers. Phosphate rock itself (tricalcium phosphate, $Ca_3(PO_4)_2$, and hydroxy-apatite) is too slightly soluble to serve as an effective source of phosphorus for plants. It is accordingly converted into the more soluble substance calcium dihydrogen phosphate, $Ca(H_2PO_4)_2$. This may be done by treatment with sulfuric acid:

$$Ca_3(PO_4)_2 + 2H_2SO_4 \rightarrow 2CaSO_4 + Ca(H_2PO_4)_2$$

Enough water is added to convert the calcium sulfate to its dihydrate, gypsum, and the mixture of gypsum and calcium dihydrogen phosphate is sold as "superphosphate of lime." Sometimes the phosphate rock is treated with phosphoric acid:

$$Ca_3(PO_4)_2 + 4H_3PO_4 \rightarrow 3Ca(H_2PO_4)_2$$

This product is much richer in phosphorus than the "superphosphate"; it is called "triple phosphate." Over ten million tons of phosphate rock is converted into phosphate fertilizer each year. Ammonium dihydrogen phosphate, $NH_4H_2PO_4$, has recently come into use.

Phosphoric acid is present in the nucleic acids, which are of fundamental importance in the biological process of reproduction of living organisms.

The Condensed Phosphoric Acids

Phosphoric acid easily undergoes the process of *condensation*. Condensation is the reaction of two or more molecules to form larger molecules, either without any other products (in which case the condensation is also called *polymerization*), or with the elimination of small molecules, such as water. Condensation of two phosphoric acid molecules occurs by the reaction of two hydroxyl groups to form water and an oxygen atom held by single bonds to two phosphorus atoms.

When orthophosphoric acid is heated it loses water and condenses to *diphosphoric acid* or *pyrophosphoric acid*, $H_4P_2O_7$:

$$2H_3PO_4 \rightleftarrows H_4P_2O_7 + H_2O$$

(The name pyrophosphoric acid is the one customarily used.) This acid is a white crystalline substance, with melting point 61°C. Its salts may be made by neutralization of the acid or by strongly heating the hydrogen orthophosphates or ammonium orthophosphates of the metals. Magnesium pyrophosphate, $Mg_2P_2O_7$, is obtained in a useful method for quantitative analysis for either magnesium or orthophosphate. A solution containing orthophosphate ion may be mixed with a solution of magnesium chloride (or sulfate), ammonium chloride, and ammonium hydroxide. The very slightly soluble substance magnesium ammonium phosphate, $MgNH_4PO_4{\cdot}6H_2O$, then slowly precipitates. The precipitate is washed with dilute ammonium hydroxide, dried, and heated to a dull red heat, causing it to form magnesium pyrophosphate, which is then weighed:

$$2MgNH_4PO_4{\cdot}6H_2O \longrightarrow Mg_2P_2O_7 + 2NH_3 + 13H_2O$$

Larger condensed phosphoric acids also occur, such as *triphosphoric acid*, $H_5P_3O_{10}$. The interconversion of triphosphates, pyrophosphates, and phosphates is important in many bodily processes, including the absorption and metabolism of sugar. These reactions occur at body temperature under the influence of special enzymes.

An important class of condensed phosphoric acids is that in which each phosphate tetrahedron is bonded by oxygen atoms to two other tetrahedra. These acids have the composition $(HPO_3)_x$, with $x = 3, 4, 5, 6, \ldots$. They are called the *metaphosphoric acids*. Among these acids are tetrametaphosphoric acid and hexametaphosphoric acid.

Metaphosphoric acid is made by heating orthophosphoric acid or pyrophosphoric acid or by adding water to phosphorus pentoxide. It is a viscous sticky mass, which contains, in addition to ring molecules such as $H_4P_4O_{12}$, long chains approaching $(HPO_3)_\infty$ in composition. It is the long chains, which may also be condensed together to form branched chains, that, by becoming entangled, make the acid viscous and sticky.

The metaphosphates are used as water softeners. Sodium hexametaphosphate, $Na_6P_6O_{18}$, is especially effective for this purpose.

Oxidation-reduction Properties of Phosphorus Compounds

The compounds of phosphorus differ in a striking way from the analogous compounds of chlorine in their oxidation-reduction properties (those of sulfur are intermediate). Thus the highest oxygen acid of phosphorus, H_3PO_4, is stable, and not an oxidizing agent, whereas that of chlorine, $HClO_4$, is a very strong oxidizing agent. The lower oxygen acids of phosphorus are strong reducing agents, and those of chlorine are strong oxidizing agents. The phosphide ion, P^{---}, is such a strong reducing agent that it cannot be obtained; sulfide ion, S^{--}, is a strong reducing agent; but chloride ion is stable.

These differences in properties may be accounted for in terms of the different electronegativities of the elements – 3.0 for Cl, 2.5 for S, and 2.1 for P – which lead to increasing stability of the bonds with oxygen in the sequence Cl, S, P. The energies of addition of four oxygen atoms in the corresponding reactions of oxidation from the lowest to the highest oxidation states are the following:*

$$Cl^-(aq) + 4O(g) \longrightarrow ClO_4^-(aq) + 4 \times 240 \text{ kJ mole}^{-1}$$

$$S^{--}(aq) + 4O(g) \longrightarrow SO_4^{--}(aq) + 4 \times 485 \text{ kJ mole}^{-1}$$

$$P^{---}(aq) + 4O(g) \longrightarrow PO_4^{---}(aq) + 4 \times 620 \text{ kJ mole}^{-1}$$

These values correspond to the average energy 240 kJ mole^{-1} for the $Cl{=}\ddot{O}\colon$ bond, 485 kJ mole^{-1} for the $S{=}\ddot{O}\colon$ bond, and 620 kJ mole^{-1} for the $P{=}\ddot{O}\colon$ bond; the increase from chlorine to phosphorus is roughly as expected from the increasing amount of ionic character of the bonds.

8-4. High-energy Molecules and High-energy Bonds

Phosphoric acid has great importance for living organisms. It is, for example, an essential part of the nucleic acids DNA and RNA that are involved in the hereditary process and the functioning of cells (Section 15-6), and also an essential part of many coenzymes and other molecules that participate in biochemical reactions (Sections 14-4, 14-6). The poly-

*The values are obtained from the standard enthalpy values given in Appendix VI, with use of the value 200 kJ mole^{-1} for P^{---}(aq) estimated from the values for PH_3, H_2S, S^{--}, HCl, and Cl^-.

phosphoric acids, especially diphosphoric acid and triphosphoric acid, are also very important. The compound ATP (adenosine triphosphate) provides the energy for most of the endergonic (usually endothermic, Chapter 14) reactions that take place in the human body and in other organisms.

ATP has the formula

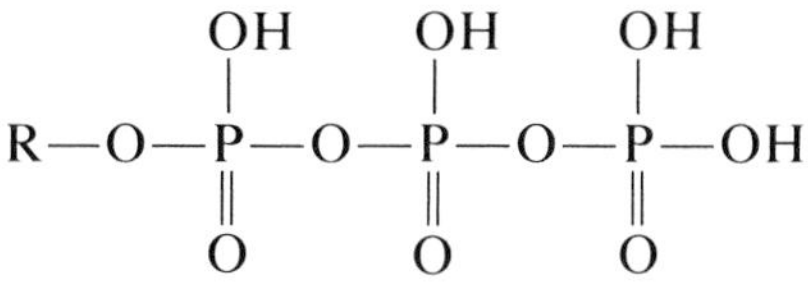

```
       OH      OH      OH
       |       |       |
R—O—P—O—P—O—P—OH
       ||      ||      ||
       O       O       O
```

in which R is the adenosine radical $C_{10}H_{12}N_5O_3$ (Section 15-6). It is formed from ADP, adenosine diphosphate, which has the formula

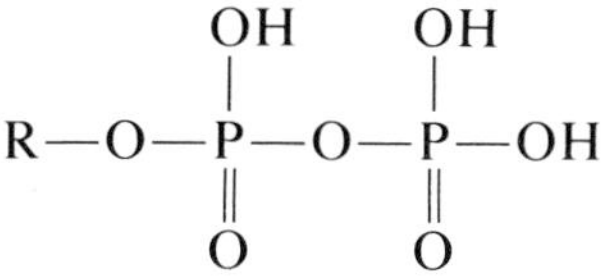

```
       OH      OH
       |       |
R—O—P—O—P—OH
       ||      ||
       O       O
```

and phosphoric acid, H_3PO_4, by the reaction

$$RP_2O_7H_3 + H_3PO_4 \longrightarrow RP_3O_{10}H_4 + H_2O \qquad (8\text{-}1)$$

or from AMP (adenosine monophosphate) and pyrophosphoric acid, $H_4P_2O_7$, by the reaction

$$RPO_4H_2 + H_4P_2O_7 \longrightarrow RP_3O_{10}H_4 + H_2O \qquad (8\text{-}2)$$

These reactions are endergonic. The energy required to make them take place comes from sunlight, in the case of plants (Section 14-6), and from the oxidation of carbohydrates, fats, and proteins, in the case of animals (Section 14-7). ATP liberates energy when it undergoes either one of the reverse reactions:

$$\text{ATP} + H_2O \longrightarrow \text{ADP} + H_3PO_4 \qquad (8\text{-}3)$$

$$\text{ATP} + H_2O \longrightarrow \text{AMP} + H_4P_2O_7 \qquad (8\text{-}4)$$

(In physiological solutions the ATP and ADP are largely ionized (Chapter 12), and the above reactions involve the ions.) These reactions are catalyzed by any one of a large number of enzymes, which also at the same time catalyze an endergonic biochemical reaction in such a way that the energy liberated by the transformation of ATP to ADP or AMP is transferred to the endergonic reaction, permitting it to take place. It is justified to say that these reactions are among the most important of all chemical reactions that take place in the human body.

It is customary to refer to ATP and to other molecules that can undergo reactions with liberation of a considerable amount of energy (free energy) as *energy-rich molecules*. Often a structural feature (a chemical bond or two or more bonds related to one another in a particular way) in an energy-rich molecule can be identified as especially important in determining the high energy content of the molecule. Such features are called *energy-rich bonds*. In ATP the energy-rich feature is two phosphorus-oxygen bonds in juxtaposition, P—O—P.

Examples of Energy-rich Molecules

It has been found by experiment that compounds containing N—N bonds are less stable than elementary nitrogen, :N≡N:. Hence the single bond between two nitrogen atoms is energy-rich relative to a triple bond between these atoms. In consequence the hydrazine molecule, $H_2N{-}NH_2$, is energy-rich relative to nitrogen and ammonia:

$$3N_2H_4 \longrightarrow N_2 + 4NH_3$$

In this reaction there is no change in the number of N—H bonds, but three N—N bonds are converted to one N≡N bond. The bond energies (Tables V-1 and V-2) are 159 kJ mole^{-1} for N—N and 946 kJ mole^{-1} for N≡N. These values lead to $946 - 3 \times 159 = 469$ kJ mole^{-1} as the expected amount of heat liberated in the reaction, in close agreement with the experimental value. The energy-rich character of the N—N bond is one of the reasons for the use of hydrazine as a rocket fuel.

The high bond energy of the triple bond in N_2 makes many compounds of nitrogen energy-rich relative to products including N_2. For example, nitrogen trichloride (Section 6-11) may explode to form elementary chlorine and nitrogen. The N—Cl single bond is not stabilized by partial ionic character (Section 6-12), because nitrogen and chlorine have the same electronegativity, and heat is evolved in the same amount, 469 kJ mole^{-1}, as in the decomposition of hydrazine. Nitrogen trifluoride, on the other hand, is not an energy-rich molecule; the stabilization of the N—F bonds by the partial ionic character overcomes the effect of the special stability of the N≡N bond.

The acetylene molecule, $HC{\equiv}CH$, is an energy-rich molecule because the carbon-carbon triple bond is less stable than carbon-carbon single bonds (Tables V-1 and V-2). Liquid acetylene sometimes explodes with violence, forming graphite and methane or other hydrocarbons.

Why ATP Is Energy-rich

The energy-rich character of ATP can be understood by consideration of the electroneutrality principle (Section 6-13) and the theory of resonance (Section 6-8).

For the phosphate ion, PO_4^{---}, we write the transargononic structure

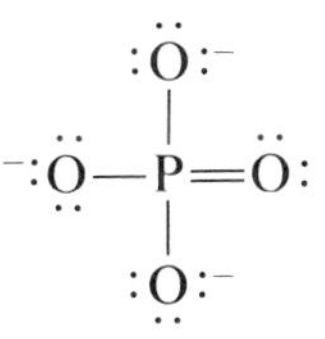

$$\begin{array}{ccccc} & & :\ddot{O}:^- & & \\ & & | & & \\ ^-:\ddot{\underset{..}{O}} & - & P & = & \ddot{O}: \\ & & | & & \\ & & :\underset{..}{O}:^- & & \end{array}$$

This structure corresponds to zero formal charge on the phosphorus atom. In aqueous solution the formation of hydrogen bonds (Section 9-6) permits the negative charges to be transferred to many oxygen atoms of surrounding water molecules, so that no atom has a charge much different from zero. Also, the four oxygen atoms may be similarly hydrated, so that the double bond can resonate among all four structures, stabilizing the hydrated ion by the resonance energy:

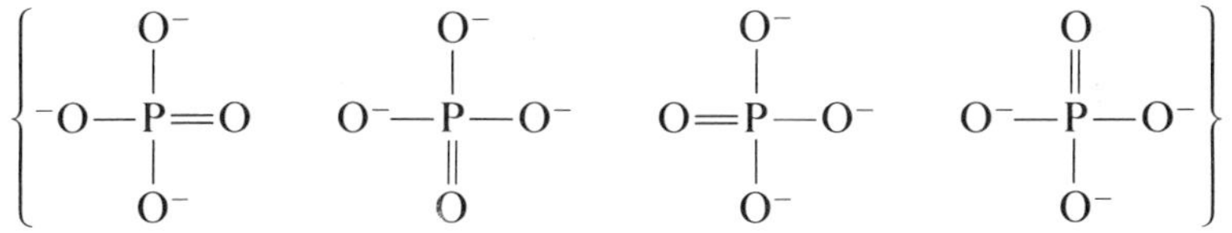

$$\left\{\begin{array}{ccc} & O^- & \\ & | & \\ ^-O- & P & =O \\ & | & \\ & O^- & \end{array}\quad \begin{array}{ccc} & O^- & \\ & | & \\ O^-- & P & -O^- \\ & \| & \\ & O & \end{array}\quad \begin{array}{ccc} & O^- & \\ & | & \\ O= & P & -O^- \\ & | & \\ & O^- & \end{array}\quad \begin{array}{ccc} & O & \\ & \| & \\ O^-- & P & -O^- \\ & | & \\ & O^- & \end{array}\right\}$$

Similarly, for H_3PO_4, $H_2PO_4^-$, and HPO_4^{--}, the hydrogen bonds can so adjust themselves (Section 9-8) that the double bond resonates equally among the four oxygen atoms. We may say that each oxygen atom forms a $1\frac{1}{4}$ bond with the phosphorus atom and a $\frac{3}{4}$ bond with one or more hydrogen atoms.

In a polyphosphate, however, the oxygen atom of a P—O—P bond is restrained from being involved in the resonance of the double bonds. Each of the two structures $P{=}\ddot{O}{-}P$ and $P{-}\ddot{O}{=}P$ gives oxygen the formal charge +1, and the structure $P{=}O{=}P$ gives it the formal charge +2. Both of these deviations from electroneutrality correspond to instability. In consequence, the double bonds are inhibited from resonating into these positions, and there is a corresponding decrease in the amount of stabilization by resonance energy, and the molecule containing the P—O—P arrangement of bond becomes an energy-rich molecule.

This effect of inhibition of resonance is seen clearly in the observed bond lengths. In salts of phosphoric acid in which the four oxygen atoms are equivalent the bond length is 154 pm, corresponding to the expected amount of double-bond character. In polyphosphates, however, the P—O—P bonds are about 162 pm long, corresponding to a much smaller amount of double-bond character, and the other bonds are 149 pm long.

We may say that for maximum stability each phosphorus-oxygen bond should have strength $1\frac{1}{4}$ corresponding to complete resonance, and that the strength of the bond reaching the oxygen atom should add up to 2, to satisfy the valence of oxygen. For P—O—P this sum is $2\frac{1}{2}$, and the deviation from 2 gives high-energy character to molecules containing this structure.

Silicon, which has valence 4 and can form four bonds with strength 1, would be expected to form Si—O—Si bonds with no excess energy. In fact, the polysilicic acids, up to the limit SiO_2, are as stable as silicic acid itself. On the other hand, the strength $1\frac{1}{2}$ for four equivalent sulfur-oxygen bonds (valence 6 for sulfur) leads to the sum 3 for S—O—S, twice as great a deviation from the value 2 required to saturate the valence of oxygen as in the case of P—O—P. The experimental values of the heat evolved in the hydrolysis of the three di-acids and the corresponding oxide of chlorine, Cl_2O_7 (with a still larger deviation from electroneutrality) are the following:*

$$H_6Si_2O_7(c) + H_2O(l) \rightarrow 2H_4SiO_4(c) + 0 \text{ kJ mole}^{-1}$$

$$H_4P_2O_7(c) + H_2O(l) \rightarrow 2H_3PO_4(c) + 21 \text{ kJ mole}^{-1}$$

$$H_2S_2O_7(c) + H_2O(l) \rightarrow 2H_2SO_4(c) + 70 \text{ kJ mole}^{-1}$$

$$Cl_2O_7(g) + H_2O(l) \rightarrow 2HClO_4(l) + 114 \text{ kJ mole}^{-1}$$

ATP contains two P—O—P groups, and ADP contains one. Hence ATP is a high-energy molecule relative to ADP and phosphoric acid, or to AMP (adenosine monophosphate) and diphosphoric acid, and ADP is a high-energy molecule relative to AMP and phosphoric acid.

A further discussion is to be found in Section 14-6.

*The deviations from electroneutrality are 0, $\frac{1}{2}$, 1, and $\frac{3}{2}$, respectively; that is, in the ratios 0:1:2:3. Since negative as well as positive deviations lead to instability, the amount of energy in the energy-rich bonds would not be proportional to the deviation. As a first approximation, it might be proportional to the square of the deviation. This relation holds reasonably well except for Cl—O—Cl.

8-5. Oxycompounds of Nitrogen

Compounds of nitrogen are known representing all oxidation levels from -3 to $+5$. Some of these compounds are shown in the following chart:

$+5$	N_2O_5, dinitrogen pentoxide	HNO_3, nitric acid
$+4$	NO_2, nitrogen dioxide N_2O_4, dinitrogen tetroxide	
$+3$	N_2O_3, dinitrogen trioxide	HNO_2, nitrous acid
$+2$	NO, nitric oxide	
$+1$	N_2O, nitrous oxide	$H_2N_2O_2$, hyponitrous acid
0	N_2, free nitrogen	
-1	NH_2OH, hydroxylamine	
-2	N_2H_4, hydrazine	
-3	NH_3, ammonia	NH_4^+, ammonium ion

The Oxides of Nitrogen

Nitrous oxide, N_2O, is made by heating ammonium nitrate:

$$NH_4NO_3 \longrightarrow 2H_2O + N_2O$$

It is a colorless, odorless gas, which has the power of supporting combustion, by giving up its atom of oxygen, leaving molecular nitrogen. When breathed for a short time the gas causes hysteria; this effect, discovered in 1799 by Humphry Davy, led to the use of the name *laughing gas* for the substance. Longer inhalation causes unconsciousness, and the gas, mixed with air or oxygen, is used as a general anesthetic for minor operations. The gas also finds use in making whipped cream; under pressure it dissolves in the cream, and when the pressure is released it fills the cream with many small bubbles, simulating ordinary whipped cream.

The electronic structure of nitrous oxide is

$$\left\{ :\ddot{N}=N=\ddot{O}: \qquad :N\equiv N-\underset{..}{\ddot{O}}: \right\}$$

The position of the oxygen atom at the end of the linear molecule explains the ease with which nitrous oxide acts as an oxidizing agent, with liberation of $:N\equiv N:$.

Nitric oxide, NO, can be made by reduction of dilute nitric acid with copper or mercury:

$$3Cu + 8H^+ + 2NO_3^- \rightarrow 3Cu^{++} + 4H_2O + 2NO$$

When made in this way the gas usually contains impurities such as nitrogen and nitrogen dioxide. If the gas is collected over water, in which it is only slightly soluble, the nitrogen dioxide is removed by solution in the water.

A metal or other reducing agent may reduce nitric acid to any lower stage of oxidation, producing nitrogen dioxide, nitrous acid, nitric oxide, nitrous oxide, nitrogen, hydroxylamine, hydrazine, or ammonia (ammonium ion), depending upon the conditions of the reduction. Conditions may be found that strongly favor one product, but usually appreciable amounts of other products are also formed. Nitric oxide is produced preferentially under the conditions mentioned above.

Nitric oxide is a colorless, difficultly condensable gas (b.p. −151.7°C, m.p. −163.6°C). It combines readily with oxygen to form the red gas nitrogen dioxide, NO_2.

Dinitrogen trioxide, N_2O_3, can be obtained as a blue liquid by cooling an equimolal mixture of nitric oxide and nitrogen dioxide. It is the anhydride of nitrous acid, and produces this acid on solution in water:

$$N_2O_3 + H_2O \rightarrow 2HNO_2$$

Nitrogen dioxide, NO_2, a red gas, and its dimer *dinitrogen tetroxide,* N_2O_4, a colorless, easily condensable gas, exist in equilibrium with one another:

$$\underset{\text{Red}}{2NO_2} \rightleftarrows \underset{\text{Colorless}}{N_2O_4}$$

The mixture of these gases may be made by adding nitric oxide to oxygen, or by reducing concentrated nitric acid with copper:

$$Cu + 4H^+ + 2NO_3^- \rightarrow Cu^{++} + 2H_2O + 2NO_2$$

It is also easily obtained by decomposing lead nitrate by heat:

$$2Pb(NO_3)_2 \rightarrow 2PbO + 4NO_2 + O_2$$

The gas dissolves readily in water or alkali, forming a mixture of nitrate ion and nitrite ion.

Dinitrogen pentoxide, N_2O_5, the anhydride of nitric acid, can be made, as white crystals, by carefully dehydrating nitric acid with tetraphosphorus decoxide or by oxidizing nitrogen dioxide with ozone. It is unstable, decomposing spontaneously at room temperature into nitrogen dioxide and oxygen.

The electronic structures of the oxides of nitrogen are shown below. Most of these molecules are resonance hybrids, and the contributing structures are not all shown; for dinitrogen pentoxide, for example, the single and double bonds may change places.*

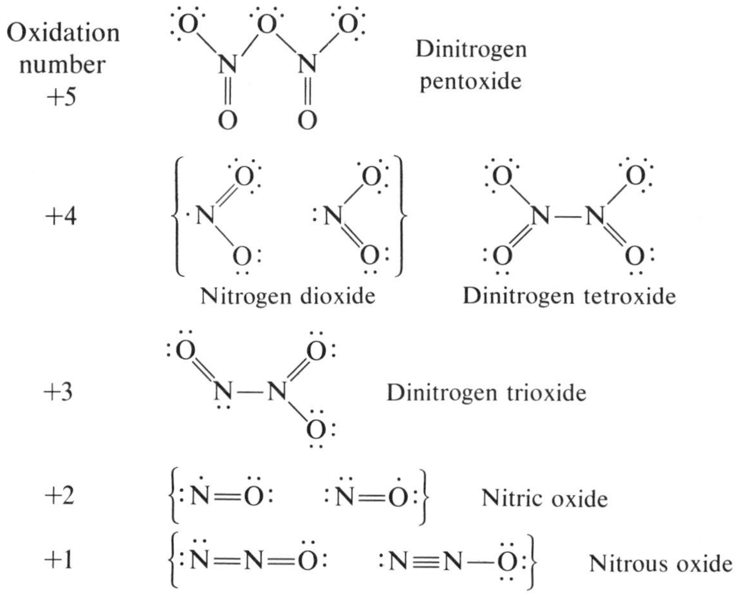

Oxidation number	Structure	Name
+5	:Ö: N(=O)—Ö—N(=O)—Ö:	Dinitrogen pentoxide
+4	{·N(=Ö:)—Ö: :N(—Ö:)=Ö:}	Nitrogen dioxide
+4	O₂N—NO₂	Dinitrogen tetroxide
+3	:Ö=N̈—N(=Ö:)—Ö:	Dinitrogen trioxide
+2	{:Ṅ=Ö: :N̈=Ȯ:}	Nitric oxide
+1	{:N̈=N=Ö: :N≡N—Ö:}	Nitrous oxide

Nitric Acid and the Nitrates

Nitric acid, HNO_3, is a colorless liquid with melting point $-42°C$, boiling point 86°C, and density 1.52 g cm^{-3}. It is a strong acid, completely ionized to hydrogen ion and nitrate ion (NO_3^-) in aqueous solution; and it is a strong oxidizing agent. It attacks the skin, and gives it a yellow color.

Nitric acid can be made in the laboratory by heating sodium nitrate with sulfuric acid in an all-glass apparatus:

$$NaNO_3 + H_2SO_4 \longrightarrow NaHSO_4 + HNO_3$$

---

*The structure shown for dinitrogen pentoxide is that for the gas molecule. The crystal contains NO_2^+ and NO_3^- ions.

The substance is also made commercially in this way, from natural sodium nitrate (Chile saltpeter). Much nitric acid is made by the oxidation of ammonia.

Sodium nitrate, $NaNO_3$, forms colorless crystals closely resembling crystals of calcite, $CaCO_3$. This resemblance is not accidental. The crystals have the same structure, with Na^+ replacing Ca^{++} and NO_3^- replacing CO_3^{--}. The crystals of sodium nitrate have the same property of birefringence (double refraction) as calcite. Sodium nitrate is used as a fertilizer and for conversion into nitric acid and other nitrates. *Potassium nitrate,* KNO_3 *(saltpeter),* is used in pickling meat (ham, corned beef), in medicine, and in the manufacture of *gunpowder,* which is an intimate mixture of potassium nitrate, charcoal, and sulfur, which explodes when ignited in a closed space.

The nitrate ion has a planar structure, with each bond a hybrid of a single bond and a double bond:

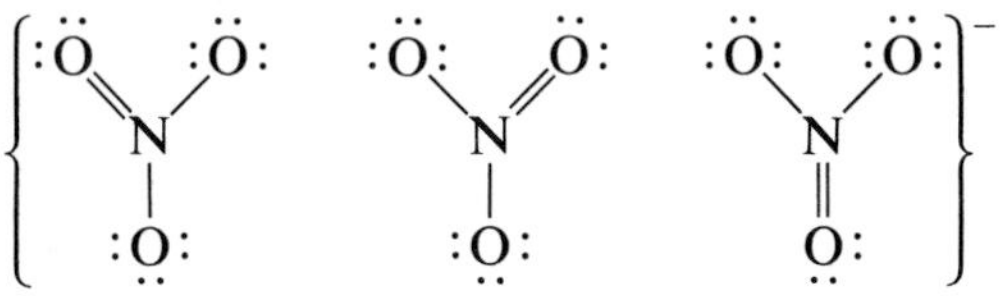

The nitrates of all metals are soluble in water.

Nitrous Acid and the Nitrites

Nitrous acid, HNO_2, forms in small quantity together with nitric acid when nitrogen dioxide is dissolved in water. Nitrite ion can be made together with nitrate ion by solution of nitrogen dioxide in alkali:

$$2NO_2 + 2OH^- \longrightarrow NO_2^- + NO_3^- + H_2O$$

Sodium nitrite, $NaNO_2$, and *potassium nitrite,* KNO_2, can be made also by decomposing the nitrates by heat:

$$2NaNO_3 \longrightarrow 2NaNO_2 + O_2$$

or by reduction with lead:

$$NaNO_3 + Pb \longrightarrow NaNO_2 + PbO$$

These nitrites are slightly yellow crystalline substances, and their solutions are yellow. They are used in the manufacture of dyes and in the chemical laboratory.

The nitrite ion is a reducing agent, being oxidized to nitrate ion by bromine, permanganate ion, chromate ion, and similar oxidizing agents. It is also itself an oxidizing agent, able to oxidize iodide ion to iodine. This property may be used, with the starch test (blue color) for iodine, to distinguish nitrite from nitrate ion, which does not oxidize iodide ion readily.

The electronic structure of the nitrite ion is

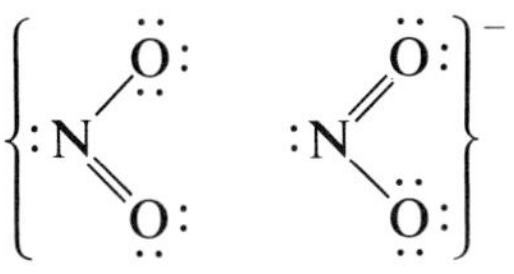

$$\left\{ :\mathrm{N}\begin{matrix} \diagup \ddot{\mathrm{O}}: \\ \diagdown\!\!\diagdown \ddot{\mathrm{O}}: \end{matrix} \quad :\mathrm{N}\begin{matrix} /\!\!/ \ddot{\mathrm{O}}: \\ \diagdown \ddot{\mathrm{O}}: \end{matrix} \right\}^-$$

Other Compounds of Nitrogen

Hydrogen cyanide, HCN (structural formula $H{-}C{\equiv}N{:}$), is a gas which dissolves in water and acts as a very weak acid. It is made by treating a cyanide, such as *potassium cyanide,* KCN, with sulfuric acid, and is used as a fumigant and rat poison. It smells like bitter almonds and crushed fruit kernels, which in fact owe their odor to it. Hydrogen cyanide and its salts are very poisonous.

Cyanides are made by action of carbon and nitrogen on metallic oxides. For example, barium cyanide is made by heating a mixture of barium oxide and carbon to a red heat in a stream of nitrogen:

$$BaO + 3C + N_2 \longrightarrow Ba(CN)_2 + CO$$

The cyanide ion, $[{:}C{\equiv}N{:}]^-$, is closely similar to a halide ion in its properties. By oxidation it can be converted into *cyanogen,* C_2N_2 $({:}N{\equiv}C{-}C{\equiv}N{:})$, which is analogous to the halogen molecules F_2, Cl_2, etc. By suitable procedures three anions can be made that are similar in structure to the carbon dioxide molecule $:\ddot{O}{=}C{=}\ddot{O}:$ and the nitrous oxide molecule $:\ddot{N}{=}N{=}\ddot{O}:$ (these structures are hybridized with other structures, such as $:O{\equiv}C{-}\underset{\cdot\cdot}{\ddot{O}}:$ and its analogues). These anions are

$$[:\ddot{N}{=}C{=}\ddot{O}:]^- \quad \text{Cyanate}$$

$$[:\ddot{C}{=}N{=}\ddot{O}:]^- \quad \text{Fulminate ion}$$

$$[:\ddot{N}{=}N{=}\ddot{N}:]^- \quad \text{Azide ion}$$

A related ion is the thiocyanate ion, $[:\ddot{N}{=}C{=}\ddot{S}:]^-$, which forms a deep red complex with ferric ion, used as a test for iron. The azide ion also forms a deep red complex with ferric ion.

The fulminates and azides of the heavy metals are very sensitive explosives. *Mercuric fulminate,* $Hg(CNO)_2$, and *lead azide,* $Pb(N_3)_2$, are used as detonators.

8-6. Oxycompounds of Carbon

Carbon burns to form the gases *carbon monoxide*, CO, and *carbon dioxide*, CO_2, the former being produced when there is a deficiency of oxygen or when the flame temperature is very high.

Carbon Monoxide

Carbon monoxide is a colorless, odorless gas with small solubility in water (35.4 ml per liter of water at 0°C and 1 atm). It is poisonous, because of its ability to combine with the hemoglobin in the blood in the same way that oxygen does; thus the carbon monoxide prevents the hemoglobin from combining with oxygen in the lungs and carrying it to the tissues. It causes death when about one-half of the hemoglobin in the blood has been converted into carbonmonoxyhemoglobin. The exhaust gas from automobile engines contains some carbon monoxide, and it is accordingly dangerous to be in a closed garage with an automobile whose engine is running. Carbon monoxide is a valuable industrial gas, for use as a fuel and as a reducing agent.

The enthalpy of formation of carbon monoxide is -110.5 kJ $mole^{-1}$ and its heat of combustion is 283.0 kJ $mole^{-1}$. The blue lambent flame seen over a charcoal fire involves the combustion of the carbon monoxide that has been formed by the surface combustion of the charcoal.

Carbon Dioxide

Carbon dioxide is a colorless, odorless gas with a weakly acid taste, owing to the formation of some carbonic acid when it is dissolved in water. It is about 50% heavier than air. It is easily soluble in water, one liter of water at 0°C dissolving 1713 ml of the gas under 1 atm pressure. Its melting point (freezing point) is higher than the point of vaporization at 1 atm of the crystalline form. When crystalline carbon dioxide is heated from a very low temperature, its vapor pressure reaches 1 atm at −79°C, at which temperature it vaporizes (sublimes) without melting. If the pressure is increased to 5.2 atm, the crystalline substance melts to a liquid at

−56.6°C. Under ordinary pressure, then, the solid substance is changed directly to a gas. This property has made solid carbon dioxide (dry ice) popular as a refrigerating agent.

The enthalpy of formation of carbon dioxide is −394 kJ mole^{-1}. Its heat of sublimation at −78.48°C (1 atm pressure) is 25 kJ mole^{-1}. The molecule is linear, with carbon-oxygen bond length 115.9 pm.

Carbon dioxide is used for the manufacture of *sodium carbonate*, $Na_2CO_3 \cdot 10H_2O$ (washing soda); *sodium hydrogen carbonate*, $NaHCO_3$ (baking soda); and carbonated water, for use as a beverage (soda water). Carbonated water is charged with carbon dioxide under a pressure of 3 or 4 atm.

Carbon dioxide can be used to extinguish fires by smothering them. One form of portable fire extinguisher is a cylinder of liquid carbon dioxide—the gas can be liquefied at ordinary temperatures under pressures of about 70 atm. Some commercial carbon dioxide (mainly solid carbon dioxide) is made from the gas emitted in nearly pure state from gas wells in the western United States. Most of the carbon dioxide used commercially is a by-product (subsidiary substance produced in the process) of cement mills, limekilns, iron blast furnaces, and breweries.

Carbonic Acid and Carbonates

When carbon dioxide dissolves in water, some of it reacts to form carbonic acid:

$$CO_2 + H_2O \longrightarrow H_2CO_3$$

The acid is diprotic; with a base such as sodium hydroxide it may form both a normal salt, Na_2CO_3, and an acid salt, $NaHCO_3$. The normal salt contains the carbonate ion, CO_3^{--}, and the acid salt contains the hydrogen carbonate ion, HCO_3^-.

The most important carbonate mineral is calcium carbonate, $CaCO_3$. This substance occurs in beautiful colorless hexagonal crystals as the mineral *calcite*. *Marble* is a microcrystalline form of calcium carbonate, and *limestone* is a rock composed mainly of this substance. Calcium carbonate is the principal constituent also of pearls, coral, and most sea shells. It also occurs in a second crystalline form, as the orthorhombic mineral *aragonite*.

When calcium carbonate is heated (as in a limekiln, where limestone is mixed with fuel, which is burned), it decomposes, forming calcium oxide (quicklime):

$$CaCO_3 \longrightarrow CaO + CO_2(g)$$

Quicklime is slaked by adding water, to form calcium hydroxide:

$$CaO + H_2O \rightarrow Ca(OH)_2$$

Slaked lime prepared in this way is a white powder that can be mixed with water and sand to form *mortar.* The mortar hardens by first forming crystals of calcium hydroxide, which cement the grains of sand together; then on exposure to air the mortar continues to get harder by taking up carbon dioxide and forming calcium carbonate.

Large amounts of limestone are used also in the manufacture of Portland cement, described in Section 18-11.

Sodium carbonate (washing soda), $Na_2CO_3 \cdot 10H_2O$, is a white, crystalline substance used as a household alkali, for washing and cleaning, and as an industrial chemical. The crystals of the decahydrate lose water readily, forming the monohydrate, $Na_2CO_3 \cdot H_2O$. The monohydrate when heated to 100° changes to anhydrous sodium carbonate (soda ash), Na_2CO_3.

Sodium hydrogen carbonate (baking soda, bicarbonate of soda), $NaHCO_3$, is a white substance usually available as a powder. It is used in cooking, in medicine, and in the manufacture of baking powder. Baking powder is a leavening agent used in making biscuits, cakes, and other food. Its purpose is to provide bubbles of gas, to make the dough "rise." The same foods can be made by use of sodium hydrogen carbonate and sour milk, instead of baking powder. In each case the reaction that occurs involves the action of an acid on sodium hydrogen carbonate, to form carbon dioxide. When sour milk is used, the acid that reacts with the sodium hydrogen carbonate is lactic acid, $HC_3H_5O_3$, the equation for the reaction being

$$NaHCO_3 + HC_3H_5O_3 \rightarrow NaC_3H_5O_3 + H_2O + CO_2(g)$$

The product $NaC_3H_5O_3$ is sodium lactate, the sodium salt of lactic acid (Section 13-5). Cream-of-tartar baking powder consists of sodium hydrogen carbonate, potassium hydrogen tartrate ($KHC_4H_4O_6$, commonly known as cream of tartar), and starch, the starch being added to keep water vapor in the air from causing the powder to form a solid cake. The reaction that occurs when water is added to a cream-of-tartar baking powder is

$$NaHCO_3 + KHC_4H_4O_6 \rightarrow NaKC_4H_4O_6 + H_2O + CO_2(g)$$

The product sodium potassium tartrate, $NaKC_4H_4O_6$, has the common name "Rochelle salt." Baking powders are also made with calcium dihydrogen phosphate, $Ca(H_2PO_4)_2$, sodium dihydrogen phosphate,

NaH_2PO_4, or sodium aluminum sulfate, $NaAl(SO_4)_2$, as the acidic constituent. The last of these substances is acidic because of the hydrolysis of the aluminum salt (Section 12-6).

The leavening agent in ordinary bread dough is yeast, a microorganism that converts sugars to ethyl alcohol and carbon dioxide.

Methyl Alcohol and Ethyl Alcohol

An *alcohol* is obtained from a hydrocarbon by replacing one hydrogen atom by a hydroxyl group, —OH. Thus methane, CH_4, gives *methyl alcohol,* CH_3OH, and ethane, C_2H_6, gives *ethyl alcohol,* C_2H_5OH. The names of the alcohols are often written by using the ending *ol;* methyl alcohol is called *methanol,* and ethyl alcohol *ethanol.* They have the following structural formulas:

```
     H                 H   H
     |                 |   |
 H—C—O—H          H—C—C—O—H
     |                 |   |
     H                 H   H
Methyl alcohol      Ethyl alcohol
```

To make methyl alcohol from methane, the methane may be converted to methyl chloride by treatment with chlorine, as described in Section 7-6, and the methyl chloride then converted to methyl alcohol by treatment with sodium hydroxide:

$$CH_3Cl + NaOH \rightarrow CH_3OH + NaCl$$

Methyl alcohol is made by the destructive distillation of wood; it is sometimes called wood alcohol. It is a poisonous substance, which on ingestion causes blindness and death. It is used as a solvent and for the preparation of other organic compounds.

The most important method of making ethyl alcohol is by the fermentation of sugars with yeast. Grains and molasses are the usual raw materials for this purpose. Yeast produces an enzyme that catalyzes the fermentation reaction. In the following equation the formula $C_6H_{12}O_6$ is that of a sugar, glucose (also called dextrose and grape sugar; Chapter 13):

$$C_6H_{12}O_6 \rightarrow 2CO_2 + 2C_2H_5OH$$

Ethyl alcohol is a colorless liquid (m.p. −117°C, b.p. 79°C) with a characteristic odor. It is used as a fuel, as a solvent, and as the starting material for preparing other compounds. Beer contains 3 to 5% alcohol, wine usually 10 to 12%, and distilled liquors such as whiskey, brandy, and gin 40 to 50%.

The *ethers* are compounds obtained by reaction of two alcohol molecules, with elimination of water. The most important ether is *diethyl ether* (ordinary ether), $(C_2H_5)_2O$. It is made by treating ethyl alcohol with concentrated sulfuric acid, which serves as a dehydrating agent:

$$2C_2H_5OH \xrightarrow[H_2SO_4]{} C_2H_5OC_2H_5 + H_2O$$

It is used as a general anesthetic and as a solvent.

The Organic Acids

Ethyl alcohol can be oxidized by oxygen of the air to *acetic acid,* $HC_2H_3O_2$ or CH_3COOH:

$$C_2H_5OH + O_2 \rightarrow CH_3COOH + H_2O$$

This reaction occurs easily in nature. If wine, containing ethyl alcohol, is allowed to stand in an open container, it undergoes acetic-acid fermentation and changes into vinegar by the above reaction. The change is brought about by microorganisms ("mother of vinegar"), which produce enzymes that catalyze the reaction.

Acetic acid has the following structural formula:

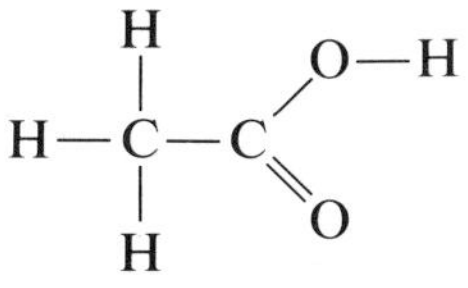

```
     H      O—H
     |     /
  H—C—C
     |     \\
     H      O
```

It contains the group

```
      O—H
     /
  —C
     \\
      O
```

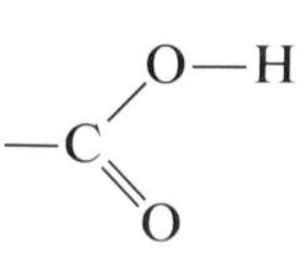

which is called the *carboxyl group.* It is this group that gives acidic properties to the organic acids.

Acetic acid melts at 17°C and boils at 118°C. It is soluble in water and alcohol. The molecule contains one hydrogen atom that ionizes from it in water, producing the *acetate ion,* $C_2H_3O_2^-$. The acid reacts with bases to form salts. An example is sodium acetate, $NaC_2H_3O_2$, a white solid:

$$HC_2H_3O_2 + NaOH \longrightarrow NaC_2H_3O_2 + H_2O$$

Formic acid, HCOOH, is the simplest of the carboxylic acids. Some others are discussed in Chapter 13.

The structure shown above for acetic acid is not completely satisfactory. The experimental value for the C—OH bond length is 136 pm, 7 pm less than that for a C—O single bond. The same value, 136 pm, is found in methyl formate, $HCOOCH_3$. The carboxylic acids are accordingly assigned the following resonance structure:

```
      ..                    ..
      O:                   :O:⁻
     //                    /
R—C                   R—C
     \                     \\
     :O—H                   O⁺—H
      ..                    ..
      A                     B
```

The bond lengths 122 pm and 136 pm indicate that structure A contributes about 80% and structure B about 20% for the acids and esters. For the carboxylate ion the two structures A′ and B′ are equivalent:

```
      ..                    ..
      O:                   :O:⁻
     //                    /
R—C                   R—C
     \                     \\
     :O:⁻                   O:
      ..                    ..
      A′                    B′
```

and each contributes 50% to the normal state of the ion. The resonance energy, relative to structure A or A′, is found to be about 117 kJ $mole^{-1}$ for acids and esters and 150 kJ $mole^{-1}$ for carboxylate ions. It is mentioned in Section 13-5 that the extra resonance energy of the carboxylate ions provides an explanation of the fact that the OH group is far more acidic in the carboxylic acids than in the alcohols.

8-7. Transargononic Compounds of the Argonons

In 1933, on the basis of arguments about the electronic structure of molecules, it was pointed out that transargononic compounds of krypton, xenon, and radon with fluorine and oxygen should be stable. The known acids H_8SnO_6, H_7SbO_6, H_6TeO_6, and H_5IO_6, for example, strongly suggest that H_4XeO_6, perxenic acid, should also exist. An effort was made to synthesize XeF_6 by reaction of xenon and fluorine, but without success. Then in 1962 and 1963 several compounds of xenon were synthesized. The first one to be reported (by the English chemist Neil Bartlett (born 1933)) was xenon hexafluoroplatinate, $XePtF_6$, a yellow crystalline substance. Later (1963) he reported the synthesis of the corresponding rhodium compound, $XeRhF_6$. Scientists in the Argonne National Laboratory, and later other investigators, prepared several xenon fluorides, including XeF_2, XeF_4, and XeF_6. Similar compounds of krypton and radon have also been synthesized, including KrF_2, KrF_4, and RnF_4.

The molecule XeF_2 is linear, with bond length 200 pm; XeF_4 has the square planar configuration, with bond length 195 pm; and XeF_6, with bond length 190 pm, is known not to be a regular octahedron (it has one unshared electron pair, as well as six shared pairs, about the xenon atom). The decrease in Xe—F bond length in this sequence reflects the increasing amount of *d* character of the xenon bond orbitals.

The xenon fluorides react vigorously with water; for example,

$$XeF_6 + H_2O \rightarrow XeOF_4 + 2HF$$

$$XeOF_4 + 2H_2O \rightarrow XeO_3 + 4HF$$

$$XeO_3 + H_2O \rightarrow H_2XeO_4$$

Xenon also forms the tetroxide, XeO_4, and the corresponding acid, H_4XeO_6. X-ray study of the crystal $Na_4XeO_6 \cdot 6H_2O$ has shown that the perxenate ion has the structure of a regular octahedron with Xe—O bond length 184 pm. Perxenic acid is a powerful oxidizing agent, which can oxidize manganous ion, Mn^{++}, to permanganate ion, MnO_4^-.

From measurement of the heat of reaction of xenon tetrafluoride with a solution of potassium iodide the heat of formation has been determined:

$$Xe(g) + 2F_2(g) \rightarrow XeF_4(g) + 188 \text{ kJ mole}^{-1}$$

The bond energy of the Xe—F bond accordingly is equal to 126 kJ mole^{-1}. This value, compared with the values for transargononic I—F, Br—F, and Cl—F bonds, which are 727, 370, and 188 kJ mole^{-1}, respectively, suggests that the electronegativity of xenon is about 3.1.

The xenon oxides are unstable: XeO_3 explodes about as violently as TNT. The heat of explosion has been measured:

$$XeO_3(c) \rightarrow Xe(g) + \tfrac{3}{2}O_2(g) + 402 \text{ kJ mole}^{-1}$$

The enthalpy of sublimation of XeO_3 is 80 kJ mole^{-1}; hence the Xe=O bond energy is 88 kJ mole^{-1}.

EXERCISES

8-1. What are the normal-valence oxycompounds of chlorine? What are their electronic structures?

8-2. What is unsatisfactory about the argononic structure of dichlorine heptoxide? What structure is better?

8-3. What are the oxidation numbers shown in stable compounds of each of the halogens?

8-4. Give the names, formulas, and electronic structures of each of the oxygen acids of chlorine.

8-5. Give reaction equations illustrating the oxidizing power of the hypochlorite ion, the chlorite ion, and the chlorate ion. What are the changes in oxidation number involved?

8-6. What is dangerous about chlorates? Give a reaction equation illustrating this danger.

8-7. What weight percentage of potassium perchlorate is required to mix with carbon black to form the most efficient rocket propellent?

8-8. Antiseptics (substances that can kill bacteria) used in minor medicine include solutions of sodium hypochlorite (Dakin's solution, Javelle water). Why is this substance effective?

8-9. Paramagnetism (attraction into a magnetic field) is caused by unpaired electron spins, as in O_2. Would you expect chlorine monoxide to be paramagnetic? Chlorine dioxide?

8-10. What are the names, formulas, and electronic structures of the stable oxyacids of bromine and their anions?

8-11. What are the names, formulas, and electronic structures of the stable oxyacids of iodine and their anions?

8-12. Using the information that iodine dioxide is paramagnetic, write an electronic structure (or structures as appropriate) for the compound. What is the oxidation number of iodine in this compound?

8-13. Compare the oxidizing power of the elements, the reducing power of their oxygen compounds, and the electronegativity of the halogens.

8-14. Give the reactions involved in the production of sulfuric acid from sulfur.

8-15. What are the electronic structures, bond lengths, and bond angles of sulfur dioxide and sulfur trioxide?

8-16. In the discussion of sulfur dioxide (Section 8-2), it is mentioned that the bond length 149 pm is expected for the S=O bond. How is this value obtained?

8-17. What are the three chief uses of sulfuric acid? Give examples.

8-18. What is the electronic structure of the thiosulfate ion? In terms of the relative ability to form double bonds of oxygen and sulfur, why do you think thiosulfuric acid is unstable?

8-19. What are the oxidation numbers of phosphorus, arsenic, antimony, and bismuth shown in stable compounds of each?

8-20. Draw electronic structures for tetraphosphorus hexoxide and tetraphosphorus decoxide.

8-21. Draw the electronic structure of phosphoric acid.

8-22. From enthalpy values in Appendix VI calculate the heat of combustion of carbon monoxide to carbon dioxide. Is more or less heat evolved in burning carbon monoxide than in burning an equal volume of hydrogen?

8-23. (*a*) From enthalpy values in Appendix VI calculate the heat evolved in the reaction of ethane with oxygen to form ethanol. (*b*) Also calculate the heat evolved in the reaction of acetaldehyde, H_3CCHO, with oxygen to form acetic acid, H_3CCOOH. (*c*) Note that the same bonds are broken and formed in these two reactions; the difference between them is that acetic acid is stabilized by the resonance of the double bond in the carboxylic acid group. From the above calculations, what is the magnitude of this resonance energy? (See Section 8-6).
(Answer: (*a*) 152 kJ $mole^{-1}$; (*b*) 269 kJ $mole^{-1}$; (*c*) 117 kJ $mole^{-1}$)

8-24. A value for the N=O bond energy can be calculated from the value of the standard enthalpy of formation of ethyl nitrite, $C_2H_5ONO(g)$, which is −104 kJ $mole^{-1}$. (*a*) Use this value to calculate the enthalpy of formation of the substance from gaseous atoms (Table V-3). (*b*) Subtract the values of the bond energies other than N=O (Table V-1). (*c*) Subtract 117 kJ $mole^{-1}$, the resonance energy in carboxylic esters, on the assumption that the value applies also to the similar resonance of the double bond in the nitrite group, to obtain a value for the N=O bond energy.
(Answer: (*a*) −3595 kJ $mole^{-1}$; (*c*) 534 kJ $mole^{-1}$)

8-25. (*a*) From the value of the standard enthalpy of formation of NO (Table VI-7) calculate the total bond energy. (*b*) By subtracting the value for the N=O bond (preceding Exercise), obtain the contribution to the bond energy made by the interchange of the odd electron and an electron pair between the two atoms (the energy of the three-electron bond).
(Answer: (*a*) 632 kJ $mole^{-1}$; (*b*) 98 kJ $mole^{-1}$.)

8-26. Make a calculation, similar to that of the preceding Exercise, of the resonance energy of the three-electron bond and that of interchange of the two kinds of bonds to the two oxygen atoms in NO_2. This resonance energy is expected to be about twice that for the three-electron bond in NO.
(Answer: 228 kJ $mole^{-1}$.)

8-27. Note that NO can be converted to NO_2 and N_2O_3 to N_2O_4 by addition of an oxygen atom to an unshared electron pair of a nitrogen atom. Calculate the bond energy of the bond that is formed (values in Appendix V apply only to normal-valence compounds, not to those in which the atoms have formal electric charges, as here: $N^+—O^-$).
(Answer: 305, 289 kJ mole^{-1}.)

9

Water and Solutions

Water is the most important of all chemical substances. It is a major constituent of living matter and of the environment in which we live. Its physical properties are strikingly different from those of other substances, in ways that determine the nature of the physical and biological world.

9-1. The Composition of Water

Water was thought by the ancients to be an element. Henry Cavendish in 1781 showed that water is formed when hydrogen is burned in air, and Lavoisier first recognized that water is a compound of the two elements hydrogen and oxygen in 1783.

The formula of water is H_2O. The relative weights of hydrogen and oxygen in the substance have been very carefully determined as 2.0160 : 16.0000. This determination has been made both by weighing the amounts of hydrogen and oxygen liberated from water by electrolysis and by determining the weights of hydrogen and oxygen which combine to form water.

Ordinary water is impure; it usually contains dissolved salts and dissolves gases, and sometimes organic matter. Hard water is water containing cations of calcium, magnesium, and iron, which are undesirable because they form a precipitate with ordinary soap and react with other substances.

9-2. Methods of Purifying Water

For chemical work water is purified by distillation. Pure tin vessels and pipes are often used for storing and transporting distilled water. Glass vessels are not satisfactory, because the alkaline constituents of glass slowly dissolve in water. Distilling apparatus and vessels made of fused silica are used in making very pure water.

The impurity which is hardest to keep out of distilled water is carbon dioxide, which dissolves readily from the air.

Removal of Ionic Impurities from Water.

Ionic impurities can be effectively and cheaply removed from water by an interesting process that involves the use of *giant molecules*—molecular structures that are so big as to constitute visible particles. A crystal of diamond is an example of such a giant molecule (Chapter 7). Some complex inorganic crystals, such as the minerals called *zeolites,* are of this nature. These minerals are used to "soften" hard water, to remove the heavy metal ions from the water. Such a process is termed *ion exchange.*

Both the positive ions and the negative ions can be removed from water by a method using two substances in two tanks. The first tank, A, contains grains that consist of giant organic resin molecules in the form of a porous framework to which acidic groups are attached. These groups are represented as *sulfonic groups,* —SO_3H:*

$$\mathrm{R{-}S({=}\ddot{O}{:})({=}\ddot{O}{:}){-}\ddot{O}{:}{-}H}$$

The reactions that occur when a solution containing salts passes through tank A may be written as

$$\mathrm{RSO_3H(c) + Na^+ \rightarrow (RSO_3^-)Na^+(c) + H^+}$$

$$\mathrm{2RSO_3H(c) + Ca^{++} \rightarrow (RSO_3^-)_2Ca^{++}(c) + 2H^+}$$

*R represents a part of the framework.

That is, sodium ions and calcium ions are removed from the solution by the acidic framework, and hydrogen ions are added to the solution. The solution is changed from a salt solution (Na^+, Cl^-, etc.) to an acid solution (H^+, Cl^-, etc.).

This acid then runs through tank B, which contains grains of giant organic resin molecules with basic groups attached. These groups are shown as *substituted ammonium hydroxide* groups, $(RNH_3^+)(OH^-)$:

$$\left[\begin{array}{c} \text{H} \\ | \\ \text{R—N—H} \\ | \\ \text{H} \end{array}\right]^+ \qquad [:\ddot{\underset{\cdot\cdot}{O}}\text{—H}]^-$$

The hydroxide ion of these groups combines with the hydrogen ion in the water:

$$OH^- + H^+ \rightarrow H_2O$$

The negative ions then remain, held by the ammonium ions of the framework. The reactions are

$$(RNH_3^+)(OH^-)(c) + Cl^- + H^+ \rightarrow (RNH_3^+)Cl^-(c) + H_2O$$

$$2(RNH_3^+)(OH^-)(c) + SO_4^{--} + 2H^+ \rightarrow (RNH_3^+)_2(SO_4^{--})(c) + 2H_2O$$

The water which passes out of the second tank contains practically no ions, and may be used in the laboratory and in industrial processes in place of distilled water.

The giant molecules in tank A may be regenerated after use by passing dilute sulfuric acid through the tank:

$$2(RSO_3^-)Na^+(c) + H_2SO_4 \rightarrow 2RSO_3H(c) + 2Na^+ + SO_4^{--}$$

Those in tank B may be regenerated by use of a moderately concentrated solution of sodium hydroxide:

$$(RNH_3^+)Cl^-(c) + OH^- \rightarrow (RNH_3^+)OH^-(c) + Cl^-$$

Other Ways of Purifying Water

Hard water may also be softened by chemical treatment. In practice the use of giant organic molecules (synthetic resins) for deionizing water, described above, is restricted to industries requiring very pure water, as in making medicinal products. Water for a city is usually treated by the addition of chemicals, followed by sedimentation when the water is allowed to stand in large reservoirs, and then by filtration through beds of sand. The settling process removes suspended matter in the water together with precipitated substances that might be produced by the added chemicals, and some living microorganisms. After filtration, the remaining living organisms may be destroyed by treatment with chlorine, bleaching powder, sodium hypochlorite or calcium hypochlorite, or ozone.

The hardness of water is due mainly to calcium ion, ferrous ion (Fe^{++}), and magnesium ion; it is these ions that form insoluble compounds with ordinary soap. Hardness is usually reported in parts per million (ppm), calculated as calcium carbonate (or sometimes in grains per gallon: 1 grain per gallon is equal to 17.1 ppm). Domestic water with hardness less than 100 ppm is good, and that with hardness between 100 and 200 ppm is fair.

Ground water in limestone regions may contain a large amount of calcium ion and hydrogen carbonate ion, HCO_3^-. Although calcium carbonate itself is insoluble, calcium hydrogen carbonate, $Ca(HCO_3)_2$, is a soluble substance. A water of this sort (which is said to have *temporary hardness*) can be softened simply by boiling, which causes the excess carbon dioxide to be driven off, and the calcium carbonate to precipitate:

$$Ca^{++} + 2HCO_3^- \rightarrow CaCO_3(c) + H_2O + CO_2(g)$$

This method of softening water cannot be applied economically in the treatment of the water supply of a city, however, because of the large fuel cost. Instead, the water is softened by the addition of calcium hydroxide, slaked lime:

$$Ca^{++} + 2HCO_3^- + Ca(OH)_2 \rightarrow 2CaCO_3(c) + 2H_2O$$

If sulfate ion or chloride ion is present in solution instead of hydrogen carbonate ion, the hardness of the water is not affected by boiling—the water is said to have *permanent hardness*. Permanently hard water can be softened by treatment with sodium carbonate:

$$Ca^{++} + CO_3^{--} \rightarrow CaCO_3(c)$$

The sodium ions of the sodium carbonate are left in solution in the water, together with the sulfate or chloride ions that were already there.

In softening water by use of calcium hydroxide or sodium carbonate, enough of the substance is used to cause magnesium ion to be precipitated as magnesium hydroxide and iron as ferrous hydroxide or ferric hydroxide. Sometimes, in addition to the softening agent, a small amount of aluminum sulfate, alum, or ferric sulfate is added as a coagulant. These substances, with the alkaline reagents, form a flocculent, gelatinous precipitate of aluminum hydroxide, $Al(OH)_3$, or ferric hydroxide, $Fe(OH)_3$, which entraps the precipitate produced in the softening reaction, and helps it to settle out. The gelatinous precipitate also tends to adsorb coloring matter and other impurities in the water.*

A water that is used in a steam boiler often deposits a scale of calcium sulfate, which is left as the water is boiled away. In order to prevent this, boiler water is sometimes treated with sodium carbonate, causing the precipitation of calcium carbonate as a sludge, and preventing the formation of the calcium sulfate scale. Sometimes trisodium phosphate, Na_3PO_4, is used, leading to the precipitation of calcium as hydroxyapatite, $Ca_5(PO_4)_3OH$, as a sludge. In either case the sludge is removed from the boiler by draining at intervals. Large modern boilers use deionized water obtained by the organic-resin ion-exchange process.

9-3. The Ionic Dissociation of Water

An acidic solution contains hydrogen ions, H^+ (actually hydronium ions, H_3O^+). A basic solution contains hydroxide ions, OH^-. A number of years ago chemists asked, and answered, the question, "Are these ions present in pure neutral water?" The answer is that they are present, in equal but very small concentrations.

Pure water contains hydrogen ions in concentration 1×10^{-7} moles per liter, and hydroxide ions in the same concentration. These ions are formed by the dissociation of water:

$$H_2O \rightleftarrows H^+ + OH^-$$

When a small amount of acid is added to pure water, the concentration of hydrogen ion is increased. The concentration of hydroxide ion then decreases, *but not to zero.* Acidic solutions contain hydrogen ion in large concentration and hydroxide ion in very small concentration. Many of the chemical properties of water come from its having both acid and base functions.

**Adsorption* is the adhesion of molecules of a gas, liquid, or dissolved substance or of particles to the surface of a solid substance. *Absorption* is the assimilation of molecules into a solid or liquid substance, with the formation of a solution or a compound. Sometimes the word *sorption* is used to include both of these phenomena. We say that a heated glass vessel *adsorbs* water vapor from the air on cooling, and becomes coated with a very thin layer of water: a dehydrating agent such as concentrated sulfuric acid *absorbs* water, forming hydrates.

9-4. Physical Properties of Water

Water is a clear, transparent liquid, colorless in thin layers. Thick layers of water have a bluish-green color.

The physical properties of water are used to define many physical constants and units. The freezing point of water (saturated with air at 1 atm pressure) is taken as 0°C, and the boiling point of water at 1 atm is taken as 100°C. The unit of volume in the metric system is chosen so that 1 ml of water at 3.98°C (the temperature of its maximum density) weighs 1.00000 gram. A similar relation holds in the English system: 1 cu ft of water weighs approximately 1000 ounces.

Most substances diminish in volume, and hence increase in density, with decrease in temperature. Water has the very unusual property of having a temperature at which its density is a maximum. This temperature is 3.98°C. With further cooling below this temperature the volume of a sample of water increases somewhat (Figure 9-1).

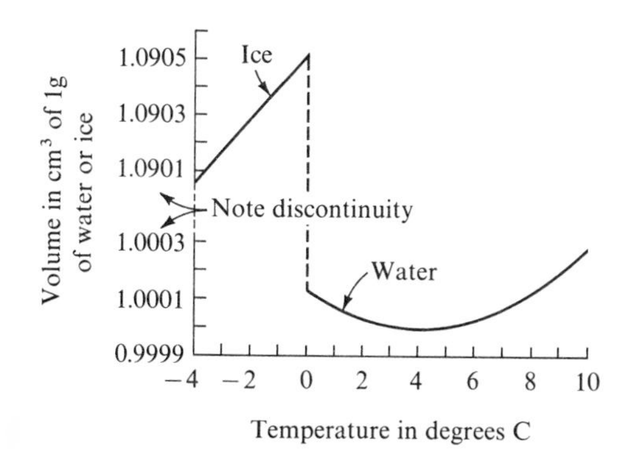

FIGURE 9-1
Dependence of the volume of ice and water on temperature.

Water is unusual (but not unique) in that the density of the solid phase, ice, is substantially lower than the density of the liquid phase, water. This property is important in that the form of life for many living organisms would have to be very different from what it is if ice sank instead of floating in freezing bodies of water.

Both the freezing point and the boiling point of water are much higher than those expected by comparison with related substances (Section 9-6). The nature of the mechanisms of life are such that life would probably be impossible without these unusual properties of water.

The dielectric constant of water is unusually high and it is this property of water that gives it its important properties as a solvent.

The number of nearest neighbors to each molecule in a simple crystalline solid is 12 (Chapter 2) and in a simple liquid approximately 10.6, as determined by various techniques. It is this difference that accounts for the usually observed decrease in density on melting. Water is unusual in that in both the solid and the liquid states the number of nearest neighbors of each water molecule is approximately four. The molecular structure of liquid water is not known; this remains one of the most interesting and important structural problems to be solved.

9-5. The Melting Points and Boiling Points of Substances

All molecules exert a weak attraction upon one another. This attraction, the *electronic van der Waals attraction,* is the result of the mutual interaction of the electrons and nuclei of the molecules; it has its origin in the electrostatic attraction of the nuclei of one molecule for the electrons of another, which is largely but not completely compensated by the repulsion of electrons by electrons and nuclei by nuclei. The van der Waals attraction is significant only when the molecules are very close together—almost in contact with one another. At small distances (about 400 pm for argon, for example) the force of attraction is balanced by a force of repulsion due to interpenetration of the outer electron shells of the molecules (Figure 9-2).

It is these intermolecular forces of electronic van der Waals attraction that cause substances such as the argonons, the halogens, etc., to condense to liquids and to freeze into solids at sufficiently low temperatures. The boiling point is a measure of the amount of molecular agitation necessary to overcome the forces of van der Waals attraction, and hence is an indication of the magnitude of these forces. In general, *the electronic van der Waals attraction between molecules increases with increase in the number of electrons per molecule.* Since the molecular weight is roughly proportional to the number of electrons in the molecule, usually about twice the number of electrons, the van der Waals attraction usually increases with increase in the molecular weight. Large molecules (containing many electrons) attract one another more strongly than small molecules (containing few electrons); hence normal molecular substances with large molecular weight have high boiling points, and those with small molecular weight have low boiling points.

This generalization is indicated in Figure 9-3, in which the boiling points of some molecular substances are shown. The steady increase in boiling point for sequences such as He, Ne, Ar, Kr, Xe, Rn, and H_2, F_2, Cl_2, Br_2, I_2 is striking.

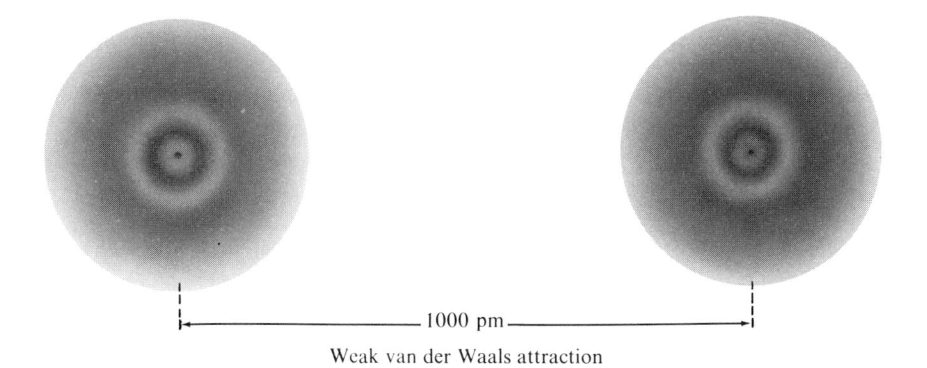

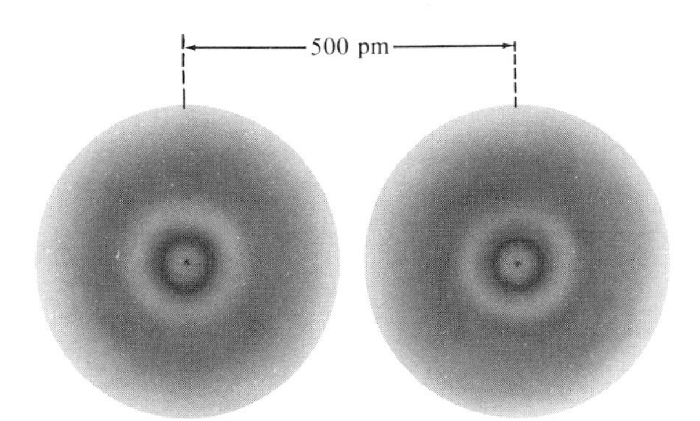

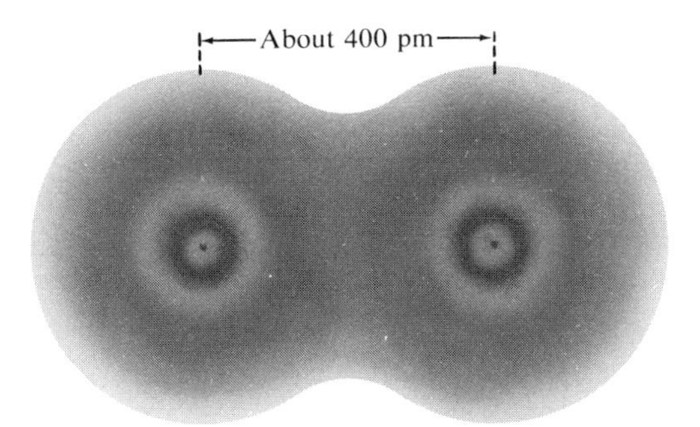

FIGURE 9-2
Diagram illustrating van der Waals attraction and repulsion in relation to electron distribution of monatomic molecules of argon.

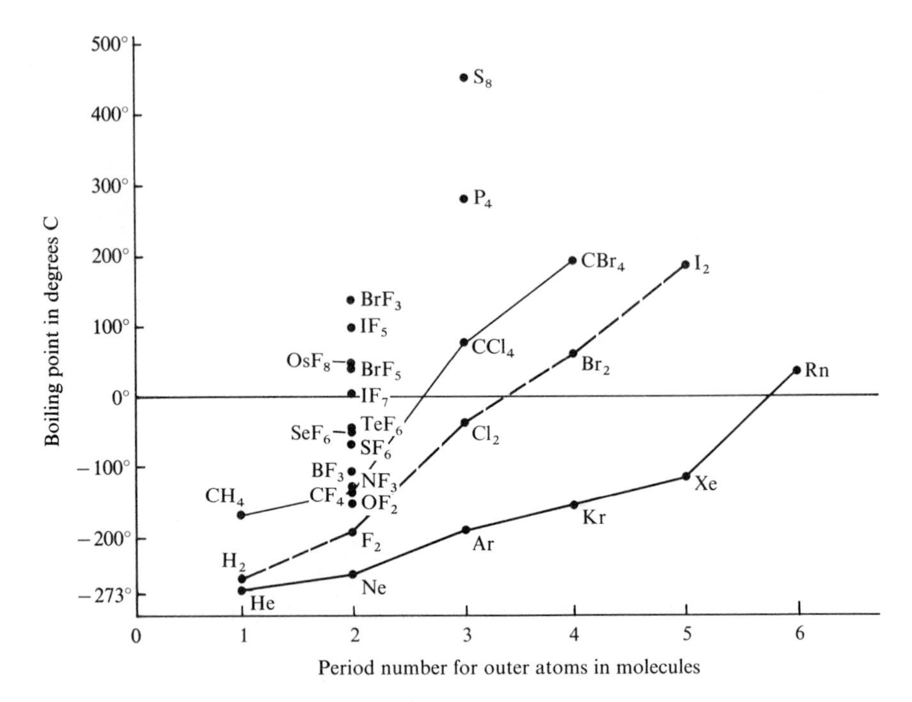

FIGURE 9-3
Diagram showing increase in boiling point with increase in molecular complexity.

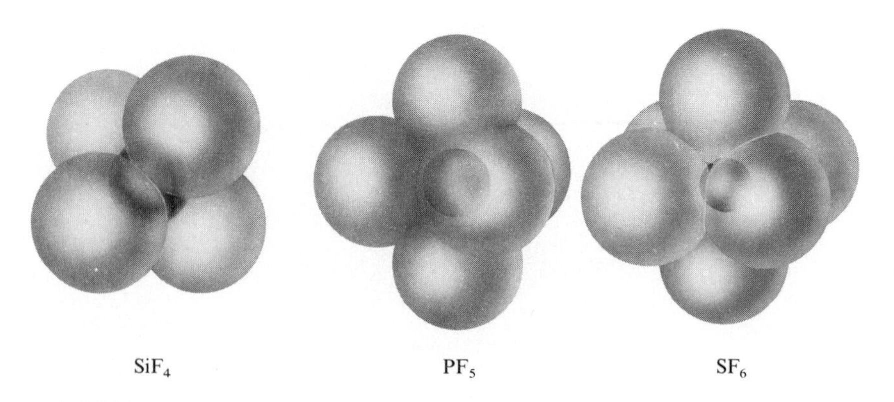

FIGURE 9-4
Molecules of silicon tetrafluoride, phosphorus pentafluoride, and sulfur hexafluoride, three very volatile substances.

The similar effect of increase in the number of atoms (with nearly the same atomic number) in the molecule is shown by the following sequences:

	Ar	Cl_2	P_4	S_8
Boiling point	−185.7°	−34.6°	280°	444.6°C

	Ne	F_2	CF_4	SF_6	IF_7	OsF_8
Boiling point	−245.9°	−184°	−161.4°	−62°	4.5°	47.5°C

Bond Type and Atomic Arrangement

Fifty years ago, before modern structural chemistry had been developed, it was thought that an abrupt change in melting point or boiling point in a series of related compounds could be accepted as proof of a change in type of bond. The fluorides of the elements of the second period, for example, have the following melting points and boiling points:

	NaF	MgF_2	AlF_3	SiF_4	PF_5	SF_6
Melting point	995°	1263°	>1257°	−90°	−94°	−51°C
Boiling point	1704°	2227°	1257°*	−95°*	−85°	−64°C*

The great change between aluminum trifluoride and silicon tetrafluoride is, however, not due to any great change in bond type—the bonds are in all cases intermediate in character between extreme ionic bonds M^+F^- and normal covalent bonds $M{:}\ddot{\underset{..}{F}}{:}$ —but rather to a *change in atomic arrangement*. The three easily volatile substances exist as discrete molecules SiF_4, PF_5, and SF_6 (with no dipole moments) in the liquid and crystalline states as well as the gaseous state (Figure 9-4), and the thermal agitation necessary for fusion or vaporization is only that needed to overcome the weak intermolecular forces, and is essentially independent of the strength or nature of the interatomic bonds within a molecule. But the other three substances in the crystalline state are giant molecules, with strong bonds between neighboring ions holding the whole crystal together (NaF, sodium chloride arrangement, Figure 6-19; MgF_2, Figure 18-2). To melt such a crystal some of these strong bonds must be broken, and to boil the liquid more must be broken; hence the melting

*Note that aluminum trifluoride, silicon tetrafluoride, and sulfur hexafluoride have the interesting property, as does carbon dioxide, of subliming at 1 atm pressure without melting. The temperatures given in the table as the boiling points of these two substances are in fact the subliming points, when the vapor pressure of the crystals becomes equal to 1 atm.

point and boiling point are high. A detailed discussion of these substances and their properties in terms of the relative sizes of the atoms (ionic radius ratio) is given in Section 18-2.

The extreme case is that in which the entire crystal is held together by very strong covalent bonds; this occurs for diamond, with sublimation point 4347°C at 1 atm and melting point at some higher value.

9-6. The Hydrogen Bond—the Cause of the Unusual Properties of Water

The unusual properties of water mentioned in Section 9-4 are due to the power of its molecules to attract one another especially strongly. This power is associated with a structural feature called the hydrogen bond. The melting points and boiling points of the hydrides of some nonmetallic elements are shown in Figure 9-5. The variation for a series of congeners is normal for the sequences. The curves through the points for H_2Te, H_2Se, and H_2S show the expected trend, but when extrapolated they indicate values of about −100°C and −80°C, respectively, for the melting point and boiling point of water. The observed value of the melting point is 100° greater, and that of the boiling point is 180° greater, than would be expected for water if it were a normal substance; and hydrogen fluoride and ammonia show similar, but smaller, deviations.

The Nature of the Hydrogen Bond

The hydrogen ion is a bare nucleus, with charge +1. If hydrogen fluoride, HF, had an extreme ionic structure, it could be represented as in A of Figure 9-6. The positive charge of the hydrogen ion could then strongly attract a negative ion, such as a fluoride ion, forming an $[F^-H^+F^-]^-$ or HF_2^- ion, as shown in B. This does indeed occur, and the stable ion HF_2^-, called the *hydrogen difluoride ion,* exists in considerable concentration in acidic fluoride solutions, and in salts such as KHF_2, potassium hydrogen difluoride. The bond holding this complex ion together, called the **hydrogen bond,** is weaker than ordinary ionic or covalent bonds, but stronger than ordinary van der Waals forces of intermolecular attraction.

Hydrogen bonds are also formed between hydrogen fluoride molecules, causing the gaseous substance to be largely polymerized into the molecular species H_2F_2, H_3F_3, H_4F_4, H_5F_5, and H_6F_6 (Figure 9-7).

In a hydrogen bond the hydrogen atom is usually attached more strongly to one of the two electronegative atoms which it holds together

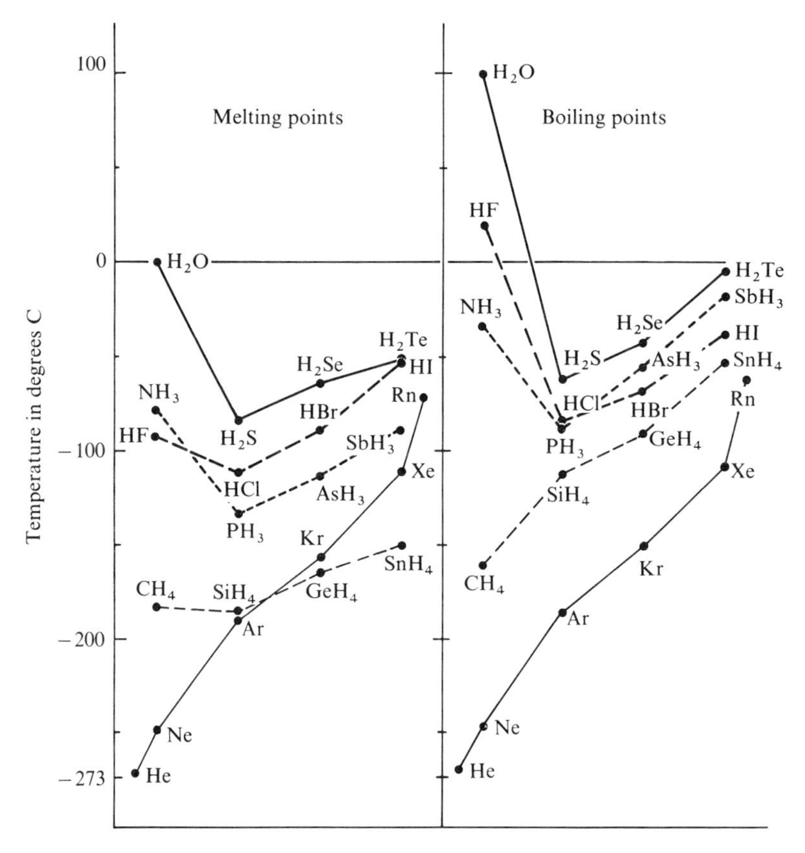

FIGURE 9-5
Melting points and boiling points of hydrides of nonmetallic elements, showing abnormally high values for hydrogen fluoride, water, and ammonia, caused by hydrogen-bond formation.

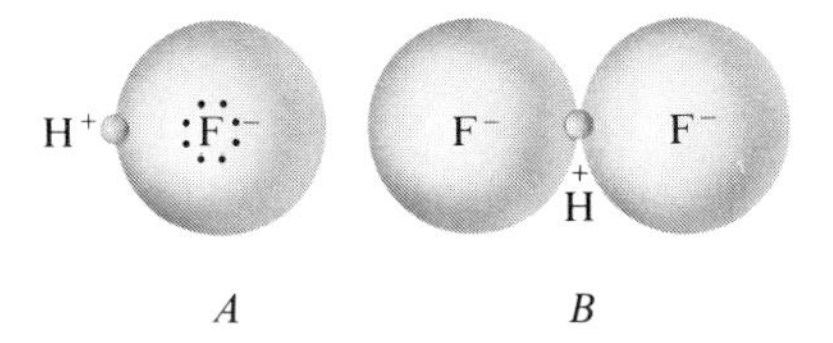

FIGURE 9-6
The hydrogen fluoride molecule (A) and the hydrogen difluoride ion, containing a hydrogen bond (B).

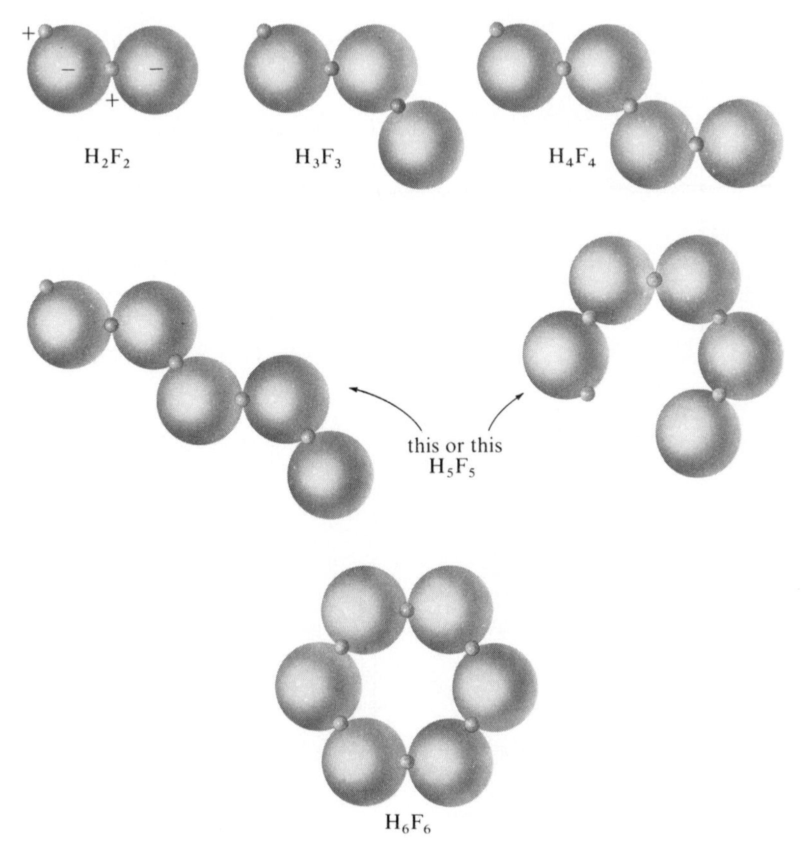

FIGURE 9-7
Some polymers of hydrogen fluoride.

than to the others.* The structure of the dimer of hydrogen fluoride may be represented by the formula

$$F^{-}—H^{+} \cdot\cdot\cdot F^{-}—H^{+}$$

in which the dotted line represents the hydrogen bonding.

Because of the electrostatic origin of the hydrogen bond, only the most electronegative atoms—fluorine, oxygen, nitrogen—form these bonds. Usually an unshared electron pair of the attracted atom approaches closely to the attracting hydrogen ion. Water is an especially suitable

*In KHF_2 and a few other exceptional substances the hydrogen atom is midway between the hydrogen-bonded atoms.

substance for hydrogen-bond formation, because each molecule has two attached hydrogen atoms and two unshared electron pairs, and hence can form four hydrogen bonds. The tetrahedral arrangement of the shared and unshared electron pairs causes these four bonds to extend in the four tetrahedral directions in space, and leads to the characteristic crystal structure of ice (Figure 9-8). This structure, in which each molecule is surrounded by only four immediate neighbors, is a very open structure, and accordingly ice is a substance with abnormally low density. When ice melts, this tetrahedral structure is partially destroyed, and the water molecules are packed more closely together, causing water to have greater density than ice. Many of the hydrogen bonds remain, however,

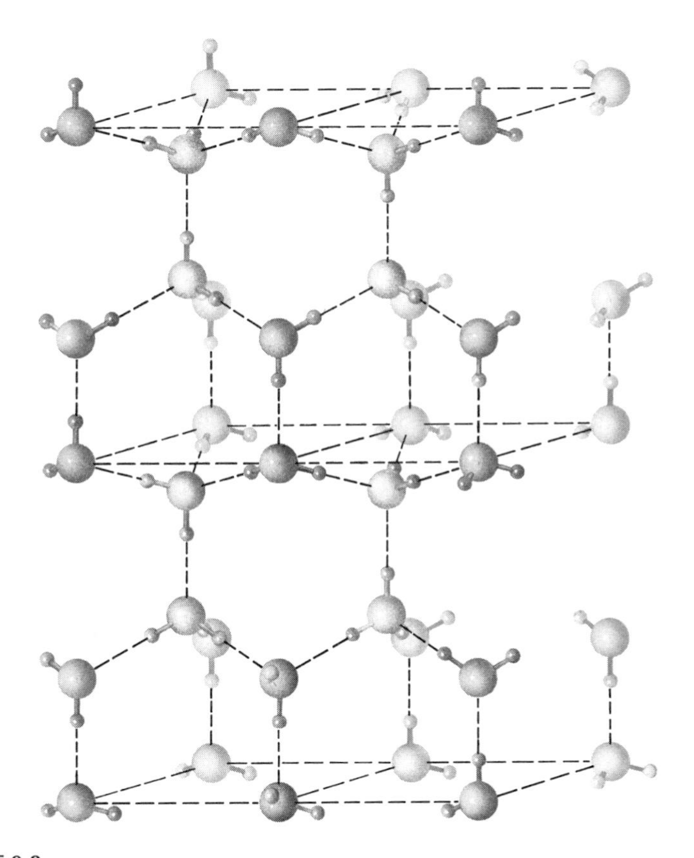

FIGURE 9-8
A small part of a crystal of ice. The molecules above are shown with approximately their correct size (relative to the interatomic distances). Note hydrogen bonds, and the open structure that gives ice its low density. The molecules are indicated diagrammatically as small spheres for oxygen atoms and still smaller spheres for hydrogen atoms.

and aggregates of molecules with a tetrahedral structure persist in water at the freezing point. With increase in temperature some of these aggregates break up, causing a further increase in density of the liquid; only at 4°C does the normal expansion due to increase in molecular agitation overcome this effect, and cause water to begin to show the usual decrease in density with increasing temperature.

Example 9-1. The heat of sublimation of ice (to form water vapor) is 51 kJ mole^{-1}. The corresponding values for CH_4, H_2Se, GeH_4, H_2Te, and SnH_4 are 8, 22, 17, 28, and 21 kJ mole^{-1}, respectively. (*a*) Use the values for CH_4, H_2Se, GeH_4, H_2Te, and SnH_4 to estimate the value for a hypothetical form of ice in which the water molecules are held together by van der Waals attraction only. (*b*) Assuming that this value represents the energy of van der Waals attraction in ice, use it to obtain from the observed heat of sublimation of ice a value for the energy of the O—H···O hydrogen bond in ice.

Solution. The molecules H_2Se and GeH_4 are isoelectronic (same number of electrons), as are also H_2Te and SnH_4. For each pair the ratio of their heats of sublimation is approximately 1.32. We assume, as a rough approximation, that the same ratio would apply also to the isoelectronic pair H_2O and CH_4 if H_2O did not form hydrogen bonds. This leads to 11 kJ mole^{-1} for the heat of vaporization of the hypothetical form of ice, and to the value $51 - 11 = 40$ kJ mole^{-1} for the energy of the hydrogen bonds in ice. There are two hydrogen bonds per H_2O molecule; hence the energy per O—H···O bond in ice is $40/2 = 20$ kJ mole^{-1}.

Example 9-2. The heat of sublimation of ammonia is 29 kJ mole^{-1}. Using information in Example 9-1, estimate the energy of the hydrogen bonds in ammonia.

Solution. To obtain a value of the expected contribution of van der Waals attraction to the heat of sublimation of ammonia we assume a linear relationship between the heat of sublimation caused by the van der Waals attraction in the isoelectronic series CH_4, NH_3, H_2O. The value calculated for NH_3 is thus the mean $(8 + 11)/2 = 10$ kJ mole^{-1} using the values for CH_4 and H_2O given and calculated, respectively, in Example 9-1. The hydrogen bond energy of ammonia is thus $29 - 10 = 19$ kJ mole^{-1} per NH_3 molecule. Ammonia has three hydrogen atoms, hydrogen bond donors, but only one electron pair, the hydrogen bond proton acceptors. The electron pair forms three approximately $\frac{1}{3}$ N—H···N hydrogen bonds, each with energy $19/3 = 6$ kJ mole^{-1}. The value 19 kJ mole^{-1} represents the maximum energy of the N—H···N hydrogen bond.

Example 9-3. The heat of fusion of ice is 6.02 kJ mole^{-1} and that of methane is 0.92 kJ mole^{-1}. Using the value in Example 9-1 for the energy of H—O···H bonds in ice, estimate the fraction of the hydrogen bonds in ice that are broken in the process of fusion.

Solution. The energy involved in breaking hydrogen bonds in the melting of ice is $6.02 - 0.92 = 5.10$ kJ mole^{-1}. The total energy of the hydrogen bonds in ice was calculated in Example 9-1 to be 40 kJ mole^{-1}. Therefore the fraction of hydrogen bonds broken in the melting of ice to form liquid water is $5.10/40 =$ approximately $\frac{1}{8}$ or 12%. Actually, it is known from other experiments that liquid water is almost completely hydrogen bonded and this energy of fusion is better interpreted as indicating an increase in strain and distortion of hydrogen bonds in liquid water rather than an actual breaking of the bonds.

9-7. The Properties of Solutions

One of the most striking properties of water is its ability to dissolve many substances, forming *aqueous solutions*. Solutions are very important kinds of matter—important for industry and for life. The ocean is an aqueous solution which contains thousands of components: ions of the metals and nonmetals, complex inorganic ions, many different organic substances. It was in this solution that the first living organisms developed, and from it that they obtained the ions and molecules needed for their growth and life. In the course of time organisms were evolved that could leave this aqueous environment, and move out onto the land and into the air. They achieved this ability by carrying the aqueous solution with them, as tissue fluid, blood plasma, and intracellular fluids containing the necessary supply of ions and molecules.

The properties of solutions have been extensively studied, and it has been found that they can be correlated in large part by some simple laws. These laws and some descriptive information about solutions are discussed in the following sections.

Types of Solutions. Nomenclature

In Chapter 1 a solution was defined as a homogeneous material that does not have a definite composition.

The most common solutions are liquids. Carbonated water, for example, is a *liquid solution* of carbon dioxide in water. Air is a *gaseous solution* of nitrogen, oxygen, carbon dioxide, water vapor, and the argonons. Coinage silver is a *solid solution* or *crystalline solution* of silver and copper. The structure of this crystalline solution is like that of crystalline copper, described in Chapter 2. The atoms are arranged in the same regular way, cubic closest packing, but atoms of silver and atoms of copper follow one another in a largely random sequence.

The concentration of a solute is often expressed as the number of grams per 100 g of solvent or the number of grams per liter of solution. It is often convenient to give the number of gram formula masses per liter of solution (the *formality*), the number of gram molecular masses (moles) per liter of solution (the *molarity*), or the number of equivalent masses (Section 11-3) per liter of solution (the *normality*). Sometimes they are referred to 1000 g of solvent; they are then called the *mass-formality*, *mass-molarity*, and *mass-normality*, respectively.

It is worth noting that a 1 M aqueous solution cannot be made up accurately by dissolving one mole of solute in 1 liter of water, because the volume of the solution is in general different from that of the solvent. Nor is it equal to the sum of the volumes of the components; for example, 1 liter of water and 1 liter of alcohol on mixing give 1.93 liters of solution; there occurs a volume contraction of 3.5%. There is no reliable way of predicting the density of a solution; tables of experimental values for important solutions are given in reference books.

Example 9-4. A solution is made by dissolving 64.11 g of $Mg(NO_3)_2 \cdot 6H_2O$ in water enough to bring the volume to 1 liter. Describe the solution.

Solution. The formula mass of $Mg(NO_3)_2 \cdot 6H_2O$ is 256.43; hence the solution is 0.25 F (0.25 formal) in this substance. The salt is, however, completely ionized in solution, to give magnesium ions Mg^{++} and nitrate ions NO_3^-. The solution is 0.25 M (0.25 molar) in Mg^{++} and 0.50 M in NO_3^-. Because magnesium is bivalent, its equivalent mass is one half its gram molecular mass. Hence the solution is also 0.50 N (0.50 normal) in Mg^{++} and 0.50 N in NO_3^-.

9-8. The Importance of Water as an Electrolytic Solvent

Salts are insoluble in most solvents. Gasoline, benzene, carbon disulfide, carbon tetrachloride, alcohol, ether—these substances are "good solvents" for grease, rubber, organic materials generally; but they do not generally dissolve salts.

The reasons that water is so effective in dissolving salts are that *it has a very high dielectric constant* (about 81 at room temperature) and *its molecules tend to combine with ions, to form hydrated ions*. Both of these properties are related to the large electric dipole moment of the water molecule.

The force of attraction or repulsion of electric charges is inversely proportional to the dielectric constant of the medium surrounding the charges. This means that two opposite electric charges in water attract each other with a force only $\frac{1}{81}$ as strong as in air (or a vacuum). It is clear that the ions of a crystal of sodium chloride placed in water could

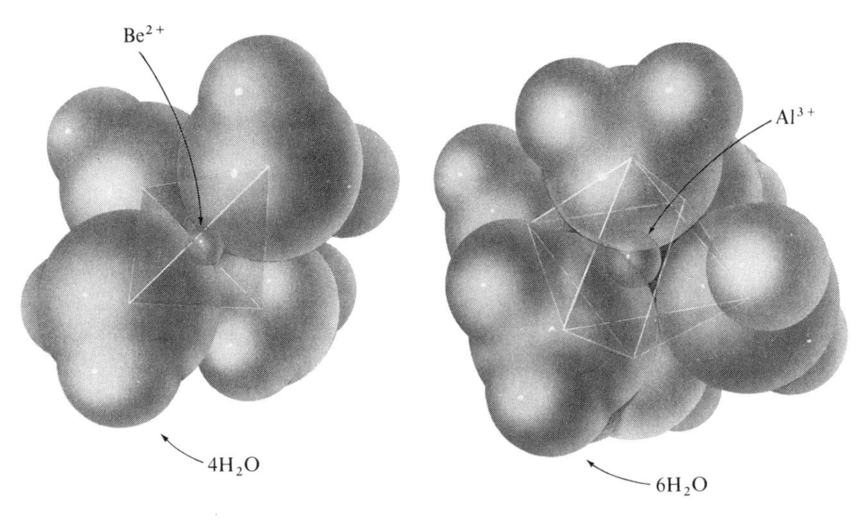

FIGURE 9-9
Diagrams showing the structure of hydrated ions.

dissociate away from the crystal far more easily than if the crystal were in air, since the electrostatic force bringing an ion back to the surface of the crystal from the aqueous solution is only $\frac{1}{81}$ as strong as from air. It is accordingly not surprising that the thermal agitation of the ions in a salt crystal at room temperature is not great enough to cause the ions to dissociate away into the air, but that it is great enough to overcome the relatively weak attraction when the crystal is surrounded by water, thus allowing large numbers of the ions to dissociate into aqueous solution.

The Hydration of an Ion

A related effect that stabilizes the dissolved ions is the formation of *hydrates* of the ions. Each negative ion attracts the positive ends of the adjacent water molecules, and tends to hold several water molecules attached to itself. The positive ions, which are usually smaller than the negative ions, show this effect still more strongly; each positive ion attracts the negative ends of the water molecules, and binds several molecules tightly about itself, forming a hydrate, which may have considerable stability, especially for the bipositive and terpositive cations.

The number of water molecules attached to a cation, its **ligancy,*** is determined by the size of the cation. The small cation Be^{++} forms the tetrahydrate $Be(OH_2)_4^{++}$. A somewhat larger ion, such as Mg^{++} or Al^{+++}, forms a hexahydrate, $Mg(OH_2)_6^{++}$ or $Al(OH_2)_6^{+++}$ (Figure 9-9).

*The ligancy of an atom is the number of atoms bonded to it or in contact with it. The ligancy was formerly called the *coordination number*.

The forces between cations and water molecules are so strong that the ions often retain a layer of water molecules in crystals. This water is called *water of crystallization.* This effect is more pronounced for bipositive and terpositive ions than for unipositive ions. The tetrahedral complex $Be(OH_2)_4^{++}$ occurs in various salts, including $BeCO_3 \cdot 4H_2O$, $BeCl_2 \cdot 4H_2O$, and $BeSO_4 \cdot 4H_2O$, and is no doubt present also in solution. The following salts contain larger ions, with six water molecules in octahedral coordination:

$MgCl_2 \cdot 6H_2O$	$AlCl_3 \cdot 6H_2O$
$Mg(ClO_3)_2 \cdot 6H_2O$	$KAl(SO_4)_2 \cdot 12H_2O$
$Mg(ClO_4)_2 \cdot 6H_2O$	$Fe(NH_4)_2(SO_4)_2 \cdot 6H_2O$
$MgSiF_6 \cdot 6H_2O$	$Fe(NO_3)_2 \cdot 6H_2O$
$NiSnCl_6 \cdot 6H_2O$	$FeCl_3 \cdot 6H_2O$

In a crystal such as $FeSO_4 \cdot 7H_2O$, six of the water molecules are attached to the iron ion, in the complex $Fe(OH_2)_6^{++}$, and the seventh occupies another position, being packed near a sulfate ion of the crystal. In alum, $KAl(SO_4)_2 \cdot 12H_2O$, six of the twelve water molecules are coordinated about the aluminum ion and the other six about the potassium ion.

Crystals also exist in which some or all of the water molecules have been removed from the cations. For example, magnesium sulfate forms the three crystalline compounds $MgSO_4 \cdot 7H_2O$, $MgSO_4 \cdot H_2O$, and $MgSO_4$.

The stability of ions in aqueous solution results from the distribution of the electric charge over a number of atoms in such a way that no atom shows a great deviation from electroneutrality (Section 6-13). Let us consider the hydrated cations $Be(OH_2)_4^{++}$ and $Al(OH_2)_6^{+++}$ shown in Figure 9-9. Both beryllium and aluminum have electronegativity 1.5 and oxygen has electronegativity 3.5. The difference corresponds to a little over 50% of ionic character, enough to transfer half an electronic charge for each bond to the central atom, leaving it approximately neutral. The O—H bonds might have 25% ionic character, which would place the entire charge of the ions on the eight hydrogen atoms of $Be(OH_2)_4^{++}$ and the twelve of $Al(OH_2)_6^{+++}$, each having charge $\frac{1}{4}+$. Moreover, each of these hydrogen atoms may be involved in a weak bond with another water molecule in such a way that its charge is neutralized by interaction with an electron pair of the oxygen atom, the entire charge of the hydrated cations $Be(OH_2)_4(OH_2)_8^{++}$ and $Al(OH_2)_6(OH_2)_{12}^{+++}$ then being distributed over the outermost hydrogen atoms, each with the charge $\frac{1}{8}+$. In fact, this sort of electric polarization of the water continues for long distances; it is responsible for the large dielectric constant of water.

In the discussion of high-energy molecules in Section 8-4 it was mentioned that formation of hydrogen bonds by a molecule such as H_3PO_4 in aqueous solution could make the four oxygen atoms nearly equivalent, permitting nearly complete resonance of the double bond among the four positions. With this resonance each oxygen atom has valence $1\frac{1}{4}$ satisfied by the bonds to phosphorus, leaving $\frac{3}{4}$ for bonds to hydrogen. If each of the three OH groups uses its hydrogen atom in a weak bond (a one-quarter bond) to the oxygen atom of a water molecule, the other three-quarters of a bond will suffice to make the phosphate oxygen atoms electrically neutral. Similarly, the phosphate oxygen without a hydrogen atom can form weak (one-quarter) bonds with hydrogen atoms of three neighboring water molecules, making it also electrically neutral.

Each of the four oxygen atoms of the phosphate ion, PO_4^{---}, may similarly form hydrogen bonds with three water molecules. The electric charge of the hydrated ion $PO_4(HOH)_{12}^{---}$, would then be distributed over the twelve outer oxygen atoms, each with charge $\frac{1}{4}-$. Similar hydrated structures are formed by the ions $(HO)_2PO_2^-$ and $HOPO_3^{--}$, which are present in about equal amounts in fluids of living organisms.

Clathrate Compounds

The argonons, simple hydrocarbons, and many other substances form crystalline hydrates; for example, xenon forms the hydrate $Xe \cdot 5\frac{3}{4}H_2O$, stable at about 2°C and partial pressure of xenon 1 atm, and methane forms a similar hydrate, $CH_4 \cdot 5\frac{3}{4}H_2O$. X-ray investigation has shown that these crystals have a structure in which the water molecules form a hydrogen-bonded framework, resembling that of ice in that each water molecule is tetrahedrally surrounded by four others, at 276 pm, but with a more open arrangement, such as to provide cavities (pentagonal dodecahedra and other polyhedra with pentagonal or hexagonal faces) that are big enough to permit occupancy by the argonon atoms or other molecules. Crystals of this sort are called clathrate crystals.

The structure of xenon hydrate and the hydrates of argon, krypton, methane, chlorine, bromine, hydrogen sulfide, and some other substances is shown in Figure 9-10. The cubic unit of structure has edge about 1200 pm and contains 46 water molecules. Chloroform hydrate, $CHCl_3 \cdot 17H_2O$, has a somewhat more complicated structure, in which the chloroform molecule is surrounded by a 16-sided polyhedron formed by 28 water molecules.

Clathrate compounds also can be made in which the hydrogen-bonded framework is formed by organic molecules such as urea, $(H_2N)_2CO$.

An interesting mechanism for the mode of action of chemically inert anesthetic agents such as halothane, $F_3CCBrClH$, and xenon has been

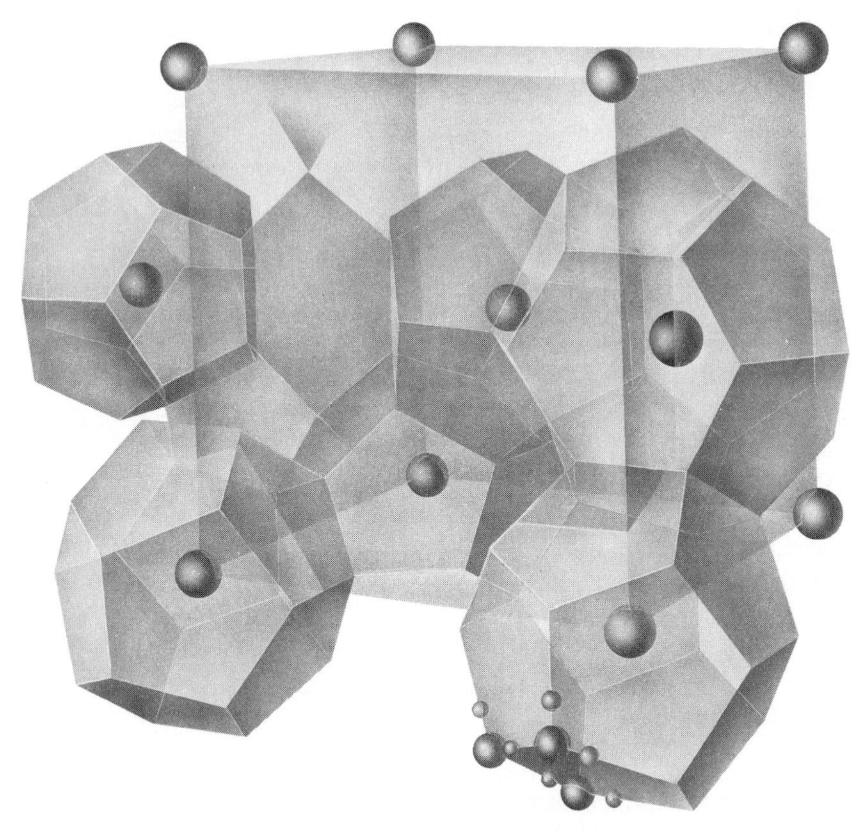

FIGURE 9-10
The structure of a clathrate crystal, xenon hydrate. The xenon atoms occupy cavities (eight per unit cube) in a hydrogen-bonded three-dimensional network formed by the water molecules (46 per unit cube). The O—H···O distance is 276 pm, as in ice. Two xenon atoms, at 0, 0, 0 and $\frac{1}{2}, \frac{1}{2}, \frac{1}{2}$, are at the centers of nearly regular pentagonal dodecahedra. The other six, at $0, \frac{1}{4}, \frac{1}{2}$; $0, \frac{3}{4}, \frac{1}{2}$; $\frac{1}{2}, 0, \frac{1}{4}$; $\frac{1}{2}, 0, \frac{3}{4}$; $\frac{1}{4}, \frac{1}{2}, 0$; and $\frac{3}{4}, \frac{1}{2}, 0$, are at the centers of tetrakaidecahedra. Each tetrakaidecahedron (one is outlined, right center) has 24 corners (water molecules), two hexagonal faces, and 12 pentagonal faces.

proposed in which the anesthetic disrupts the water structure of intracellular or intercellular fluid by forming clathrate structures that interfere with the normal intercellular communication systems. Local anesthetic agents are different in their action. Their molecules can form hydrogen bonds, and they probably exert their action by combining with protein molecules or other constituents of nerves.

Other Electrolytic Solvents

Some liquids other than water can serve as ionizing solvents, with the power of dissolving electrolytes to give electrically conducting solutions. These liquids include hydrogen peroxide, hydrogen fluoride, liquid ammonia, and hydrogen cyanide. All of these liquids, like water, have large dielectric constants. Liquids with low dielectric constants, such as benzene and carbon disulfide, do not act as ionizing solvents.

Liquids with large dielectric constants are sometimes called *polar liquids*.

The large dielectric constant of water, which is responsible for the striking power of water to dissolve ionic substances, is due in part to its power to form hydrogen bonds. The hydrogen bonds help the water molecules to line up in such a direction as to neutralize part of the electric field. Hydrogen bonds are also formed in the other liquids [hydrogen peroxide, hydrogen fluoride, ammonia (boiling point −33.4°C), and hydrogen cyanide] that can dissolve ionic substances.

9-9. Solubility

An isolated system is in *equilibrium* when its properties, in particular the distribution of components among phases, remain constant with the passage of time.

If the system in equilibrium contains a solution and another phase that is one of the components of the solution in the form of a pure substance, the concentration of that substance in the solution is called the *solubility* of the substance. The solution is called a *saturated solution*.

For example, at 0°C a solution of borax containing 1.3 g of anhydrous sodium tetraborate, $Na_2B_4O_7$, in 100 g of water is in equilibrium with the solid phase $Na_2B_4O_7 \cdot 10H_2O$, sodium tetraborate decahydrate; on standing the system does not change, the composition of the solution remaining constant. The solubility of $Na_2B_4O_7 \cdot 10H_2O$ in water is hence 1.3 g of $Na_2B_4O_7$ per 100 g, or, correcting for the water of hydration, 2.5 g of $Na_2B_4O_7 \cdot 10H_2O$ per 100 g.

Change in the Solid Phase

The solubility of $Na_2B_4O_7 \cdot 10H_2O$ increases rapidly with increasing temperature; at 60°C it is 20.3 g of $Na_2B_4O_7$ per 100 g (Figure 9-11). If the system is heated to 70° and held there for some time, a new phenomenon occurs. A third phase appears, a crystalline phase with composition $Na_2B_4O_7 \cdot 5H_2O$, and the other solid phase disappears. At this temperature the solubility of the decahydrate is greater than that of the pentahydrate; a solution saturated with the decahydrate is supersaturated with respect to the pentahydrate, and will deposit crystals of the pentahydrate.*

*The addition of "seeds" (small crystals of the substance) is sometimes necessary to cause the process of crystallization to begin.

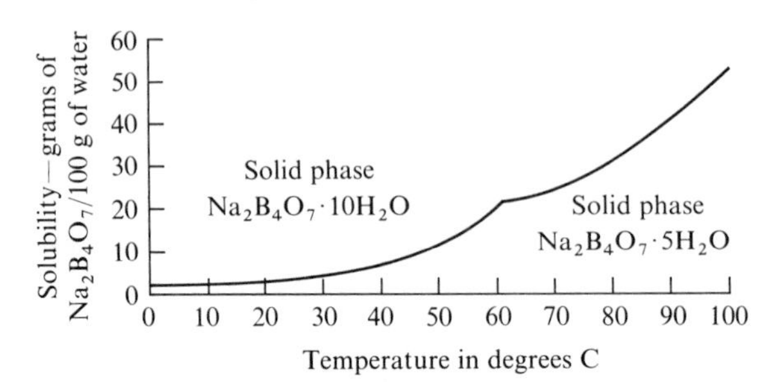

FIGURE 9-11
Solubility of sodium tetraborate in water.

The process of solution of the unstable phase and crystallization of the stable phase will then continue until none of the unstable phase remains.*

In this case the decahydrate is less soluble than the pentahydrate below 61°C, and is hence the stable phase below this temperature. The solubility curves of the two hydrates cross at 61°, the pentahydrate being stable in contact with solution above this temperature.

Change other than solvation may occur in the stable solid phase. Thus rhombic sulfur is less soluble in suitable solvents than is monoclinic sulfur at temperatures below 95.5°C, the transition temperature between the two forms; above this temperature the monoclinic form is the less soluble. The principles of thermodynamics require that the temperature at which the solubility curves of the two forms cross be the same for all solvents, and be also the temperature at which the vapor-pressure curves intersect.

The Dependence of Solubility on Temperature

The solubility of a substance may either increase or decrease with increasing temperature. An interesting case is provided by sodium sulfate. The solubility of $Na_2SO_4 \cdot 10H_2O$ (the stable solid phase below 32.4°C) increases very rapidly with increasing temperature, from 5 g of Na_2SO_4 per 100 g of water at 0° to 55 g at 32.4°C. Above 32.4° the stable solid phase is Na_2SO_4; the solubility of this phase decreases rapidly with increasing temperature, from 55 g at 32.4° to 42 g at 100° (Figure 9-12).

Most salts show increased solubility with increase in temperature; a good number (NaCl, K_2CrO_4) change only slightly in solubility with increase in temperature; and a few, such as Na_2SO_4, $FeSO_4 \cdot H_2O$, and $Na_2CO_3 \cdot H_2O$, show decreased solubility (Figures 9-12 and 9-13).

---

*The third hydrate of sodium tetraborate, kernite, $Na_2B_4O_7 \cdot 4H_2O$, is more soluble than the other phases.

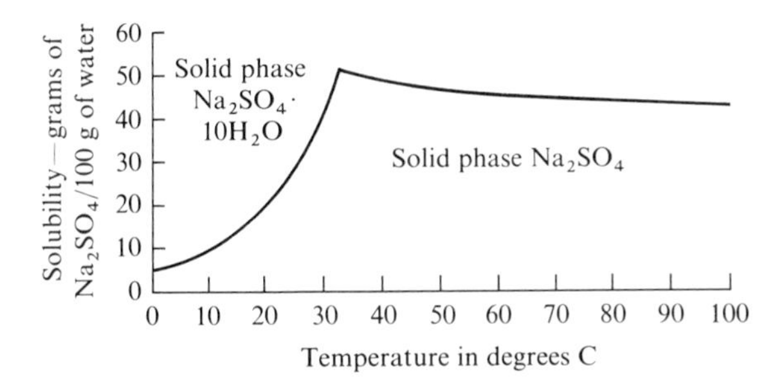

FIGURE 9-12
Solubility of sodium sulfate in water.

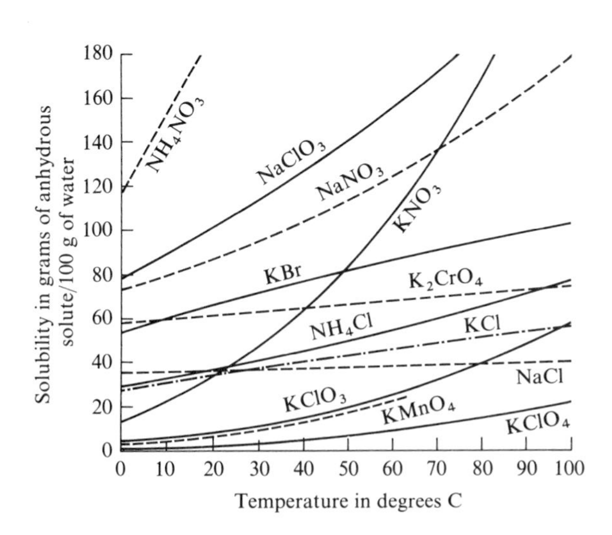

FIGURE 9-13
Solubility curves for some salts in water.

9-10. The Dependence of Solubility on the Nature of Solute and Solvent

Substances vary greatly in their solubilities in various solvents. There are a few general rules about solubility, which, however, apply in the main to organic compounds.

One of these rules is that *a substance tends to dissolve in solvents that are chemically similar to it*. For example, the hydrocarbon naphthalene, $C_{10}H_8$, has a high solubility in gasoline, which is a mixture of hydrocarbons; it has a somewhat smaller solubility in ethyl alcohol, C_2H_5OH, whose molecules consist of short hydrocarbon chains with hydroxide groups attached, and a very small solubility in water, which is much different from a hydrocarbon. On the other hand, boric acid, $B(OH)_3$, a hydroxide compound, is moderately soluble in both water and alcohol, which themselves contain hydroxide groups, and is insoluble in gasoline. In fact, the three solvents themselves show the same phenomenon—both gasoline and water are soluble in alcohol, whereas gasoline and water dissolve in each other only in very small amounts.

The explanation of these facts is the following. Hydrocarbon groups (involving only carbon and hydrogen atoms) attract hydrocarbon groups only weakly, as is shown by the low melting and boiling points of hydrocarbons, relative to other substances with similar molecular masses. But hydroxide groups and water molecules show very strong intermolecular attraction; the melting point and boiling point of water are higher than those of any other substance with low molecular mass. This strong attraction is due to the partial ionic character of the O—H bonds, which places electric charges on the atoms. The positively charged hydrogen atoms are then attracted to the negative oxygen atoms of other molecules, forming hydrogen bonds and holding the molecules firmly together.

The term *hydrophilic* is often said to describe substances or groups that attract water, with the term *hydrophobic* used for substances or groups that repel water and attract hydrocarbons. In fact, the molecules of a hydrophobic substance exert an electronic van der Waals attraction on water molecules as well as on hydrocarbon molecules. For example, the solubility at 25°C of water vapor at 0.0313 atm pressure (the vapor pressure of liquid water at this temperature) in kerosene (a mixture of hydrocarbons) is 72 mg per kg of solvent, whereas the solubility of methane at the same partial pressure is somewhat less, 10 mg per kg of kerosene. The water molecules are attracted somewhat more strongly by the kerosene molecules than are the methane molecules. The difference between water and methane is that at a higher partial pressure water vapor condenses to liquid water, which is stabilized by the intramolecular hydrogen bonds, whereas methane remains a gas.

The solubility of methane in polar solvents is nearly the same as in nonpolar solvents; in alcohols from methanol, CH_3OH, to pentanol, $C_5H_{11}OH$, it varies from 72 percent to 80 percent of the value in kerosene. The forces of van der Waals attraction of the solvent molecules for the methane molecule are nearly the same for the various solvents. On the other hand, the solubility of water vapor at 0.313 atm in pentanol is 1400 times that in kerosene, and water is miscible in all proportions with the lighter alcohols.

The solubility of substances with small nonpolar molecules, such as oxygen, nitrogen, and methane, in water is less than that in nonpolar solvents, about one-tenth as great. Substances with larger nonpolar molecules are essentially insoluble in water, and usually easily soluble in nonpolar solvents. The action of water in excluding these molecules is attributed to the breaking or bending of water-water hydrogen bonds that would be needed to provide a cavity for them. The reason that substances such as gasoline and naphthalene do not dissolve in water is that their molecules in solution would prevent water molecules from forming as many strong hydrogen bonds as in pure water; on the other hand, boric acid is soluble in water because the decrease in the number of water-water bonds is compensated by the formation of strong hydrogen bonds between the water molecules and the hydroxide groups of the boric acid molecules.

9-11. Solubility of Salts and Hydroxides in Water

In the study of inorganic chemistry, especially qualitative analysis, it is useful to know the approximate solubility of common substances. The simple rules of solubility are given below. These rules apply to compounds of the common cations Na^+, K^+, NH_4^+, Mg^{++}, Ca^{++}, Sr^{++}, Ba^{++}, Al^{+++}, Cr^{+++}, Mn^{++}, Fe^{++}, Fe^{+++}, Co^{++}, Ni^{++}, Cu^{++}, Zn^{++}, Ag^+, Cd^{++}, Sn^{++}, Hg_2^{++}, Hg^{++}, and Pb^{++}. By "soluble" it is meant that the solubility is more than about 1 g per 100 ml (roughly 0.1 *M* in the cation), and by "insoluble" that the solubility is less than about 0.1 g per 100 ml (roughly 0.01 *M*); substances with solubilities within or close to these limits are described as *sparingly soluble*.

Class of mainly soluble substances:

All *nitrates* are soluble.

All *acetates* are soluble.

All *chlorides, bromides,* and *iodides* are soluble except those of silver, mercurous mercury (mercury with oxidation number +1), and lead. $PbCl_2$ and $PbBr_2$ are sparingly soluble in cold water (1 g per 100 ml at 20°C) and more soluble in hot water (3 g, 5 g, respectively, per 100 ml at 100°C).

All *sulfates* are soluble except those of barium, strontium, and lead. $CaSO_4$, Ag_2SO_4, and Hg_2SO_4 (mercurous sulfate) are sparingly soluble.

All salts of *sodium, potassium,* and *ammonium* are soluble except $NaSb(OH)_6$ (sodium antimonate), K_2PtCl_6 (potassium hexachloroplatinate), $(NH_4)_2PtCl_6$, $K_3Co(NO_2)_6$ $(NH_4)_3Co(NO_2)_6$, and $KClO_4$.

Class of mainly insoluble substances:

All *hydroxides* are insoluble except those of the alkali metals, ammonium, and barium. $Ca(OH)_2$ and $Sr(OH)_2$ are sparingly soluble.

All normal *carbonates* and *phosphates* are insoluble except those of the alkali metals and ammonium. Many hydrogen carbonates and phosphates, such as $Ca(HCO_3)_2$, $Ca(H_2PO_4)_2$, etc., are soluble.

All *sulfides* except those of the alkali metals, ammonium, and the alkaline-earth metals are insoluble.*

9-12. The Solubility of Gases in Liquids

Air is somewhat soluble in water: at room temperature (20°C) one liter of water dissolves 19.0 ml of air at 1 atm pressure. (The amount of dissolved air decreases with increasing temperature.) If the pressure is doubled, the solubility of air is doubled. This proportionality of the solubility of air to its pressure illustrates *Henry's law*,† which may be stated in the following way: *At constant temperature, the partial pressure in the gas phase of one component of a solution is, at equilibrium, proportional to the concentration of the component in the solution, in the region of low concentration.*

This is equivalent to saying that *the solubility of a gas in a liquid is proportional to the partial pressure of the gas.*

Example 9-5. The solubility of atmospheric nitrogen‡ in water at 0°C and 1 atm is 23.54 ml l^{-1}, and that of oxygen is 48.89 ml l^{-1}. Air contains 79% N_2 and 21% O_2 by volume. What is the composition of the dissolved air?

Solution. The solubilities of nitrogen and oxygen at partial pressures 0.79 and 0.21 atm respectively are $0.79 \times 23.54 = 18.60$ ml l^{-1} and $0.21 \times 48.89 = 10.27$ ml l^{-1} respectively. The composition of the dissolved air is hence $\frac{18.60}{18.60 + 10.27} = 64.4\%$ nitrogen and $\frac{10.27}{18.60 + 10.27} = 35.6\%$ oxygen.

The solubilities of most gases in water are of the order of magnitude of that of air. Exceptions are those gases which combine chemically with water or which dissociate largely into ions, including CO_2 (solubility 1,713 ml l^{-1} at 0°C), H_2S (4,670), and SO_2 and NH_3, which are extremely soluble.

---

*The sulfides of aluminum and chromium are hydrolyzed by water, precipitating $Al(OH)_3$ and $Cr(OH)_3$.

†Discovered by the English chemist William Henry (1775–1836).

‡98.8% N_2 and 1.2% A.

The Partition of a Solute Between Two Solvents

If a solution of iodine in water is shaken with chloroform, most of the iodine is transferred to the chloroform. The ratio of concentrations of iodine in the two phases, called the *distribution ratio*, is a constant in the range of small concentrations of the solute in each solution. For iodine in chloroform and water the value of this ratio at room temperature is 250; hence whenever a solution of iodine in chloroform is shaken with water or a solution in water is shaken with chloroform until equilibrium is reached, the iodine concentration in the chloroform phase is 250 times that in the water phase.

It is seen on consideration of the various equilibria that the distribution ratio of a solute between two solvents is equal to the ratio of the solubilities of the solute (as a crystalline, liquid, or gaseous phase) in the two solvents, provided that the solubilities are small.

The method of shaking a solution with an immiscible solvent is of great use in organic chemistry, especially the chemistry of natural products, for removing one of several solutes from a solution. In inorganic chemistry it is useful in another way—for determining concentrations of particular molecular species. Thus iodine combines with iodide ion to form the tri-iodide ion: $I_2 + I^- \longrightarrow I_3^-$. The concentration of molecular iodine, I_2, in a solution containing both I_2 and I_3^- can be determined by shaking with chloroform, analyzing the chloroform solution, and dividing by the distribution ratio. (Tri-iodide ion is not soluble in chloroform.)

9-13. The Freezing Point and Boiling Point of Solutions

It is well known that the freezing point of a solution is lower than that of the pure solvent; for example, in cold climates it is customary to add a solute such as alcohol or glycerol or ethylene glycol to the radiator water of automobiles to keep it from freezing. Freezing-point lowering by the solute also underlies the use of a salt-ice mixture for cooling, as in freezing ice cream; the salt dissolves in the water, making a solution, which is in equilibrium with ice at a temperature below the freezing point of water.

Let us consider what happens as a salt solution (concentration 1 molar, say) is cooled, with solid carbon dioxide, for example. The temperature falls a little below the freezing point of the solution, −3.4°C, and then, as ice begins to form, it rises to this value and remains constant. As ice continues to form, however, the salt concentration of the solution slowly increases, and its freezing point drops. When half of the water has frozen to ice the solution is 2 M in NaCl, and the temperature is −6.9°C. Ice forms, the solution increases in concentration, and the temperature drops until it reaches the value −21.1°C. At this temperature the solution becomes saturated with respect to the solute, which begins to crystallize out as the solid phase $NaCl \cdot 2H_2O$ (sodium chloride dihydrate). The system then remains at this temperature, called the *eutectic temperature,* as the fine-grained mixture of two solid phases, ice and $NaCl \cdot 2H_2O$. This mixture is called the *eutectic mixture* or *eutectic*.

It is found by experiment that the freezing-point lowering of a dilute solution is proportional to the concentration of the solute. In 1883 the French chemist F. M. Raoult made the very interesting discovery that *the mass-molar freezing-point lowering produced by different solutes is the same for a given solvent.* Thus the following freezing points are observed for 0.1 *M* solutions of the following solutes in water:

Hydrogen peroxide	H_2O_2	−0.186°C
Methanol	CH_3OH	−0.181
Ethanol	C_2H_5OH	−0.183
Dextrose	$C_6H_{12}O_6$	−0.186
Sucrose	$C_{12}H_{22}O_{11}$	−0.188

The *molar freezing-point constant* for water has the value 1.86°C, the freezing point of a solution containing *c* moles of solute per 1000 g of water being −1.86 *c* in degrees C. For other solvents the values of this constant are shown in Table 9-1.

TABLE 9-1

Solvent	Freezing Point	Molar* Freezing-Point Constant
Benzene	5.6°C	4.90°C
Acetic acid	17	3.90
Phenol	40	7.27
Camphor	180	40

*Molar = mass-molar, moles per 1000 g of solvent.

The Determination of Molecular Mass by the Freezing-Point Method

The freezing-point method is a very useful way of determining the molecular masses of substances in solution. Camphor, with its very large constant, is of particular value for the study of organic substances.

Example 9-6. The freezing point of a solution of 0.244 g of benzoic acid in 20 g of benzene was observed to be 5.232°C, and that of pure benzene to be 5.478°. What is the molecular mass of benzoic acid in this solution?

Solution. The solution contains $\frac{0.244 \times 1{,}000}{20} = 12.2$ g of benzoic acid per 1,000 g of solvent. The number of moles of solute per 1,000 g of solvent is found from the observed freezing-point lowering 0.246° to be $\frac{0.246}{4.90} = 0.0502$. Hence the molecular mass is $\frac{12.2}{0.0502} = 243$. The explanation of this high value (the formula mass for benzoic acid, C_6H_5COOH, being 122.05) is that in this solvent the substance forms double molecules, $(C_6H_5COOH)_2$.

Evidence for Electrolytic Dissociation

One of the strongest arguments advanced by the Swedish chemist Svante Arrhenius in 1887 in support of the theory of electrolytic dissociation of salts into ions in aqueous solution was the fact that the freezing-point lowering of salt solutions is much larger than that calculated for undissociated molecules, the observed lowering for a salt such as NaCl or $MgSO_4$ in very dilute solution being just twice as great and for a salt such as Na_2SO_4 or $CaCl_2$ just three times as great as expected.

Elevation of Boiling Point

The boiling point of a solution is higher than that of the pure solvent by an amount proportional to the molar concentration of the solute. Values of the proportionality factor, the *molar boiling-point constant,* are given in Table 9-2 for some important solvents.

Boiling-point data for a solution can be used to obtain the molecular weight of the solute in the same way as freezing-point data.

TABLE 9-2

Solvent	Boiling Point	Molar Boiling-Point Constant
Water	100°C	0.52°C
Ethyl alcohol	78.5	1.19
Ethyl ether	34.5	2.11
Benzene	79.6	2.65

The Vapor Pressure of Solutions

It was found experimentally by Raoult in 1887 that the partial pressure of solvent vapor in equilibrium with a dilute solution is directly proportional to the mole fraction of solvent in the solution. It can be expressed by the equation

$$P = P_0 x$$

in which P is the partial pressure of the solvent above the solution, P_0 is the vapor pressure of the pure solvent, and x is the mole fraction of solvent in the solution.

Determination of Molecular Mass from Vapor Pressure

Raoult's law permits the direct calculation of the effective molecular mass of a solute from data on the solvent vapor pressure of the solution, as shown by the following example.

Example 9-7. A 10-g sample of an unknown nonvolatile substance is dissolved in 100 g of benzene, C_6H_6. A stream of air is then bubbled through the solution, and the loss in weight of the solution (through saturation of the air with benzene vapor) is determined as 1.205 g. The same volume of air passed through pure benzene at the same temperature caused a loss of 1.273 g. What is the molecular mass of the solute?

Solution. The loss in weight by evaporation of benzene is proportional to the vapor pressure. Hence the mole fraction of benzene in the solution is 1.205/1.273 = 0.947, and the mole fraction of solute is 0.053. The molecular mass of benzene, C_6H_6, is 78; and the number of moles of benzene in 100 g is 100/78. Let x be the molecular mass of the solute; the number of moles of solute in the solution (containing 10 g of solute) is $10/x$. Hence

$$\frac{10/x}{100/78} = \frac{0.053}{0.947}$$

or

$$x = \frac{78}{100} \times \frac{0.947}{0.053} \times 10 = 139$$

This is the molecular mass of the substance.

The Osmotic Pressure of Solutions

If red blood corpuscles are placed in pure water they swell, become round, and finally burst. This is the result of the fact that the cell wall is permeable to water but not to some of the solutes of the cell solution (hemoglobin, other proteins); in the effort to reach a condition of equilibrium (equality of water vapor pressure) between the two liquids water enters the cell. If the cell wall were sufficiently strong, equilibrium would be reached when the hydrostatic pressure in the cell had reached a certain value, at which the water vapor pressure of the solution equals the vapor pressure of the pure water outside the cell. This equilibrium hydrostatic pressure is called the *osmotic pressure* of the solution.

A semipermeable membrane is a membrane with very small holes in it, of such a size that molecules of the solvent are able to pass through but molecules of the solute are not. A useful semipermeable membrane for measurement of osmotic pressure is made by precipitating cupric ferrocyanide, $Cu_2Fe(CN)_6$, in the pores of an unglazed porcelain cup, which gives the membrane mechanical support to enable it to withstand high pressures. Accurate measurements have been made in this way to over 250 atm. Cellophane membranes may also be used, if the osmotic pressure is not large.

It is found experimentally that the osmotic pressure of a dilute solution satisfies the equation

$$\pi V = n_1 RT$$

with n_1 the number of moles of solute (to which the membrane is impermeable) in volume V, π the osmotic pressure, R the gas constant, and T the absolute temperature. This relation was discovered by van't Hoff in 1887. It is striking that the equation is identical in form with the perfect-gas equation; van't Hoff emphasized the similarity of a dissolved substance and a gas.

For inorganic substances and simple organic substances the osmotic-pressure method of determining molecular mass offers no advantages over other methods, such as the measurement of freezing-point lowering. It has, however, been found useful for substances of very high molecular mass; the molecular mass of hemoglobin was first reliably determined in this way by Adair in 1925. The value found by Adair, 68,000, has been verified by measurements made with the ultracentrifuge.

Example 9-8. (*a*) Horse hemoglobin is a protein in the red cells of the blood that is found on analysis of the dehydrated substance to contain 0.328% iron. What is the minimum molecular mass of horse hemoglobin?

(*b*) Adair in one experiment found that a solution with 80 g of hemoglobin per liter gave the osmotic pressure $\pi = 0.026$ atm at 4°C. What is the correct molecular mass?

Solution. (*a*) The smallest possible molecule would contain one iron atom, atomic weight 55.85, and would have molecular mass 55.85/0.00328 = 17,027.

(*b*) From the above osmotic-pressure equation with $\pi = 0.026$ atm, $V = 1$ l, $R = 0.082$ l atm deg^{-1} mole^{-1}, and $T = 277$°K, we obtain

$$n_1 = \frac{0.026}{0.082 \times 277} = 0.00114$$

Hence the molecular mass is $80/0.00114 \cong 70{,}000$. The number of significant figures suggests that the iron analysis is about ten times as accurate as the osmotic-pressure measurement. We conclude from the latter that the molecule contains four atoms of iron, and that the molecular mass is $4 \times 17{,}027 = 68{,}100$.

9-14. Colloidal Solutions and Dispersions

It was found by the British chemist Thomas Graham (1805–1869) in the years around 1860 that substances such as glue, gelatin, albumin, and starch diffuse very slowly in solution, their diffusion rates being as small as one-hundredth of those for ordinary solutes (such as salt or sugar). Graham also found that substances of these two types differ markedly in their ability to pass through a membrane such as parchment paper or collodion; if a solution of sugar and glue is put into a collodion bag and the bag is placed in a stream of running water, the sugar soon dialyzes through the bag into the water, and the glue remains behind. This process of dialysis gives a useful method of separating substances of these two kinds.

We now recognize that these differences in ability to pass through the pores of a membrane and in rates of diffusion are due to differences in size of the solute molecules. Graham thought that there was a deeper difference between ordinary, easily crystallizable substances and the slowly diffusing, nondialyzing substances, which he was unable to crystallize. He named the substances of the latter class *colloids* (Greek *kolla,* glue), in contradistinction to ordinary *crystalloids*. It is now known that there is not a sharp line of demarcation between the two classes (many substances of large molecular mass have been crystallized), but it has been found convenient to retain the name "colloid" for substances of large molecular mass.

Some colloids consist of well-defined molecules, with constant molecular weight and definite molecular shape, permitting them to be ordered in a crystalline array. Proteins have molecular masses ranging from about 10,000 to several hundred thousand.

Graham introduced the words *sol* for a colloidal solution (a dispersion of a solid substance in a fluid medium) and *gel* for a dispersion that has developed a structure that prevents it from being mobile. A solution of gelatin in water at high temperatures is a sol and at low temperatures a gel. A *hydrosol* is a dispersion in water, and an *aerosol* is a dispersion of a solid substance in air.

Inorganic sols may be made by dispersing a solid substance that is normally insoluble, such as gold, ferric oxide, and arsenious sulfide, in water. Gold sols, made by adding a reducing agent to a dilute solution of gold chloride, were known to the alchemists of the seventeenth century, and were studied by Michael Faraday. They often have striking colors—ruby red, blue, green, and others—that are the result of diffraction of light by the gold sol particles with dimensions approaching the wavelengths of light. The sols are stabilized by the presence of electric charge on the surface of the particles, a negative charge in the case of gold sols. Faraday found that on addition of a small amount of a salt the ruby gold sols turned blue. This is the result of the aggregation of smaller particles to form larger ones, which scatter light of longer wavelengths. With more salt the particles coagulate. The coagulation results from the action of small ions with opposite charge (Na^+, Mg^{++}, Al^{+++}) attaching themselves to the negative charges on the surface of the gold particles and neutralizing the charge, so as to permit the particles to approach closely enough to form the coagulum. The coagulating power of the cations is approximately proportional to the sixth power of their charge: an aluminum salt is effective as a coagulant at a concentration 700 times less than a sodium salt.

An *emulsion* is a colloidal dispersion of one liquid in another. The most common emulsions are those of oil in water or of water in oil. Emulsions are stabilized by the presence of emulsifying agents, such as soap, proteins, bile salts, gums, or carbohydrates. A molecule of an effective emulsifying agent may usually be described as having one end soluble in oil and the other end soluble in water. The oil-soluble end may be an alkyl chain, and the water-soluble end an ionic group (carboxylate ion, ammonium ion) or a hydrogen-bond-forming group, such as hydroxyl. Emulsifying agents such as soap are used for dispersing solid fatty substances in water, as well as liquid oils.

Soaps and Other Surfactants

Ordinary soap is made by heating fat in an iron kettle with a strong aqueous solution of sodium hydroxide until the fat is completely hydrolyzed. Fats are the glyceryl esters of long-chain fatty acids, such as palmitic acid (Section 13-5). For example, glyceryl tripalmitate has the structural formula

$$
\begin{array}{l}
H_2C-O-\overset{\displaystyle O}{\overset{\|}{C}}-(CH_2)_{14}CH_3 \\
\;\;| \\
HC-O-\overset{\displaystyle O}{\overset{\|}{C}}-(CH_2)_{14}CH_3 \\
\;\;| \\
H_2C-O-\overset{\displaystyle O}{\overset{\|}{C}}-(CH_2)_{14}CH_3
\end{array}
$$

A soap made by the reaction of this fat with sodium hydroxide consists of sodium palmitate, $C_{15}H_{31}COONa$. The sodium soaps are solid at room temperature, and the potassium soaps are liquid. Potassium soaps (soft soap) used to be made by heating wood ashes, fat, and water (wood ashes contain potassium hydroxide or potassium carbonate).

Soap is effective as a cleaning agent because of its ability to emulsify fats and oil in water. The fatty acid anions, such as palmitate ion, form a layer around an oil droplet, as shown in Figure 9-14. The sodium ions dissolve in the water, and the negatively charged carboxylate ends of the fatty acid anions remain in the aqueous phase, at its interface with the oil droplet. The hydrocarbon chains of the anions interact more strongly with the fats, however, than with water (a hydrocarbon is much more soluble in oil than in water), and they extend into the oil droplet, constituting the oil side of the interface. We might say that the ionic (polar) ends of the soap molecules are dissolved in the aqueous phase and the hydrocarbon (nonpolar) ends are dissolved in the oil phase; because these ends are attached to one another an interface is produced.

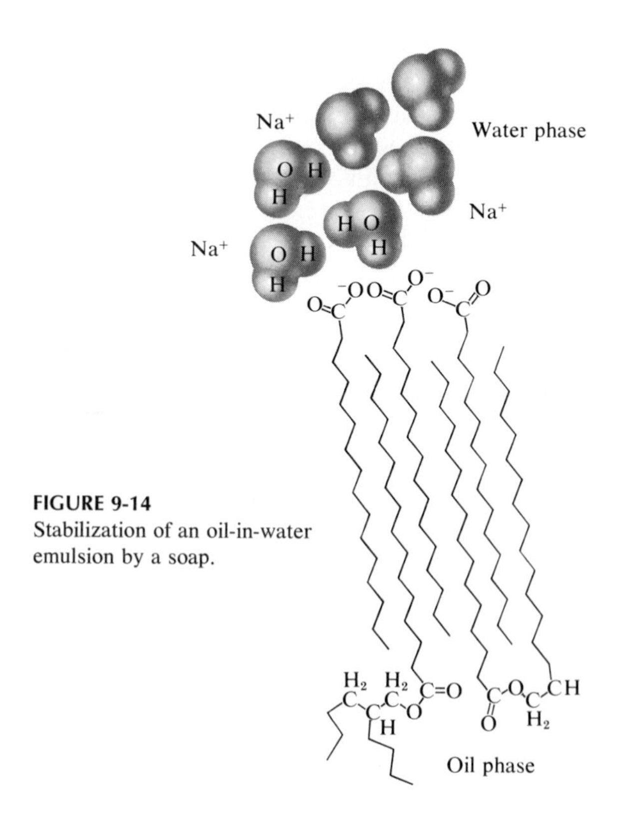

FIGURE 9-14
Stabilization of an oil-in-water emulsion by a soap.

Substances that affect the properties of surfaces (interfaces between two phases) are called *surface-active agents* or *surfactants*. In practice the term is restricted to substances effective when one of the phases is aqueous. A surfactant used for cleansing is called a *detergent*.

There are many kinds of synthetic detergents. These detergents have displaced soap in most household and industrial uses. A common detergent has the sulfonate ion group, $—SO_3^-$, in place of the carboxylate group, $—CO_2^-$. Ordinary soap forms with hard water a greasy precipitate of the calcium salt of the fatty acid. The sulfonate detergents have the advantage of not forming this precipitate, because of the greater solubility of the calcium salts.

Soaps and other detergents have valuable antiseptic properties. The cationic surfactants are especially valuable for their bacteriostatic and bactericidal properties. An example is cetylpyridinium chloride, which contains the hexadecylpyridinium cation:

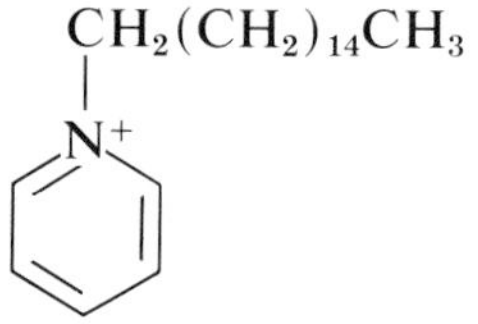

$CH_2(CH_2)_{14}CH_3$

N^+

It is a constituent of many oral antibacterial mouthwashes and throat lozenges. It is probably effective in bacteriostasis by interfering with the functioning of cell membranes. Another surfactant used in oral antibacterial preparations is 4-hexylresorcinol:

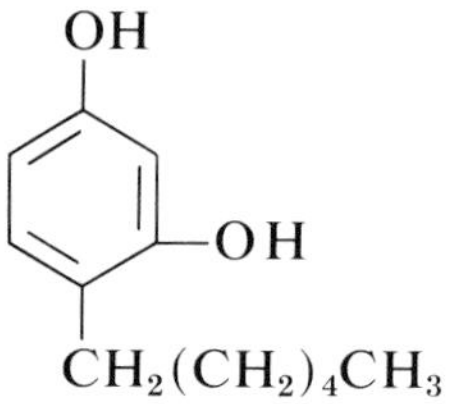

OH

—OH

$CH_2(CH_2)_4CH_3$

Resorcinol, $C_6H_4(OH)_2$, is soluble in water but not in oils, whereas hexane is insoluble in water and soluble in oils. The solubility of resorcinol in water and the surfactant properties of hexylresorcinol are the result of hydrogen-bond formation by the hydroxyl groups with water molecules.

EXERCISES

9-1. A zeolite is a mineral containing sodium ions, which can be replaced by calcium ions. It can be regenerated by passing a concentrated solution of sodium chloride over it. Write the fundamental chemical equations for the softening of water by zeolite and for the regeneration of the zeolite.

9-2. Write the fundamental chemical equations for the removal of most of the ionic impurities in water by the "ion-exchange" process. Why do you suppose this process is sometimes preferred to distillation for the preparation of moderately pure water for industrial use? What do you think is the simplest method of determining when the absorbers in tanks A and B are saturated with ions and should be regenerated?

9-3. In softening water, aluminum sulfate or ferric sulfate is often added as well as calcium hydroxide, with the formation of a flocculent precipitate of aluminum hydroxide or ferric hydroxide. Write equations for the formation of these two hydroxides. Why are these hydroxides useful in the process of purifying water?

9-4. What explanation can you give of the fact that calcium fluoride, CaF_2 (the mineral fluorite), is a crystalline substance with high melting point, whereas stannic chloride, $SnCl_4$, is an easily volatile liquid?

9-5. Describe the structure of ice. Explain why ice floats, and mention some ways in which this property affects our lives.

9-6. By reference to Figure 9-5, estimate the melting points and boiling points that hydrogen fluoride, water, and ammonia would be expected to have if these substances did not form hydrogen bonds. What would you expect the relative density of ice and water to be if hydrogen bonds were not formed?

9-7. Why are there no strong hydrogen bonds in liquid H_2S?

9-8. Give an example of a gaseous solution, a liquid solution, and a crystalline solution.

9-9. A solution contains 10.00 g of anhydrous cupric sulfate, $CuSO_4$, in 1000 ml of solution. What is the formality of this solution in $CuSO_4$?

9-10. Saturated salt solution (20°C) contains 35.1 g NaCl per 100 g of water. What is its mass-formality? The density of the solution is 1.197 g ml^{-1}. What is its formality?

9-11. A 0.2 mass F solution of HCl is neutralized with 0.2 mass F NaOH. What is the mass-formality of NaCl in the resulting solution?

9-12. The only substance known to form a crystalline solution with water (that is, to be present in ice) is ammonium fluoride. Can you explain this fact? What is the structure of the crystalline solution?

9-13. Explain why sodium chloride crystallizes from solution as unhydrated NaCl, beryllium chloride as $BeCl_2 \cdot 4H_2O$, and magnesium chloride as $MgCl_2 \cdot 6H_2O$.

9-14. What is the difference in orientation between the water molecules of hydration surrounding an anion and those surrounding a cation?

9-15. The unit of structure of the clathrate crystal shown in Figure 9-10 contains two small polyhedra, each with 20 water molecules at its corners, and six larger polyhedra, each with 24 water molecules at its corners. The edges are O—H···O hydrogen bonds, with length 276 pm. The small polyhedron is the pentagonal dodecahedron. The distance from the center to a corner of a regular pentagonal dodecahedron is 1.40 times the length of the edge. Can you explain why xenon hydrate and methane hydrate have composition $8Xe \cdot 46H_2O$ and $8CH_4 \cdot 46H_2O$, whereas chlorine hydrate and bromine hydrate have composition $6Cl_2 \cdot 46H_2O$ and $6Br_2 \cdot 46H_2O$? What would you predict for the composition of crystals formed by water, chlorine, and xenon?

9-16. What is a polar solvent? Give some examples.

9-17. Define solubility and saturated solution.

9-18. Make qualitative predictions about the solubility of the following:
(*a*) Ethyl ether, $C_2H_5OC_2H_5$, in water, alcohol, and benzene.
(*b*) Hydrogen chloride in water and gasoline.
(*c*) Ice in liquid hydrogen fluoride and in cooled gasoline.
(*d*) Sodium tetraborate in water, in ether, and in carbon tetrachloride.
(*e*) Iodoform, HCI_3, in water and in carbon tetrachloride.
(*f*) Decane, $C_{10}H_{22}$, in water and in gasoline.

9-19. What can you say about the solubility in water of the substances $AgNO_3$, $PbCl_2$, PbI_2, Hg_2SO_4, $BaSO_4$, $Mg(OH)_2$, $Ba(OH)_2$, PbS, $NaSb(OH)_6$, K_2PtCl_6, KCl?

9-20. Plot (make a graph) of the solubilities of nitrogen and oxygen in water at 0°C as the partial pressure of each varies from 0 to 2 atm. Indicate the units.

9-21. How would you separate I_2 from I_3^- in aqueous solution?

9-22. The solubility of nitrogen at 1 atm partial pressure in water at 0°C is 23.54 ml l^{-1}, and that of oxygen is 48.89 ml l^{-1}. Calculate the amount by which the freezing points of air-saturated water and air-free water differ.

9-23. A solution containing 2.30 g of glycerol in 100 ml of water was found to freeze at −0.465°C. What is the approximate molecular mass of glycerol dissolved in water? The formula of glycerol is $C_3H_5(OH)_3$. What would you predict as to the miscibility of this substance with water (its solubility)?

9-24. When 0.412 g of naphthalene ($C_{10}H_8$) was dissolved in 10.0 g of camphor, the freezing point was found to be 13.0° below that of pure camphor. What is the mass-molar freezing-point constant for camphor, calculated from this observation? Can you explain why camphor is frequently used in molecular mass determinations? (Answer: 40.4°.)

9-25. The freezing point of ethylene glycol is −17°C, yet when mixed with water in a car radiator it can prevent freezing at even lower temperatures. How is this explained?

9-26. Adding alcohol to water lowers the freezing point of water, but also lowers the boiling point. Explain.

9-27. If the vapor pressure of some solvent is 0.080 atm at 25°C, what is the vapor pressure of the solvent at 25°C when a solute is added to the extent that the mole fraction of the solvent in the solution is 0.96?

9-28. Why do red blood cells burst when placed in distilled water?

9-29. Considering that the cell wall is an osmotic membrane, explain why a lettuce salad containing salt and vinegar becomes limp in a few hours.

9-30. Find the osmotic pressure at 17°C of a solution containing 17.5 g of sucrose ($C_{12}H_{22}O_{11}$) in 150 ml of solution (Answer: 8.1 atm.)

9-31. An aqueous solution of amygdalin (a sugarlike substance obtained from almonds) containing 96 g of solute per liter was found to have osmotic pressure 4.74 atm at 0°C. What is the molecular mass of the solute?

9-32. A 3% aqueous solution of gum arabic (simplest formula $C_6H_{10}O_5$) was found to have an osmotic pressure of 0.0272 atm at 25°C. What are the average molecular mass and degree of polymerization of the solute?

9-33. Differentiate between ordinary aqueous solutions and colloidal dispersions in water.

9-34. Assuming that each soap molecule in the surface of an emulsified fat droplet occupies the area 1 nm^2 and that the droplets are 0.1 mm in diameter, calculate the minimum weight of soap (sodium palmitate) needed to emulsify 1 g of oil.

9-35. In 1935 the German scientist Gerhard Domack reported that the trimethylalkylammonium chlorides, which contain the cation $(H_3C)_3NR^+$, with $R{=}C_nH_{2n+1}$, are effective as germicides only when n is between 8 and 18. Can you suggest a plausible explanation of this fact?

10

Chemical Equilibrium and the Rate of Chemical Reaction

Two questions may be asked in the consideration of a proposed chemical process, such as the preparation of a useful substance. One of these questions is "Are the stability relations of the reactants and the expected products such that it is possible for the reaction to occur?" The second question is equally important: it is "Under what conditions will the reaction proceed sufficiently rapidly for the method of preparation to be practicable?"

10-1. Factors Influencing the Rate of Reaction

Every chemical reaction requires some time for its completion, but some reactions are fast and some are slow. Reactions between ions in solution without change in oxidation state are usually extremely fast. An example is the neutralization of a strong acid by a strong base, which proceeds as fast as the solutions can be mixed. Presumably nearly every time a hydronium ion collides with a hydroxide ion reaction occurs, and the number of collisions is very great, so that there is little delay in the reaction.

The formation of a precipitate, such as that of silver chloride when a solution containing silver ions is mixed with a solution containing chloride ion, may require a few seconds, to permit the ions to diffuse together to form the crystalline grains of the precipitate:

$$Ag^+ + Cl^- \rightarrow AgCl(c)$$

On the other hand, ionic oxidation-reduction reactions are sometimes very slow. An example is the reduction of permanganate ion by hydrogen peroxide in sulfuric acid solution. When a drop of permanganate solution is added to a solution of hydrogen peroxide and sulfuric acid, the solution is colored pink, and this pink color may remain for several minutes, indicating that very little reaction has taken place. When, after several minutes, the solution has become colorless, another drop of permanganate is found to produce a pink color that remains for a shorter time, and a third and fourth drop are found to be decolorized still more rapidly. Finally, after a considerable amount of permanganate solution has been added and has undergone reaction, with the formation of manganous ion and the liberation of free oxygen, it is found that the permanganate solution poured in a steady stream into the container is decolorized as rapidly as it can be stirred into the hydrogen peroxide solution. The explanation of this interesting phenomenon is that a product of the reaction, manganese in a lower state of oxidation, acts as a catalyst for the reaction; the first drop of permanganate reacts slowly, in the absence of any catalyst, but the reaction undergone by subsequent drops is the catalyzed reaction. The presence of a catalyst may be one of the most important factors influencing the rate of a reaction.

An example of a reaction which is extremely slow at room temperature is that between hydrogen and oxygen:

$$2H_2 + O_2 \rightarrow 2H_2O$$

A mixture of hydrogen and oxygen can be kept for years without appreciable reaction. If the gas is ignited, however, a very rapid reaction—an explosion— occurs, because of the energy liberated.

Homogeneous and Heterogeneous Reactions

A reaction that takes place in a homogeneous system (consisting of a single phase) is called a **homogeneous reaction.** The most important of these reactions are those in gases (such as the formation of nitric oxide in the electric arc, $N_2 + O_2 \rightleftarrows 2NO$) and those in liquid solutions.

A **heterogeneous reaction** is a reaction involving two or more phases. An example is the oxidation of carbon by potassium perchlorate:

$$KClO_4(c) + 2C(c) \rightarrow KCl(c) + 2CO_2(g)$$

This is a reaction of two solid phases. This reaction and similar reactions occur when perchlorate propellants are burned. (These propellants, which are used for assisted takeoff of airplanes and for propulsion of rockets, consist of intimate mixtures of very fine grains of carbon black and potassium perchlorate held together by a plastic binder.) Another example is the solution of zinc in acid:

$$Zn(c) + 2H^+(aq) \rightarrow Zn^{++}(aq) + H_2(g)$$

In this reaction three phases are involved: the solid zinc phase, the aqueous solution, and the gaseous phase formed by the evolved hydrogen.

Most actual chemical processes are very complicated, and the analysis of their rates is difficult. As a reaction proceeds, the reacting substances are used up and new ones are formed; the temperature of the system is changed by the heat evolved or absorbed by the reaction; and other effects may occur that influence the reaction in a complex way. In order to obtain an understanding of the rates of reaction, chemists have attempted to simplify the problem as much as possible. A good understanding has been obtained of homogeneous reactions (in a gaseous or liquid solution) that take place at constant temperature. Experimental studies are made by placing the reaction vessel in a thermostat, which is held at a fixed temperature. For example, hydrogen gas and iodine vapor might be mixed, at room temperature, and their conversion into hydrogen iodide followed by observing the change in color of the gas, the iodine vapor having a violet color and the other substances involved in the reaction being colorless.

The *explosion* of a gaseous mixture, such as hydrogen and oxygen, and the *detonation* of a high explosive, such as glyceryl trinitrate (nitroglycerin), are interesting chemical reactions; but the analysis of the rates of these reactions is made difficult by the great changes in temperature and pressure that accompany them, and we shall not attempt it in this book.

The detonation of a high explosive such as glyceryl trinitrate illustrates the great rate of some chemical reactions. The rate at which a detonation wave moves along a sample of glyceryl trinitrate is about 20,000 feet per second. A specimen of high explosive weighing several grams may accordingly be completely decomposed within a millionth of a second, the time required for the detonation wave to move one quarter of an inch.

Another reaction that can occur very rapidly is the fission of the nuclei of heavy atoms. The nuclear fission of several kilograms of U^{235} or Pu^{239} may take place in a few millionths of a second in the explosion of a nuclear bomb (Chapter 20).

A heterogeneous reaction takes place at the surfaces (the *interfaces*) of the reacting phases, and it can be made to go faster by *increasing the extent of the surfaces.* Thus finely divided zinc reacts more rapidly with acid than does coarse zinc, and the rate of burning of a perchlorate propellant is increased by grinding the potassium perchlorate to a finer crystalline powder.

Sometimes a *reactant is exhausted* in the neighborhood of the interface, and the reaction is slowed down. Stirring the mixture then accelerates the reaction, by bringing fresh supplies of the reactant into the reaction region.

Catalysts may accelerate heterogeneous as well as homogeneous reactions.

The rates of nearly all chemical reactions depend greatly on the *temperature.* The effect of temperature is discussed in a later section of this chapter.

Special devices may be utilized to accelerate certain chemical reactions. The formation of a zinc amalgam on the surface of the grains of zinc by treatment with a small amount of mercury increases the speed of the reduction reactions of zinc.

The solution of zinc in acid is retarded somewhat by the bubbles of liberated hydrogen, which prevent the acid from achieving contact with the zinc over its entire surface. This effect can be avoided by bringing a plate of unreactive metal, such as copper or platinum, into electric contact with the zinc (Figure 10-1). The reaction then proceeds as two separate electron reactions. Hydrogen is liberated at the surface of the

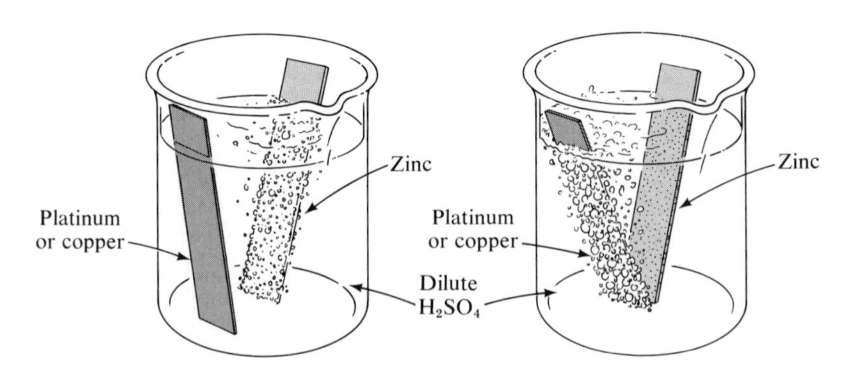

FIGURE 10-1
The interaction of an inert metal plate and a zinc plate with sulfuric acid, when the plates are not in contact (left) and when the plates are in contact (right).

copper or platinum, and zinc dissolves at the surface of the zinc plate:

$$2H^+ + 2e^- \longrightarrow H_2(g) \text{ at copper surface}$$

$$Zn \longrightarrow Zn^{++} + 2e^- \text{ at zinc surface}$$

The electrons flow from the zinc plate to the copper plate through the electric contact, and electric neutrality in the different regions of the solution is maintained by the migration of ions.

The solution of zinc in acid can be accelerated by adding a small amount of cupric ion to the acid. The probable mechanism of this effect is that zinc replaces cupric ion from the solution, depositing small particles of metallic copper on the surface of the zinc, and these small particles then act in the way described above.

10-2. The Rate of a First-order Reaction at Constant Temperature

If a molecule, which we represent by the general symbol A, has a tendency to decompose spontaneously

$$A \longrightarrow \text{products}$$

at a rate that is not influenced by the presence of other molecules, we expect that *the number of molecules that decompose by such a unimolecular process in unit time will be proportional to the number present.* If the volume of the system remains constant, the concentration of A will decrease at a rate proportional to this concentration. Let us use the symbol [A] for the concentration of A (in moles per liter). The rate of decrease in concentration with time is, in the language of the calculus, $-d[A]/dt$. For a unimolecular decomposition we accordingly may write the equation

$$-\frac{d[A]}{dt} = k[A] \qquad (10\text{-}1)$$

as *the differential equation determining the rate of the reaction.** The factor k is called the *first-order rate constant.* A reaction of this kind is

*Do not be discouraged by this equation even if you have not studied the calculus and are not familiar with equations of this sort. The expression on the left of Equation 10-1 is the rate of the reaction—the amount of decrease of concentration of the reacting substance in unit time. The expression on the right shows that this rate of decrease is proportional to the concentration itself.

called a *first-order reaction;* **the order of a reaction is the sum of the powers of the concentration factors in the rate expression** (on the right side of the rate equation).

The dimensions of k for a first-order reaction are seen to be t^{-1}; with time in seconds, s^{-1}.

For example, the rate constant k may have the value 0.001 s^{-1}. The equation would then state that during each second 1/1,000 of the molecules present would decompose. Suppose that at the time $t = 0$ there were 1,000,000,000 molecules per milliliter in the reaction vessel. During the first second 1,000,000 of these molecules would decompose, and there would remain at $t = 1$ sec only 999,000,000 molecules undecomposed. During the next second 999,000 molecules would decompose, and there would remain 998,001,000 molecules.* After some time (about 693 seconds) half of the molecules would have decomposed, and there would remain only 500,000,000 undecomposed molecules per milliliter. Of these about 500,000 would decompose during the next second, and so on.

The foregoing statement about the time required for half of the molecules to decompose can be verified by integrating Equation 10-1, to obtain the *first-order reaction-rate equation in the integrated form.* We rewrite Equation 10-1 as

$$\frac{d[A]}{[A]} = -k\,dt$$

We see that the expression on the left side is the derivative of ln [A] (plus a constant) and that on the right side is the derivative of $-kt$ (plus a constant). If we write $-\ln\,[A]_0$ for the first constant and kt_0 for the second constant, we obtain

$$\ln\,[A] - \ln\,[A]_0 = \ln\,([A]/[A]_0) = -kt + kt_0 \qquad (10\text{-}2)$$

On raising each side to the power e, this equation becomes

$$[A] = [A]_0 e^{-k(t-t_0)} \qquad (10\text{-}3)$$

We see that the constant $[A]_0$ is the concentration of A at the time $t = t_0$.

*These numbers are not precise. The molecules decompose at random, at the *average* rate given by Equation 10-1, and a statistical fluctuation from this rate is to be expected. The fluctuation is measured by the square root of the number of molecules that have decomposed. In the theory of probability it is shown that for expectation number n of independent events the standard deviation (root mean square error) is $\sqrt{n/2} = 0.7071\sqrt{n}$, the probable error (including half the observations) is $0.4769\sqrt{n}$, and the average error is $\sqrt{n/\pi}$ $= 0.5642\sqrt{n}$. The fractional error (ratio of error to expectation number) is seen to be proportional to $n^{-1/2}$. To halve the fractional error the number of observations must be quadrupled.

It is customary to place t_0 equal to 0 (that is, to measure the time from the moment when the concentration is $[A]_0$). The equation then becomes

$$[A] = [A]_0 e^{-kt} \tag{10-4}$$

The ratio of the concentration of the reactant to the initial concentration decreases exponentially with time, as shown in Figure 10-2. The first-order character of a reaction may be tested by observing the rate of disappearance of the reactant at various concentrations and comparing with Equation 10-1, or by measuring the concentration of one component of the system at several different times and comparing with Equation 10-4.

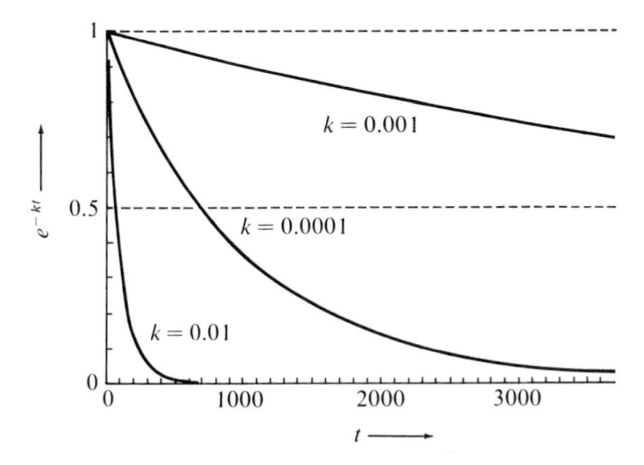

FIGURE 10-2
Curves showing decrease with time of the amount remaining of a substance decomposing by a first-order reaction, with indicated values of the reaction-rate constant.

A representative first-order gas-phase chemical reaction is the decomposition of azomethane, $CH_3—N{=}N—CH_3$, into ethane and nitrogen:*

$$CH_3NNCH_3 \rightarrow C_2H_6 + N_2$$

The molecular mechanism of this reaction is indicated in Figure 10-3. Most of the azomethane molecules have the configuration shown as *A*, with the methyl groups at opposite sides of the N=N axis. This is called the *trans* configuration. A few molecules have the *cis* configuration, *B*. If

*Some other products also are formed, by other reactions of azomethane.

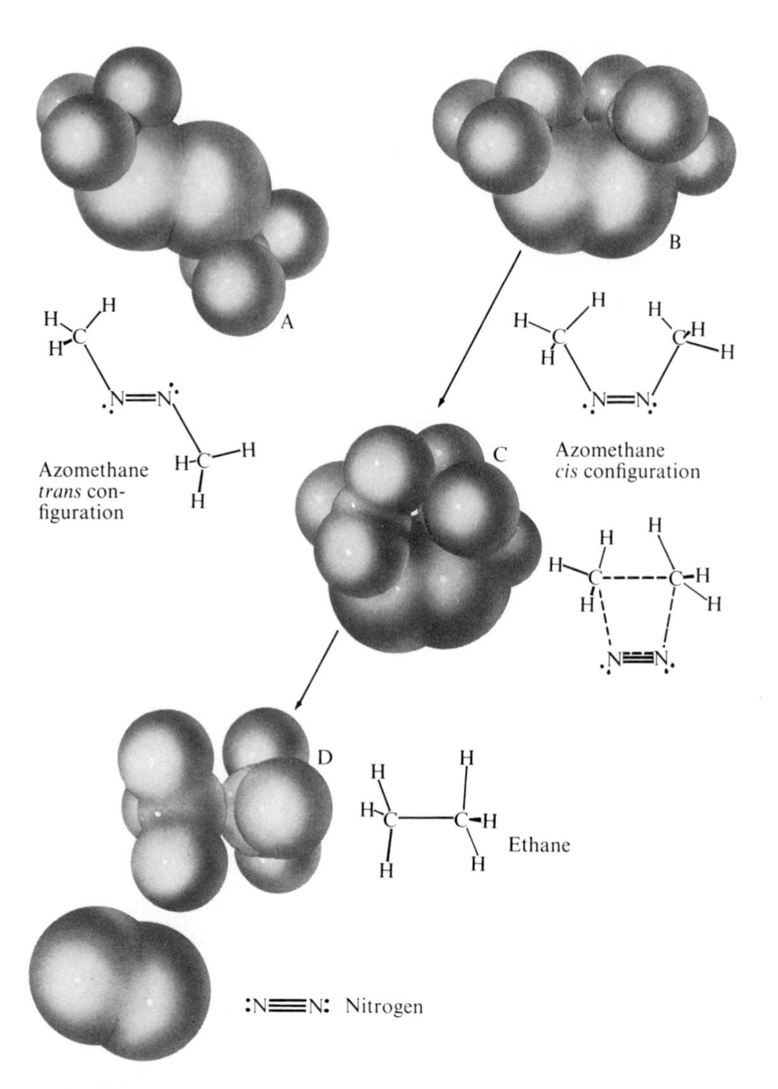

FIGURE 10-3
The unimolecular decomposition of azomethane into nitrogen and ethane.

a molecule with the *cis* configuration collides very vigorously with another molecule, it may be set into very violent vibration, sufficient to bring the two carbon atoms close together, as shown in *C*. In this configuration the two N—C bonds tend to break, forming a C—C bond and another N—N bond. This tendency is indicated by the dashed valence-bond lines in *C*. The molecule *C* may either return to configuration *B*, or break the C—N bonds and separate into two molecules, *D*.

Two other first-order chemical reactions are the decomposition of dinitrogen pentoxide, N_2O_5, into nitrogen dioxide and oxygen, and the decomposition of dimethyl ether, $CH_3—O—CH_3$, into methane, carbon monoxide, and hydrogen:

$$2N_2O_5 \longrightarrow 4NO_2 + O_2$$

$$CH_3OCH_3 \longrightarrow CH_4 + CO + H_2$$

It is to be noted that *the order of a reaction cannot be predicted from the stoichiometric over-all equation.* The equation for the N_2O_5 decomposition makes use of two molecules of the reactant, but the reaction is in fact a first-order reaction. This shows that the reaction takes place in steps; first there occurs a first-order decomposition, probably

$$N_2O_5 \longrightarrow NO_3 + NO_2$$

This is followed by other reactions, such as

$$2NO_3 \longrightarrow 2NO_2 + O_2$$

A simple method of studying the rate of a gaseous reaction in which the number of molecules of product is either greater or smaller than the number of molecules of reactant is by measuring the change in pressure during the reaction. The reactant is placed in the reaction vessel in the thermostat, and the gas pressure is then measured from time to time by a gauge.

Half-life

Let $t = nt'$. The exponential expression e^{-kt} is seen to be equal to $e^{-knt'} = (e^{-kt'})^n$. Hence in each succeeding period of time t' the concentration of undecomposed molecules decreases by the same factor.

In the example discussed above, with $k = 0.001\ s^{-1}$, it was mentioned that at $t' = 693$ s the value of $[A]/[A]_0$ is $\frac{1}{2}$. We see that after another 693 s (at $t = 2t'$) it is $\frac{1}{4}$, at $t = 3t'$ it is $\frac{1}{8}$, and so on.

The time required for the concentration of a first-order reactant to decrease by one-half is called its *half-life*. We see from Equation 10-4 that the half-life is the value of t' at which $e^{-kt'} = \frac{1}{2}$. To evaluate t' we write

$$\ln (e^{-kt'}) = -kt' = \ln (\tfrac{1}{2})$$

$$kt' = \ln 2 = 2.30259 \log 2 = 0.69315$$

$$\text{Half-life} = t' = \frac{0.69315}{k} \qquad (10\text{-}5)$$

From this equation, relating the half-life and the rate constant, we see that the statement that "For $k = 0.001\ s^{-1}$ the half-life is 693 s" is verified.

Radioactive Decomposition of Nuclei

The most important class of first-order reactions is the radioactive decomposition of atomic nuclei. Each nucleus of radium 226 or other radionucleide has a probability of decomposition in unit time that is independent of the concentration (in general, of the presence of other particles), and in consequence the process of radioactive decay is represented by Equations 10-1 and 10-4.

Example 10-1. The half-life of radium ($_{88}Ra^{226}$) is 1590 years. What is the value of the decay constant? What fraction decays in one year?

Solution. From Equation 10-5 we write

$$k = \frac{0.693}{1590} = 0.000436\ \text{year}^{-1}$$

Hence the decay constant has the value $4.36 \times 10^{-4}\ \text{year}^{-1}$.

On expanding the exponential term in Equation 10-4 we obtain

$$e^{-kt} = 1 - kt + \tfrac{1}{2}k^2t^2 - \cdots$$

For t small only the linear term above needs to be retained. We see that in unit time the fraction k decays. Hence in one year the fraction 4.36×10^{-4}, which is 0.0436%, of the radium decays.

Example 10-2. In Section 20-18 it is pointed out that the age of a piece of wood can be determined by measuring its carbon-14 activity. Carbon 14 has half-life 5,760 years. The carbon 14 in fresh wood decomposes at the rate 15.3 atoms per minute per gram of carbon (this is the number of carbon-14 β-rays emitted per minute, as determined with a beta counter). Wood from trees that were buried by ash in the volcanic eruption of Mt. Mazama in southern Oregon has been found to give 6.90 carbon-14 beta counts per minute per gram of carbon. When did the eruption occur?

Solution. By the method used in the preceding example we find for k the value

$$k = \frac{0.693}{5760\ \text{Y}} = 1.204 \times 10^{-4}\ \text{Y}^{-1}$$

The fraction of C^{14} undecomposed is $6.90/15.3 = 0.451$. Hence we write

$$e^{-kt} = 0.451$$

$$kt = -\ln 0.451 = -2.303 \log 0.451 = 2.303 \times 0.347 = 0.800$$

$$t = 0.800/k = 0.800/(1.204 \times 10^{-4}\ \text{year}^{-1}) = 6640\ \text{years}$$

Hence the eruption of Mt. Mazama occurred about 6640 years ago.

Example 10-3. Skeletons assigned to a form of early man called *Zinjanthropus* (East African man) have been found in the Olduvai Canyon in Tanganyika in association with volcanic ash containing potassium minerals. By use of a mass spectrograph the amount of argon 40 in the ash was measured; it was found to be 0.078% of the amount of potassium 40 present. (Potassium 40, half-life 15×10^8 years, is a radioactive isotope of potassium constituting 0.011% of natural potassium.) This argon 40 had been formed by beta decay of potassium 40, since the ash was deposited in a volcanic eruption; the older argon 40 would have been boiled out of the molten lava at the time of the eruption. How old are the skeletons?

Solution. The rate constant k for decay of K^{40} is $0.693/15.10^8$ Y $= 4.6 \times 10^{-10}$ Y^{-1}. The time t required for 0.078% of the K^{40} to decompose is given by the equation

$$kt = 7.8 \times 10^{-4}$$

$$t = 7.8 \times 10^{-4}/k = 7.8 \times 10^{-4}/4.6 \times 10^{-10}\ Y^{-1} = 1.7 \times 10^6\ Y$$

Hence the ash was laid down about 1,700,000 years ago; this is the presumable age of the skeletons found in association with it.

10-3. Reactions of Higher Order

If reaction occurs by collision and interaction of two molecules A and B, the rate of reaction will be proportional to the number of collisions. The number of collisions in unit volume is seen from simple kinetic considerations to be proportional to the product of the concentrations of A and B. Hence we may write as the differential rate expression for this second-order reaction

$$\text{rate of reaction} = -\frac{d[A]}{dt} = -\frac{d[B]}{dt} = k[A][B] \tag{10-6}$$

Here $-d[A]/dt$ is the rate of decrease in concentration of [A] and $-d[B]/dt$ is that of [B]; they are equal, the reaction being $A + B \rightarrow$ products. The factor k is the *second-order rate constant.* Its dimensions are seen to be liter $mole^{-1}$, the reciprocal of concentration, times s^{-1}.

It must be emphasized that *the stoichiometric equation for a reaction does not determine its rate equation.* Thus the oxidation of iodide ion by persulfate ion

$$S_2O_8^{--} + 2I^- \rightarrow 2SO_4^{--} + I_2$$

might be a third-order reaction, with rate proportional to $[S_2O_8^{--}][I^-]^2$;

it is in fact second-order, with rate proportional to $[S_2O_8^{--}][I^-]$. In this case the slow, rate-determining reaction is the reaction between one persulfate ion and one iodide ion,

$$S_2O_8^{--} + I^- \longrightarrow \text{products}$$

This is followed by a rapid reaction between the products and another iodide ion.*

For gases it is convenient to use partial pressures instead of concentrations of substances. For a bimolecular gas reaction we may write

$$\text{rate of reaction} = -\frac{dP_A}{dt} = -\frac{dP_B}{dt} = kP_A P_B \quad (10\text{-}7)$$

The second-order reaction-rate equation can be integrated. Let the concentrations of reactants be equal:

$$x = [A] = [B] \text{ (or } x = P_A = P_B)$$

Our equation in differential form is

$$-\frac{dx}{dt} = kx^2$$

Its solution by the method of integral calculus is found to be

$$x = \frac{c}{ckt + 1}$$

By placing $t = 0$, we see that c is the initial value of x.

Example 10-4. Hydrogen and iodine react by a bimolecular mechanism:

$$H_2(g) + I_2(g) \longrightarrow 2HI(g)$$

If at a certain temperature and pressure 1% of the substance present in the smaller amount had undergone reaction in 1 minute, how long would it take for 1% to react, at the same temperature, if the volume of the sample of gas were doubled?

Solution. We see from Equation 10-7 that

$$-\frac{1}{P_A}\frac{dP_A}{dt} = kP_B$$

*The slow reaction probably produces the hypoiodite ion, IO^-.

The fractional rate of removal of A is accordingly proportional to P_B. Since doubling the volume halves P_B, and hence the rate, it would take 2 minutes, twice as long, for the reaction to take place to the same extent.

10-4. Mechanism of Reactions. Dependence of Reaction Rate on Temperature

It is everyday experience that chemical reactions are accelerated by increased temperature: the rate of many reactions at room temperature is approximately doubled for every 10°C increase in temperature.

This is a useful rule. It is only a rough rule. Reactions of large molecules, such as proteins, may have very large temperature factors; the rate of denaturation of ovalbumin (the process that occurs when an egg is boiled) increases about fiftyfold for a 10° rise in temperature.

We may obtain further insight into this question by considering the mechanism of reactions. For the reaction

$$H_2(g) + I_2(g) \longrightarrow 2HI(g)$$

at temperature T we may write the following rate equation:

$$-\frac{dP_{H_2}}{dt} = -\frac{dP_{I_2}}{dt} = kP_{H_2}P_{I_2}$$

The rate constant k is determined in part by the number of collisions at $P_{H_2} = P_{I_2} = 1$ atm. This number can be calculated by use of the kinetic theory of gases and certain assumptions about the sizes of the molecules. Only a small fraction of the collisions, however, are effective. If the molecules collide with small relative velocity, the van der Waals force of repulsion causes them to rebound elastically from one another.

The Arrhenius Reaction-rate Equations

In 1889 the Swedish scientist Svante Arrhenius, who also discovered the existence of ions in electrolytic solutions, formulated the following equation for the change in the reaction-rate parameter k with change in temperature:

$$\frac{d \ln k}{dT} = \frac{E_a}{RT^2} \tag{10-8}$$

This equation can be transformed into an equation giving the change in k with respect to change in the reciprocal of the temperature, $1/T$:

$$\frac{d(\ln k)}{d(1/T)} = -\frac{E_a}{R} \tag{10-9}$$

The integrated form of these equations, with E_a taken to be constant, is

$$k = A \exp\left(\frac{-E_a}{RT}\right) \tag{10-10}$$

or

$$\ln k = \frac{-E_a}{RT} + \ln A \tag{10-11}$$

Equations 10-8 to 10-11 are called the Arrhenius reaction-rate equations.

Activation Energy

In order to react the pair of molecules must pass through a configuration intermediate between the initial one and the final one. This configuration is called the *activated complex.* In general this intermediate configuration is one in which some of the bonds in the reactant molecule are stretched, in the process of being broken, while new bonds are in the process of being formed.

In Equations 10-8 to 10-11 E_a is the activation energy per mole and A is a coefficient representing the collision number, the chance that the activated complex will break up to form the reactants rather than the products, and other factors affecting the rate.

The factors depend to some extent on the temperature, but as a good approximation they may usually be taken to be constants.

The value of the activation energy for a reaction can be determined by measuring the reaction-rate parameter at two or more temperatures, as described in Example 10-5. This value can then be used to calculate the rate at other temperatures.

Example 10-5. The observed reaction-rate curve for H_2 and I_2, Figure 10-4, has slope −21.6 in the units indicated. What is the value of the activation energy for the reaction?

Solution. With x axis $1/T$ in place of 1000 K/T the slope would be

$$\frac{d \ln k}{d(1/T)} = -21.6 \times 1000 = -21{,}600 \text{ K}$$

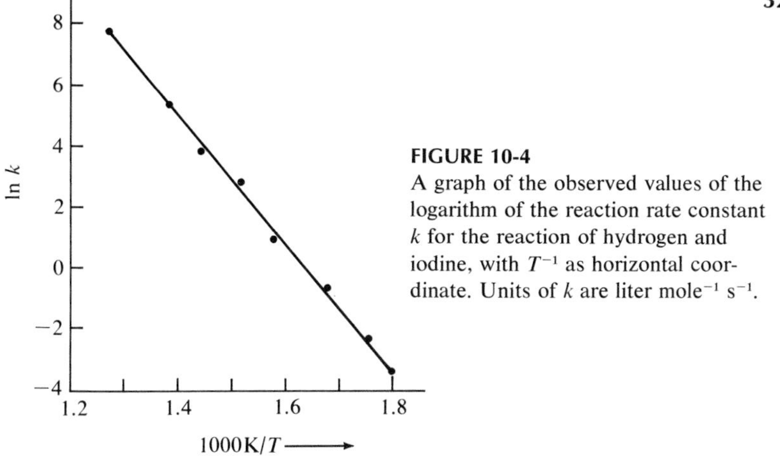

FIGURE 10-4
A graph of the observed values of the logarithm of the reaction rate constant k for the reaction of hydrogen and iodine, with T^{-1} as horizontal coordinate. Units of k are liter mole^{-1} s^{-1}.

From Equation 10-9 we see that

$$-\frac{E_a}{R} = -21{,}600 \text{ K}$$

and hence

$$E_a = 21{,}600 \text{ K} \times R = 21{,}600 \text{ K} \times 8.315 \text{ J K}^{-1} \text{ mole}^{-1}$$
$$= 179.6 \text{ kJ mole}^{-1}$$

The activation energy for the reaction is thus found to have the value 179.6 kJ mole^{-1}.

Activation Energy for the Reverse Reaction

In Figure 10-5 there is shown diagrammatically the change in energy accompanying a reaction. In order for the activated complex to be formed from the reactant molecules the activation energy E_a (forward) is needed.

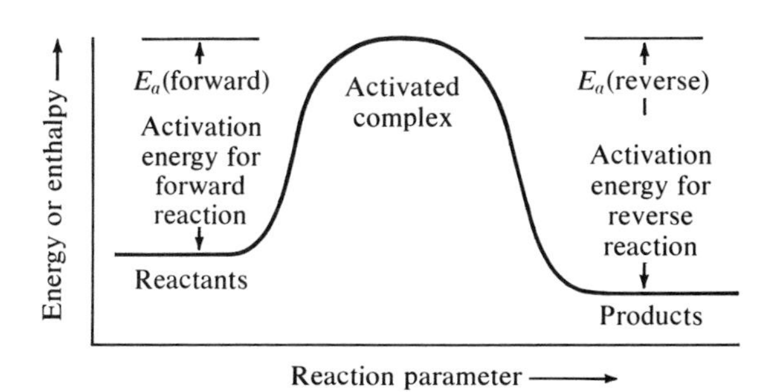

FIGURE 10-5
Diagrammatic representation of the change in energy accompanying a chemical reaction. The top of the curve corresponds to the activated complex for the reaction.

A detailed analysis shows that the activation energy E_a (reverse) differs from E_a (forward) by the difference in enthalpy of the reactants and the products. For the reactions $H_2 + I_2 \rightleftarrows 2HI$ this enthalpy difference is 10 kJ mole^{-1} (Table VI-10); hence, with the result from Example 10-5, we can state that the value of E_a (reverse) for the reaction $2HI \rightarrow H_2 + I_2$ is $180 + 10 = 190$ kJ mole^{-1}.

The Relation between Activation Energy and Bond Energy

Values of the activation energy can often be estimated from bond-energy values (Table V-1).

For example, the activation energy for the combination of two atoms, such as

$$I + I \rightarrow I_2$$

is zero. Reaction may occur whenever the two atoms collide, with electron spins opposed, even at low relative velocity. (The bond energy is converted to kinetic energy by collision with a third body.) The activation energy for the reverse reaction, the dissociation of a diatomic molecule, is just equal to the bond energy (enthalpy). For the dissociation of I_2 it is 151 kJ mole^{-1} (Table V-1).

A rough empirical rule can be formulated for reactions in which two bonds are broken and two are formed, such as

$$H_2 + I_2 \rightarrow 2HI$$

and its reverse reaction

$$2HI \rightarrow H_2 + I_2$$

It is found that the activation energy for the exothermic reaction (the first one of this pair) is approximately 30% of the sum of the bond energies of the two bonds that are broken (H—H and I—I, in this case).

For an exothermic reaction in which one bond is broken and another is formed, such as

$$F + H_2 \rightarrow HF + H$$

or

$$D + H_2 \rightarrow HD + H$$

the activation energy is approximately 8% of the energy of the bond that is broken.

10-5. Catalysis

The study of the factors that affect the rate of reactions has become more and more important with the continued great development of chemical industry. A modern method of manufacturing toluene, used for making the explosive trinitrotoluene and for other purposes, may be quoted as an example. The substance methylcyclohexane, C_7H_{14}, occurs in large quantities in petroleum. At high temperature and low pressure this substance should decompose into toluene, C_7H_8, and hydrogen. The reaction is so slow, however, that the process could not be carried out commercially until the discovery was made that a certain mixture of oxides increases the rate of reaction enough for the process to be put into practice. Many examples of catalysis (the process of accelerating a reaction by use of a catalyst) have already been mentioned, and others are mentioned in later chapters (see Section 14-4).

It is thought that catalysts speed up reactions by bringing the reacting molecules together and holding them in configurations favorable to reaction. Unfortunately so little is known about the fundamental nature of catalytic activity that the search for suitable catalysts is largely empirical. The test of a catalytic reaction, as of any proposed process, is made by trying to see if it works.

Catalysis and Activation Energy

There is evidence that some catalysts have a structure that leads to strong interaction with the activated complex and only weak interaction with the reactants and products. The strong interaction with the activated complex leads to a decrease in the activation energy, as indicated in Figure 10-6, and hence to increase in the rate of the reaction.

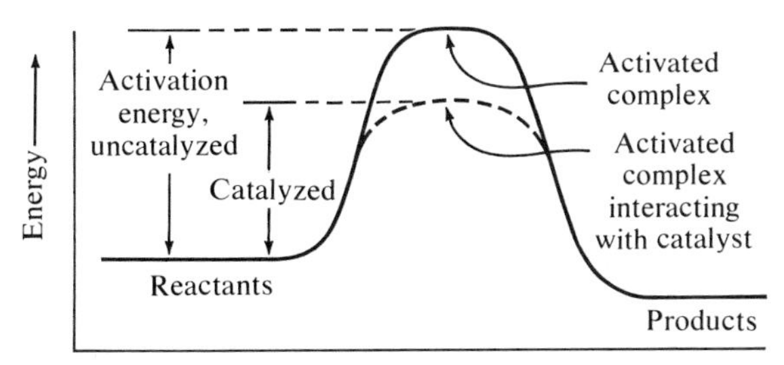

FIGURE 10-6
A diagram showing the possible effect of a catalyst in decreasing the value of the activation energy for a reaction, and thus increasing the rate of both the forward reaction and the reverse reaction.

10-6. Chain Reactions

When the reaction

$$H_2(g) + Br_2(g) \longrightarrow 2HBr(g)$$

was carefully investigated about 50 years ago, it was found that its rate is not proportional to the product $[H_2][Br_2]$, as expected (Equation 10-6). Instead, in its initial stages, with [HBr] small, the rate was observed to be proportional to $[H_2][Br_2]^{1/2}$:

$$-\frac{d[H_2]}{dt} = -\frac{d[Br_2]}{dt} = k[H_2][Br_2]^{1/2}$$

This observation was accounted for by the assumption that the reaction occurs in a series of steps. The first is the dissociation of a bromine molecule

$$Br_2 \longrightarrow 2Br$$

This is followed by a sequence of reactions, called a *chain:*

$$Br + H_2 \longrightarrow HBr + H$$

$$H + Br_2 \longrightarrow HBr + Br$$

$$Br + H_2 \longrightarrow HBr + H$$

$$H + Br_2 \longrightarrow HBr + Br$$

$$\cdots$$

The chain is ultimately broken by the reaction

$$2Br \longrightarrow Br_2$$

Explosions

A mixture of hydrogen and chlorine explodes when ignited. The overall reaction $H_2 + Cl_2 \longrightarrow 2HCl$, which takes place by the chain mechanism analogous to that for $H_2 + Br_2$, liberates so much heat that the gas may begin to increase in temperature, instead of dissipating the heat to the environment. The reaction then proceeds more rapidly, the temperature increases more rapidly, and a very rapid reaction, called *a thermal explosion,* results.

The explosion of a mixture of hydrogen and oxygen is different in kind; it is called a *branched-chain explosion*. An initial reaction with an atom of oxygen (or of hydrogen) is followed by two reactions, that one by three, then five, and so on:*

$$O + H_2 \rightarrow OH + H$$

$$OH + H_2 \rightarrow H_2O + O \quad H + O_2 \rightarrow OH + O$$

$$O + H_2 \rightarrow OH + H \quad OH + H_2 \rightarrow H_2O + O \quad O + H_2 \rightarrow OH + H$$

$$\cdots$$

Important chain reactions, which under some conditions lead to explosion, are the fission and fusion of atomic nuclei (Section 20-22).

10-7. Chemical Equilibrium—a Dynamic Steady State

Sometimes a chemical reaction begins, continues for a while, and then appears to stop before any one of the reactants is used up: the reaction is said to have reached *equilibrium*. The reaction between nitrogen dioxide, NO_2, and dinitrogen tetroxide, N_2O_4, provides an interesting example. The gas that is obtained by heating concentrated nitric acid with copper is found to have a density at high temperatures corresponding to the formula NO_2, and a density at low temperatures and high pressures approximating the formula N_2O_4. At high temperatures the gas is deep red in color, and at low temperatures it becomes lighter in color, the crystals formed when the gas is solidified being colorless. The change in the color of the gas and in its other properties with change in temperature and change in pressure can be accounted for by assuming that the gas is a mixture of the two molecular species NO_2 and N_2O_4, in equilibrium with one another according to the equation

$$\underset{\text{Colorless}}{N_2O_4} \rightleftarrows \underset{\text{Red}}{2NO_2} \tag{10-12}$$

It has been found by experiment that the amounts of nitrogen dioxide and dinitrogen tetroxide in the gas mixture are determined by a simple equation:

$$\frac{[NO_2]^2}{[N_2O_4]} = K \tag{10-13}$$

*Other reactions, involving the radical HO_2, also occur.

This equation, which is called the *equilibrium equation* for the reaction, is seen to involve in the numerator the concentration of the substance on the right side of the chemical equation (Equation 10-12), with the exponent 2, which is the coefficient shown in the chemical equation. The denominator contains the concentration of the substance on the left side. Its exponent is 1, because in the equation as written the coefficient of N_2O_4 is 1.

The quantity K is called the *equilibrium constant* of the reaction of dissociation of dinitrogen tetroxide to nitrogen dioxide. The equilibrium constant is independent of the pressure of the system, or of the concentration of the reacting substances. It is, however, dependent on the temperature.

The Relation between the Equilibrium Equation and Rates of Reaction

The reaction $N_2O_4 \rightarrow 2NO_2$ is a first-order reaction. Its rate equation is

$$-\frac{d[N_2O_4]}{dt} = k_1[N_2O_4]$$

The reverse reaction, $2NO_2 \rightarrow N_2O_4$, is a second-order reaction, with rate equation (expressed as rate of formation of the product)

$$\frac{d[N_2O_4]}{dt} = k_2[NO_2]^2$$

In a gas in which both reactions are taking place, the over-all rate of formation of N_2O_4 is given by

$$\frac{d[N_2O_4]}{dt} = k_2[NO_2]^2 - k_1[N_2O_4]$$

When the steady state with $d[N_2O_4]/dt = 0$ is reached, this equation becomes

$$k_2[NO_2]^2 = k_1[N_2O_4]$$

or

$$\frac{[NO_2]^2}{[N_2O_4]} = \frac{k_1}{k_2} = K \qquad (10\text{-}14)$$

We see that the expression for k_1/k_2, the ratio of the two reaction rate constants, is exactly the same as the equilibrium expression of Equation

10-13, and hence that the equilibrium constant K is the ratio of the rate constants for the two opposing reactions.

We repeat the statement of the general principle: *chemical equilibrium is a steady state, in which opposing chemical reactions occur at equal rates.*

In some cases it has been found possible to determine the rates of the opposing reactions, and to show experimentally that the ratio of the two rate constants is indeed equal to the equilibrium constant. This has not been done for the nitrogen dioxide-dinitrogen tetroxide equilibrium, however, because the individual chemical reactions take place so rapidly that experiments have not been able to determine their rates.

Equilibria of this sort are very important in chemistry. Many industrial processes have been made practicable by the discovery of a way of shifting an equilibrium so as to produce a satisfactory amount of a desired product. In this chapter and later chapters we shall discuss quantitatively the principles of chemical equilibrium and the methods of shifting the equilibrium of a system in one direction or the other.

The Effect of Catalysts on Chemical Equilibrium

It is a consequence of the laws of thermodynamics—the impossibility of perpetual motion—that *a system in equilibrium is not changed by the addition of a catalyst.* The catalyst may increase the rate at which the system approaches its final equilibrium state, but it cannot change the value of the equilibrium constant. Under equilibrium conditions a catalyst has the same effect on the rate of the backward reaction as on that of the corresponding forward reaction.

It is true that a system that has stood unchanged for a long period of time, apparently in equilibrium, may undergo reaction when a small amount of a catalyst is added. Thus a mixture of hydrogen and oxygen at room temperature remains apparently unchanged for a very long period of time; however, if even a minute amount of finely divided platinum (platinum black) is placed in the gas, chemical reaction begins and continues until very little of one of the reacting gases remains. In this case the system in the absence of the catalyst is not in equilibrium with respect to the reaction $2H_2 + O_2 \rightleftarrows 2H_2O$, but only in *metastable equilibrium,* the rate of formation of water being so small that true equilibrium would not be approached in a millenium.

Because of the possibility of metastable equilibrium it is necessary in practice to apply the following **equilibrium criterion:** *a system is considered to have reached equilibrium with respect to a certain reaction when the same final state is reached by approach by the reverse reaction as by the forward reaction.* This true equilibrium is called *stable equilibrium.*

The General Equation for the Equilibrium Constant

The chemical equation for a general reaction can be written in the following form:

$$a\text{A} + b\text{B} + \cdots \rightleftarrows d\text{D} + e\text{E} + \cdots \qquad (10\text{-}15)$$

Here the capital letters A, B, D, E are used to represent different molecular species, the reactants and the products, and the small letters *a, b, d, e* are the numerical coefficients that tell how many molecules of the different sorts are involved in the reaction.

The equilibrium equation for this reaction is

$$\frac{[\text{D}]^d[\text{E}]^e \cdots}{[\text{A}]^a[\text{B}]^b \cdots} = K \qquad (10\text{-}16)$$

Here K is the equilibrium constant for the reaction.

It is customary to write the concentration ratio in the way given in Equation 10-16 for a chemical equation such as Equation 10-15; that is, the *concentrations of the products (to the appropriate powers) are written in the numerator and the concentrations of the reactants in the denominator.* This is a convention that has been accepted by all chemists.

The equilibrium constant for gases is often written with partial pressure replacing concentration:

$$\frac{P_{\text{D}}{}^d P_{\text{E}}{}^e \cdots}{P_{\text{A}}{}^a P_{\text{B}}{}^b \cdots} = K_P \qquad (10\text{-}17)$$

Unless the reaction involves no change in number of molecules, the numerical value of K_P differs from that of K.

The validity of the equilibrium equation, with K a constant at constant temperature, is a consequence of the laws of thermodynamics, if the reactants and the products are gases obeying the perfect gas laws or are solutes in dilute solution. In gases under high pressure and in concentrated solutions there occur some deviations from this equation, similar in magnitude to the deviations from the perfect gas laws.

Examples of the use of the general equilibrium equation will be given in the following chapters of this book. This simple equation permits the chemist to answer many important questions that arise in his work.

Example 10-6. Hydrogen iodide, HI, is not a very stable substance. The pure gas is colorless, but when it is made in the laboratory the gas may have a violet color, indicating the presence of free iodine. In fact, hydrogen iodide decomposes to an appreciable amount according to the equation

$$2HI \rightleftarrows H_2 + I_2(g)$$

The equilibrium constant for this decomposition reaction has been found by experiment to have the value 0.0190 at 300°C. To what extent does hydrogen iodide decompose at this temperature?

Solution. In this example the value of the equilibrium constant is given without a statement as to its dimensions. Let us write the expression for the equilibrium constant:

$$K = \frac{[H_2][I_2]}{[HI]^2} = 0.0190$$

Each of the concentrations $[H_2]$, $[I_2]$, and $[HI]$ has the dimensions moles liter^{-1}. Hence we see that for this reaction the dimensions of K are those of a pure number:

$$\text{Dimensions of } K = \frac{(\text{moles liter}^{-1})(\text{moles liter}^{-1})}{(\text{moles liter}^{-1})^2} = 1$$

When hydrogen iodide decomposes, equal numbers of molecules of hydrogen and iodine are formed. Accordingly the concentrations of hydrogen and iodine present in the gas formed when hydrogen iodide undergoes some decomposition are equal. Let us use the symbol x to represent the concentration of hydrogen and also the concentration of iodine:

$$[H_2] = [I_2] = x$$

Then we have

$$\frac{x^2}{[HI]^2} = 0.0190$$

or

$$x^2 = 0.0190[HI]^2$$

$$x = \sqrt{0.0190} \times [HI] = 0.138[HI]$$

By solving this equation we have found that after the hydrogen iodide has decomposed enough to produce the equilibrium state at 300°C the molar concentration of hydrogen is equal to 13.8% of the molar concentration of HI. The molar concentration of iodine is also equal to 13.8% of the molar concentration of HI. The question "To what extent does hydrogen iodide decompose at this temperature?" is to be interpreted as meaning "What percentage of pure hydrogen iodide originally produced decomposes to give hydrogen and iodine?" The equation for the chemical reaction shows that two molecules of HI on reaction form only one molecule of H_2 and one of I_2. Accordingly, there must have been 27.6% more moles of HI present initially than when equilibrium is reached. Hence the extent of decomposition of the hydrogen iodide is 27.6/127.6 = 0.216, or 21.6%.

Example 10-7. At 500°C the equilibrium constant for the formation of ammonia by the reaction

$$N_2 + 3H_2 \rightleftarrows 2NH_3$$

has a value such that at 1 atm total pressure the partial pressure of ammonia in a stoichiometric mixture (1 mole of N_2 to 3 moles of H_2) is only 0.001 atm. Would the fraction of nitrogen converted to ammonia be increased or decreased if the gas were compressed to 1/100 its volume, at this temperature.

Solution. The equilibrium equation for this reaction is

$$\frac{P_{NH_3}^2}{P_{N_2}P_{H_2}^3} = K$$

The problem states that

$$P_{H_2} = 3P_{N_2}$$

because the stoichiometric ration N_2 to $3H_2$ is specified for these gases. We have the values $P_{NH_3} = 0.001$ atm, $P_{N_2} = 0.25$ atm, and $P_{H_2} = 0.75$ atm, giving

$$K = \frac{(0.001)^2}{0.25 \times (0.75)^3} = 9.5 \times 10^{-6} \text{ atm}^{-2}$$

If the hundred-fold compression did not change the composition of the gas, the partial pressures would be 0.1 atm, 25 atm, and 75 atm, respectively, and the ratio $P_{NH_3}^2/P_{N_2}P_{H_2}^3$ would have the value 9.5×10^{-10} atm^{-2}. This value is 10,000 times smaller than the known value of K. Accordingly the numerator of the equilibrium expression must be increased (and the denominator decreased). We see that if we ignore the change in the denominator, the value of P_{NH_3} is increased 100-fold, to 10 atm, and the fractional conversion to ammonia is about 100 times as great as at 1 atm total pressure.

10-8. Le Chatelier's Principle

An important and interesting qualitative principle about equilibrium is the *principle of Le Chatelier.* This principle, which is named after the French chemist Henry Louis Le Chatelier (1850–1936), may be expressed in the following way: *if the conditions of a system, initially at equilibrium, are changed, the equilibrium will shift in such a direction as to tend to restore the original conditions, if such a shift is possible.*

From the equilibrium equation 10-16 at constant temperature, we see that increasing the partial pressure of a reactant A or B causes the equilibrium to shift in such a way as to decrease the partial pressure of that reactant toward its initial value (with changes also in the other partial pressures). Increasing the partial pressure of a product, D or E, causes the reaction to go in the opposite direction, decreasing the partial pressure of that product toward its original value.

Increasing the total pressure causes the equilibrium to shift in such a way as to decrease the total pressure, if the equilibrium is dependent on the total pressure. We have seen that increase in total pressure of the system shifts the ammonia equilibrium in the direction of the product, NH_3; this shift decreases the total pressure, because the volume of $2NH_3(g)$ is only half that of $N_2(g) + 3H_2(g)$. The reaction $H_2(g) + I_2(g) \rightleftarrows 2HI(g)$ is an example of one in which change in total pressure causes no shift in the equilibrium; it is the existence of systems of this sort that requires the final qualifying clause in our statement of Le Chatelier's principle.

10-9. The Effect of Change of Temperature on Chemical Equilibrium

From the principle of Le Chatelier we can predict that *increase in temperature will drive a reaction further toward completion (by increasing the equilibrium constant) if the reaction is endothermic and will drive it back (by decreasing the equilibrium constant) if the reaction is exothermic.*

For example, let us consider the NO_2-N_2O_4 equilibrium mixture at room temperature. Heat is absorbed when an N_2O_4 molecule dissociates into two NO_2 molecules:

$$N_2O_4(g) \rightarrow 2NO_2(g) \qquad \Delta H^\circ = 62.8 \text{ kJ mole}^{-1}$$

If the reaction mixture were to be increased in temperature by a few degrees the equilibrium would be changed, according to the principle of Le Chatelier, in such a way as to tend to restore the original temperature—that is, in such a way as to lower the temperature of the system, by using up some of the heat energy. This would be achieved by the decomposition of some additional molecules of dinitrogen tetroxide. Accordingly, in agreement with the statement above, the equilibrium constant would change in such a way as to correspond to the dissociation of more of the N_2O_4 molecules.

This principle is of great practical importance. For example, the synthesis of ammonia from nitrogen and hydrogen is exothermic (the heat evolved is 11.0 kcal per mole of ammonia formed); hence the yield of ammonia is made a maximum by keeping the temperature as low as possible. The commercial process of manufacturing ammonia from the elements became practicable when catalysts were found that caused the reaction to proceed rapidly at low temperatures.

In Section 10-7 it was pointed out (Equation 10-14) that the equilibrium constant K for a reaction is equal to k_1/k_2—that is, to the ratio of the rate constant for the forward reaction to that of the reverse reaction. We may accordingly write

$$\ln K = \ln (k_1/k_2) = \ln k_1 - \ln k_2$$

and

$$\frac{d \ln K}{dT} = \frac{d \ln k_1}{dT} - \frac{d \ln k_2}{dT} \tag{10-18}$$

One of the Arrhenius equations, Equation 10-8, is

$$\frac{d \ln k}{dT} = \frac{E_a}{RT^2}$$

We may accordingly rewrite Equation 10-18 as

$$\frac{d \ln K}{dT} = \frac{E_a \text{ (forward)} - E_a \text{ (reverse)}}{RT^2}$$

By an argument illustrated in Figure 10-5 we have indicated that E_a (forward) $- E_a$ (reverse) is equal to $-\Delta H°$, the standard enthalpy change accompanying the reaction. Thus we are led to the important equation

$$\frac{d \ln K}{dT} = \frac{\Delta H°}{RT^2} \tag{10-19}$$

This equation expresses the dependence of the equilibrium constant for a chemical reaction on the temperature. Our development of the equation has not been a rigorous one, but the equation can be rigorously derived from fundamental principles by the methods of either statistical mechanics or chemical thermodynamics.

Over a small range of temperature, $\Delta H°$ can be considered to be a constant. When Equation 10-19 is integrated on this assumption, the following equation is obtained:

$$\ln \left(\frac{K(T_2)}{K(T_1)}\right) = \frac{\Delta H°}{R}\left(\frac{1}{T_1} - \frac{1}{T_2}\right) \tag{10-20}$$

Example 10-8. The vapor pressure of water at 0°C is 0.0060 atm. What is the heat of vaporization of water?

Solution. The equilibrium constant for the reaction $H_2O(l) \longrightarrow H_2O(g)$ is

$$\frac{[H_2O(g)]}{[H_2O(l)]} = K_{\text{concentration}}$$

We may replace $[H_2O(l)]$ by a constant, and incorporate it into the equilibrium constant:

$$[H_2O(g)] = K'_{\text{concentration}}$$

or, using pressure instead of concentration,

$$P_{H_2O(g)} = K$$

We know that $P_{H_2O(g)}$ has the values $K(T_1) = 0.0060$ atm at $T_1 = 273.16$ K and $K(T_2) = 1$ atm at $T_2 = 373.16$ K. From Equation 10-20 we obtain the equation

$$\ln\left(\frac{1}{0.0060}\right) = \frac{\Delta H^\circ}{R}\left(\frac{1}{273.16} - \frac{1}{373.16}\right)$$

Hence

$$2.303 \log 166 = \frac{\Delta H^\circ}{R}(0.00366 - 0.00268)$$

$$2.303 \times 2.220 = \frac{\Delta H^\circ}{R} \times 0.00098$$

$$\Delta H^\circ = \frac{2.303 \times 2.220R}{0.00098} = 43{,}400 \text{ J mole}^{-1}$$

Thus we have obtained 45.3 kJ mole^{-1} as the heat of vaporization of water over the temperature range 0°C to 100°C, with the assumption that it is constant over this range.

By use of values of the vapor pressure over smaller ranges of temperature the same procedure leads to 45.3 kJ mole^{-1} at 0°C, 42.8 kJ mole^{-1} at 50°C, and 40.7 kJ mole^{-1} at 100°C. These values are verified by calorimetric measurements.

10-10. The Phase Rule—a Method of Classifying All Systems in Equilibrium

It was discovered by J. Willard Gibbs in the course of his early work on chemical thermodynamics that *a simple, unifying principle holds for all systems in equilibrium.* This principle is called the *phase rule.*

The phase rule is a relation among the number of independent components, the number of phases, and the variance of a system in equilibrium. The independent components (or, briefly, the components) of a system are the substances that must be added to realize the system. The phases have been defined earlier (Section 1-3). Thus a system containing ice, water, and water vapor consists of three phases but only one component (water-substance), since any two of the phases can be formed from the third. The *variance* of the system is the number of independent ways in which the system can be varied; these ways may include varying the temperature and the pressure, and also varying the composition of any solutions (gaseous, liquid, or crystalline) that exist as phases in the system.

The nature of the phase rule can be induced from some simple examples. Consider the system represented in Figure 10-7. It is made of water-substance (water in its various forms), in a cylinder with movable piston (to permit the pressure to be changed), placed in a thermostat with changeable temperature. If only one phase is present, both the pressure

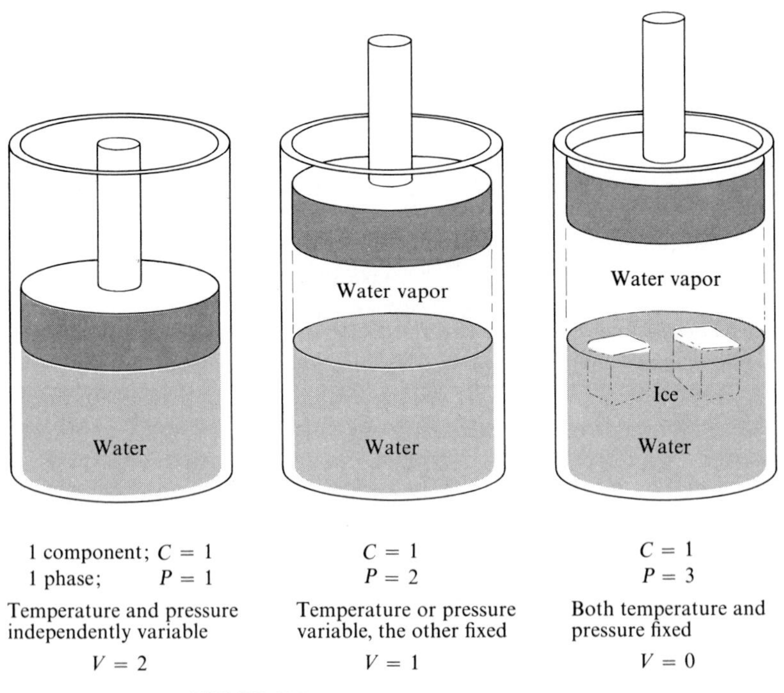

FIGURE 10-7
A simple system illustrating the phase rule.

and the temperature can be arbitrarily varied over wide ranges: the variance is 2. For example, liquid water can be held at any temperature from its freezing point to its boiling point under any applied pressure. But if two phases are present, the pressure is automatically determined by the temperature: the variance is reduced to 1. For example, pure water vapor in equilibrium with water at a given temperature has a definite pressure, the vapor pressure of water at that temperature. And if three phases are present in equilibrium, ice, water, and water vapor, both the temperature and the pressure are exactly fixed: the variance is 0. This condition is called the triple point of ice, water, and water vapor. It occurs at temperature +0.0099°C and pressure 0.0060 atm.

We see that for this simple system, with one component, the sum of the number of phases and the variance is equal to 3. It was discovered by Gibbs that *for every system in equilibrium the sum of the number of phases and the variance is 2 greater than the number of components:*

$$\textit{Number of phases} + \textit{variance} = \textit{number of components} + 2$$

or, using the abbreviations P, V, and C,

$$P + V = C + 2$$

This is the phase rule.

Let us now consider its application. We ask: Is it ever possible for four phases to exist together in equilibrium? The answer is seen to be that it is, provided that there are at least two components. If there are only two components the four phases can coexist only at exactly fixed temperature and pressure. For example, we might add sodium chloride to the water in our system. The components would then be two in number ($C = 2$). When ice, liquid solution, and water vapor were present ($P = 3$) the temperature could still be varied somewhat, by varying the concentration of sodium chloride in the liquid solution. The variance would then be 1, with three phases. On lowering the temperature there would ultimately be formed crystals of sodium chloride dihydrate, $NaCl \cdot 2H_2O$. The system would then be at fixed temperature (−21.2°C), fixed composition of the liquid phase (22.42 g NaCl per 100 g of solution), and fixed pressure (the vapor pressure of ice at that temperature, 0.00091 atm). The variance is zero, as required for $C = 2$ and $P = 4$. There is another invariant point for the system, with the four phases $H_2O(g)$, salt solution, $NaCl \cdot 2H_2O(c)$, and NaCl(c), at temperature 0.15°C, concentration of solution 26.3 g NaCl per 100 g, and pressure 0.0055 atm. A derivation of the phase rule is given at the end of the following section.

10-11. The Driving Force of Chemical Reactions

What makes a chemical reaction go? This is a question that chemists and students have asked ever since chemical reactions began to be investigated. At the beginning of the nineteenth century the question was answered by saying that two substances react if they have a "chemical affinity" for each other. The answer, of course, had no real value until some quantitative meaning was given to "chemical affinity" and some way was found for measuring or predicting it.

It might be thought that the heat of a reaction is its driving force, and that a reaction will proceed if it evolves heat, and not proceed if it would absorb heat. In fact, however, many reactions take place with absorption of heat. We have mentioned some of these reactions in the preceding sections of this chapter; another example is that when mercuric oxide is heated it decomposes into mercury and oxygen, with absorption of heat.

Let us first ask why a liquid evaporates, even though the reaction of evaporation is endothermic. The explanation of this phenomenon is given by the examination of *probability*. Let us consider a large flask, with volume 10 liters, into which some water vapor is introduced. We might well think that it would be equally probable that a particular water molecule would be in any place in the flask—that the probability would be 1 in 10,000 that the molecule would be within any particular milliliter of volume within the flask. If enough water vapor has been introduced into the flask, however, some of it will liquefy, the rest remaining as water vapor. Let us suppose that there is 1 ml of liquid water present in a little puddle at the bottom of the flask. At room temperature most of the water-substance present in the flask will be in this puddle of liquid water, only a small fraction of the water molecules being present as water vapor. Now, although it seems very improbable that a water molecule should stay in the small volume, 1 ml, occupied by the liquid water, instead of occupying the remaining 9,999 ml of space, we know that the reason that the water vapor condenses to liquid water is that liquid water is the more stable state, and that condensation proceeds until the rate at which gas molecules strike the surface of the liquid and stick is just equal to the rate at which molecules of the liquid leave the surface and escape into the gas. This is the equilibrium condition. We see that the equilibrium condition involves a balance between the effect of *energy*, which tends to concentrate the molecules into the liquid phase, and the effect of *probability*, which tends to change the liquid into the gas. If the volume of the flask were five times as great, making the probability for the gas phase 49,999 to 1 instead of 9,999 to 1, five times as many molecules would leave the liquid phase and move to the gaseous phase.

Accordingly we see that this effect of probability can be made to cause more of the liquid to evaporate, simply by increasing the volume of the system.

In the branch of science called thermodynamic chemistry a more detailed consideration is given to the relative effects of energy and probability. It has been found that the effect of probability can be described quantitatively by a new property of substances. This new property, which represents the probability of a substance in various states, is called *entropy*.

Whereas the energy change that accompanies a chemical reaction does not depend very much on the pressures of the gases or the concentrations of the solutes involved in the reaction, the entropy change does depend on these partial pressures and concentrations. In general, a system held at constant temperature will reach a steady state, called the state of equilibrium. In this state of the system the reaction has no preferential tendency to proceed either forward or backward; it has no driving force in either direction. If, however, the concentration of one of the reactants (a solute or a gas) is increased, a driving force comes into existence, which causes the reaction to go in the forward direction, until the equilibrium expression, involving the concentrations or partial pressures of reactants and products, again becomes equal to the equilibrium constant for the reaction.

It is clear from these considerations that *the driving force of a reaction depends not only on the chemical formulas of the reactants and the structure of their molecules, but also on the concentrations of the reactants and of the products.*

A great step forward was made during the second half of the last century when it was found that an energy quantity called its *free energy* can be assigned to each substance, such that a reaction in a system held at constant temperature tends to proceed if it is accompanied by a decrease in free energy; that is, if the free energy of the reactants is greater than that of the products. *The free energy of a substance is a property that expresses the resultant of the enthalpy of the substance and its inherent probability (entropy).* If the substances whose formulas are written on the left of the double arrow in a chemical equation and those whose formulas are written on the right have the same entropy (probability), the reaction will proceed in the direction that leads to the evolution of heat, that is, in the exothermic direction. If the substances on the left and those on the right have the same enthalpy, the reaction will proceed from the substances with the smaller probability (entropy) toward the substances with the greater probability (entropy). At equilibrium, when a reaction has no preferential tendency to go in either the forward or backward direction, the free energy of the substances on the left side is exactly equal to that of the substances on the right side. *At equilibrium the driving force of the enthalpy change accompanying a reaction is exactly balanced by the driving force of the probability change (entropy change).*

The relation between free energy, enthalpy, and entropy is

$$G = H - TS \tag{10-22}$$

For a reaction at constant temperature the change in free energy is

$$\Delta G = \Delta H - T\Delta S \tag{10-23}$$

A substance in the form of a perfect crystal at the absolute zero is in its lowest quantum state, which may be a single state. The value of the entropy for such a substance is zero; that is, $S = k \ln 1 = 0$. This is the *third law of thermodynamics*. It was discovered by the German chemist W. Nernst (1864–1941) in 1906.

The Relation of the Equilibrium Constant to Standard Free-energy Change

It can be shown by the methods of statistical mechanics or chemical thermodynamics that the equilibrium constant K for a reaction is related to the change in standard free energy $\Delta G°$ accompanying the reaction by the equation

$$RT \ln K = -\Delta G° \tag{10-24}$$

Values of $\Delta G°$ of formation of thousands of compounds, at 0 K and 298.16 K (25°C), are given in the U.S. Bureau of Standards Circular 500, "Selected Values of Chemical Thermodynamic Properties." These values may be combined to give the values of $\Delta G°$ for many other reactions, for which the equilibrium constants can then be calculated by use of Equation 10-24.

The study of the free energy of substances constitutes a complex subject and only a bare introduction to it can be given in a course in general chemistry. The following chapter deals with free-energy changes accompanying oxidation-reduction reactions; a similar treatment can also be given to other reactions.

We can derive the phase rule by considering the free energy of the system. Let us assume, for generality, that each of the P phases is a solution (solid, liquid, or gas) of all C components. The composition of each phase can be specified by the mole fractions $x_1, x_2, \ldots, x_C$ of the components, with their sum equal to 1. Hence $C - 1$ values are required to specify the composition of each phase. The state of the system can be specified by $P(C - 1)$ compositional parameters, plus two more, the temperature and pressure:

$$\text{Total number of variables} = PC - P + 2$$

The equilibrium state of the system at fixed temperature and pressure is the state of minimum free energy. Let us consider a variation from the stable state, consisting in removing a small amount of the first component from the first phase and adding it to the second phase. Because G is at its minimum, this variation leaves G unchanged. Hence there is a restraint on the compositions of the first and second phases: the change in free energy of the first phase on removing a small amount of the first component is the negative of that of the second phase on adding this small amount of the first component. (This is equivalent to saying that it equals that of the second phase on removal of the same small amount.) Similarly, the third phase and other phases are restrained in composition, with respect to the first component; $P - 1$ restraints for each component, and a total of $C(P - 1)$:

$$\text{Number of restraints} = CP - C$$

The variance of the system, V, is the total number of variables minus the number of restraints:

$$V = PC - P + 2 - PC + C$$

or

$$P + V = C + 2$$

The foregoing argument is essentially identical with that used by Gibbs in his discovery of the phase rule in 1876.

EXERCISES

10-1. What two properties of a chemical reaction are important for determining whether the reaction is useful?

10-2. List as many factors as possible that might affect the rates of chemical reactions.

10-3. Differentiate between homogeneous and heterogeneous reactions, with examples.

10-4. Define a catalyst and give examples.

10-5. What quantities do the differential form of the rate equation and the integrated form of the rate equation relate to time?

10-6. What is the order of a reaction? Illustrate with a general equation.

10-7. Azomethane, $CH_3—N=N—CH_3$, decomposes in the gas phase by a first-order mechanism, to produce ethane and nitrogen: $CH_3NNCH_3 \rightarrow C_2H_6 + N_2$. At 267°C with initial pressure 320 mm Hg (azomethane the only substance present), the pressure reaches 323.2 mm Hg after 400 seconds.
(*a*) What fraction of the azomethane has decomposed?
(*b*) What is the value of the reaction-rate constant?
(*c*) What is the half-life of azomethane at this temperature?
(*d*) How long would it take for 99.22% of the substance to decompose, at this temperature?

10-8. A sample of tritium, H^3, was observed to produce beta rays at the rate of 1200 per minute. Ten months later the beta rate of the sample was 1144 counts per minute. To what value of the half-life of tritium does this measurement lead?

10-9. Carbon dioxide in the air contains some carbon 14, which is produced at a steady rate in the upper atmosphere by the reaction of cosmic-ray neutrons with nitrogen nuclei. The half-life of C^{14} is 5760 years. Atmospheric carbon dioxide gives a C^{14} beta-ray count of 184 counts per minute per mole. How many atoms of $C^{12} + C^{13}$ are there in atmospheric CO_2 per atom of C^{14}?
(Answer: 0.74×10^{12}.)

10-10. What is the relationship between the stoichiometric equation for a reaction and its rate equation?

10-11. What is the time required for half the reactants to undergo second-order reaction, when they are present in equal amounts? How does this compare with the expression for the half-life of a first-order reaction?

10-12. For a reaction A + B $\longrightarrow$ products that proceeds by a second-order mechanism, the rate equation is

$$\frac{-d[A]}{dt} = k_2[A][B]$$

Show that the decrease in concentration of A can be described by the first-order equation

$$\frac{-d[A]}{dt} = k_1[A]$$

if [B] is initially very large compared with [A]. What is the relation between k_1 and k_2? (Answer: $k_1 = k_2[B]_0$.)

10-13. The reaction

$$CF_3 + H_2 \longrightarrow CF_3H + H$$

is a second-order reaction with rate constant $k_2 = 4.5 \times 10^3$ liter mole^{-1} s^{-1} at 400 K. What is the electronic structure of the radical CF_3? If a small amount of this radical were to be introduced into hydrogen at 1 atm pressure and 400 K, how long would it take for half of it to be converted to methyl fluoride?

10-14. Why do foods cook faster in a pressure cooker than in an ordinary cooking pot?

10-15. What value of the activation energy for a reaction corresponds to doubling the reaction rate on increase of temperature by 10°C at room temperature? At 300°C? At 600°C? (Answer: About 50 kJ mole^{-1}; 190; 440.)

10-16. What is the relation between the activation energy for a forward reaction, the activation energy for the reverse reaction, and the total change in enthalpy for the reaction?

10-17. What value would you predict for the activation energy of the reaction of Exercise 10-13 from the rough rule of Section 10-4? (The experimental value is 40 kJ mole^{-1}.) At what temperature would the reaction proceed twice as fast as at 400 K?

10-18. It was found by the American chemists D. P. Stevenson and D. O. Schissler, by use of a mass spectrograph to determine the composition of the reacting gas, that the reaction

$$Kr^+ + H_2 \longrightarrow KrH^+ + H$$

occurs at every collision between the reacting molecules, and that the activation energy for the reaction is zero.

(*a*) What electronic structure would you assign to KrH^+?

(*b*) Is the reaction exothermic or endothermic?

(*c*) What is the minimum value of the bond energy between Kr^+ and H in KrH^+? (Answer: (*c*) 436 kJ mole^{-1}.)

10-19. An enzyme in the human body may increase the rate of a chemical reaction a millionfold. What energy of binding of the activated complex to the enzyme would be needed to account for this effect? (Answer: 36 kJ mole^{-1}.)

10-20. How does the equilibrium of a chemical reaction change in the presence of a catalyst? Why?

10-21. One mole of hydrogen iodide was introduced into a vessel and held at 350°C for some time. The gas was then cooled rapidly enough to prevent change in its composition. The quenched equilibrium mixture was found to contain 0.117 mole of I_2. What is the equilibrium constant for the dissociation of hydrogen iodide at 350°C? (Answer: 0.0233.)

10-22. The density of the equilibrium mixture of N_2O_4 and NO_2 at 1 atm and 25°C is 3.18 g liter^{-1}. What is the average molecular mass of the gas? What are the partial pressures of NO_2 and N_2O_4? What is the value of the equilibrium constant for the reaction $N_2O_4 \rightleftarrows 2NO_2$? To what extent is N_2O_4 dissociated under these conditions?

10-23. In the contact process for making sulfuric acid, sulfur dioxide is oxidized to sulfur trioxide.
(*a*) Write the equation for the reaction between sulfur dioxide and oxygen, and write the equilibrium expression.
(*b*) At a certain temperature and concentration of oxygen the value of the equilibrium constant is such that 50% of the SO_2 is converted to SO_3 when the total pressure is 10 atm. Would the fraction converted be increased or decreased by doubling the total pressure?

10-24. It is found by experiment that when hydrogen iodide is heated the degree of dissociation increases. Is the dissociation of hydrogen iodide an exothermic or an endothermic reaction?

11

Oxidation-Reduction Reactions. Electrolysis

It was mentioned in Section 6-9 that in the period from 1884 to 1887 Svante Arrhenius developed the theory that electrolytes (salts, acids, bases) in aqueous solution are dissociated into electrically charged atoms or groups of atoms, called cations and anions. The present chapter is devoted in part to the phenomena involved in the interaction of molten salts and ionic solutions with an electric current. It is found that the electron reactions that take place at electrodes can be described as involving oxidation or reduction of atoms or groups of atoms, and that the chemical reactions called oxidation-reduction reactions (sometimes shortened to redox reactions) can often be conveniently described in terms of two electrode reactions.

11-1. The Electrolytic Decomposition of Molten Salts

The discovery of ions resulted from the experimental investigations of the interaction of an electric current with chemical substances. These investigations were begun early in the nineteenth century, and were carried on effectively by Michael Faraday (1791–1867) in the period around 1830.

The Electrolysis of Molten Sodium Chloride

Molten sodium chloride (the salt melts at 801°C) conducts an electric current, as do other molten salts. During the process of conducting the current a chemical reaction occurs; the salt is decomposed. If two electrodes (carbon rods) are dipped into a crucible containing molten sodium chloride and an electric potential, from a battery or generator, is applied, metallic sodium is produced at the cathode and chlorine gas at the anode. Such electric decomposition of a substance is called electrolysis.

The Mechanism of Ionic Conduction

Molten sodium chloride, like the crystalline substance, consists of equal numbers of sodium ions and chloride ions. These ions are very stable, and do not gain electrons or lose electrons easily. Whereas the ions in the crystal are firmly held in place by their neighbors, those in the molten salt move about with considerable freedom.

An electric generator or battery forces electrons into the cathode and pumps them away from the anode—electrons move freely in a metal or a semimetallic conductor such as graphite. But electrons cannot ordinarily get into a substance such as salt; the crystalline substance is an insulator, and the electric conductivity shown by the molten salt is not electronic conductivity (metallic conductivity), but is conductivity of a different kind, called ionic conductivity or electrolytic conductivity. This sort of conductivity results from the motion of the ions in the liquid; the cations, Na^+, move toward the negatively charged cathode, and the anions, Cl^-, move toward the anode (Figure 11-1).

The Electrode Reactions

The preceding statement describes the mechanism of the conduction of the current through the liquid. We must now consider the way in which the current passes between the electrodes and the liquid; that is, we consider the electrode reactions.

The process that occurs at the cathode is this: sodium ions, attracted to the cathode, combine with the electrons carried by the cathode to form sodium atoms; that is, to form sodium metal. The cathode reaction accordingly is

$$Na^+ + e^- \longrightarrow Na \qquad (11\text{-}1)$$

The symbol e^- represents an electron, which in this case comes from the cathode. Similarly, at the anode chloride ions give up their extra electrons

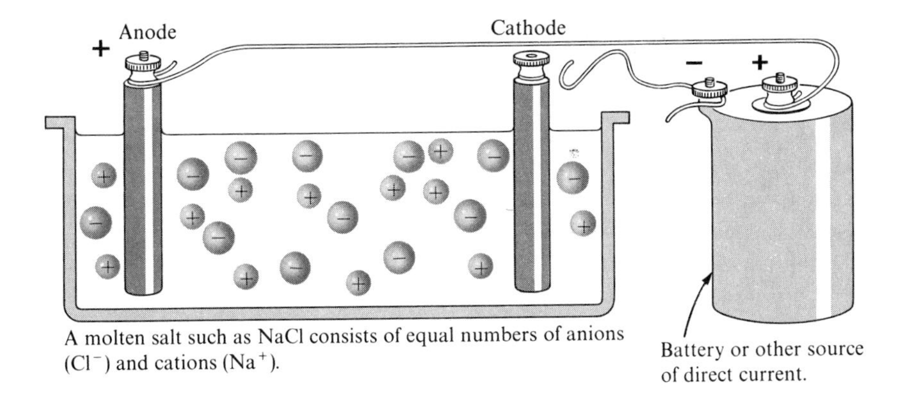

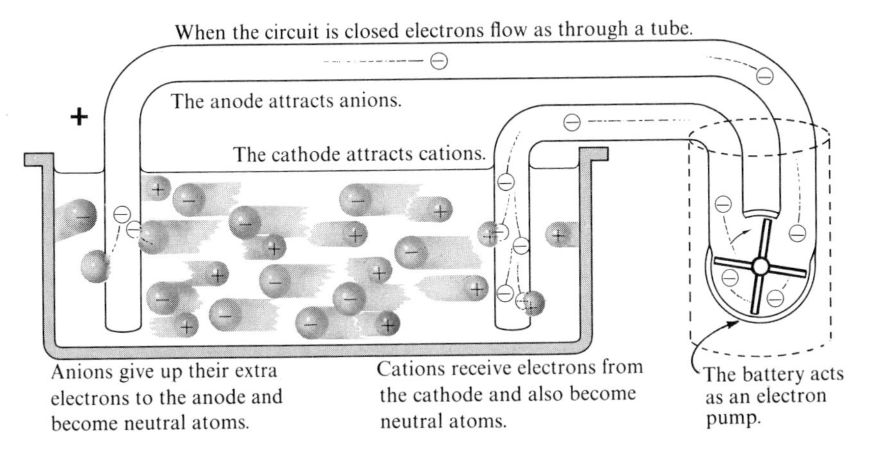

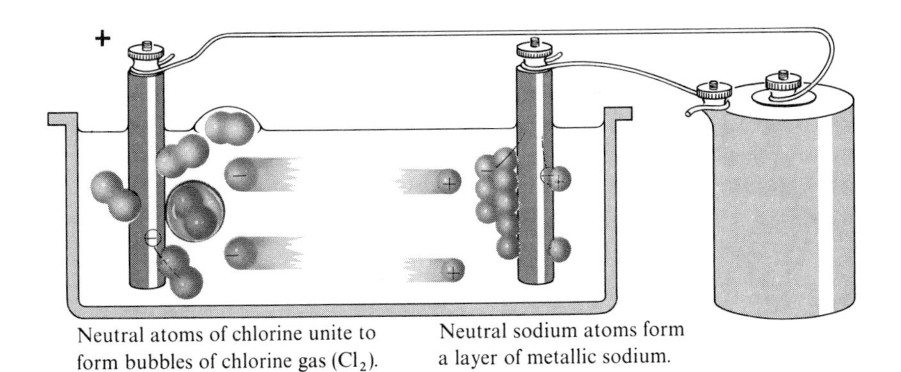

FIGURE 11-1
Electrolysis of molten sodium chloride.

to the anode, and become chlorine atoms, which are combined as the molecules of a chlorine gas. The anode reaction is

$$2Cl^- \rightarrow Cl_2 + 2e^- \qquad (11\text{-}2)$$

The Over-all Reaction

The whole process of electric conduction in this system thus occurs in the following steps.

1. An electron is pumped into the cathode.
2. The electron moves out of the cathode onto an adjacent sodium ion, converting it into an atom of sodium metal.
3. The charge of the electron is conducted across the liquid by the motion of the ions.
4. A chloride ion gives its extra electron to the anode, and becomes half of a molecule of chlorine gas.
5. The electron moves out of the anode toward the generator or battery.

The over-all reaction for the electrolytic decomposition is the sum of the two electrode reactions. Since two electrons are shown on their way around the circuit in Equation 11-2, we must double Equation 11-1:

$$\begin{array}{rcl} 2Na^+ + 2e^- & \longrightarrow & 2Na \\ 2Cl^- & \longrightarrow & Cl_2 + 2e^- \\ \hline 2Na^+ + 2Cl^- & \xrightarrow[\text{electr.}]{} & 2Na + Cl_2 \end{array} \qquad (11\text{-}3)$$

$$2NaCl \xrightarrow[\text{electr.}]{} 2Na + Cl_2 \qquad (11\text{-}4)$$

The equations 11-3 and 11-4 are equivalent; they both represent the decomposition of sodium chloride into its elementary constituents. The abbreviation "electr." (for electrolysis) is written beneath the arrow to indicate that the reaction occurs on the passage of an electric current.

Ionic Conduction by Crystals

Metals conduct electricity by the motion of electrons from atom to atom within the crystal. The conductivity decreases with increasing temperature; the electrons are scattered by the disorder of the atoms associated with thermal agitation. Metalloids and other semiconductors, which also conduct electricity by motion of electrons, have smaller conductivity and a positive, rather than negative, temperature coefficient of conductivity. Some crystalline substances have large ionic conductivity, with positive temperature coefficient.

The relation between the current, I (in amperes), the applied voltage, E

(in volts), and the resistance, R (in ohms), of a conductor is given by Ohm's law, $E = IR$. For a wire or other conductor with cross section A (in cm^2) and length l (in cm) the resistivity ρ (in ohm cm) is equal to RA/l. The reciprocal of the resistivity, $\sigma = \rho^{-1}$, is called the conductivity. It is usually reported in the units $ohm^{-1}\ cm^{-1}$, and is equal to the current in amperes that flows through a conductor with area 1 cm^2 under the potential 1 V per centimeter of length.

The conductivity of metals at 20°C ranges from about $1 \times 10^4\ ohm^{-1}\ cm^{-1}$ for the poorer conductors, such as barium ($\sigma = 1.7 \times 10^4$) and gadolinium ($\sigma = 0.7 \times 10^4$), to 0.7×10^6 for the best conductor, silver.

For ionic crystals, such as sodium chloride, the increase in conductivity with increasing temperature can be represented by an exponential factor:

$$\sigma(T) = \sigma_0 \exp(E^*/RT) \tag{11-5}$$

The value of E^*, which can be interpreted as the excitation energy required to move a sodium ion from its normal position in the crystal, is 190 kJ $mole^{-1}$. The conductivity remains very low, about $1 \times 10^{-4}\ ohm^{-1}\ cm^{-1}$, even at 800°C, only a degree below the melting point.

Silver iodide is an example of a crystal with large ionic conductivity, which reaches the value 2.5 $ohm^{-1}\ cm^{-1}$ at 555°C, 3° below the melting point. At the melting point the conductivity of the crystal is greater than that of the liquid.

The very large ionic conductivity of silver iodide crystal is explained by its structure. The crystal is cubic, with the four iodide ions in the unit cell in the close-packed positions $0\ 0\ 0$, $0\ \frac{1}{2}\ \frac{1}{2}$, $\frac{1}{2}\ 0\ \frac{1}{2}$, $\frac{1}{2}\ \frac{1}{2}\ 0$ (Figure 2-7). The silver ions might be in the octahedral positions $\frac{1}{2}\ \frac{1}{2}\ \frac{1}{2}$, and so on, which would give the sodium chloride structure (Figure 6-19), or the tetrahedral positions $\frac{1}{4}\ \frac{1}{4}\ \frac{1}{4}$, and so on, or in positions midway between adjacent iodide ions (ligancy 2 for silver, as found in the ion AgI_2^-). In fact, the x-ray diffraction pattern shows that the silver ions are distributed among all of these positions. They move with nearly complete freedom from one position to an adjacent (unoccupied) position. The potential barrier associated with this motion is small; the observed temperature coefficient of the conductivity corresponds to the value 5.1 kJ $mole^{-1}$ for the excitation energy E^*.

11-2. The Electrolysis of an Aqueous Salt Solution

Although pure water does not conduct electricity very well (conductivity $4.4 \times 10^{-4}\ ohm^{-1}\ cm^{-1}$ at 20°C), a solution of salt (or acid or base) is a good conductor. During electrolysis chemical reactions take place at the electrodes, as described below.

The phenomena that occur when a current of electricity is passed through such a solution are analogous to those described in the preceding section for molten salt. The five steps are the following.

1. Electrons are pumped into the cathode.
2. Electrons jump from the cathode to adjacent ions or molecules, producing the cathode reaction.
3. The current is conducted across the liquid by the motion of the dissolved ions.
4. Electrons jump from ions or molecules in the solution to the anode, producing the anode reaction.
5. The electrons move out of the anode toward the generator or battery.

Let us consider a dilute solution of sodium chloride (Figure 11-2). The process of conduction through this solution (step 3) is closely similar to that for molten sodium chloride. Here it is the dissolved sodium ions that move toward the cathode and the dissolved chloride ions that move toward the anode. By the motion of the ions in this way, negative electric charge is carried toward the anode and away from the cathode.

But the electrode reactions for dilute salt solutions are entirely different from those for molten salts. Electrolysis of dilute salt solution produces hydrogen at the cathode and oxygen at the anode, whereas electrolysis of molten salt produces sodium and chlorine.

The *cathode reaction* for a dilute salt solution is

$$2e^- + 2H_2O \rightarrow H_2 + 2OH^- \qquad (11\text{-}6a)$$

Two electrons from the cathode react with two water molecules to produce a molecule of hydrogen and two hydroxide ions. The molecular hydrogen bubbles off as hydrogen gas (after the solution near the cathode has become saturated with hydrogen) and the hydroxide ions stay in the solution. The *anode reaction* is

$$2H_2O \rightarrow O_2 + 4H^+ + 4e^- \qquad (11\text{-}6b)$$

Four electrons enter the anode from two water molecules, which decompose to form an oxygen molecule and four hydrogen ions.

These electrode reactions, like other chemical reactions, may occur in steps; the description given in the preceding sentence of the course of the anode reaction is not to be interpreted as giving the necessary sequence of events.

The *over-all reaction* is obtained by multiplying Equation 11-6a by 2 and adding Equation 11-6b:

$$6H_2O \xrightarrow[\text{electr.}]{} \underset{\text{Cathode}}{2H_2} + \underset{\text{Anode}}{O_2} + \underset{\text{Anode}}{4H^+} + \underset{\text{Cathode}}{4OH^-} \qquad (11\text{-}7)$$

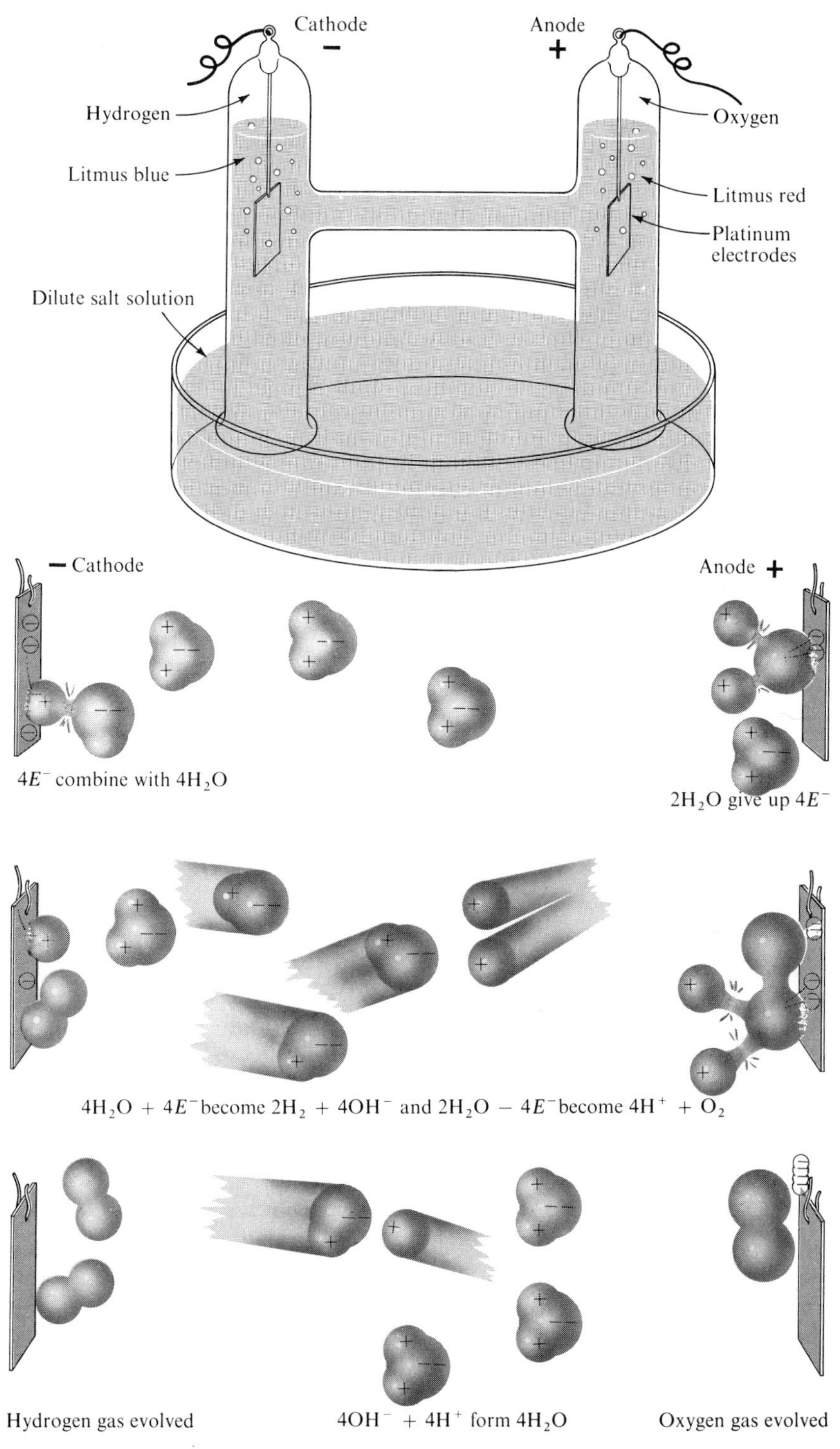

FIGURE 11-2
Electrolysis of dilute aqueous salt solution.

It is seen that in the electrolysis of the salt solution the solution around the anode becomes acidic, because of the production of hydrogen ions, and that around the cathode becomes basic, because of the production of hydroxide ions. This reaction could accordingly be used for the manufacture of acids such as hydrochloric acid and bases such as sodium hydroxide.

In the course of time, if the system were allowed to stand, the hydrogen ions produced near the anode and the hydroxide ions produced near the cathode would diffuse together and combine to form water:

$$H^+ + OH^- \rightarrow H_2O$$

This reaction would, in particular, occur if the solution of electrolytes were to be stirred during the electrolysis. If this reaction of neutralization of hydrogen ion by hydroxide ion occurs completely, the over-all electrolysis reaction is

$$\underset{}{2H_2O} \xrightarrow[\text{electr.}]{} \underset{\text{Cathode}}{2H_2} + \underset{\text{Anode}}{O_2} \qquad (11\text{-}8)$$

In discussing the electrode reactions we have made little use of the fact that the electrolyte is sodium chloride. Indeed, *the electrode reactions are the same for almost all dilute aqueous electrolytic solution,* and even for pure water as well. When electrodes are placed in pure water and an electric potential is applied, the electrode reactions shown in Equations 11-6a and 11-6b begin to take place. Very soon, however, a large enough concentration of hydroxide ions is built up near the cathode and of hydrogen ions near the anode to produce a back electric potential that tends to stop the reactions. Even in pure water there are a few ions (hydrogen ions and hydroxide ions); these ions move slowly toward the electrodes, and neutralize the ions (OH^- and H^+, respectively) formed by the electrode reactions. It is the smallness of the current that the very few ions that are present in pure water can carry through the region between the electrodes that causes the electrolysis of pure water to proceed only very slowly.

Equations 11-6a and 11-6b show water molecules undergoing decomposition at the electrodes. These equations probably represent the usual molecular reactions in neutral salt solutions. In acidic solutions, however, in which there is a high concentration of hydrogen ions, the cathode reaction may well be simply the reaction

$$2H^+ + 2e^- \rightarrow H_2$$

and in basic solutions, in which there is a high concentration of hydroxide ions, the anode reaction may be

$$4OH^- \rightarrow O_2 + 2H_2O + 4e^-$$

The ions in an electrolytic solution can carry a larger current between the electrodes than can the very few ions in pure water. In a sodium chloride solution undergoing electrolysis, sodium ions move to the cathode region, where their positive electric charges compensate the negative charges of the hydroxide ions that have been formed by the cathode reaction. Similarly, the chloride ions that move toward the anode compensate electrically the hydrogen ions that have been formed by the acid reaction.

Production of hydroxide ions at the cathode and of hydrogen ions at the anode during the electrolysis can be demonstrated by means of litmus or a similar indicator.

The electrolysis of dilute aqueous solutions of other electrolytes is closely similar to that of sodium chloride, producing hydrogen and oxygen gases at the electrodes. Concentrated electrolytic solutions may behave differently; concentrated brine (sodium chloride solution) on electrolysis produces chlorine at the anode, as well as oxygen. We may understand this fact by remembering that in concentrated brine there are a great many chloride ions near the anode, and some of these give up electrons to the anode, and form chlorine molecules.

11-3. Oxidation-Reduction Reactions

From the rules for assigning oxidation numbers given in Section 6-16 we see that the anode reaction 11-2 involves an increase in oxidation number for chlorine, from -1 to 0, accompanied by loss of an electron to the electrode (the anode). Increase in oxidation number is described as oxidation. The anode reaction is an oxidation reaction. Similarly, the cathode reaction, Equation 11-1, in which an electron from the cathode causes the decrease in oxidation number of sodium from $+1$ to 0, is described as reduction: the cathode reaction is a reduction reaction.

We see that oxidation can be described as de-electronation, and reduction as electronation. In ordinary oxidation-reduction reactions the two processes take place simultaneously, sometimes by direct transfer of electrons from the atoms that are oxidized to those that are reduced.

It is often convenient in writing the equation for an oxidation-reduction reaction to write and balance equations for two electrode reactions (which may be hypothetical), and then to add these reactions in such a way that the electrons cancel.

The first step is to be sure that you know what the reactants are and what the products are. Then you identify the reducing agent and the oxidizing agent, write the equations for the de-electronation and electronation reactions, and combine these equations, as illustrated in the following example.

Example 11-1. The oxidizing agent permanganate ion, MnO_4^-, on reduction in acid solution forms manganous ion, Mn^{++}. Ferrous ion, Fe^{++}, can accomplish this reduction. Write the equation for the reaction between permanganate ion and ferrous ion in acid solution.

Solution. The oxidation number of manganese in permanganate ion is +7, $[Mn^{+7}(O^{-2})_4]^-$. That of manganous ion is +2. Hence five electrons are involved in the reduction of permanganate ion. The electron reaction is

$$[Mn^{+7}(O^{-2})_4]^- + 5e^- + \text{other reactants} \longrightarrow Mn^{++} + \text{other products} \qquad (11\text{-}9)$$

In reactions in aqueous solutions water, hydrogen ion, and hydroxide ion may come into action as reactants or products. For example, in an acid solution hydrogen ion may be either a reactant or a product, and water may also be either a reactant or a product in the same reaction. In acid solutions hydroxide ion exists only in extremely low concentration, and would hardly be expected to enter into the reaction. Hence water and hydrogen ion may enter into the reaction now under consideration.

Reaction 11-9 is not balanced electrically; there are six negative charges on the left side and two positive charges on the right side. The only other ion that can enter into the reaction is hydrogen ion, and the number needed to give conservation of electric charge is 8. Thus we obtain

$$MnO_4^- + 5e^- + 8H^+ \longrightarrow Mn^{++} + \text{other products} \qquad (11\text{-}10)$$

Oxygen and hydrogen occur here on the left side and not the right side of the reaction; conservation of atoms is satisfied if $4H_2O$ is written in as the "other products":

$$MnO_4^- + 5e^- + 8H^+ \longrightarrow Mn^{++} + 4H_2O \qquad (11\text{-}11)$$

We check this equation on three points—*proper change in oxidation number* (5 electrons used, with change of −5 in oxidation number of manganese, from Mn^{+7} to Mn^{+2}), *conservation of electric charge* (from −1 −5 +8 to +2), and *conservation of atoms of each kind*—and convince ourselves that it is correct.

The electron reaction for the oxidation of ferrous ion is now written:

$$Fe^{++} \longrightarrow Fe^{+++} + e^- \qquad (11\text{-}12)$$

This equation checks on all three points.

The equation for the oxidation-reduction reaction is obtained by combining the two electron reactions in such a way that the electrons liberated in one are used up in the other. We see that this is achieved by multiplying Equation 11-12 by 5 and adding it to Equation 11-11:

$$5Fe^{++} \longrightarrow 5Fe^{+++} + 5e^-$$
$$MnO_4^- + 5e^- + 8H^+ \longrightarrow Mn^{++} + 4H_2O$$
$$MnO_4^- + 5Fe^{++} + 8H^+ \longrightarrow Mn^{++} + 5Fe^{+++} + 4H_2O$$

It is good practice to check this final equation also on all three points, to be sure that no mistake has been made.

It is not always necessary to carry through this entire procedure. Sometimes an equation is so simple that it can be written at once and verified by inspection. An example is the reduction of silver ion, Ag^+, by metallic zinc:

$$Zn(c) + 2Ag^+(aq) \longrightarrow 2Ag(c) + Zn^{++}(aq)$$

Sometimes, too, the conditions determine the reaction, as when a single substance decomposes. Thus ammonium nitrite decomposes to give water and nitrogen:

$$NH_4NO_2 \longrightarrow N_2 + 2H_2O$$

Here N^{+3} (of NO_2^-) oxidizes N^{-3} (of NH_4^+), both going to N^0 (of N_2).

Oxidation Equivalents and Reduction Equivalents

The *oxidizing capacity* or *reducing capacity* of an oxidizing agent or reducing agent is equal to the number of electrons involved in its reduction or oxidation. An oxidation-reduction equation is balanced when the amounts of oxidizing agent and reducing agent indicated as reacting have the same capacities.

An *oxidation equivalent* or *reduction equivalent* of a substance is the amount that takes up or gives up one electron (one mole of electrons). Thus the gram equivalent mass (equivalent mass expressed in grams) of potassium permanganate as an oxidizing agent in acid solution (see Equation 11-11) is one-fifth of the gram formula mass, whereas the gram equivalent mass of ferrous ion as a reducing agent is just the gram atomic mass.

Equivalent masses of oxidizing agents and reducing agents react exactly with one another, since they involve the taking up or giving up of the same number of electrons.

Normal Solutions of Oxidizing and Reducing Agents

A solution of an oxidizing agent or reducing agent containing 1 gram equivalent mass per liter of solution is called a 1 normal (1 N) solution. In general the *normality* of the solution is the number of gram equivalent masses of the oxidizing or reducing agent present per liter.

It is seen from the definition that equal volumes of an oxidizing solution and a reducing solution of the same normality react exactly with each other.

Example 11-2. What is the normality, for use as an oxidizing agent in acid solution, of a permanganate solution made by dissolving one-tenth of a gram formula mass of $KMnO_4$ ($\frac{1}{10} \times 158.03$ g) in water and diluting to a volume of 1 l?

Solution. The reduction of permanganate ion in acid solution involves 5 electrons (Equation 11-11). Hence 1 gram formula mass is 5 equivalents. The solution is accordingly 0.5 *N*.

It is necessary to take care that the conditions of the use of a reagent are known in stating its normality. Thus permanganate ion is sometimes used as an oxidizing agent in neutral or basic solution, in which it is reduced by only three steps to manganese dioxide, MnO_2, in which manganese has oxidation number 4. The above solution would have normality 0.3 for this use.

11-4. Quantitative Relations in Electrolysis

In 1832 and 1833 Michael Faraday reported his discovery by experiment of the fundamental laws of electrolysis. These are

1. *The mass of a substance produced by a cathode or anode reaction in electrolysis is directly proportional to the quantity of electricity passed through the cell.*

2. *The masses of different substances produced by the same quantity of electricity are proportional to the equivalent masses of the substances.*

These laws are now known to be the result of the fact that electricity is composed of individual particles, the electrons. Quantity of electricity can be expressed as number of electrons; the number of molecules of a substance produced by an electrode reaction is related in a simple way to the number of electrons involved; hence the amount of substance produced is proportional to the quantity of electricity, in a way that involves the molecular mass or chemical equivalent mass of the substance.

The amounts of substances produced by a given quantity of current can be calculated from knowledge of the electrode reactions. The magnitude of the charge of one mole of electrons is 96,490 coulombs. This is called a *faraday:* 1 F = 1 faraday = 96,490 C.

It is customary to define the faraday as this quantity of positive electricity. The quantitative treatment of electrochemical reactions is made in the same way as the calculation of mass relations in ordinary chemical reactions, with use of the faraday to represent one mole of electrons.

Example 11-3. For how long a time would a current of 10 A have to be passed through a cell containing fused sodium chloride to produce 23 g of metallic sodium at the cathode? How much chlorine would be produced at the anode?

Solution. The cathode reaction is

$$Na^+ + e^- \longrightarrow Na$$

Hence 1 mole of electrons passing through the cell would produce 1 mole of sodium atoms. One mole of electrons is 1 faraday, and 1 mole of sodium atoms is a gram atom of sodium, 23.00 g. Hence the amount of electricity required is 96,490 coulombs. One coulomb is 1 ampere second. Hence 96,490 coulombs of electricity passes through the cell if 10 A flows for 9649 seconds. The answer is thus 9649 s.

The anode reaction is

$$2Cl^- \longrightarrow Cl_2 + 2e^-$$

To produce 1 mole of molecular chlorine, Cl_2, 2 F must pass through the cell. One faraday will produce $\frac{1}{2}$ mole of Cl_2, or 1 gram atom of chlorine, which is 35.46 g. Note that two moles of electrons are required to produce one mole of molecular chlorine.

Example 11-4. Two cells are set up in series, and a current is passed through them. Cell A contains an aqueous solution of silver sulfate, and has platinum electrodes (which are unreactive). Cell B contains a copper sulfate solution, and has copper electrodes. The current is passed through until 1.6 g of oxygen has been liberated at the anode of cell A. What has occurred at the other electrodes? (See Figure 11-3.)

Solution. At the anode of cell A the reaction is

$$2H_2O \longrightarrow O_2(g) + 4H^+ + 4e^-$$

Hence 4 F of electricity liberates 32 g of oxygen, and 0.2 F liberates 1.6 g.

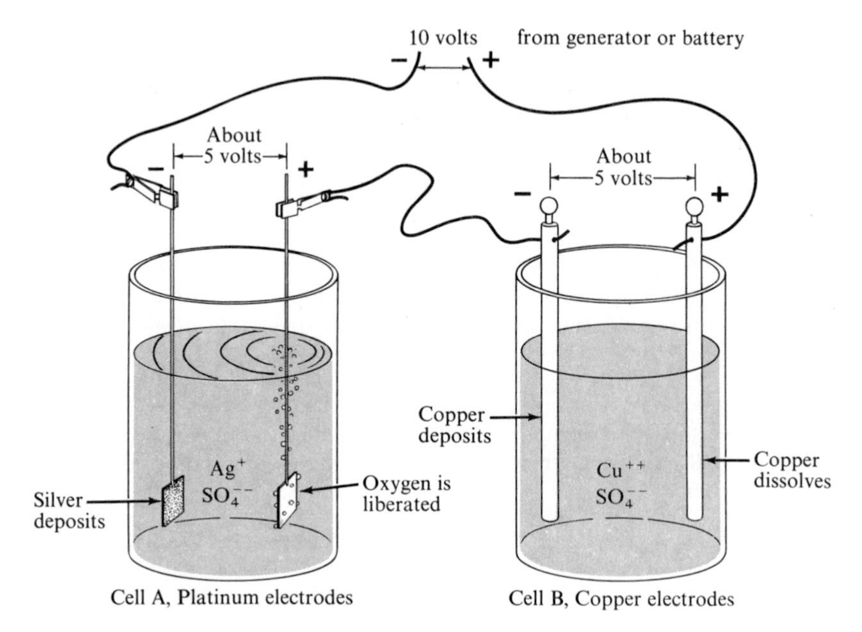

FIGURE 11-3
Two electrolytic cells in series.

At the cathode of cell A the reaction is

$$Ag^+ + e^- \rightarrow Ag(c)$$

One gram atom of silver, 107.880 g, would be deposited by 1 F, and the passage of 0.2 F through the cell would hence deposit 21.576 g of silver on the platinum cathode.

At the cathode in cell B the reaction is

$$Cu^{++} + 2e^- \rightarrow Cu(c)$$

One gram atom of copper, 63.57 g, would hence be deposited on the cathode by 2 F of electricity, and 6.357 g by 0.2 F.

The same amount of copper, 6.357 g, would be dissolved from the anode of cell B, since the same number of electrons flow through the anode as through the cathode. The anode reaction is

$$Cu(c) \rightarrow Cu^{++} + 2e^-$$

Note that the total voltage difference supplied by the generator or battery (shown as 10 volts) is divided between the two cells in series. The division need not be equal, as indicated, but is determined by the properties of the two cells.

11-5. The Electromotive-force Series of the Elements

It is found that if a piece of one metal is put into a solution containing ions of another metallic element the first metal may dissolve, with the deposition of the second metal from its ions. Thus a strip of zinc placed in a solution of a copper salt causes a layer of metallic copper to deposit on the zinc, as the zinc goes into solution:

$$\begin{array}{r} Zn(c) \rightarrow Zn^{++} + 2e^- \\ Cu^{++} + 2e^- \rightarrow Cu(c) \\ \hline Zn(c) + Cu^{++} \rightarrow Zn^{++} + Cu(c) \end{array}$$

On the other hand, a strip of copper placed in a solution of a zinc salt does not cause metallic zinc to deposit.

It is not strictly correct to say that zinc can replace copper in solution and that copper cannot replace zinc. If a piece of metallic copper is placed in a solution containing zinc ion in appreciable concentration (1 mole per liter, say) and no cupric ion at all, the reaction

$$Cu(c) + Zn^{++} \rightarrow Cu^{++} + Zn(c)$$

will occur to a very small extent, stopping when a certain very small concentration of copper ion has been produced. If metallic zinc is added to a 1-molar solution of cupric ion, the reaction

$$Zn(c) + Cu^{++} \rightarrow Zn^{++} + Cu(c)$$

the reverse of the preceding reaction, will take place almost to completion, stopping when the concentration of copper ion has become very small. The principles of thermodynamics require that the ratio of concentrations of the two ions Cu^{++} and Zn^{++} in equilibrium with solid copper and solid zinc be the same whether the equilibrium is approached from the Cu, Zn^{++} side or the Zn, Cu^{++} side. The statement "zinc replaces copper from solution" means that at equilibrium the ratio of cupric-ion concentration to zinc-ion concentration is small.

By experiments of this sort, the elements can be arranged in a table showing their ability to reduce ions of other metals. This table is given as Table 11-1. The metal with the greatest reducing power is at the head of the list. This series is called the electromotive-force series because the tendency of one metal to reduce ions of another can be measured by setting up an electric cell and measuring the voltage that it produces. ("Electromotive force" is here a synonym for "voltage.") It is conventional to tabulate the voltage corresponding to unit activity for each ion;

TABLE 11-1
Standard Oxidation-reduction Potentials and Equilibrium Constants (The values apply to temperature 25°C, with standard concentration for aqueous solutions 1 *M* and standard pressure of gases 1 atm.)

	$E°$	K
$Li \rightleftarrows Li^+ + e^-$	−3.05	4×10^{50}
$Cs \rightleftarrows Cs^+ + e^-$	−2.92	1×10^{49}
$Rb \rightleftarrows Rb^+ + e^-$	−2.92	1×10^{49}
$K \rightleftarrows K^+ + e^-$	−2.92	1×10^{49}
$\frac{1}{2}Ba \rightleftarrows \frac{1}{2}Ba^{++} + e^-$	−2.90	5×10^{48}
$\frac{1}{2}Sr \rightleftarrows \frac{1}{2}Sr^{++} + e^-$	−2.89	4×10^{48}
$\frac{1}{2}Ca \rightleftarrows \frac{1}{2}Ca^{++} + e^-$	−2.87	2×10^{48}
$Na \rightleftarrows Na^+ + e^-$	−2.712	4.0×10^{45}
$\frac{1}{3}Al + \frac{4}{3}OH^- \rightleftarrows \frac{1}{3}Al(OH)_4^- + e^-$	−2.35	3×10^{39}
$\frac{1}{2}Mg \rightleftarrows \frac{1}{2}Mg^{++} + e^-$	−2.34	2×10^{39}
$\frac{1}{2}Be \rightleftarrows \frac{1}{2}Be^{++} + e^-$	−1.85	1×10^{31}
$\frac{1}{3}Al \rightleftarrows \frac{1}{3}Al^{+++} + e^-$	−1.67	1×10^{28}
$\frac{1}{2}Zn + 2OH^- \rightleftarrows \frac{1}{2}Zn(OH)_4^{--} + e^-$	−1.216	2.7×10^{20}
$\frac{1}{2}Mn \rightleftarrows \frac{1}{2}Mn^{++} + e^-$	−1.18	7×10^{19}
$\frac{1}{2}Zn + 2NH_3 \rightleftarrows \frac{1}{2}Zn(NH_3)_4^{++} + e^-$	−1.03	2×10^{17}
$Co(CN)_6^{----} \rightleftarrows Co(CN)_6^{---} + e^-$	−0.83	1×10^{14}
$\frac{1}{2}Zn \rightleftarrows \frac{1}{2}Zn^{++} + e^-$	−0.762	6.5×10^{12}
$\frac{1}{3}Cr \rightleftarrows \frac{1}{3}Cr^{+++} + e^-$	−0.74	3×10^{12}
$\frac{1}{2}H_2C_2O_4(aq) \rightleftarrows CO_2 + H^+ + e^-$	−0.49	2×10^{8}
$\frac{1}{2}Fe \rightleftarrows \frac{1}{2}Fe^{++} + e^-$	−0.440	2.5×10^{7}
$\frac{1}{2}Cd \rightleftarrows \frac{1}{2}Cd^{++} + e^-$	−0.402	5.7×10^{6}
$\frac{1}{2}Co \rightleftarrows \frac{1}{2}Co^{++} + e^-$	−0.277	4.5×10^{4}
$\frac{1}{2}Ni \rightleftarrows \frac{1}{2}Ni^{++} + e^-$	−0.250	1.6×10^{4}
$I^- + Cu \rightleftarrows CuI(s) + e^-$	−0.187	1.4×10^{3}
$\frac{1}{2}Sn \rightleftarrows \frac{1}{2}Sn^{++} + e^-$	−0.136	1.9×10^{2}
$\frac{1}{2}Pb \rightleftarrows \frac{1}{2}Pb^{++} + e^-$	−0.126	1.3×10^{2}
$\frac{1}{2}H_2 \rightleftarrows H^+ + e^-$	0.000	1
$\frac{1}{2}H_2S \rightleftarrows \frac{1}{2}S + H^+ + e^-$	0.141	4.3×10^{-3}
$Cu^+ \rightleftarrows Cu^{++} + e^-$	0.153	2.7×10^{-3}
$\frac{1}{2}H_2O + \frac{1}{2}H_2SO_3 \rightleftarrows \frac{1}{2}SO_4^{--} + 2H^+ + e^-$	0.17	1×10^{-3}
$\frac{1}{2}Cu \rightleftarrows \frac{1}{2}Cu^{++} + e^-$	0.345	1.6×10^{-6}
$Fe(CN)_6^{----} \rightleftarrows Fe(CN)_6^{---} + e^-$	0.36	9×10^{-7}
$I^- \rightleftarrows \frac{1}{2}I_2(s) + e^-$	0.53	1×10^{-9}
$MnO_4^{--} \rightleftarrows MnO_4^- + e^-$	0.54	1×10^{-9}
$\frac{4}{3}OH^- + \frac{1}{3}MnO_2 \rightleftarrows \frac{1}{3}MnO_4^- + \frac{2}{3}H_2O + e^-$	0.57	3×10^{-10}
$\frac{1}{2}H_2O \rightleftarrows \frac{1}{4}O_2 + H^+ + e^-$	0.682	3.5×10^{-12}
$Fe^{++} \rightleftarrows Fe^{+++} + e^-$	0.771	1.1×10^{-13}
$Hg \rightleftarrows \frac{1}{2}Hg_2^{++} + e^-$	0.799	3.7×10^{-14}

TABLE 11-1 (continued)

	$E°$	K
$Ag \rightleftarrows Ag^+ + e^-$	0.800	3.5×10^{-14}
$H_2O + NO_2 \rightleftarrows NO_3^- + 2H^+ + e^-$	0.81	3×10^{-14}
$\frac{1}{2}Hg \rightleftarrows \frac{1}{2}Hg^{++} + e^-$	0.854	4.5×10^{-15}
$\frac{1}{2}Hg_2^{++} \rightleftarrows Hg^{++} + e^-$	0.910	5.0×10^{-16}
$\frac{1}{2}HNO_2 + \frac{1}{2}H_2O \rightleftarrows \frac{1}{2}NO_3^- + \frac{3}{2}H^+ + e^-$	0.94	2×10^{-16}
$NO + H_2O \rightleftarrows HNO_2 + H^+ + e^-$	0.99	2×10^{-17}
$\frac{1}{2}ClO_3^- + \frac{1}{2}H_2O \rightleftarrows \frac{1}{2}ClO_4^- + H^+ + e^-$	1.00	2×10^{-17}
$Br^- \rightleftarrows \frac{1}{2}Br_2(l) + e^-$	1.065	1.3×10^{-18}
$H_2O + \frac{1}{2}Mn^{++} \rightleftarrows \frac{1}{2}MnO_2 + 2H^+ + e^-$	1.23	2×10^{-21}
$Cl^- \rightleftarrows \frac{1}{2}Cl_2 + e^-$	1.358	1.5×10^{-23}
$\frac{7}{6}H_2O + \frac{1}{3}Cr^{+++} \rightleftarrows \frac{1}{6}Cr_2O_7^{--} + \frac{7}{3}H^+ + e^-$	1.36	1×10^{-23}
$\frac{1}{2}H_2O + \frac{1}{6}Cl^- \rightleftarrows \frac{1}{6}ClO_3^- + H^+ + e^-$	1.45	4×10^{-25}
$\frac{1}{3}Au \rightleftarrows \frac{1}{3}Au^{+++} + e^-$	1.50	6×10^{-26}
$\frac{4}{5}H_2O + \frac{1}{5}Mn^{++} \rightleftarrows \frac{1}{5}MnO_4^- + \frac{8}{5}H^+ + e^-$	1.52	3×10^{-26}
$\frac{1}{2}Cl_2 + H_2O \rightleftarrows HClO + H^+ + e^-$	1.63	4×10^{-28}
$H_2O \rightleftarrows \frac{1}{2}H_2O_2 + H^+ + e^-$	1.77	2×10^{-30}
$Co^{++} \rightleftarrows Co^{+++} + e^-$	1.84	1×10^{-31}
$F^- \rightleftarrows \frac{1}{2}F_2 + e^-$	2.65	4×10^{-44}

that is, the concentration of the ion multiplied by a factor correcting for interionic interactions in concentrated solutions. Thus the cell shown in Figure 11-4 would be used to measure the voltage between the electrodes at which occur the electrode reactions

$$Zn(c) \rightleftarrows Zn^{++}(aq, a = 1) + 2e^-$$

and

$$Cu^{++}(aq, a = 1) + 2e^- \rightleftarrows Cu(c)$$

This cell produces a voltage* of about 1.107 V, the difference of the values of $E°$ shown in the table. The cell is used to some extent in practice; it is called the gravity cell when made as shown in Figure 11-5.

The standard reference point in the EMF series is the *hydrogen electrode,* which consists of gaseous hydrogen at 1 atm bubbling over a platinum electrode in an acidic solution with activity 1 for the hydrogen ion (Figure 11-6). Similar electrodes can be made for some other nonmetallic elements, and a few of these elements are included in the table, as well as some other oxidation-reduction pairs.

*Correction by a few millivolts must be made for the *liquid-junction potential,* which results from effects in the connecting arm where the two solutions meet.

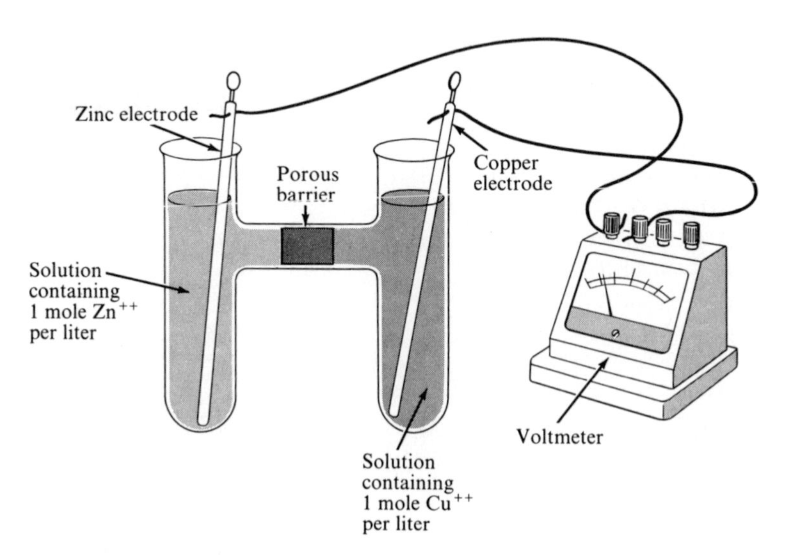

FIGURE 11-4
A cell involving the Zn, Zn^{++} electrode and the Cu, Cu^{++} electrode.

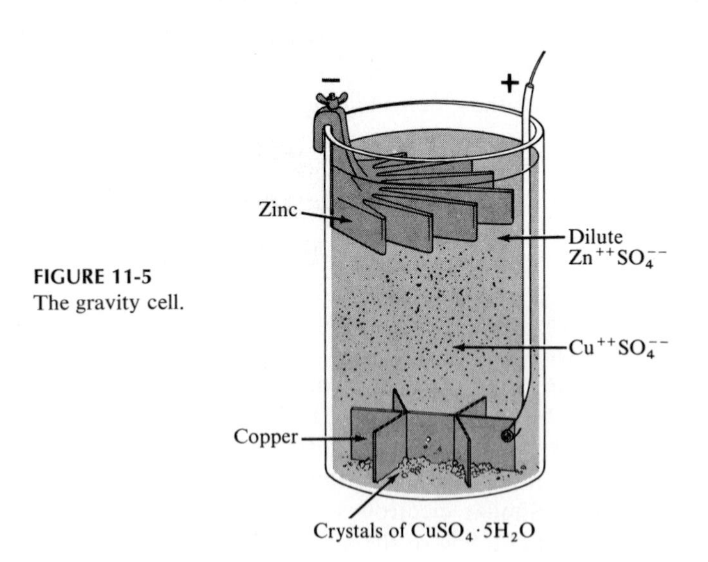

FIGURE 11-5
The gravity cell.

11-6. Equilibrium Constants for Oxidation-Reduction Couples

The electric cell shown in Figure 11-6 is conventionally written as

$$\mathrm{Zn(c)}|\mathrm{Zn^{++}}(a=1)|\mathrm{H^+}(a=1)|\mathrm{H_2}(1\ \mathrm{atm})$$

Here the vertical lines denote contact between two phases. The electromotive force is taken as $E = E(\text{left}) - E(\text{right})$. For this cell $E(\text{right})$ is equal to zero, since the right half-cell in this example is taken as the standard, with $E° = 0$, and the value of E is $E(\text{left})$. From Table 11-1 we find that $E(\text{left})$ is -0.762 V, and hence that the EMF of the cell is -0.762 V. The negative value means that there is a greater electron pressure at the left electrode than at the right electrode.

If two moles of electrons (2 F) were to flow through the wire from the left electrode to the right electrode, with the corresponding electron reactions taking place at the electrodes, the change in the system would correspond to the equation

$$\mathrm{Zn(c)} + 2\mathrm{H^+}(a=1) \rightarrow \mathrm{Zn^{++}}(a=1) + \mathrm{H_2}(1\ \mathrm{atm})$$

The electric current could be used completely in doing work, the amount of work being $-E \times 2F$ (or, in general, $-nEF$, with n the number

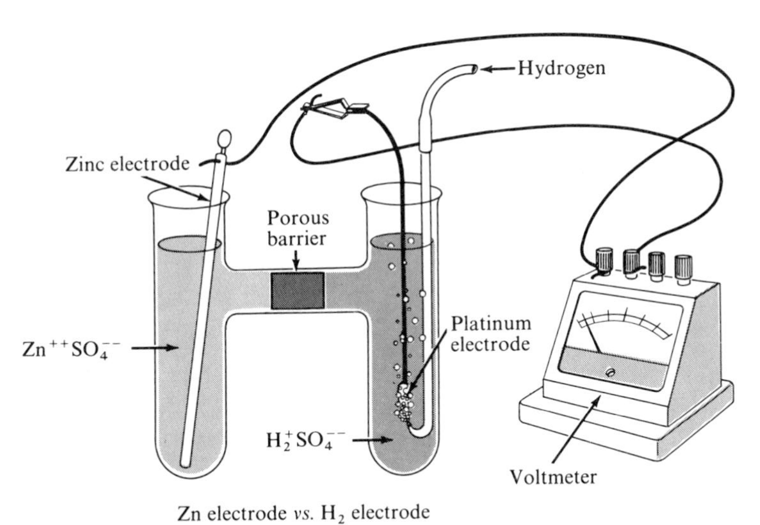

FIGURE 11-6
A cell involving the zinc electrode and the hydrogen electrode.

of electrons involved in the electrode reactions). The amount of work done by a system when it undergoes a reversible change in state at constant pressure and temperature is $-\Delta G$, the decrease in the free energy of the system. Hence we write

$$nEF = \Delta G \tag{11-13}$$

If the reactants and products in the cell are in their standard states the decrease in free energy is $-\Delta G^\circ$:

$$nE^\circ F = \Delta G^\circ \tag{11-14}$$

We have seen, however, that ΔG° is related to the equilibrium constant K of the reaction (Equation 10-24):

$$-\Delta G^\circ = RT \ln K \tag{11-15}$$

We may rewrite this equation as

$$K = \exp(-\Delta G^\circ/RT) \tag{11-16}$$

or

$$K = \exp(-nE^\circ F/RT) \tag{11-17}$$

Many equilibrium constants (and free-energy values) have been determined experimentally by EMF measurements of electric cells. One difficulty that has not been overcome for many possible half-cells is that of finding an electrode surface that permits the half-cell (electron) reaction to take place in a reversible manner. A platinum electrode covered with finely divided platinum (platinum black) is effective for many half-cells.

The electron reactions are all written in Table 11-1 so as to produce one electron. This is done for convenience; with this convention the ratio of two values of K gives the equilibrium constant for the reaction obtained by subtracting the equation for one couple from that for another. It is sometimes desirable to clear the equation of fractions by multiplying by a suitable factor; this involves raising the equilibrium constant to the power equal to this factor.

Many questions about chemical reactions can be answered by reference to a table of standard oxidation-reduction potentials. In particular it can be determined whether or not a given oxidizing agent and a given reducing agent can possibly react to an appreciable extent, and the extent of possible reactions can be predicted. It cannot be said, however, that the reaction will necessarily proceed at a significant rate under given conditions; the table gives information only about the state of chemical equilibrium and not about the rate at which equilibrium is approached. For this reason the most valuable use of the table is in connection with reactions that are known to take place, to answer questions as to the extent of reaction; but the table is also valuable in telling whether or not it is worth while to try to make a reaction go by changing conditions.

The great simplification introduced by this procedure can be seen by examining Table 11-1. This table contains only 56 entries, which correspond to 56 different electron reactions. By combining any two of these electron reactions the equation for an ordinary oxidation-reduction reaction can be written. There are 1540 ($56 \times \frac{55}{2}$) of these oxidation-reduction reactions that can be formed from the 56 electron reactions. The 56 numbers in the table can be combined in such a way as to give the 1540 values of their equilibrium constants; accordingly, this small table permits a prediction to be made as to whether any one of these 1540 reactions will tend to go in a forward direction or the reverse direction.

A more extensive table in the book *The Oxidation States of the Elements and Their Potentials in Aqueous Solutions,* by W. M. Latimer, occupies eight pages; the information given on these eight pages permits one to calculate values of the equilibrium constants for about 85,000 reactions, which, if written out, would occupy 1750 pages of the same size as the pages in Latimer's book; and, moreover, it is evident that if the equilibrium constants were independent of one another, and had to be determined by separate experiments, 85,000 experiments would have had to be carried out, instead of only about 400.

Electrode Potentials and Electronegativity

There is a rough general correlation between standard electrode potentials and electronegativity values, illustrated in Figure 11-7. It is seen that to within the uncertainty in the electronegativity values, ± 0.05, the points for chlorine, bromine, and iodine lie on the straight line connecting the points for fluorine and hydrogen. The contribution of the entropy term to the standard free energy change (which determines $E°$) is different for the different kinds of oxidation-reduction couples, and a close correlation between $E°$ and x is not to be expected.

Some ways in which the values in Table 11-1 can be used are illustrated below.

Example 11-5. Would you expect reaction to occur on mixing solutions of ferrous sulfate and mercuric sulfate?

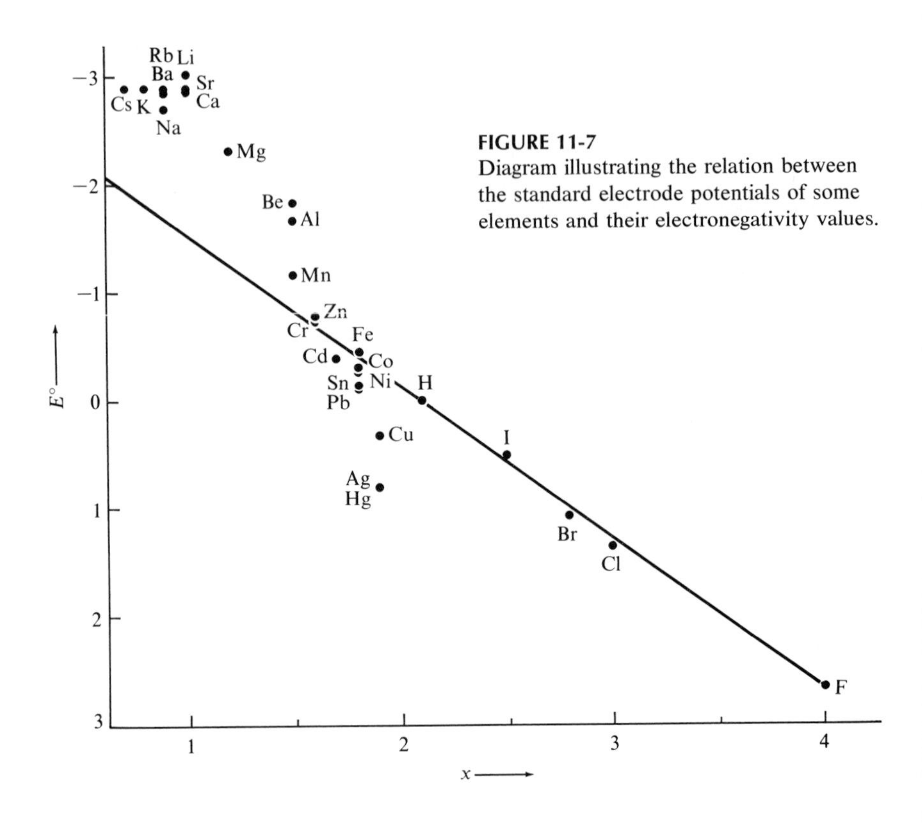

FIGURE 11-7
Diagram illustrating the relation between the standard electrode potentials of some elements and their electronegativity values.

Solution. The ferrous-ferric couple has potential 0.771 V and the mercurous-mercuric couple 0.910 V; hence the latter couple is the stronger oxidizing of the two, and the reaction

$$2Fe^{++} + 2Hg^{++} \longrightarrow 2Fe^{+++} + Hg_2^{++}$$

would occur, and proceed well toward completion.

Example 11-6. In the manufacture of potassium permanganate a solution containing manganate ion is oxidized by chlorine. Would bromine or iodine be as good?

Solution. From the table we see that the values of $E°$ and K are the following:

	$E°$	K
$MnO_4^{--} \rightleftarrows MnO_4^- + e^-$	0.54	1×10^{-9}
$Cl^- \rightleftarrows \frac{1}{2}Cl_2 + e^-$	1.358	2×10^{-23}
$Br^- \rightleftarrows \frac{1}{2}Br_2(l) + e^-$	1.065	1.3×10^{-18}
$I^- \rightleftarrows \frac{1}{2}I_2(s) + e^-$	0.53	1×10^{-9}

The values for iodine is so close to that for manganate-permanganate that effective oxidation by iodine (approaching completion) would not occur; hence iodine would be unsatisfactory. Bromine would produce essentially complete reaction, and in this respect would be as good as chlorine; but it costs ten times as much, and so should not be used.

Example 11-7. What equilibrium composition would you expect to obtain by mixing equal volumes of a 0.2 F solution of $K_4Co(CN)_6$ and a 0.2 F solution of $K_3Fe(CN)_6$?

Solution. From Table 11-1 we obtain the following information:

	$E°$	K
$Co(CN)_6^{----} \rightleftarrows Co(CN)_6^{---} + e^-$	-0.83	1×10^{14}
$Fe(CN)_6^{----} \rightleftarrows Fe(CN)_6^{---} + e^-$	0.36	9×10^{-7}

The electron-reaction equilibrium equations are

$$\frac{[Co(CN)_6^{---}][e^-]}{[Co(CN)_6^{----}]} = 1 \times 10^{14}$$

$$\frac{[Fe(CN)_6^{---}][e^-]}{[Fe(CN)_6^{----}]} = 9 \times 10^{-7}$$

By dividing the first by the second we obtain the equilibrium equation for the over-all reaction:

$$\frac{[Co(CN)_6^{---}][Fe(CN)_6^{----}]}{[Co(CN)_6^{----}][Fe(CN)_6^{---}]} = \frac{1 \times 10^{14}}{9 \times 10^{-7}} = 1 \times 10^{20}$$

The stated conditions of the reaction are such that the equilibrium concentrations of the two reactants are equal to each other, and those of the two products are equal to each other. We write

$$x = [Co(CN)_6^{----}] = [Fe(CN)_6^{---}]$$

$$0.1 - x = [Co(CN)_6^{---}] = [Fe(CN)_6^{----}]$$

Hence

$$\frac{(0.1 - x)^2}{x^2} = 1 \times 10^{20}$$

$$x^2 = 1 \times 10^{-20} \times (0.1 - x)^2 \cong 1 \times 10^{-22}$$

$$x = 1 \times 10^{-11} \text{ mole l}^{-1}$$

Hence the reaction proceeds essentially to completion, with the concentration of each of the products 10^{10} times that of each of the reactants.

11-7. Primary Cells and Storage Cells

The production of an electric current through chemical reaction is achieved in *primary cells* and *storage cells.*

Primary cells are cells in which an oxidation-reduction reaction can be carried out in such a way that its driving force produces an electric potential. This is achieved by having the oxidizing agent and the reducing agent separated; the oxidizing agent then removes electrons from one electrode and the reducing agent gives electrons to another electrode, the flow of current through the cell itself being carried by ions.

Storage cells are similar cells, which, however, can be returned to their original state after current has been drawn from them (can be *charged*) by applying an impressed electrical potential between the electrodes, and thus reversing the oxidation-reduction reaction.

The Common Dry Cell

One primary cell, the gravity cell, has been described in Section 11-5. This cell is called a *wet cell,* because it contains a liquid electrolyte. A very useful primary cell is the *common dry cell,* shown in Figure 11-8. The common dry cell consists of a zinc cylinder that contains as electrolyte a paste of ammonium chloride (NH_4Cl), a little zinc chloride ($ZnCl_2$), water, and diatomaceous earth or other filler. The dry cell is not dry; water must be present in the paste that serves as electrolyte. The

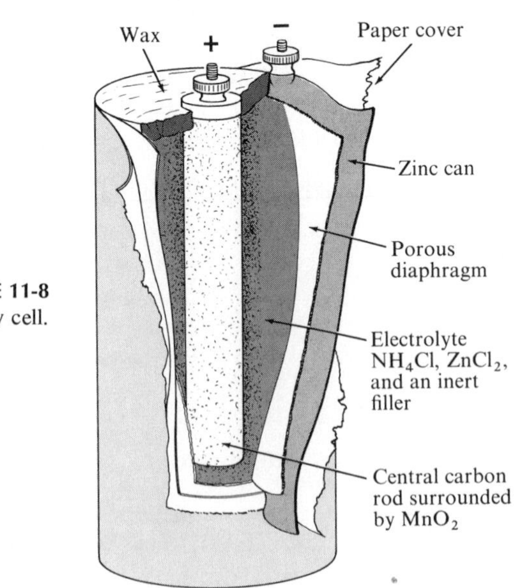

FIGURE 11-8
The dry cell.

central electrode is a mixture of carbon and manganese dioxide, embedded in a paste of these substances. The electrode reactions are

$$Zn \rightarrow Zn^{++} + 2e^-$$

$$2NH_4^+ + 2MnO_2 + 2e^- \rightarrow 2MnO(OH) + 2NH_3$$

(The zinc ion combines to some extent with ammonia to form the zinc-ammonia complex ion, $Zn(NH_3)_4^{++}$.) This cell produces a potential of about 1.48 V.

The Lead Storage Battery

The most common storage cell is that in the *lead storage battery* (Figure 11-9). The electrolyte in this cell is a mixture of water and sulfuric acid with density about 1.290 g cm^{-3} in the charged cell (38% H_2SO_4 by weight). The plates are lattices made of a lead alloy, the pores of one plate being filled with spongy metallic lead, and those of the other with lead dioxide, PbO_2. The spongy lead is the reducing agent and the lead

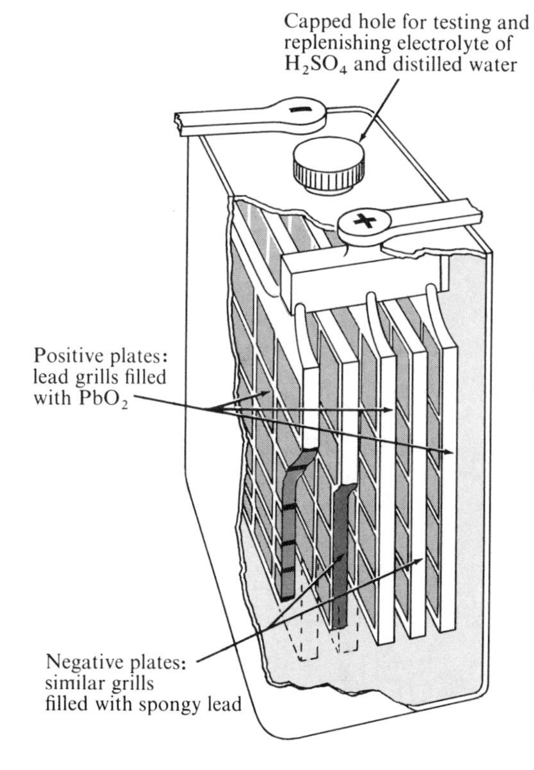

FIGURE 11-9
The lead storage cell.

dioxide the oxidizing agent in the chemical reaction that takes place in the cell. The electrode reactions that occur as the cell is being discharged are

$$Pb + SO_4^{--} \longrightarrow PbSO_4 + 2e^-$$

$$PbO_2 + SO_4^{--} + 4H^+ + 2e^- \longrightarrow PbSO_4 + 2H_2O$$

Each of these reactions produces the insoluble substance $PbSO_4$, lead sulfate, which adheres to the plates. As the cell is discharged, sulfuric acid is removed from the electrolyte, which decreases in density. The state of charge or discharge of the cell can accordingly be determined with use of a hydrometer, by measuring the density of the electrolyte.

The cell can be charged again by applying an electric potential across the terminals, and causing the above electrode reactions to take place in the opposite directions. The charged cell produces an electromotive force of slightly over 2 volts. A 6-V battery consists of three cells in series, and a 12-V battery of six cells.

It is interesting that in this cell the same element changes its oxidation state in the two plates: the oxidizing agent is PbO_2 (containing lead with oxidation number +4, which changes to +2 as the cell discharges), and the reducing agent is Pb (lead with oxidation number 0, which changes to +2).

11-8. Electrolytic Production of Elements

Many metals and some nonmetals are made by electrolytic methods. Hydrogen and oxygen are produced by the electrolysis of water containing an electrolyte. The alkali metals, alkaline-earth metals, magnesium, aluminum, and many other metals are manufactured either entirely or for special uses by electrochemical reduction of their compounds.

All commercial aluminum is made electrolytically, by a process discovered in 1886 by a young American, Charles M. Hall (1863–1914), and independently, in the same year, by a young Frenchman, P. L. T. Heroult (1863–1914). A carbon-lined iron box, which serves as cathode, contains the electrolyte, which is the molten mineral cryolite, Na_3AlF_6 (or a mixture of AlF_3, NaF, and sometimes CaF_2, to lower the melting point), in which aluminum oxide, Al_2O_3, is dissolved. The aluminum oxide is obtained from the ore *bauxite* by a process of purification, which is described below. The anodes in the cell are made of carbon. The passage of the current provides heat enought to keep the electrolyte molten, at about 1000°C. The aluminum metal that is produced by the process of electrolysis sinks to the bottom of the cell, and is tapped off. The cathode reaction is

$$Al^{+++} + 3e^- \longrightarrow Al$$

The anode reaction involves the carbon of the electrodes, which is converted into carbon dioxide:

$$C + 2O^{--} \longrightarrow CO_2 + 4e^-$$

The cells operate at about 5 V potential difference between the electrodes. Bauxite is a mixture of aluminum minerals ($AlHO_2$, $Al(OH)_3$), which contains some iron oxide. It is purified by treatment with sodium hydroxide solution, which dissolves hydrated aluminum oxide, as the aluminate ion, $Al(OH)_4^-$, but does not dissolve iron oxide:

$$Al(OH)_3 + OH^- \longrightarrow Al(OH)_4^-$$

The solution is filtered, and is then acidified with carbon dioxide, which reverses the above reaction by forming hydrogen carbonate ion, HCO_3^-:

$$Al(OH)_4^- + CO_2 \longrightarrow HCO_3^- + Al(OH)_3$$

The precipitated aluminum hydroxide is then dehydrated by ignition (heating to a high temperature), and the purified aluminum oxide is ready for addition to the electrolyte.

The Electrolytic Refining of Metals

Several metals, won from their ores by either chemical or electrochemical processes, are further refined by electrolytic methods.

Metallic copper is sometimes obtained by leaching a copper ore with sulfuric acid and then depositing the metal by electrolysis of the copper sulfate solution. Most copper ores, however, are converted into crude copper by chemical reduction. This crude copper is cast into anode plates about 2 cm thick, and is then refined electrolytically. In this process the anodes of crude copper alternate with cathodes of thin sheets of pure copper coated with graphite, which makes it possible to strip off the deposit. The electrolyte is copper sulfate. As the current passes through, the crude copper dissolves from the anodes and a purer copper deposits on the cathodes. Metals below copper in the EMF series, such as gold, silver, and platinum, remain undissolved, and fall to the bottom of the tank as sludge, from which they can be recovered. More active metals, such as iron, remain in the solution.

11-9. The Reduction of Ores. Metallurgy

Metals are obtained from ores. An *ore* is a mineral or other natural material that may be profitably treated for the extraction of one or more metals.

The process of extracting a metal from the ore is called *winning* the metal. *Refining* is the purification of the metal that has been extracted from the ore. *Metallurgy* is the science and art of winning and refining metals, and preparing them for use.

The simplest processes for winning metals are those used to obtain the metals that occur in nature in the elementary state. Thus nuggets of gold and of the platinum metals may be picked up by hand, in some deposits, or may be separated by a hydraulic process (use of a stream of water) when the nuggets occur mixed with lighter materials in a placer deposit.* A quartz vein containing native gold may be treated by mining it, pulverizing the quartz in a stamp mill, and then mixing the rock powder with mercury. The gold dissolves in the mercury, which is easily separated from the rock powder because of its great density, and the gold can be recovered from the amalgam by distilling off the mercury.

The chemical processes involved in the winning of metals are mainly the reduction of a compound of the metal, usually oxide or sulfide. The principal reducing agent that is used is carbon, often in the form of coke. An example is the reduction of iron oxide with coke in a blast furnace (Chapter 19). Occasionally reducing agents other than carbon are used; thus antimony is won from stibnite, Sb_2S_3, by heating it with iron:

$$Sb_2S_3 + 3Fe \longrightarrow 3FeS + 2Sb$$

Whether or not a reaction may be used for winning a metal depends upon the free energies of the reactants and the products. For some reactions, especially those in which the products are similar to the reactants, the change in free energy for the reaction is nearly equal to the enthalpy change. An example is the reaction of stibnite and iron given above. Whether such a reaction is exothermic or endothermic can be predicted with reasonable confidence by use of electronegativity values. In the above reaction six Sb—S bonds are broken and six Fe—S bonds are formed. The electronegativity values are 1.9 for Sb, 2.5 for S, and 1.8 for Fe. The bond energy per metal-sulfur bond is $100(x_A - x_B)^2$ kJ mole^{-1}; its values are $100(2.5-1.9)^2 = 36$ kJ mole^{-1} for Sb—S and $100(2.5-1.8)^2 = 49$ kJ mole^{-1} for Fe—S. We conclude that the reaction as written is

*A placer deposit is a glacial deposit or alluvial deposit (made by a river, lake, or arm of the sea), as of sand or gravel, containing gold or other valuable material.

exothermic and probably also exergonic (accompanied by the evolution of free energy).

From the foregoing argument we would conclude that any element more electropositive than antimony could be used to prepare antimony from stibnite. Reference to Table 11-1 shows that many metals are in this class. Iron is used in practice rather than some other metal because it is the cheapest metal.

Carbon is more effective for reducing metal oxides than might be expected from its electronegativity, 2.5, for two reasons: first, the special stability of CO (low enthalpy) corresponding to the great strength of multiple bonds between first-row elements, and second, the large entropy of the gaseous product. It can be used to reduce oxides of metals with electronegativity as low as 1.6.

The Metallurgy of Copper

Copper occurs in nature as *native copper;* that is, in the free state. Other ores of copper include *cuprite,* Cu_2O; *chalcocite,* Cu_2S; *chalcopyrite,* $CuFeS_2$; *malachite,* $Cu_2CO_3(OH)_2$; and *azurite,* $Cu_3(CO_3)_2(OH)_2$. Malachite, a beautiful green mineral, is sometimes polished and used in jewelry.

An ore containing native copper may be treated by grinding it and then washing away the gangue (the associated rock or earthy material), and melting and casting the copper. Oxide or carbonate ores may be leached with dilute sulfuric acid, to produce a cupric solution from which the copper can be deposited by electrolysis. High-grade oxide and carbonate ores may be reduced by heating with coke mixed with a suitable flux. (A flux is a material, such as limestone, that combines with the silicate minerals of the gangue to form a slag that is liquid at the temperature of the furnace, and can be easily separated from the metal.)

Sulfide ores are smelted by a complex process. Low-grade ores are first concentrated, by a process such as *flotation.* The finely ground ore is treated with a mixture of water and a suitable oil. The oil wets the sulfide minerals, and the water wets the silicate minerals of the gangue. Air is then blown through to produce a froth, which contains the oil and the sulfide minerals; the silicate minerals sink to the bottom.

The concentrate or the rich sulfide ore is then roasted in a furnace through which air is passing. This removes some of the sulfur as sulfur dioxide, and leaves a mixture of Cu_2S, FeO, SiO_2, and other substances. This roasted ore is then mixed with limestone to serve as flux, and is heated in a furnace. The iron oxide and silica combine with the limestone to form a slag, and the cuprous sulfide melts and can be drawn off. This impure cuprous sulfide is called *matte*. It is then reduced by blowing air through the molten material:

$$Cu_2S + O_2 \rightarrow SO_2 + 2Cu$$

Some copper oxide is also formed by the blast of air, and this is reduced by stirring the molten metal with poles of green wood. The copper obtained in this way has a characteristic appearance, and is called *blister copper*. It contains about 1% of iron, gold, silver, and other impurities, and is usually refined electrolytically, as described in the last part of Section 11-8.

Reduction of Metal Oxides or Halides by Strongly Electropositive Metals

Some metals, including titanium, zirconium, hafnium, lanthanum, and the lanthanons, are most conveniently obtained by reaction of their oxides or halides with a more electropositive metal. Sodium, potassium, calcium, and aluminum are often used for this purpose. Thus titanium may be made by reduction of titanium tetrachloride by calcium:

$$TiCl_4 + 2Ca \rightarrow Ti + 2CaCl_2$$

Titanium, zirconium, and hafnium are purified by the decomposition of their tetraiodides on a hot wire. The impure metal is heated with iodine in an evacuated flask, to produce the tetraiodide as a gas:

$$Zr + 2I_2 \rightarrow ZrI_4$$

The gas comes into contact with a hot filament, where it is decomposed, forming a wire of the purified metal:

$$ZrI_4 \rightarrow Zr + 2I_2$$

The process of preparing a metal by reduction of its oxide by aluminum is called the *aluminothermic process*. For example, chromium can be prepared by igniting a mixture of powdered chromium(III) oxide and powdered aluminum:

$$Cr_2O_3 + 2Al \rightarrow Al_2O_3 + 2Cr$$

The heat liberated by this reaction is so great as to produce molten chromium. The aluminothermic process is a convenient way of obtaining a small amount of liquid metal, such as iron for welding.

EXERCISES

11-1. Define electrolysis. What is the mechanism of electrolytic conductivity? How does it differ from electronic conductivity?

11-2. Write electrode equations for the electrolytic production of (*a*) magmesium metal from molten magnesium chloride; (*b*) perchlorate ion, ClO_4^-, from chlorate ion, ClO_3^-, in aqueous solution; (*c*) fluorine from fluoride ion in a molten salt. State in each case whether the reaction occurs at the anode or at the cathode.

11-3. Ionic crystals generally have very low conductivity. Account for the relatively high conductivity of crystalline silver iodide and the positive temperature coefficient of conductivity.

11-4. How does electrolysis of a dilute aqueous salt solution differ from that of the molten salt? Write electrode equations for the electrolysis of a dilute aqueous solution of sodium chloride.

11-5. Describe oxidation and reduction in terms of gain or loss of electrons and in terms of change in oxidation number.

11-6. What are the three conservation rules for any oxidation-reduction equation?

11-7. Complete and balance the following oxidation-reduction equations:
$Cl_2 + I^- \rightarrow I_2 + Cl^-$
$Sn + I_2 \rightarrow SnI_4$
$KClO_3 \rightarrow KClO_4 + KCl$
$MnO_2 + H^+ + Cl^- \rightarrow Mn^{++} + Cl_2$
$ClO_4^- + Sn^{++} \rightarrow Cl^- + Sn^{++++}$

11-8. Indicate which substances are reducing agents (are oxidized) and which substances are oxidizing agents (are reduced) in the equations above.

11-9. What fraction of the formula mass is the oxidation equivalent mass or reduction equivalent mass of each of the oxidizing or reducing agents in the equations of Exercise 11-7?

11-10. What are Faraday's laws of electrolysis? What is a faraday?

11-11. How much of what products are obtained by passing a current of 10 A through a cell of (*a*) molten silver iodide; (*b*) a dilute aqueous solution of silver iodide?

11-12. What mass of 3.00% hydrogen peroxide solution would be required to oxidize 2.00 g of lead sulfide, PbS, to lead sulfate, $PbSO_4$?

11-13. Would you expect zinc to reduce cadmium ion? (Refer to the electromotive-force series.) Iron to reduce mercuric ion? Zinc to reduce lead ion? Potassium to reduce magnesium ion?

11-14. Which metal ions would you expect gold to reduce? Suggest a reason for calling gold and platinum noble metals.

11-15. What would you expect to happen if a large piece of lead were put in a beaker containing a solution of stannous salt?

11-16. What would happen if chlorine gas were bubbled into a solution containing fluoride ion and bromide ion? If chlorine were bubbled into a solution containing both bromide ion and iodide ion?

11-17. Why are hydrogen peroxide and potassium permanganate both antiseptics?

11-18. Explain each of the terms in the equation $K = \exp(-nE°F/RT)$.

11-19. Why is there a correlation between EMF and electronegativity? Would you expect a correlation between EMF and ionization potential?

11-20. Define and determine the equilibrium constant in a solution obtained by adding 0.1 mole ferrous sulfate and 0.1 mole mercuric sulfate to water and bringing the total volume to 1 l.

11-21. The following electric cell is set up:

$$Hg(l), Hg_2Cl_2(c)|Cl^-(1\ F)|Cl_2(g, 1\ atm)$$

The aqueous phase (1 F KCl in water) is saturated with Hg_2Cl_2 in the neighborhood of the mixture of calomel crystals and liquid mercury. There are platinum electrodes at each end of the cell.
(*a*) Write an equation for the reaction accompanying the passage of one faraday through the cell.
(*b*) The EMF of the cell at 25°C is −1.091 V. What is the standard free-energy change for the reaction at this temperature?

11-22. Assuming that the reaction goes to completion and then stops, how much of what constituents would a common dry cell have to contain to light a 2.96-watt flash light for 10 hours?

11-23. A current of 10 A is drawn from a lead storage cell for 1 hour. How much $PbSO_4$ is formed from Pb at one electrode plate? How much is formed from PbO_2 at the plate of opposite polarity?

11-24. At $0.03 per kilowatt hour, how much does the electricity cost to produce a ton of aluminum? At $0.01 a pound, how much does the carbon cost?

11-25. The process of winning an elemental metal usually involves the reduction of an ore. Why? Give an example with equations.

11-26. Thermite is a mixture of Al and Fe_2O_3 used for welding iron. Describe the reaction.

12

Acids, Bases, and Buffers

An acid may be defined as a hydrogen-containing substance that dissociates on solution in water to produce hydrogen ions, and a base may be defined as a substance containing the hydroxide ion or the hydroxyl group that dissociates in aqueous solution as the hydroxide ion. Acidic solutions have a characteristic sharp taste, that of the hydronium ion, H_3O^+, and basic solutions have a characteristic brackish taste, that of the hydroxide ion, OH^-. The ordinary mineral acids (hydrochloric acid, nitric acid, sulfuric acid) are completely ionized (dissociated) in solution, producing one hydrogen ion for every acidic hydrogen atom in the formula of the acid, whereas other acids, such as acetic acid, produce only a smaller number of hydrogen ions. Acids such as acetic acid are called weak acids. A 1 *F* solution of acetic acid does not have nearly so sharp a taste and does not react nearly so vigorously with an active metal, such as zinc, as does a 1 *F* solution of hydrochloric acid, because the 1 *F* solution of acetic acid contains a great number of undissociated molecules $HC_2H_3O_2$, and only a relatively small number of ions H_3O^+ and $C_2H_3O_2^-$. There exists in a solution of acetic acid a steady state, corresponding to the equation

$$HC_2H_3O_2 + H_2O \rightleftarrows H_3O^+ + C_2H_3O_2^-$$

In order to understand the properties of acetic acid it is necessary to formulate the equilibrium expression for this steady state; by use of this equilibrium expression the properties of acetic acid solutions of different concentrations can be predicted.

The general principles of chemical equilibrium can similarly be used in the discussion of a weak base, such as ammonium hydroxide, and also of salts formed by weak acids and weak bases. In addition, these principles are important in providing an understanding of the behavior of indicators, which are useful for determining whether a solution is acidic, neutral, or basic. These principles are of further importance in permitting a discussion of the relation between the concentrations of hydronium ion and hydroxide ion in the same solution.

The definitions of an acid and a base given in the first sentence of this chapter are satisfactory for most purposes. More general definitions, which are sometimes convenient, have been suggested. One of these definitions is that an acid is a molecule or ion that can give a proton to another molecule or ion, and a base is a molecule or ion that can accept a proton. This definition is useful, for example, in discussing reactions in liquid ammonia. Potassium amide, KNH_2, functions as a base: it can react with the acid hydrogen chloride to form ammonia and the salt potassium chloride (or potassium ion and chloride ion). In liquid ammonia the ammonium ion, NH_4^+, and the amide ion, NH_2^-, are analogous to the hydronium ion and the hydroxide ion in aqueous systems. Some other nonaqueous systems are discussed in Section 12-9.

A still more general theory of acids and bases than the proton donor-acceptor theory was introduced by G. N. Lewis. He called a base anything that has available an unshared pair of electrons such as NH_3,

```
   H
  /
:N—H
  \
   H
```

and an acid anything that might attach itself to such a pair of electrons, such as H^+, to form NH_4^+, or BF_3, to form $F_3B—NH_3$.

This concept explains many phenomena. An example is the effect of certain substances other than hydrogen ion in changing the color of indicators. Another interesting application of the concept is its explanation of salt formation by reaction of acidic oxides and basic oxides.

12-1. Hydronium-ion (Hydrogen-ion) Concentration

It was mentioned in the chapter on water (Chapter 9) that pure water does not consist simply of H_2O molecules, but also contains hydronium ions in concentration about 1×10^{-7} moles per liter (at 25°C), and hydroxide ions in the same concentration. These ions are formed by the *autoprotolysis* of water (reaction of one molecule of water acting as an acid with another molecule of water acting as a base; this reaction is also called the ionization of water):

$$2H_2O \rightleftarrows H_3O^+ + OH^-$$

A molecule, such as the water molecule, that can both lose a proton and add a proton is called an *amphiprotic* molecule (Greek *amphi,* both). Only amphiprotic molecules or ions can undergo autoprotolysis.

The means by which it has been found that pure water contains hydronium ions and hydroxide ions is the measurement of the electric conductivity of water. The mechanism of the electric conductivity of a solution was discussed in Section 11-2. According to this discussion, electric charge is transferred through the body of the solution by the motion of cations from the region around the anode to the region around the cathode, and anions from the region around the cathode to the region around the anode. If pure water contained no ions its electric conductivity would be zero. When investigators made water as pure as possible, by distilling it over and over again, it was found that the electric conductivity approached a certain small value, about one ten-millionth that of a 1 F solution of hydrochloric acid or sodium hydroxide. This indicates that the autoprotolysis of water occurs to such an extent as to give hydronium ions and hydroxide ions in concentration about one ten-millionth mole per liter. Refined measurements have provided the value 1.00×10^{-7} M for $[H_3O^+]$ and $[OH^-]$ in pure water at 25°C.*

Although the hydronium ion, H_3O^+, is present in water and confers acidic properties upon aqueous solutions, it is customary to use the symbol H^+ in place of H_3O^+ and to speak of hydrogen ion in place of hydronium. In the later sections of this chapter we shall follow this custom except when the discussion is based upon the proton donor-acceptor theory (the Brønsted-Lowry theory). Accordingly, we shall use the symbol H^+ with two meanings: to represent the unhydrated proton when the Brønsted-Lowry theory is being applied, and to represent the hydronium ion, H_3O^+, when this theory is not being applied. We shall

*The extent of autoprotolysis depends somewhat on the temperature. At 0°C $[H_3O^+]$ and $[OH^-]$ are 0.83×10^{-7} M, and at 100°C they are 6.9×10^{-7} M. When a solution of a strong acid and a solution of a strong base are mixed, a large amount of heat is given off. This shows that the reaction gives off heat, and accordingly that the reaction of dissociation of water absorbs heat. In accordance with Le Chatelier's principle, increase in the temperature would shift the equilibrium of dissociation of water in such a way as to tend to restore the original temperature; that is, the reaction would take place in the direction that absorbs heat. This direction is the dissociation of water to hydrogen ions and hydroxide ions, and accordingly the principle requires that increase in temperature cause an increased amount of dissociation of water, as is found experimentally.

refer to H^+ in the Brønsted-Lowry applications as the proton and to H^+ (representing H_3O^+) in other discussions as the hydrogen ion (representing the hydronium ion).

The *p*H

Instead of saying that the concentration of hydrogen ion in pure water is 1.00×10^{-7} *M*, it is customary to say that the *p*H of pure water is 7. This symbol, *p*H, is defined in the following way: *the* pH *is the negative common logarithm of the hydrogen-ion concentration:*

$$pH = -\log [H^+]$$

or

$$[H^+] = 10^{-pH} = \text{antilog}\ (-pH)$$

We see from this definition of *p*H that a solution containing 1 mole of hydrogen ions per liter, that is, with a concentration 10^{-0} in H^+, has *p*H zero. A solution only one-tenth as strong in hydrogen ion, containing 0.1 mole of hydrogen ions per liter, has $[H^+] = 10^{-1}$, and hence has *p*H 1. The relation between the hydrogen-ion concentration and the *p*H is shown for simple concentrations along the left side of Figure 12-1.

In science and medicine it is customary to describe the acidity of a solution by saying "The *p*H of the solution is 3," for example, instead of saying "The hydrogen-ion concentration of the solution is 10^{-3}." It is

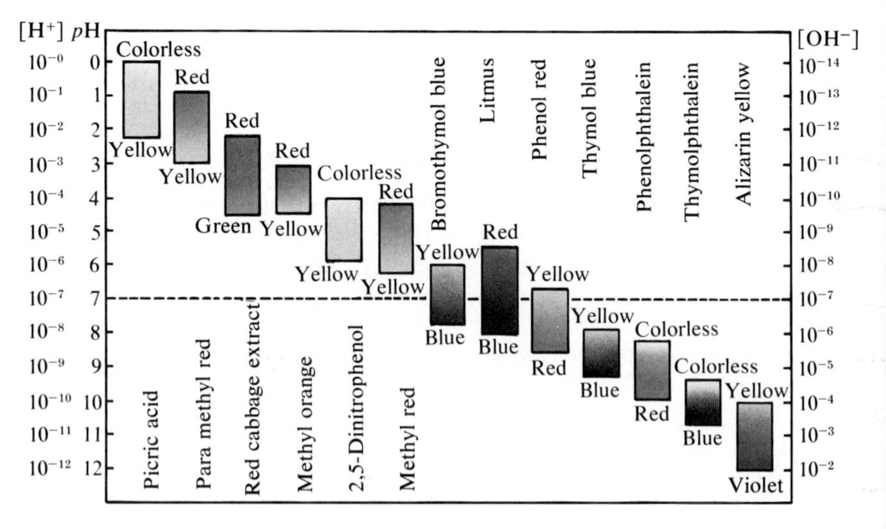

FIGURE 12-1
Color changes of indicators.

evident that the quantity *p*H is useful, in permitting the exponential expression to be avoided.

The chemical reactions involved in biological processes are often very sensitive to the hydrogen-ion concentration of the medium. In industries such as the fermentation industry the control of the *p*H of the materials being processed is very important. It is not surprising that the symbol *p*H was introduced by a Danish biochemist, S. P. L. Sørensen, while he was working on problems connected with the brewing of beer.

Example 12-1. What is the *p*H of a solution with $[H^+] = 0.0200$?

Solution. The log of 0.0200 is equal to the log of 2×10^{-2}, which is $0.301 - 2 = -1.699$. The *p*H of the solution is the negative of the logarithm of the hydrogen-ion concentration. Hence the *p*H of this solution is 1.699.

Example 12-2. What is the hydrogen-ion concentration of a solution with *p*H 4.30?

Solution. A solution with *p*H 4.30 has $\log [H^+] = -4.30$, or $0.70 - 5$. The antilog of 0.70 is 5.0, and the antilog of -5 is 10^{-5}. Hence the hydrogen-ion concentration in this solution is 5.0×10^{-5}.

12-2. The Equilibrium between Hydrogen Ion and Hydroxide Ion in Aqueous Solution

The equation for the autoprotolysis of water is

$$2H_2O \rightleftarrows H_3O^+ + OH^-$$

The expression for the equilibrium constant, in accordance with the principle developed in the preceding chapter, is

$$\frac{[H_3O^+][OH^-]}{[H_2O]^2} = K_1$$

In this expression the symbol $[H_2O]$ represents the activity (concentration) of water in the solution. Since the activity of water in a dilute aqueous solution is nearly the same as that for pure water, it is customary to omit the activity of water in the equilibrium expression for dilute solutions. Accordingly, the product of K_1 and $[H_2O]^2$ may be taken as another constant K_w, and we may write

$$[H_3O^+] \times [OH^-] = K_w = 1.00 \times 10^{-14} \text{ mole}^2 \text{ liter}^{-2}$$

This expression states that the product of the hydronium-ion concentration and the hydroxide-ion concentration in water and in dilute aqueous solutions is a constant, at given temperature. The value of K_w is 1.00×10^{-14} mole^2 liter^{-2} at 25°C. *Hence in pure water both H_3O^+ and OH^- have the concentration 1.00×10^{-7} moles per liter at 25°C, and in acidic or basic solutions the product of the concentrations of these ions equals 1.00×10^{-14}.*

Thus a neutral solution contains both hydrogen ions (hydronium ions) and hydroxide ions at the same concentration, 1.00×10^{-7}. A slightly acidic solution, containing 10 times as many hydrogen ions (concentration 10^{-6}, *p*H 6), also contains some hydroxide ions, one-tenth as many as a neutral solution. A solution containing 100 times as much hydrogen ion as a neutral solution (concentration 10^{-5}, *p*H 5) contains a smaller amount of hydroxide ion, one one-hundredth as much as a neutral solution; and so on. A solution containing 1 mole of strong acid per liter has hydrogen-ion concentration 1, and *p*H 0; such a strongly acidic solution also contains some hydroxide ion, the concentration of hydroxide ion being 1×10^{-14}. Although this is a very small number, it still represents a large number of actual ions in a macroscopic volume. Avogadro's number is 0.602×10^{24}, and accordingly a concentration of 10^{-14} moles per liter corresponds to 0.602×10^{10} ions per liter, or 0.602×10^7 ions per milliliter.

12-3. Indicators

Indicators such as litmus may be used to tell whether a solution is acidic, neutral, or basic. The change in color of an indicator as the *p*H of the solution changes is not sharp, but extends over a range of one or two *p*H units. This is the result of the existence of chemical equilibrium between the two differently colored forms of the indicator, and the dependence of the color on the hydrogen-ion concentration is due to the participation of hydrogen ion in the equilibrium.

Thus the red form of litmus may be represented by the formula HIn and the blue form by In^-, resulting from the dissociation reaction

$$\underset{\substack{\text{Red} \\ \text{acidic form}}}{\text{HIn}} \rightleftarrows \underset{\substack{\text{Blue} \\ \text{basic form}}}{H^+ + In^-}$$

In alkaline solutions, with $[H^+]$ very small, the equilibrium is shifted to the right, and the indicator is converted almost entirely into the basic form (blue for litmus). In acidic solutions, with $[H^+]$ large, the equilibrium is shifted to the left, and the indicator assumes the acidic form.

Let us calculate the relative amount of the two forms as a function of $[H^+]$. The equilibrium expression for the indicator reaction written above is

$$\frac{[H^+][In^-]}{[HIn]} = K_{In}$$

in which K_{In} is the *equilibrium constant for the indicator.* We rewrite this as

$$\frac{[HIn]}{[In^-]} = \frac{[H^+]}{K_{In}}$$

This equation shows how the ratio of the two forms of the indicator depends on $[H^+]$. When the two forms are present in equal amounts, the ratio of acidic form to alkaline form, $[HIn]/[In^-]$, has the value 1, and hence $[H^+] = K_{In}$. *The indicator constant K_{In} is thus the value of the hydrogen-ion concentration at which the change in color of the indicator is half completed.* The corresponding pH value is called the pK of the indicator.

When the pH is decreased by one unit the value of $[H^+]$ becomes ten times K_{In} and the ratio $[HIn]/[In^-]$ then equals 10. Thus at a pH value 1 less than the pK of the indicator (its midpoint) the acidic form of the indicator predominates over the basic form in the ratio 10:1. In this solution 91% of the indicator is in the acidic form, and 9% in the basic form. Over a range of 2 pH units the indicator accordingly changes from 91% acidic form to 91% basic form. For most indicators the color change detectable by the eye occurs over a range of about 1.2 to 1.8 pH units.

Indicators differ in their pK values; pure water, with pH 7, is neutral to litmus (which has pK equal to 6.8), acidic to phenolphthalein (pK 8.8), and basic to methyl orange (pK 3.7).

A chart showing the color changes and effective pH ranges of several indicators is given in Figure 12-1. The approximate pH of a solution can be determined by finding by test the indicator toward which the solution shows a neutral reaction. Test paper, made with a mixture of indicators and showing several color changes, is now available with which the pH of a solution can be estimated to within about 1 unit over the pH range 1 to 13.

In titrating a weak acid or a weak base the indicator must be chosen with care. The way of choosing the proper indicator is described in Section 12-6.

It is seen that an indicator behaves as a weak organic acid; the equilibrium expression for an indicator is the same as that for an ordinary weak acid.

By the use of color standards for the indicator, the *p*H of a solution may be estimated to about 0.1 unit by the indicator method. A more satisfactory general method of determining the *p*H of a solution is by use of an instrument that measures the hydrogen-ion concentration by measuring the electric potential of a cell with cell reaction involving hydrogen ions. Modern glass-electrode *p*H meters are now available that cover the *p*H range 0 to 14 with an accuracy approaching 0.01 (Figure 12-2).

The structure of the glass electrode is shown at the right in Figure 12-2. The Ag-AgCl electrode provides a reversible electric connection between the terminal wire and the HCl solution. The glass bulb at the bottom is

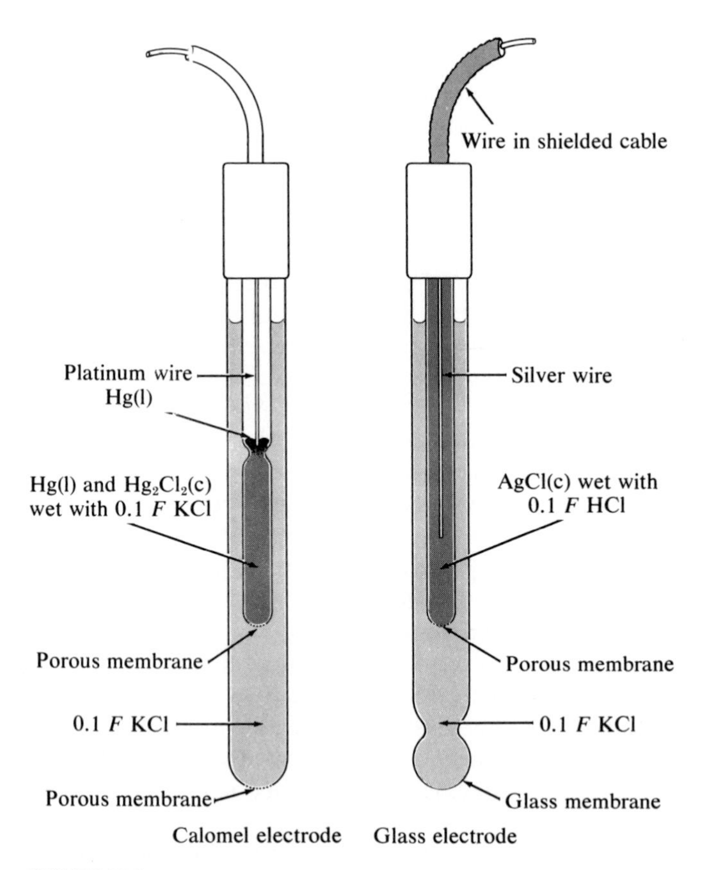

FIGURE 12-2
A representation of the two electrodes of a glass-electrode *p*H meter. The glass membrane is made of a special glass that has an effective permeability to hydrogen ions. The potential of this electrode is determined by the concentration of hydrogen ions in the medium surrounding the glass membrane.

made of a special glass that conducts the electric current by accepting protons, passing them from one oxygen atom to another, and liberating protons on the other side—this glass is not permeable to other ions. The mercury-calomel electrode shown at the left permits a second reversible electric contact, not dependent on the hydrogen-ion concentration, to be made with the solution. In measuring the *p*H of a solution the ends of the two electrodes are placed in the solution and the electromotive force that is developed is measured with a high-impedance voltmeter. Since the conduction of current through the cell involves the transfer of hydrogen ions from a solution with one hydrogen-ion activity to a solution with another hydrogen-ion activity (the solutions on the two sides of the glass membrane) and the hydrogen-ion activities are not significantly involved in the other conduction steps, the electromotive force depends upon the *p*H of the solution being tested. It is linear in the *p*H, changing by 0.059 volt per *p*H unit at 25°C.

12-4. Equivalent Masses of Acids and Bases

A solution containing one gram formula mass of hydrochloric acid, HCl, per liter is 1 *F* in hydrogen ion. Similarly a solution containing 0.5 gram formula mass of sulfuric acid, H_2SO_4, per liter is 1 *F* in replaceable hydrogen. Each of these solutions is neutralized* by an equal volume of a solution containing one gram formula mass of sodium hydroxide, NaOH, per liter, and the masses of the acids are hence equivalent to one gram formula mass of the alkali.

The quotient of the gram formula mass of an acid by the number of hydrogen atoms which are replaceable for the reaction under consideration is called the *equivalent mass of the acid.* Likewise, the quotient of the gram formula mass of a base by the number of hydroxyl groups which are replaceable for the reaction under consideration is called the *equivalent mass of the base.*

One equivalent mass of an acid neutralizes one equivalent mass of a base. It is important to note that the equivalent mass of a polyprotic acid is not invariant; for H_3PO_4 it may be the gram formula mass, one half this, or one third, depending on whether one, two, or three hydrogens are effective in the reaction under consideration.

The *normality* of a solution of an acid or base is the number of equivalents of acid or base per liter; a 1 *N* solution contains 1 equivalent per liter of solution. By determining, with use of an indicator such as litmus,

*The meaning of "neutralized" in the case of weak acids or bases is discussed in a later section of this chapter.

the relative volumes of acidic and alkaline solutions that are equivalent, the normality of one solution can be calculated from the known value of the other. This process of *acid-base titration* (the determination of the *titer* or strength of an unknown solution) with use of special apparatus such as graduated burets and pipets is an important method of volumetric quantitative analysis.

Example 12-3. It is found by experiment that 25.0 ml of a solution of sodium hydroxide is neutralized by 20.0 ml of a 0.100 N acid solution. What are the normality of the alkaline solution and the mass of NaOH per liter?

Solution. The unknown normality x of the alkaline solution is found by solving the equation that expresses the equivalence of the portions of the two solutions:

$$25.0x = 20.0 \times 0.100$$

$$x = \frac{20.0 \times 0.100}{25.0} = 0.080$$

The mass of NaOH per liter is 0.080 times the equivalent mass, 40.0, or 3.20 g.

You may find it useful to fix in your mind the following equation:

$$V_1N_1 = V_2N_2$$

Here V_1 is the volume of a solution with normality N_1 and V_2 is the equivalent volume (containing the same number of replaceable hydrogens or hydroxyls) of a solution with normality N_2. In solving the above exercise we began by writing this equation; $25.0x$ is V_1N_1, and 20.0×0.100 is V_2N_2, in this case.

12-5. Weak Acids and Bases

Ionization of a Weak Acid

A 0.1 N solution of a strong acid such as hydrochloric acid is 0.1 N in hydrogen ion, since this acid is very nearly completely dissociated into ions except in very concentrated solutions. But a 0.1 N solution of acetic acid contains hydrogen ions in much smaller concentration, as is seen by testing with indicators, observing the rate of attack of metals, or simply by tasting. Acetic acid is a weak acid; the acetic acid molecules hold their protons so firmly that not all of them are transferred to water molecules to form hydronium ions. Instead, there is an equilibrium reaction,

$$HC_2H_3O_2 + H_2O \rightleftarrows H_3O^+ + C_2H_3O_2^-$$

or, ignoring the hydration of the proton,

$$HC_2H_3O_2 \rightleftharpoons H^+ + C_2H_3O_2^-$$

The equilibrium expression for this reaction is

$$\frac{[H^+][C_2H_3O_2^-]}{[HC_2H_3O_2]} = K$$

In general, for an acid HA in equilibrium with ions H^+ and A^- the equilibrium expression is

$$\frac{[H^+][A^-]}{[HA]} = K_a$$

The constant K_a, characteristic of the acid, is called its *acid constant* or *ionization constant.*

Values of acid constants are found experimentally by measuring the *p*H of solutions of the acids. A table of values is given later in this chapter (Section 12-8).

The hydrogen-ion concentration of a weak acid (containing no other electrolytes that react with it or its ions) in 1 *N* concentration is approximately equal to the square root of its acid constant, as is seen from the Example 12-5.

Example 12-4. The *p*H of a 0.100 *N* solution of acetic acid is found by experiment to be 2.874. What is the acid constant, K_a, of this acid?

Solution. To calculate the acid constant we note that acetic acid added to pure water ionizes to produce hydrogen ions and acetate ions in equal quantities. Moreover, since the amount of hydrogen ion resulting from the dissociation of water is negligible compared with the total amount present, we have

$$[H^+] = [C_2H_3O_2^-] = \text{antilog}\ (-2.874) = 1.34 \times 10^{-3}$$

The concentration $[HC_2H_3O_2]$ is hence $0.100 - 0.001 = 0.099$, and the acid constant has the value

$$K_a = (1.34 \times 10^{-3})^2/0.099 = 1.80 \times 10^{-5}$$

Example 12-5. What is $[H^+]$ of a 1 *N* solution of HCN, hydrocyanic acid, which has $K_a = 4 \times 10^{-10}$?

Solution. Let $x = [H^+]$. Then we can write $[CN^-] = x$ (neglecting the amount of hydrogen ion due to ionization of the water), and $[HCN] = 1 - x$. The equilibrium equation is

$$\frac{x^2}{1-x} = K_a = 4 \times 10^{-10}$$

We know that x is going to be much smaller than 1, since this weak acid is only very slightly ionized. Hence we replace $1-x$ by 1 (neglecting the small difference between unionized hydrocyanic acid and the total cyanide concentration), obtaining

$$x^2 = 4 \times 10^{-10}$$

$$x = 2 \times 10^{-5} = [H^+]$$

The neglect of the ionization of water is also seen to be justified, since even in this very slightly acidic solution the value of $[H^+]$ is 200 times the value for pure water.

Successive Ionization of a Polyprotic Acid

A polyprotic acid has several acid constants, corresponding to dissociation of successive hydrogen ions. For phosphoric acid, H_3PO_4, there are three equilibrium expressions:

$$H_3PO_4 \rightleftarrows H^+ + H_2PO_4^-$$

$$K_1 = \frac{[H^+][H_2PO_4^-]}{[H_3PO_4]} = 7.5 \times 10^{-3} = K_{H_3PO_4}$$

$$H_2PO_4^- \rightleftarrows H^+ + HPO_4^{--}$$

$$K_2 = \frac{[H^+][HPO_4^{--}]}{[H_2PO_4^-]} = 6.2 \times 10^{-8} = K_{H_2PO_4^-}$$

$$HPO_4^{--} \rightleftarrows H^+ + PO_4^{---}$$

$$K_3 = \frac{[H^+][PO_4^{---}]}{[HPO_4^{--}]} = 10^{-12} = K_{HPO_4^{--}}$$

Note that these constants have the dimensions of concentration, mole liter^{-1}.

The ratio of successive ionization constants for a polybasic acid is usually about 10^{-5}, as in this case. We see that with respect to its first hydrogen phosphoric acid is a moderately strong acid—considerably stronger than acetic acid. With respect to its second hydrogen it is weak, and to its third very weak.

Ionization of a Weak Base

A weak base dissociates in part to produce hydroxide ions:

$$MOH \rightleftarrows M^+ + OH^-$$

The corresponding equilibrium expression is

$$\frac{[M^+][OH^-]}{[MOH]} = K_b$$

The constant K_b is called the *basic constant* of the base.

Ammonium hydroxide is the only common weak base. Its basic constant has the value 1.81×10^{-5} at 25°C. The hydroxides of the alkali metals and the alkaline-earth metals are strong bases.

Very many problems in solution chemistry are solved with use of the acid and base equilibrium equations. The uses of these equations in discussing the titration of weak acids and bases, the hydrolysis of salts, and the properties of buffered solutions are illustrated in the following sections of this chapter.

The student while working a problem should not substitute numbers in the equations in a routine way, but should think carefully about the chemical reactions and equilibria involved and the magnitudes of the concentrations of the different molecular species. Every problem solved should add to his understanding of solution chemistry. The ultimate goal is such an understanding of the subject that the student can estimate the orders of magnitude of concentrations of the various ionic and molecular species in a solution without having to solve the equilibrium equations.

Example 12-6. What is the pH of a 0.1 F solution of ammonium hydroxide?

Solution. Our fundamental equation is

$$\frac{[NH_4^+][OH^-]}{[NH_4OH]} = K_b = 1.81 \times 10^{-5}$$

Since the ions NH_4^+ and OH^- are produced in equal amounts by the dissociation of the base and the amount of OH^- from dissociation of water is negligible, we put

$$[NH_4^+] = [OH^-] = x$$

The concentration of NH_4OH is accordingly $0.1 - x$, and we obtain the equation

$$\frac{x^2}{0.1 - x} = 1.81 \times 10^{-5}$$

(Here we have made the calculation as though all the undissociated solute were NH_4OH. Actually there is dissolved NH_3 present; however, since the equilibrium $NH_3 + H_2O \rightleftarrows NH_4OH$ is of such a nature that the ratio $[NH_4OH]/[NH_3]$ is constant, we can write the equilibrium expression for the base as shown above, with the symbol $[NH_4OH]$

representing the total concentration of the undissociated solute, including the molecular species NH_3 as well as NH_4OH.)

Solving this equation, we obtain the result

$$x = [OH^-] = [NH_4^+] = 1.34 \times 10^{-3}$$

The solution is hence only slightly alkaline—its hydroxide-ion concentration is the same as that of a 0.00134 *N* solution of sodium hydroxide.

This value of $[OH^-]$ corresponds to $[H^+] = (1.00 \times 10^{-14})/(1.34 \times 10^{-3}) = 7.46 \times 10^{-12}$, as calculated from the water equilibrium equation

$$[H^+][OH^-] = 1.00 \times 10^{-14}$$

The corresponding *p*H is 11.13, which is the answer to the problem.

12-6. The Titration of Weak Acids and Bases

A liter of solution containing 0.2 mole of a strong acid such as hydrochloric acid has $[H^+] = 0.2$ and $pH = 0.7$. The addition of a strong base, such as 0.2 *N* NaOH, causes the hydrogen-ion concentration to diminish through neutralization by the added hydroxide ion. When 990 ml of strong base has been added, the excess of acid over base is $0.2 \times 10/1000 = 0.002$ mole, and since the total volume is very close to 2 liters the value of $[H^+]$ is 0.001, and the *p*H is 3. When 999 ml has been added, and the neutralization reaction is within 0.1% of completion, the values are $[H^+] = 0.0001$ and $pH = 4$. At $pH = 5$ the reaction is within 0.01% of completion, and at *p*H 6 within 0.001%. Finally *p*H 7, neutrality, is reached when an amount of strong base has been added exactly equivalent to the amount of strong acid present. A very small excess of strong base causes the *p*H to increase beyond 7.

We see that to obtain the most accurate results in titrating a strong acid and a strong base an indicator with indicator constant about 10^{-7} ($pK = 7$) should be chosen, such as litmus or bromthymol blue. The titration curve calculated above, and given in Figure 12-3, shows, however, that the choice of an indicator is in this case not crucial; any indicator with *pK* between 4 (methyl orange) and 10 (thymolphthalein) could be used with error less than 0.2%.

In titrating a weak acid (with a strong base) or a weak base (with a strong acid) greater care is needed in the selection of an indicator. Let us consider the titration of 0.2 *N* acetic acid, a moderately weak acid with $K_a = 1.80 \times 10^{-5}$, with 0.2 *N* sodium hydroxide. When an amount of the alkali equivalent to that of the acid has been added, the resultant solution is the same as would be obtained by dissolving 0.1 mole of the

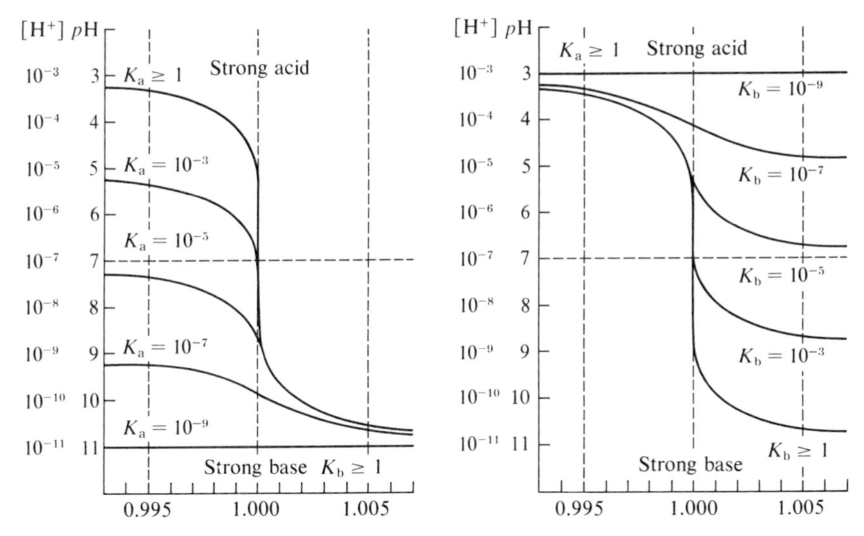

Ratio of equivalents of base to acid in titration of a 0.2 N acid with a 0.2 N base, either acid or base being strong

FIGURE 12-3
Acid-base titration curves.

salt $NaC_2H_3O_2$ in a liter of water. The solution of this salt is not neutral, with pH 7, however, but is alkaline, as can be seen from the following argument, based on the Brønsted-Lowry theory.

The salt $NaC_2H_3O_2$ is completely dissociated into ions, Na^+ and $C_2H_3O_2^+$, when it is dissolved in water. The ion Na^+ has no protons, and hence is not an acid. The acetate anion, $C_2H_3O_2^+$, is, however, a base—it is the base conjugate to the acid $HC_2H_3O_2$, and it can accept a proton from an acid, such as H_2O:

$$H_2O + C_2H_3O_2^- \rightleftharpoons HC_2H_3O_2 + OH^-$$

This reaction takes place to the extent determined by the value of its equilibrium constant:

$$\frac{[HC_2H_3O_2][OH^-]}{[H_2O][C_2H_3O_2^-]} = K_b$$

We may say that an aqueous solution of the neutral salt sodium acetate is alkaline, with pH greater than 7, because it contains the base (proton acceptor) acetate ion.

The value of K_b for a base is closely related to the value of the acid constant K_a of the homologous acid. Let us consider these constants for an acid HA and its homologous base A^-:

$$K_a = \frac{[H_3O^+][A^-]}{[HA][H_2O]}$$

$$K_b = \frac{[HA][OH^-]}{[H_2O][A^-]}$$

On multiplying the left sides and the right sides of these equations, we obtain

$$K_aK_b = \frac{[H_3O^+][A^-]}{[HA][H_2O]} \frac{[HA][OH^-]}{[H_2O][A^-]} = \frac{[H_3O^+][OH^-]}{[H_2O]^2}$$

The expression on the right side is seen to be the autoprotolysis constant for water, with value 1.00×10^{-14} (for $[H_2O]$ taken as 1; see Section 12-2); hence we have obtained the relation

$$K_b = \frac{K_w}{K_a} = \frac{1.00 \times 10^{-14}}{K_a}$$

between the acid constant and the base constant of a conjugate pair.

The acid constant of acetic acid is 1.80×10^{-5} (Section 12-5); the base constant of acetate ion accordingly has the value $1.00 \times 10^{-14}/1.80 \times 10^{-5} = 5.56 \times 10^{-10}$.

Let x be the number of acetate ions (mole liter^{-1}) that have undergone reaction with water to produce $HC_2H_3O_2$ and OH^-. The concentrations of solutes are seen to be

$$[HC_2H_3O_2] = [OH^-] = x$$

$$[C_2H_3O_2^-] = 0.1 - x$$

and the equilibrium equation is

$$\frac{x^2}{0.1 - x} = K_b = 5.56 \times 10^{-10}$$

Solution of this equation gives

$$x = 0.75 \times 10^{-5} \text{ mole liter}^{-1}$$

Hence $[OH^-] = 0.75 \times 10^{-5}$ and $[H^+] = 1.34 \times 10^{-9}$.

The pH of the solution of sodium acetate is hence 8.87. By reference to Figure 12-1 we see that *phenolphthalein,* with $pK = 9$, *is the best indicator to use for titrating a moderately weak acid such as acetic acid.*

The complete titration curve, showing the pH of the solution as a function of the amount of strong base added, can be calculated in essentially this way. Its course is shown in Figure 12-3 ($K_a = 10^{-5}$). We see that the solution has pH 7 when there is about 1% excess of acid; hence if litmus were used as the indicator an error of about 1% would be made in the titration.

The basic constant of ammonium hydroxide has about the same value as the acid constant of acetic acid. Hence to *titrate a weak base such as ammonium hydroxide with a strong acid methyl orange* (pK 3.8) *may be used as the indicator.*

It is possible by suitable selection of indicators to titrate separately a strong acid and a weak acid or a strong base and a weak base in a mixture of the two. Let us consider, for example, a solution of sodium hydroxide and ammonium hydroxide (ammonia). If strong acid is added until the pH is 11.1, which is that of 0.1 N ammonium hydroxide solution, the strong base will be within 1% of neutralization (Figure 12-3). Hence by using alizarine yellow (pK 11) as indicator the concentration of the strong base can be determined, and then by a second titration with methyl orange the concentration of ammonium hydroxide can be determined.

The Acidic Properties of Hydrated Ions of Metals other than the Alkalis and Alkaline Earths

Metal salts of strong acids, such as $FeCl_3$, $CuSO_4$, and $KAl(SO_4)_2 \cdot 12H_2O$ (alum), produce acidic solutions; the sour taste of these salts is characteristic.

It will be recalled from the discussion in Chapter 9 that the aluminum ion in aqueous solution is hydrated, having the formula $Al(H_2O)_6{}^{+++}$, with the six water molecules arranged octahedrally about the aluminum ion. The hydrolysis of aluminum salts may be represented by the equations

$$Al(H_2O)_6{}^{+++} + H_2O \rightleftarrows H_3O^+ + Al(H_2O)_5OH^{++}$$

$$Al(H_2O)_5OH^{++} + H_2O \rightleftarrows H_3O^+ + Al(H_2O)_4(OH)_2{}^+$$

$$Al(H_2O)_4(OH)_2{}^+ + H_2O \rightleftarrows H_3O^+ + Al(H_2O)_3(OH)_3$$

$$\rightleftarrows Al(OH)_3(c) + 3H_2O + H_3O^+$$

In these reactions the hydrated ions of aluminum lose protons, forming successive hydroxide complexes; the final neutral complex then loses water to form the insoluble hydroxide $Al(OH)_3$.

The complex ions $Al(H_2O)_5(OH)^{++}$ and $Al(H_2O)_4(OH)_2^+$ remain in solution, whereas the hydroxide $Al(OH)_3$ is very slightly soluble and precipitates if more than a very small amount is formed; precipitation occurs when the *p*H is greater than 3.

The protolysis of hydrated ferric ion occurs to such an extent that the color of ferric ion itself, $Fe(H_2O)_6^{+++}$, is usually masked by that of the hydroxide complexes. Ferric ion is nearly colorless; it seems to have a very pale violet color, seen in crystals of ferric alum, $KFe(SO_4)_2 \cdot 12H_2O$, and ferric nitrate, $Fe(NO_3)_3 \cdot 9H_2O$, and in ferric solutions strongly acidified with nitric or perchloric acid. Solutions of ferric salts ordinarily have the characteristic yellow to brown color of the hydroxide complexes $Fe(H_2O)_5OH^{++}$ and $Fe(H_2O)_4(OH)_2^+$, or even the red-brown color of colloidal particles of hydrated ferric hydroxide.

12-7. Buffered Solutions

Very small amounts of strong acid or base suffice to change the hydrogen-ion concentration of water in the slightly acidic to slightly basic region; one drop of strong concentrated acid added to a liter of water makes it appreciably acidic, increasing the hydrogen-ion concentration by a factor of 5000, and two drops of strong alkali would then make it basic, decreasing the hydrogen-ion concentration by a factor of over a million. Yet there are solutions to which large amounts of strong acid or base can be added with only very small resultant change in hydrogen-ion concentration. Such solutions are called *buffered solutions*.

Blood and other physiological solutions are buffered; the *p*H of blood changes only slowly from its normal value (about 7.4) on addition of acid or base. Important among the buffering substances in blood are the serum proteins (Chapter 14), which contain basic and acidic groups that can combine with the added acid or base.

A drop of concentrated acid, which when added to a liter of pure water increases $[H^+]$ 5000-fold (from 10^{-7} to 5×10^{-4}], produces an increase of $[H^+]$ of less than 1% (from 1.00×10^{-7} to 1.01×10^{-7}, for example) when added to a liter of buffered solution such as the phosphate buffer made by dissolving 0.2 moles of phosphoric acid in a liter of water and adding 0.3 moles of sodium hydroxide.

This is a half-neutralized phosphoric acid solution; its principal ionic constituents and their concentrations are Na^+, 0.3 *M;* HPO_4^{--}, 0.1 *M;* $H_2PO_4^-$, 0.1 *M;* H^+, about 10^{-7} *M*. From the titration curve of Figure 12-4 we see that this solution is a good buffer; to change its *p*H from 7 to 6.5 or 7.5 (tripling the hydrogen ion or hydroxide ion concentration) about

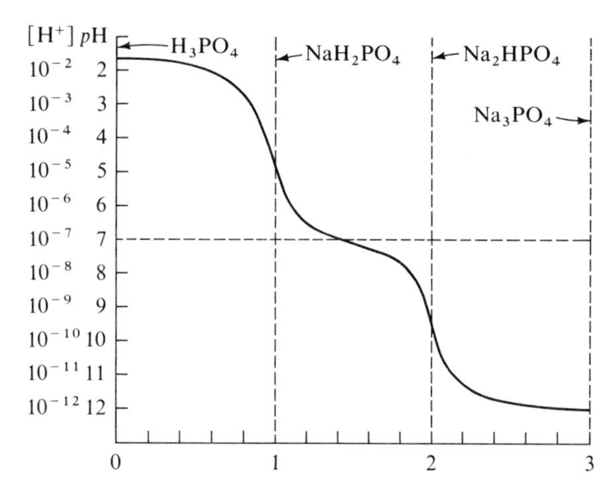

FIGURE 12-4
Titration curve for phosphoric acid (0.1 *F*) and a strong base.

one-twentieth of an equivalent of strong acid or base is needed per liter, whereas this amount of acid or base in water would cause a change of 5.7 *p*H units (an increase or decrease of $[H^+]$ by the factor 500,000). Such a solution, usually made by dissolving the two well-crystallized salts KH_2PO_4 and $Na_2HPO_4 \cdot 2H_2O$ in water, is widely used for buffering in the neutral region (*p*H 5.3 to 8.0).* Other useful buffers are sodium citrate-hydrochloric acid (*p*H 1 to 3.5), acetic acid-sodium acetate (*p*H 3.6 to 5.6), boric acid-sodium hydroxide (*p*H 7.8 to 10.0), and glycine-sodium hydroxide (*p*H 8.5 to 13).

The behavior of a buffer can be understood from the equilibrium equation for the acid dissociation. Let us consider the case of acetic acid-sodium acetate. The solution contains $HC_2H_3O_2$ and $C_2H_3O_2^-$ in equal or comparable concentrations. The equilibrium expression

$$\frac{[H^+][C_2H_3O_2^-]}{[HC_2H_3O_2]} = K_a$$

may be written as

$$[H^+] = \frac{[HC_2H_3O_2]}{[C_2H_3O_2^-]} K_a$$

*A concentrated neutral buffer solution containing one-half mole of each salt per liter may be kept in the laboratory to neutralize either acid or base spilled on the body.

This shows that when $[C_2H_3O_2^-]$ and $[HC_2H_3O_2]$ are equal, as in an equimolal mixed solution of $HC_2H_3O_2$ and $NaC_2H_3O_2$, the value of $[H^+]$ is just that of K_a, 1.80×10^{-5}, and hence the pH is 4.7. A 1:5 mixture of $HC_2H_3O_2$ and $NaC_2H_3O_2$ has $[H^+] = \frac{1}{5}K_a$ and pH 5.4, and a 5:1 mixture has $[H^+] = 5K_a$ and pH 4.0. By choosing a suitable ratio of $HC_2H_3O_2$ to $NaC_2H_3O_2$, any desired hydrogen-ion concentration in this neighborhood can be obtained.

It is seen from the equilibrium expressions that *the effectiveness of a buffer depends on the concentrations of the buffering substances;* a tenfold dilution of the buffer decreases by the factor 10 the amount of acid or base per liter that can be added without causing the pH to change more than the desired amount.

For the phosphate buffer in the pH 7 region the equilibrium constant of interest is that for the reaction

$$H_2PO_4^- \rightleftharpoons HPO_4^{--} + H^+$$

The value of $K_{H_2PO_4^-}$ is 6.2×10^{-8}; this is accordingly the value of $[H^+]$ expected for a solution with $[H_2PO_4^-] = [HPO_4^{--}]$.

If the buffered solution is dilute, this is its hydrogen-ion concentration. Because the activities of ions are affected by other ions, however, there is appreciable deviation from the calculated values in salt solutions as concentrated as 0.1 *M*. This fact accounts for the small discrepancies between the pH values calculated from equilibrium constants and those given in the buffer tables.

12-8. The Strengths of the Oxygen Acids

The oxygen acids, which consist of oxygen atoms O and hydroxide groups OH attached to a central atom ($HClO_4 = ClO_3(OH)$, $H_2SO_4 = SO_2(OH)_2$, and so on), vary widely in strength, from very strong acids such as perchloric acid, $HClO_4$, to very weak ones such as boric acid, H_3BO_3. It is often useful to know the approximate strengths of these acids. Fortunately there have been formulated some simple and easily remembered rules about these acid strengths.

The Rules Expressing the Strengths of the Oxygen Acids

The strengths of the oxygen acids are expressed approximately by the following two rules:

Rule 1. The successive acid constants K_1, K_2, K_3, . . . *are in the ratios* $1{:}10^{-5}{:}10^{-10}{:}$ . . . .

We note the examples of phosphoric acid

$$K_{H_3PO_4} = 7.5 \times 10^{-3} \qquad K_{H_2PO_4^-} = 6.2 \times 10^{-8} \qquad K_{HPO_4^{--}} = 1 \times 10^{-12}$$

and sulfurous acid

$$K_{H_2SO_3} = 1.2 \times 10^{-2} \qquad K_{HSO_3^-} = 1 \times 10^{-7}$$

The rule holds well for all the acids of the class under consideration.

Rule 2. The value of the first ionization constant is determined by the value of m in the formula $XO_m(OH)_n$*: if m is zero (no excess of oxygen atoms over hydrogen atoms, as in* $B(OH)_3$*) the acid is very weak, with* $K_1 \leqq 10^{-7}$*; for* $m = 1$ *the acid is weak, with* $K_1 \cong 10^{-2}$*; for* $m = 2$ $(K_1 \cong 10^3)$ *or* $m = 3$ $(K_1 \cong 10^8)$ *the acid is strong.*

Note the occurrence of the factor 10^{-5} in both this rule and the first one. The applicability of this rule is shown by the tables at the end of this section.

The second rule can be understood in the following way. The force attracting H^+ to ClO^- to form ClOH (hypochlorous acid) is that of an O—H valence bond. But the force between H^+ and either one of the two oxygen atoms of the ion ClO_2^- to form ClOOH (chlorous acid) may be smaller than that for an O—H valence bond because the total attraction for the proton is divided between the two oxygen atoms, and hence this acid (of the second class) may well be expected to be more highly dissociated than hypochlorous acid. An acid of the third class would be still more highly dissociated, since the total attraction for the proton would be divided among three oxygen atoms.

With use of these rules we can answer questions about the choice of indicators for titration without referring to tables of acid constants.

Example 12-7. What reaction to litmus would be expected of solutions of the following salts: NaClO, $NaClO_2$, $NaClO_3$, $NaClO_4$?

Solution. The corresponding acids are shown by the rule to be very weak, weak, strong, and very strong, respectively. Hence NaClO and $NaClO_2$ would through hydrolysis give basic solutions, and the other two salts would give neutral solutions.

Example 12-8. What indicator could be used for titrating periodic acid, H_5IO_6?

Solution. This acid has one extra oxygen atom, and is hence of the second class, as is phosphoric acid. We accordingly refer to Figures 12-4 and 12-1, and see that methyl orange should be satisfactory for titrating the first hydrogen, or phenolphthalein for titrating the first two hydrogens.

Experimental Values of Acid Constants

Some values of acid constants that have been determined by experiment are given in the following tabulation.

First class; Very weak acids $X(OH)_n$ or H_nXO_n

First acid constant about 10^{-7} or less

	K_1
Hypochlorous acid, $HClO$	3.2×10^{-8}
Hypobromous acid, $HBrO$	2×10^{-9}
Hypoiodous acid, HIO	1×10^{-11}
Silicic acid, H_4SiO_4	1×10^{-10}
Germanic acid, H_4GeO_4	3×10^{-9}
Boric acid, H_3BO_3	5.8×10^{-10}
Arsenious acid, H_3AsO_3	6×10^{-10}
Antimonous acid, H_3SbO_3	1×10^{-11}

Second Class; Weak acids $XO(OH)_n$ or H_nXO_{n+1}

First acid constant about 10^{-2}

	K_1
Chlorous acid, $HClO_2$	1.1×10^{-2}
Sulfurous acid, H_2SO_3	1.2×10^{-2}
Selenious acid, H_2SeO_3	0.3×10^{-2}
Phosphoric acid, H_3PO_4	0.75×10^{-2}
Phosphorous acid,* H_2HPO_3	1.6×10^{-2}
Hypophosphorous acid,* HH_2PO_2	1×10^{-2}
Arsenic acid, H_3AsO_4	0.5×10^{-2}
Periodic acid, H_5IO_6	1×10^{-3}
Nitrous acid, HNO_2	0.45×10^{-3}
Acetic acid, $HC_2H_3O_2$	1.80×10^{-5}
Carbonic acid,† H_2CO_3	0.45×10^{-6}

---

*It is known that phosphorous acid has the structure H—P(=O)(—OH)—OH and hypophosphorus acid the structure H—P(=O)(—H)—OH ; the hydrogen atoms that are bonded to the phosphorus atom are not counted in applying the rule.

†The low value for carbonic acid is due in part to the existence of some of the un-ionized acid in the form of dissolved CO_2 molecules rather than H_2CO_3. The proton dissociation constant for the molecular species H_2CO_3 is about 2×10^{-4}.

Third Class; Strong acids $XO_2(OH)_n$ or H_nXO_{n+2}

First acid constant about 10^3
Second acid constant about 10^{-2}

	K_1	K_2
Chloric acid, $HClO_3$	Large	
Sulfuric acid, H_2SO_4	Large	1.2×10^{-2}
Selenic acid, H_2SeO_4	Large	1×10^{-2}

Fourth Class; Very strong acids $XO_3(OH)_n$ or H_nXO_{n+3}

First acid constant about 10^8

Perchloric acid, $HClO_4$	Very strong
Permanganic acid, $HMnO_4$	Very strong

There is no simple way of remembering the strengths of acids other than those discussed above. HCl, HBr, and HI are strong, but HF is weak, with $K_a = 7.2 \times 10^{-4}$. The homologues of water are weak acids, with the following reported acid constants:

	K_1	K_2
Hydrosulfuric acid, H_2S	1.1×10^{-7}	1.0×10^{-14}
Hydroselenic acid, H_2Se	1.7×10^{-4}	1×10^{-12}
Hydrotelluric acid, H_2Te	2.3×10^{-3}	1×10^{-11}

The hydrides NH_3 and PH_3 function as bases by adding protons rather than as acids by losing them.

Oxygen acids that do not contain a single central atom have strengths corresponding to reasonable extensions of our rules, as shown by the following examples.

Very weak acids: $K_1 = 10^{-7}$ or less

	K_1	K_2
Hydrogen peroxide, HO—OH	2.4×10^{-12}	
Hyponitrous acid, HON—NOH	9×10^{-8}	1×10^{-11}

Weak acids: $K_1 = 10^{-2}$

	K_1	K_2
Oxalic acid, HOOC—COOH	5.9×10^{-2}	6.4×10^{-5}

The following acids are not easily classified:

	K_1
Hydrocyanic acid, HCN	4×10^{-10}
Cyanic acid, HOCN	Strong
Thiocyanic acid, HSCN	Strong
Hydrazoic acid, HN_3	1.8×10^{-5}

Organic Phosphates and Other Acids of Biological Importance

Many organic phosphates are involved in biochemical reactions in the cells of living organisms. The high-energy molecules ATP and ADP have been mentioned in Section 8-4 and are discussed also in Section 14-6. Glucose phosphate and many other phosphates are involved in metabolic processes. These molecules are ionized in solution in body fluids to an extent determined by their ionization constants.

The three ionization constants of phosphoric acid have the values 7.5×10^{-3}, 6.2×10^{-8}, and 1×10^{-12}. When one or two hydrogen atoms are replaced by carbon atoms of organic radicals the first two acid constants become somewhat larger, as shown by the following examples:

	K_1	K_2
Methylphosphoric acid	3×10^{-2}	5×10^{-7}
n-Butylphosphoric acid	2×10^{-2}	2×10^{-7}
Glucose-1-phosphate	8×10^{-2}	7×10^{-7}
Dimethylphosphoric acid	5×10^{-2}	
Di-*n*-butylphosphoric acid	2×10^{-2}	

These values show that monoesters of phosphoric acid are present in neutral solutions mainly as the ions $ROPO_2OH^-$ and $ROPO_3^{--}$, with the latter predominating. Diesters are completely ionized to $(RO)_2PO_2^-$.

Pyrophosphoric acid has acid constants 7×10^{-1}, 3×10^{-2}, 2×10^{-6}, and 6×10^{-9}. It is probable that organic pyrophosphates and triphosphates such as ADP and ATP are mainly present in cells as the trinegative anions.

Many polycarboxylic acids, such as citric acid, are involved in metabolic processes (Section 14-7). Values of acid constants for some of them are the following:

		K_1	K_2
Succinic acid	$H_2C—COOH$ | $H_2C—COOH$	7×10^{-5}	3×10^{-6}
Malic acid	H $HOC—COOH$ | $H_2C—COOH$	4×10^{-4}	8×10^{-6}

		K_1	K_2	K_3
Fumaric acid	HC—COOH ‖ HOOC—CH	9×10^{-4}	3×10^{-5}	
Citric acid	H_2C—COOH | HOC—COOH | H_2C—COOH	8×10^{-4}	2×10^{-5}	4×10^{-6}

These acids are seen to be essentially completely ionized in physiological solutions.

Acid Strength and Condensation

It is observed that the tendency of oxygen acids to condense to larger molecules is correlated with their acid strengths. Very strong acids, such as $HClO_4$ and $HMnO_4$, condense only with difficulty, and the substances formed, Cl_2O_7 and Mn_2O_7, are very unstable. Less strong acids, such as H_2SO_4, form condensation products such as $H_2S_2O_7$, disulfuric acid, on strong heating, but these products are not stable in aqueous solution. Phosphoric acid forms pyrophosphate ion and other condensed ions in aqueous solution, but these ions easily hydrolyze to the orthophosphate ion; other weak acids behave similarly. The very weak oxygen acids, including silicic acid (Chapter 18) and boric acid, condense very readily, and their condensation products are very stable substances.

This correlation is reasonable. The un-ionized acids contain oxygen atoms bonded to hydrogen atoms, and the condensed acids contain oxygen atoms bonded to two central atoms:

```
H—Ö:  :Ö—H
    \  /
     Si
    /  \
H—Ö:  :Ö—H

H—Ö:             :Ö:             :Ö—H
    \          /     \          /
     Si ------         ------ Si
    /  \                     /  \
H—Ö:  :Ö—H            H—Ö:  :Ö—H
```

It is hence not surprising that stability of the un-ionized acid (low acid strength) should be correlated with stability of the condensed molecules.

12-9. Nonaqueous Amphiprotic Solvents

In Section 12-1 the autoprotolysis of the amphiprotic substance water was discussed. Every solvent whose molecules contain one or more protons and one or more unshared electron pairs in outer shells can act as an amphiprotic solvent. One of its molecules can donate a proton to a sufficiently strong base or accept a proton from a sufficiently strong acid. For example, perchloric acid dissolved in pure sulfuric acid undergoes the following reaction, in which H_2SO_4 acts as a base:

$$HClO_4 + H_2SO_4 \rightleftharpoons H_3SO_4^+ + ClO_4^-$$

A prediction of the probable behavior of a solute in an amphiprotic solvent can be made by comparing the acid constant of the solute when dissolved in water with the acid constant of the solvent when dissolved in water. For example, in water perchloric acid is about 10^5 times as strong as sulfuric acid. Its great proton-donating power probably applies also when sulfuric acid is the solvent.

Phosphoric acid, on the other hand, is in aqueous solution a weaker acid than sulfuric acid; it reacts as a base when dissolved in pure sulfuric acid:

$$H_2SO_4 + H_3PO_4 \rightleftharpoons H_4PO_4^+ + HSO_4^-$$

Sulfuric acid also undergoes autoprotolysis:

$$2H_2SO_4 \rightleftharpoons H_3SO_4^+ + HSO_4^-$$

Values of the autoprotolysis constants of sulfuric acid and some other amphiprotic solvents, as determined by measuring the electric conductivity of the solvent and some of its solutions, are given in Table 12-1.

TABLE 12-1
Values of the Autoprotolysis Constant

Solvent	Autoprotolysis Constant*
H_2O	$[H_3O^+][OH^-] = 1.0 \times 10^{-14}$ (mole liter^{-1})^2
NH_3	$[NH_4^+][NH_2^-] = 1 \times 10^{-33}$
H_2SO_4	$[H_3SO_4^+][HSO_4^-] = 2 \times 10^{-4}$
HCOOH (formic acid)	$[HC(OH)_2^+][HCOO^-] = 6 \times 10^{-7}$
CH_3COOH (acetic acid)	$[CH_3C(OH)_2^+][CH_3COO^-] = 1 \times 10^{-13}$
CH_3OH (methyl alcohol)	$[CH_3OH_2^+][CH_3O^-] = 2 \times 10^{-17}$
C_2H_5OH (ethyl alcohol)	$[C_2H_5OH_2^+][C_2H_5O^-] = 3 \times 10^{-20}$

*All values are for 25°C except that for ammonia, which is for −33°C. The value for water at 100°C is 0.5×10^{-12}.

EXERCISES

12-1. Give three definitions of acids and bases, in the order of increasing generality.

12-2. Differentiate between a strong acid and a weak acid.

12-3. What is autoprotolysis? Give the equation for the autoprotolysis of water. What is the concentration of H_3O^+ and that of OH^- in pure water at 25°C? At 0°C? At 100°C?

12-4. Which of these oxides form acids in water and which form bases? Write an equation for each representing its reaction with water.

P_2O_3	N_2O_5	Na_2O	Mn_2O_7
Cl_2O	B_2O_3	I_2O_5	SO_2
Cl_2O_7	CO_2	SO_3	

12-5. Draw a graph of the H_3O^+ concentration and the OH^- concentration of water and dilute solutions as a function of pH.

12-6. Define the K_{In} and pK of an acid-base indicator.

12-7. Describe three methods of measuring the pH of a solution, giving the expected accuracy of the method.

12-8. What is the pH, to the nearest pH unit, of 1 N HCl? of 0.1 N HCl? of 10 N HCl? of 0.1 N NaOH? of 10 N NaOH?

12-9. What is the normality of a solution of a strong acid 25.00 ml of which is rendered neutral by 28.75 ml of 0.1063 N NaOH solution?

12-10. The poisonous *botulinus* organism does not grow in canned vegetables if the pH is less than 4.5. Some investigators [*Journal of Chemical Education, 22,* 409 (1945)] have recommended that in home canning of nonacid foods, such as beans, without a pressure canner a quantity of hydrochloric acid be added. The amount of hydrochloric acid recommended is 25 ml of 0.5 N hydrochloric acid per half-liter jar. Calculate the pH that this solution would have, assuming it originally to be neutral, and neglecting the buffering action of the organic material. Also calculate the amount of baking soda ($NaHCO_3$), measured in teaspoonfuls, that would be required to neutralize the acid after the jar is open. One teaspoon equals 4 grams of baking soda.

12-11. Calculate the hydrogen-ion concentration in the following solutions:
(*a*) 1 F $HC_2H_3O_2$, $K = 1.8 \times 10^{-5}$
(*b*) 0.06 F HNO_2, $K = 0.45 \times 10^{-3}$
(*c*) 0.004 F NH_4OH, $K_b = 1.8 \times 10^{-5}$
(*d*) 0.1 F HF, $K = 6.7 \times 10^{-4}$
What are the pH values of the solutions?

12-12. Calculate the concentrations of the various ionic and molecular species in a solution prepared by mixing equal volumes of 1 N NaOH and 0.5 N NH_4OH.

12-13. Calculate the pH of a solution that is 0.1 F in HNO_2 and 0.1 F in HCl.

12-14. Calculate the concentrations of the various ionic and molecular species in the following solutions:
(*a*) 0.1 F H_2Se ($K_1 = 1.7 \times 10^{-4}$, $K_2 = 1 \times 10^{-11}$)
(*b*) 0.01 F H_2CO_3 ($K_1 = 4.5 \times 10^{-7}$, $K_2 = 6 \times 10^{-11}$)
(*c*) 1 F H_2CrO_4 ($K_1 = 0.18$, $K_2 = 3.2 \times 10^{-7}$)
(*d*) 0.5 F H_3PO_4 ($K_1 = 7.5 \times 10^{-3}$, $K_2 = 0.6 \times 10^{-7}$, $K_3 = 1 \times 10^{-12}$)
(*e*) 1 F H_2SO_4 ($K_2 = 1.20 \times 10^{-2}$)
(*f*) 0.01 F H_2SO_4

12-15. Boric acid loses only one hydrogen ion. In 0.1 M H_3BO_3, $[H^+] = 1.05 \times 10^{-5}$. Calculate the ionization constant for boric acid.

12-16. A patent medicine for stomach ulcers contains 2.1 g of $Al(OH)_3$ per 100 ml. How far wrong is the statement on the label that the preparation is "capable of combining with 16 times its volume of N/10 HCl"?

12-17. Which of these substances form acidic solutions, which neutral, and which basic? Write equations for the reactions which give H^+ or OH^-.

NaCl	$(NH_4)_2SO_4$	$CuSO_4 \cdot 5H_2O$
NaCN	$NaHSO_4$	$FeCl_2$
Na_3PO_4	NaH_2PO_4	$KAlSO_4 \cdot 12H_2O$
NH_4Cl	Na_2HPO_4	$Zn(ClO_4)_2$
NH_4CN	$KClO_4$	BaO

12-18. Approximately how much acetic acid must be added to a 0.1 N solution of sodium acetate to make the solution neutral (pH 7)?

12-19. What indicators should be used in titrating the following acids?

	K_a
HNO_2	4.5×10^{-4}
H_2S (first hydrogen)	1.1×10^{-7}
HCN	4×10^{-10}

With what indicators could you titrate separately for HCl and $HC_2H_3O_2$ in a solution containing both acids?

12-20. Calculate the pH of a solution that is
(*a*) 0.1 F in NH_4Cl, 0.1 F in NH_4OH
(*b*) 0.05 F in NH_4Cl, 0.15 F in NH_4OH
(*c*) 1.0 F in $HC_2H_3O_2$, 0.3 F in $NaC_2H_3O_2$
(*d*) prepared by mixing 10 ml 1 F $HC_2H_3O_2$ with 90 ml 0.05 F NaOH

12-21. Calculate the pH of a solution that is prepared from
(*a*) 10 ml 1 F HCN, 10 ml 1 F NaOH
(*b*) 10 ml 1 F NH_4OH, 10 ml 1 F HCl
(*c*) 10 ml 1 F NH_4OH, 10 ml 1 F NH_4Cl

12-22. Calculate the concentration of the various ionic and molecular species in
(*a*) 0.4 F NH_4Cl
(*b*) 0.1 F $NH_4C_2H_3O_2$
(*c*) 0.1 F $NaHCO_3$
(*d*) 0.1 F Na_2CO_3

12-23. Calculate the concentration of the various ionic and molecular species in a solution that is
(*a*) 0.3 *F* in HCl, and 0.1 *F* in H_2S
(*b*) buffered to a *p*H of 4, and 0.1 *F* in H_2S
(*c*) 0.2 *F* in KHS
(*d*(0.2 *F* in K_2S

12-24. What relative masses of KH_2PO_4 and $Na_2HPO_4 \cdot 2H_2O$ should be taken to make a buffered solution with *p*H 6.0?

12-25. Carbon dioxide, produced by oxidation of substances in the tissues, is carried by the blood to the lungs. Part of it is in solution as carbonic acid, and part as hydrogen carbonate ion, HCO_3^-. If the *p*H of the blood is 7.4, what fraction is carried as the ion?

12-26. The value of K_1 for H_2S is 1.1×10^{-7}. What is the ratio $[H_2S]/[HS^-]$ at *p*H 8? If hydrogen sulfide at 1 atm pressure is 0.1 *F* soluble in acid solution, what would be its solubility at *p*H 8?

12-27. Would water act as an acid or a base when dissolved in liquid H_2S? Would H_2Se act as an acid or a base?

12-28. Hydrogen cyanide, H—C≡N:, is amphiprotic. What is its conjugate acid? Its conjugate base?

12-29. What reaction would you expect to take place when HCN is dissolved in pure sulfuric acid?

12-30. What is the concentration of the cation $H_3SO_4^+$ in pure sulfuric acid? Of the anion HSO_4^-? (See Table 12-1.) (Answer: 0.014 mole liter^{-1}, 0.014 mole liter^{-1}.)

12-31. What is the concentration of the cation $H_2C_2H_3O_2^+$ in pure acetic acid? Of the anion $C_2H_3O_2^-$?

12-32. Explain why a metal hydroxide such as ferric hydroxide, $Fe(OH)_3$, is much more soluble in acidic solution than it is in a basic solution.

12-33. Discuss the reaction $F^- + BF_3 \longrightarrow BF_4^-$ in terms of the Lewis theory of acids and bases.

13

Organic Chemistry

13-1. The Nature and Extent of Organic Chemistry

Organic chemistry is the chemistry of the compounds of carbon. It is a very great subject—over a million different organic compounds have already been reported and described in the chemical literature. Many of these substances have been isolated from living matter, and many more have been synthesized by chemists in the laboratory.

The occurrence in nature, methods of preparation, composition, structure, properties, and uses of some organic compounds (hydrocarbons, alcohols, chlorine derivatives of hydrocarbons, and organic acids) were discussed in Chapters 7 and 8. This discussion is continued in the following sections, with emphasis on natural products, especially the valuable substances obtained from plants, and on synthetic substances useful to man. Several large parts of organic chemistry will not be discussed at all; these include the methods of isolation and purification of naturally occurring compounds, the methods of analysis and determination of structure, and the methods of synthesis used in organic chemistry, except to the extent that they have been described in Chapters 7 and 8.

There are two principal ways in which organic chemists work. One of these ways is to begin the investigation of some natural material, such as a plant, that is known to have special properties. This plant might, for example, have been found by the natives of a tropical region to be beneficial in the treatment of malaria. The chemist then proceeds to make an extract from the plant, with use of a solvent such as alcohol or ether, and, by various methods of separation, to divide the extract into fractions. After each fractionation a study is made to see which fraction still contains the active substance. Finally this process may be carried so far that a pure crystalline active substance is obtained. The chemist then analyzes the substance, and determines its molecular weight, in order to find out what atoms are contained in the molecule of the substance. He next investigates the chemical properties of the substance, splitting its molecules into smaller molecules of known substances, in order to determine its molecular structure. When the structure has been determined, he attempts to synthesize the substance; if he is successful, the active material may be made available in large quantity and at low cost.

During recent decades the organic chemist, especially if he is working with natural products (from plants or animals), has made great use of physical methods of determining the chemical formulas and molecular structures of substances. Sometimes, as with penicillin and cyanocobalamin (vitamin B_{12}), the correct structural formula has been discovered through a complete determination of the crystal structure by the x-ray diffraction method. Mass spectrometry is also a valuable method of determining the structural formulas of substances. The molecules are ionized by electron bombardment, which usually splits the molecules into fragments. The masses of the fragments often permit their identification, and the structural formula can then be deduced as the one that would be expected to give the observed fragments. Various spectroscopic methods, especially the nuclear-spin magnetic-resonance method, are also useful.

The other way in which organic chemists work involves the synthesis and study of a large number of organic compounds, and the continued effort to correlate the empirical facts by means of theoretical principles. Often a knowledge of the structure and properties of natural substances is valuable in indicating the general nature of the compounds that are worth investigation. The ultimate goal of this branch of organic chemistry is the complete understanding of the physical and chemical properties, and also the physiological properties, of substances in terms of their molecular structure. At the present time chemists have obtained a remarkable insight into the dependence of the physical and chemical properties of substances on the structure of their molecules. So far, however, only a small beginning has been made in attacking the great problem of the relation between structure and physiological activity. This problem remains one of the greatest and most important problems of science, challenging the new generation of scientists.

13-2. Petroleum and the Hydrocarbons

One of the most important sources of organic compounds is petroleum (crude oil). Petroleum, which is obtained from underground deposits that have been tapped by drilling oil wells, is a dark-colored, viscous liquid that is in the main a mixture of hydrocarbons (compounds of hydrogen and carbon; see Section 7-2). A very great amount of it, approximately one billion tons, is produced and used each year. Much of it is burned, for direct use as a fuel, but much is separated or converted into other materials.

The Refining of Petroleum

Petroleum may be separated into especially useful materials by a process of distillation, called *refining*. It was mentioned in Section 7-2 that petroleum ether, obtained in this way, is an easily volatile pentane-hexane-heptane (C_5H_{12} to C_7H_{16}) mixture that is used as a solvent and in the dry cleaning of clothes, gasoline is the heptane-to-nonane (C_7H_{16} to C_9H_{20}) mixture used in internal-combustion engines, kerosene is the decane-to-hexadecane ($C_{10}H_{22}$ to $C_{16}H_{34}$) mixture used as a fuel, and heavy fuel oil is a mixture of still larger hydrocarbon molecules.

The residue from distillation is a black, tarry material called *petroleum asphalt*. It is used in making roads, for asphalt composition roofing materials, for stabilizing loose soil, and as a binder for coal dust in the manufacture of briquets for use as a fuel. A similar material, *bitumen* or *rock asphalt,* is found in Trinidad, Texas, Oklahoma, and other parts of the world, where it presumably has been formed as the residue from the slow distillation of pools of oil.

It is thought that petroleum, like coal, is the result of the decomposition of the remains of plants that grew on the earth about 250 million years ago.

Cracking and Polymerizing Processes

As the demand for gasoline became greater, methods were devised for increasing the yield of gasoline from petroleum. The simple "cracking" process consists in the use of high temperature to break the larger molecules into smaller ones; for example, a molecule of $C_{12}H_{26}$ might be broken into a molecule of C_6H_{14} (hexane) and a molecule of C_6H_{12} (hexene, containing one double bond). There are now several rather complicated cracking processes in use. Some involve heating liquid petroleum, under pressure of about 50 atm, to about 500°C, perhaps with a catalyst such as aluminum chloride, $AlCl_3$. Others involve heating petroleum vapor with a catalyst such as clay containing some zirconium dioxide.

Polymerization is also used to make gasoline from the lighter hydrocarbons containing double bonds. For example, two molecules of ethylene, C_2H_4 can react to form one molecule of butylene, C_4H_8 (structural formula $CH_3—CH=CH—CH_3$).

Some gasoline is also made by the hydrogenation (reaction with hydrogen) of petroleum and coal. Many organic chemicals are prepared in great quantities from these important raw materials.

Hydrocarbons Containing Several Double Bonds

The structure and properties of ethylene, a substance whose molecules contain a double bond, were discussed in Section 7-2. Some important natural products are hydrocarbons containing several double bonds. For example, the red coloring matter of tomatoes, called *lycopene,* is an unsaturated hydrocarbon, $C_{40}H_{56}$, with the structure shown in Figure 13-1.

The molecule of this substance contains thirteen double bonds. It is seen that eleven of these double bonds are related to one another in a special way—they alternate regularly with single bonds. A regular alternation of double bonds and single bonds in a hydrocarbon chain is called a *conjugated system of double bonds.* The existence of this structural feature in a molecule confers upon the molecule special properties, such as the power of absorbing visible light, causing the substance to be colored.

Other yellow and red substances, isomers of lycopene, with the same formula $C_{40}H_{56}$, are called α-carotene, β-carotene, and similar names. These substances occur in butter, milk, green leafy vegetables, eggs, cod liver oil, halibut liver oil, carrots, tomatoes, and other vegetables and fruits. They are important substances because they serve in the human body as a source of vitamin A (see Chapter 14).

Polycyclic Substances

Many important substances exists whose molecules contain two or more rings of atoms: these substances are called *polycyclic substances;* naphthalene, anthracene, and phenanthrene are examples of polycyclic aromatic hydrocarbons (Section 7-3). An example of a polycyclic aliphatic hydrocarbon is *pinene,* $C_{10}H_{16}$, which is the principal constituent of *turpentine.* Turpentine is an oil obtained by distilling a semifluid

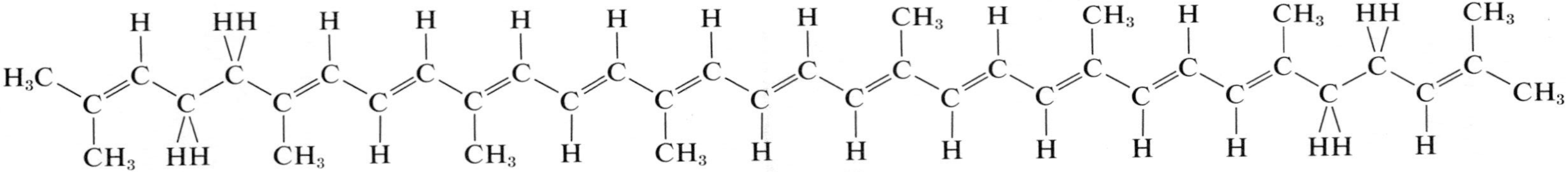

Lycopene, $C_{40}H_{56}$

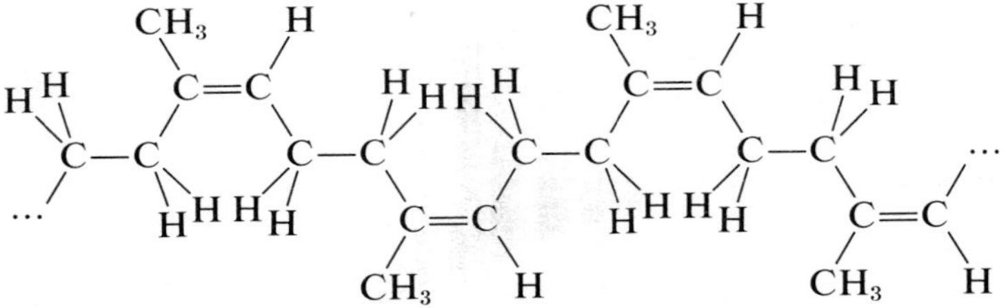

Portion of rubber polymer $(C_5H_8)_\infty$

FIGURE 13-1
Structural formulas of lycopene and rubber.

resinous material that exudes from pine trees. The pinene molecule has the following structure:

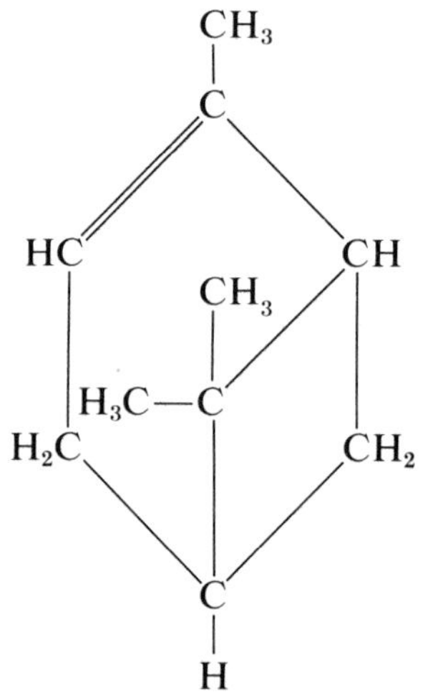

CH_3
C
HC CH
CH_3
H_3C—C
H_2C CH_2
C
H

Another interesting polycyclic substance is *camphor*, obtained by steam distillation of the wood of the camphor tree, or, in recent years, by a synthetic process starting with pinene. The molecule of camphor is roughly spherical in shape—it is a sort of "cage" molecule:

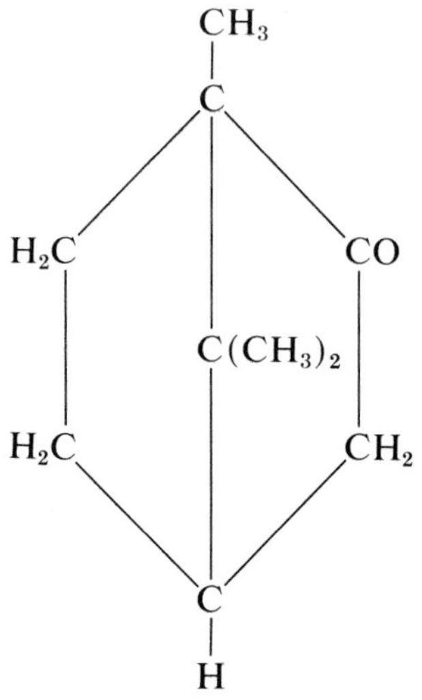

CH_3
C
H_2C CO
$C(CH_3)_2$
H_2C CH_2
C
H

Camphor contains one oxygen atom, its formula being $C_{10}H_{16}O$. A hydrocarbon is obtained by replacing the oxygen atom by two hydrogen atoms, producing the substance called *camphane*. Camphor is used in medicine and in the manufacture of plastics. Ordinary *celluloid* consists of nitrocellulose plasticized with camphor.

Rubber

Rubber is an organic substance, obtained mainly from the sap of the rubber tree, *Hevea brasiliensis*. Rubber consists of very long molecules, which are polymers of *isoprene,* C_5H_8. The structure of isoprene is

```
      H     H
      |     |
   H—C     C
      \\  /  \\
        C      C—H
        |      |
       CH3     H
```

and that of the rubber polymer, as produced in the plant, is shown in Figure 13-1.

The characteristic properties of rubber are due to the fact that it is an aggregate of very long molecules, intertwined with one another in a rather random way. The structure of the molecules is such that they do not tend to align themselves side by side in a regular way—that is, to crystallize—but instead tend to retain an irregular arrangement.

It is interesting to note that the rubber molecule contains a large number of double bonds, one for each C_5H_8 residue. In natural rubber the configuration about the double bonds is the *cis* configuration, as shown in the structural formula in Figure 13-1. *Gutta percha,* a similar plant product that does not have the elasticity of rubber, contains the same molecules, with, however, the *trans* configuration around the double bonds. This difference in configuration permits the molecules of gutta percha to crystallize more readily than those of rubber.

Ordinary unvulcanized rubber is sticky, as a result of a tendency for the molecules to pull away from one another, a portion of the rubber thus adhering to any material with which it comes in contact. The stickiness is eliminated by the process of *vulcanization,* which consists in heating rubber with sulfur. During this process sulfur molecules, S_8, open up and combine with the double bonds of rubber molecules, forming bridges of sulfur chains from one rubber molecule to another rubber molecule. These sulfur bridges bind the aggregate of rubber molecules together into a large molecular framework, extending through the whole sample of rubber. Vulcanization with a small amount of sulfur leads to a soft product, such as that in rubber bands or (with a filler of carbon black or zinc oxide) in automobile tires. A much harder material, called vulcanite, is obtained by using a larger amount of sulfur.

The materials called *synthetic rubber* are not really synthetic rubber, since they are not identical with the natural product. They are, rather, substitutes for rubber—materials with properties and structure similar to but not identical with those of natural rubber. For example, the substance *chloroprene,* C_4H_5Cl, with the structure

$$\begin{array}{ccccccc} & H & & H & & & \\ & | & & | & & & \\ H-C & & & C & & & \\ & \diagdown\!\!\diagdown & & / & \diagdown\!\!\diagdown & & \\ & & C & & & C-H & \\ & & | & & & | & \\ & & Cl & & & H & \end{array}$$

is similar to isoprene except for the replacement of a methyl group by a chlorine atom. Chloroprene polymerizes to a rubber called *chloroprene rubber.* It and other synthetic rubbers have found extensive uses, and are superior to natural rubber for some purposes.

13-3. Alcohols and Phenols

The aliphatic alcohols have a hydroxyl group, —OH, attached to a carbon atom in place of one of the hydrogen atoms of an aliphatic hydrocarbon. The two simplest alcohols, methyl alcohol (methanol) and ethyl alcohol (ethanol), have been discussed in Section 8-6. The melting points, boiling points, and densities of some alcohols are given in Table 13-1.

Some of the heavier alcohols are made from the olefines that are obtained as by-products in the refining of petroleum. For example, propylene, $CH_2{=}CH{-}CH_3$, can be hydrated by addition of water vapor at high temperature and pressure in the presence of a catalyst:

$$CH_2{=}CH{-}CH_3 + H_2O \rightarrow CH_3{-}\underset{\displaystyle OH}{\underset{|}{C}H}{-}CH_3$$

The product is called *isopropyl alcohol* or *2-propanol* (the number 2 means that the substituent is on the second carbon atom in the chain, and the suffix *ol* means that the substituent is the hydroxyl group). An alcohol

of this kind, with formula

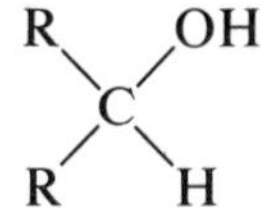

$$\begin{array}{ccc} R & & OH \\ & \diagdown\diagup & \\ & C & \\ & \diagup\diagdown & \\ R & & H \end{array}$$

(R being a radical with a carbon atom forming the bond) is called a secondary alcohol. Isopropyl alcohol may be called *sec*-propanol.

An alcohol

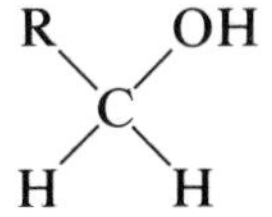

$$\begin{array}{ccc} R & & OH \\ & \diagdown\diagup & \\ & C & \\ & \diagup\diagdown & \\ H & & H \end{array}$$

is called a *primary alcohol;* examples are ethanol and 1-propanol, $CH_3CH_2CH_2OH$. *Tertiary alcohols* have the formula

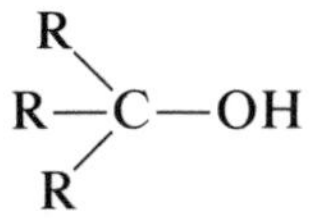

$$\begin{array}{l} R \\ \;\diagdown \\ R—C—OH \\ \;\diagup \\ R \end{array}$$

TABLE 13-1
Physical Properties of Some Alcohols and Phenols

	Melting Point	Boiling Point	Density of Liquid
Methyl alcohol, CH_3OH	−97.8°C	64.7°C	0.796 g ml^{-1}
Ethyl alcohol, CH_3CH_2OH	−117.3°	78.5°	.789
Propyl alcohol, $CH_3(CH_2)_2OH$	−127°	97.2°	.804
Isopropyl alcohol, $CH_3CHOHCH_3$	−89°	82.3°	.785
Butyl alcohol, $CH_3(CH_2)_3OH$	−89°	117.7°	.810
sec-Butyl alcohol, $CH_3CHOHCH_2CH_3$	−89°	100°	.808
tert-Butyl alcohol, $(CH_3)_3COH$	25°	83°	.789
1-Pentanol, $CH_3(CH_2)_4OH$	−78°	138°	.814
Glycol, CH_2OHCH_2OH	−17°	197°	1.116
1,2-Propanediol, $CH_2OHCHOHCH_3$		189°	1.040
1,3-Propanediol, $CH_2OHCH_2CH_2OH$		214°	1.053
Glycerol, $CH_2OHCHOHCH_2OH$	17.9°	290°	1.260
Benzyl alcohol, $C_6H_5CH_2OH$	−15.3°	205°	1.050
Phenol, C_6H_5OH	41°	182°	1.072*
o-Cresol, $CH_3C_6H_4OH$	30°	192°	1.047*
m-Cresol, $CH_3C_6H_4OH$	11°	203°	1.034
p-Cresol, $CH_3C_6H_4OH$	36°	203°	1.035*

*Density of crystalline substance.

The simplest example is *tert*-butyl alcohol $(CH_3)_3COH$. The propyl and butyl alcohols are used as solvents for lacquers and other materials.

The formation of hydrogen bonds by the hydroxyl groups causes the alcohols to have higher melting and boiling points and larger solubility in water than other organic compounds with corresponding molecular mass. The lower alcohols, including *tert*-butanol, are soluble in water in all proportions. The other butanols have limited solubility in water, presumably because their less compact $—C_4H_9$ groups fit less readily than the *tert*-butyl group into the water structure (see the discussion of crystalline hydrates in Chapter 9).

The Polyhydroxy Alcohols

Alcohols containing two or more hydroxyl groups attached to different carbon atoms can be made. They are called the *polyhydroxy alcohols* or *polyhydric* alcohols. *Glycol,*

$$\begin{array}{l} CH_2OH \\ | \\ CH_2OH \end{array}$$

is used as a solvent and as an antifreeze material for automobile radiators. *Glycerol* (glycerine), $C_3H_5(OH)_3$, is a trihydroxypropane, with the structure

$$\begin{array}{c} \quad H \\ \quad | \\ H—C—OH \\ \quad | \\ H—C—OH \\ \quad | \\ H—C—OH \\ \quad | \\ \quad H \end{array}$$

Glycerol is a viscous liquid that is used as an antifreeze material, as a humectant (moistening agent) for tobacco, and especially for use in manufacturing explosives. It reacts with a mixture of nitric acid and sulfuric acid to form the viscous liquid *glyceryl trinitrate* (common name, *nitroglycerine*):

$$\begin{array}{l} CH_2OH \\ | \\ CHOH \\ | \\ CH_2OH \end{array} + 3HONO_2 \xrightarrow[H_2SO_4]{} \begin{array}{l} CH_2ONO_2 \\ | \\ CHONO_2 \\ | \\ CH_2ONO_2 \end{array} + 3H_2O$$

Glyceryl trinitrate is a powerful and treacherous explosive. It was used extensively for blasting and mining in the decades about 1860, despite numerous fatal accidents. Then in 1867 the Swedish industrial chemist Alfred Nobel (1833–1896) discovered that the hazards of handling it would be greatly reduced by mixing it with an absorbent material such as diatomaceous earth, to form the product called *dynamite.* In the year 1876 Nobel also discovered the powerful detonating explosive blasting gelatin, which consists of cellulose nitrate (guncotton) that has soaked up glyceryl trinitrate, and in 1889 he developed the propellant ballistite, a plasticized mixture of cellulose nitrate and glyceryl trinitrate with composition such that it burns smoothly and rapidly and does not detonate.

The Aromatic Alcohols

An example of an aromatic alcohol is *benzyl alcohol,* $C_6H_5—CH_2OH$. In this substance the hydroxyl group is attached to the carbon atom of the alkyl group (methyl group) that is itself attached to the benzene ring. The properties of benzyl alchol and other aromatic alcohols resemble those of the aliphatic alcohols.

The Phenols

A compound in which a hydroxyl group is attached directly to the carbon atom of a benzene ring (or of naphthalene or other aromatic ring system) is called a *phenol.* The simplest phenol is phenol (hydroxybenzene), C_6H_5OH. The three *cresols* (ortho, meta, and para) are 1-hydroxy-2-methylbenzene, 1-hydroxy-3-methylbenzene, and 1-hydroxy-4-methylbenzene, respectively:

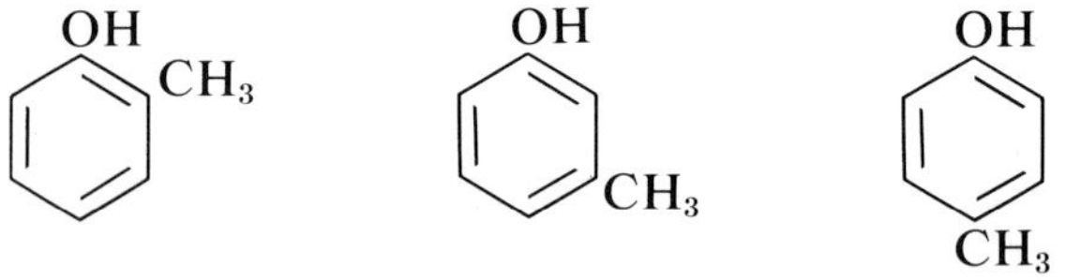

o-Cresol *m*-Cresol *p*-Cresol

They are obtained in the refining of coal tar, and are used as disinfectants and in the manufacture of plastics.

The properties of phenols differ considerably from those of the aliphatic and aromatic alcohols in ways that can be accounted for by the theory of resonance. The main difference is in acid strength: the alcohols (in aqueous solution) have acid constants about 1×10^{-16}, whereas the phenols are about a million times stronger, with acid constants about 1×10^{-10}.

The acid dissociation corresponds to the equilibrium reaction

$$ROH \rightleftarrows RO^- + H^+$$

For an alcohol, such as methanol, the anion RO^- has the electronic structure $H_3C—\ddot{\underset{..}{O}}:^-$. For phenol, however, the phenolate ion can be assigned a structure that is the hybrid of several valence-bond structures:

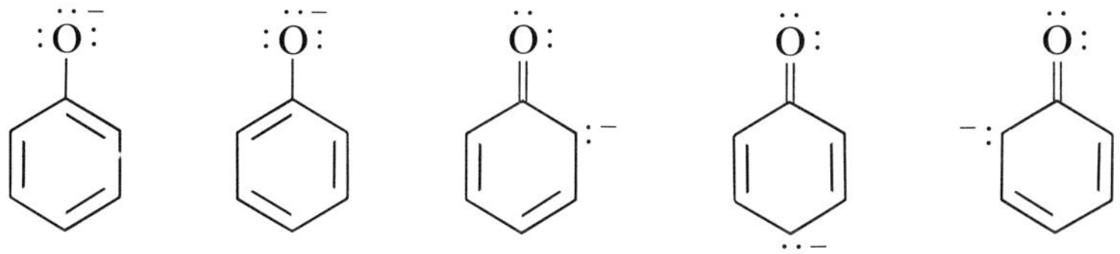

I II III IV V

The resonance energy for these five structures stabilizes the phenolate ion more than the amount by which the undissociated phenol molecule is stabilized by resonance between the two Kekulé structures (with only small contributions by the other three, which involve a separation of charges). The extra stabilization of the anion increases the acid constant; the observed factor 10^6 corresponds to the reasonable value 33 kJ mole^{-1} for the extra resonance energy of the phenolate ion.

13-4. Aldehydes and Ketones

The alcohols and ethers (Section 8-6) represent the first stage of oxidation of hydrocarbons. Further oxidation leads to substances called *aldehydes* and *ketones*. The aldehydes have the formula

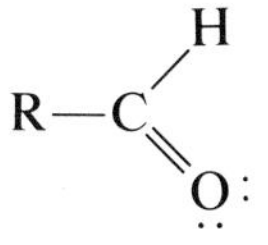

$$R—C\begin{smallmatrix} H \\ \\ \ddot{\underset{..}{O}}: \end{smallmatrix}$$

and the ketones the formula

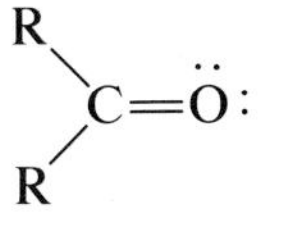

$$\begin{smallmatrix} R \\ \\ R \end{smallmatrix} C=\ddot{O}:$$

The group

$$\rangle C{=}O$$

is called the *carbonyl group*. The substance *formaldehyde,*

$$\begin{array}{l} H \\ \quad \rangle C{=}\ddot{O}: \\ H \end{array}$$

is also classed as an aldehyde. It can be made by passing methyl alcohol vapor and air over a heated metal catalyst:

$$2CH_3OH + O_2 \rightarrow 2HCHO + 2H_2O$$

Formaldehyde is a gas with a sharp irritating odor. It is used as a disinfectant and antiseptic, and in the manufacture of plastics and of leather and artificial silk.

Acetaldehyde, CH_3CHO, is a similar substance made from ethyl alcohol.

The ketones are effective solvents for organic compounds, and are extensively used in chemical industry for this purpose. *Acetone,* $(CH_3)_2CO$, which is dimethyl ketone, is the simplest and most important of these substances. It is a good solvent for nitrocellulose.

Acrolein, $CH_2{=}CHCHO$, is the simplest unsaturated aldehyde. It is a liquid with the characteristic pungent odor of burning fat. It is produced when fats or oils are heated above 300°C, and it can be made by heating glycerol with a dehydrating agent:

$$C_3H_5(OH)_3 \xrightarrow[KHSO_4]{} CH_2CHCHO + 2H_2O$$

Many of the higher aldehydes and ketones have pleasant odors, and some of the aromatic aldehydes are used as flavors. An example is *vanillin,* the fragrant principle of the vanilla bean; its structural formula is

$$\begin{array}{c} OCH_3 \quad \\ | \quad OH \\ \text{(benzene ring)} \\ | \\ HC{=}O \end{array}$$

Vanillin is seen to be a phenol and an aromatic ether as well as an aldehyde. An example of a strongly fragrant ketone is *muscone,* which is

TABLE 13-2
Physical Properties of Some Aldehydes and Ketones

	Melting Point	Boiling Point	Density of Liquid
Formaldehyde, HCHO	−92°C	−21°C	0.82 g ml^{-1}
Acetaldehyde, CH_3CHO	−124°	21°	.782
Propionaldehyde, CH_3CH_2CHO	−81°	49°	.807
n-Butyraldehyde, $CH_3(CH_2)_2CHO$	−98°	76°	.817
Isobutyraldehyde, $(CH_3)_2CHCHO$	−66°	62°	.794
Glyoxal, OHCCHO	15°	50°	1.14
Acrolein, $CH_2{=}CHCHO$	−88°	53°	0.841
Benzaldehyde, C_6H_5CHO	−26°	180°	1.050
Acetone, CH_3COCH_3	−95°	57°	0.792
Methyl ethyl ketone, $CH_3COCH_2CH_3$	−86°	80°	.805
Methyl *n*-propylketone, $CH_3CO(CH_3)_2CH_3$	−79°	102°	.812
Diethyl ketone, $CH_3CH_2COCH_2CH_3$	−42°	103°	.815
Biacetyl, $CH_3COCOCH_3$		88°	.978
Acetylacetone, $CH_3COCH_2COCH_3$	−23°	137°	.976
Acetophenone, $CH_3COC_6H_5$	20°	202°	1.026
Benzophenone, $C_6H_5COC_6H_5$	49°	306°	1.098*

*Density of crystalline substance.

obtained from the scent glands of the male musk deer and is used in perfumes. Its formula is

$$H_3C-CH-CH_2-C{=}O \quad \text{(CH and C joined through } (CH_2)_{12})$$

It contains an unusually large ring (15 carbon atoms).

Physical properties of some aldehydes and ketones are given in Table 13-2.

13-5. The Organic Acids and Their Esters

Acetic acid, CH_3COOH, was mentioned in Section 8-6 as an example of an organic acid. The simplest organic acid is *formic acid,* HCOOH. It can be made by distilling ants, and its name is from the Latin word for ants.

Properties of some of the organic acids are given in Table 13-3. It is seen that the acid constants for the monocarboxylic acids lie in the range

TABLE 13-3
Properties of Some Carboxylic Acids

	Melting Point	Boiling Point	Density of Liquid	pK_a
Formic, $HCOOH$	8°C	101°C	1.226	3.77
Acetic, CH_3COOH	17°	118°	1.049	4.76
Propionic, CH_3CH_2COOH	−22°	141°	0.992	4.88
Butyric, $CH_3(CH_2)_2COOH$	−6°	164°	.959	4.82
Isobutyric, $(CH_3)_2CHCOOH$	−47°	154°	.949	4.85
Valeric, $CH_3(CH_2)_3COOH$	−35°	187°	.942	4.81
Caproic, $CH_3(CH_2)_4COOH$	−1°	205°	.945	4.81
Palmitic, $CH_3(CH_2)_{14}COOH$	64°	380°	.853	
Stearic, $CH_3(CH_2)_{16}COOH$	69°	383°	.847	
Acrylic, $CH_2{=}CHCOOH$	12°	142°	1.062	4.26
Oleic, $CH_3(CH_2)_7CH{=}CH(CH_2)_7COOH$	14°	300°	0.895	
Lactic, $CH_3CHOHCOOH$	18°		1.248	3.87
Oxalic, $HOOCCOOH$	189°			1.46*
Malonic, $HOOCCH_2COOH$	136°		1.631†	2.80*
Succinic, $HOOC(CH_2)_2COOH$	185°		1.564†	4.17*
Benzoic, C_6H_5COOH	122°	249°	1.266†	4.17
Salicylic, *o*-HOC_6H_4COOH	159°		1.443†	3.00

*For first dissociation.
†Density of crystalline substance.

2×10^{-4} to 1×10^{-5} (pK 3.7 to 5). The explanation of the greater acid strength of the —OH group in the carboxylic acids than in the alcohols is given by the theory of resonance; it is similar to that already given (Section 13-3) of the acid strength of the phenols. The dissociation of a carboxylic acid is represented by the equation

$$RCOOH \rightleftarrows RCOO^- + H^+$$

The anion $RCOO^-$ can be assigned two electronic structures:

$$\text{A: } R{-}C(={\ddot{O}}{:}){-}\ddot{\underset{..}{O}}{:}^- \qquad \text{B: } R{-}C(-\ddot{\underset{..}{O}}{:}^-)={\ddot{O}}{:}$$

A B

These two structures are equivalent, and the normal state of the anion can be described as a hybrid structure to which the two valence-bond structures A and B contribute equally. The anion is stabilized by the maximum amount of resonance energy, corresponding to complete resonance between the two valence-bond structures. For the undissociated acid the two valence-bond structures are A′ and B′:

$$\underset{A'}{R{-}C({=}\ddot{O}:){-}\ddot{O}{-}H} \qquad \underset{B'}{R{-}C({-}\ddot{O}:^-){=}O^+{-}H}$$

Structure B′ is less stable than structure A′ because it involves the separation of electric charge, and accordingly the normal state of the acid is a hybrid involving mainly A′, with only a small contribution of B′, and only a small amount of resonance stabilization. The anion is accordingly stabilized by resonance relative to the undissociated acid; this stabilization energy shifts the equilibrium to favor the ion, and thus increases the acid strength. The change in acid constant from about 1×10^{-16} (for alcohols) to 1×10^{-4} corresponds to about 67 kJ mole^{-1} greater resonance energy in the carboxylate anion than in the undissociated acid.

Formic acid and acetic acid are the first two members of a series of carboxylic acids, the *fatty acids*. The next two acids in the series are *propionic acid*, CH_3CH_2COOH, and *butyric acid*, $CH_3CH_2CH_2COOH$. Butyric acid is the principal odorous substance in rancid butter.

Some of the important organic acids occurring in nature are those in which there is a carboxyl group at the end of a long hydrocarbon chain. *Palmitic acid*, $CH_3(CH_2)_{14}COOH$, and *stearic acid*, $CH_3(CH_2)_{16}COOH$, have structures of this sort. *Oleic acid* is similar to stearic acid except that it contains a double bond between two of the carbon atoms in the chain: $CH_3(CH_2)_7CH{=}CH(CH_2)_7COOH$.

Oxalic acid, $(COOH)_2$, is a poisonous substance that occurs in some plants. Its molecule consists of two carboxyl groups bonded together:

$$HO{-}C({=}O){-}C({=}O){-}OH$$

Lactic acid, having the structural formula

$$H_3C{-}\overset{OH}{\underset{H}{C}}{-}COOH$$

contains a hydroxyl group as well as a carboxyl group; it is a hydroxypropionic acid. It is formed when milk sours and when cabbage ferments, and it gives the sour taste to sour milk and sauerkraut. Tartaric acid, which occurs in grapes, is a dihydroxydicarboxylic acid, with the structural formula

```
        H
        |
  HO—C—COOH
        |
  HO—C—COOH
        |
        H
```

Citric acid, which occurs in citrus fruits, is a hydroxytricarboxylic acid, with the formula

```
        H
       HC—COOH
        |
  HO—C—COOH
        |
       HC—COOH
        H
```

Benzoic acid, C_6H_5COOH, is the simplest aromatic acid. It is used in medicine as an antiseptic (in benzoated lard). *Salicylic acid,* which is *o*-hydroxybenzoic acid, *o*-HOC_6H_4COOH, is also used in medicine.

Esters

Esters are the products of reaction of acids and alcohols or phenols. For example, ethyl alcohol and acetic acid react with the elimination of water to produce *ethyl acetate:*

$$C_2H_5OH + CH_3COOH \rightarrow H_2O + CH_3COOC_2H_5$$

Ethyl acetate is a volatile liquid with a pleasing, fruity odor. It is used as a solvent, especially in lacquers.

Many of the esters have pleasant odors, and are used in perfumes and flavorings. The esters are the principal flavorful and odorous constituents of fruits and flowers. Butyl acetate, $CH_3COO(CH_2)_3CH_3$, and amyl acetate, $CH_3COO(CH_2)_4CH_3$, have the odor characteristic of bananas, methyl butyrate, $CH_3(CH_2)_2COOCH_3$, has the odor characteristic of pineapples, and amyl butyrate, $CH_3(CH_2)_2COO(CH_2)_4CH_3$, has that of apricots. Methyl salicylate, *o*-$OHC_6H_4COOCH_3$, is oil of wintergreen.

Fats and Waxes

Fats are the glyceryl esters of fatty acids. Glycerol is the trihydroxy alcohol $CH_2OH—CHOH—CH_2OH$ (Section 13-3). It combines with three molecules of a fatty acid, such as palmitic acid (Table 13-3), to produce a fat:

$$\begin{array}{l} H_2C—OH \\ \quad | \\ HC—OH \\ \quad | \\ H_2C—OH \end{array} + 3HOOC(CH_2)_{14}CH_3 \rightarrow \begin{array}{l} H_2C—OOC(CH_2)_{14}CH_3 \\ \quad | \\ HC—OOC(CH_2)_{14}CH_3 \\ \quad | \\ H_2C—OOC(CH_2)_{14}CH_3 \end{array} + 3H_2O$$

Glycerol   Palmitic acid   Glyceryl tripalmitate

A fat with three fatty acid groups is usually called a triglyceride. The amount of triglycerides in blood plasma is one of the measures of state of health.

The most common saturated fatty acids in fats are the following:

Myristic acid: $H_3C(CH_2)_{12}COOH$

Palmitic acid: $H_3C(CH_2)_{14}COOH$

Stearic acid: $H_3C(CH_2)_{16}COOH$

The most important unsaturated acids in fats are

Oleic acid: $H_3C(CH_2)_7CH=CH(CH_2)_7COOH$

Linoleic acid: $H_3C(CH_2)_4CH=CHCH_2CH=CH(CH_2)_7COOH$

Linolenic acid:

$$H_3CCH_2CH=CHCH_2CH=CHCH_2CH=CH(CH_2)_7COOH$$

Arachidonic acid:

$$H_3C(CH_2)_4CH=CHCH_2CH=CHCH_2CH=CHCH_2CH=CH(CH_2)_3COOH$$

Palmitic acid and oleic acid are important constituents of most fats; the triglycerides often contain one residue of each, with the third position occupied by one of the other fatty acids. The unsaturated fatty acids have the *cis* configuration about the double bonds, causing the hydrocarbon chain to be bent. In consequence the chain cannot fit properly into a crystalline array of the saturated chains, which are most stable in the extended *trans* configuration, and the melting point of the fat is lowered. Fats with a high content of unsaturated fatty acids (polyunsaturated fats) are oils at room temperature. Vegetable oils are converted into solid fats (margarine) by hydrogenation, an addition reaction of hydrogen to the double-bonded carbon atoms.

The fats are important foods (Section 14-8). The use of fats in making soap has been discussed in Section 9-14.

Waxes are esters of fatty acids with alcohols other than glycerol, such as myricyl alcohol (1-triacontanol, $C_{30}H_{61}OH$). Beeswax, from the honeycomb of bees, is chiefly myricyl palmitate. Carnauba wax, used in shoe polish and floor waxes, is a mixture of myricyl alcohol, cerotic acid ($C_{26}H_{53}COOH$), and myricyl cerotate.

13-6. Sugars

The carbohydrates are organic substances in which most of the carbon atoms have an H and an OH group attached. Their formulas usually approximate $(CH_2O)_n$. They occur widely in nature. The simpler carbohydrates are called sugars (monosaccharides and disaccharides), and the complex ones, consisting of very large molecules, are called polysaccharides.

A common simple sugar is D-glucose (also called dextrose and grape sugar), $C_6H_{12}O_6$. It occurs in many fruits, in plants, and in the blood of animals. Its structural formula may be written in three ways:

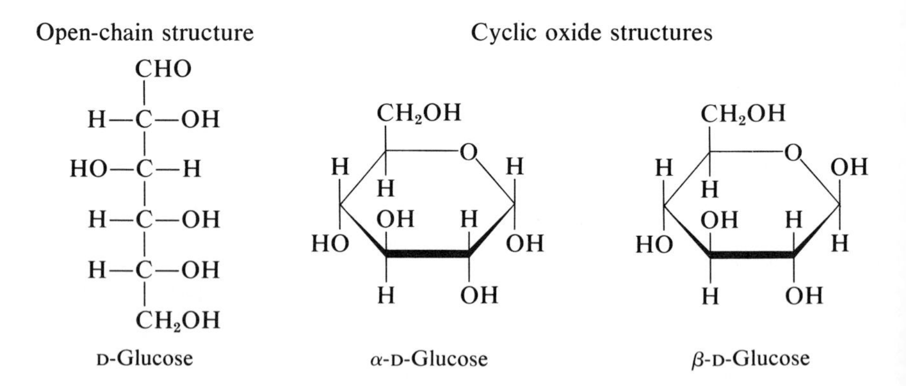

D-Glucose α-D-Glucose β-D-Glucose

The first carbon atom in the open-chain structure is part of an aldehyde group, $-\mathrm{C}\begin{array}{l}\diagup\!\!\!\diagup \mathrm{O}\\ \diagdown \mathrm{H}\end{array}$. The four middle carbon atoms of the chains are asymmetric in the way characteristic of glucose. Other hexoses (monosaccharides with six carbon atoms), such as fructose, mannose, and galactose, have different arrangements of the groups around these carbon atoms:

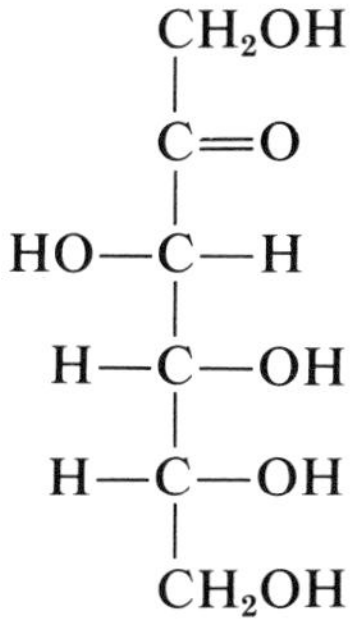

```
    CH2OH              CHO                 CHO
     |                  |                   |
     C=O           HO—C—H               H—C—OH
     |                  |                   |
HO—C—H             HO—C—H              HO—C—H
     |                  |                   |
 H—C—OH              H—C—OH            HO—C—H
     |                  |                   |
 H—C—OH              H—C—OH             H—C—OH
     |                  |                   |
    CH2OH              CH2OH               CH2OH
```

D-Fructose 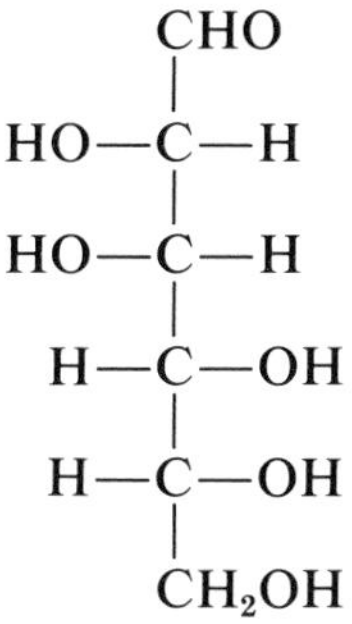 D-Mannose 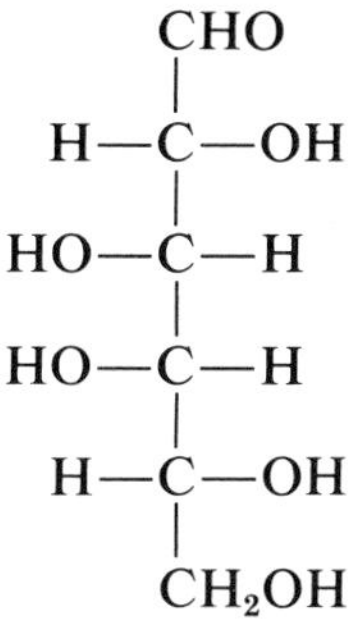 D-Galactose

Fructose also differs from the others in having a ketonic carbonyl group instead of an aldehyde group.

The molecules can easily assume ring structures, as shown above for glucose. The rings are, of course, not planar, but are puckered, as required by the tetrahedral bond angle 110°. The puckered rings have the chair conformations (see also Figure 13-2):

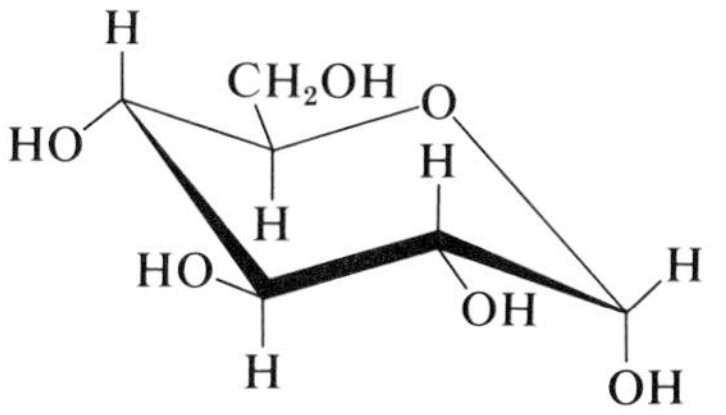

H CH₂OH O HO H H HO H OH H OH

α-D-Glucose

H CH₂OH O HO H H HO OH OH H H

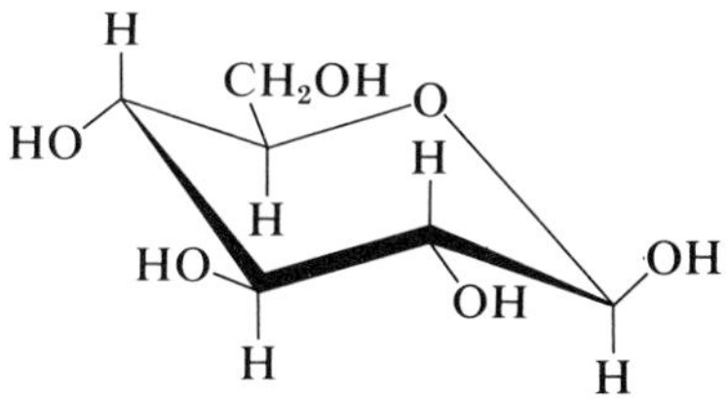

β-D-Glucose

At equilibrium an aqueous solution of glucose at room temperature contains 64 percent of the β ring, 36 percent of the α ring, and about 0.02 percent of the open-chain molecules.

Because of the presence of the aldehyde group or keto group, sugars are reducing agents. A simple test for the disease diabetes mellitus, in which the concentration of glucose in the blood is so large that some of it appears in the urine, is to boil a mixture of the urine with Benedict's solution, which contains a complex of cupric ion, Cu^{++}. The cupric ion is reduced to the univalent state, and appears as a yellow or brick red precipitate of cuprous oxide, Cu_2O. The glucose is oxidized to gluconic acid, $C_6H_{12}O_7$, which has the same structure as glucose except that the aldehyde group, —CHO, has been converted to the carboxyl group,

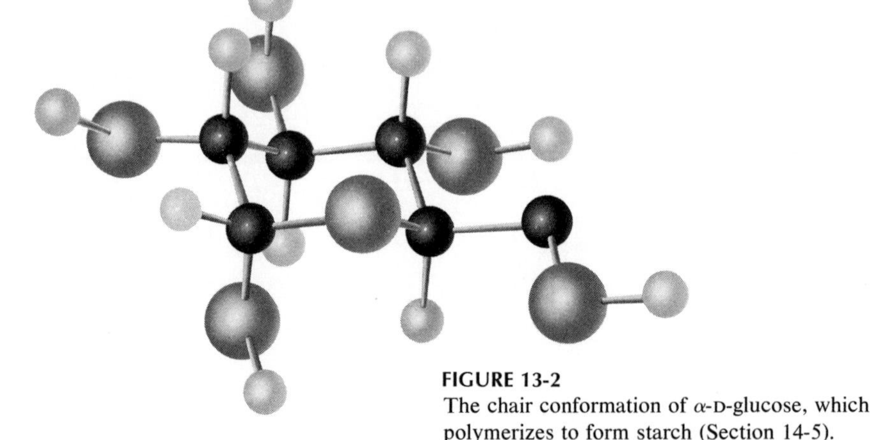

FIGURE 13-2
The chair conformation of α-D-glucose, which polymerizes to form starch (Section 14-5).

—COOH. A strong oxidizing agent, such as nitric acid, converts glucose to glucaric acid, $C_6H_{12}O_8$, which has a carboxyl group at each end.

Some monosaccharides, called *pentoses,* contain only five carbon atoms. The most important pentoses are D-ribose and 2-deoxy-D-ribose, which are present in nucleic acids, D-ribose in the ribonucleic acids (RNA), and 2-deoxy-D-ribose in the deoxyribonucleic acids (DNA—see Section 15-6). These pentoses have the structural formulas shown below. In the nucleic acids they are cyclized, with the α structure.

CHO
H—C—OH
H—C—OH
H—C—OH
CH_2OH
D-Ribose

HOH_2C O H
H H
H OH
OH OH
α-D-Ribose

CHO
CH_2
H—C—OH
H—C—OH
CH_2OH
2-Deoxy-D-ribose

HOH_2C O H
H H
H OH
OH H
α-2-Deoxy-D-ribose

Disaccharides are sugars formed by condensation of two monosaccharides, with elimination of water. The most important disaccharides are sucrose (table sugar), maltose, and lactose, all of which are dihexoses, with formula $C_{12}H_{22}O_{11}$:

Sucrose: D-glucose + D-fructose
Maltose: D-glucose + D-glucose
Lactose: D-galactose + D-glucose

The hexoses have the ring structure, and are linked together by an oxygen atom, as shown in the following diagrams:

Sucrose

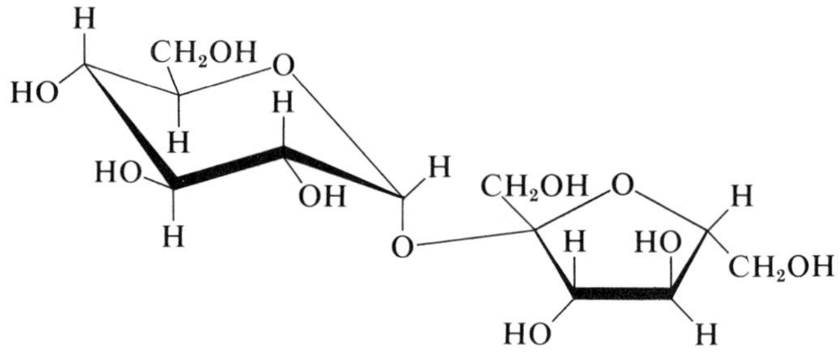

Maltose

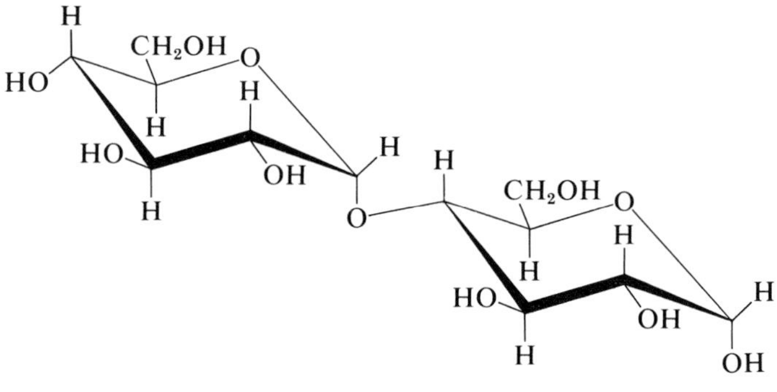

Lactose

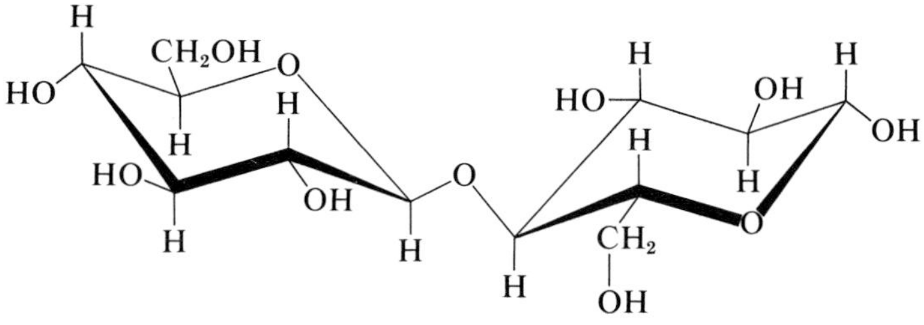

The disaccharides, like the other sugars, have the properties of dissolving readily in water and of crystallizing in rather hard crystals. These properties are attributed to the presence of a number of hydroxyl groups in these molecules, which form hydrogen bonds with water molecules and (in the crystals) with each other.

The sugars have a sweet taste. Fructose is sweeter than sucrose, and glucose is less sweet.

Sucrose is extracted from sugar cane and sugar beets and is purified and sold in very large amounts for use in foods. Maltose (malt sugar) is made by the enzymatic hydrolysis of starch, which is a polyglucose. Lactose is the sugar in milk. It is not as sweet as sucrose.

Polysaccharides, including starch and cellulose, are discussed in Section 14-5.

13.7 Amines and Other Organic Compounds of Nitrogen

The amines are derivatives of ammonia, NH_3, obtained by replacing one or more of the hydrogen atoms by organic radicals. The lighter amines, such as *methylamine,* CH_3NH_2, *dimethylamine,* $(CH_3)_2NH$, and *trimethylamine,* $(CH_3)_3N$, are gases.

Aniline is aminobenzene, $C_6H_5NH_2$. It is a colorless oily liquid, which on standing becomes dark in color because of oxidation to highly colored derivatives. It is used in the manufacture of dyes and other chemicals.

Many substances that occur in plant and animal tissues are compounds of nitrogen. Especially important are the proteins and nucleic acids, which are discussed in the following chapters. The principal product of the metabolism of proteins in the human body is *urea,* $(NH_2)_2CO$. It is the main nitrogenous constituent of urine.

Heterocyclic Nitrogen Compounds. Purines and Pyrimidines

Heterocyclic compounds are cyclic compounds in which one or more atoms other than carbon (usually nitrogen, oxygen, or sulfur) are present in the ring. An example is *pyridine,* C_5H_5N, a colorless liquid with an unpleasant odor, which is among the products of distilling coal. The electronic structure of pyridine can be described as a hybrid of several valence-bond structures:

```
    H              H              H              H              H
    C              C              C+             C              C
  /   \\         //   \         /   \          /   \\         //   \
HC      CH     HC      CH     HC      CH     HC      CH     HC      CH
||      |      |       ||     ||      ||     ||      |      |       ||
HC      CH     HC      CH     HC      CH     HC      +CH    HC+     CH
  \   //         \\   /         \ - /          \ - /          \ - /
    N              N             :N:            :N:            :N:
    ..             ..
```

The resonance energy of pyridine, relative to one of the Kekulé-like structures, is 180 kJ mole^{-1}. Pyridine is a base; in acidic solution it adds a proton to the unshared electron pair of the nitrogen atom, forming the pyridonium ion, $C_5H_5NH^+$.

Six-membered rings containing two or more nitrogen atoms also exist. *Pyrimidine,* $C_4H_4N_2$, is an important example. It is a colorless substance with melting point 22°C and boiling point 124°C. The two nitrogen atoms are in the meta position in the ring. Its electronic structure is a hybrid of

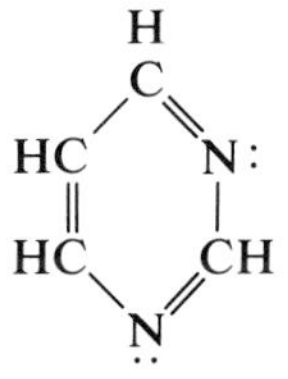

and other valence-bond structures similar to those shown above for pyridine. The partial double-bond character of all of the bonds in the ring requires that the molecules of pyridine and pyrimidine be planar.

The derivatives of pyrimidine, called the pyrimidines, include three substances, thymine, uracil, and cytosine, that are of great importance in the chemistry of heredity. They will be discussed in the following two chapters.

The purines constitute another important class of nitrogen heterocycles. They are the derivatives of the substance *purine,* $C_5H_4N_4$, a colorless crystalline substance with melting point 217°C. The purine molecule is planar; its electronic structure is a hybrid of

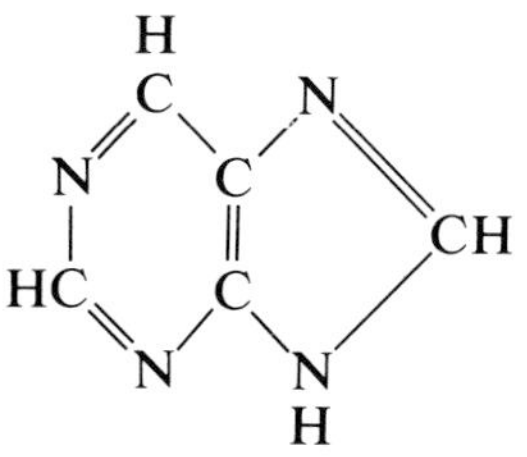

and several other valence-bond structures.

Two of the purines, adenine and guanine, are important in the chemistry of heredity, and will be discussed in the following two chapters.

Caffeine, a stimulant found in coffee, tea, maté, and cola drinks, is a purine. It is a colorless, odorless substance with melting point 236°C. Its structural formula (showing only one of the several valence-bond distributions) is

```
        O     CH3
        ‖     |
  H3C   C     N
     \ / \   / \
      N   C     \
      |   ‖      CH
      C   C     //
    // \ / \   //
   O    N     N
        |
        CH3
```

Pyrrole, Protoporphyrin, Heme

Pyrrole, C_4H_5N, is a colorless liquid, boiling point 130°C, with an unpleasant odor. It is present in coal tar and in the oil obtained by distilling bones. Its structural formula,

```
       H
       N
   HC / ·· \ CH
   ‖         ‖
   HC ——————— CH
```

describes the molecule only partially; resonance with other structures, such as

```
        H
        N+
   HC: /  \\ CH
   |         |
   HC ═══════ CH
```

and

```
        H
        N+
   HC /   \\ CH
   ‖          |
   HC ———————  CH
               ··
```

stabilizes the molecule by 130 kJ mole^{-1} of resonance energy.

The pyrrole ring is especially important because of its presence in the heme group of hemoglobin, in chlorophyll (Section 14-6), and in other substances found in living organisms. Four pyrrole rings are attached to one another by CH groups in the porphyrins. Protoporphyrin has the structural formula

```
               H3C             CH=CH2

               HC—    ..     —CH
H2C=HC                N                CH3
                      H
                 N:        :N

                      H
H3C                   N                CH2CH2COOH
               HC=    ..     =CH

               H3C             CH2CH2COOH
```

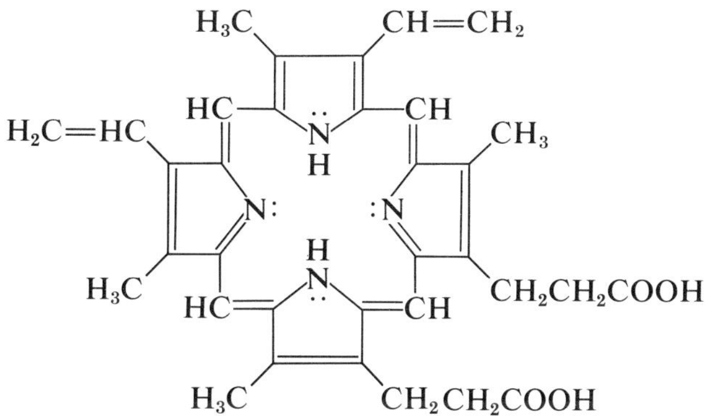

The double bonds resonate among the various positions, requiring the molecule to be planar, and the two central hydrogen atoms may be distributed in various ways among the four nitrogen atoms.

Ferroheme, also called heme, is iron(II) protoporphyrin, with a bipositive iron atom replacing the two central hydrogen atoms. The iron atom is located in or near the plane of the molecule, and it forms bonds with all four nitrogen atoms. Ferroheme is a constituent of hemoglobin and of some other proteins.

Ferroheme is easily oxidized to ferriheme, containing iron(III). Ferriheme chloride (old name hemin chloride) forms black crystals. Protoporphyrin is purple; its color, the black color of hemin salts, and the red color of blood are related to the system of resonating double bonds in the porphyrin structure.

There are many porphyrins, which differ from one another in the nature and disposition of the attached groups (vinyl, propionic acid, and methyl in protoporphyrin). Coproporphyrin I, the principal porphyrin in human feces, has four methyl groups and four propionic acid groups in alternation around the molecule, and uroporphyrin I, sometimes present in human urine, has four acetic acid groups and four propionic acid groups in alternation. In the disease congenital porphyria there is a defect in the enzyme that normally catalyzes the conversion of uroporphyrin to coproporphyrin.

The *barbiturates,* which include several important drugs used as sedatives (tranquilizers) and hypnotics (sleep-producers), are closely related to the pyrimidines. The structural formulas of barbituric acid and two of its derivatives are given below; in these formulas the distribution of the hydrogen atoms between oxygen and nitrogen is uncertain, and only one of several pertinent valence-bond structures is indicated. The mechanism of the physiological action of the drugs is not known in detail. There is, however, little doubt that they act by fitting into receptor sites of molecules in the nerve ends, and forming hydrogen bonds with them.

Barbituric acid

Barbital

Phenobarbital

Alkaloids

Alkaloids are basic (that is, alkali-like) substances of plant origin that contain at least one nitrogen atom, usually in a heterocyclic ring. Most of the alkaloids are physiologically active, and many are useful in medicine. An example is *cocaine,* a powerful local anesthetic and stimulant obtained from coca leaves. Its formula is

Nicotine, $C_{10}H_{14}N_2$, is the principal alkaloid in the tobacco plant. Its formula is

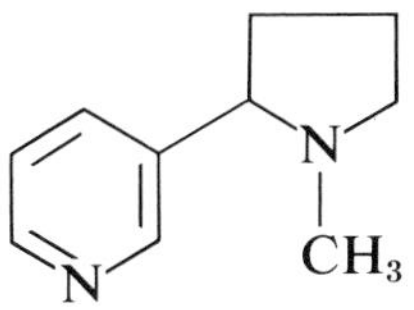

N CH3 N

It is highly toxic and is used as an insecticide. In small quantities it acts as a stimulant and raises the blood pressure. The decreased life expectancy of cigarette smokers is thought to be due in some part to the effect of the inhaled nicotine, which is absorbed into the blood stream, but for the most part to carcinogenic hydrocarbons and other harmful substances in the smoke.

13-8. Fibers and Plastics

Silk and wool are protein fibers, consisting of long polypeptide chains (see Chapter 15). Cotton and linen are polysaccharides (carbohydrates), with composition $(C_6H_{10}O_5)_x$. These fibers consist of long chains made from carbon, hydrogen, and oxygen atoms, with no nitrogen atoms present.

In recent years synthetic fibers have been made, by synthesizing long molecules in the laboratory. One of these, which has valuable properties, is *nylon*. It is the product of condensation of adipic acid and diaminohexane. These two substances have the following structures:

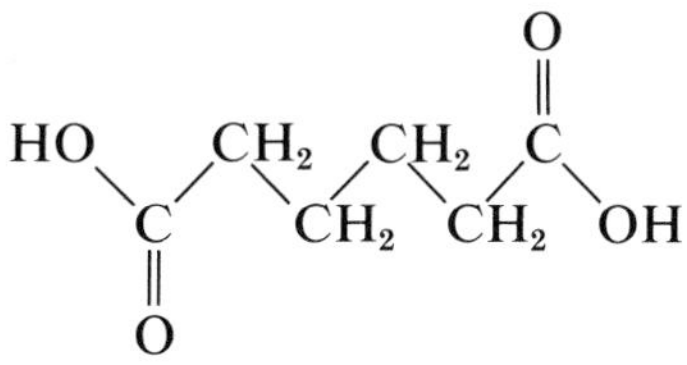

$$HO-C(=O)-CH_2-CH_2-CH_2-CH_2-C(=O)-OH$$

Adipic acid

$$H_2N-CH_2-CH_2-CH_2-CH_2-CH_2-CH_2-NH_2$$

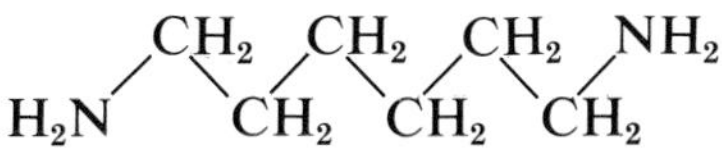

Diaminohexane

Adipic acid is a chain of four methylene groups with a carboxyl group at each end, and diaminohexane is a similar chain of six methylene groups with an amino group at each end. A molecule of adipic acid can react with a molecule of diaminohexane in the following way:

$$
\begin{array}{ccccccccccccc}
 & & & & & & O & & & & & & \\
 & & & & & & \| & & & & & & \\
HO & & CH_2 & & CH_2 & & C & & CH_2 & & CH_2 & & CH_2 \quad NH_2 \\
 & C & & CH_2 & & CH_2 & & NH & & CH_2 & & CH_2 & \quad CH_2 \\
 & \| & & & & & & & & & & & \\
 & O & & & & & & & & & & &
\end{array}
$$

If this process is continued, a very long molecule can be made, in which the adipic acid residues alternate with the diaminohexane residues. Nylon is a fibrous material that consists of these long molecules in approximately parallel orientation.

Other artificial fibers and plastics are made by similar condensation reactions. A *thermolabile plastic* usually is an aggregate of long molecules of this sort that softens upon heating, and can be molded into shape. A *thermosetting plastic* is an aggregate of long molecules containing some reactive groups, capable of further condensation. When this material is molded and heated, these groups react in such a way as to tie the molecules together into a three-dimensional framework, producing a plastic material that cannot be further molded.

With a great number of substances available for use as his starting materials, the chemist has succeeded in making fibers and plastics that are for many purposes superior to natural materials. This field of chemistry, that of synthetic giant molecules, is now advancing rapidly, and we may look forward to further progress in it in the coming years.

EXERCISES

13-1. Define organic chemistry.

13-2. Describe petroleum ether, gasoline, and kerosene in terms of constituents.

13-3. What is a conjugated system of double bonds? As an example, hexadiene contains a chain of six carbon atoms. (*a*) How many hydrogen atoms does it contain? (*b*) The position of a double bond is indicated by the number of the first carbon atom; hexadiene-(1,5) has a double bond at each end. What are the formulas of the other hexadienes? Which ones are conjugated?

13-4. Pentadiene-(1,3)(g) has standard enthalpy of formation 78.0 kJ $mole^{-1}$ (average of 78.2 and 77.8 for *cis* and *trans* isomers), smaller than the value 105.4 kJ $mole^{-1}$ for pentadiene-(1,4)(g). To what do you attribute the greater stability of the 1,3 isomers?

13-5. Adjacent double bonds, as in allene, $H_2C{=}C{=}CH_2$, are unstable. Evaluate the amount of instability, relative to unconjugated double bonds, from the standard enthalpy of formation of pentadiene-(1,2)(g), 129.7 kJ $mole^{-1}$. (Answer: 24.3 kJ $mole^{-1}$)

13-6. Vulcanization is an addition reaction (Section 7-2). Draw an example of part of the chemical structure of vulcanized chloroprene rubber.

13-7. What is the characteristic chemical group of all alcohols? Is it possible to have more than one such group in a molecule?

13-8. Primary and tertiary alcohols have been described. What is a secondary alcohol?

13-9. What is the difference between phenol and benzyl alcohol? Draw chemical structures for them. Estimate their relative acid strengths.

13-10. What are the characteristic chemical groups of aldehydes and ketones? How do they differ?

13-11. Draw structural formulas for acetaldehyde and acetone.

13-12. What is the characteristic chemical group of the organic acids? Compare the acid strengths of aliphatic alcohols, phenols, and organic acids.

13-13. Draw the structural formula of ethyl acetate. What is the characteristic chemical group of an ester?

13-14. Write the equation for the conversion of tristearylglyceride into soap.

13-15. What are the structural differences between waxes and fats?

13-16. Characterize sugars (saccharides) in terms of chemical structure with respect to other substances.

13-17. Draw structural formulas of cyclic fructose, mannose, and galactose.

13-18. From the equilibrium quantities of α, β, and open chain glucose in solution, list them in order of increasing stability.

13-19. What bonds are different in α and β glucose? From values of the bond energy in Appendix V, calculate the expected difference in energy of the two forms. (Note that some difference in energy of hydrogen bonds with water may be expected for the substances in solution.)

14

Biochemistry

In 1806, in his book *Lectures in Animal Chemistry,* the great Swedish chemist Jöns Jakob Berzelius defined organic chemistry as the part of physiology that describes the composition of living bodies (organisms) and the chemical processes that occur in them. It was then thought that organic compounds were the result of "vital forces," and could not be synthesized artificially from inorganic substances. After Wöhler synthesized urea, $(NH_2)_2CO$, from inorganic substances in 1828 this view was given up, and organic chemistry was defined as the chemistry of compounds of carbon. In the course of time the terms biochemistry and physiological chemistry came into use to describe the study of the substances found in living organisms, especially in human beings in good health or suffering from disease, and the chemical reactions that take place in these organisms. During the period since 1940 great progress has been made in determining the detailed molecular structure of many substances present in living organisms and the molecular mechanisms of the processes characteristic of life. This new field of science has become so important that it has been given a name, molecular biology. Both biochemistry and molecular biology have become very large branches of science.

Some aspects of biochemistry are discussed in the following sections of this chapter, and some aspects of molecular biology in the next chapter.

14-1. The Nature of Life

All of our ideas about life involve chemical reactions. What is it that distinguishes a living organism,* such as a man or some other animal or a plant, from an inanimate object, such as a piece of granite? We recognize that the plant or animal may have several attributes that are not possessed by the rock. The plant or animal has, in general, the power of *reproduction* —the power of having progeny, which are sufficiently similar to itself to be recognized as belonging to the same species of living organisms. The process of reproduction involves chemical reactions, the reactions that take place during the growth of the progeny. The growth of the new organism may occur only during a small fraction of the total lifetime of the animal, or may continue throughout its lifetime.

A plant or animal in general has the ability of ingesting certain materials, foods, subjecting them to chemical reactions, involving the release of energy, and secreting some of the products of the reactions. This process, by which the organism makes use of the food that it ingests by subjecting it to chemical reaction, is called *metabolism.*

Most plants and animals have the ability to respond to their environment. A plant may grow toward the direction from which a beam of light is coming, in response to the stimulus of the beam of light, and an animal may walk or run in a direction indicated by increasing intensity of the odor of a palatable food.

To illustrate the difficulty of defining a living organism, let us consider the simplest kinds of matter that have been thought to be alive. These are the *viruses,* such as the tomato bushy stunt virus, of which an electron micrograph has been shown as Figure 2-14. These viruses have the power of reproducing themselves when in the appropriate environment. A single particle (individual organism) of tomato bushy stunt virus, when placed on the leaf of a tomato plant, can cause the material in the cells of the leaf to be in large part converted into replicas of itself. This power of reproduction seems, however, to be the only characteristic of living organisms possessed by the virus. After the particles are formed, they do not grow. They do not ingest food nor carry on any metabolic processes. So far as can be told by use of the electron microscope and by other methods of investigation, the individual particles of the virus are identical with one another, and show no change with time—there is no phenomenon of aging, of growing old. The virus particles seem to have no means of locomotion, and seem not to respond to external stimuli in the way that large living organisms do. But they do have the power of reproducing themselves.

*The word *organism* is used to refer to anything that lives or has ever been living—we speak of dead organisms, as well as of living organisms.

Considering these facts, should we say that a virus is a living organism, or that it is not? At the present time scientists do not agree about the answer to this question—indeed, the question may not be a scientific one at all, but simply a matter of the definition of words. If we were to define a living organism as a material structure with the power of reproducing itself, then we would include the plant viruses among the living organisms. If, however, we require that living organisms also have the property of carrying on some metabolic reactions, then the plant viruses would be described simply as molecules (with molecular mass of the order of magnitude of 10,000,000) that have such a molecular structure as to permit them to catalyze a chemical reaction, in a proper medium, leading to the synthesis of molecules identical with themselves.

14-2. The Structure of Living Organisms

Chemical investigation of the plant viruses has shown that they consist of the materials called *proteins* and *nucleic acids,* the nature of which is discussed in this chapter and the following chapter. The giant virus particles or molecules, with molecular mass of the order of magnitude of 10,000,000, may be described as aggregates of smaller molecules, tied together in a definite way.

Many microorganisms, such as molds and bacteria, consist of single *cells.* These cells may be just big enough to be seen with an ordinary microscope, having diameter around 1 μm (10^{-6} m), or they may be much bigger—as large as a millimeter or more in diameter. The cells have a well-organized structure, consisting of a *cell wall,* a few dozen nanometers in thickness, within which is enclosed a semifluid material called *cytoplasm,* and often other structures that can be seen with the microscope. Other plants and animals consist largely of aggregates of cells, which may be of many different kinds in one organism. The muscles, blood vessel and lymph vessel walls, tendons, connective tissues, nerves, skin, and other parts of the body of a man consist of cells attached to one another to constitute a well-defined structure. In addition there are many cells that are not attached to this structure, but float around in the body fluids. Most numerous among these cells are the *red corpuscles* of the blood. The red corpuscles in man are flattened disks, about 7.5 μm in diameter and 2 μm thick. The number of red cells in a human adult is very large. There are about 5 million red cells per cubic millimeter of blood, and a man contains about 5 liters of blood, that is, 5 million cubic millimeters of blood. Accordingly there are 25×10^{12} red cells in his body. In addition, there are many other cells, some of them small, like the red cells, and some somewhat larger—a single nerve cell may be about 1 μm in diameter and 1 m long, extending from the toe to the spinal cord. The total number of cells in the human body is about 5×10^{14}. The amount of *organization* in the human organism is accordingly very great.

The human body does not consist of cells alone. In addition there are the *bones*, which have been laid down as excretions of bone-making cells. The bones consist of inorganic constituents, calcium hydroxyphosphate, $Ca_5(PO_4)_3OH$, and calcium carbonate, and an organic constituent, *collagen,* which is a protein. The body also contains the body fluids blood and lymph, as well as fluids that are secreted by special organs, such as saliva and the digestive juices. Very many different chemical substances are present in these fluids.

The structure of cells is determined by their framework materials, which constitute the cell walls and, in some cases, reinforcing frameworks within the cells. In plants the carbohydrate cellulose is the most important constituent of the cell walls. In animals the framework materials are proteins. Moreover, the cell contents consist largely of proteins. For example, a red cell is a thin membrane enclosing a medium that consists of 60% water, 5% miscellaneous materials, and 35% *hemoglobin,* an iron-containing protein, which has molecular mass 68,000, and has the power of combining reversibly with oxygen. It is this power that permits the blood to combine with a large amount of oxygen in the lungs, and to carry it to the tissues, making it available there for oxidation of foodstuffs and body constituents. It has been mentioned earlier in this section that the simplest forms of matter with the power of reproducing themselves, the viruses, consist in part of proteins, as do also the most complex living organisms.

14-3. Amino Acids and Proteins

Proteins may well be considered the most important of all the substances present in plants and animals. Proteins occur either as separate molecules, usually with very large molecular mass, ranging from about 10,000 to many millions, or as reticular constituents of cells, constituting their structural framework. The human body contains many thousands of different proteins, which have special structures that permit them to carry out specific tasks.

All proteins are nitrogenous substances, containing approximately 16% of nitrogen, together with carbon, hydrogen, oxygen, and often other elements such as sulfur, phosphorus, iron (four atoms of iron are present in each molecule of hemoglobin), and copper.

The framework proteins of the body are called *fibrous proteins*. The principal fibrous proteins are *keratin,* a protein found in hair, fingernail, and muscle, and also in the horns, quills, and feathers of animals, and *collagen,* the protein of tendon, skin, bone, and the connective tissue between cells. When collagen is boiled it hydrolyzes to a soluble protein called gelatin. The soluble proteins in the body are called *globular proteins*. The *albumins,* such as serum albumin in the blood, ovalbumin in egg white, and lactalbumin in milk, are soluble in cold water as well as in dilute salt solution. The *globulins,* such as serum globulin and fibrinogen in blood, egg globulin in egg white, and lactoglobulins in milk, are soluble in dilute salt solution but not in cold water.

Amino Acids

When proteins are heated in acidic or basic solution they undergo hydrolysis, producing substances called amino acids. Amino acids are carboxylic acids in which one hydrogen atom has been replaced by an amino group, $—NH_2$. The amino acids that are obtained from proteins are *alpha* amino acids, with the amino group attached to the carbon atom next to the carboxyl group (this carbon atom is called the alpha carbon atom). The simplest of these amino acids is *glycine,* $CH_2(NH_2)COOH$. The other natural amino acids contain another group, usually called R, in place of one of the hydrogen atoms on the alpha carbon atom, their general formula thus being $CHR(NH_2)COOH$.

The amino group is sufficiently basic and the carboxyl group is sufficiently acidic that in solution in water the proton is transferred from the carboxyl group to the amino group. The carboxyl group is thus converted into a carboxyl ion, and the amino group into a substituted ammonium ion. The structure of glycine and of the other amino acids in aqueous solution is accordingly the following:

```
 H  H     O
  \ |     ‖
H—N+     C—O-
    \   /
      C
    /   \
   H     R
```

The amino groups and carboxyl groups of most amino acids dissolved in animal or plant liquids, which usually have pH about 7, are internally ionized in this way, to form an ammonium ion group and a carboxyl ion group within the same molecule.

There are twenty-three amino acids that have been recognized as important constituents of proteins. Their names are given in Table 14-1, together with the formulas of the characteristic group R. Some of the amino acids have an extra carboxyl group or an extra amino group. There is one double amino acid, *cystine,* which is closely related to a simple amino acid, *cysteine.* Four of the amino acids contain heterocyclic rings —rings of carbon atoms and one or more other atoms, in this case nitrogen atoms. Two of the amino acids given in the table, *asparagine* and *glutamine,* are closely related to two others, *aspartic acid* and *glutamic acid,* differing from them only in having the extra carboxyl group changed into an amide group,

$$-C\begin{matrix} \nearrow O \\ \searrow NH_2 \end{matrix}$$

Proteins are important constituents of food. They are digested by the digestive juices in the stomach and intestines, being split in the process of digestion into small molecules, probably mainly the amino acids themselves. These small molecules are able to pass through the walls of the stomach and intestines into the blood stream, by which they are carried around into the tissues, where they may then serve as building stones for the manufacture of the body proteins. Sometimes people who are ill and cannot digest foods satisfactorily are fed by the injection of a solution of amino acids directly into the blood stream. A solution of amino acids for this purpose is usually obtained by hydrolyzing proteins.

Right-handed and Left-handed Amino-acid Molecules

It was pointed out in Section 6-4 that some substances exist in two isomeric (enantiomeric) forms, called L (levo) and D (dextro) forms, with molecules that are mirror images of one another. These two forms exist for every amino acid except glycine; they differ from one another in the arrangement in space of the four groups attached to the α-carbon atom. Figure 14-1 shows the two enantiomers of the amino acid alanine, in which R is the methyl group, CH_3.

A most extraordinary fact is that only one of the two enantiomers of each of the amino acids has been found to occur in plant and animal proteins, and that this enantiomer has the same configuration for all of these amino acids; that is, the hydrogen atom, carboxyl ion group, and ammonium ion group occupy the same position relative to the group R around the alpha carbon atom. This configuration is called the L configuration—*proteins are built entirely of* L*-amino acids.*

This is a very puzzling fact. Nobody knows why it is that we are built of L-amino acid molecules, rather than of D-amino acid molecules. All the proteins that have been investigated, obtained from animals and from plants, from higher organisms and from very simple organisms—bacteria, molds, even viruses—are found to have been made of L-amino acids.*

*Residues of D-amino acids are found in a few simple peptides in living organisms.

TABLE 14-1
The Principal Amino Acids Occurring in Proteins

	MONOAMINOMONOCARBOXYLIC ACIDS	
Gly	Glycine, aminoacetic acid	$-R = -H$
Ala	Alanine, α-aminopropionic acid	$-CH_3$
Ser	Serine, α-amino-β-hydroxypropionic acid	$-CH_2OH$
Thr	Threonine, α-amino-β-hydroxybutyric acid	$-CH(CH_3)(OH)$
Met	Methionine, α-amino-γ-methylmercapto-butyric acid	$-CH_2-CH_2-S-CH_3$
Val	Valine, α-amino-isovaleric acid	$-CH(CH_3)(CH_3)$
Leu	Leucine, α-amino-isocaproic acid	$-CH_2-CH(CH_3)(CH_3)$
Ile	Isoleucine, α-amino-β-methylvaleric acid	$-CH(CH_2-CH_3)(CH_3)$
Phe	Phenylalanine, α-amino-β-phenylpropionic acid	$-CH_2-C_6H_5$ (benzene ring with H, H, H, H, H)
Tyr	Tyrosine, α-amino-β-(para-hydroxy-phenyl)-propionic acid	$-CH_2-C_6H_4-OH$ (benzene ring with H, H, H, H, OH)
Cys	Cysteine, α-amino-β-sulfhydrylpropionic acid	$-CH_2-SH$
	MONOAMINODICARBOXYLIC ACIDS	
Asp	Aspartic acid, aminosuccinic acid	$-CH_2-COOH$
Glu	Glutamic acid, α-aminoglutaric acid	$-CH_2-CH_2-COOH$
	Hydroxyglutamic acid, α-amino-β-hydroxyglutaric acid	$-CH(CH_2-COOH)(OH)$

	DIAMINOMONOCARBOXYLIC ACIDS	
Arg	Arginine, α-amino-δ-guanidinovaleric acid	$-CH_2-CH_2-CH_2-NH-C(=NH)-NH_2$
Lys	Lysine, α,ϵ-diaminocaproic acid	$-CH_2-CH_2-CH_2-CH_2-NH_2$
	A DIAMINODICARBOXYLIC ACID	
Cys	Cystine, di-β-thio-α-aminopropionic acid	$-CH_2-S-S-CH_2-$
	AMINO ACIDS CONTAINING HETEROCYCLIC RINGS	
His	Histidine, α-amino-β-imidazolepropionic acid	$-CH_2-C$ (ring: C=CH–NH–CH=N–C)
Pro	Proline, 2-pyrrolidinecarboxylic acid*	H_2N^+ ring: H_2C, H_2C, CH_2, $C(H)-C(=O)O^-$
Hyp	Hydroxyproline, 4-hydroxy-2-pyrrolidinecarboxylic acid*	H_2N^+ ring: H_2C, $HC(OH)$, CH_2, $C(H)-C(=O)O^-$
Trp	Tryptophan, α-amino-β-indolepropionic acid†	benzene ring fused: $C(CH_2-)=CH-N(H)$
	AMINO ACIDS CONTAINING AN AMIDE GROUP	
Asn	Asparagine, aminosuccinic acid monoamide	$-CH_2-C(=O)-NH_2$
Gln	Glutamine, α-aminoglutaric acid monoamide	$-CH_2-CH_2-C(=O)-NH_2$

*The formulas given for proline and hydroxyproline are those of the complete molecules, and not just of the groups R.

†The hexagon represents a benzene ring.

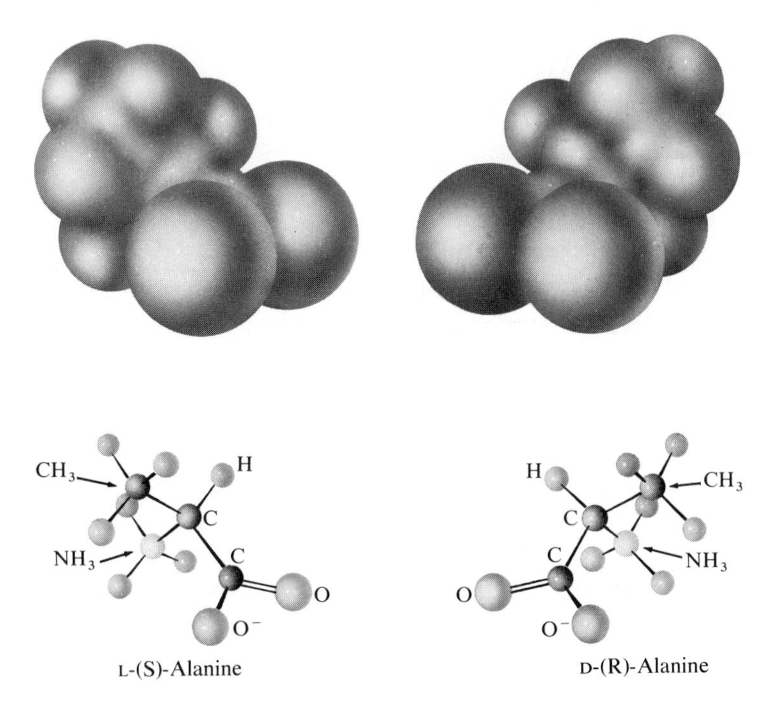

FIGURE 14-1
The two enantiomers of the amino acid alanine.

Right-handed molecules and left-handed molecules have exactly the same properties, so far as their interaction with ordinary substances is concerned—they differ in their properties only when they interact with other right-handed or left-handed molecules. The earth might just as well be populated with living organisms made of D-amino acids as with those made of L-amino acids. A man who was suddenly converted into an exact mirror image of himself would not at first know that anything had changed about him, except that he would write with his left hand, instead of his right, his hair would be parted on the right side instead of the left, his heartbeat would show his heart to be on the right side, and so on; he could drink water, inhale air and use the oxygen in it for combustion, exhale carbon dioxide, and carry on other bodily functions just as well as ever—so long as he did not eat any ordinary food. If he were to eat ordinary plant or animal food, he would find that he could not digest it.* He could be kept alive only on a diet containing synthetic D-amino acids, made in

*Alice: "Perhaps Looking-glass milk isn't good to drink." In *Through the Looking-Glass*, by Lewis Carroll (Charles Lutwidge Dodgson), 1872.

the chemical laboratory. He could not have any children, unless he could find a wife who had been subjected to the same process of reflection into a mirror image of her original self. We see that there is the possibility that the earth might have been populated with two completely independent kinds of life—plants, animals, human beings of two kinds, who could not use one another's food, could not produce hybrid progeny.

No one knows why living organisms are constructed of L-amino acids. We have no strong reason to believe that molecules resembling proteins could not be built up of equal numbers of right-handed and left-handed amino acid molecules. Perhaps the protein molecules that are made of amino acid molecules of one sort only are especially suited to the construction of a living organism—but if this is so, we do not know why.*

Nor do we know why it is that living organisms have evolved in the L-system rather than in the D-system. The suggestion has been made that the first living organism happened by chance to make use of a few molecules with the L configuration, which were present with D molecules in equal number; and that all succeeding forms of life that have evolved have continued to use L-amino acid molecules through inheritance of the character from the original form of life. Perhaps a better explanation than this can be found—but we do not know what it is.

The Essential Amino Acids

Although all the amino acids listed in Table 14-1 are present in the proteins of the human body, not all of them need to be in the food. Experiments have been carried out that show that nine of the amino acids are essential to man. These *nine essential amino* acids are histidine, isoleucine, leucine, lysine, methionine, phenylalanine, threonine, tryptophan, and valine. The human body is able to manufacture the others, which are called the nonessential amino acids. The minimum amounts of amino acids required by young men were determined by the American biochemist W. C. Rose. He found that when the daily intake of any one of the eight amino acids listed above (not including histidine) falls below a certain value the subject begins to excrete more nitrogen compounds than he ingests; his body proteins are being broken down faster than they are being synthesized. The young men differed from one another in their requirements over about a two-fold range; for example, between 0.4 and 0.8 g per day for lysine. The minimum daily requirements listed by Rose are the highest values for any of his subjects. There is no doubt that human beings differ from one another in their genetic character and

*A possible reason is that there would be serious overcrowding of the side chains for a mixture of D and L residues in the alpha helix (Chapter 15).

accordingly also in their biochemical characteristics. The values given in Table 14-2 are twice the values given by Rose. Presumably these amounts are enough to prevent the deterioration of body proteins in most persons (99 percent). The requirements of women are about two-thirds as great.

The ninth amino acid listed above, histidine, seems not to be required for adults but is required for the growth of infants in amount more than 30 mg per kg body weight per day.

Protein foods for man may be classed as *good protein foods,* those that contain all of the essential amino acids, and *poor protein foods,* those that are lacking in one or more of the essential amino acids. *Casein,* the principal protein in milk, is a good protein, from this point of view, whereas *gelatin,* a protein obtained by boiling bones and tendons (partial hydrolysis of the insoluble protein collagen produces gelatin) is a poor protein. Gelatin contains no tryptophan, no valine, and little or no threonine.

The amounts of the essential amino acids in eight representative foods are given in Table 14-3. It is seen that plant foods contain a smaller amount (25 to 30 g per 100 g of protein) of the essential amino acids than animal foods (36 to 42 percent). Comparison with Table 14-2 shows that 100 g of mixed protein per day provides about twice the recommended daily intake of all of the essential amino acids except methionine. A deficiency in methionine is partially met by the intake of cystine, which is present in amount about 2 g per 100 g of animal protein and 1 g per 100

TABLE 14-2
Requirements of Amino Acids for Adult Men and Amino-acid Contents of American Diets

	Recommended daily intake*	Amount in average daily diets†
Isoleucine	1.40 g	0.7 to 5.7 g
Leucine	2.20	1.3 to 7.8
Lysine	1.60	1.3 to 8.6
Methionine‡	2.20	0.7 to 3.0
Phenylalanine§	2.20	0.9 to 5.0
Threonine	1.00	0.9 to 3.8
Tryptophan	0.50	0.4 to 1.3
Valine	1.60	0.8 to 5.4

*Values are twice the minimum values given by Rose for men. Values for women are somewhat smaller.

†Ranges reported in four investigations in 1952 and 1953.

‡Three-quarters of the methionine can be replaced by four times as much cystine.

§Three quarters of the phenylalanine can be replaced by tyrosine.

TABLE 14-3
Essential Amino Acids in Proteins of Some Foods*

	Raw Lean Beef	Whole Eggs	Fish	Cow's Milk	Pecans	Whole Grain Flour	Raw Potatoes	Peas	Average
His	3.1	2.1	2.8	2.4	2.2	1.7	1.3	1.4	2.1
Ile	4.6	5.8	4.4	5.7	4.4	3.5	3.8	4.0	4.5
Leu	7.2	7.7	6.6	8.7	6.1	5.5	4.4	5.5	6.5
Lys	7.6	5.6	7.8	6.9	3.4	2.2	4.7	4.1	5.3
Met	2.2	2.7	2.5	2.2	1.2	1.2	1.1	0.7	1.7
Phe	3.6	5.1	3.2	4.3	4.6	4.0	3.9	3.4	4.0
Thr	3.9	4.4	3.8	4.1	3.1	2.4	3.4	3.2	3.5
Try	1.0	1.4	0.8	1.3	1.1	1.0	0.9	0.7	1.0
Val	4.9	6.5	4.7	6.1	4.1	3.8	4.7	3.6	4.8
Sum	38.1	41.3	36.6	41.7	30.2	25.3	28.2	26.6	33.4
Protein†	21.6%	12.9%	20%	3.5%	9.3%	13.5%	2.1%	3.4%	

*The amounts are in grams per 100 g of protein (about 115 g of all amino acids).
†The amount of protein in the food.

g of plant protein. It is likely that a daily intake of 1 g of protein (somewhat less of high-quality protein) per kg body weight is sufficient for good health, but somewhat more may be desirable for optimum health.

Some organisms that we usually consider to be simpler than man have greater powers than the human organism in that they are able to manufacture all of the amino acids from inorganic constituents. The red bread mold, Neurospora, has this power. There is an evolutionary advantage for the organism in getting rid of the chemical machinery (enzymes) for manufacturing vital substances that are available in the food supply. The larger vertebrates, such as the dog and the rat, require not only the essential amino acids required by man, but also another one, arginine.

The Primary Structure of Proteins

During the past century much effort has been devoted by scientists to the problem of the structure of proteins. This is a very important problem, and much progress has been made in attacking it. We now have a much better understanding than formerly of the nature of physiological reactions, and the new knowledge about the structure of protein molecules is helping in the attack on important medical problems, such as the control of heart disease and cancer.

In the period between 1900 and 1910 strong evidence was obtained by the German chemist Emil Fischer (1852–1919) to show that the amino acids in proteins are combined into long chains, called *polypeptide chains.* For example, two molecules of glycine can be condensed together, with elimination of water, to form the double molecule glycyglycine, shown in Figure 14-2. The bond formed in this way is called a *peptide bond.* The process of forming these bonds can be continued, resulting in the production of a long chain containing many amino-acid residues, as shown in Figure 14-2.

Chemical methods have been developed to determine how many polypeptide chains there are in a protein molecule. These methods involve the use of a reagent (fluorodinitrobenzene) that combines with the free amino group of the amino acid residue at the end of the polypeptide chain to form a colored complex, which can be isolated and identified after the protein has been hydrolyzed into its constituent amino acids, including the end amino acid with the colored group attached. The amino acids are then separated from one another by the chromatographic technique (Figure 14-3), and the end amino acid, with its colored group attached, is identified. The protein may be partially hydrolyzed into peptides, each containing several amino-acid residues. These peptides are then isolated by chromatography, and their amino-acid sequences are determined. For example, in his work on insulin the English biochemist Fred Sanger (born 1918) obtained the following dipeptides, among others:

Ile-Val
Val-Glu
Glu-Glu
Glu-Cys

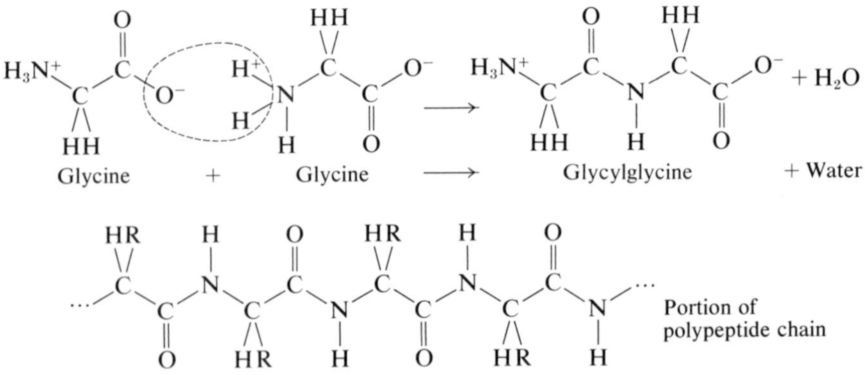

Glycine + Glycine ⟶ Glycylglycine + Water

Portion of polypeptide chain

FIGURE 14-2
The condensation of glycine to a dipeptide, glycylglycine; the structural formula of a portion of a polypeptide chain.

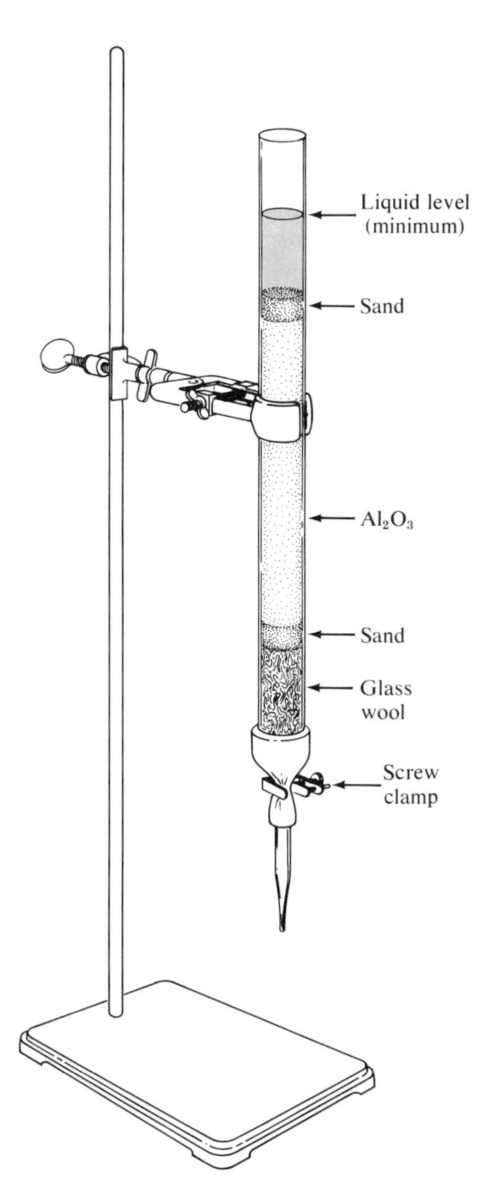

FIGURE 14-3
Apparatus for column chromatography. The vertical glass tube contains a column of fine particles of a substance that attaches solute molecules to its surface. Different solute molecules (amino acids, for example) are adsorbed more or less strongly. A solution of the solutes is added at the top of the column. A solvent is then added. As it slowly runs through the column the different adsorbed substances move downward at different rates. If they are colored they may be seen as separated bands (hence the name chromatography). Fractions of solvent containing the different solutes may be collected at the bottom. Filter-paper chromatography and gas-liquid chromatography are also important techniques.

These dipeptides overlap to indicate the sequence Ile-Val-Glu-Glu-Cys. He also obtained the tetrapeptide with Gly as the labeled residue (amino end of a chain) and Ile, Val, and Glu, in undetermined order. He concluded that insulin contains a chain with initial sequence Gly-Ile-Val-Glu-Glu-Cys.

The order of amino acid residues in the polypeptide chains (called the *primary structure*) was first determined (in this way) for the protein insulin. The insulin molecule has molecular mass 5,733 d. It consists of two polypeptide chains, of which one contains 21 amino acid residues and the

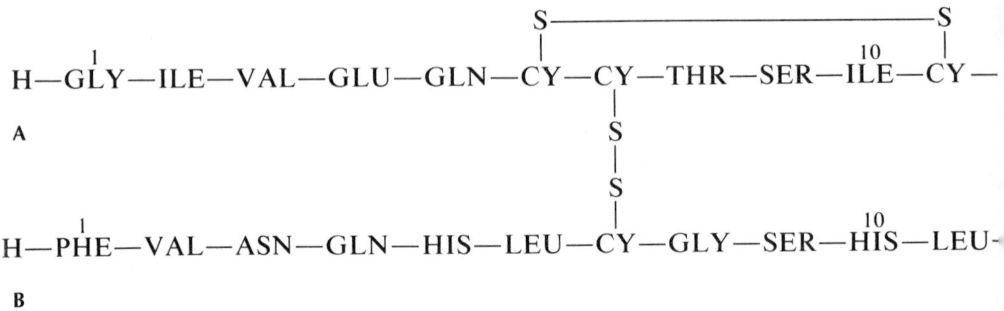

```
                                          S————————————————————S
         1                                |               10   |
H—GLY—ILE—VAL—GLU—GLN—CY—CY—THR—SER—ILE—CY—
                                             |
A                                            S
                                             |
                                             S
       1                                     |             10
H—PHE—VAL—ASN—GLN—HIS—LEU—CY—GLY—SER—HIS—LEU-

B
```

FIGURE 14-4
The primary structure of human insulin. H indicates the free amino end and OH the carboxylate end of the chains.

other contains 30. The sequences of amino acids in the short chain and the long chain were determined in the years between 1945 and 1952 by Sanger and his collaborators. The chains are attached to one another by sulfur-sulfur bonds, between the halves of cystine residues.

The sequence for human insulin is shown in Figure 14-4. Note that the symbol Cy—S is used for the half-molecule of cystine. There is a sulfur-sulfur bond between the sixth residue and the eleventh residue in the A chain, producing a ring, and there are two sulfur-sulfur bonds between the A chain and the B chain.

Animal insulins have somewhat different primary structure from human insulin. The animal insulins used for injection to control diabetes are pig insulin and ox insulin. The A chain of pig insulin is identical with the human A chain, and the B chain differs only in one residue (Ala in place of Thr in the 30th position). Ox insulin has the same B chain as pig insulin, and the A chain differs from the human A in having Ala-Ser-Val in place of Thr-Ser-Ile in positions 8 to 10. Elephant insulin has the same B chain as human insulin, and the A chain differs in two positions, 9 and 10, with Gly-Val in place of Ser-Ile. Dog insulin is identical with pig insulin, differing from human in only the last position in the B chain.

Insulin is a hormone. Its physiological activity is discussed in Section 14-10.

The Denaturation of Proteins

Proteins such as insulin and hemoglobin have certain special properties that make them valuable to the organism. Insulin is a hormone that assists in the process of oxidation of sugar in the body. Hemoglobin has the power of combining reversibly with oxygen, permitting it to attach oxygen molecules to itself in the lungs, and to liberate them in the tissues. These well-defined properties show that the protein molecules have very definite structures.

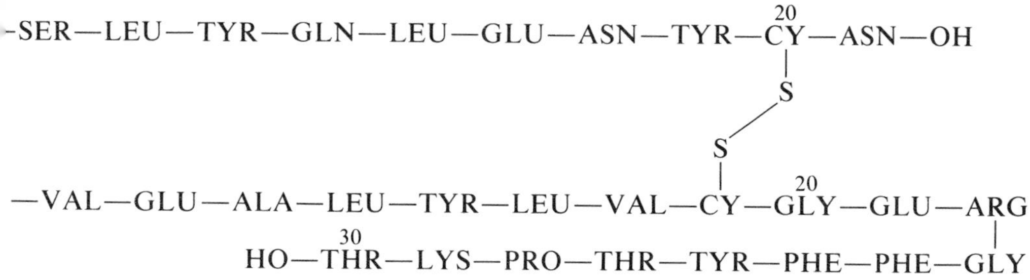

```
                                                         20
-SER—LEU—TYR—GLN—LEU—GLU—ASN—TYR—CY—ASN—OH
                                                         |
                                                         S
                                                        /
                                                      S
                                                      |     20
—VAL—GLU—ALA—LEU—TYR—LEU—VAL—CY—GLY—GLU—ARG
                    30                                            |
              HO—THR—LYS—PRO—THR—TYR—PHE—PHE—GLY
```

A protein that retains its characteristic properties is called a *native protein:* hemoglobin as it exists in the red cell or in a carefully prepared hemoglobin solution, in which it still has the power of combining reversibly with oxygen, is called native hemoglobin. Many proteins lose their characteristic properties very easily. They are then said to have been *denatured.* Hemoglobin can be denatured simply by heating its solution to 65°C. It then coagulates, to form a brick-red insoluble coagulum of denatured hemoglobin. Most other proteins are also denatured by heating to approximately this temperature. Egg white, for example, is a solution consisting mainly of the protein *ovalbumin,* with molecular mass 43,000. Ovalbumin is a soluble protein. When its solution is heated for a little while at about 65°C the ovalbumin is denatured, forming an insoluble white coagulum of denatured ovalbumin. This phenomenon is observed when an egg is cooked.

It is believed that the process of denaturation involves uncoiling the polypeptide chains from the characteristic structure of the native protein. In the coagulum of denatured hemoglobin or denatured ovalbumin the uncoiled polypeptide chains of different molecules of the protein have become tangled up with one another in such a way that they cannot be separated; hence the denatured protein is insoluble. Some chemical agents, including strong acid, strong alkali, and alcohol, are good denaturing agents.

14-4. Enzymes

Enzymes are substances formed by living organisms that have catalytic power; that is, the power to increase the rate of certain chemical reactions. Enzyme action is involved in the processes of fermentation in the preparation of wine, vinegar, beer, and bread, which have been known and practised since prehistoric times. In 1680 Antony Leeuwenhoek used his microscope to observe yeast cells and bacteria, but he did not recognize them as living organisms. In 1857 Louis Pasteur showed that yeast is a living organism and recognized that fermentation is a physiological process. In 1897 E. Buchner showed that yeast cells are not needed for fermentation. He extracted the cells and obtained a cell-free solution that had enzymic activity. The word enzyme means "in yeast".

It was not until 1926 that enzymes were shown to be proteins. In that year James B. Sumner (1887–1955) of Cornell University purified and crystallized the enzyme urease from jack beans. Urease is a protein that catalyzes the hydrolysis of urea:

$$CO(NH_2)_2 + H_2O \rightarrow CO_2 + 2NH_3$$

Its molecular mass is 480,000 d, the molecule being composed of six subunits. Its amino-acid sequence has not yet been determined.

The Mechanism of Catalysis by Enzymes

In Section 10-5 it was pointed out that a catalyst increases the rate of a reaction (and also that of the reverse reaction) by decreasing the activation energy of the reaction. It accomplishes this end by interacting more strongly with the complex molecule representing the structure half way between the reactant molecules and the product molecules than it does with either the reactant molecules or the product molecules. Let us consider the hydrolysis of urea, catalyzed by urease. The urea molecule is planar; its electronic structure is a hybrid of the following three structures:

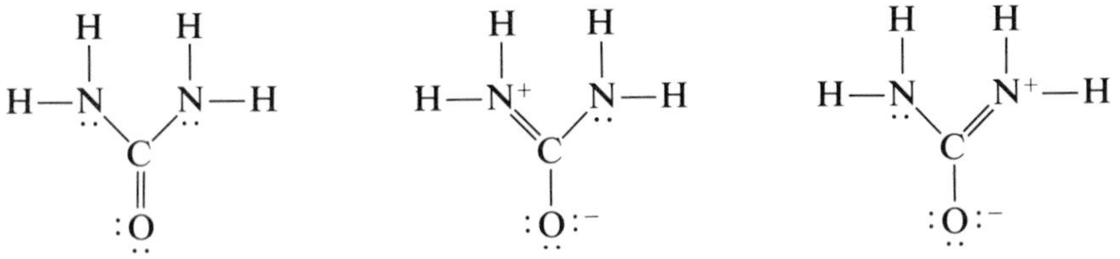

```
   H   H              H   H              H   H
   |   |              |   |              |   |
 H—N   N—H         H—N⁺   N—H         H—N    N⁺—H
   ..\ /..            \\ /..            ..\ //
     C                  C                  C
     ||                 |                  |
    :O                 :O:⁻               :O:⁻
     ..                 ..                 ..
```

In order to form the products $CO_2 + 2NH_3$ a water molecule must approach it in such a way that its oxygen atom is attached to the carbon atom of urea on the side nearly opposite the oxygen of urea and each of the two hydrogen atoms must attach themselves to a nitrogen atom. The intermediate complex probably involves a bent urea molecule, with the carbon atom pushed up from the NNO plane and the four hydrogen atoms pulled down from this plane. This bending could be facilitated by the use of NH or OH groups of the enzyme to form hydrogen bonds with the urea oxygen atom and of oxygen atoms of the enzyme to form hydrogen bonds with the two urea nitrogen atoms, while a non-hydrogen-bond-forming group (CH_2, for example) of the enzyme pushed the urea carbon atom up. The water molecule might also be held by the enzyme in a suitable position adjacent to the strained urea molecule. At body temperature a decrease in activation energy by 5.93 kJ mole^{-1} leads to a tenfold increase in rate of reaction (Section 10-4). The energy of an O—H···O or N—H···O hydrogen bond is about 20 kJ mole^{-1}. It is accordingly not unreasonable that the strain produced by several hydrogen bonds could amount to 25 or 30 kJ mole^{-1}, causing a 10,000- or 100,000-fold increase in reaction rate, as is observed for some enzyme-catalyzed reactions.

The region to which the reactant molecules (forming the intermediate complex) attach themselves is called the active region of the enzyme. For urease it need not be very large—only a little larger than the urea molecule. We may ask why the enzyme itself is so large, containing thousands of atoms. Part of the answer probably is that the atoms constituting the active region need to be held tightly in the proper positions, and the other atoms of the enzyme molecule provide reinforcement for them. Furthermore, the local atomic environment of many enzyme reactions is very different (hydrophobic, for example) from the aqueous environment in which enzymes operate and the large enzyme molecule provides this change in local environment.

Conjugated Proteins. Coenzymes

Some proteins consist solely of polypeptide chains. They are called simple proteins. Many, however, consist of polypeptide chains and other molecules. The human hemoglobin molecule, for example, consists of four polypeptide chains and four hemes, the heme being an organic ion compound. Such proteins are called *conjugated proteins,* and the non-protein part is called the *prosthetic group.*

Many enzymes are conjugated proteins, formed by combination of a simple protein, called the apoenzyme (apo means off, separated from), and one or more other molecules or ions, called coenzymes. Some coenzymes are metal ions. An example is the carbonic anhydrase in human erythrocytes. The apoenzyme has molecular mass 28,000 d. It combines with one zinc ion, Zn^{++}, to form the active enzyme, which catalyzes the decomposition of carbonic acid to water and carbon dioxide. Another example is the amylase in human saliva, which helps digest starch. It has molecular mass 50,000 d, and requires one calcium ion, Ca^{++}, as coenzyme. Many enzymes have vitamins or derivatives of vitamins as coenzymes.

Kinetics of Enzyme Reactions

Many enzyme-catalyzed reactions take place at rates that can be described by a simple equation, the Michaelis-Menten equation (Equation 14-1). A reactant in such a reaction is usually called the substrate. Let us assume that reaction occurs through combination of the enzyme E and the substrate S to form the complex ES, which can then decompose in two ways, forming either E and S again, or E and the products P:

$$\mathrm{E} + \mathrm{S} \underset{}{\overset{K}{\rightleftarrows}} \mathrm{ES} \xrightarrow{k} \mathrm{E} + \mathrm{P}$$

If the rate-constant k is small, the concentration of ES approximates closely the equilibrium value:

$$\frac{[\mathrm{ES}]}{[\mathrm{E}][\mathrm{S}]} = K \tag{14-1}$$

The rate of reaction is proportional to [ES], and hence to [E][S]:

$$-\frac{d[\mathrm{S}]}{dt} = k[\mathrm{ES}] = kK[\mathrm{E}][\mathrm{S}] \tag{14-2}$$

The total enzyme concentration is $[\mathrm{E}]_{\mathrm{total}} = [\mathrm{E}] + [\mathrm{ES}]$. From Equation 14-1 we see that

$$[\mathrm{E}] = \frac{[\mathrm{E}]_{\mathrm{total}}}{K[\mathrm{S}] + 1}$$

and hence Equation 14-2 may be rewritten as

$$\mathrm{Rate} = -\frac{d[\mathrm{S}]}{dt} = \frac{k[\mathrm{S}][\mathrm{E}]_{\mathrm{total}}}{[\mathrm{S}] + (1/K)} \tag{14-3}$$

From this equation we see that when [S] is small compared with K^{-1} the rate is proportional to [S], being $kK[\mathrm{S}][\mathrm{E}]_{\mathrm{total}}$, and when [S] is so large as to saturate the enzyme the rate is independent of [S], being just $k[\mathrm{E}]_{\mathrm{total}}$. Some calculated curves are shown in Figure 14-5.

It is interesting to compare the curve for $K = 0.5$ with the experimental curve of Figure 14-6, which shows the rate of growth of a mutant of red bread mold along a glass tube containing the nutrient solution as a function of the concentration of a growth substance, para-aminobenzoic acid, in the solution. The similarity in shape indicates that an equilibrium corresponding to Equation 14-1 determines the rate of growth of the organism.

It is possible that the equilibrium involved is not that between an

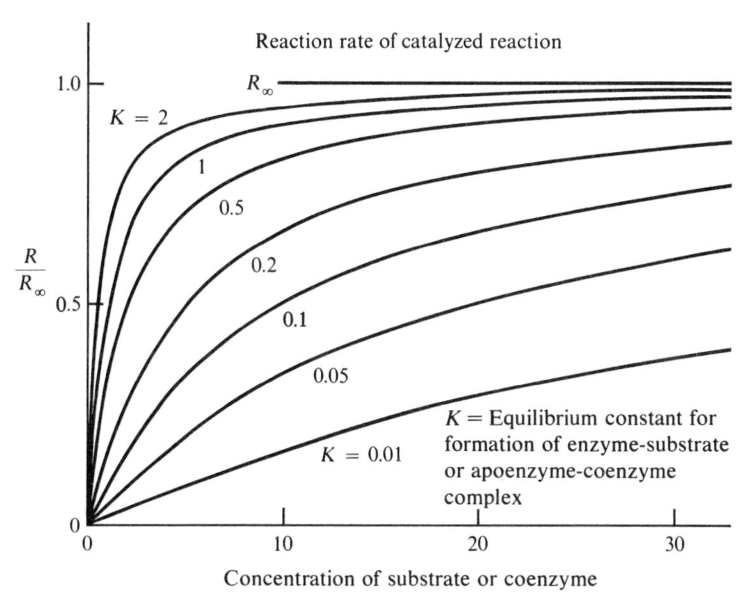

FIGURE 14-5
Curves showing calculated reaction rate R/R_∞ of a catalyzed reaction as function of the concentration of the substrate or coenzyme, for different values of the equilibrium constant K for formation of the enzyme-substrate or apoenzyme-coenzyme complex.

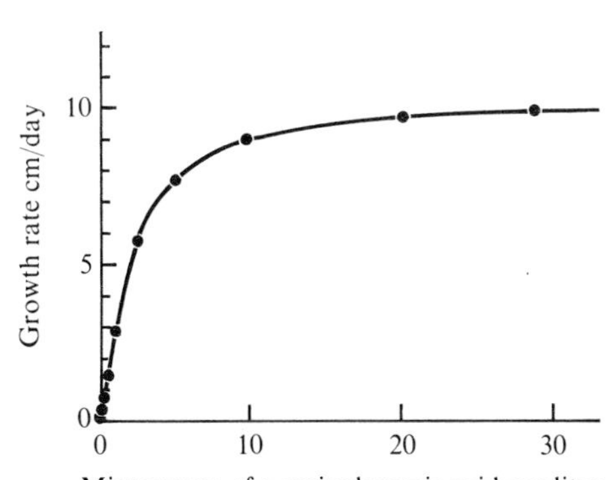

FIGURE 14-6
The observed rate of growth of a *p*-aminobenzoic-acid-requiring *Neurospora* mutant (Tatum and Beadle, 1942), as function of the concentration of the growth substance in the medium.

enzyme and a substrate but rather that between an apoenzyme (a protein without enzymatic activity) and a coenzyme (often a vitamin) to form the active enzyme:

$$A + C \rightleftarrows E$$

$$\frac{[E]}{[A][C]} = K' \qquad (14\text{-}4)$$

You can verify that the rate of the reaction is affected by the concentration of the coenzyme in the same way as by the concentration of the substrate.

Many diseases are known to involve gene mutations that decrease the activity of an important enzyme in the human body. One way in which this occurs is through the formation of an altered apoenzyme with a greatly decreased value of the combining constant K' with its coenzyme. One disease of this sort is cystathioninuria, which is recognized by the appearance of cystathionine, a derivative of the amino acid methionine (see Table 14-1), in the urine of the patient. The disease causes mental retardation and unpleasant physical manifestations. The enzyme involved in the metabolism of cystathionine has pyridoxine (vitamin B_6) as its coenzyme, which usually is ingested in the amount of about 1 mg per day. It has been found that ingestion of 100 times this amount leads to the disappearance of cystathionine from the urine and alleviation of other manifestations of the disease, presumably by shifting the equilibrium of Equation 14-4 in the direction of the active enzyme. This example illustrates the use of the principles of chemical equilibrium and chemical kinetics in the treatment of disease.

Enzyme Inhibitors

An enzyme inhibitor is a substance that combines with an enzyme and makes it inactive. An inhibitor may be competitive with the substrate or noncompetitive. If it is competitive, the amount of inhibition is determined by the ratio of the concentrations of inhibitor and substrate. The molecules of inhibitor and substrate compete in attaching themselves to the active region of the enzyme.

An example of competitive inhibition is the inhibition of succinate dehydrogenase by malonate ion. The catalyzed reaction is the oxidation of succinate ion to fumarate ion:

$$^{-}OOC{-}CH_2{-}CH_2{-}COO^{-} + \tfrac{1}{2}O_2 \rightarrow {}^{-}OOC{-}CH{=}CH{-}COO^{-} + H_2O$$

Malonate ion, $^{-}OOC—CH_2—COO^{-}$, has its two carboxylate groups closer together than succinate ion. It has been found by experiment that the equilibrium constant for binding malonate ion is nearly the same as that for binding succinate ion (about one third, corresponding to a difference in free energy of combination of 2.8 kJ mole^{-1}, which is small compared with the total free energy of combination with either ion). We conclude that the binding sites of the enzyme for the two carboxylate groups are at a distance intermediate between the distances for the unstrained succinate and malonate ions.

Most powerful poisons are lethal because they inhibit enzymes required for life. Hydrogen cyanide and hydrogen sulfide liberate ions, CN^- and HS^-, that combine with ferric iron atoms in the cytochromes, which are heme enzymes that catalyze the cellular oxidation reactions required for life. Many drugs are enzyme inhibitors.

Often the products of an enzyme-catalyzed reaction are themselves inhibitors of the enzyme. This provides a valuable feedback mechanism, preventing the concentration of the product from becoming too large.

Sometimes the action of an enzyme is controlled by a substance other than the substrate or products. These enzymes are called *allosteric* enzymes. The site on the enzyme molecule that reacts with the allosteric control molecule is different from the active reaction site of the enzyme. The separate allosteric control sites and active sites have been observed by x-ray diffraction in the enzyme aspartate transcarbamylase, an enzyme of molecular mass 310,000 d which consists of six subunits.

14-5. Polysaccharides

The simpler carbohydrates, called sugars, have been discussed in Section 13-6. Those with larger molecules are called polysaccharides.

Important polysaccharides include *starch, glycogen,* and *cellulose.* Starch, $(C_6H_{10}O_5)_x$, occurs in plants, mainly in their seeds or tubers. It is an important constituent of foods. Glycogen, $(C_6H_{10}O_5)_x$, is a substance similar to starch, which occurs in the blood and the internal organs, especially the liver, of animals. Glycogen serves as a reservoir of readily available food for the body; whenever the concentration of glucose in the blood becomes low, glycogen is rapidly hydrolyzed into glucose.

Cellulose, which also has the formula $(C_6H_{10}O_5)_x$, is a stable polysaccharide that serves as a structural element for plants, forming the walls of cells.

These three polysaccharides consist of glucose rings, thousands in a chain, held together in different ways. In starch the residues are oriented similarly, held by oxygen bonds between carbon atoms 1 and 4, on opposite sides of the ring, as shown in Section 13-6 for maltose (Figure 14-7). In glycogen some of the glucose residues are attached to one another by oxygen bonds between carbon atoms 1 and 6. In both starch and glycogen the rings are α rings. Cellulose has a structure like that of starch, with 1,4 bonds between the rings, but the rings have the β structure (Figure 14-7). Man produces enzymes that digest starch, but not cellulose.

Dextrins are polysaccharides of intermediate molecular mass produced by partial hydrolysis of starch. They are used in making tablets of drugs, in sizing paper and fabrics, in printing inks and glues, and many other ways. The mucilage on postage stamps is dextrin.

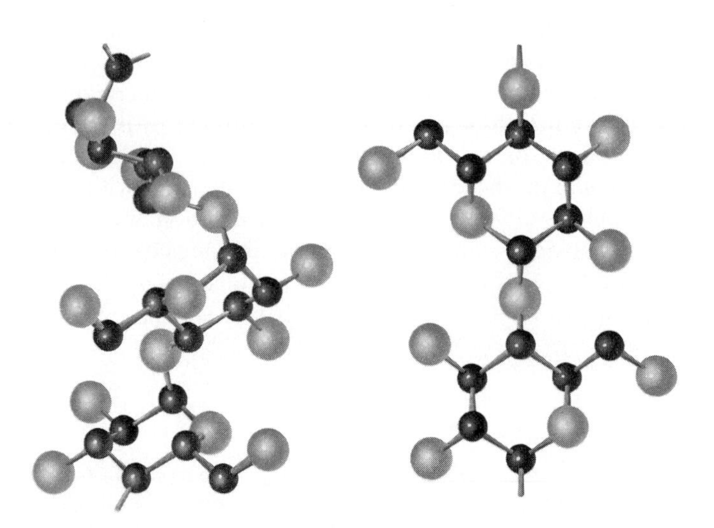

FIGURE 14-7
Starch and cellulose *(left and right)* are giant molecules formed from α-D-glucose and β-D-glucose, respectively. In this illustration hydrogen atoms have been omitted for clarity.

14-6. Photosynthesis and ATP

The carbohydrates (sugars and starch) are important foods, providing most of the energy required by the human body (Section 14-8). The energy for the synthesis of sugars, starch, and cellulose by plants is obtained from sunlight. The process, called *photosynthesis,* involves the green substance chlorophyll, a compound of magnesium. The formula of chlorophyll a is

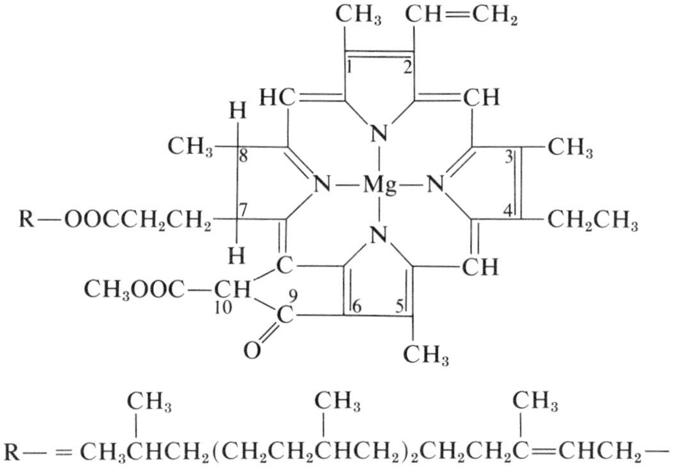

```
                         CH3  CH=CH2
                          |     |
                          |1   2|
              H   HC======        =====CH
              |           \      /       |
     CH3 -----8-----|       N          |-----3------- CH3
              |      \\     |          //   ||
              |        N — Mg — N           ||
              |      /      |          \    ||
R—OOCCH2CH2 --7-----|        N          |----4------- CH2CH3
              |     ||     /     \      ||
              H      C-----        -----CH
                   /        ||    ||
   CH3OOC — CH    9         ||6  5||
            10   \ C--------        
                  //              |
                 O               CH3

              CH3                 CH3                     CH3
               |                   |                       |
R— = CH3CHCH2(CH2CH2CHCH2)2CH2CH2C=CHCH2—
```

$$R— = CH_3\overset{\displaystyle CH_3}{\overset{|}{C}}HCH_2(CH_2CH_2\overset{\displaystyle CH_3}{\overset{|}{C}}HCH_2)_2CH_2CH_2\overset{\displaystyle CH_3}{\overset{|}{C}}{=}CHCH_2—$$

Chlorophyll a

Chlorophyll b has a similar structure, differing in that the methyl group in the 3 position is replaced by a CHO group.

The overall reaction of photosynthesis is

$$6CO_2 + 6H_2O \longrightarrow \underset{\text{Glucose}}{C_6H_{12}O_6} + 6O_2$$

The enthalpy increase for this reaction is 2816 kJ per mole of glucose (the heat of combustion of glucose). Chlorophyll has principal absorption bands in the wave-length regions around 425 and 660 nm, corresponding to energy of about 280 and 180 kJ per mole of photons, respectively. Several photons are accordingly required for each mole of carbon dioxide converted to glucose. The process of photosynthesis is a very complicated one.

This important process produces about 10^{12} kg of starch and cellulose per year, and liberates about 1.1×10^{12} kg of oxygen into the atmosphere per year, replacing the oxygen used in respiration and combustion.

It has been found that one primary step in photosynthesis is the conversion of ADP (adenosine diphosphate) to ATP (adenosine triphosphate):

$$ADP + H_3PO_4 + \text{energy} \longrightarrow ATP + H_2O$$

The amount of energy required for this reaction is about 30 kJ $mole^{-1}$. ATP is a high-energy molecule, relative to ADP (Section 8-4). After its formation, it can hydrolyze to ADP and phosphoric acid, by the reverse of the above reaction, releasing its energy to permit a sequence of endergonic reactions to take place that leads ultimately to the synthesis of glucose.

ATP is the triphosphate of the nucleotide adenosine, which is composed of the purine adenine and the sugar D-ribose (Section 13-6). Adenine is one of the units of DNA (Section 15-6). The formula of ATP is

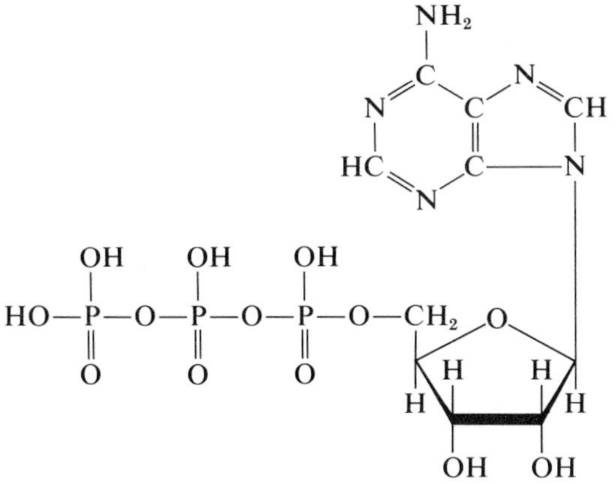

The reactions of photosynthesis can be carried out in the dark by cells of green plants if they are provided with a supply both of ATP and of another substance, NADPH (reduced nicotinamide-adenine dinucleotide phosphate), which has the formula

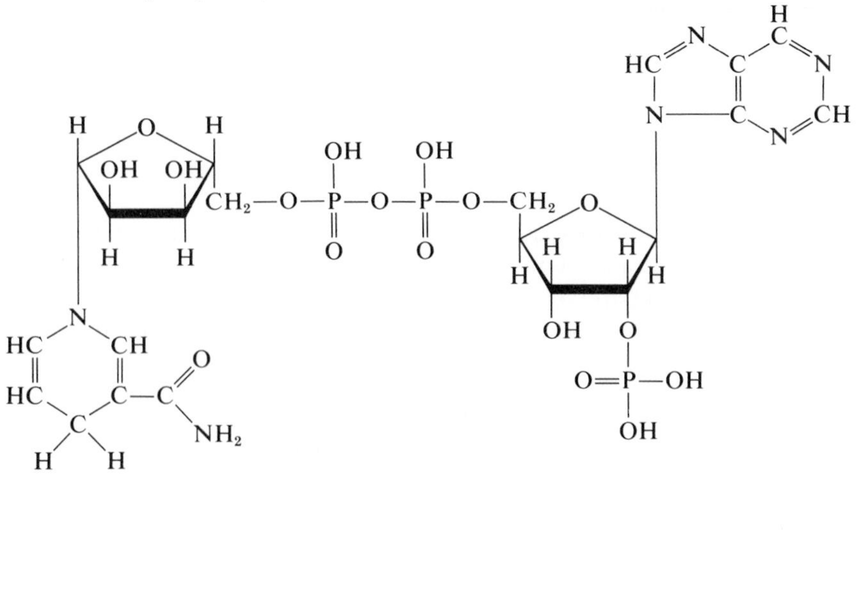

NADPH is a reducing agent; it can be oxidized to $NADP^+$ (nicotinamide-adenine dinucleotide cation, with change in the ring at the left (the reduced nicotinamide ring):

```
        |                               |
        N                               N+
      /   \                           //  \
   HC       CH                      HC      CH
   ||       ||     O                 |      ||     O
   HC       C — C//    + ½O2 →      HC      C — C//     + OH-
     \     /      \                   \\   /      \
        C          NH2                  C          NH2
      /   \                             H
     H     H
```

Apparently the light absorbed by the green plant converts ADP to ATP and $NADP^+$ to NADPH. ATP, by serving as a source of energy, and NADPH, as a reducing agent (a source of hydrogen atoms), can then carry on the reactions that lead to the conversion of water and carbon dioxide to glucose:

$$2NADP^+OH^- + \text{light} \longrightarrow 2NADPH + O_2$$

$$2NADPH + H_2O + CO_2 \xrightarrow[ATP]{} 2NADP^+OH^- + \frac{1}{n}(CH_2O)_n$$

In the cells of the human body and of other animals a reaction occurs that is the reverse of photosynthesis, the oxidation of glucose. In this reaction the liberated energy is used to convert ADP to ATP, and probably also to favor the reduction of $NADP^+$ to NADPH. These energy-rich molecules then serve to fuel the many physiological mechanisms of the body.

14-7. The Citric Acid Cycle

In the period between 1935 and 1950 there was discovered a principal way in which the oxidation of carbohydrates to water and carbon dioxide is carried out with production of a number of high-energy molecules for each molecule of carbon dioxide formed. This biochemical mechanism is called the *citric acid cycle* or the *Krebs cycle*. It was in large part formulated by 1943 by the British biochemist Hans Adolf Krebs (born 1900), after Albert Szent-Györgyi in 1935 had discovered that enzymes from muscle could catalyze the oxidation of dicarboxylic four-carbon acids (succinic, fumaric, malic, and oxaloacetic acid).

The cycle involves the addition of two carbon atoms (acetic acid,

CH_3COOH) to oxaloacetic acid, $HOOCCH_2COCOOH$, to form citric acid, a six-carbon tricarboxylic acid:

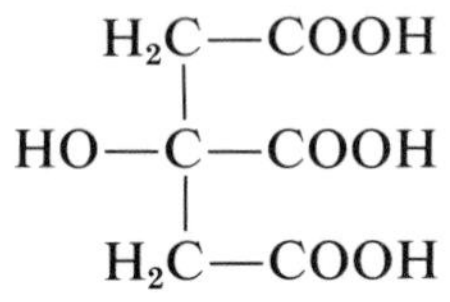

$$\begin{array}{c} H_2C\text{—}COOH \\ | \\ HO\text{—}C\text{—}COOH \\ | \\ H_2C\text{—}COOH \end{array}$$

The citric acid is then converted to a four-carbon dicarboxylic acid by a series of six steps, in two of which a molecule of carbon dioxide is released. The four-carbon acid is converted to oxaloacetic acid in three additional steps. All of these steps are catalyzed by specific enzymes, and in each of several of them a high-energy molecule is produced. In this way a considerable fraction of the very large amount of energy (470 kJ per mole of CO_2) released in the oxidation of glucose is made available for various purposes.

Acetic acid itself (acetate ion) cannot enter the citric acid cycle. It was discovered in 1950 by the American biochemist Fritz Lipmann (born 1899) that the form in which it does enter the cycle is as a compound with a complex substance called coenzyme A, abbreviated as CoASH (it contains the sulfhydryl group SH). The compound is acetyl-SCoA, $H_3C\text{—}C(\text{=O})\text{—}SCoA$. CoASH has the structure

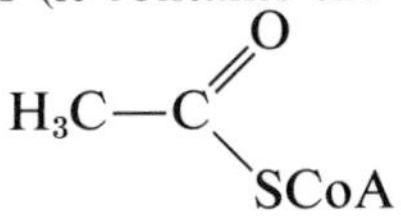

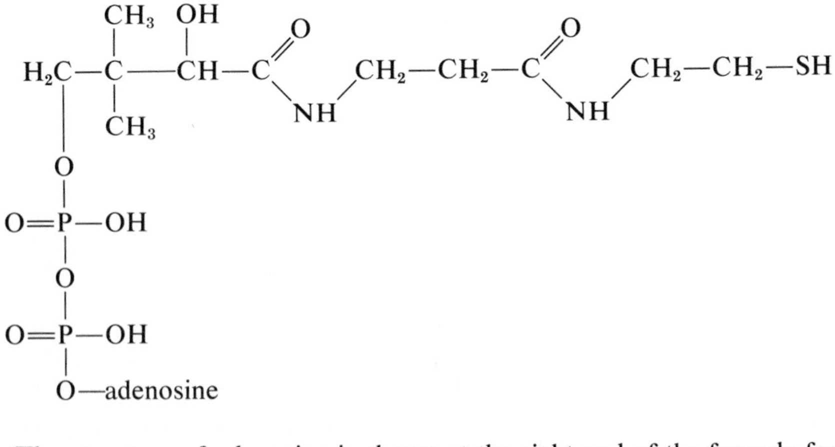

$$H_2C(\text{—}O\text{—}P(\text{=O})(OH)\text{—}O\text{—}P(\text{=O})(OH)\text{—}O\text{—adenosine})\text{—}C(CH_3)_2\text{—}CH(OH)\text{—}C(\text{=O})\text{—}NH\text{—}CH_2\text{—}CH_2\text{—}C(\text{=O})\text{—}NH\text{—}CH_2\text{—}CH_2\text{—}SH$$

(The structure of adenosine is shown at the right end of the formula for NAPDH in the preceding section.)

The citric acid cycle in its present form is shown in Figure 14-8. (There is the possibility that some refinements remain to be made.) Each step

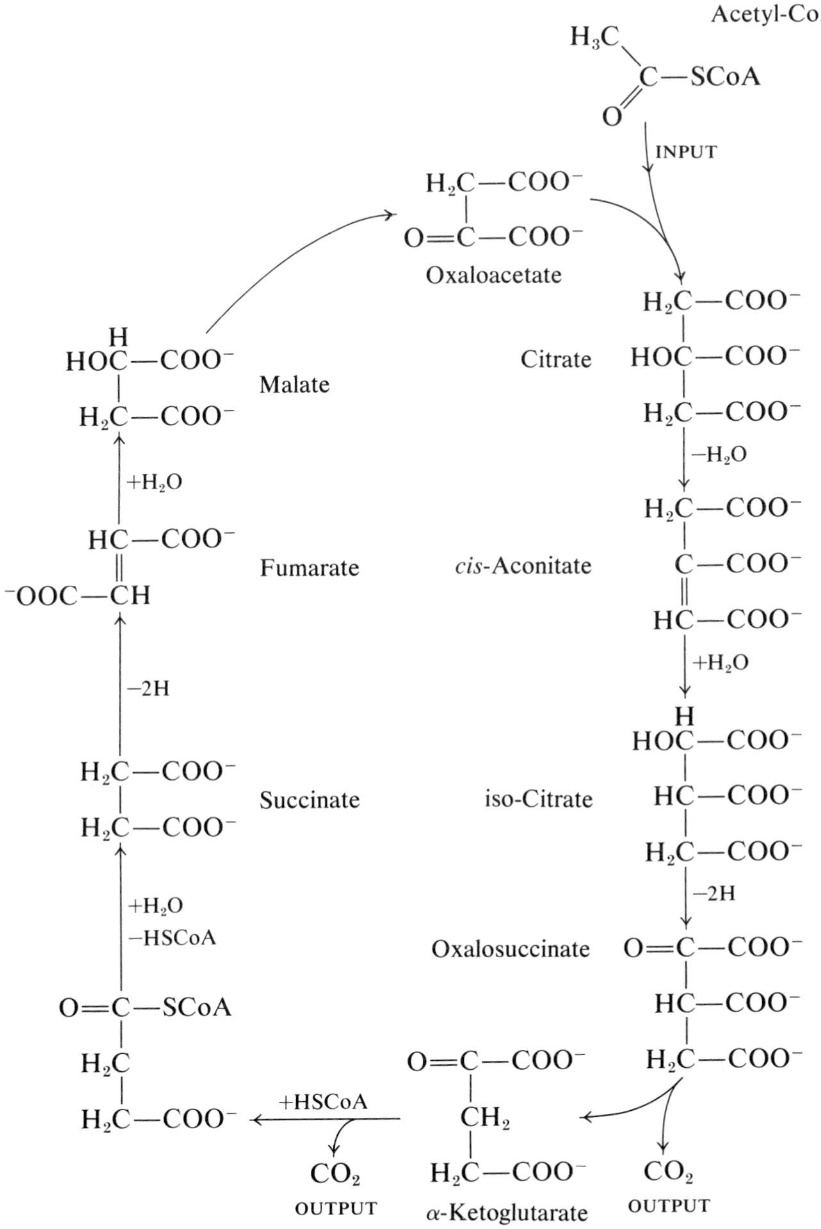

Acetyl-CoA

$H_3C{-}C({=}O){-}SCoA$

INPUT

$H_2C{-}COO^-$
$O{=}C{-}COO^-$
Oxaloacetate

Citrate
$H_2C{-}COO^-$
$HOC{-}COO^-$
$H_2C{-}COO^-$

$-H_2O$

cis-Aconitate
$H_2C{-}COO^-$
$C{-}COO^-$
$HC{-}COO^-$

$+H_2O$

iso-Citrate
$H\,HOC{-}COO^-$
$HC{-}COO^-$
$H_2C{-}COO^-$

$-2H$

Oxalosuccinate
$O{=}C{-}COO^-$
$HC{-}COO^-$
$H_2C{-}COO^-$

CO_2
OUTPUT

$O{=}C{-}COO^-$
CH_2
$H_2C{-}COO^-$
α-Ketoglutarate

$+HSCoA$

CO_2
OUTPUT

$O{=}C{-}SCoA$
H_2C
$H_2C{-}COO^-$

$+H_2O$
$-HSCoA$

Succinate
$H_2C{-}COO^-$
$H_2C{-}COO^-$

$-2H$

Fumarate
$HC{-}COO^-$
$^-OOC{-}CH$

$+H_2O$

Malate
$H\,HOC{-}COO^-$
$H_2C{-}COO^-$

FIGURE 14-8
The citric acid cycle.

is catalyzed by specific enzymes, and some of the steps are accompanied by other reactions, a few of which are indicated. Some of the steps lead to the conversion of ADP to ATP. The acetylSCoA that enters the cycle may come from polysaccharides, fatty acids, or amino acids.

The citric acid cycle has been found to take place in microorganisms and plant seedlings as well as in the cells of animals. The existence of this common feature, as well as others, indicates a common origin, as is assumed in the theory of evolution. There is evidence that for some microorganisms the cycle serves mainly to produce molecules with special structure for special purposes (such as α-ketoglutaric acid for the synthesis of glutamic acid and some other amino acids). For man and other animals it supplies both these special substances and energy.

14-8. Lipids

Lipids are substances that are soluble in many organic solvents (not containing groups that form hydrogen bonds) but are insoluble or only very slightly soluble in water. Many lipids are important constituents of plants and animals. They are either compounds of fatty acids or are complex alcohols.

Fats

The fats (glyceryl esters of fatty acids, discussed in Section 13-5) are important lipids in plants and animals and important constituents of food. Their value as a source of energy is discussed below.

The polyunsaturated fatty acids (linoleic, linolenic, and arachidonic acid, Section 13-5) are called *essential fatty acids*. Small amounts have been shown to be required for the growth and well-being of rats, and probably are required also by man. A large intake, however, is harmful, unless it is accompanied by an increased intake of vitamin E (Section 14-9).

The Fats in Nutrition. Heat Values of Foods

The fats are digested in the intestines, with enzymes called lipases as catalysts. They constitute an important part of the food of most people: an average daily diet for a healthy young man may include 80 g of protein, 385 g of carbohydrate, and 100 g of fat.

One important use of foods is to serve as a source of energy, permitting work to be done, and of heat, keeping the body warm. Foods serve in this way through their oxidation within the body by oxygen that is extracted from the air in the lungs and is carried to the tissues by the hemoglobin of the blood. The ultimate products of oxidation of most of the hydrogen and carbon in foods are water and carbon dioxide.

Heats of combustion of foods and their relation to dietary requirements have been thoroughly studied. The food ingested daily by a healthy man of average size doing a moderate amount of muscular work should have a total heat of combustion of about 12000 kJ. About 90% of this is made available as work and heat by digestion and metabolism of the food.

Fats and carbohydrates are the principal sources of energy in foods. Pure fat has a caloric value (heat of combustion) of 37.6 kJ per g, and pure carbohydrate (sugar) a caloric value of about 17 kJ per g (17.5 for starch, 16.5 for sucrose, and 15.6 for glucose). The caloric values of foods are obtained by use of a bomb calorimeter, as described in Appendix VI. The third main constituent of food, protein, is needed primarily for growth and for the repair of tissues. About 50 g of protein is the daily requirement for an adult of average size. Usually somewhat more, 80 g, is ingested, with caloric value about 1400 kJ, the heat of combustion of protein being about 18 kJ per g. Accordingly, fat and carbohydrate must provide about 10600 of the 12000 kJ required daily. Usually fat provides about one third of the total (100 g provides 3760 kJ), and carbohydrates about 60 percent. Persons doing very hard physical work, such as lumberjacks and arctic explorers, with very high needs, may increase their fat intake to as much as 250 g per day, fat being a more concentrated source of energy than carbohydrate.

Fats are oxidized (catabolized) in the body by splitting off two carbon atoms as acetic acid and forming a shorter chain; for example

$$H_3C(CH_2)_{16}COOH + O_2 \longrightarrow H_3C(CH_2)_{14}COOH + H_3CCOOH$$

If the ratio of fat to carbohydrate is too large the catabolism is incomplete, and a high concentration of acetone, acetoacetic acid, and β-hydroxybutyric acid is reached in the blood and urine:

$(CH_3)_2CO$	H_3CCOCH_2COOH	$H_3CCHOHCH_2COOH$
Acetone	Acetoacetic acid	β-Hydroxybutyric acid

This condition is called ketosis or acidosis. The acidosis, a decrease in pH of the blood, results from the elimination of the two acids in the urine as the ammonium or sodium salts. Ketosis and acidosis occur in diabetes, liver disease, starvation, alcoholism, and ingestion of a ketogenic diet (with large fat-carbohydrate ratio).

Phospholipids

Phospholipids are important substances that are present in every tissue in the body, especially in cell membranes and the sheaths on nerves (Section 15-7). On hydrolysis they yield fatty acids, an alcohol, phosphoric acid, and a nitrogen-containing compound.

The *lecithins* have the formula

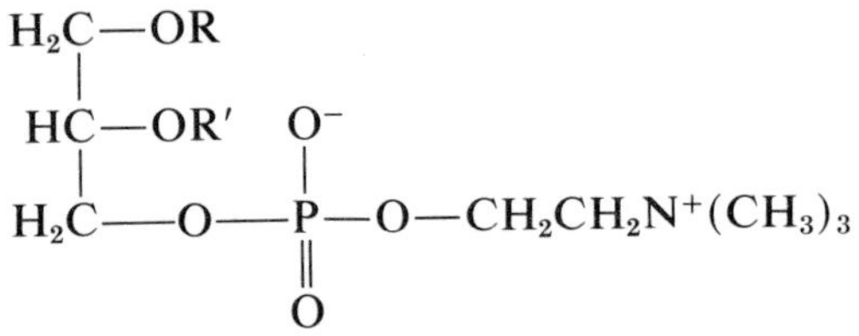

$$
\begin{array}{l}
H_2C\text{—}OR \\
\quad | \\
\;HC\text{—}OR' \qquad O^- \\
\quad | \qquad\qquad\quad\;\, | \\
H_2C\text{——}O\text{——}P\text{—}O\text{—}CH_2CH_2N^+(CH_3)_3 \\
\qquad\qquad\qquad\;\; \| \\
\qquad\qquad\qquad\;\; O
\end{array}
$$

The nitrogen atom in lecithin is in the cholyl radical, $—O(CH_2)_2N(CH_3)_3^+$. The radicals R and R′ are those of different fatty acids (palmitic, stearic, oleic, etc.), in different ratios for different plants and animals. Lecithin is formed in the body from fats. It is a good emulsifying agent, and probably helps in carrying fats to the tissues.

The *cephalins* have structure similar to that of the lecithins but with the radical $—OCH_2CH_2NH_2$ in place of the cholyl radical; that is, they are the glycerophosphate esters of the alcohol β-aminoethanol, $HO(CH_2)_2NH_2$. The *sphingomyelins* are the phosphate diesters of choline and the complex alcohol sphingosine, $H_3C(CH_2)_{12}CH{=}CHCH(OH)CH(NH_2)CH_2OH$. The cephalins and sphingomyelins are found in the brain and nerves.*

Sterols and Steroids

The sterols (from Greek stereos, solid) are crystalline alcohols with a structure involving three six-membered rings, condensed together as in phenanthrene, and one five-membered ring:

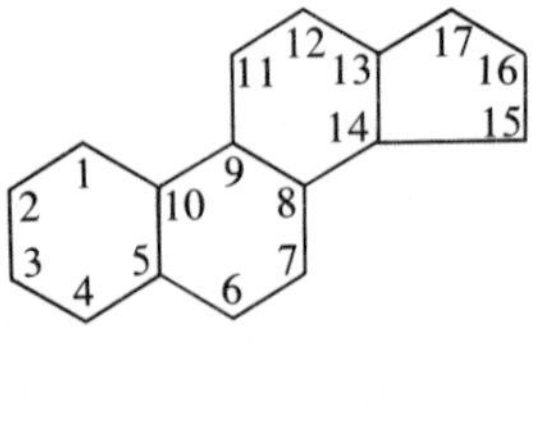

*The cerebrosides, also found in these tissues, are compounds (glycolipids) of sphingosine and the sugar galactose.

The rings are not benzenoid, but are saturated except for one or two double bonds. A hydroxyl group is attached to carbon 3, methyl groups to C10 and C13, and a chain, usually C_8H_{17}, to C17. The steroids are substances related to the sterols.

Sterols are present in all plant and animal tissues. The chief animal sterol is cholesterol, $C_{27}H_{46}O$, with structural formula

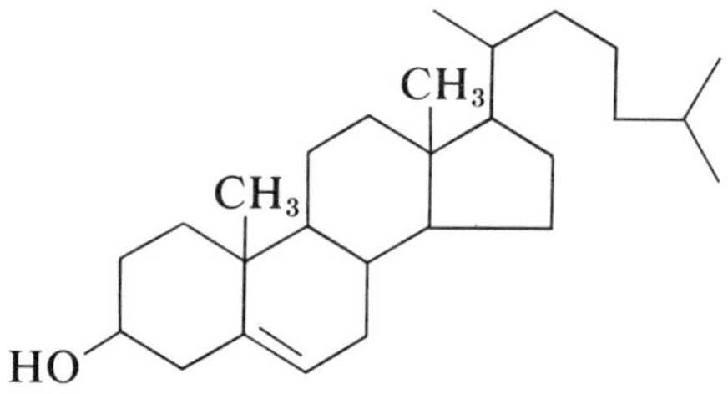

CH3 CH3 HO

(Hydrogen atoms not shown. The large side chain is C_8H_{17}.) It is insoluble in water, but soluble in ether, benzene, and other organic solvents—the definition of lipids.

Cholesterol and its fatty-acid esters enter the intestinal cells and are combined with proteins to form lipoproteins that are carried to the tissues, especially the brain, by the blood. In addition, cholesterol is synthesized from acetate ion by human beings, in amount about 1000 mg per day. The daily food intake may provide 500 to 1000 mg (an egg, a high-cholesterol food, contains about 250 mg). Cholesterol is broken down at a rate equal to the intake, and is eliminated in the bile as *bile acids*. The bile acids have a carboxyl group at the end of the side chain; for example, cholic acid, $C_{24}H_{40}O_5$, differs from cholesterol in having the side chain $—CH(CH_3)CH_2CH_2COOH$ at C17, as well as hydroxyl groups at C7 and C13. The bile acids are steroids; other important steroids are hormones (Section 14-10).

The rate of conversion of cholesterol to bile acids is proportional to the concentration in the blood, and a steady state is reached at a certain concentration, dependent on the intake and the genotype of the person (activity of the enzymes controlling synthesis and breakdown). For most persons this concentration is in the range 150 to 250 mg per deciliter of blood. There is a correlation between incidence of coronary heart disease and the serum concentration of cholesterol. In one study of men 50 to 60 years old the average incidence of coronary heart disease was 1.5 times as great for those in the range 200 to 240 as for those below 200 mg dl^{-1}; the factor for range 240 to 260 was 3.0, and that for more than 260 mg dl^{-1} was 4.2. A similar correlation is found with the concentration of triglycerides in the blood serum.

Decreasing the intake of fats, especially animal fats (saturated fats), decreases the cholesterol level somewhat. The cholesterol content of eggs, however, may not be harmful. For most people even ten eggs per day does not raise the cholesterol concentration by more than a few percent. The other nutrient substances, such as the phospholipid lecithin, in this valuable food may assist in controlling the cholesterol.

The most important way of avoiding heart disease by keeping the cholesterol level low is probably the restriction of the intake of sucrose, ordinary sugar. At the present time the average daily intake of carbohydrate in the United States and other affluent countries is about 175 g of starch, 140 g of sucrose, 20 g of lactose, 10 g of fructose, and 5 g of other sugars. (One hundred fifty years ago the intake of sucrose was only one sixth as great.) It has been shown that subjects receiving 100 g of sucrose per day have cholesterol concentration in the serum 50 mg dl^{-1} greater than when they ingest only glucose polysaccharides (starch). The explanation of this effect is that the fructose half of sucrose undergoes reactions leading to the synthesis of extra cholesterol. John Yudkin, British biochemist and nutritionist, has shown that the incidence of coronary heart disease increases with increase in the amount of sugar (sucrose) ingested; for an intake of 150 g or more per day it is six times as great, at a given age, as for an intake of 75 g or less per day. A high intake of sucrose is also correlated with increased incidence of other diseases. A good way to improve one's health is to decrease one's intake of sucrose by avoiding sweet desserts, soft drinks, and the sugar in the sugar bowl.

14-9. Vitamins

Man requires the essential amino acids in his diet, in order to keep in good health. It is not enough, however, that the diet contain proteins that provide these amino acids, and a sufficient supply of carbohydrates and fats to provide energy. Other substances, both inorganic and organic, are also essential to health.

The organic compounds other than the essential amino acids that are required in small amounts for health are called *vitamins*. Man is known to require at least thirteen vitamins: vitamins A, B_1 (thiamine), B_2 (riboflavin), B_6 (pyridoxine), B_{12}, C (ascorbic acid), D, K, niacin, pantothenic acid, inositol, para-aminobenzoic acid, and biotin.

Although it has been recognized for over a century that certain diseases occur when the diet is restricted, and can be prevented by additions to the diet (such as lime juice for the prevention of scurvy), the identification of the essential food factors as chemical substances was not made until a few decades ago. Progress in the isolation of these substances and in the determination of their structure has been rapid in recent years, and many of the vitamins are now being made synthetically, for use as dietary supplements. It is possible for a diet to be obtained that provides all of the essential food substances in satisfactory amounts, but it is usually wise to have the diet supplemented by vitamin preparations, in order to achieve the best of health.

Vitamin A has the formula $C_{20}H_{29}OH$, and the structure

```
                                 CH3               CH3
  H3C     CH3                     |                 |
     \   /                        |                 |
       C       CH                 C       CH        C        CH2
     /   \    /  \\             /  \\    /  \\     /  \\     /  \
  H2C     C        CH        CH      CH      CH       CH       OH
  |       ||
  H2C     C
     \   /  \
      CH2    CH3
```

It is a yellow, oily substance, which occurs in nature in butterfat and fish oils. Lack of vitamin A in the diet causes a scaly condition of the eyes, and similar abnormality of the skin in general, together with a decreased resistance to infection of the eyes and skin. In addition there occurs a decreased ability to see at night, called *night blindness*. There are two mechanisms for vision, one situated in the cones of the retina of the eye, which are especially concentrated in the neighborhood of the fovea (the center of vision), and the other situated in the rods of the retina. Color vision, which is the ordinary vision, used when the intensity of light is normal, involves the retinal cones. Night vision, which operates when the intensity of light is very small, involves the rods; it is not associated with a recognition of color. It has been found that a certain protein, *visual purple,* which occurs in the rods, takes part in the process of night vision—it absorbs light, and activates the visual nerve. There are three other colored substances, in the cones, that absorb light in the three wavelength ranges that provide color vision. All four of these substances are conjugated proteins involving vitamin A or one of its derivatives.

It is not essential that vitamin A itself be present in food in order to prevent the vitamin A deficiency symptoms. Certain hydrocarbons, the *carotenes,* with formula $C_{40}H_{56}$ (similar in structure to lycopene, Figure 13-1), can be converted into vitamin A in the body. These substances, which are designated by the name *provitamin A,* are red and yellow substances that are found in carrots, tomatoes, and other vegetables and fruits, as well as in butter, milk, green leafy vegetables, and eggs.

Thiamine, vitamin B_1, has the following formula (that shown is for thiamine chloride):

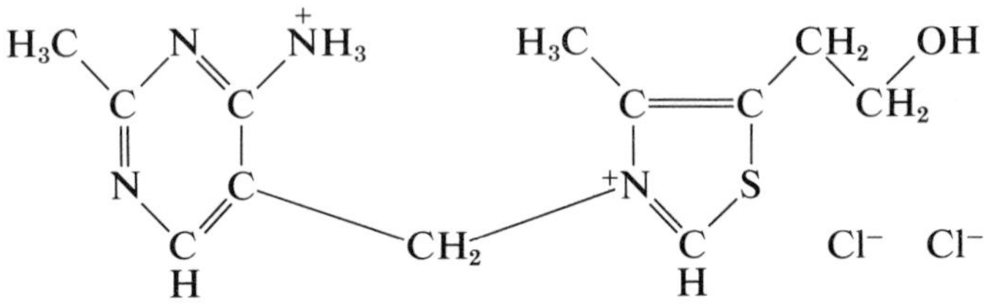

```
H3C     N     +NH3          H3C              CH2    OH
   \  /   \\  /                \            /   \  /
    C       C                   C=========C      CH2
    ||      |                   |         |
    N       C                  +N         S
     \    //  \               /   \\     /
      C         \__ CH2 ___/        C             Cl-   Cl-
      H                             H
```

A lack of thiamine in the diet causes the disease beri-beri, a nerve disease that in past years was common in the Orient. Just before 1900 it was found by Christiaan Eijkman (1858–1930) in Java that beri-beri occurred as a consequence of a diet consisting largely of polished rice, and that it could be cured by adding the rice polishings to the diet. In 1911 Casimir Funk assumed that beri-beri and similar diseases were due to lack of a substance present in a satisfactory diet and missing from a deficient diet, and he attempted to isolate the substance whose lack was responsible for beri-beri. He coined the name vitamin for substances of this sort (he spelled it vitamine because he thought that the substances were amines). The structure of vitamin B_1, thiamine, was determined by R. R. Williams, E. R. Buchman, and their collaborators in 1936.

Thiamine seems to be important for metabolic processes in the cells of the body, but the exact way in which it operates is not known. There is some evidence that it is the prosthetic group for an enzyme involved in the oxidation of carbohydrates. The vitamin is present in potatoes, whole cereals, milk, pork, eggs, and other vegetables and meats.

Riboflavin, vitamin B_2, has the following structure:

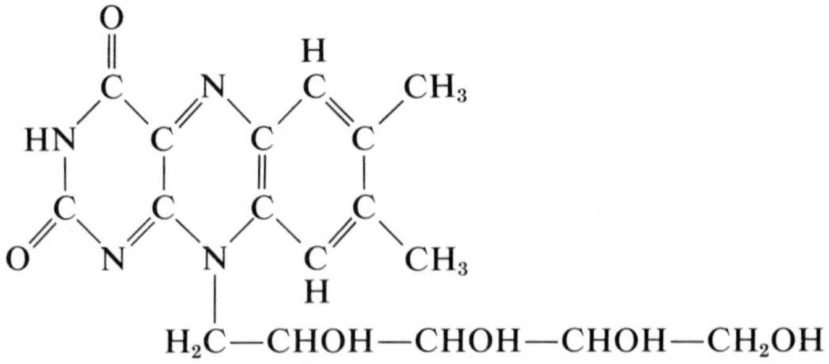

```
        O
        ||              H
        C       N       C       CH3
      /   \   //  \   /   \\   /
   HN       C       C       C
   |        |       ||      |
    C       C       C       C
  //  \   //  \   /   \   //  \
 O      N       N       C       CH3
                |       H
                |
           H2C—CHOH—CHOH—CHOH—CH2OH
```

It is essential for growth and for a healthy condition of the skin. A derivative of riboflavin is the prosthetic group of an enzyme, called *yellow en-*

zyme, that catalyzes the oxidation of glucose and certain other substances in the animal body, and riboflavin is present in many other enzymes.

Vitamin B_6 (pyridoxine) has the formula

```
 H         N         CH3
   \     /   \\     /
     C           C
     ||          |
     C           C
   /   \       //  \
H2C       C         OH
 |        |
 OH    H2C—OH
```

It is present in yeast, liver, rice polishings, and other plant and animal foods, and is also produced synthetically. It has the power of stimulating growth, and of preventing skin eruptions (dermatitis). Pyridoxine and its derivatives serve as coenzymes for scores of enzyme systems.

Vitamin B_{12} (cyanocobalamin, Figure 14-9) is involved in the manufacture of the red corpuscles of the blood. It is used for the treatment of pernicious anemia, and it is perhaps the most potent substance known in its physiological activity: 2 micrograms per day (2×10^{-6} g) of vitamin B_{12} is effective in the control of the disease. The vitamin can be isolated from liver tissue, and is also produced by molds and other microorganisms. Each molecule of vitamin B_{12} contains one cobalt atom. This is the only compound of cobalt that is known to be present in the human body.

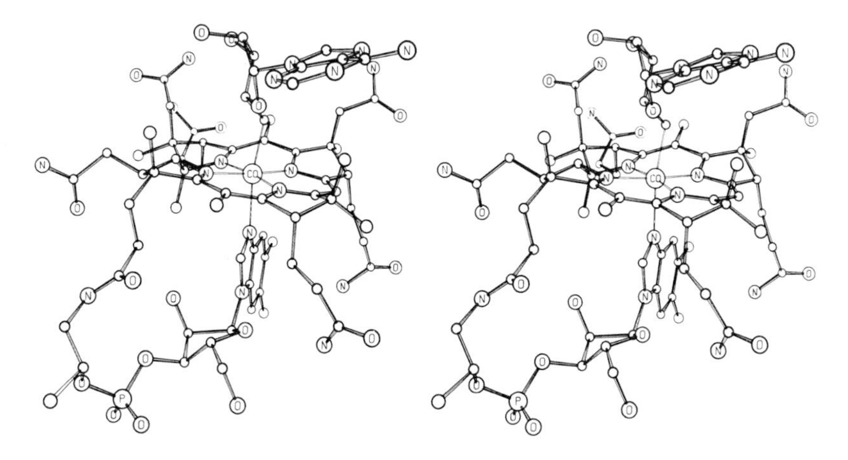

FIGURE 14-9
A stereoscopic drawing of 5′-deoxyadenosylcobalamin (coenzyme B_{12}). [From T. G. Lenhert, *Proceedings of the Royal Society of London* **A303,** 45–84 (1968).]

Vitamin B_{12} serves as the coenzyme for some enzymes. One of these is the enzyme that catalyzes the reaction of isomerization of methylmalonic acid, $HOOCCH(CH_3)COOH$, to succinic acid. Some children suffer from the disease methylmalonicaciduria, in which the concentration of methylmalonic acid in the blood and urine is very large because of a genetic defect that causes an abnormality in this enzyme and the consequent failure to convert the methylmalonic acid to succinic acid. For about half of the patients the disease can be kept under control by the daily intake of about 1 mg of B_{12}, 500 times the usually recommended intake. The probable explanation of this effect is that the patient manufactures a defective apoenzyme with a very small combination constant with the coenzyme, so that at equilibrium under ordinary nutrition only a very small fraction of the apoenzyme is converted to the active enzyme. It can be seen from Equation 14-4, however, that the amount converted to active enzyme is increased by an increase in the concentration of the coenzyme.

Ascorbic acid, vitamin C, is a water-soluble vitamin of great importance. A deficiency of vitamin C in the diet leads to scurvy, a disease characterized by loss of weight, general weakness, hemorrhagic condition of the gums and skin, loosening of the teeth, and other symptoms. Sound tooth development seems to depend upon a satisfactory supply of this vitamin, and a deficiency is thought to cause a tendency to incidence of a number of diseases.

The formula of ascorbic acid is the following:

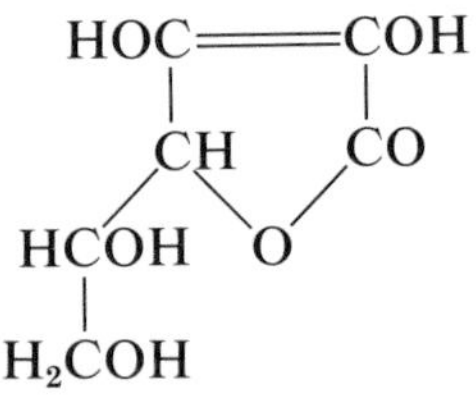

```
      HOC=====COH
        |       |
        CH      CO
       /  \    /
   HCOH    O
     |
   H2COH
```

It is seen that it is not a carboxylic acid; instead of the carboxyl resonance between adjacent single and double bonds to oxygen atoms (Section 13-5), the acidic character results from resonance between the structures

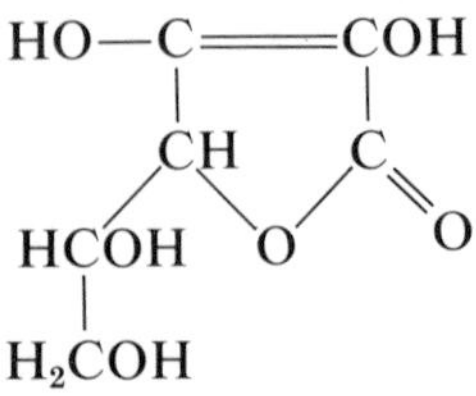

```
HO—C=====COH                    HO+=C——————COH
   |       |                        |       ||
   CH      C                        CH      C
  /  \    / \\                     /  \    / \
HCOH   O     O          and     HCOH   O     O-
 |                               |
H2COH                           H2COH
```

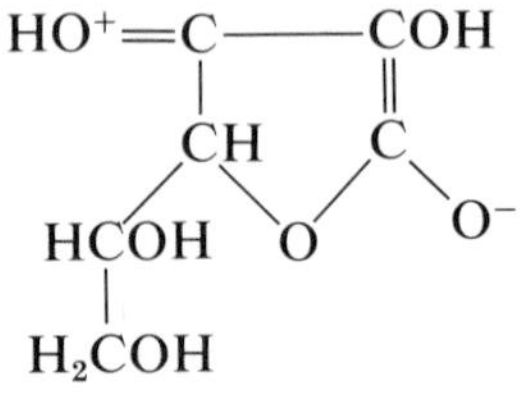

The acid ionization constant of ascorbic acid is 6.76×10^{-5}, about the same as for the carboxylic acids.

Ascorbic acid is optically active. The vitamin is L-ascorbic acid. It is a sugar acid, related to glucose. It is a reducing agent; oxidation converts it to dehydroascorbic acid,

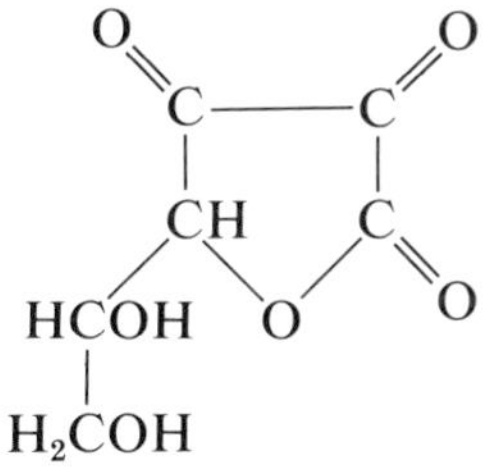

```
  O         O
   \\      //
    C------C
    |      |
    CH     C
   /  \   / \\
HCOH   O     O
 |
H2COH
```

Its action as a reducing agent may be part of its physiological function. It is known to be required for the synthesis of the protein of connective tissue, collagen, in particular for the conversion of prolyl residues to hydroxyprolyl residues, which constitute one-seventh of the amino acids of this protein. It probably has other physiological functions, but it is not known to serve as the coenzyme in any enzyme system. This vitamin is present in many foods, especially fresh green peppers, turnip greens, parsnip greens, spinach, orange juice, tomato juice, and potatoes. The daily intake of vitamin C that suffices to prevent scurvy in most people is about 45 mg, but larger amounts, 1000 to 5000 mg per day, are valuable in preventing or ameliorating the common cold and other diseases.

Most vitamins required by man are required also by all other animals. Vitamin C is an exception: most animals manufacture ascorbic acid from glucose, by the following enzyme-catalyzed oxidation reactions:

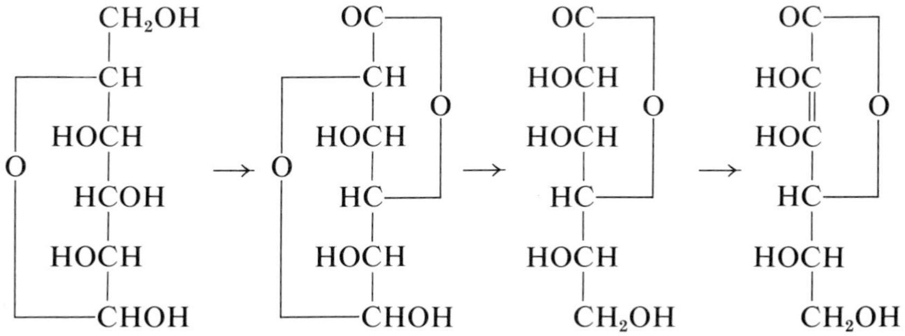

```
   CH2OH            OC---+           OC---+           OC---+
   |                |    |           |    |           |    |
+--CH          +----CH   |         HOCH   |          HOC   |
|  |           |    |    O           |    O          ||    O
| HOCH         |  HOCH   |         HOCH   |          HOC   |
O  |      ->   O    |    |    ->     |    |    ->    |     |
|  HCOH        |    HC---+           HC---+          HC----+
|  |           |    |                |               |
| HOCH         |  HOCH             HOCH             HOCH
|  |           |    |                |               |
+--CHOH        +----CHOH            CH2OH           CH2OH
```

$C_6H_{12}O_6$	→	$C_6H_8O_6$	→	$C_6H_{10}O_6$	→	$C_6H_8O_6$
D-Glucose		D-Glucurono-γ-lactone		L-Gulono-γ-lactone		L-Ascorbic acid

Man and his close relatives (other primates), however, have lost the ability to make the enzyme, gulonolactone dehydrogenase, that catalyzes the last step, and are accordingly dependent on exogenous sources. The rate at which various animals synthesize ascorbic acid is proportional to body weight; it is, for rat, mouse, cat, dog, goat, and others, between 2 g and 20 g per day, calculated to 70 kg body weight. This observation suggests that the optimum intake for man might be in this range, as is suggested also by the observed greater resistance to disease associated with a much larger intake than that needed to prevent scurvy.

Vitamin D is necessary in the diet for the prevention of rickets, a disease involving malformation of the bones and unsatisfactory development of the teeth. There are several substances with antirachitic activity. The form that occurs in oils from fish livers is called vitamin D_3; it has the following chemical structure:

```
                                    H3C
                                       \
                        H2C      CH2     CH—CH2
                   H2C      C———CH            |
                   |        |    |            CH2        CH3
                   H2C      C    CH2          |         /
                        C   |  CH2            CH2—CH
              CH2       ||  H                         \
     H2C—C              CH                             CH3
  H2C        C=CH
     C—CH2
 HO   H
```

Only a very small amount of vitamin D is necessary for health—approximately 0.01 mg (400 international units, IU) per day. The vitamin is a fat-soluble vitamin, occurring in cod-liver oil, egg yolks, milk, and in very small amounts in other foods. Cereals, yeast, and milk acquire an added vitamin D potency when irradiated with ultraviolet light. The radiation converts the sterol *ergosterol* into *calciferol* (vitamin D_2), which has vitamin-D activity. Calciferol differs from D_3 in having the side chain $—CH(CH_3)CH{=}CHCH(CH_3)CH(CH_3)_2$ in place of $—CH(CH_3)CH_2CH_2CH_2CH(CH_3)_2$.

Vitamin E activity is shown by several closely related substances, called tocopherols. The most potent is α-tocopherol, with formula

```
       CH3           CH3      CH3                CH3                 CH3
H3C          O                |                  |                   |
                    —(CH2)3—CH—(CH2)3—CH—(CH2)3—CH—CH3
HO                H2
             H2
       CH3
```

The others (β, γ, δ, ϵ, ζ) have slightly different formulas. The tocopherols are yellow oils. Vitamin E is required for reproduction and lactation in rats, and probably is required for good health by man. One IU is equal to 1 mg of D,L-α-tocopheryl acetate (0.74 mg of the D acetate, 1.36 mg of the L acetate). Vitamin E is the principal fat-soluble antioxidant; it protects the tissues, especially the unsaturated fatty acids, against damage by oxidation. Animals deficient in vitamin E develop muscular dystrophy and heart disease.

Vitamin E is found in seed germ oils, alfalfa, lettuce, and other plant foods. Much processed food is stripped of most of its vitamin E. The recommended daily allowance, 15 IU per day, may be found in a good diet, but the best of health, especially of the heart and blood vessels, may require much more, several hundred IU per day.

Pantothenic acid, inositol, *p*-amino-benzoic acid, and biotin are substances involved in the process of normal growth. Vitamin K is a vitamin that prevents bleeding, by assisting in the process of clotting of the blood.

Nutritional Requirements of Vitamins and Minerals

In Figure 14-10 there are shown the recommended daily amounts (RDA) of vitamins, as estimated by the Food and Nutrition Board of the U.S. National Academy of Sciences—National Research Council. These amounts are described as enough to prevent manifestations of vitamin deficiency in most healthy persons, and it is said that the amounts are contained in a good diet, so that there is no need for use of vitamin supplements by healthy persons ingesting a good diet.

We might ask, however, what the optimum daily amounts of these important substances are, what amounts should be ingested to lead to the best health. The Food and Nutrition Board has not attempted to answer this question, and in fact there is little evidence available for human beings. The recommended diet for laboratory monkeys contains about 100 times the amount of ascorbic acid that is recommended for man, and there is experimental evidence showing that the general health and resistance to disease of the monkeys are better on this diet than on one containing less ascorbic acid. There is also some evidence for man, as mentioned above. It is likely that the optimum intake of vitamin C for most people is much greater than the RDA.

For vitamin E, too, the optimum daily intake may be much greater than the RDA. It is possible that an intake of the other vitamins somewhat larger than the RDA's approaches the optimum, but little direct evidence is available. Care should be taken in increasing the intake of vitamin A and vitamin D, because of their toxicity when taken in large amounts.

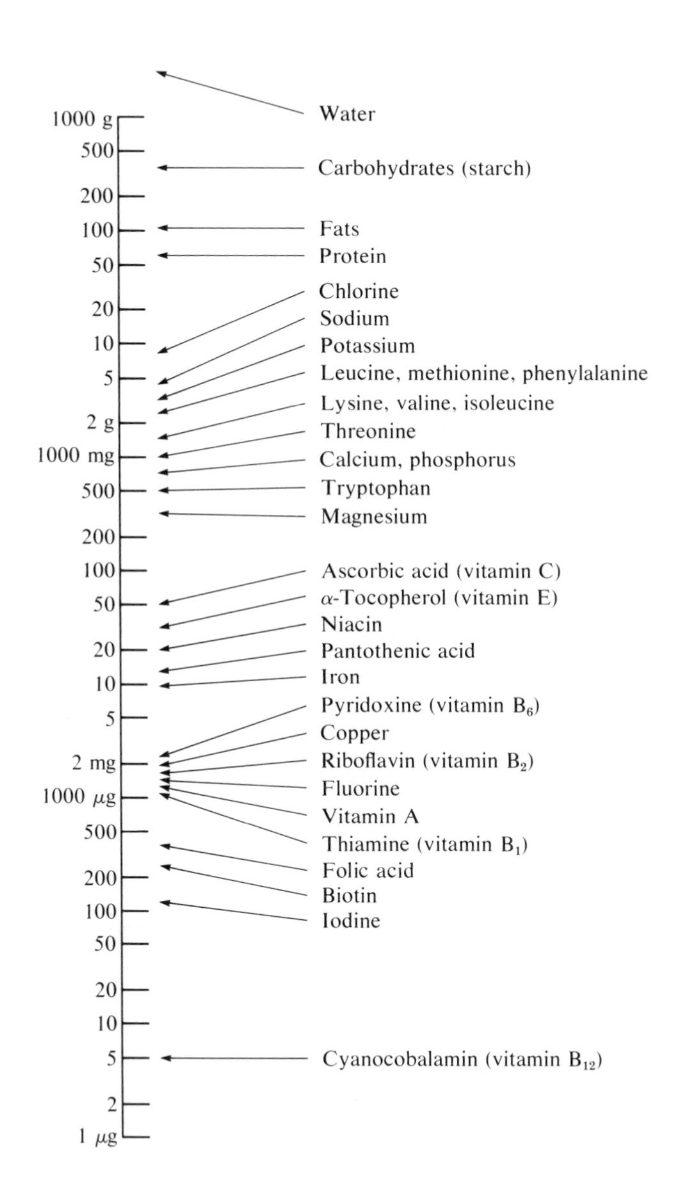

FIGURE 14-10
Recommended dietary allowances for a male adult (daily intake, in foods and food supplements) of some nutrients, usually the amounts estimated as needed to prevent overt manifestation of deficiency disease in most persons. For the substances listed in smaller amounts the optimum intake, leading to the best of health, may be somewhat greater. Not shown, but probably or possibly required, are the essential fatty acids, *p*-aminobenzoic acid, choline, vitamin D, vitamin K, chromium, manganese, cobalt, nickel, zinc, selenium, molybdenum, vanadium, tin, and silicon.

Essential Elements

Twenty of the first thirty elements, and also four heavier ones, are required for life. Hydrogen, carbon, nitrogen, and oxygen are present in many compounds in the body. Sodium, potassium, magnesium, calcium, and chlorine are present as ions in the blood and intercellular fluids. Phosphorus is found as phosphate ion in the blood, phosphate esters in the phospholipids and other compounds, and calcium hydroxyapatite in the bones and teeth. Sulfur is an important part of insulin and other proteins. Fluorine, as fluoride ion in the drinking water, is required for the formation of strong teeth and bones, and is required by the rat for growth. Silicon, vanadium, chromium, manganese, iron, cobalt, copper, zinc, selenium, molybdenum, tin, and iodine (the nutritional trace elements or minerals) are all required in small amounts for life. The evidence about some of them has been obtained only with animals, especially the rat, but it is likely that it applies to man also.

Many enzymes incorporate one or more metal ions as essential parts of their structure. Different metalloenzymes make use of ions of magnesium, calcium, manganese, iron, cobalt, copper, zinc, or molybdenum. For example, the molecule of alcohol dehydrogenase (molecular mass 87000 d), which catalyzes the oxidation of ethanol to acetic acid in the human liver, contains two atoms of zinc, and the amylase in human saliva contains an atom of calcium (Ca^{++}). Some enzyme molecules contain several metal atoms, which may be of different kinds. An example is cysteamine oxidase, which catalyzes the oxidation of cysteamine, $HSCH_2CH_2NH_2$; this enzyme contains an atom of iron, an atom of copper, and an atom of zinc.

14-10. Hormones

Another class of substances of importance in the activity of the human body consists of the *hormones,* which are substances that serve as messengers from one part of the body to another, moving by way of the body fluids. The hormones control various physiological processes. For example, when a man is suddenly frightened, a substance called *epinephrine* (also called adrenalin) is secreted by the suprarenal glands, small glands that lie just above the kidneys. The formula of epinephrine is

OH
CH NH
CH$_2$ CH$_3$
HO
OH

When epinephrine is introduced into the blood stream it speeds up the action of the heart, causes the blood vessels to contract, thus increasing the blood pressure, and causes the liberation of glucose inside the cells, providing an immediate source of extra energy.

Hormones are characteristic of large organisms. The single-cell organisms do not need them. They provide a mechanism by means of which the cells in different parts of the body can work harmoniously together.

Thyroxine is a secretion of the thyroid gland that controls metabolism. Its structural formula is

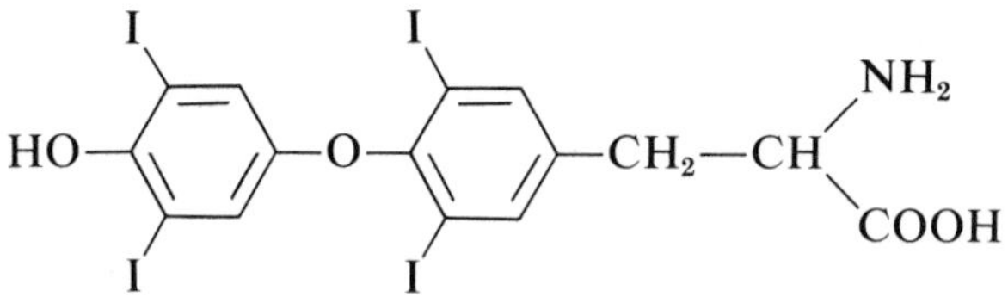

I I NH₂ HO—O—CH₂—CH COOH I I

(It is an amino acid.) It is synthesized in the thyroid gland, and when liberated into the blood stream it combines with a protein (thyroxine-binding globulin, molecular mass about 45,000 d) and is carried to the cells, where it accelerates their metabolism. In regions where soil and drinking water are low in iodine people may develop hypothyroidism or cretinism (severe hypothyroidism), with retarded growth and mental development. It was observed in 1820, only nine years after the element iodine was discovered, that intake of a few milligrams of iodine per day prevents cretinism from developing, and the suggestion was soon made that a small amount of sodium iodide be added to table salt to achieve this end. The medical profession objected, however, with the argument that it might not be safe, because iodine in large amounts is poisonous (2 to 3 g may be fatal), and the use of iodized salt was postponed for 100 years.

The protein *insulin* (Section 14-3) is a hormone, manufactured in the pancreas and liberated into the blood. It stimulates the transfer of glucose and some other sugars, as well as vitamin C, across cell membranes.

The *sex hormones* are sterols. The principal estrogenic hormones (stimulating estrus) are *estradiol, estrone,* and *estriol.* Estradiol has the structure

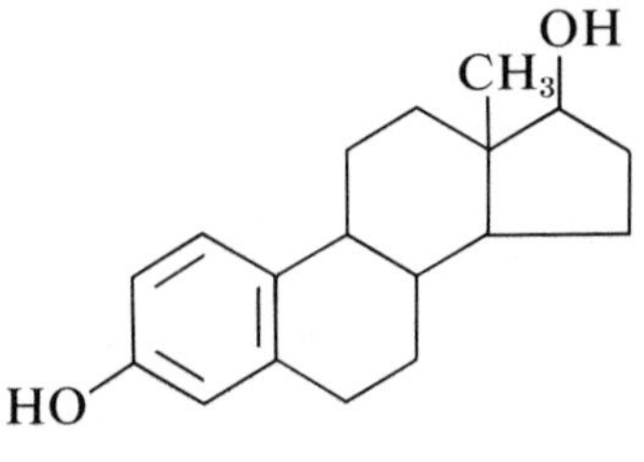

OH CH_3 HO

and the others are similar, estrone having O and estriol OH OH in place of OH .

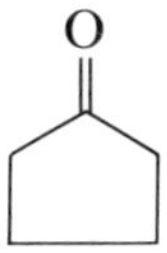

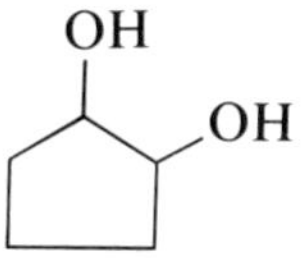

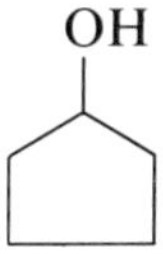

Stilbestrol is a synthetic compound (not a sterol) that has estrogenic properties. Its formula is

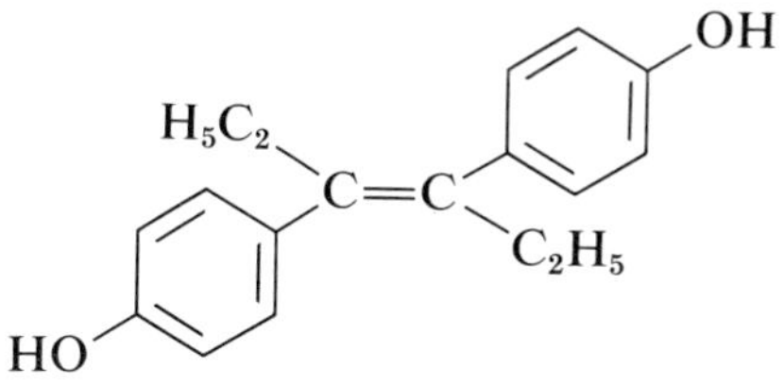

H_5C_2 C=C C_2H_5 OH HO

It seems likely that the estrogenic activity of the hormones and stilbestrol involves having two hydroxyl groups about 1200 pm apart, as in these molecules; presumably they form hydrogen bonds with a specific protein.

Estradiol is made by the cells of a follicle in the ovary containing an ovum that is ready to be expelled. In the male the interstitial cells produce *testosterone,* one of the male sex hormones. It is converted into *androsterone*. Their formulas are

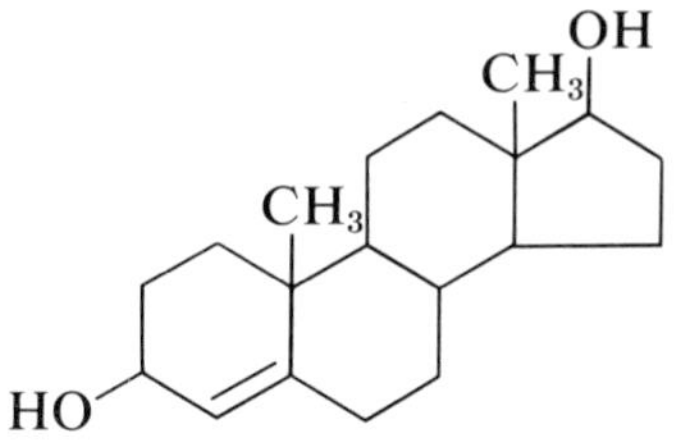

OH CH_3 CH_3 HO

Testosterone

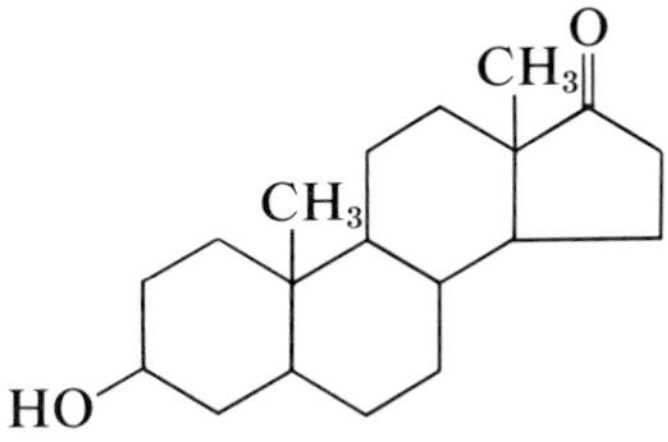

O CH_3 CH_3 HO

Androsterone

Their liberation leads to the development of the secondary sex characters at puberty.

Cortisone is an important steroid hormone synthesized by the cortex of the adrenal gland. Its formula is

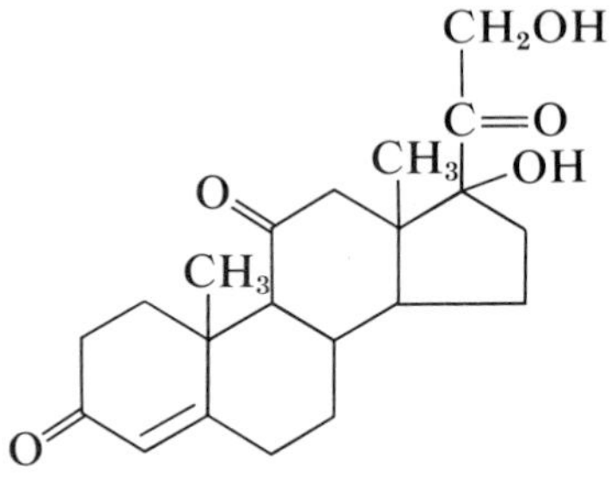

It is used in medicine as a powerful anti-inflammatory and anti-allergic agent, but it has serious side effects on continued use. Many related compounds (corticosteroids) are used in medicine. ACTH is a protein hormone that stimulates the adrenal cortex to produce the corticosteroids. It has molecular mass about 3,500 d; ox ACTH contains 39 amino-acid residues. It is used to some extent in medicine.

Cyclic AMP

A significant contribution to the understanding to the mechanism of hormone action was made by the American biochemist Earl Wilbur Sutherland (1915–1973) through his study of *cyclic AMP* (*cyclic adenosine monophosphate*). While he was investigating the effect of the hormone epinephrine on liver and muscle cells he found a new chemical substance that acts as an intermediate between the hormone and the cell, transmitting the message from the hormone to the machinery of the cell. He called it a "second messenger." He identified the substance as cyclic AMP, with the structure

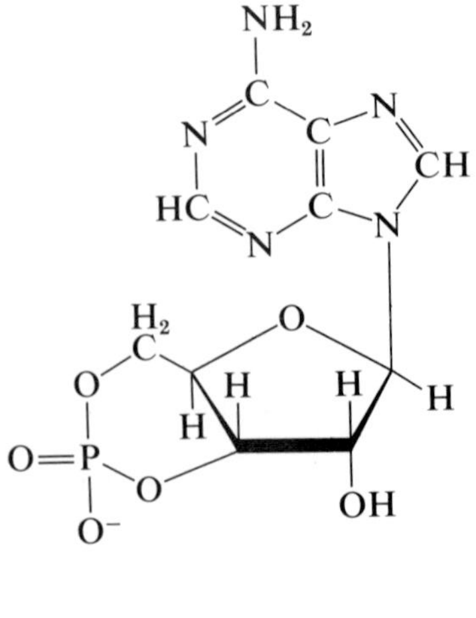

It is closely similar to AMP, but involves a second bond between the phosphate group and the sugar, forming a ring. It is formed from ATP by catalytic action of a specific enzyme, adenylate cyclase. In 1965 Sutherland was able to show that this enzyme is neither inside the cell nor outside it, but is in the cell membrane. It is normally inactive, but becomes active on combination with a molecule of the hormone epinephrine. It then converts ATP to cyclic AMP, which is released inside the cell, stimulating it in various ways. Thus the hormone itself, epinephrine, never enters the cell. Later the cyclic AMP is destroyed by enzymatic conversion to AMP.

Cyclic AMP has been shown to be involved, presumably as a second messenger, in more than forty cellular processes, in addition to the release of glucose. Among its effects are the increased contractility of heart muscle, increased secretion of hydrochloric acid by the gastric mucosa, decreased aggregation of blood platelets, and increased or decreased formation of several enzymes. Further study of cyclic AMP and perhaps of other second messengers should provide much more information about how the cells of an organism collaborate with one another.

14-11. Chemistry and Medicine

From the earliest times chemicals have been used in the treatment of disease. The substances that were first used as drugs are natural products such as in the leaves, branches, and roots of plants. Man (perhaps more often women than men) slowly learned by trial and error that the ingestion of certain plants had certain effects. It was discovered, for example, that ingestion of certain plants of the family Solanaceae causes mydriasis, the dilation of the pupil of the eye. One of these plants is named *Atropa belladonna*. Belladonna means beautiful woman and it was, and perhaps still is, thought that women with large pupils are more beautiful than those with small pupils.

Beginning in the nineteenth century, chemists and physicians began extracting and isolating the active principles of these plants. In 1881 the alkaloid hyoscyamine (atropine, belladonna) was isolated from the plants mentioned above. It is now used in medicine.

Inorganic substances are also used in medicine. As the alchemists discovered or made new chemical substances, these substances were tried out to see if they had physiological activity, and many of them were introduced into early medical practice. For example, both mercuric chloride, $HgCl_2$, and mercurous chloride, Hg_2Cl_2, were used in medicine, mercuric chloride as an antiseptic, and mercurous chloride, taken internally, as a cathartic and general medicament.

The modern period of *chemotherapy,* the treatment of disease by use of chemical substances, began with the work of Paul Ehrlich (1854–1915). It was known at the beginning of the present century that certain organic compounds of arsenic would kill protozoa, parasitic microorganisms responsible for certain diseases, and Ehrlich set himself the tast of synthesizing a large number of arsenic compounds, in an effort to find one that would be at the same time toxic (poisonous) to protozoa in the human body and nontoxic to the human host of the microorganism. After preparing many compounds he synthesized *arsphenamine,* which has the structure of a linear high polymer:

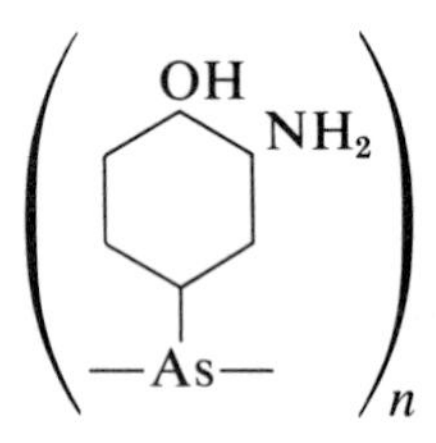

Arsphenamine has been found to be extremely valuable. Its greatest use is in the treatment of syphilis; the drug attacks the microorganism responsible for this disease, *Spirocheta pallida.* It has also been useful in the treatment of some other diseases. Now it has been superseded by penicillin (which we shall discuss below) in the treatment of syphilis.

Since Ehrlich's time there has been continual progress in the development of new chemotherapeutic agents. Thirty years ago the infectious diseases constituted the principal cause of death; now most of these diseases are under effective control by chemotherapeutic agents, some of which have been synthesized in the laboratory and some of which have been isolated from microorganisms. At the present time only a few of the infectious diseases constitute major hazards to the health of man, and we may confidently anticipate that the control of these diseases by chemotherapeutic agents will be achieved in a few years.

The recent period of rapid progress began with the discovery of the *sulfa drugs* by G. Domagk. In 1935 Domagk discovered that the compound prontosil, a derivative of sulfanilamide, was effective in the control of streptococcus infections. It was soon found by other workers that sulfanilamide itself is just as effective in the treatment of these diseases, and that it could be administered by mouth. The formula of sulfanilamide is given in Figure 14-11. Sulfanilamide is effective against hemolytic streptococcic infections and meningococcic infections. As soon as the value of sulfanilamide was recognized chemists synthesized hundreds of related substances, and investigations were made of their usefulness as bacteriostatic agents (agents with the power of controlling the spread

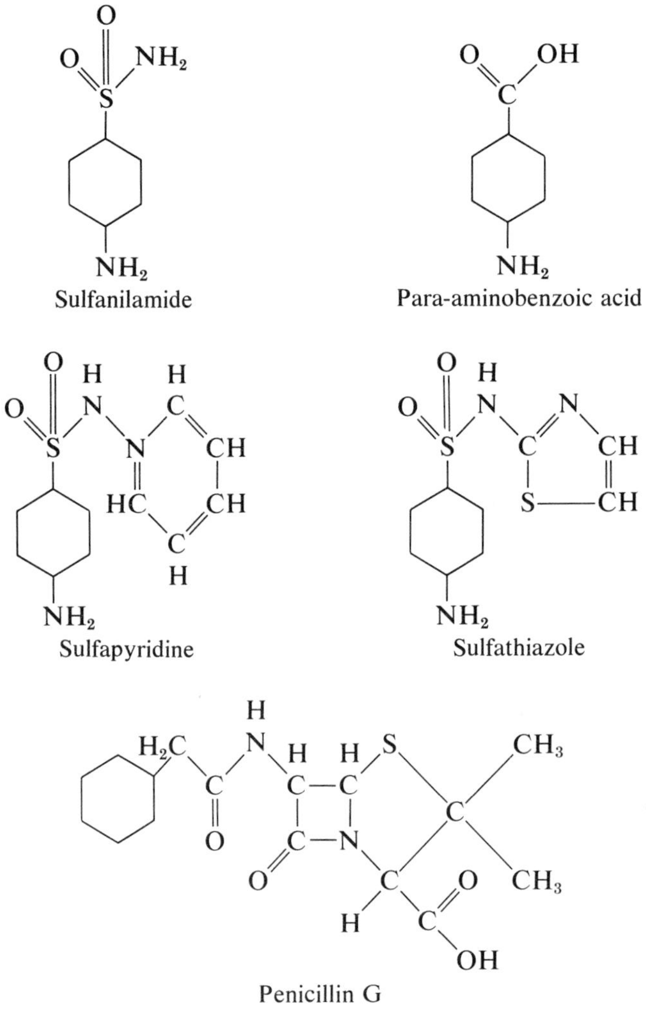

Sulfanilamide

Para-aminobenzoic acid

Sulfapyridine

Sulfathiazole

Penicillin G

FIGURE 14-11
Structural formulas of sulfa drugs and penicillin.

of bacterial infections). It was found that many of these related substances are valuable, and their use is now an important part of medical practice. Sulfapyridine has been found valuable for the control of pneumococcic pneumonia (pneumonia due to the *Pneumococcus* microorganisms), as well as of other pneumococcic infections and gonorrhea. Sulfathiazole is used for these infections and also for the control of staphylococcic infections, which occur especially in carbuncles and eruptions of the skin. These and other sulfa drugs are all derivatives of sulfanilamide itself, obtained by replacing one of the hydrogen atoms of the amide group (the NH_2 bonded to the sulfur atom) by some other group (Figure 14-11).

The introduction of *penicillin* into medical treatment was the next great step forward. In 1929 Professor Alexander Fleming, a bacteriologist working in the University of London, noticed that bacteria that he was growing in a dish in his laboratory were not able to grow in the region immediately surrounding a bit of mold that had accidentally begun to develop. He surmised that the mold was able to produce a chemical substance that had bacteriostatic action, the power of preventing the bacteria from growing, and he made a preliminary investigation of the nature of this substance. Ten years later, perhaps spurred on by the successful use of the sulfa drugs in medicine, Professor Howard Florey and Dr. E. B. Chain of the University of Oxford decided to make a careful study of the antibacterial substances that had been reported in order to see whether they would be similarly useful in the treatment of disease. When they tested the bacteriostatic power of the liquid in which the mold *Penicillium notatum* that had been observed by Fleming was growing, they found it to be very great, and within a few months the new antibiotic substance penicillin was being used in the treatment of patients. Through a cooperative effort of many investigators in the United States and England rapid progress was made during the next two or three years in the determination of the structure of penicillin, the development of methods of manufacturing it in large quantities, and the investigation of the diseases that could be effectively treated by use of it. Within less than a decade this new antibiotic agent had become the most valuable of all drugs. It provides an effective therapeutic treatment of many diseases.

The structure of penicillin is shown in Figure 14-11. The substance has been synthesized, but no cheap method of synthesizing it has been developed, and the large amount of penicillin that is being manufactured and used in the treatment of disease is made by growing the mold penicillium in a suitable medium and then extracting the penicillin from the medium. Important forward steps in the introduction of penicillin into medical treatment were the development of strains of the mold that produced the desired penicillin in large quantities, and the discovery of the best medium on which to grow the mold.

It is interesting that a number of slightly different penicillins are formed in nature by different strains of the mold. The formula in Figure 14-11 represents benzyl penicillin (penicillin G), which is the product that is now manufactured and used.

The spectacular success of penicillin as a chemotherapeutic agent has led to the search for other antibiotic products of living organisms. *Streptomycin,* which is produced by the mold *Actinomyces griseus,* has been found to be valuable in the treatment of diseases that are not effectively controlled by penicillin, and some other bacteriostatic agents also have been found to have significant value.

Another step forward has been made since 1955 by the discovery of substances that can control the development of viral infections. Penicillin, streptomycin, and the sulfa drugs are effective against bacteria but not against viruses. It has been found, however, that *chloramphenicol* (Chloromycetin) and *aureomycin,* both of which are substances manufactured by molds (the molds *Streptomyces venezuele* and *Streptomyces aureofaciens* respectively), have the power of controlling certain viral infections. The value of vitamin C in combatting viral diseases as well as bacterial diseases has been recognized recently.

The Relation between the Molecular Structure of Substances and Their Physiological Activity

It has been found by trial that some drugs have valuable chemotherapeutic effects without many serious side effects. All drugs are toxic to some extent. The value of a drug in relation to the danger in using it may be estimated from the therapeutic index, which is the ratio of the effective dose to the lethal dose.

Although the empirical basis for using many drugs is sound, it would be valuable to understand the detailed mechanisms of their actions. We know the structural formulas of many drugs, and also of vitamins and hormones—some of these formulas have been given in the preceding sections. In general, however, we do not know how the molecules of these substances interact with the molecules of the human body.

In recent years some progress has been made toward understanding mechanisms of drug action. An example is the way in which the sulfa drugs exercise their bacteriostatic action. It has been found that a concentration of sulfanilamide or other sulfa drug that would prevent bacterial cultures from growing under ordinary circumstances loses this power when some para-aminobenzoic acid is added. The amount of para-aminobenzoic acid required to permit the bacteria to increase in number is approximately proportional to the excess of the amount of the sulfa drug over the minimum that would produce bacteriostatic action. This competition between the sulfa drug and para-aminobenzoic acid can be given a reasonable explanation. Let us assume that the bacteria need to have some para-aminobenzoic acid in order to grow; that is, that para-aminobenzoic acid is a vitamin for the bacteria. Probably it serves as a vitamin by combining with a protein to form an essential enzyme; presumably it serves as the prosthetic group of this enzyme. It is likely that the bacterium synthesizes a protein molecule that has a small region, a cavity, on one side of itself into which the para-aminobenzoic acid molecule just fits.

The sulfanilamide molecule is closely similar in structure to the para-aminobenzoic molecule (see Figure 14-11). Each of the molecules contains a benzene ring, an amino group ($—NH_2$) attached to one of the carbon atoms of the benzene ring, and another group attached to the opposite carbon atom. It seems not unlikely that the sulfanilamide molecule can fit into the cavity on the protein, thus preventing the para-aminobenzoic molecule from getting into this place. If it is further assumed that the sulfanilamide molecule is not able to function in such a way as to make the complex with the protein able to act as an enzyme, then the explanation of the action of sulfanilamide is complete. It is thought that the protein fits tightly around the benzene ring and the amino group, but not around the other end of the molecule. The evidence for this is that derivatives of sulfanilamide in which various other groups are attached to the sulfur atom are effective as bacteriostatic agents, whereas compounds in which other groups are attached to the benzene ring or the amino group are not effective.

EXERCISES

14-1. After 1828 the definition of organic chemistry (Exercise 13-1) was changed. How was it changed and what experiment required the change?

14-2. What are the general characteristics of living organisms? Which of these characteristics are possessed by viruses and which are not?

14-3. What are the chemical constituents of viruses? Give some of the additional constituents of the human body.

14-4. Distinguish between fibrous proteins and globular proteins.

14-5. Define an amino acid. How many principal amino acids occur in proteins?

14-6. If you sit on the R group of an L-(S)-amino acid in the plane of R—C—H looking toward the α-carbon atom with the hydrogen atom sticking out toward you above your head:

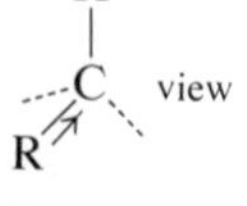

H
|
C view
R

on which side, right or left, is the amino group and on which side the carboxyl group? (See Section 6-4.)

14-7. Which are the essential amino acids?

14-8. Which amino acid is essential for infants but not adults? Which essential amino acids can be partially substituted for by which other amino acids in what relative amounts?

14-9. Write the chemical reaction equation for the hydrolysis of the dipeptide glycylalanine into the constituent amino acids.

14-10. How does ox insulin differ from human insulin in primary structure?

14-11. What is the denaturation of proteins and what is believed to be the process?

14-12. What is an enzyme? A prosthetic group? A conjugated protein? A coenzyme? A substrate?

14-13. Assuming that all the energy of hydrogen-bond formation of an enzyme-substrate complex directly lowers the activation energy of the enzyme-catalyzed reaction, approximately how many hydrogen bonds would have to be formed to increase the rate of reaction by a factor of 100,000?

14-14. What is the coenzyme of carbonic anhydrase? Of amylase?

14-15. Assuming that the rate constant for the formation of products from the enzyme-substrate complex is small compared to the rate constant for the formation of the complex from the reactants, how does the overall rate constant for the formation of products vary with the concentration of the substrate?

14-16. What is the likely mechanism of the treatment of the genetic disease cystathioninuria with large amounts of the vitamin pyridoxine?

14-17. The reaction between an enzyme and a competitive inhibitor is necessarily reversible. Would you expect hydrogen cyanide and hydrogen sulfide ions to be competitive or non-competitive inhibitors of the cytochromes?

14-18. What are monosaccharides? Disaccharides? Polysaccharides? Give examples.

14-19. Draw a structural formula for glycogen.

14-20. If you homogenize living tissue and extract it with water and an immiscible organic solvent, in which phase do you find the globular proteins? In which phase do you find the lipids?

14-21. Compare the average daily ingestion of fats, carbohydrates, and proteins with the amount of energy obtained from them.

14-22. The alcohol found in phospholipids is usually glycerine, 1,2,3-trihydroxypropane (Section 13-3). The phosphoric acid group links the alcohol to the nitrogen group. The phosphoric acid is in the 1 position of the glycerine and the two fatty acids are in the 2 and 3 positions. Draw a structural formula of a typical lecithin.

14-23. Draw the structural formula of cholic acid.

14-24. Define vitamins.

14-25. List the vitamins given in the text and their known physiological functions.

14-26. Compare the vitamin treatment of methylmalonicaciduria with that of cystathioninuria. Why are these treatments used?

14-27. Draw the necessary portion of the periodic table and indicate which elements are essential to human life.

14-28. What are hormones? What purpose do they serve?

14-29. Compare the chemical structures of estradiol (a female sex hormone) and testosterone (a male sex hormone).

14-30. Describe the probable mechanism of action of the sulfa drugs.

15

Molecular Biology

Molecular biology is the study of the structure of biological materials and the mechanism of their functions at the detailed molecular and atomic level. This branch of science began to develop between 1930 and 1940, when a penetrating understanding of the structure and properties of smaller molecules had been obtained by spectroscopic and magnetic techniques and especially x-ray diffraction by crystals and electron diffraction by gas molecules, aided by the theoretical insight provided by quantum mechanics. The first x-ray diffraction patterns of fibrous proteins and cellulose were made in 1918 and of crystals of globular proteins in 1934; only many years later, however, were the structures of proteins determined.

Molecular biology has developed very rapidly during the last quarter century; only a few aspects of the subject can be described in this chapter.

15-1. The Alpha Helix and the Pleated Sheets

The primary structure of a protein, the sequence of amino-acid residues in the polypeptide chains, has been discussed in Section 14-3. The term *secondary structure* is used to refer to certain simple ways in which the chains are coiled or folded and arranged together, the most important being the alpha helix and the two pleated sheets. (Tertiary structure is the aggregate of secondary structures for a polypeptide chain, together with the regions where one secondary structure changes to another, and quaternary structure is the mutual arrangement of several chains.)

The alpha helix and the pleated sheets were discovered not by direct deduction from experimental observations on proteins but rather by theoretical considerations based on the study of simpler substances. In amino acids one carbon atom, called the alpha carbon atom, C_α, lies between the amino group and the carboxyl group. When amino acids combine with one another to form a peptide or polypeptide, with elimination of water, the atoms between two alpha carbon atoms have the following structure:

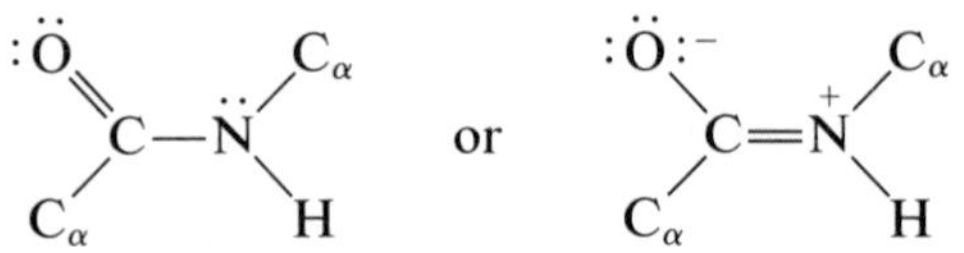

$$\begin{array}{ccc} :\ddot{O} & & C_\alpha \\ \| & & / \\ C & - \ddot{N} & \\ / & & \backslash \\ C_\alpha & & H \end{array} \quad \text{or} \quad \begin{array}{ccc} :\ddot{O}:^- & & C_\alpha \\ \backslash & & / \\ C & = \overset{+}{N} & \\ / & & \backslash \\ C_\alpha & & H \end{array}$$

There is resonance between these two valence-bond structures, with the structure on the left (the normal-valence structure) contributing about 60 percent and the other structure (the separated-charge structure) contributing about 40 percent to the normal state of the peptide group. The contribution of the second structure, which gives 40-percent double-bond character to the central bond, is great enough to hold the group of six atoms closely to a common plane, and to reduce the C—N bond length from the single-bond value, 147 pm, to 132 pm. These planar units can be rotated rather freely about the C_α—C and N—C_α single bonds. The problem was to rotate them in such a way as to permit the formation of hydrogen bonds, with N—H···O length about 280 pm, between the NH of one amide group and the O of another farther along the chain or in an adjacent chain, without distorting the planar amide groups by more than a small amount.

Several solutions to this problem were reported in 1950. In the alpha helix, shown in Figures 15-1 and 15-2, the chain has the conformation of a helix and the hydrogen bonds are parallel to the axis of the helix, linking adjacent turns together. There are about 3.6 amino-acid residues per turn (18 in 5 turns), and the pitch (distance along the axis) is 150 pm per residue, 540 pm per turn.

The alpha helix has been found to be present in many fibrous proteins and globular proteins. It might occur as either a right-handed or a left-handed helix, but in fact only the right-handed helix has been found in proteins and in synthetic polypeptides of L-amino acids. The probable

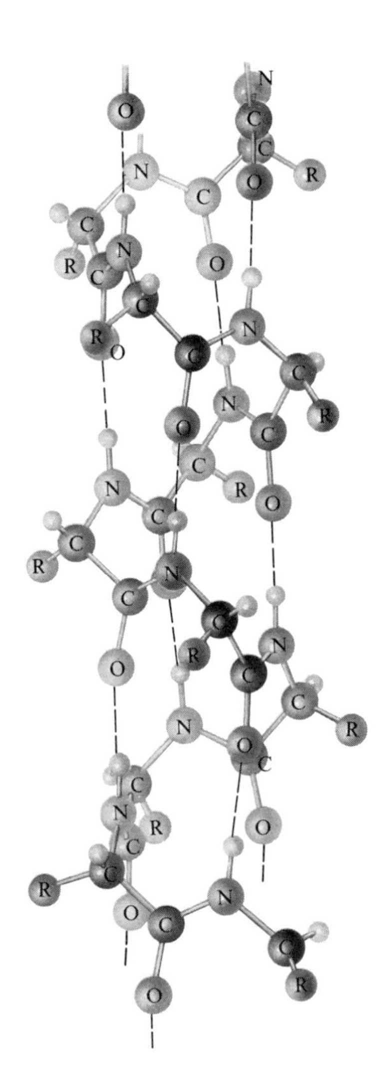

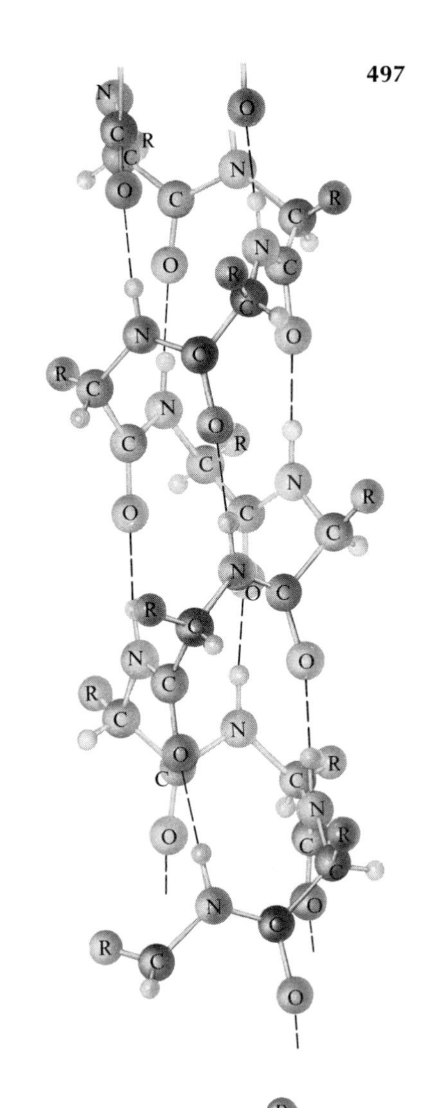

FIGURE 15-1
A drawing showing two possible forms of the α helix; the one on the left is a left-handed helix, and the one on the right is a right-handed helix. The right-handed helix of polypeptide chains is found in many proteins. The amino-acid residues have the L configuration in each case. The circles labeled R represent the side chains of the various residues.

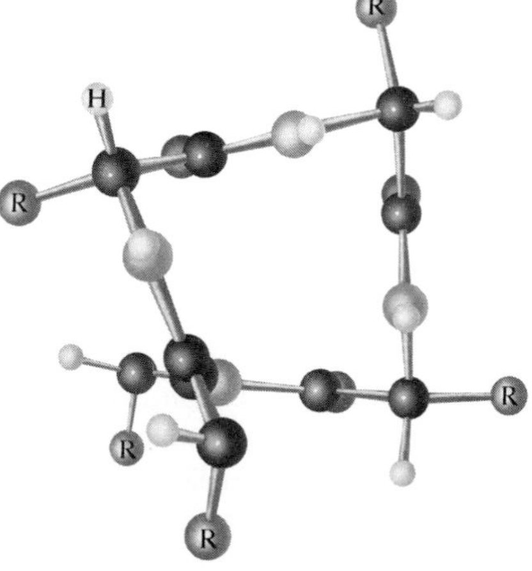

FIGURE 15-2
View down axis of the right-handed α helix.

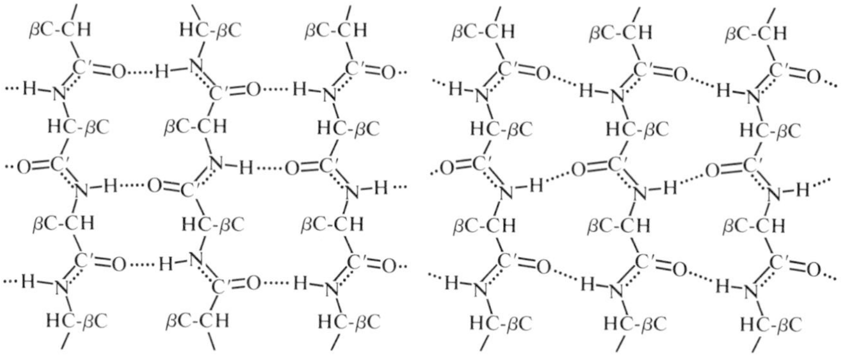

FIGURE 15-3
Diagrammatic representation of the antiparallel-chain pleated-sheet structure *(left)* and the parallel-chain pleated-sheet structure *(right)*.

explanation is that more room is provided for the bulky side chains by the right-handed than by the left-handed helix.

The pleated sheets have lateral hydrogen bonds between adjacent chains in a layer, as shown in Figure 15-3. From the figure it looks as though the chains could be fully extended, with the amide groups in the plane of the sheet. It is found, however, by calculation or construction of a model that the bond lengths and bond angles do not permit such a planar sheet to be formed. Satisfactory structures can be made by bending the chains at the alpha carbon atoms to form pleated sheets, as shown in Figures 15-4 and 15-5. The pleated sheets are found in silk, stretched hair, and globular proteins.

15-2. The Fibrous Proteins

One of the important fibrous proteins is *keratin,* which composes hair, fingernail, horn, porcupine quill, and other parts of animals. From your own experience you can conclude that keratin (from Greek *keros,* horn) serves as a valuable protective interface between an animal and its environment. It was found by the British molecular biologist W. T. Astbury (1898–1961) in 1931 that hair and other keratins normally produce a characteristic x-ray pattern, which he named the α-keratin pattern, and that stretched hair gives a different pattern, the β-keratin pattern. The α-keratin pattern is that of the alpha helix (the prefix α in α-helix was selected for this reason). The x-ray pattern of hair shows that the α-helixes are not simply aligned side by side, but instead are twisted about in groups of three or seven, as shown in Figure 15-6.

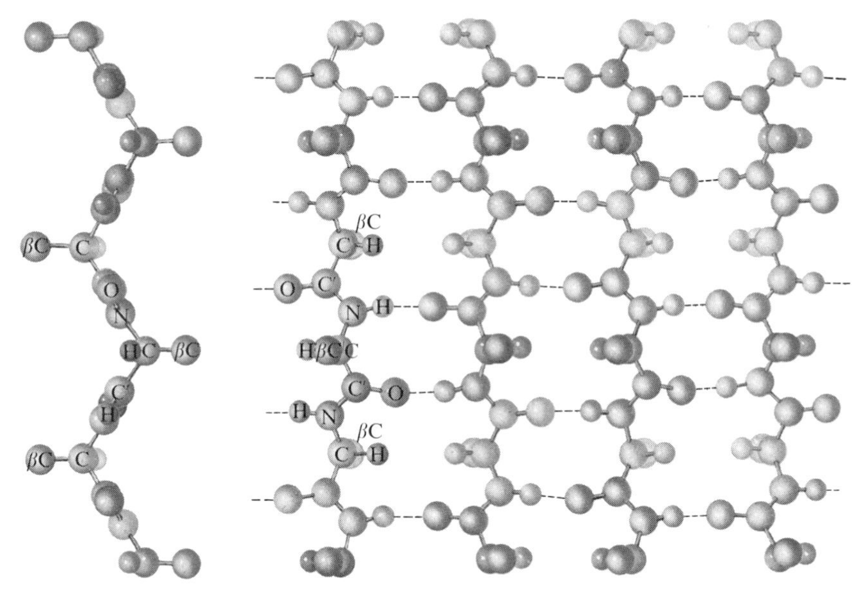

FIGURE 15-4
Drawing representing the antiparallel-chain pleated-sheet structure.

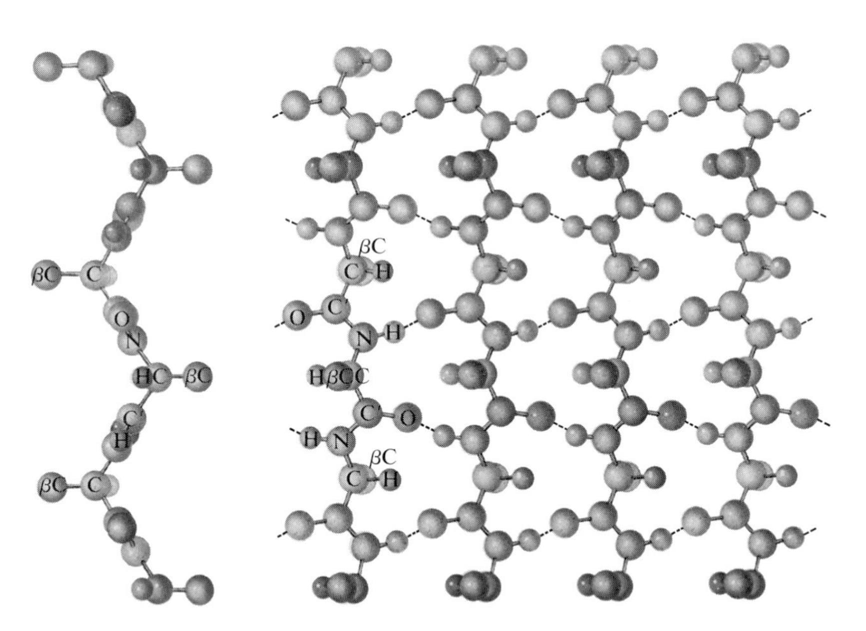

FIGURE 15-5
Drawing representing the parallel-chain pleated-sheet structure.

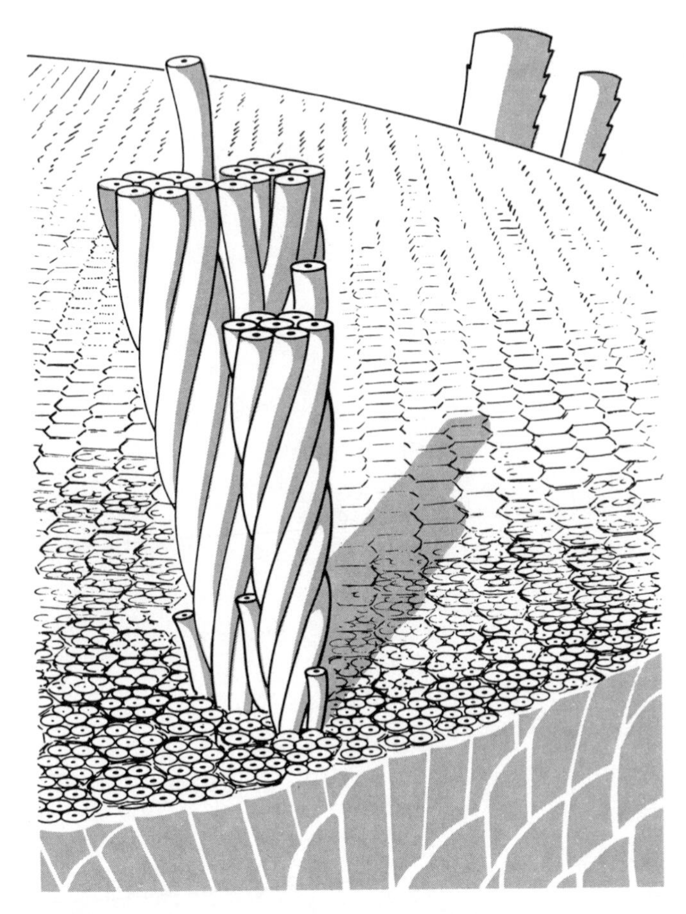

FIGURE 15-6
A drawing representing the molecular structure of hair, fingernail, muscle, and related fibrous proteins. The protein molecules have the configuration of the α helix (Figure 15-1); each molecule is represented in this drawing as a rod with circular cross section. These fibrous proteins contain seven-stranded cables, consisting of a central α helix and six others that are twisted about it. The spaces between these cables are filled with additional α helixes. [From J. A. Campbell, *Chemical Systems*, W. H. Freeman and Company. Copyright © 1970.]

Human hair contains about 12 percent of cystine, the double amino acid with two amino-acid groups connected by an S—S bond. For many cystine residues the two ends are in adjacent polypeptide chains, rather than in the same chain. This extensive cross-linking binds the entire hair together, explaining the insolubility of the protein in water, salt solution, and other solvents. If, however, the hair is treated with a reducing agent, such as sodium sulfide or an alkyl hydrosulfide, the cystine molecule is reduced to two cysteine molecules:

$$\underset{\text{cystine}}{R—S—S—R} + 2HSR' \rightleftarrows \underset{\text{cysteine}}{2R—SH} + R'—S—S—R'$$

The hair treated in this way becomes soluble and pliable. It can, for example, be curled, and then set by application of an oxidizing agent. Because of the insolubility of keratin most animals are unable to digest wool. An exception is the clothes moth, which has a high concentration of hydrosulfides in its digestive system.

After some of the disulfide bonds are broken hair can be stretched to a little over twice its normal length. It then gives a pleated-sheet x-ray pattern. The structure seems to be that of the parallel-chain pleated sheet (length per residue along the chain axis 325 pm, 2.17 times that for the α-helix), rather than of the antiparallel-chain pleated sheet (length per residue 350 pm, 2.33 times that for the α-helix).

Silk fibroin, the silk spun by silk worms and spiders, has the antiparallel-chain pleated-sheet structure. The observed length per residue along the fiber axis is 348 pm, in good agreement with the predicted value 350 pm. *Bombyx mori* silkworm silk contains 50 percent glycine residues, which alternate with other residues, mainly alanine and serine. As a result there are only hydrogen atoms on one side of the sheet, with the larger side chains, methyl and hydroxymethyl, on the other side. Alternate sheets are turned over, as shown in Figure 15-7, so that the interlayer distances are alternately small and large. Another kind of silk,

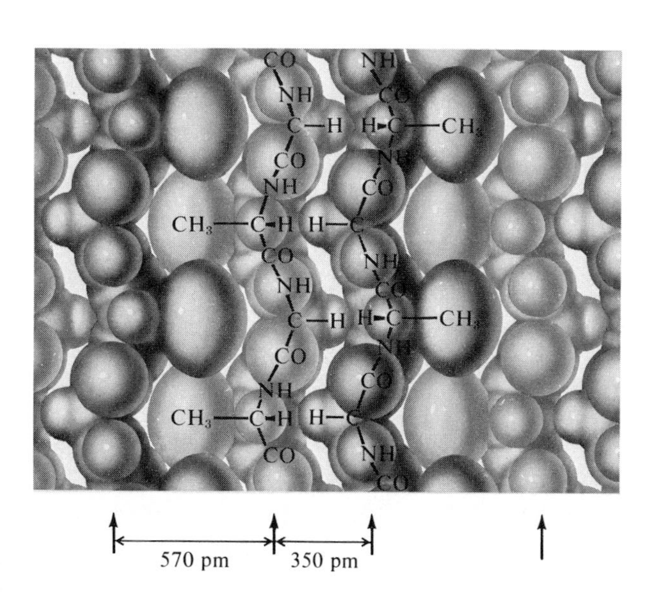

FIGURE 15-7
A cross section of *Bombyx mori* silk fibroin, showing four pleated sheets (indicated by arrows).

Tussah silk, contains only 27 percent glycine residues, and the pleated sheets are separated by the same distance, 530 pm, rather than alternating as in *Bombyx mori* silk.

Collagen is the major fibrous protein in skin, tendon, cartilage, bone, the cornea of the eye, the walls of arteries, and other tissues. It is also an important constituent, as collagen fibrils, of the intercellular cement that cements cells together in tissues (other important substances in the intercellular cement are hyaluronic acid and other long-chain mucopolysaccharides). Collagen differs from most other proteins in its high content of prolyl and hydroxyprolyl residues, which constitute about 25 percent of the amino-acid residues, and of glycyl residues, 34 percent. In the process of synthesis of collagen, the protein procollagen is first made. It contains no hydroxyproline, and collagen is formed from it by the hydroxylation of about half of the prolyl residues. Vitamin C is required for this hydroxylation reaction.

The polypeptide chains in collagen are about 280 nm long, and they consist of about 980 amino-acid residues. When skin, bones, and other tissues containing collagen are boiled the collagen hydrolyzes to shorter chains, forming the substance called *gelatin.* Gelatin is a good food, except that it contains no tryptophan and is rather low in methionine and some other essential amino acids.

Collagen differs in structure from the other fibrous proteins. Each polypeptide chain has the conformation of a left-handed helix, and three of these chains are held together by hydrogen bonds in a right-handed triple helix. The structure explains the fact that every third residue in the polypeptide chain of collagen is a glycyl residue: the three chains are held so closely together by the interchain hydrogen bonds that there is no room for a side chain on these residues larger than a hydrogen atom. The nearly extended chains of atoms linked together by covalent bonds give great strength to this fibrous protein—a fiber of tendon has about the same tensile strength as a wire of low-carbon steel. In tendon the chains extend along the axis of the tissue, whereas in the cornea of the eye there are alternating layers with the chains at right angles to one another.

Chirality of Organisms

It was mentioned in Section 14-3 that the reason that all proteins are composed of L-amino acids, rather than a mixture of L and D, is not known. It is, however, probable that the nature of the pleated sheets and the alpha helix, which are the principal secondary structures of proteins, provides the explanation. In both pleated sheets the structure is such that one of the two side-chain bonds of the alpha carbon atom extends out nearly at right angles to the plane of the sheet, giving plenty of room for the side chain, whereas the other lies nearly in the plane of the sheet, where there is room only for a hydrogen atom. Also, in the alpha helix with all L (or all D) residues the side chains (first carbon atoms) are more than 500 pm apart, whereas with L and D mixed they are only 350 pm apart. The structures with all L or all D amino-acid residues are accordingly more stable, because of less crowding of large side chains, than those with L,D mixtures. Organisms based entirely on L (or D) amino acids (and corresponding carbohydrates and other substances) are, moreover, far simpler than those with both L and D. Enzymes are in general stereospecific; an enzyme that catalyzes a reaction involving an L substrate will not catalyze the same reaction for the corresponding D substrate. Accordingly the existing organisms now get along with only half as many enzymes as they would need if they were composed of both L and D isomers. The choice of L amino acids rather than D was probably accidental.

15-3. The Structure of Muscle and the Mechanism of Muscular Contraction

Animals differ from plants in their ability to move voluntarily. This motion is accomplished by the use of muscles. A muscle fiber may be 10 to 1,000 μm in diameter, and, in man, as much as 35 cm long (the sartorius muscle).

The contraction of muscle results from the interaction of protein molecules. A unit of muscle (a sarcomere) has the structure shown in Figure 15-8, as determined by use of the electron microscope. In the center of each sarcomere there is an array of filaments of the protein *myosin,* each filament being about 16 nm in diameter and about 1,500 nm long. In the interstices between these filaments, at each end, there are smaller filaments of another protein, *actin,* about 7 nm in diameter and 1,000 nm long. The actin filaments are attached to plates that mark the end of one sarcomere and the beginning of another, as shown in Figure 15-8. In the muscle of vertebrates there are four times as many actin filaments as myosin filaments. Figure 15-9 shows an electron micrograph, magnification 100,000 diameters, of a transverse section of the leg muscle of a frog, showing twice as many thin filaments as thick ones. (There is another net of thin filaments at the other end of the thick ones.) The centers of the thick filaments are about 45 nm apart, and the smallest distance between the surfaces of a thin filament and a thick one is about 15 nm.

The actin filaments can be dissociated into globular molecules with molecular mass about 58,000 d. In the filaments themselves these molecules are aggregated into two chains that are twisted about one another

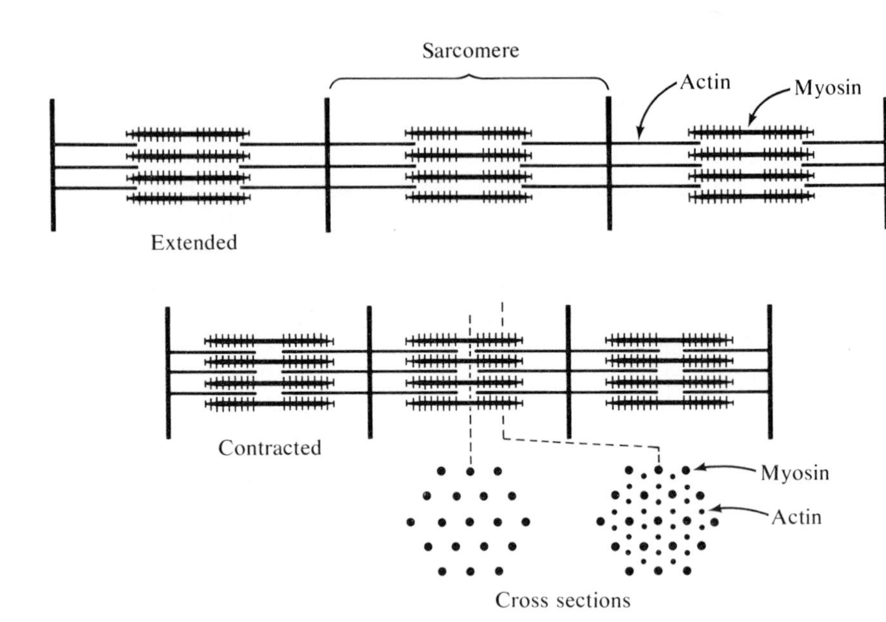

FIGURE 15-8
Diagram illustrating the structure of muscle.

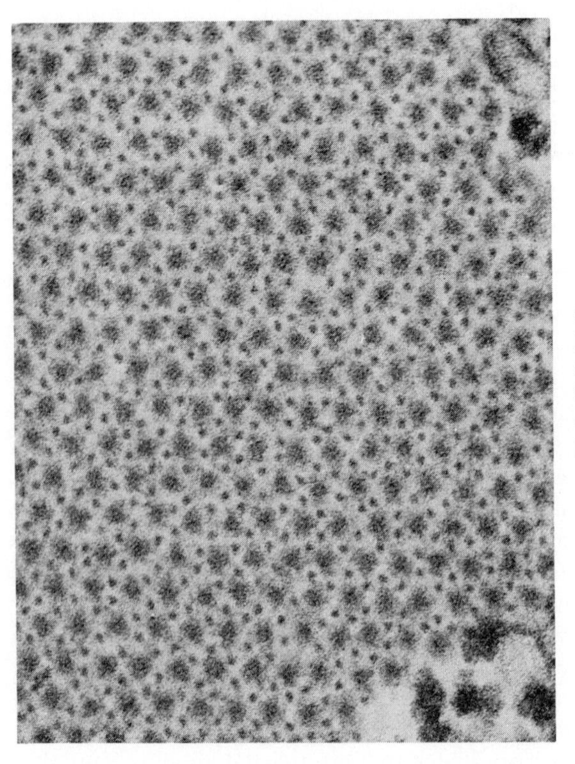

FIGURE 15-9
Electron micrograph showing a transverse section of the leg muscle of a frog, magnification 100,000 diameters. [Courtesy of H. E. Huxley.]

into a double helix, as shown in Figure 15-10. The myosin filaments can be dissociated into myosin molecules, which have molecular mass about 525,000 d. They are seen in the electron microscope to have the shape of a rod 200 nm long and 2.0 nm in diameter, with a knob 20 nm long and 4 nm in diameter at one end. The x-ray pattern shows the secondary structure to be that of the alpha helix, and the rod probably consists of three alpha-helix chains twisted about one another. There are about 600 molecules in each myosin filament. Electron micrographs and x-ray patterns show that the molecules are approximately parallel to one another in the filament, their tails being intertwined and their heads projecting above the surface. The heads are directed toward the closer end of the filament, and the middle part of the filament, about 200 nm long, is bare of heads. The heads consist of about 300 amino-acid residues in each of the three chains. With the alpha-helix structure each chain extends 45 nm (more for the extended structure), easily bridging the distance to an adjacent actin filament.

Electron micrographs of an actin filament in a solution of myosin molecules show that the myosin molecules are attached to the actin

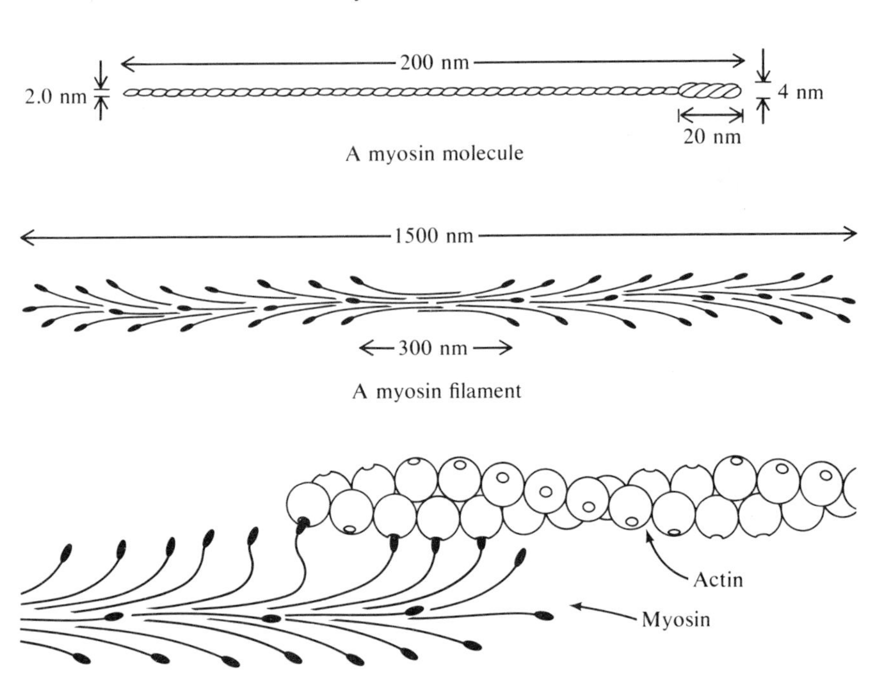

FIGURE 15-10
A myosin molecule and a myosin filament are represented above. The diagram below shows a myosin filament creeping along an actin filament by the interaction of the combining groups in the myosin head with complementary regions of the actin molecules.

filament in such a way that their tails all extend toward one end of the filament.

These observations are compatible with a theory of muscular contraction illustrated in Figure 15-10. Each myosin filament has about 1800 chain ends projecting from it. There are regions on each actin molecule in the actin filaments that are complementary in structure to a region on the myosin chain ends, and can combine with it to form a weak bond (see Section 15-5). The complementariness can be destroyed by a change in structure of one of the combining groups, which can be achieved by interaction with a source of energy. When the muscle is stimulated to contract the mutually complementary regions begin to combine with one another, if they can be brought together. The end chains of myosin reach along the actin filament, and each chain moves to the next actin combining site, as the actin filament is drawn further into the myosin region, thus achieving the contraction of the sarcomere.

During the relaxation phase the reaction with ATP changes the structure of one of the combining regions in such a way as to destroy the complementariness, and the actin filaments then withdraw from their interstitial positions.

No detailed information has been obtained as yet about the nature of the combining groups of actin and myosin. There are about 0.1 g of myosin and 0.04 g of actin in 1 cm^3 of muscle, corresponding to 0.6 x 10^{-6} mole of myosin chains (three per molecule) and 0.7×10^{-6} mole of actin molecules. The approximate agreement of these numbers supports the reasonable assumption of one combining region per myosin chain and per actin molecule.

Muscular force is about 3.5 kg per cm^2 of cross section. The amount of contraction is about 33 percent, and hence the work done by 1 cm^3 in contracting is about 0.1 J, which corresponds to 143 kJ per mole of actin. This large value, several times the energy provided by one ATP molecule, suggests that the reaction occurs in several steps.

15-4. The Structure and Properties of Globular Proteins

The first nearly complete determination of the structure of a globular protein was made by the English scientist John C. Kendrew (born 1917) and many collaborators during the period 1946 to 1960. They made and interpreted the x-ray diffraction photographs of a crystal of sperm-whale *myoglobin.* Myoglobin, which is present in muscle, is a protein rather similar to hemoglobin, but with only one polypeptide chain in the molecule (about 150 amino-acid residues, molecular mass about 17,000 d), and one heme group (Section 13-7). The x-ray investigation of crystals of hemoglobin, carried out by the English scientist Max Perutz (born 1914) and his collaborators, has shown that its molecule is an aggregate of four subunits, each of which has a structure closely resembling that of myoglobin.

The principal secondary structure found in myoglobin and hemoglobin is the alpha helix. There are eight segments of the polypeptide chain that are coiled into the conformation of the alpha helix; these segments constitute 80 percent of the residues, with the remaining 20 percent being involved in turning the corner from one segment of alpha helix to the next. The resulting tertiary structure forms a cleft, within which the heme group is contained.

An approximate representation of the structure is shown in Figure 15-11, a stereodrawing of the hemoglobin of the bloodworm (a polychaete annelid, *Glycera dibranchiata*). Each peptide group, CCONHC, is represented by a rod between two small circles, which show the positions in space of the alpha-carbon atoms of the 150 or 146 amino-acid residues. Several segments of alpha helix can be clearly seen. The iron atom, represented by the large circle in the center of the nearly planar heme group, forms six bonds: four with nitrogen atoms of the heme group, one with a nitrogen atom of the imidazole ring in the side chain of the histidine residue (residue 90, measured from the free amino end of the chain), and one with an oxygen atom of the attached oxygen molecule (shown by a circle to the left) or the carbon atom of an attached carbon monoxide molecule.

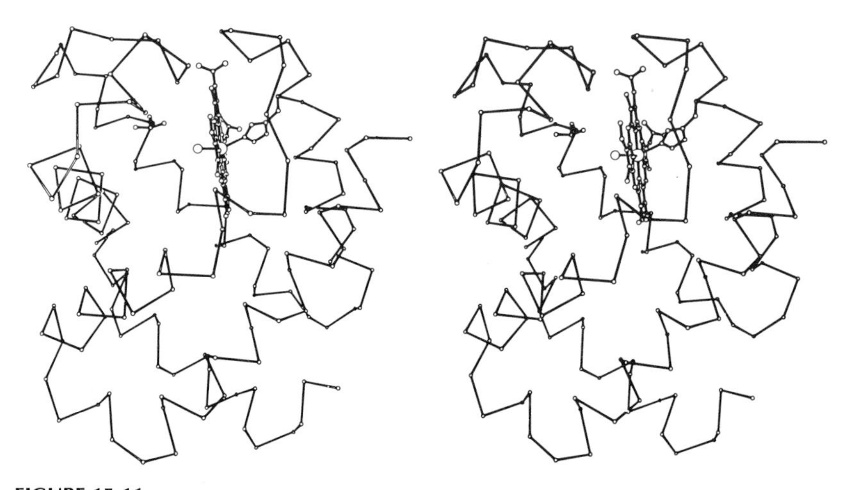

FIGURE 15-11
Stereodrawing of the molecule of *Glycera* hemoglobin. The positions of the alpha carbon atoms are shown, connected by straight rods. The heme group is also shown. [From E. A. Padlan and W. E. Love, *Journal of Biological Chemistry* **249**, 4067 (1974).]

The Alpha and Beta Chains of Hemoglobin

By Sanger's methods (Section 14-3) it has been found that a molecule of mammalian hemoglobin contains four polypeptide chains, each with its attached heme group. In most mammalian hemoglobins the chains are of two kinds, called the alpha chain and the beta chain, with two alpha chains and two beta chains in each molecule. Normal adult human hemoglobin contains alpha chains with 140 amino-acid residues and beta chains with 146 residues; for other mammalian species the numbers of residues are nearly the same. The sequences of amino-acid residues are completely known for the normal human hemoglobin chains, for many abnormal human hemoglobins (see Section 15-8), and for many hemoglobins of other species of animals. The first few residues for normal adult human hemoglobin are the following:

ALPHA

```
    1   2   3   4   5   6   7   8   9   10  11  12
H2N-Val-Leu-Ser-Pro-Ala-Asp-Lys-Thr-Asn-Val-Lys-Ala-
```

BETA

```
H2N-Val-His-Leu-Thr-Pro-Glu-Glu-Lys-Ser-Ala-Val-Thr-
```

The two kinds of chains are rather similar; about half of the residues are the same in corresponding loci (note deletion of His between loci 1 and 2 in the alpha chain). Some invertebrate hemoglobins consist of molecules with only one kind of chain.

The Oxygen Equilibrium of Hemoglobin. Allosteric Structures of Proteins

The molecule of human hemoglobin, like those of other mammalian hemoglobins, consists of four polypeptide chains, each with a heme group, and it can combine reversibly with four oxygen molecules. It was found many years ago that the oxygen equilibrium curve for hemoglobin has a sigmoid (S-like) shape, as shown in Figure 15-12, differing from that for myoglobin. Myoglobin, with one heme group in the molecule, would be expected to show an equilibrium curve with oxygen corresponding to the reaction

$$\text{My} + \text{O}_2 \rightleftharpoons \text{MyO}_2$$

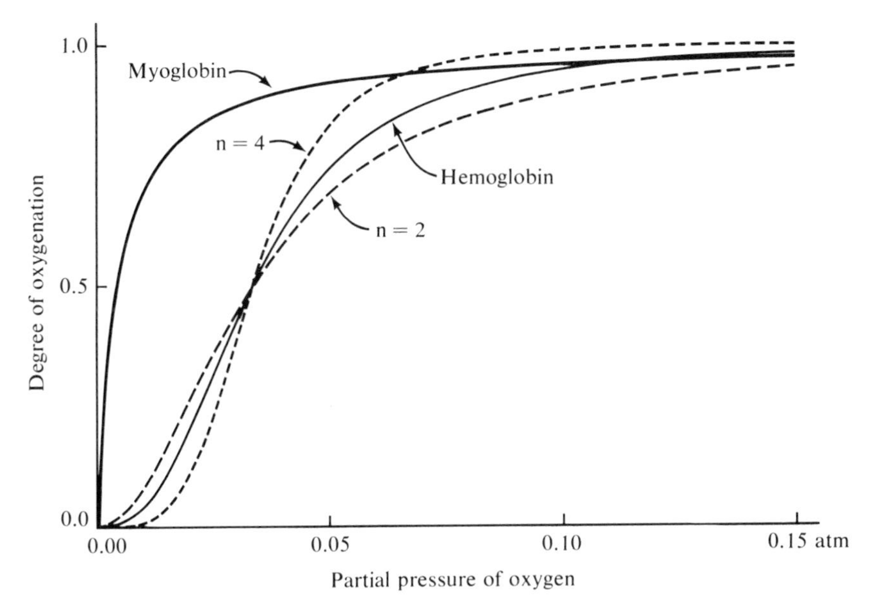

FIGURE 15-12
Diagram showing the observed degree of oxygenation in relation to the partial pressure of oxygen for myoglobin and hemoglobin and calculated curves for addition of two or four oxygen molecules simultaneously (very strong heme-heme interactions).

The corresponding equilibrium expression is

$$\frac{[\mathrm{MyO_2}]}{[\mathrm{My}]P_{\mathrm{O_2}}} = K$$

or

$$\text{Fraction oxygenated} = \frac{[\mathrm{MyO_2}]}{[\mathrm{My}] + [\mathrm{MyO_2}]} = \frac{K_{P_{\mathrm{O_2}}}}{1 + K_{P_{\mathrm{O_2}}}}$$

The experimental values agree well with this expression.

If the four hemes of the hemoglobin molecule combined with oxygen independently of one another the same expression would apply also to hemoglobin. The observed sigmoid curve shows that there is an interaction between the heme groups, such that when one is oxygenated one or more of the others is changed in such a way as to increase its oxygen equilibrium constant.

One possibility is that the heme-heme interactions are so strong that there are no intermediates in appreciable concentration between Hb_4 and

$Hb_4(O_2)_4$. The equilibrium expression for the corresponding reaction,

$$Hb_4 + 4O_2 \rightleftarrows Hb_4(O_2)_4$$

is

$$\frac{[Hb_4(O_2)_4]}{[Hb_4]} = K'P_{O_2}{}^4$$

The curve given by this relation is shown in Figure 15-12. We see that it does not fit the experimental curve, and hence that the assumption of very strong heme-heme interactions is wrong.

Another simple assumption is that the hemes interact very strongly with one another in pairs, with no interaction between the pairs. The corresponding equilibrium expression

$$\frac{[Hb_2(O_2)_2]}{[Hb_2]} = K''P_{O_2}{}^2$$

gives the curve also shown in Figure 15-12. We see that it also does not fit, being less sigmoid than the experimental curve.

Good fit with experiment is obtained by the assumption that a heme increases its oxygen combination constant by the factor 12 for each adjacent HbO_2, the hemes being arranged in a square. The factor 12 corresponds to a heme-heme interaction energy of $RT \ln 12 = 6$ kJ mole^{-1}.

The hemes are far apart (3000 pm) in the molecule, and no known direct interaction is large enough to account for this effect. It has been found, however, that the addition of oxygen or other ligand changes the shape of the coiled polypeptide chain, and the interaction with the adjacent chains transmits the effect.

The first evidence that a ligand combining with a heme of myoglobin or hemoglobin is crowded into a cavity was provided by studies of combination with alkyl isocyanides, RNC, with R = ethyl, isopropyl, and tertiary butyl. These groups, $—CH_2CH_3$, $—CH(CH_3)_2$, and $—C(CH_3)_3$, become successively larger in this order. They combine with ferroheme itself, and their combination constants are very nearly the same: the strength of the Fe—CNR bond is independent of the size of the alkyl group. With both myoglobin and hemoglobin, however, the combination constants decrease with increasing size of the group (divisor about 14 for each CH_3 replacing H). A reasonable explanation of this observation is that the ligand is being crowded into its position in the cleft in the myoglobin molecule and the hemoglobin quarter-molecule. This introduction of the ligand into a too-small hole can be achieved by an expansion of the hole through a change in structure of the molecule. Additional evidence

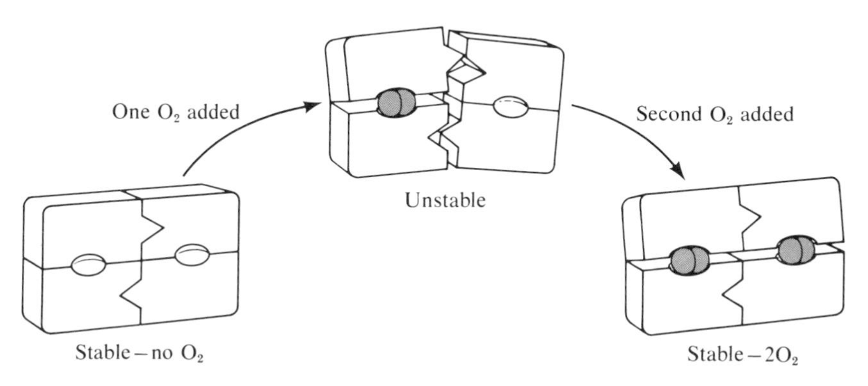

FIGURE 15-13
Diagram representing dimer with heme-heme interaction through allosterism.

that a change in structure occurs on oxygenation of hemoglobin is provided by the fact that human oxyhemoglobin crystals are different from human hemoglobin crystals (also for hemoglobins of other species); if no change in the molecules occurred the structures of the crystals would be the same.

The four subunits of hemoglobin are held to one another by weak interactions between complementary surface structures, as discussed in Section 15-5 and illustrated (for two subunits) in Figure 15-13. The complementariness is partially destroyed on the addition of one oxygen molecule, and regained on the addition of the second. The changes in interaction energy operate to decrease the value of the combination constant for the first oxygen molecule and increase it for the second.

This mechanism has value for the organism, in that it increases the effective load of oxygen carried from the lungs to the tissues. In the lungs, with partial pressure of oxygen about 0.15 atm, somewhat less than that in the air, the hemoglobin is about 95 percent oxygenated, and in the tissues, with partial pressure about 0.04 atm, it is 20 percent oxygenated.* Its effective load is accordingly 75 percent of the maximum possible, as compared with only 32 percent for independent hemes.

The change in conformation of the polypeptide chains of hemoglobin on addition of oxygen is an example of *allosterism.* Allosteric forms of other proteins, especially enzymes, have also been recognized. Allosterism provides a mechanism whereby various properties of a large protein molecule can be changed by combination with a small molecule. For hemoglobin the value of the phenomenon is that it permits the hemes to interact in such a way as to lead to a sigmoid oxygen equilibrium curve.

*In fact, the partial pressure is somewhat larger than 0.04 atm, and the release of oxygen from hemoglobin is aided by the increased acidity of the blood resulting from transfer of carbon dioxide from the tissues to the blood.

Enzymes and Other Globular Proteins

The method of x-ray diffraction by crystals has been used to determine the detailed structures of a number of globular proteins. The alpha helix and the two pleated sheets have been found to be the main types of secondary structure in these proteins. The location of the catalytically active region of an enzyme can be discovered by the x-ray study of crystals of the enzyme combined with an inhibitor.

Many of the enzymes that catalyze oxidation-reduction reactions contain iron atoms. Examples are the cytochromes, present in every living organism, which contain heme groups bound in a different way than in myoglobin and hemoglobin. An interesting non-heme iron-containing protein is HiPIP (high-potential iron protein). This protein has been extracted from the cells of several species of purple bacteria. It can be reversibly oxidized by one step (extraction of one electron) by ferricyanide ion and other oxidizing agents, and presumably catalyzes some physiological oxidation reactions. The structure of the molecule is shown in Figure 15-14. Some short segments of alpha helix can be seen, and a couple of pieces of antiparallel-chain pleated sheet. In the middle of the molecule there is a group of four iron atoms and eight sulfur atoms. The four iron atoms can be described as lying at four non-adjacent corners of a cube, with four sulfur atoms a little way out from the four other corners

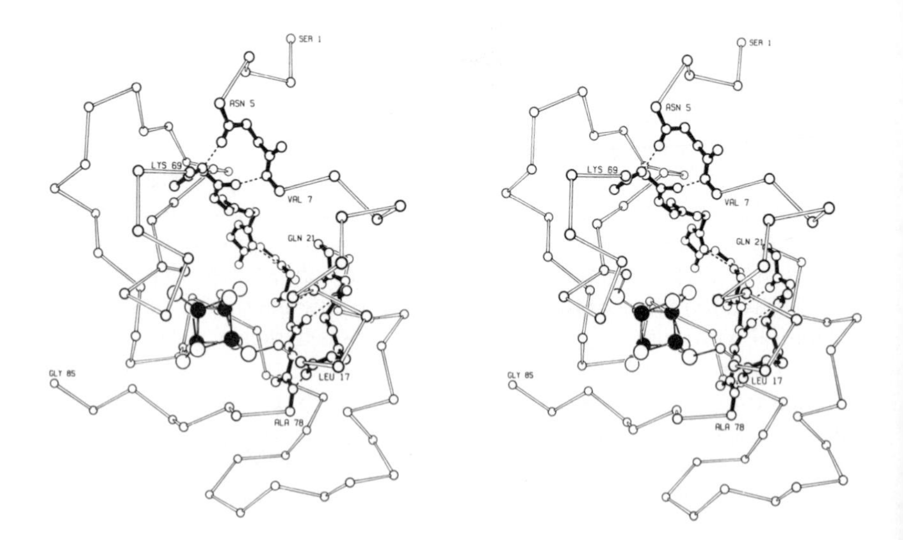

FIGURE 15-14
A stereodrawing showing the alpha carbon atoms of the 85 amino-acid residues and the Fe_4S_8 group of the protein HiPIP. [From C. W. Carter, Jr., J. Kraut, S. T. Freer, N. Xuong, R. A. Alden, and R. G. Bartsch, *Journal of Biological Chemistry* **249,** 4212 (1974).]

of the cube. In addition, there are four more sulfur atoms, which lie radially out from the iron atoms. These outer sulfur atoms represent cysteine residues in positions 43, 46, 63, and 77 of the 85-residue chain. The hydrogen atom of each cysteine sulfhydryl group is replaced by an iron atom. An iron(III) atom in the group can be described as forming a single covalent bond with the adjacent cysteine sulfur atom and two-thirds bonds with each of the other three adjacent sulfur atoms (two covalent bonds resonating among three positions). The tetrahedral arrangement of four sulfur atoms around an iron atom is found also in some sulfide minerals, such as chalcopyrite, $CuFeS_2$.

The structure of a proteolytic enzyme, carboxypeptidase A, is shown in Figure 15-15. The polypeptide chain contains 307 amino-acid residues and one zinc ion. There are several sequences of alpha helix, as well as a somewhat twisted pleated sheet (near the center). The catalytically active region is near the zinc atom. A stereoview of a portion of an enzyme (lysozyme, found in tears and in eggwhite, providing protection against infection by catalyzing the hydrolysis of polysaccharides of bacteria), together with a molecule of substrate, is shown in Figure 15-16. Hydrogen bonds with the enzyme strain the substrate molecule in such a way as to decrease the activation energy for its hydrolysis into two smaller molecules.

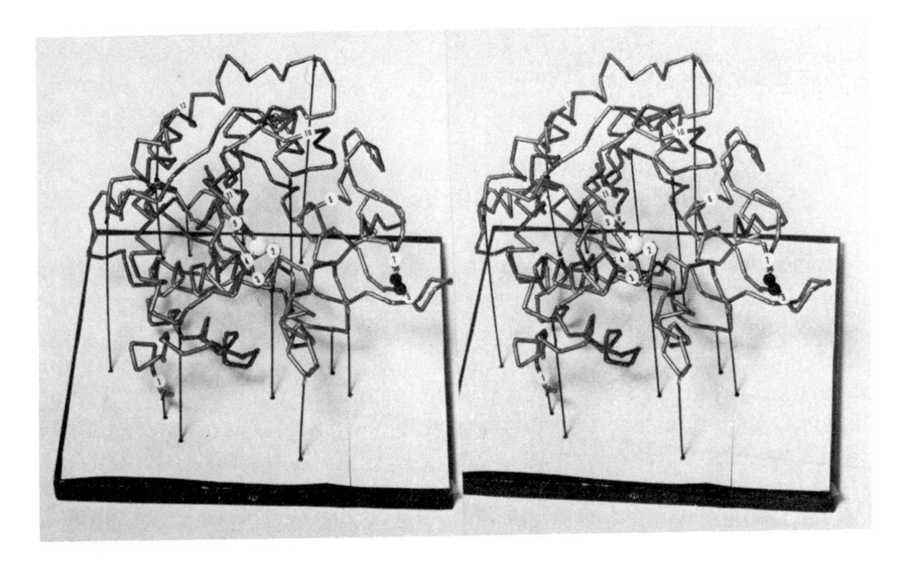

FIGURE 15-15
A stereographic view of the molecule of the proteolytic enzyme carboxypeptidase A. Each amino-acid residue is represented by a rod between the alpha carbon atoms. There are 307 residues in the polypeptide chain, and a zinc atom, shown as a white sphere between the numbers 2 and 4. [From W. N. Lipscomb et al., *Structure, Function, and Evolution in Proteins,* Brookhaven Symposia in Biology, No. 21 (1968).]

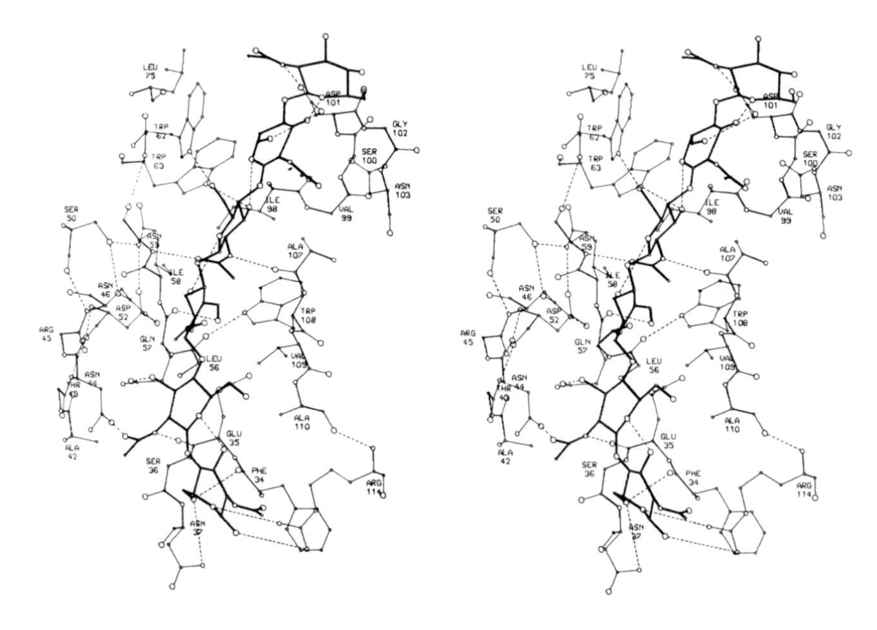

FIGURE 15-16
A stereodrawing showing portions of the polypeptide chain of the enzyme lysozyme and a substrate molecule with six glucose rings, held in a deformed position by hydrogen bonds to the enzyme. The molecule is cleaved by hydrolysis between the third and the fourth ring. [From Professor D. C. Phillips, Oxford University.]

15-5. Antigens and Antibodies. The Molecular Basis of Biological Specificity

A person who contracts smallpox may die, as a result of the damage done to his body by the smallpox virus multiplying in his cells. If he survives he has immunity against smallpox for the rest of his life; but this immunity does not extend to other viral diseases. The immunity results from the induced manufacture of specific protein molecules, called *antibodies;* the molecules (in this case the smallpox virus particles) that induce antibody formation are called *antigens.*

Human antibodies are the proteins called gamma globulin in the serum of the blood. They have molecular mass about 270,000 d. Proteins and polysaccharides foreign to the individual or the species of animals, when injected, may serve as antigens. Also, small molecules may be attached to proteins and injected to induce formation of antibodies with the power of combining specifically with them. For example, if *p*-aminobenzoic acid, $H_2NC_6H_4COOH$, is treated with hydrochloric acid and sodium nitrate it forms the diazonium ion $:N{\equiv}^{+}N{-}C_6H_4COOH$, which can react with some side-chain groups of proteins to give azoproteins, having the azobenzoate group $—NNC_6H_4COO^-$ attached. Antibodies produced by injecting this azoprotein into an animal may have small combining power with the original protein and large combining power with the attached azobenzoate group (which is called the *haptenic group* or *hapten,* from the Greek word *haptein,* to grasp). The antiserum forms a precipitate with any protein to which azobenzoate groups are attached, and it agglutinates (clumps) cells to which these groups are attached. The antiserum does not form a precipitate with benzoate ion itself. Instead, benzoate ion in high enough concentration prevents precipitation of homologous azoprotein and agglutination of azocells by the antiserum.

These facts are explained by the assumption that the antibody molecule has two or more combining regions. The azoprotein molecule has several haptenic groups on its surface, and the antibody molecules bind the azoprotein molecules into a framework, which constitutes the serological precipitate (Figure 15-17). With the hapten, the antibody molecule remains in solution, even though the haptens are attached to their combining regions.

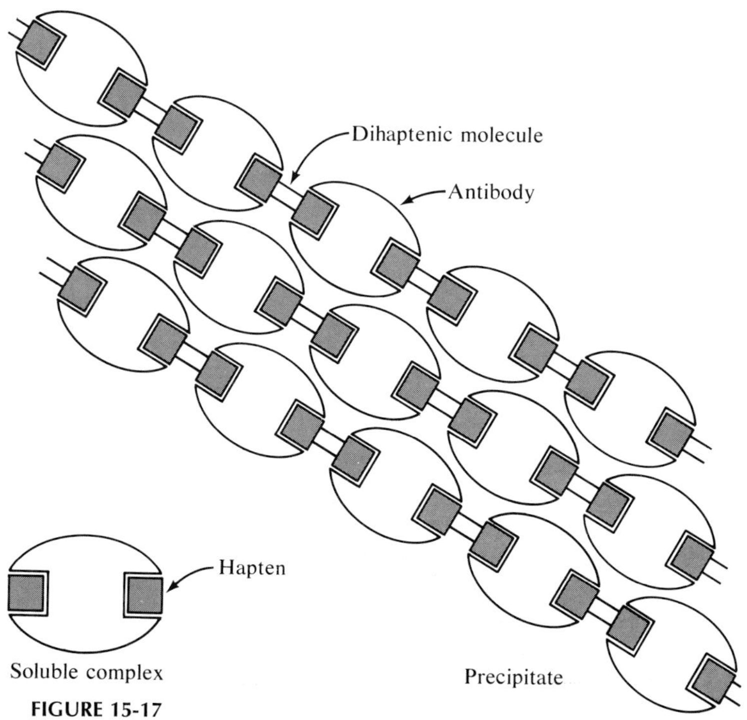

FIGURE 15-17
Diagram of soluble complex of bivalent antibody with two haptens and of precipitate of antibody with a dihaptenic substance.

It was in fact found that a substance such as diazobenzoate *o*-cresol,

$$^{-}OOCC_6H_4NN-C_6H_2(OH)(CH_3)-NNC_6H_4COO^{-}$$

containing two haptenic groups in the molecule, forms a precipitate with the anti-azobenzoate antiserum. Moreover, the precipitate contains equal numbers of molecules of this dihaptenic substance and of antibody, showing that the precipitating and agglutinating antibody molecules have two combining regions.

The Structure of Antibodies

It has been shown, largely through the work of R. R. Porter in England and G. M. Edelman in the United States, that the molecule of human antibody consists of four polypeptide chains, held together by the sulfur-sulfur bonds of cystine, as shown in Figure 15-18. Each of the two long chains contains 446 amino-acid residues and each of the two short chains contains 214 residues. The primary structure (nature of the 660 residues) has been completely determined. The first 108 residues in each chain are different for different antibodies; it is these variable parts of the antibody that give it the power of combining selectively with the homologous antigen or hapten. The ways in which the polypeptide chains are coiled have not yet been determined.

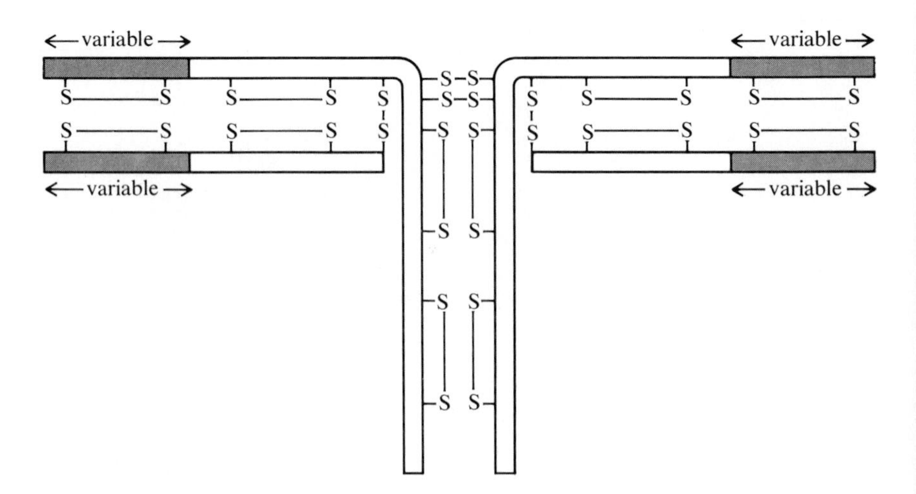

FIGURE 15-18
The arrangement of the four polypeptide chains in a molecule of human antibody.

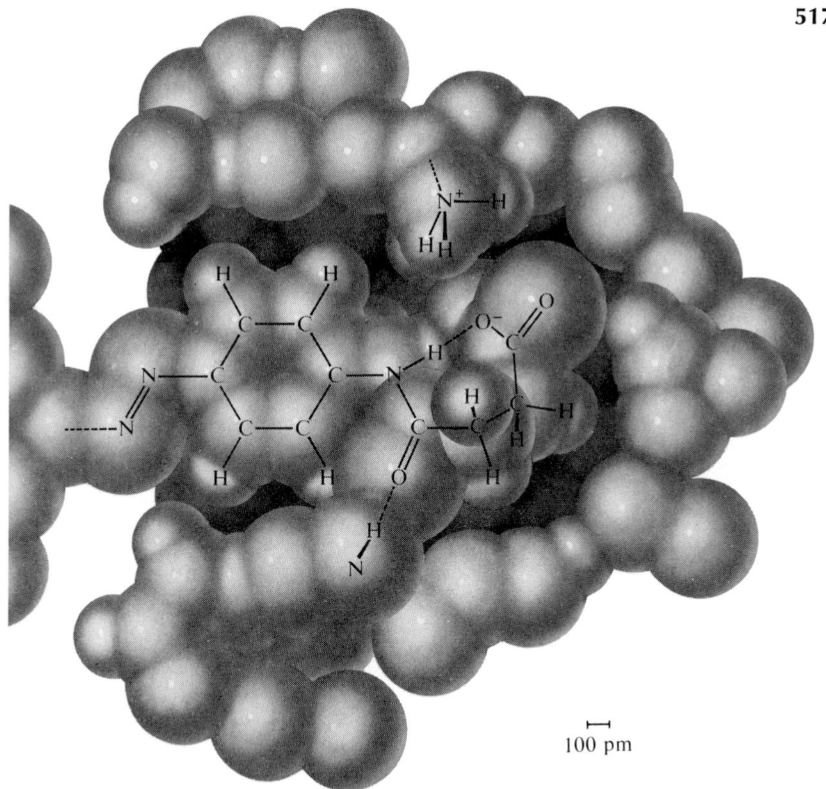

FIGURE 15-19
Drawing of the *p*-azosuccinanilate ion and the combining region of a homologous antibody, showing the close complementariness in structure.

The Antigen-antibody Bond

The bond between the combining region of an antibody and a hapten or the haptenic group of an antigen is not a covalent bond, but is the sum of a number of weak interactions—electronic van der Waals attraction, formation of hydrogen bonds, attraction of electrically charged groups. Several interactions cooperate to give a resultant attraction strong enough to resist the disrupting influence of thermal agitation. These forces fall off rapidly with increasing distance between the groups. Accordingly to form an effective bond the combining region must be closely complementary in shape and size and in position of corresponding groups to the hapten, as shown for the hapten *p*-azosuccinanilate ion in Figure 15-19.

The closeness of fit has been determined by measuring the equilibrium constants of combination of various haptens with antibodies. This can be done by determining the concentration of the hapten needed to inhibit by 50 percent the precipitation of the antibody by a dihaptenic substance or the azoprotein. The antibody against the *p*-azobenzoate ion, for example, has combination constant with *o*-chlorobenzoate ion only one fifth as great as with benzoate ion. The larger size of the chlorine atom (van der Waals radius 181 pm) than of the hydrogen atom (110 pm) makes it difficult for the chlorine-substituted hapten to fit into the region occupied by a hydrogen atom in the haptenic group of the azoprotein that stimulated the rabbit to produce the antibody. On the other hand, *p*-chlorobenzoate ion combines 3.5 times more strongly with the antibody than benzoate ion. In this para position, occupied by the azo group in the stimulating haptenic group, there is no steric interference with the chlorine atom, and the greater electric polarizability of the chlorine atom than of the hydrogen atom leads to a greater value of the van der Waals attraction for the adjacent parts of the combining region of the antibody.

For the succinanilate system there is evidence that a hydrogen bond is formed between the CO group and the antibody. The combining constant decreases to 0.01 of the value for succinanilate ion when the CO group is replaced by CH_2. There is similar evidence for a positive electric charge in the antibody, close to the negative charge of the haptenic group.

Quantitative studies of the combination of hundreds of haptenic groups of known structure with antibodies have yielded convincing evidence that the striking specificity of antibodies results from their having structures that are closely complementary to those of the stimulating antigens. The combining region of the antibody extends over the whole of the hapten. The fit is close, to within about 100 pm, less than the diameter of an atom. Electric charges and hydrogen-bond-forming groups complementary to those in the hapten and fixed in the suitable positions for maximum attraction are present in the antibody. The combining region of the antibody is rigid, and does not adjust itself to the structure of the hapten.

It is likely that biological specificity in general is the result of a detailed complementariness in structure of the interacting molecules. Enzymes are complementary in structure to a strained configuration of the reacting molecules. Genes are complementary nucleic acid chains, as discussed in Section 15-6.

Complementariness in Crystallization

There is an analog to biological specificity in the inanimate world. It is the phenomenon of crystallization. A growing crystal can remove certain molecules from a complex solution or melt, rejecting all others. An example is provided by the crystals of potassium hydrogen tartrate that sometimes form in a glass of grape jelly. Even though the jelly contains hundreds of different kinds of molecules and ions, the crystals, when washed and analyzed, are found to be 99.99-percent pure.

The specificity of crystallization results from the complementariness in structure of each constituent molecule or ion and the rest of the crystal. We can imagine removing a single tartrate ion from the crystal. We can see that the cavity has just the size and shape to permit the tartrate ion to fit neatly into it. Moreover, there are positively charged ions in the rest of the crystal close to the places to be occupied by the negative charges of the tartrate ion, and also complementary hydrogen-bond-forming groups.

The same failures to achieve complete specificity are found in serological systems and in crystals. For example, the antibody to the *p*-azo-*m*-bromobenzoate ion group and that to the *p*-azo-*m*-methylbenzoate ion group react with the two haptens nearly equally well; the antibodies are unable to distinguish clearly between the methyl group and the bromine atom, which have nearly the same size and shape. In the same way, crystals that form in a solution of *m*-bromobenzoic acid and *m*-methylbenzoic acid are not pure crystals of one substance or the other; instead, they are crystalline solutions, containing the two kinds of molecules in ratio determined by the concentrations in the liquid.

The Process of Formation of Antibodies

An old idea (1940) about the structure and process of formation of antibodies started with the assumption that a precipitating or agglutinating antibody molecule consists of a central part, with well-defined structure, and two ends, consisting of polypeptide chains that could fold into conformations complementary in structure to the haptenic groups of the antigen, forming the two combining regions of the antibody molecule. This assumption is in agreement with later discoveries about the structure of antibodies, as shown in Figure 15-18. It was also assumed that the end chains had such an amino-acid composition as to permit them to fold in many different ways, all with about the same energy. In the presence of an antigen these chains would fold into a structure complementary to that of a haptenic group of the antigen, stabilized by the interaction energy with this group. The antibody molecule would then dissociate away, leaving the antigen to serve as the template for another antibody molecule.

This picture has now been replaced by another one. The ends of the four chains in Figure 15-18 have been found to have different amino-acid sequences for antibodies homologous to different antigens. The sequences presumably are such as to cause the antibody to fold into a conformation that provides complementariness to the haptenic group. Under the stimulus of the antigen a clone of cells develops, with each cell synthesizing the antibodies complementary to the stimulating antigen.

The observed specificity of antibodies against thousands of different antigens indicates that there are cells waiting to be stimulated into multiplication and production of one or another of thousands of different polypeptide chains. The production of these chains is under the control of genes. The stimulation of production of a particular antibody by an antigen results with little doubt from the combination of the antigen with its complementary antibody in the cell, but the detailed mechanism has not yet been discovered.

15-6. Nucleic Acids. The Chemistry of Heredity

One of the most amazing and interesting aspects of the world is the existence of human beings and other living organisms who are able to have progeny, to whom they transmit many of their own characteristics. The mechanism by means of which a child develops in such a way as to resemble his parents has been under intensive study for a century, and the progress in understanding this phenomenon has been especially rapid during the last 25 years.

In 1866 the Abbot Gregor Johann Mendel (1822–1884) developed a simple theory of inheritance on the basis of experiments that he had carried out with peas in the garden of the Augustinian monastery at Brno, in Moravia (now Czechoslovakia). He found that his experimental results could be accounted for by assuming that each of the plants of the second generation receives from each of the two parent plants a determiner or factor (now called a *gene*) for each inherited character. The genes are now described as being arranged linearly in a larger structure, one of the chromosomes, which can be seen in the nuclei of cells.

Different genes that may occur at the same locus in a chromosome are called *alleles* or *allelomorphic genes*. For example, Mendel hybridized two strains of peas that differed from one another in that the seeds were round in one strain and wrinkled in the other. The first-generation hybrid progeny had round seeds. However, when they were allowed to become self-fertilized he found that about three-quarters of the second-generation progeny had round seeds and about one-quarter had wrinkled seeds. His explanation of this observation, and of many others like it, is that the peas of the first strain carry two alleles for roundness, and those of the second strain two alleles for wrinkledness. The hybrids of these two strains inherit one of each of these two alleles (one from each parent), and Mendel assumed that the allele for roundness is the *dominant* gene and that for wrinkledness is *recessive,* so that the possession of one each of the two allelomorphic genes leads to roundness (as does the possession of two genes for roundness). In the next generation, obtained by self-fertilization of the first-generation progeny, the allele for roundness or the allele for wrinkledness is inherited at random from the one parent, and also at random from the other parent. About one-quarter of the progeny would then be expected to have the genic composition RR (with R representing the dominant allele), one-half to have the genic constitution Rr or rR, and one-quarter to have the genic constitution rr. The progeny RR would have round seeds, the heterozygotes Rr and rR would also have round seeds, because of the assumed dominance of R, and the recessive homozygotes rr would have wrinkled seeds.

The theory of the gene was greatly developed in the years following 1910 as the result of work on the fruit fly, *Drosophila,* carried out by Thomas Hunt Morgan and his collaborators (especially A. H. Sturtevant, Calvin Bridges, and H. J. Muller), who were able to determine the order in which many genes are located in the chromosomes of this organism. Further progress was made by other investigators (G. W. Beadle and E. L. Tatum, in particular) with use of the red bread mold, *Neurospora,* and by J. Lederberg and others who have studied the genetics of bacteria.

An example of the relation between genes and protein molecules is provided by the different kinds of hemoglobin that have been found in the red cells of human beings. In 1949 it was discovered that some human beings, patients with the disease sickle-cell anemia, have in their red cells a form of hemoglobin (hemoglobin S) that is different from that in the red cells of most people (hemoglobin A). The difference is not great: the two alpha chains of the hemoglobin-S molecule are identical with those of the hemoglobin-A molecule, and each beta chain has one amino-acid residue that is different. The beta chain of hemoglobin A has a residue of glutamic acid in the sixth position from the free amino end, whereas the beta chain of hemoglobin S has in this position a residue of valine; all of the other amino-acid residues are the same.

The abnormal hemoglobin in the red cells of the sickle-cell-anemia patients causes a very serious disease. Each of the two parents of a patient with this disease is found by experiment to have in his red cells a fifty-fifty mixture of hemoglobin A and hemoglobin S, and one-quarter of the children of such marriages are found, on the average, to be sickle-cell homozygotes, with the genic constitution SS and the disease sickle-cell anemia. It is evident that the two genes A and S carry out their functions essentially independently of one another; in a heterozygote, with genic constitution AS, each of the genes manufactures its own kind of hemoglobin, and each red cell contains a mixture of hemoglobin A and hemoglobin S.

About 25 years ago evidence was obtained showing that a gene is a molecule of *deoxyribonucleic acid* (usually abbreviated as *DNA*). The chemical nature of DNA has now been determined, and its molecular structure is known. The nature of this structure is such as to permit considerable insight to be obtained about the mechanism by means of which these molecules duplicate themselves, in order that the duplicates may be passed on to the progeny, or in order that the living organism may grow, through cell division, with each cell having its complement of genes.

DNA consists of units, called nucleotides (several hundred), that are held together by chemical bonds in a linear array, called a polynucleotide chain or a nucleic acid molecule. Each nucleotide consists of three parts: a molecule of phosphoric acid, a molecule of a sugar, *deoxyribose* (Section 13-6), and a molecule of a nitrogen compound, called a nitrogen base. The molecules of sugar and molecules of phosphoric acid are condensed together to form long chains:

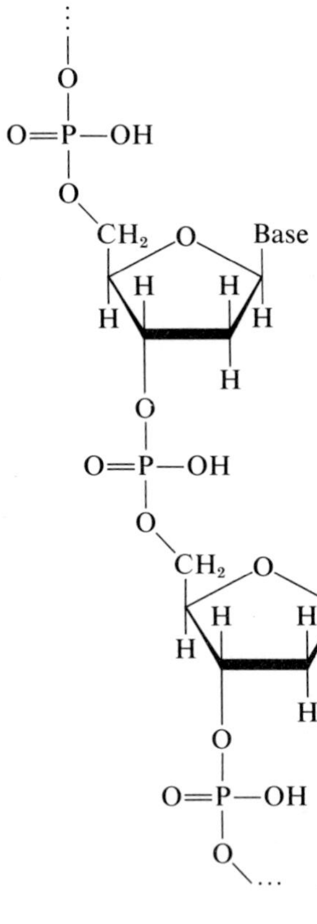

Deoxyribose is a pentose (sugar with formula $C_5H_{10}O_5$) that has lost one oxygen atom, giving it the formula $C_5H_{10}O_4$; its structural formula (see Section 13-6) is

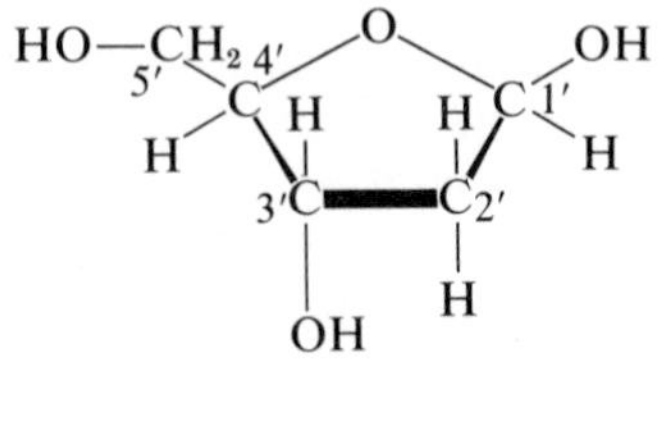

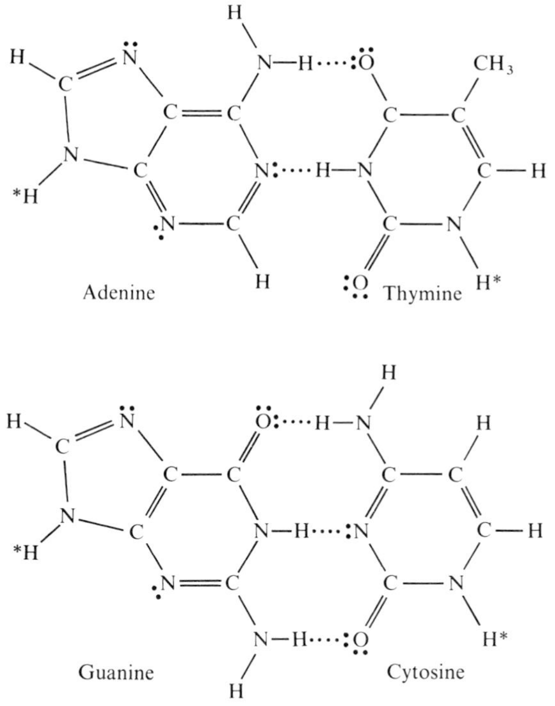

Adenine Thymine

Guanine Cytosine

FIGURE 15-20
Specific hydrogen bonding between adenine and thymine and between cytosine and guanine.

In DNA the two hydroxyl groups attached to carbon atoms 3′ and 5′ condense with hydroxyl groups of separate molecules of phosphoric acid, $OP(OH)_3$, to form the DNA chain. The nitrogen atom of the nitrogen base replaces the hydroxyl group attached to carbon atom 1′.

The nitrogen bases found in DNA comprise the two purines *adenine* and *guanine* and the two pyrimidines *thymine* and *cytosine;* in the formulas shown in Figure 15-20 the asterisk indicates the hydrogen atom that is replaced by the carbon atom of the sugar ring in DNA, and the double bonds correspond to only one of the several valence-bond structures for each molecule. The molecules are planar, because each of the bonds in the purine and pyrimidine rings has some double-bond character.

Chemical analysis of DNA from the nuclei of cells showed that, although the relative number of molecules of the two purines adenine and guanine varies from species to species, the molecular ratio adenine/thymine is unity and the ratio guanine/cytosine is unity. For example, the percentages in human sperm are 31% adenine, 19% guanine, 31% thymine, and 19% cytosine.

This experimental result was interpreted only when a theory of the structure of DNA had been developed. In 1953, making use of excellent x-ray diffraction patterns of DNA that had been made by M. H. F. Wilkins, the American biologist J. D. Watson and the British biophysicist F. H. C. Crick proposed that molecules of DNA consist of two chains wrapped about one another in a helical configuration, in such a way that at every level, 330 pm apart along the axis of the double helix, there occurs a residue of either adenine or guanine and one of either thymine or cytosine, and that these residues occur in complementary pairs: either as an adenine-thymine pair or as a guanine-cytosine pair (Figure 15-21). The explanation of this complementary pairing is shown in Figure 15-20. It is seen that adenine and thymine can form two hydrogen bonds with one another, whereas cytosine and guanine can form three.

According to the Watson-Crick proposal, the four bases adenine, thymine, guanine, and cytosine, which may be represented by the letters A, T, G, and C, occur in a characteristic sequence in one of the two

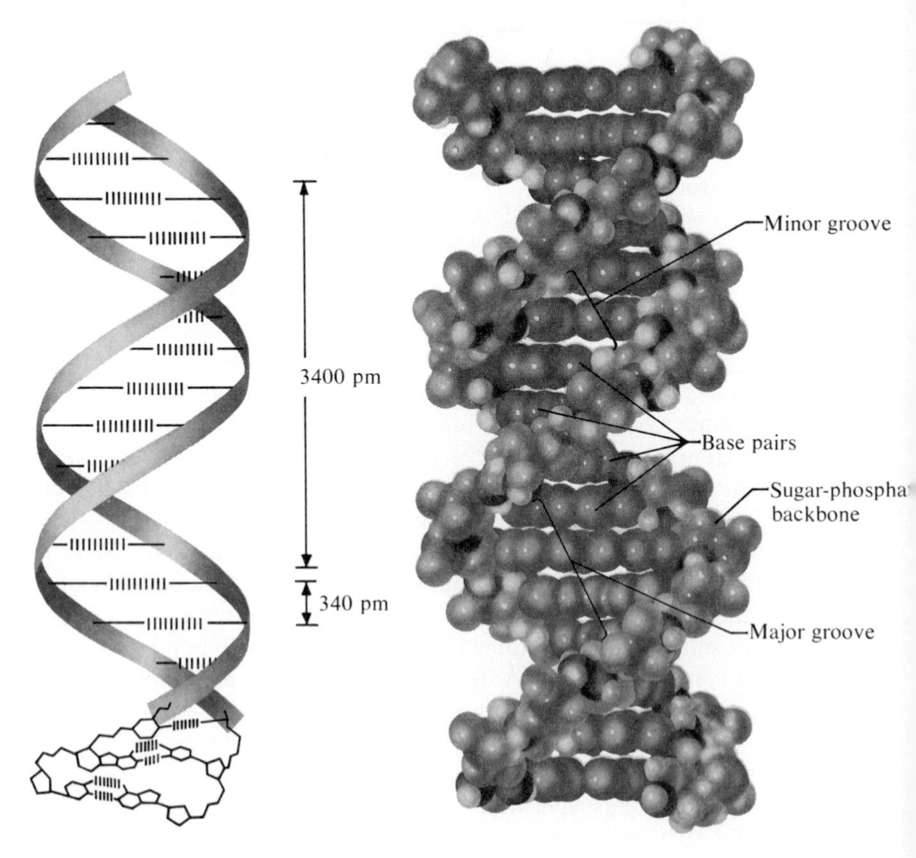

FIGURE 15-21
The double helix of DNA, as formulated by Watson and Crick.

polynucleotide chains of a gene and in the complementary sequence in the other polynucleotide chain. At each level there is one of the following four pairs of nitrogen bases: —A=T—, —T=A—, —G≡C—, —C≡G—. The dashes indicate either two or three hydrogen bonds, as shown in Figure 15-20.

The Replication of DNA

In addition to controlling the manufacture of other molecules, as discussed below, the DNA replicates itself. The Watson-Crick mechanism of reduplication of DNA molecules in the course of cell division is postulated to be the following: a double helix of two complementary polynucleotides begins to uncoil into the separate chains, and new polynucleotide chains begin to be synthesized, with the aid of enzymes as catalysts and with the old chains as the templates. The new chain that is being synthesized in approximation to each of the old chains is identical with the other old chain, in order to preserve the complementariness. Thus when the process is completed there are two identical double helixes, each consisting of one old chain and one newly synthesized chain (Figure 15-22).

This process represents the most striking and remarkable known example of the general phenomenon of achieving biological specificity through the interaction of complementary structures, as discussed for antibodies and antigens (Section 15-5). Each of the two purine-pyrimidine pairs involves complementariness in relation to the formation of two or three hydrogen bonds. The sugar-phosphate backbones of the two chains hold the nitrogen bases in such relative positions that only a purine in one chain and a pyrimidine in the other can form hydrogen bonds with one another. The wrong pairs that might be formed are AC and GT. By reference to Figure 15-20 we see that A and C form no hydrogen bonds with one another when juxtaposed in the way determined by the positions of the backbones. G and T might form one hydrogen bond, but in fact the hydrogen atoms in the middle positions of both G and T introduce steric hindrance that keeps G and T too far away to permit this bond to form. The energy of a hydrogen bond is about 20 kJ $mole^{-1}$, and hence the introduction into the growing chain of the right rather than the wrong purine or pyrimidine will be favored by 40 or 60 kJ $mole^{-1}$.

We can make a rough but nevertheless significant calculation of the probability that an error at a particular locus will be made in the synthesis of a polynucleotide chain in juxtaposition to the complementary template chain. The entropy change associated with introducing the wrong nucleotide is nearly the same as that for introducing the right one, and accordingly the difference in free energy is closely equal to the difference in

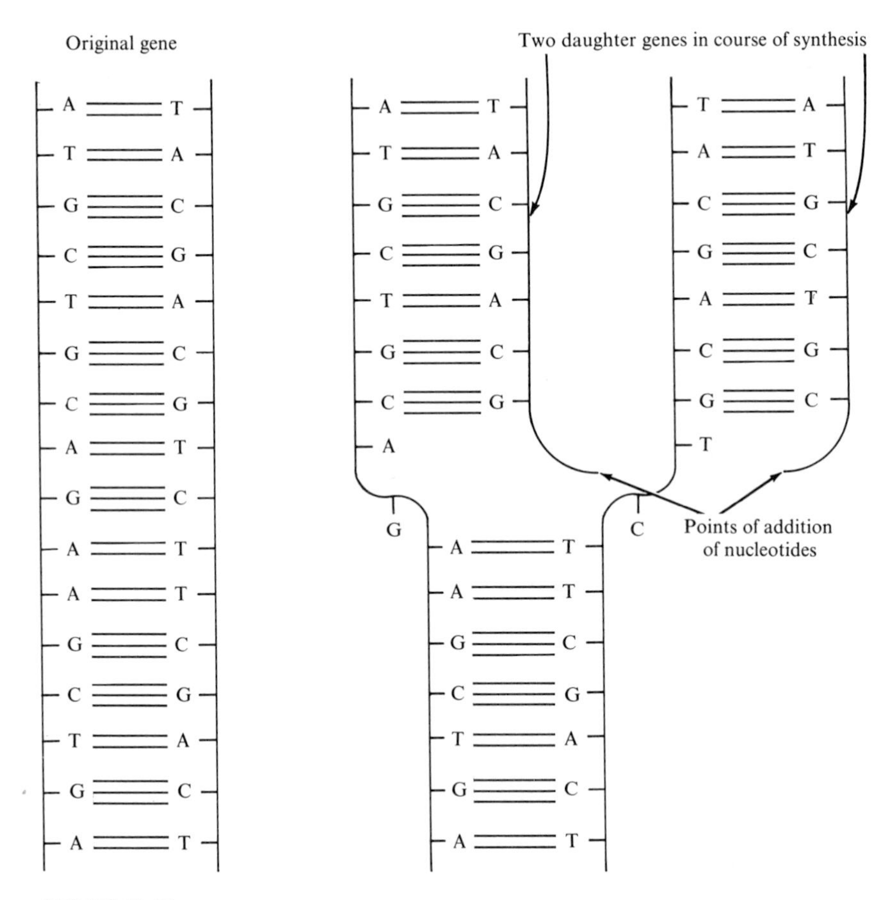

FIGURE 15-22
A diagram showing the postulated method of reduplication of the gene through formation of a polynucleotide chain complementary to each of the two mutually complementary chains of the original gene. The helical arrangement of the two chains is not indicated in this diagram.

enthalpy, 40 kJ mole^{-1} (or 60 kJ mole^{-1}). From the discussion of chemical equilibrium we see that we may write

$$RT \ln(N^*/N) = \Delta G^\circ - \Delta G^{\circ *}$$
$$= -40 \text{ kJ mole}^{-1}$$

Here ΔG° is the standard free energy change for the reaction of introducing the right nucleotide into the growing chain, $\Delta G^{\circ *}$ is that for introducing the wrong one, N is the number of right nucleotides introduced (at equilibrium), and N^* is the number of wrong ones. This equation leads to

$$\ln(N^*/N) = \frac{-40 \text{ kJ mole}^{-1}}{RT} = -15.52$$

$$N^*/N = 1.8 \times 10^{-7}$$

This calculation applies to the base pair A=T; for G≡C the ratio is far smaller. The average of the two is 0.9×10^{-7}; that is, about one chance in 10^7. The same result is obtained by considering the relative rates of the two reactions, with the difference in activation energy equal to 40 kJ mole^{-1}.

In the following discussion of protein synthesis it is pointed out that the gene that directs the synthesis of one of the polypeptide chains of hemoglobin contains about 500 base pairs. The above rough calculation accordingly leads to the conclusion that the probability that an error has been made in the synthesis of a gene is less than about 1 in 20,000; that is, 5×10^{-5}. It has been found by experiment that the number of errors is much less than this, perhaps only one percent as great. The explanation is that there are enzymes that recognize errors, and cut the polynucleotide chain to delete the incorrect nucleotide, and other enzymes that introduce the correct nucleotide, thus rectifying the error.

The half-old half-new character of first-generation daughter molecules of DNA in bacterial cultures has been verified by a striking experiment with tracer isotopes (^{15}N; Section 20-17). This experiment was reported in 1958 by M. Meselson and F. W. Stahl. They grew the bacterium E. coli for several generations on a medium containing a nitrogen compound with a high content of the heavy isotope ^{15}N. The DNA separated from the organism then has a larger molecular mass (with the same atoms in the molecule) and a larger density than that from the organism grown on the ordinary medium. The difference in density can be detected by the technique called density-gradient ultracentrifugation. A solution of cesium chloride is put in the sample tube of an ultracentrifuge, and rotated so rapidly as to produce a centrifugal force as much as 100,000 times the force of gravity. The cesium ions, which have high density, increase in concentration at the outer end of the sample tube (also carrying chloride ions along, to give electrical neutrality), thus producing a density gradient along the tube. Large molecules, such as those of DNA, are concentrated into a band at the radius where their density equals the density of the solution. Meselson and Stahl transferred the ^{15}N organisms to the normal medium. After one cell division had occurred, the DNA molecules were found to form a band midway between the ^{15}N molecules and the normal (^{14}N) molecules (Figure 15-23). Moreover, by suitable treatment the DNA molecules can be split into the two

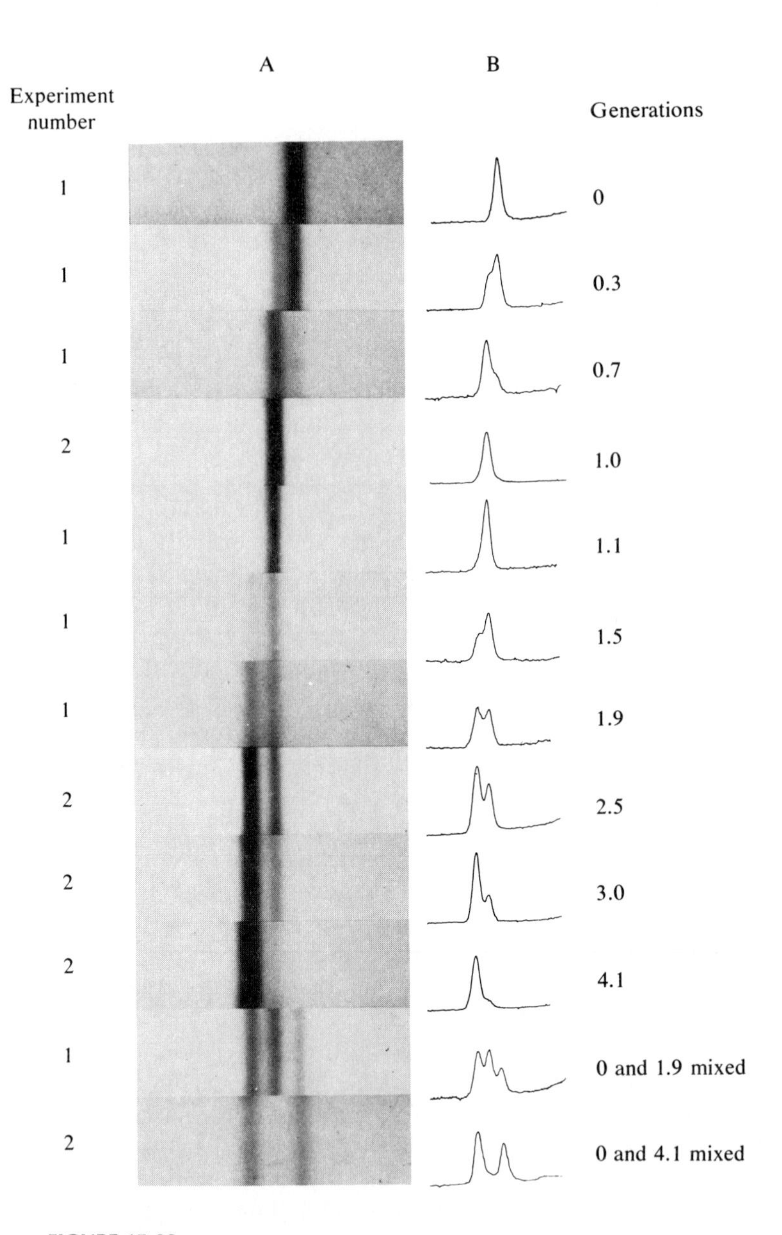

FIGURE 15-23
Density-gradient ultracentrifuge tubes showing bands of DNA, as observed by Meselson and Stahl. The top tube shows DNA made dense by its content of ^{15}N. After one generation (one cell division) in normal medium, only DNA molecules containing half the amount of ^{15}N are observed. [M. Meselson and F. W. Stahl, *Proceedings of the National Academy of Sciences*, U. S. A. **44**, 675 (1957).]

separate chains, and half of those chains were shown by density-gradient ultracentrifugation to be all-heavy, with the other half all-light. Accordingly it was verified, as had been postulated, that the first-generation DNA progeny consist of an old strand and a newly synthesized strand. After the next cell division half the DNA molecules are of this kind, and half contain no ^{15}N.

The Genetic Code and the Synthesis of Proteins

The mechanism of protein synthesis involves the transfer of information from one of the chains of the DNA helix to a molecule of RNA (ribonucleic acid) that is a complement of the DNA chain. RNA contains the sugar ribose

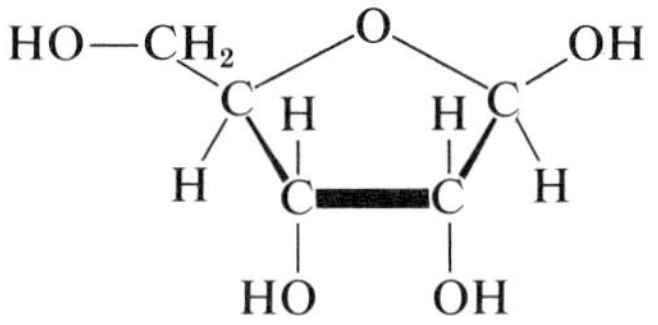

```
HO—CH2       O        OH
       \   /   \     /
        C  H   H  C
       /  \|    |/  \
      H    C----C    H
           |    |
          HO    OH
```

in place of the deoxyribose of DNA, and the pyrimidine base uracil (U)

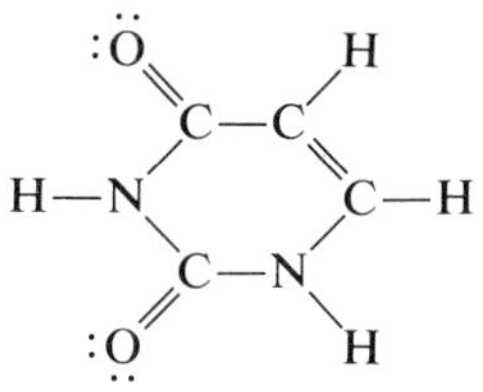

```
      :Ö        H
         \\    /
          C—C
         /    \\
    H—N        C—H
         \    /
          C—N
         //    \
      :O        H
       ..
```

in place of thymine. The power of uracil to form hydrogen bonds (two to adenine) is the same as for thymine, from which it differs only in having hydrogen in place of a methyl group (see Figure 15-20). Each gene (molecule of DNA) can serve as the template for the synthesis of many molecules of messenger RNA, each of which carries the information stored in the gene. This information is then used, with the aid of other molecules, especially certain enzymes, in the synthesis of the polypeptide chains of proteins.

It has been found that three nucleotides select an amino acid for incorporation in the chain; we may say that the gene is a sequence of three-letter words (called *codons*) formed with a four-letter alphabet, A, T, G, C for DNA and the equivalent A, U, G, C for RNA. Thus 146 codons, 438 letters (plus a few to carry the messages to start and to stop

TABLE 15-1
The Genetic Code

SECOND LETTER

FIRST LETTER	U		C		A		G		THIRD LETTER
U	UUU	Phe	UCU	Ser	UAU	Tyr	UGU	Cys	U
	UUC		UCC		UAC		UGC		C
	UUA	Leu	UCA		UAA	*	UGA	*	A
	UUG		UCG		UAG	*	UGG	Trp	G
C	CUU	Leu	CCU	Pro	CAU	His	CGU	Arg	U
	CUC		CCC		CAC		CGC		C
	CUA		CCA		CAA	Gln	CGA		A
	CUG		CCG		CAG		CGG		G
A	AUU	Ile	ACU	Thr	AAU	Asn	AGU	Ser	U
	AUC		ACC		AAC		AGC		C
	AUA		ACA		AAA	Lys	AGA	Arg	A
	AUG	Met	ACG		AAG		AGG		G
G	GUU	Val	GCU	Ala	GAU	Asp	GGU	Gly	U
	GUC		GCC		GAC		GGC		C
	GUA		GCA		GAA	Glu	GGA		A
	GUG		GCG		GAG		GGG		G

*May act as signals for terminating polypeptide chains.

the synthesis) are needed in the gene for the beta chain of hemoglobin, containing 146 amino-acid residues. Each RNA molecule manufactures hundreds of beta chains; there are about 100,000,000 hemoglobin molecules in the mature red cell.

The genetic code seems to be essentially the same in all organisms. It is given in Table 15-1. The code is redundant, in selecting among 20 amino acids; there are 64 three-letter words in the alphabet. The redundancy involves primarily the third letter.

How the code has been worked out is illustrated by the following experiment. An enzyme solution obtained from bacterial cells and added to a solution of all twenty amino acids produces a polypeptide chain consisting only of residues of the amino acid phenylalanine when provided with a synthetic RNA consisting of poly-uracil (that is, U-U-U-U- . . .). Hence UUU is the codon for phenylalanine, as shown in the table. Much of this work was done by the American scientists M. W. Nirenberg, H. G. Khorana, and R. H. Holley, and their collaborators, with use of enzymes that had been discovered by A. Kornberg and S. Ochoa.

The process of synthesis of polypeptide chains is indicated in Figure 15-24. One of the two DNA polynucleotides of a gene transfers its information to a complementary RNA molecule, called messenger RNA (mRNA). Ribosomes then travel along this mRNA molecule, and assist in the process of adding successive amino-acid residues to the growing chain until the entire chain has been synthesized. Molecules of transfer RNA (tRNA) assist in the process, as well as enzymes of several different kinds. Several ribosomes may be moving at the same time along one

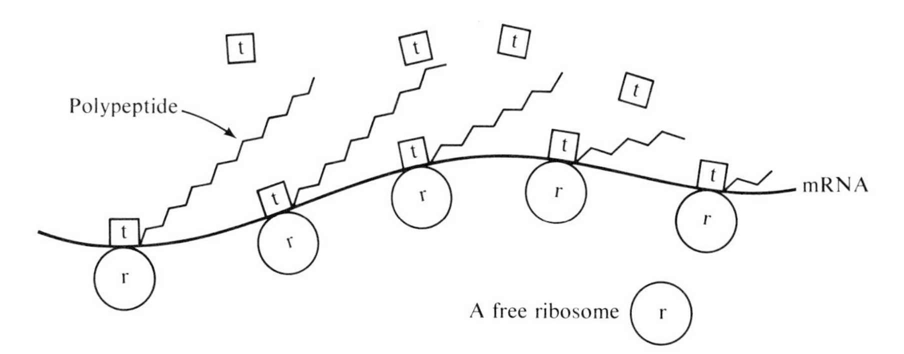

FIGURE 15-24
A diagram representing the process of synthesis of polypeptide chains. The long horizontal strand is a molecule of messenger RNA. The circles labeled r are ribosomes, which move along the strand. An appropriate molecule of transfer RNA (t) adds the proper amino acid to the growing chain.

molecule of mRNA. A single gene may produce several thousand molecules of mRNA during the lifetime of its cell, and each molecule of mRNA may direct the synthesis of hundreds of molecules of the protein.

Each mRNA molecule has a chain of several hundred adenyl residues (poly A) attached at one end, and this chain is observed to become shorter as the mRNA is used as a template for more and more protein chains. Presumably as each ribosome passes along the mRNA it snips off an A residue or two, or more likely three. This process might provide a mechanism for limiting the amount of protein synthesized, so that the cell is not overwhelmed with protein of one kind.

Ribosomes have molecular mass about 2.7×10^8 d, consisting of about 65 percent ribosomal RNA (rRNA) and 35 percent ribosomal protein. The rRNA is two kinds, a small molecule (about 500,000 d) and a large molecule (about 1,000,000 d). One of these may be involved in attachment to the mRNA molecule, and the other in the binding of the molecules of tRNA.

Transfer RNA (tRNA) has molecular mass about 25,000 d. There are about 40 different kinds of tRNA molecules, about two for each of the twenty amino acids (probably corresponding to the two or more codons (Table 15-1). A specific molecule of tRNA has the ability to recognize and combine with a particular amino acid, with the aid of a specific enzyme (amino-acylsynthetase) that helps in the recognition of the amino acid. The amino acid is attached to the tRNA by a covalent bond to its carboxyl group. This bond is a high-energy bond, and its high energy aids in accelerating the reaction of attachment of the amino-acid residue to the growing polypeptide chain. The high energy of the bond is provided by the hydrolysis of a molecule of GTP (guanine triphosphate, analogous to ATP, Section 14-6).

The sequence of the approximately 75 nucleotide residues has been determined for many different tRNA molecules. In all of them there are the same four sequences of complementary pairs A═U or G≡C, as shown in Figure 15-25. The existence of these sequences permitted the inference to be drawn that the molecule has the clover-leaf structure shown in Figure 15-25. This structure has been verified by an x-ray-diffraction study of crystals of yeast phenylalanine tRNA, with the result shown in Figure 15-26.

The residues that are encircled in Figure 15-25 are found in most tRNA molecules. It seems likely that some of these residues form hydrogen bonds with complementary residues in rRNA. The rRNA molecules have some mutually complementary sequences, indicating that they have hydrogen-bonded loops similar to those shown in Figures 15-25 and 15-26. The residues at the bottom of Figure 15-25 represent the codon for the specific amino acid. Selection of the appropriate tRNA involves the formation of hydrogen bonds with the complementary

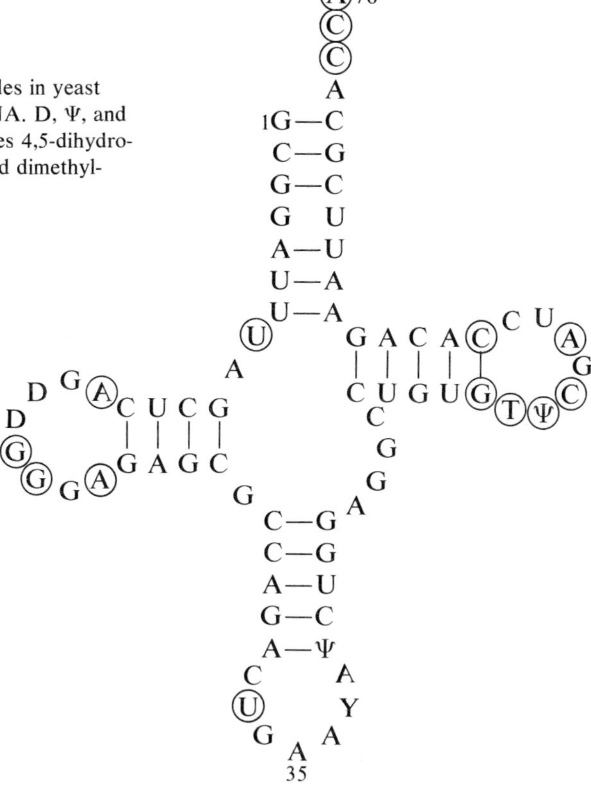

FIGURE 15-25
The sequence of nucleotides in yeast phenylalanine transfer RNA. D, Ψ, and Y represent the nucleotides 4,5-dihydrouridine, pseudouridine, and dimethylguanosine, respectively.

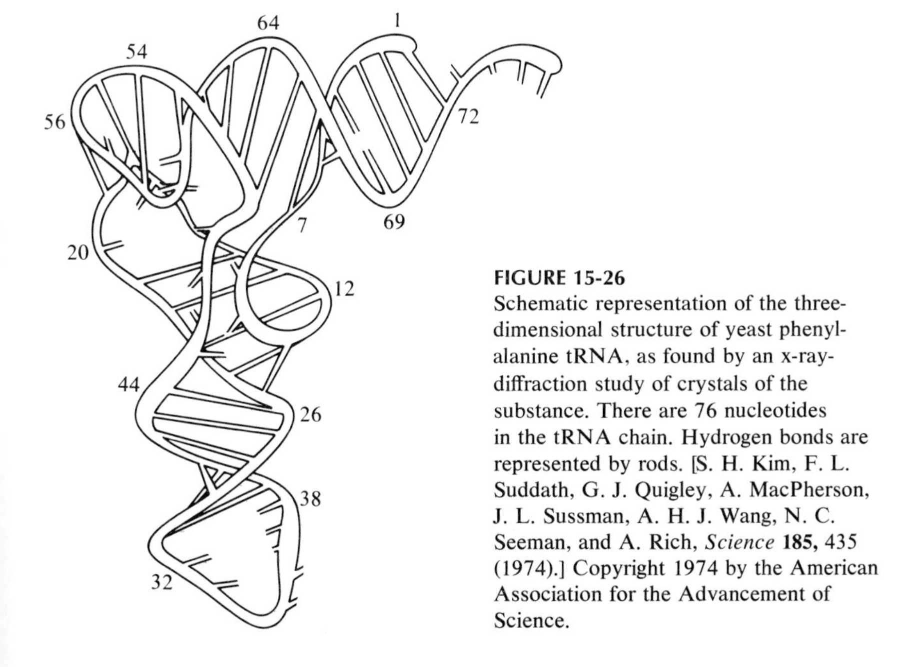

FIGURE 15-26
Schematic representation of the three-dimensional structure of yeast phenylalanine tRNA, as found by an x-ray-diffraction study of crystals of the substance. There are 76 nucleotides in the tRNA chain. Hydrogen bonds are represented by rods. [S. H. Kim, F. L. Suddath, G. J. Quigley, A. MacPherson, J. L. Sussman, A. H. J. Wang, N. C. Seeman, and A. Rich, *Science* **185,** 435 (1974).] Copyright 1974 by the American Association for the Advancement of Science.

sequence in the mRNA strand. After the amino-acid residue is attached to the growing polypeptide chain the tRNA molecule is released, the ribosome moves on to the next codon, and the appropriate tRNA attaches itself. The processes leading to addition of one amino-acid residue to the polypeptide chain take place in about 0.05 s.

15-7. The Structure of Biological Membranes

The first step in the evolution of life on earth probably was the development of molecules, in an aqueous solution that contained many small molecules that had been formed at random through the action of sunlight, lightning, and other sources of energy, that had the property of catalyzing reactions that led to the synthesis of duplicates of themselves. It is likely that this process, in fact, took part in two stages: first, the formation under the catalytic activity (as a template) of a molecule complementary in structure to the original molecule, and then, with use of the second molecule as a template, the formation of a molecule identical with the original one. The fact that this two-step process of replication (or the equivalent one-step process of replication of a molecule consisting of two complementary parts) occurs at the present time with nucleic acid when replication of genes takes place suggests that the first self-duplicating molecules on earth were in fact molecules of nucleic acid. Because of the importance of proteins in living organisms it has been suggested that proteins themselves might have been the first self-duplicating molecules, but the existing evidence favors nucleic acids.

The next step in the evolution of life might have been the development of a process for synthesizing proteins, under the guidance of nucleic acids. So long, however, as the molecules were free to escape in the surrounding aqueous medium, there would be little advantage to the development of processes for synthesizing molecules of different kinds. Only when cells began to form, consisting of a quantity of water with various dissolved substances that were restrained from escaping into the surrounding medium by a cell membrane, did it become possible for the process of molecular evolution to progress in such a way as ultimately to involve tens of thousands of substances and tens of thousands of catalyzed reactions.

During recent years much information has been obtained about the structure of biological membranes. This information has come in part from biochemical studies (the isolation of different chemical compounds from the cell membranes), x-ray diffraction, electron-spin and nuclear-spin magnetic resonance, spectroscopy, and especially the use of the electron microscope. Cell membranes, such as the membrane of the red blood cell, consist of about equal amounts of lipids and proteins. There is

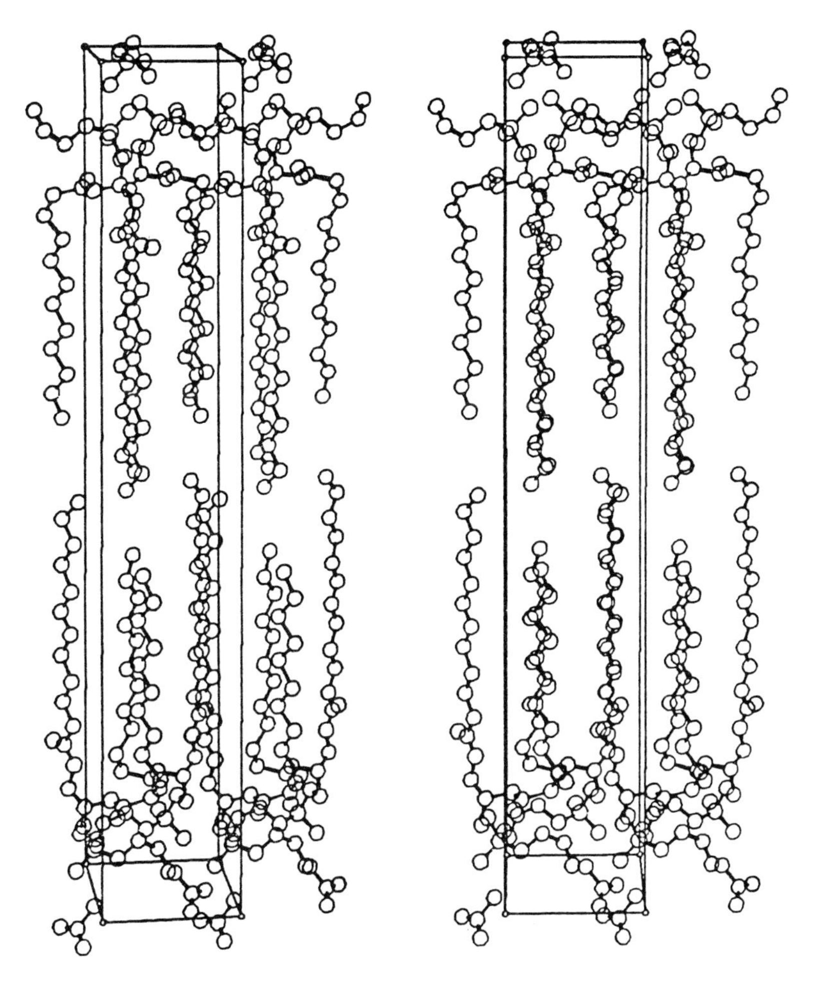

FIGURE 15-27
Stereodrawing of the structure of crystalline 1,2-dilaurylphosphotidylethanol-aminoacetic acid. [From P. B. Hitchcock, R. Mason, and K. M. Thomas, *Journal of the Chemical Society of London, Communications* **1974,** 539.]

also a small amount, a few percent, of polysaccharide, which is combined with polypeptide chains in the form of glycoproteins.

The principal properties of a membrane are largely determined by the nature of the phospholipids in it. The molecules of these substances carry electric charges and hydrogen-bond-forming groups at one end, and consist of hydrocarbon chains at the other end. The polar ends are hydrophilic, and form the surfaces of the membrane, whereas the hydrocarbon ends, which are rejected by the aqueous phase, extend toward other hydrocarbons. A double layer is formed, about 8 nm thick, with the structure indicated in Figure 15-27.

The phospholipid whose structure is shown in Figure 15-27 has lauryl side chains, containing no double bonds. In biological membranes the hydrocarbon side chains are of a number of sorts, as discussed in Section 13-5. Hydrocarbon side chains with a double bond and with the *cis* configuration about the double bond are bent, and there is some evidence that in biological membranes the hydrocarbon chains extend perpendicularly from each of the two surfaces inward for some distance and then bend through about 30°. Moreover, it has been found that the structure is well defined close to the two surfaces, essentially as shown in Figure 15-27, but that the ends of the chains move rather freely, so that the structure approximates that of a liquid in the middle portion of the membrane, and of a crystal toward the surfaces.

In addition, protein molecules are attached to the membrane in various ways. It is likely that some of the protein molecules extend from one surface to the other, and that these molecules can function to transport molecules and ions from outside the cell to the inside or from the inside out (Figure 15-28).

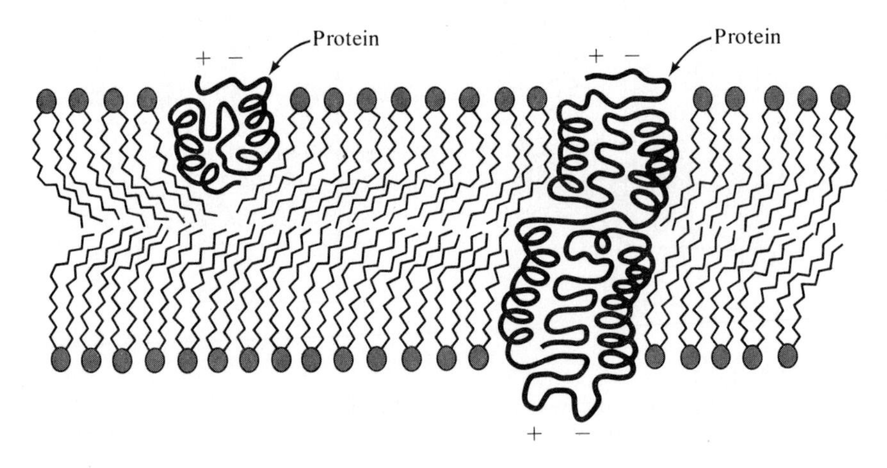

FIGURE 15-28
Diagrammatic representation of a cell membrane, showing lipid and protein molecules.

15-8. Molecular Diseases

In 1902 the English physician A. E. Garrod (1857–1936) made a study of persons whose urine turns black on exposure to the air, and found that the change in color is caused by the presence in the urine of the substance homogentisic acid, which is 2,5-dihydroxybenzeneacetic acid. He described this condition as "an inborn error of metabolism," and it was recognized later that it results from a genetic mutation such that the gene that normally converts homogentisic acid into other substances in the human body either was not synthesized or possibly was synthesized in a changed form lacking enzyme activity. In 1949 it was discovered that another genetic disease, sickle-cell anemia, results from the presence in the person of a mutated gene that synthesizes an abnormal form of polypeptide chain in the hemoglobin molecule. The beta chain of the hemoglobin molecule in patients with sickle-cell anemia has one amino-acid residue altered, from glutamic acid to valine, as discussed in Section 15-6. Because of the discovery that an abnormal hemoglobin molecule is involved in the disease, sickle-cell anemia was called a molecular disease. Since 1949 hundreds of molecular diseases have been recognized, and for many of them the nature of the gene mutation and the corresponding change in structure of the related protein molecule have been identified. For some of these molecular diseases the identification of the nature of the molecular abnormality provides an essentially complete explanation of the manifestations of the disease.

An example of a molecular disease that is now thoroughly understood is *ferrihemoglobinemia.* Ferrihemoglobinemia is characterized by the fact that two of the four iron atoms in the abnormal hemoglobin molecule are easily oxidized to the tripositive state, whereas the other two remain in the normal bipositive state. Combination with oxygen molecules in the lungs, and their release in the tissues, can occur only when the iron atom is in the bipositive (ferrous) state (iron(II)). When the iron atoms are oxidized to the ferric state they lose the power of reversible combination with oxygen, and the blood of the patient has only half the carrying power for oxygen that the blood of a normal person has.

The first fact that needs explanation is the stability of the bipositive oxidation state of iron in normal hemoglobin. Most iron(II) compounds are easily oxidized by oxygen of the air to the corresponding iron(III) compounds, but normal hemoglobin is stable to this oxidation, unless the hemoglobin molecule is denatured. The determination of the structure of myoglobin and hemoglobin by the x-ray diffraction method provided an explanation of the special stability of the bipositive state of iron in hemoglobin. It was found that there is in locus 58 in the alpha chain of hemoglobin and locus 63 in the beta chain a residue of the amino acid histidine, with the imidazole ring held firmly in a position rather close to the iron atom of the heme group. The imidazole ring is basic, and at normal pH of about 7 it picks up a proton from the solution, becoming the imidazolium ion, $C_3N_2H_4^+$, as indicated in Figure 15-29. The positive electric charge of this ionic group produces an electrostatic field in the region of the iron atom, such as to tend to hold electrons in this region, and accordingly it decreases the ease with which an electron can be removed from the iron atom, converting it to the tripositive state. This electrostatic

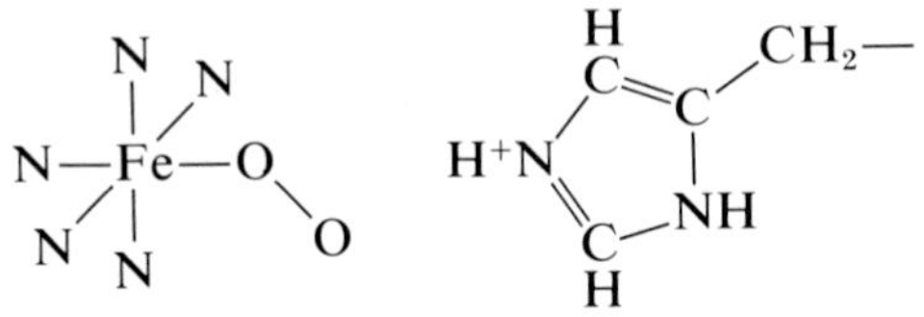

FIGURE 15-29
Drawing showing juxtaposition of an imidazolium ion of histidine residue to the iron atom in hemoglobin. The positive charge helps to keep the iron atom in the bipositive state.

effect of the imidazolium ion provides accordingly the explanation of the unusual stability of the bipositive state of iron in normal hemoglobin.

There are, however, several known mutations in human beings affecting the amino-acid sequence in the alpha chain or the beta chain in such a way as to lead to easy oxidation of the iron atom, and accordingly to the disease ferrihemoglobinemia. In one of these diseases the residue of histidine in the 58th position of the alpha chain is replaced by a residue of tyrosine. The side chain of tyrosine contains a hydroxybenzene ring, which is not acidic, and does not pick up a proton to assume a positive charge. Accordingly there is no electrostatic field in the region of the iron atom to prevent it from being oxidized to the tripositive state, and this oxidation takes place for both of the two alpha heme groups in the molecule, leading to ferrihemoglobinemia. This form of the disease is called alpha-chain ferrihemoglobinemia.

Persons are also known who have the disease beta-chain ferrihemoglobinemia, in which the histidine residue in the 63rd locus of the beta chain has been replaced by tyrosine.

This knowledge about the nature of the amino-acid substitution in the alpha chain or the beta chain provides an explanation of the nature of the disease, which manifests itself through permitting the iron atoms to be oxidized to the tripositive (ferric) state. This suggests a possible method of controlling the manifestations of the disease. If some way could be designed to introduce a positive charge into the correct part of the hemoglobin molecule, decreasing the tendency for the iron atom to be converted to the tripositive state, the disease might be controlled. Another way of controlling the disease, which, of course, does not require detailed understanding of the mechanism, but only the knowledge that the disease involves oxidation of the iron atom, is to increase the concentration in the body fluids of a reducing agent, such as ascorbic acid (vitamin C), so that the oxidation-reduction conditions are such as to favor reduction of iron to the bipositive state.

Verification of the explanation given above has been provided by observations on another molecular disease. If the histidine residue in locus 58 of the alpha chain or 63 of the beta chain is replaced by a residue of arginine, another genetic disease, rather mild in manifestations, results. This disease, however, does not have ferrihemoglobinemia as one of its manifestations; the iron atoms remain in the normal bipositive state. This fact is to be expected, in that the side chain of arginine contains a guanidine group, which at *p*H 7 picks up a proton to become the guanidinium ions, carrying a positive charge. Accordingly the electrostatic effect continues to operate to stabilize the bipositive oxidation state of the iron atom.

The discovery of molecular diseases was made only about 25 years ago, and with each passing year more information becomes available about the molecular basis of disease. In the course of time investigations along these lines should lead to important advances in controlling disease and decreasing the amount of human suffering.

EXERCISES

15-1. What is the length of an amino-acid residue along the axis of the alpha helix? Along the chain direction of the parallel-chain pleated sheet? From these numbers, calculate the increase in length that might be achieved by stretching a hair that has been treated with steam or a reducing agent.

15-2. Why can hair that has been treated with a reducing agent be stretched more easily than before this treatment?

15-3. What structural feature of *Bombyx mori* silkworm silk is associated with the fact that half the amino-acid residues for this silk are glycine residues?

15-4. Explain why the two pleated sheets can be constructed of L amino-acid residues or of D, but not of a mixture of L and D.

15-5. Make a diagram showing the structure of a sarcomere, and discuss the process of contraction of a muscle fiber.

15-6. In Section 15-4 it is stated that the effective load of oxygen that could be carried by hemoglobin with independent hemes (partial pressure of oxygen in lungs 0.15 atm, in tissues 0.04 atm) is only 32 percent of the maximum. (*a*) Verify by using the equilibrium equation. (*b*) What effective load could be carried by a hemoglobin with very large heme-heme interaction? (*c*) Why, in your opinion, has such a hemoglobin not been found in nature? (Answer: (*b*) 87 percent.)

15-7. Why, from the physiological point of view, is it useful for the partial pressure of oxygen for 50-percent saturation of myglobin to be smaller than that of hemoglobin? (See Figure 15-12.)

15-8. What sort of experimental evidence leads to the conclusion that an antibody molecule has two combining regions, each able to combine with a haptenic group?

15-9. When a solution of ovalbumin is added to rabbit anti-ovalbumin serum (obtained from a rabbit that had been injected with ovalbumin) a precipitate forms. When more ovalbumin solution is added the precipitate dissolves. Can you explain these facts?

15-10. The succinanilate ion is shown in Figure 15-19 as having the succinate chain bent, instead of having its maximum extension. Evidence for this bent conformation is provided by the observation that maleanilate ion combines much more strongly with antisuccinanilate antibody than does fumaranilate ion. (Maleanilate, $C_6H_5HNCOCH{=}CHCO_2^-$, has the *cis* conformation about the double bond, and its isomer fumaranilate has the *trans* conformation.) Can you formulate the argument that is involved?

15-11. It was found by Chargaff that DNA from different organisms contain different amounts of the four nucleotides A, T, G, and C, but that in each DNA the amounts of A and T are equal, as are also the amounts of G and C. How is this fact explained?

15-12. In what way did the experiment of Meselson and Stahl provide support for the double helix and the postulated mechanism of replication of genes?

15-13. The distances between pairs of atoms (oxygen, nitrogen) shown in Figure 15-20 as connected by hydrogen bonds in DNA lie between 280 and 300 pm. The energy of the hydrogen bond decreases by half for each increase of 20 pm in distance. By about how much would you expect the van der Waals contact of the two central NH groups to increase the separation of guanine and thymine, for the unstable GT base pair, and by about how much would the hydrogen bonds be weakened? (Answer: about 100 pm.)

15-14. Describe the structure of a biological membrane.

15-15. Why does ferrihemoglobinemia result from replacement of the histidine residue in locus 58 of the alpha chain or 63 of the beta chain of hemoglobin by tyrosine? Why does it not result from replacement by arginine?

16

Inorganic Complexes and Coordination Compounds

16-1. The Nature of Inorganic Complexes

An inorganic molecule that contains several atoms, including one or more metal atoms, is called an *inorganic complex* or *coordination compound.* An example is nickel tetracarbonyl, $Ni(CO)_4$. An inorganic complex with an electric charge is called a *complex ion.* Familiar examples of complex ions are the ferrocyanide ion, $Fe(CN)_6^{----}$, the ferricyanide ion, $Fe(CN)_6^{---}$, the hydrated aluminum ion, $Al(H_2O)_6^{+++}$, and the deep blue cupric ammonia complex ion, $Cu(NH_3)_4^{++}$, which is formed by adding ammonium hydroxide to a solution of cupric salt. Complex ions are important in the methods of separation used in qualitative and quantitative chemical analysis and in various industrial processes.

The formation of complexes constitutes an especially important part of the chemistry of the transition metals. The special feature of the electronic structure of the transition metals that leads to their formation of stable complexes is the availability of *d* orbitals, as well as *s* and *p* orbitals, for bond formation, as discussed in the following section. The transition metals have in their outer shells electrons occupying *d, s,* and *p* orbitals. Thus for the elements from potassium to krypton the outer-shell electrons may occupy the five 3*d* orbitals, the 4*s* orbital, and the three $4p$ orbitals, and in the succeeding sequences of transition metals the available orbitals are similar, but with increase of the total quantum number by 1 or 2.

The different transition metals have different numbers of d orbitals available for hybridization with the s orbitals and the three p orbitals of the valence shell, to form bond orbitals, and the nature of the bonds formed by the metal atom depends upon the number of d orbitals available. With no d orbitals available, tetrahedral sp^3 bond orbitals (Figure 6-8) of the type described in Chapter 6 may be formed. An example is provided by the zinc ion, Zn^{++}. The zinc ion has ten electrons outside of the argon shell. These ten electrons can occupy the five $3d$ orbitals in pairs, leaving the $4s$ orbital and the three $4p$ orbitals available for hybridization to form four tetrahedral bond orbitals. It is in fact found by experiment that bipositive zinc has ligancy four, forming complexes in which four atoms or groups of atoms are tetrahedrally bonded to it. Among the complexes of this sort that are discussed in following sections of the chapter and later chapters are $Zn(NH_3)_4^{++}$, $Zn(OH)_4^{--}$, and $Zn(CN)_4^{--}$.

16-2. The Nature of the Transition Elements

The long periods of the periodic system can be described as short periods with ten additional elements inserted. The first three elements of the long period between argon and krypton, which are the metals potassium, calcium, and scandium, resemble their congeners of the preceding short period, sodium, magnesium, and aluminum, respectively. Similarly the last four elements in the sequence, germanium, arsenic, selenium, and bromine, resemble their preceding congeners, silicon, phosphorus, sulfur, and chlorine, respectively. The remaining elements of the long period, titanium, vanadium, chromium, manganese, iron, cobalt, nickel, copper, zinc, and gallium, have no lighter congeners; they are not closely similar in their properties to any lighter elements.

The properties of these elements accordingly suggest that the long period can be described as involving the introduction of ten elements in the center of the series. The introduction of these elements is correlated with the insertion of ten additional electrons into the five $3d$ orbitals of the M shell, converting it from a shell of 8 electrons, as in the argon atom, to a shell of 18 electrons. It is convenient to describe the long period as involving ten transition metals, corresponding to the ten electrons. We shall consider the ten elements from titanium, group IVa, to gallium, group IIIb, as constituting the ten transition elements in the first long period, and shall take the heavier congeners of these elements as the transition elements in the later series.

The chemical properties of the transition elements do not change so strikingly with change in atomic number as do those of the other elements.

In the series potassium, calcium, scandium, the normal salts of the elements correspond to the maximum oxidation numbers given by the positions of the elements in the periodic system, 1 for potassium, 2 for calcium, and 3 for scandium; the sulfates, for example, of these elements are K_2SO_4, $CaSO_4$, and $Sc_2(SO_4)_3$. The fourth element, titanium, tends to form salts representing a lower oxidation number than its maximum, 4; although compounds such as titanium dioxide, TiO_2, and titanium tetrachloride, $TiCl_4$, can be prepared, most of the compounds of titanium represent lower oxidation states, +2 or +3. The same tendency is shown by the succeeding elements. The compounds of vanadium, chromium, and manganese that represent the maximum oxidation numbers +5, +6, and +7, respectively, are strong oxidizing agents, and are easily reduced to compounds in which these elements have oxidation numbers +2 or +3. The oxidation numbers +2 and +3 continue to be the important ones for the succeeding elements, iron, cobalt, and nickel.

A striking characteristic of most of the compounds of the transition metals is their color. Nearly every compound formed by vanadium, chromium, manganese, iron, cobalt, nickel, and copper is strongly colored, the color depending not only on the atomic number of the metallic element but also on its state of oxidation, and, to some extent, on the nature of the nonmetallic element or anion with which the metal is combined. It seems clear that the color of these compounds is associated with the presence of an incomplete M shell of electrons; that is, with an M shell containing less than its maximum number of electrons, 18. When the M shell is completed, as in the compounds of bipositive zinc ($ZnSO_4$ and others) and of unipositive copper (CuCl and others), the substances are in general colorless. Another property characteristic of incompleted inner shells is paramagnetism, the property of a substance of being attracted into a strong magnetic field. Nearly all the compounds of the transition elements in oxidation states corresponding to the presence of incompleted inner shells are strongly paramagnetic.

Octahedral Orbitals

The doubly charged iron cation, Fe^{++}, has six electrons outside of the argon shell. These six electrons can be placed in three of the five $3d$ orbitals, in pairs. The ion would then have two $3d$ orbitals available to hybridize with the $4s$ orbital and the three $4p$ orbitals, to form six bond orbitals. These d^2sp^3 hybrid bond orbitals have been found to constitute a set of six orbitals with their maxima directed in the six octahedral directions (along the $+x$, $-x$, $+y$, $-y$, $+z$, and $-z$ directions in a set of Cartesian coordinates); that is, toward the corners of a regular octahedron. Bipositive iron might accordingly be expected to use these orbitals in forming an octahedral complex, and in fact the complex ion $Fe(CN)_6^{----}$ has been shown by x-ray diffraction of ferrocyanide crystals to have the octahedral structure. Other examples of octahedral complexes are described in later sections of this chapter.

The electronic structure that would be expected for an isolated Fe^{++} ion is the one in which four of the 3*d* orbitals are occupied by single electrons, with parallel spin, and one is occupied by a pair of electrons. The ion with this structure would have a magnetic moment corresponding to four unpaired electron spins in parallel orientation. It is found by experiment that the hydrated ferrous ion, $Fe(H_2O)_6^{++}$, has a magnetic moment with this value, whereas the ferrocyanide ion has no magnetic moment. The conclusion can be drawn that the bonds in these two complex ions are different in character: in the hydrated ferrous ion the bonds, which have a large amount of ionic character, are formed with use of the 4*s* orbital and the three 4*p* orbitals, whereas in the ferrocyanide ion the orbitals are d^2sp^3 covalent bonds. Investigation of the magnetic properties of a complex can in many cases permit a decision to be made as to the nature of the bond orbitals used by the metal atom. It has been found by use of this magnetic criterion that complexes of metals with strongly electronegative atoms or groups are usually essentially ionic in character (without the 3*d* orbitals used in bonding), whereas those with less electronegative atoms or groups are covalent in character (with use of 3*d* orbitals in the hybrid bond orbitals).

Square Bond Orbitals

The bipositive nickel ion, Ni^{++}, has eight electrons outside of the argon shell. These electrons may be introduced in the five 3*d* orbitals in two ways: either by placing three electron pairs in three of the 3*d* orbitals and an odd electron in each of the other two, with their spins parallel, or by placing four electron pairs in four of the 3*d* orbitals, leaving one 3*d* orbital available for bond formation. Complexes in which bipositive nickel has the first electronic structure would have a magnetic moment, leading to paramagnetism, whereas those in which bipositive nickel has the second structure would have zero magnetic moment.

It has been found by study of the magnetic properties of different compounds of bipositive nickel that some of them, such as the hydrated nickel ion, are paramagnetic, and accordingly form bonds in which the 3*d* orbitals do not participate. Others, such as the nickel tetracyanide ion, $Ni(CN)_4^{--}$, have no magnetic moment, and the bonds may be considered to be formed by bond orbitals involving one 3*d* orbital.

The hybrid bond orbitals that can be formed by one 3*d* orbital, one 4*s* orbital, and the set of 3*p* orbitals are four bond orbitals that lie in a plane and are directed toward the corners of a square. (The third *p* orbital is not involved in this set of bond orbitals.) X-ray examination of crystals has shown that bipositive nickel, palladium, and platinum form complexes of this square planar type.

The Discovery of Octahedral and Square Complexes

The concept of the coordination of ions or groups of atoms in a definite geometric arrangement about a central metal atom was developed shortly after the beginning of the present century by the Swiss chemist A. Werner (1866–1919) to account for the existence and properties of compounds such as K_2SnCl_6, $Co(NH_3)_6I_3$, and so on. Before Werner's work was carried out these compounds had been assigned formulas such as $SnCl_4 \cdot 2KCl$ and $CoI_3 \cdot 6NH_3$, and had been classed as "molecular compounds," of unknown nature. Werner showed that the properties of many complexes formed by various transition metals could be explained by the postulate that the metal atoms have ligancy 6, with the six attached groups arranged about the central atom at the corners of a circumscribed regular octahedron.

One important property that Werner explained in this way is the existence of *isomers of inorganic complexes*. For example, there are two complexes with the formula $Co(NH_3)_4Cl_2^+$, one of which is violet in color and one green. Werner identified these two complexes with the *cis* and *trans* structures shown in Figure 16-1. In the *cis* form the chloride ions are in adjacent positions, and in the *trans* form in opposite positions. Werner identified the violet complex with the *cis* configuration through the observation that it could be made easily from the carbonate-ammonia complex $Co(NH_3)_4CO_3^+$, for which only the *cis* form is possible. Werner also discovered square coordination and identified the square *cis* and *trans* isomers.

In recent years a great amount of information about the structure of inorganic complexes has been gathered by the methods of x-ray diffraction, measurement of magnetic susceptibility, magnetic resonance spectroscopy, Mössbauer spectroscopy, and other techniques. This information about the structure of complexes has been correlated with their chemical properties in such a way as to bring reasonable order into this field of chemistry.

16-3. Ammonia Complexes

A solution of a cupric salt is blue in color. This blue color is due to the absorption of yellow and red light, and consequent preferential transmission of blue light. The molecular species that absorbs the light is the

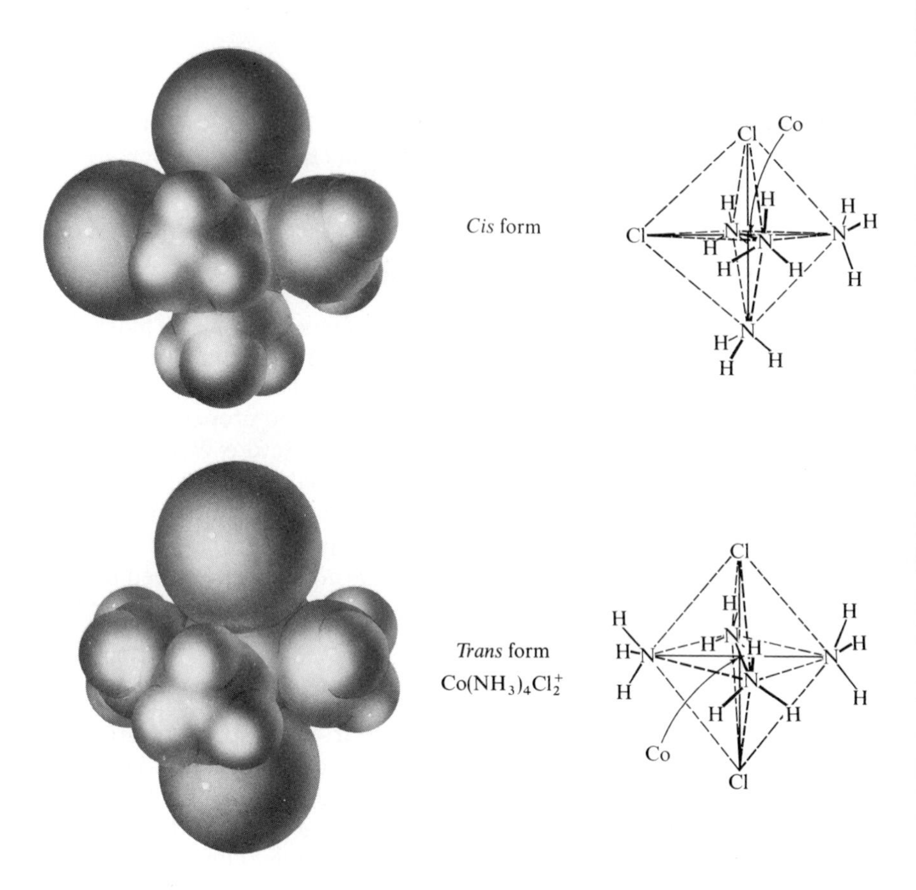

FIGURE 16-1
The *cis* and *trans* isomers of the cobaltic tetrammine dichloride ion, $Co(NH_3)_4Cl_2^+$. In the *cis* form the two chlorines occupy adjacent corners of the coordination octahedron about the cobalt atom, and in the *trans* form the two chlorine atoms occupy opposite corners.

hydrated copper ion, probably $Cu(H_2O)_4^{++}$. Crystalline hydrated cupric salts such as $CuSO_4 \cdot 5H_2O$ are blue, like the aqueous solution, whereas anhydrous $CuSO_4$ is white.*

When a few drops of sodium hydroxide solution are added to a cupric solution a blue precipitate is formed. This is cupric hydroxide, $Cu(OH)_2$, which precipitates when the ion concentration product $[Cu^{++}][OH^-]^2$ reaches the value corresponding to a saturated solution of the hydroxide. (Here the symbol Cu^{++} is used, as is conventional, for the ion species $Cu(H_2O)_4^{++}$.) Addition of more sodium hydroxide solution leads to no further change, other than the formation of more precipitate.

---

*The crystal structure of $CuSO_4 \cdot 5H_2O$ shows that in the crystal four water molecules are attached closely to the cupric ion, and the fifth is more distant.

If ammonium hydroxide is added in place of sodium hydroxide the same precipitate of $Cu(OH)_2$ is formed. On addition of more ammonium hydroxide, however, the precipitate dissolves, giving a clear solution with a deeper and more intense blue color than the original cupric solution.*

The solution of the precipitate cannot be attributed to increase in hydroxide-ion concentration, because sodium hydroxide does not cause it, nor to ammonium ion, because ammonium salts do not cause it. There remains undissociated NH_4OH or NH_3, which might combine with the cupric ion. It has in fact been found that the new deep blue ion species formed by addition of an excess of ammonium hydroxide is the *cupric ammonia complex* $Cu(NH_3)_4^{++}$, similar to the hydrated cupric ion except that the four water molecules have been replaced by ammonia molecules. This complex is sometimes called the *cupric tetrammine complex*, the word *ammine* meaning an attached ammonia molecule.

Salts of this complex ion can be crystallized from ammonia solution. The best known one is *cupric tetrammine sulfate monohydrate*, $Cu(NH_3)_4SO_4 \cdot H_2O$, which has the same deep blue color as the solution.

The reason that the precipitate of cupric hydroxide dissolves in an excess of ammonium hydroxide can be given in the following way. A precipitate of cupric hydroxide is formed because the concentration of cupric ion and the concentration of hydroxide ion are greater than the values corresponding to the solubility product of cupric hydroxide. If there were some way for copper to be present in the solution without exceeding the solubility product of cupric hydroxide, then precipitation would not occur. In the presence of ammonia, copper exists in the solution not as the cupric ion (that is, the hydrated cupric ion), but principally as the cupric ammonia complex, $Cu(NH_3)_4^{++}$. This complex is far more stable than the hydrated cupric ion. The reaction of formation of the cupric ammonia complex is

$$Cu^{++} + 4NH_3 \rightleftarrows Cu(NH_3)_4^{++}$$

We see from the equation for the reaction that the addition of ammonia to the solution causes the equilibrium to shift to the right, more of the cupric ion being converted into cupric ammonia complex as more and more ammonia is added to the solution. When sufficient ammonia is present a large amount of copper may exist in the solution as cupric ammonia complex, at the same time that the cupric ion concentration is less than that required to cause precipitation of cupric hydroxide. When ammonia is added to a solution in contact with the precipitate of cupric hydroxide, the cupric ion in the solution is converted to cupric ammonia

*In describing color the adjective deep refers not to intensity but to shade; deep blue tends toward indigo.

complex, causing the solution to be unsaturated with respect to cupric hydroxide. The cupric hydroxide precipitate then dissolves, and if enough ammonia is present the process continues until the precipitate has dissolved completely.

The process of *solution of a slightly soluble substance through formation of a complex by one of its ions* is the basis of some of the most important practical applications of complex formation. Several examples are mentioned later in this chapter.

The nickel ion forms two rather stable ammonia complexes. When a small amount of ammonium hydroxide solution is added to a solution of a nickel salt (green in color) a pale green precipitate of nickel hydroxide, $Ni(OH)_2$, is formed. On addition of more ammonium hydroxide solution this dissolves to give a blue solution, which with still more ammonium hydroxide changes color to light blue-violet.

The light blue-violet complex is shown to be the *nickel hexammine ion,* $Ni(NH_3)_6^{++}$, by the facts that the same color is shown by crystalline $Ni(NH_3)_6Cl_2$ and other crystals containing six ammonia molecules per nickel ion, and that x-ray studies have revealed the presence in these crystals of octahedral complexes in which the six ammonia molecules are situated about the nickel ion at the corners of a regular octahedron. The structure of crystalline $Ni(NH_3)_6Cl_2$ is shown in Figure 16-2.

The blue complex is probably the *nickel tetramminedihydrate ion,* $Ni(NH_3)_4(H_2O)_2^{++}$. Careful studies of the change in color with increasing ammonia concentration indicate that the ammonia molecules are added one by one and that all the complexes $Ni(H_2O)_6^{++}$, $Ni(H_2O)_5NH_3^{++}$, $Ni(H_2O)_4(NH_3)_2^{++}$, $Ni(H_2O)_3(NH_3)_3^{++}$, $Ni(H_2O)_2(NH_3)_4^{++}$, $Ni(H_2O)(NH_3)_5^{++}$, and $Ni(NH_3)_6^{++}$ exist.

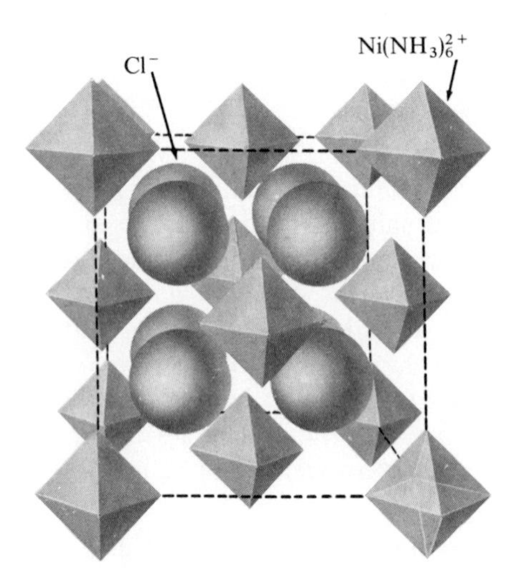

FIGURE 16-2
The structure of crystalline nickel hexammine chloride, $Ni(NH_3)_6Cl_2$. The crystal contains octahedral nickel hexammine ions and chloride ions.

Several metal ions form ammonia complexes with sufficient stability to put the hydroxides into solution. Others, such as aluminum and iron, do not. The formulas of the stable complexes are given below. There is no great apparent order about the stability or composition of the complexes, except that often the unipositive ions add two, the bipositive ions four, and the terpositive ions six ammonia molecules.

The *silver ammonia complex,* $Ag(NH_3)_2^+$, is sufficiently stable for ammonium hydroxide to dissolve precipitated silver chloride by reducing the concentration of silver ion, $[Ag^+]$, below the value required for precipitation by the solubility product of AgCl. A satisfactory test for silver ion is the formation with chloride ion of a precipitate that is soluble in ammonium hydroxide. Ammonia complexes in general are decomposed by acid, because of formation of ammonium ion; for example, as in the reaction

$$Ag(NH_3)_2^+ + Cl^- + 2H^+ \longrightarrow AgCl + 2NH_4^+$$

The common stable ammonia complexes are

$Cu(NH_3)_2^+$	$Cu(NH_3)_4^{++}$	$Co(NH_3)_6^{+++}$
$Ag(NH_3)_2^+$	$Zn(NH_3)_4^{++}$	$Cr(NH_3)_6^{+++}$
$Au(NH_3)_2^+$	$Cd(NH_3)_4^{++}$	
	$Hg(NH_3)_2^{++}$	
	$Hg(NH_3)_4^{++}$	
	$Ni(NH_3)_4^{++}$	
	$Ni(NH_3)_6^{++}$	
	$Co(NH_3)_6^{++}$	

NOTES: 1. Cobaltous ammonia ion is easily oxidized by air to cobaltic ammonia ion.
2. Chromic ammonia ion forms only slowly, and is decomposed by boiling, to give chromium hydroxide precipitate.

16-4. Cyanide Complexes

Another important class of complex ions includes those formed by the metal ions with cyanide ion. The common cyanide complexes are

$Cu(CN)_2^-$	$Zn(CN)_4^{--}$	$Fe(CN)_6^{---}$	$Au(CN)_4^-$
$Ag(CN)_2^-$	$Cd(CN)_4^{--}$	$Co(CN)_6^{---}$	
$Au(CN)_2^-$	$Hg(CN)_4^{--}$	$Mn(CN)_6^{----}$	
		$Fe(CN)_6^{----}$	
		$Co(CN)_6^{----}$	

Some of these complexes are very stable—the stability of the *argento-cyanide ion,* $Ag(CN)_2^-$, for example, is so great that addition of iodide ion does not cause silver iodide to precipitate, even though the solubility product of silver iodide is very small. The *ferrocyanide ion,* $Fe(CN)_6^{----}$, *ferricyanide ion,* $Fe(CN)_6^{---}$, and *cobalticyanide ion,* $Co(CN)_6^{---}$, are so stable that they are not appreciably decomposed by strong acid. The others are decomposed by strong acid, with the formation of hydrocyanic acid, HCN.

An illustration of the stability of the ferrocyanide complex is provided by the old method of making potassium ferrocyanide, $K_4Fe(CN)_6$, by strongly heating nitrogenous organic material (such as dried blood and hides) with potassium carbonate and iron filings.

The *cobaltocyanide ion,* $Co(CN)_6^{----}$, like the cobaltous ammonia complex, is a very strong reducing agent; it is able to decompose water, liberating hydrogen, as it changes into cobalticyanide ion.

Cyanide solutions are used in the *electroplating* of gold, silver, zinc, cadmium, and other metals. In these solutions the concentrations of uncomplexed metal ions are very small, and this favors the production of a uniform fine-grained deposit. Other complex-forming anions (tartrate, citrate, chloride, hydroxide) are also used in plating solutions.

16-5. Complex Halides and Other Complex Ions

Nearly all anions can enter into complex formation with metal ions. Thus stannic chloride, $SnCl_4$, forms with chloride ion the stable *hexachloro-stannic ion,* $SnCl_6^{--}$, which with cations crystallizes in an extensive series of salts. Various complexes of this kind are discussed below.

Many chloride complexes are known; representative are the following:

$CuCl_2(H_2O)_2$, $CuCl_3(H_2O)^-$, $CuCl_4^{--}$
$AgCl_2^-$, $AuCl_2^-$
$HgCl_4^{--}$
$CdCl_4^{--}$, $CdCl_6^{----}$
$SnCl_6^{--}$
$PtCl_6^{--}$
$AuCl_4^-$

The cupric chloride complexes are recognizable in strong hydrochloric acid solutions by their green color. The crystal $CuCl_2 \cdot 2H_2O$ is bright green, and x-ray studies have shown that it contains the complex molecule $CuCl_2(H_2O)_2$. The ion $CuCl_3(H_2O)^-$ is usually written $CuCl_3^-$; it is highly probable that the indicated water molecule is present, and, indeed, the ion $Cu(H_2O)_3Cl^+$ very probably also exists in solution.

The stability of the *tetrachloroaurate ion,* $AuCl_4^-$, is responsible for the ability of *aqua regia,* a mixture of nitric and hydrochloric acids, to dissolve gold, which is not significantly soluble in the acids separately. Nitric acid serves as the oxidizing agent that oxidizes gold to the terpositive state, and the chloride ions provided by the hydrochloric acid further the reaction by combining with the auric ion to form the stable complex:

$$Au + 6H^+ + 4Cl^- + 3NO_3^- \longrightarrow AuCl_4^- + 3NO_2 + 3H_2O$$

The solution of platinum in aqua regia likewise results in the stable *hexachloroplatinate ion,* $PtCl_6^{--}$.

The bromide and iodide complexes closely resemble the chloride complexes, and usually have similar formulas.

Fluoride ion is more effective than the other halide ions in forming complexes. Important examples are the *tetrafluoroborate ion,* BF_4^-, the *hexafluorosilicate ion,* SiF_6^{--}, the *hexafluoroaluminate ion,* AlF_6^{---}, and the *ferric hexafluoride ion,* FeF_6^{---}.

A useful complex is that formed by thiosulfate ion, $S_2O_3^{--}$, and silver ion. Its formula is $Ag(S_2O_3)_2^{---}$, and its structure is

$$\left[\begin{array}{ccccccc} & :\ddot{O} & & & & \ddot{O}: & \\ & \| & & & & \| & \\ :\ddot{O}- & S & -\ddot{S}- & Ag & -\ddot{S}- & S & -\ddot{O}: \\ & \| & & & & \| & \\ & :\underset{..}{O} & & & & \underset{..}{O}: & \end{array}\right]^{---}$$

This complex ion is sufficiently stable to cause silver chloride and bromide to be soluble in thiosulfate solutions, and this is the reason that sodium thiosulfate solution ("hypo") is used after development of a photographic film or paper to dissolve away the unreduced silver halide, which if allowed to remain in the emulsion would in the course of time darken through long exposure to light.

Of the nitrite complexes that with cobaltic ion, $Co(NO_2)_6^{---}$, called the *cobaltinitrite ion* or *hexanitrocobaltic ion,* is the most familiar. *Potassium cobaltinitrite,* $K_3Co(NO_2)_6$, is one of the least soluble potassium salts, and its precipitation on addition of sodium cobaltinitrite reagent is commonly used as a test for potassium ion.

Ferric ion and thiocyanate ion combine to give a product with an intense red color; this reaction is used as a test for ferric ion. The red color seems to be due to various complexes, ranging from $Fe(H_2O)_5NCS^{++}$ to $Fe(NCS)_6^{---}$. The azide ion, NNN^-, gives a similar color with ferric ion.

The Chromic and Cobaltic Complexes

Terpositive chromium and cobalt combine with cyanide ion, nitrite ion, chloride ion, sulfate ion, oxalate ion, water, ammonia, and many other ions and molecules to form a very great number of complexes, with a wide range of colors that are nearly the same for corresponding chromic and cobaltic complexes. Most of these complexes are stable, and are formed and decomposed slowly. Representative are the members of the series

$Cr(NH_3)_6^{+++}$	$Cr(NH_3)_5Cl^{++}$	$Cr(NH_3)_4Cl_2^{+}$
Yellow	Purple	Green

$Cr(NH_3)_3Cl_3$	$Cr(NH_3)_2Cl_4^{-}$
Violet	Orange-red

and

$Co(NH_3)_6^{+++}$	$Co(NH_3)_5H_2O^{+++}$	...	$Co(H_2O)_6^{+++}$
Yellow	Rose-red		Purple

A group such as oxalate ion, $C_2O_4^{--}$, or carbonate ion, CO_3^{--}, may occupy two of the six coordination places in an octahedral complex; examples are $Co(NH_3)_4CO_3^{+}$ and $Cr(C_2O_4)_3^{---}$.

The often puzzling color changes shown by chromic solutions are caused by reactions of these complexes. Solutions containing chromic ion, $Cr(H_2O)_6^{+++}$, are purple in color; on heating they become green, because of the formation of complexes such as $Cr(H_2O)_4Cl_2^{+}$ and $Cr(H_2O)_5SO_4^{+}$. At room temperature these green complexes slowly decompose, again forming the purple solution.

16-6. Hydroxide Complexes

If sodium hydroxide is added to a solution containing zinc ion a precipitate of zinc hydroxide is formed:

$$Zn^{++} + 2OH^{-} \rightleftarrows Zn(OH)_2$$

This hydroxide precipitate is of course soluble in acid; it is also soluble in alkali. On addition of more sodium hydroxide the precipitate goes back into solution, this process occurring at hydroxide-ion concentrations about 0.1 *M* to 1 *M*.

To explain this phenomenon we postulate the formation of a complex ion. The complex ion that is formed is the zincate ion, $Zn(OH)_4^{--}$, by the reaction

$$Zn(OH)_2 + 2OH^- \rightleftarrows Zn(OH)_4^{--}$$

The ion is closely similar to other complexes of zinc, such as $Zn(H_2O)_4^{++}$, $Zn(NH_3)_4^{++}$, and $Zn(CN)_4^{--}$, with hydroxide ions in place of water or ammonia molecules or cyanide ions. The ion $Zn(H_2O)(OH)_3^-$ is also formed to some extent.

The molecular species that exist in zinc solutions of different pH values are the following:

Acidic solution $\begin{cases} Zn(H_2O)_4^{++} \\ Zn(H_2O)_3(OH)^+ \end{cases}$

Neutral solution $Zn(H_2O)_2(OH)_2 \rightleftarrows Zn(OH)_2(c)$

Basic solution $\begin{cases} Zn(H_2O)(OH)_3^- \\ Zn(OH)_4^{--} \end{cases}$

The conversion of each complex into the following one occurs by removal of a proton from one of the four water molecules of the tetrahydrated zinc ion. Each complex except $Zn(H_2O)_4^{++}$ and $Zn(OH)_4^{--}$ is amphiprotic.

The principal common amphiprotic hydroxides and their anions are the following:

$Zn(OH)_2$	$Zn(OH)_4^{--}$,	zincate ion
$Al(OH)_3$	$Al(OH_2)_2(OH)_4^-$,	aluminate ion
$Cr(OH)_3$	$Cr(OH_2)_2(OH)_4^-$,	chromite ion
$Pb(OH)_2$	$Pb(OH)_3^-$,	plumbite ion
$Sn(OH)_2$	$Sn(OH)_3^-$,	stannite ion

In addition, the following hydroxides evidence acidic properties by combining with hydroxide ion to form complex anions:

$Sn(OH)_4$	$Sn(OH)_6^{--}$,	stannate ion
$As(OH)_3$	$As(OH)_4^-$,	arsenite ion
$Sb(OH)_3$	$Sb(OH)_4^-$,	antimonite ion
$Sb(OH)_5$	$Sb(OH)_6^-$,	antimonate ion

The hydroxides listed above form hydroxide complex anions to a sufficient extent to make them soluble in moderately strong alkali. Other common hydroxides have weaker acidic properties: $Cu(OH)_2$ and

$Co(OH)_2$ are only slightly soluble in very strong alkali, and $Cd(OH)_2$, $Fe(OH)_3$, $Mn(OH)_2$, and $Ni(OH)_2$ are effectively insoluble. The common analytical method of separation of Al^{+++}, Cr^{+++}, and Zn^{++} from Fe^{+++}, Mn^{++}, Co^{++}, and Ni^{++} with use of sodium hydroxide is based on these facts.

16-7. Sulfide Complexes

Sulfur, which is directly below oxygen in the periodic table of the elements, has many properties similar to those of oxygen. One of these is the property of combining with another atom to form complexes; there exist *sulfo acids* (thio acids) of many elements similar to the oxygen acids. An example is *sulfophosphoric acid,* H_3PS_4, which corresponds exactly in formula to phosphoric acid, H_3PO_4. This sulfo acid is not of much importance; it is unstable, and hydrolyzes in water to phosphoric acid and hydrogen sulfide:

$$H_3PS_4 + 4H_2O \rightarrow H_3PO_4 + 4H_2S$$

But other sulfo acids, such as *sulfarsenic acid,* H_3AsS_4, are stable, and are of use in analytical chemistry and in chemical industry.

All of the following arsenic acids are known:

$$H_3AsO_4 \qquad H_3AsO_3S \qquad H_3AsO_2S_2 \qquad H_3AsOS_3 \qquad H_3AsS_4$$

The structure of the five complex anions AsO_4^{---}, AsO_3S^{---}, $AsO_2S_2^{---}$, $AsOS_3^{---}$, and AsS_4^{---} is the same: an arsenic atom surrounded tetrahedrally by four other atoms, oxygen or sulfur.

Some metal sulfides are soluble in solutions of sodium sulfide or ammonium sulfide because of formation of a complex sulfo anion. The important members of this class are HgS, As_2S_3, Sb_2S_3, As_2S_5, Sb_2S_5, and SnS_2, which react with sulfide ion in the following ways:

$$HgS + S^{--} \rightleftarrows HgS_2^{--}$$

$$As_2S_3 + 3S^{--} \rightleftarrows 2AsS_3^{---}$$

$$Sb_2S_3 + 3S^{--} \rightleftarrows 2SbS_3^{---}$$

$$As_2S_5 + 3S^{--} \rightleftarrows 2AsS_4^{---}$$

$$Sb_2S_5 + 3S^{--} \rightleftarrows 2SbS_4^{---}$$

$$SnS_2 + S^{--} \rightleftarrows SnS_3^{--}$$

Mercuric sulfide is soluble in a solution of sodium sulfide and sodium hydroxide (to repress hydrolysis of the sulfide, which would decrease the sulfide-ion concentration), but not in a solution of ammonium sulfide and ammonium hydroxide, in which the sulfide-ion concentration is smaller. The other sulfides listed are soluble in both solutions. CuS, Ag_2S, Bi_2S_3, CdS, PbS, ZnS, CoS, NiS, FeS, MnS, and SnS are not soluble in sulfide solutions, but most of these form complex sulfides by fusion with Na_2S or K_2S. Although SnS is not soluble in Na_2S or $(NH_4)_2S$ solutions, it dissolves in solutions containing both sulfide and disulfide, Na_2S_2 or $(NH_4)_2S_2$, or sulfide and peroxide. The disulfide ion, S_2^{--}, or peroxide oxidizes the tin to the stannic level, and the sulfostannate ion is then formed:

$$SnS + S_2^{--} \rightleftarrows SnS_3^{--}$$

Many schemes of qualitative analysis involve separation of the copper-group sulfides (PbS, Bi_2S_3, CuS, CdS) from the tin-group sulfides (HgS, As_2S_3, As_2S_5, Sb_2S_3, Sb_2S_5, SnS, SnS_2) by treatment with Na_2S-Na_2S_2 solution, which dissolves only the tin-group sulfides.

16-8. Polydentate Complexing Agents

In analytical chemistry and industrial chemistry extensive use is made of complexing agents with more than one atom capable of attachment to the central metal atom in a complex. Such a complexing agent is called a *polydentate ligand* or a *chelating agent* (pronounced kee'lating; from the Greek *chēlē*, claw).

An example is triaminotriethylamine (tren), with the following structural formula:

$$\mathrm{N}\begin{cases}\mathrm{CH_2{-}CH_2{-}NH_2}\\ \mathrm{CH_2{-}CH_2{-}NH_2}\\ \mathrm{CH_2{-}CH_2{-}NH_2}\end{cases}$$

All four nitrogen atoms of this molecule can coordinate with a metal atom. Thus the Zn^{++} ion forms a complex with tren in which each of the four nitrogen atoms uses its unshared pair of electrons to form a bond with the zinc atom and the nitrogen atoms are arranged approximately tetrahedrally about the central atom. The formation constant $[Zn(tren)^{++}]/[Zn^{++}][tren]$ for the complex between tren and Zn^{++}, 4.5×10^{14}, is over 400,000 times larger than that $[Zn(NH_3)_4^{++}]/[Zn^{++}][NH_3]^4$ of the reaction between the zinc ion and four ammonia molecules. The large value of the formation constant for the $Zn(tren)^{++}$ complex is primarily the result of the entropy factor (the fact that the four nitrogen atoms are not free to move about in the solution independently of one another, but are bonded to one another at approximately the same distance apart as in the complex).

Another polydentate complexing agent that forms complexes with many metal ions is EDTA (ethylenediaminetetra-acetic acid), with formula

```
            CH2COOH
        ..  /
         N
        / \
    H2C    CH2COOH
     |
    H2C    CH2COOH
        \ /
         N
        ..  \
            CH2COOH
```

The anion of EDTA has a quadruple negative charge. The four carboxylate ion groups and also the two nitrogen atoms can form bonds with a metal atom; the anion is accordingly a hexadentate complexing agent. In the stable complexes that it forms with many metal ions the two nitrogen atoms and four oxygen atoms, one of each carboxylate ion group, are approximately octahedrally arranged around the central ion. The structure of this complex with tripositive cobalt as determined by x-ray diffraction of a crystal containing the complex is shown in Figure 16-3.

EDTA (also called versene) is used in analytical chemistry and also in chemical industries. In many industrial processes, such as dyeing and the manufacture of soaps and detergents, even very small concentrations of heavy metal ions in the water interfere with the reactions. EDTA and

FIGURE 16-3
The structure of the complex between tripositive cobalt and the anion of EDTA.

similar agents (sequestering agents) convert the metal ions into complexes, which may not have the harmful properties of the metal ions.

16-9. The Structure and Stability of Carbonyls and Other Covalent Complexes of the Transition Metals

The problem of the stability of the complexes of the transition metals was for many years a puzzling one. Why is the cyanide group so facile in the formation of complexes with these elements, whereas the carbon atom in other groups, such as the methyl group, does not form bonds with them? Why do the transition metals and not other metals (beryllium, aluminum, and so on) form cyanide complexes? In the ferrocyanide ion, $Fe(CN)_6^{----}$, for example, the iron atom has a formal charge of 4^-, on the assumption that it forms six covalent bonds with the six ligands; how can this large negative charge be made compatible with the tendency of metals to lose electrons and form positive ions?

The answers to these questions and other questions about the cyanide and carbonyl transition-metal complexes can be derived from the idea that the cyanide and carbonyl groups form double bonds with the transition metal atom.

Nickel tetracarbonyl, $Ni(CO)_4$, is a volatile liquid with freezing point −25°C and boiling point 43°C. It plays an important part in the refining of nickel (Section 19-6). The electron-diffraction pattern of the gas has shown that the molecule has a tetrahedral configuration with bond lengths Ni—C = 182 pm and C═O = 116 pm. These bond lengths show that the Ni—C bond has a large amount of double-bond character and the C═O bond a considerable amount of triple-bond character.* The structures A and B are suggested for the molecule by the electroneutrality principle (plus the structures that correspond to the partial ionic character of the Ni—C bonds):

```
           :Ö                                 :Ö
            ‖                                  ‖
            C                                  C
            ‖                                  ‖
   :O=C=Ni=C=Ö:                       :O=C=Ni=C=Ö:
    ¨       ‖                          ¨       |
            C                                  C
            ‖                                  ⦀
 A          O:                        B        O
            ¨                                  ¨
                                      (four of this type)
```

*Section 6-15. The observed bond length in NiH(g) is 147 pm, which leads to 117 pm for the single-bond radius of Ni.

For each of these structures all nine $3d^5 4s4p^3$ orbitals of the nickel atom are used, either for bond formation or for occupancy by an unshared pair. The electronegativity difference 0.6 of C and Ni corresponds to 9% ionic character (Table 6-4). Hence structure A places the charge −0.72 on the nickel atom, and structure B the charge +0.37. Electroneutrality would result from twice the contribution of B as of A.

Iron forms the pentacarbonyl $Fe(CO)_5$, a liquid with freezing point −21°C and boiling point 103°C. It has the trigonal bipyramidal configuration, with Fe—C bond length 184 pm, and principal electronic structure

```
         O
         C
   OC    |
      \\ |
        Fe=CO
      // ||
   OC    C
         O
```

and its equivalents with the single bond to each carbon atom. Chromium hexacarbonyl is a crystalline substance. It is less stable than nickel carbonyl and iron carbonyl, and decomposes at about 110°C. Its bond length, Cr—C = 192 pm, is compatible with the electronic structure

```
         O
         C
   OC    |    CO
      \  |  /
        Cr
      // || \\
   OC    C    CO
         O
```

and its equivalents.

The Cyanide Complexes of the Transition Elements

The structural formula usually written for the ferrocyanide ion,

```
 [        N          ]----
 [        C          ]
 [        |   CN     ]
 [        | /        ]
 [ NC—Fe—CN          ]
 [      / |          ]
 [   NC   C          ]
 [        N          ]
```

with single covalent bonds from the iron atom to each of the six carbon atoms, is seen to be surprising in that it places a charge of 4− on the iron atom, whereas iron tends to assume a positive charge, as in the ferrous ion, and not a negative charge. As suggested by the foregoing discussion of the carbonyl compounds, we assign to the complex a structure involving some iron-carbon double bonds. The structure

```
[            ..−            ]----
             N:
             ‖        N:
             C      ///
            −‖    C
             ‖  /        ..−
  :N≡C—Fe=C=N:
            //|
         C    C
   −..  //    |||
   :N         N
              ..
]
```

and its equivalents places the formal charge −1 on the iron atom (obtained by dividing the bond electrons equally between the bonded atoms), which becomes +0.08 on correction for the 12% ionic character for iron-carbon bonds indicated by the electronegativity difference. This structure thus agrees with the electroneutrality principle.

16-10. Polynuclear Complexes

The transition metals form polynuclear complexes, containing two or more transition-metal atoms, as well as the simple complexes discussed in the first part of this chapter. Cobalt(III), for example, forms many octahedral complexes, including $[Co(NH_3)_6]^{+++}$, which is yellow, and $[Co(NH_3)_5Cl]^{++}$, which is purple-red in color. It also forms the bright-blue binuclear ion $[(NH_3)_5CoNH_2Co(NH_3)_5]^{5+}$, in which each atom is octahedrally ligated to five ammino groups and one immino group, NH_2, which occupies the shared corner of each of the two octahedra. This ion in a solution containing hydrochloric acid reacts with H^+ and Cl^- to form the hexammino complex and pentamminochloro complex mentioned above. The OH group can also serve to link two octahedra together.

In some polynuclear transition-metal complexes the metal atoms are bonded directly to one another. An example is provided by the yellow substance with composition corresponding to the simple formula $MoCl_2$. This substance was discovered in 1859 by C. W. Blomstrand, who pointed out that it has the surprising property that, when it is dissolved in water and a solution of silver nitrate is added, only one third of the chlorine is precipitated as silver chloride. X-ray studies of the crystals has shown that they contain the complex ion $[Mo_6Cl_8]^{4+}$, with the structure shown in Figure 16-4. The six molybdenum atoms form an octahedron; each atom is bonded to four molybdenum atoms by single bonds (Mo—Mo distance 263 pm, which is less than in the metal, 273 pm for ligancy 8), as well as to four bridging chlorine atoms. The cationic complex can also add six anions, such as chloride or hydroxide, in the six positions directly out from each metal atom. $MoBr_2$, WCl_2, and WBr_2 also contain similar polynuclear complexes.

Many metal carbonyls and related substances are polynuclear. A representative example is dicobalt hexacarbonyl diphenylacetylene, the structure of which, determined by x-ray diffraction, is shown in Figure 16-5. The carbon-carbon triple bond has been replaced by a carbon-carbon single bond and four carbon-cobalt single bonds. Each cobalt atom forms a single bond to the other cobalt atom, two single bonds to the acetylenic carbon atoms, and a double bond to each of its attached carbonyl groups, thus using all of its nine outer electrons and nine outer orbitals in bond formation. In some polynuclear carbonyl complexes there are bridging carbonyl groups, in which the carbonyl carbon atom forms single bonds with two metal atoms, in addition to the double bond to the oxygen atom.

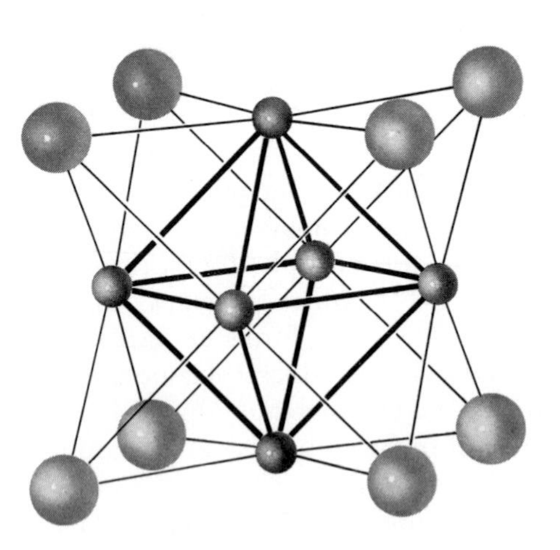

FIGURE 16-4
The structure of the complex ion $(Mo_6Cl_8)^{++++}$.

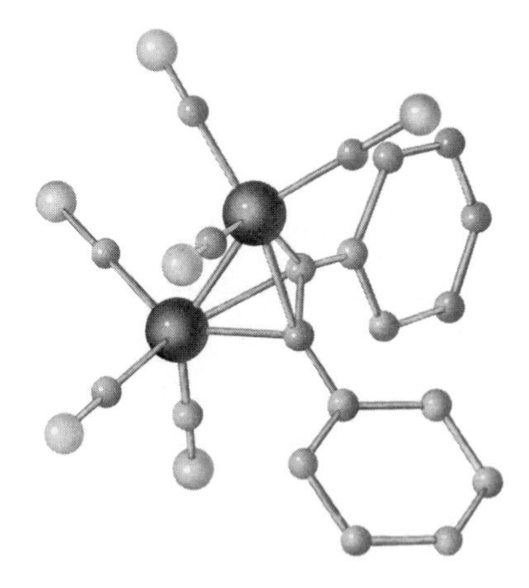

FIGURE 16-5
The structure of dicobalt hexacarbonyl diphenylacetylene, $Co_2(CO)_6C_2(C_6H_5)_2$. Large spheres represent cobalt atoms, small spheres carbon atoms, and spheres of intermediate size oxygen atoms.

EXERCISES

16-1. Define inorganic complex, coordination compound, and complex ion.

16-2. In what ways does the chemistry of inorganic complexes or coordination compounds differ from the chemistry of the other substances you have studied so far?

16-3. Characterize the transition elements in terms of their electronic structure and available bonding orbitals.

16-4. What oxidation state would you consider to be most stable of scandium? Zinc? Copper? Why?

16-5. Why should most of the transition element coordination compounds be colored (absorb light in the visible region) and some, such as those of Zn(II), be white (no absorption)?

16-6. What bond orbitals are used to form octahedral bonds? Square bonds? Tetrahedral bonds?

16-7. Why is tetrahedral four-coordinated Ni(II) paramagnetic and square four-coordinated Ni(II) diamagnetic?

16-8. Which requires more space (solid angle) around a coordinated atom, a bonding electron pair or an unshared electron pair?

16-9. On the basis of a simple electrostatic repulsion theory, what coordination polyhedra would you expect to find when the total number of bonded and non-bonded electron pairs equals 4? 5? 6? 7? 8? 12?

16-10. Discuss the effects of adding to three portions of a cupric solution (*a*) NH_4OH, (*b*) NaOH, (*c*) NH_4Cl. Write equations for the reactions.

16-11. To three portions of a solution containing Ni^{++} and Al^{+++} there are added (*a*) NaOH, (*b*) NH_4OH, (*c*) NaOH + NH_4OH. What happens in each case?

16-12. What happens when a solution of sodium chloride is added to a solution of silver nitrate? What happens when a solution of ammonium hydroxide is then added in excess?

16-13. Write the equation for the principal chemical reaction involved in fixing a photographic film.

16-14. Write the chemical equation for the solution of platinum in aqua regia. Explain why platinum dissolves in aqua regia but not in either hydrochloric acid or nitric acid alone.

16-15. Would sodium cyanide be an effective substitute for sodium thiosulfate as a photographic fixer?

16-16. What molecular species are involved when a purple chromic solution changes to green upon heating and becomes purple again upon cooling?

16-17. What happens when NaOH is added to a solution of Zn(II)? When acid is added to this system?

16-18. How many structural isomers of the octahedral complex $Co(NH_3)_3Cl_3$ are there?

16-19. How many isomers of the tetrahedral complex $Zn(NH_3)_2Cl_2$ are there? Of the square planar complex $Pt(NH_3)_2Cl_2$?

16-20. The silicon hexafluoride ion, SiF_6^{--}, is octahedral. What orbitals of the silicon atom are occupied by unshared electron pairs? By bonding electron pairs? What is the electric charge on the silicon atom and each fluorine atom, as calculated from electronegativity values?

16-21. The compounds K_2SiF_6, K_2SnF_6, and K_2SnCl_6 are known, but not K_2SiCl_6. Can you explain this fact?

16-22. The aqueous solution made by dissolving crystals of ferric nitrate, $Fe(NO_3)_3 \cdot 6H_2O$ (which has a violet color), is yellow. When an equal volume of strong nitric acid is added to the solution, it assumes a pale violet color. What reaction do you suggest to be responsible for the change in color?

16-23. What metal sulfides are soluble in sulfide solutions and what complex ions are formed? What metal sulfides are not soluble in sulfide solutions?

16-24. What are sequestering agents and give an example. What are they used for and how do they work?

16-25. Mond made a fortune by discovering a process for refining nickel using nickel tetracarbonyl. What might the process be?

17

The Nature of Metals and Alloys

About eighty of the more than one hundred elementary substances are metals. A metal may be defined as a substance that has large conductivity of electricity and of heat, has a characteristic luster, called metallic luster, and can be hammered into sheets (is malleable) and drawn into wire (is ductile); in addition, the electric conductivity increases with decrease in temperature.*

The metallic elements may be taken to include lithium and beryllium in the first short period of the periodic table, sodium, magnesium, and aluminum in the second short period, the thirteen elements from potassium to gallium in the first long period, the fourteen from rubidium to tin in the second long period, the twenty-nine from cesium to bismuth in the first very long period (including the fourteen rare-earth metals), and the eighteen from francium to khurchatovium in the second very long period.

The metals themselves and their alloys are of great usefulness to man, because of the properties characteristic of metals. Our modern civilization is based upon iron and steel and valuable alloy steels are made that involve the incorporation with iron of vanadium, chromium, manganese, cobalt, nickel, molybdenum, tungsten, and other metals. The importance of these alloys is due primarily to their hardness and strength. These

---

*Sometimes there is difficulty in classifying an element as a metal, a metalloid, or a nonmetal. For example, the element tin can exist in two forms, one of which, the common form, called white tin, is metallic, whereas the other, gray tin, has the properties of a metalloid. The next element in the periodic table, antimony, exists in only one crystalline form, with metallic luster but with the electric properties of a metalloid, and it is brittle, rather than malleable and ductile. We shall consider tin to be a metal and antimony a metalloid.

properties are a consequence of the presence in the metals of very strong bonds between the atoms. For this reason it is of especial interest to us to understand the nature of the forces that hold the metal atoms together in metals and alloys.

17-1. The Structure of Metals

In a nonmetal or metalloid the number of atoms that each atom has as its nearest neighbors is determined by its covalence. For example, the iodine atom, which is univalent, has only one other iodine atom close to it in a crystal of iodine: the crystal, like liquid iodine and iodine vapor, is composed of diatomic molecules. In a crystal of sulfur there are S_8 molecules, in which each sulfur atom has two nearest neighbors, to each of which it is attached by one of its two covalent bonds. In diamond the quadrivalent carbon atom has four nearest neighbors. On the other hand, the potassium atom in potassium metal, the calcium atom in calcium metal, and the titanium atom in titanium metal, which have one, two, and four outer electrons, respectively, do not have only one, two, and four nearest neighbors, but have, instead, eight or twelve nearest neighbors. We may state that one of the characteristic features of a metal is that each atom has a large number of neighbors; the number of small interatomic distances is greater than the number of valence electrons.

Most metals crystallize with an atomic arrangement in which each atom has surrounded itself with the maximum number of atoms that is geometrically possible. There are two common metallic structures that correspond to the closest possible packing of spheres of constant size. One of these structures, called the cubic closest-packed structure, has been described in Chapter 2. The other structure, called hexagonal closest packing, is represented in Figure 17-1. It is closely similar to the cubic closest-packed structure; each atom is surrounded by twelve equidistant neighbors, with, however, the arrangement of these neighbors slightly different from that in cubic closest packing. About fifty metals have the cubic closest-packed structure or the hexagonal closest-packed structure, or both.

Another common structure, assumed by about twenty metals, is the body-centered cubic structure. In this structure, shown as Figure 2-9, each atom has eight nearest neighbors, and six next-nearest neighbors. These six next-nearest neighbors are 15% more distant than the eight nearest neighbors; in discussing the structure it is difficult to decide whether to describe each atom as having ligancy 8 or ligancy 14.

The periodicity of properties of the elements, as functions of the atomic number, is illustrated by the observed values of the interatomic distances in the metals, as shown in Figure 17-2. These values are half of the

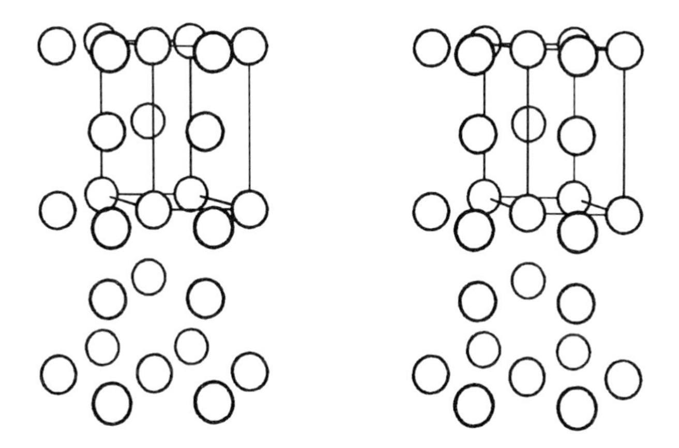

FIGURE 17-1
The hexagonal close-packed arrangement of spheres. Many metals crystallize with this structure.

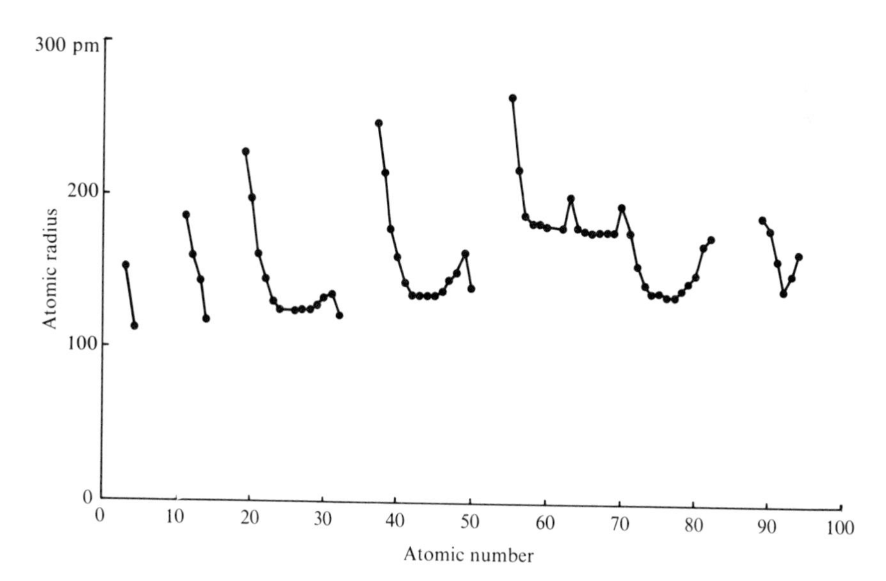

FIGURE 17-2
The atomic radii of metals, plotted against atomic number.

directly determined interatomic distances for the metals with a cubic closest-packed or hexagonal closest-packed structure. For other metals a small correction has been made; it has been observed, for example, that a metal such as iron, which crystallizes in a form with a closest-packed structure and also a form with the body-centered cubic structure, has contact interatomic distances about 3% less in the latter structure than in the former, and accordingly a correction of 3% can be made for body-centered cubic structures, to convert the interatomic distances to ligancy 12.

We may well expect that the strongest bonds would have the shortest interatomic distances, and it is accordingly not surprising that the large interatomic distances shown in Figure 17-2 are those for soft metals, such as potassium; the smallest ones, for chromium, iron, nickel, and others, refer to the strong, hard metals.

17-2. The Metallic State

Let us consider the first six metals of the first long period, potassium, calcium, scandium, titanium, vanadium, and chromium. The first of these metals, potassium, is a soft, light metal, with low melting point. The second metal, calcium, is much harder and denser, and has a much higher melting point. Similarly, the third metal, scandium, is still harder, still denser, and melts at a still higher temperature, and this change in properties continues through titanium, vanadium, and chromium. These properties are illustrated in Figure 17-3, which shows a quantity called the ideal density, equal to 50/gram-atomic volume. This ideal density, which is inversely proportional to the gram-atomic volume of the metal, is the density that these metals would have if they all had the same atomic weight, 50. It is an inverse measure of the cube of the interatomic distances in the metals. We see that the ideal density increases steadily from its minimum value of about 1 for potassium to a value of about 7 for chromium, and many other properties of the metals, including hardness and tensile strength, show a similar steady increase through this series of six metals.

There is a simple explanation of this change in properties in terms of the electronic structure of the metals. The potassium atom has only one electron outside of its completed argon shell. It could use this electron to form a single covalent bond with another potassium atom, as in the diatomic molecules K_2 that are present, together with monatomic molecules K, in potassium vapor. In the crystal of metallic potassium each potassium atom has a number of neighboring atoms, at the same distance. It is held to these neighbors by its single covalent bond, which resonates

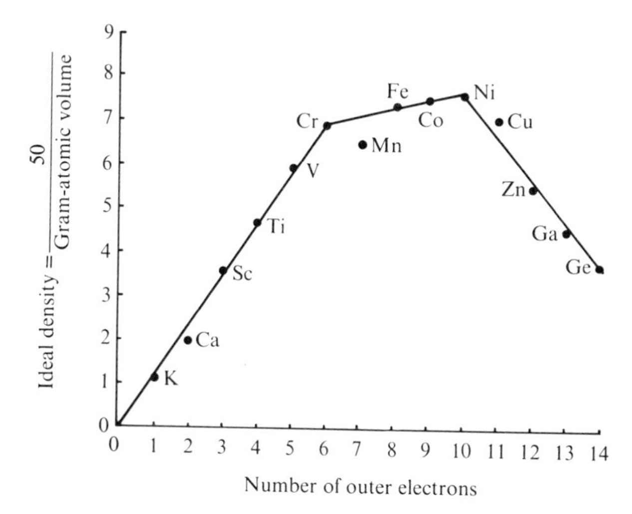

FIGURE 17-3
A graph of the ideal density of the metals of the first long period. The ideal density is defined here as the density that these metals would have if their atomic masses were all equal to 50.

among the neighbors. In metallic calcium there are *two* valence electrons per calcium atom, permitting each atom to form two bonds with its neighbors. These two bonds resonate among the calcium-calcium positions, giving a total bonding power in the metal twice as great as that in potassium. Similarly in scandium, with *three* valence electrons, the bonding is three times as great as in potassium, and so on to chromium, where, with *six* valence electrons, the bonding is six times as great.

This increase does not continue in the same way beyond chromium. Instead, the strength, hardness, and other properties of the transition metals remain essentially constant for the five elements chromium, manganese, iron, cobalt, and nickel, as is indicated by the small change in ideal density in Figure 17-3. (The low value for manganese results from the existence of this metal with an unusual crystal structure shown by no other element.) We can conclude that the metallic valence does not continue to increase, but remains at the value six for these elements. Then, after nickel, the metallic valence again decreases, through the series copper, zinc, gallium, and germanium, as is indicated by the rapid decrease in ideal density in Figure 17-3, and by a corresponding decrease in hardness, melting point, and other properties.

It is interesting to note that in the metallic state chromium has metallic valence 6, corresponding to the oxidation number +6 characteristic of the chromates and dichromates, rather than to the lower oxidation number

+3 shown in the chromium salts, and that the metals manganese, iron, cobalt, and nickel also have metallic valence 6, although nearly all of their compounds represent the oxidation state +2 or +3. *The valuable physical properties of the transition metals are the result of the high metallic valence of the elements.*

Unsynchronized Resonance of Bonds in Metals

It is mentioned above that in potassium metal each atom, with one valence electron, can form one covalent bond, and that this bond is not between the atom and a single neighboring atom, but instead resonates among several positions. For four potassium atoms in a square, we might write two valence-bond structures:

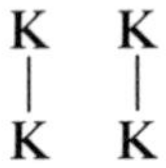

$$\begin{array}{cc} \begin{array}{cc} K & K \\ | & | \\ K & K \end{array} & \begin{array}{c} K—K \\ \\ K—K \end{array} \\ \text{I} & \text{II} \end{array}$$

These structures are analogous to the two Kekulé structures for the benzene molecule. Resonance between the two structures would stabilize the metal relative to a crystal composed of K_2 molecules, each with a fixed covalent bond.

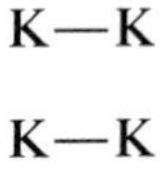

There are other structures, involving a transfer of an electron from one atom to another, that might be considered:

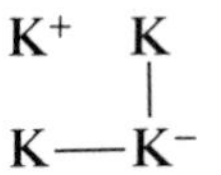

$$\begin{array}{cccc} \begin{array}{cc} K^+ & K \\ & | \\ K & \!\!—\!\! K^- \end{array} & \begin{array}{cc} K & K^+ \\ | & \\ K^- & \!\!—\!\! K \end{array} & \begin{array}{cc} K^- & \!\!—\!\! K \\ | & \\ K & K^+ \end{array} & \begin{array}{cc} K & \!\!—\!\! K^- \\ & | \\ K^+ & K \end{array} \\ \text{III} & \text{IV} & \text{V} & \text{VI} \end{array}$$

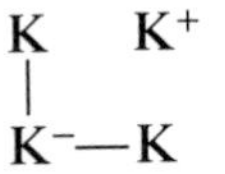

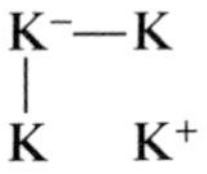

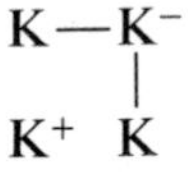

Resonance among all six structures would lead to greater stabilization than resonance between only I and II above. Moreover, this unsynchronized resonance gives a simple explanation of the characteristic properties of metals—the large electric conductivity and the negative temperature coefficient of electric conductivity.

Let us consider a row of potassium atoms:

$$K—K \quad K—K \quad K^+ \quad K—K^-—K \quad K—K \quad K—K$$

In the presence of an electric field, produced by a cathode at the left and an anode at the right, the bonds would tend to shift in such a way as to move the positive charge toward the cathode and the negative charge toward the anode:

$$\text{K—K} \quad \text{K}^+ \quad \text{K—K} \quad \text{K—K} \quad \text{K—K}^-\text{—K} \quad \text{K—K}$$

$$\text{K}^+ \ \text{K—K} \quad \text{K—K} \quad \text{K—K} \quad \text{K—K} \quad \text{K—K}^-\text{—K}$$

etc.

The unsynchronized resonance of the bonds corresponds to the transfer of electric charge (electrons) that leads to high electric conductivity. This conductivity is characteristic of the structure of the metal, and hence takes place most readily at very low temperatures, when the atoms are quite regularly arranged in the crystal. At higher temperatures the thermal oscillation of the atoms introduces some disorder in their arrangement, which interferes with the resonance of the bonds, and hence causes a decrease in conductivity (negative temperature coefficient).

The Metallic Orbital

In the valence-bond structure

$$\begin{array}{cc} \text{K}^+ & \text{K} \\ & | \\ \text{K} & \!\!\!\!\text{—K}^- \end{array}$$

one atom, K^-, has assumed a second valence electron, permitting it to form two covalent bonds, rather than just one. Two bond orbitals are accordingly being utilized by this atom, one more than by the neutral (unicovalent) atoms. This extra orbital, called the *metallic orbital,* is needed to permit the unsynchronized valence-bond resonance characteristic of metals: *the metallic orbital, an extra orbital not occupied by an electron or electron pair in the neutral atom, is the characteristic structural feature of metals.*

Potassium has nine reasonably stable orbitals in its outer shell: one 4*s* orbital, three 4*p* orbitals, and five 3*d* orbitals. Only one (an *spd* hybrid) is used as a bond orbital by the unicovalent atom, and others are available to serve as the metallic orbital; hence potassium is a metal. In diamond, on the other hand, the four stable orbitals of the valence shell (the tetrahedral orbitals formed by hybridization of the 2*s* orbital and the three 2*p* orbitals, Chapter 6) are all occupied by bond electrons; there is no metallic orbital, and hence diamond is not a metal.

17-3. Metallic Valence

It is mentioned in the preceding section that for the elements potassium, calcium, scandium, titanium, vanadium, and chromium the physical properties indicate that all of the electrons outside of the argon shell are used in forming bonds, and that the metallic valences for these elements are 1, 2, 3, 4, 5, and 6, respectively.

There are nine stable orbitals available for the transition elements (one $4s$, three $4p$, five $3d$), and, with one required as the metallic orbital, the metallic valence might be expected to continue to increase, and have the value 7 for manganese and 8 for iron. However, as mentioned above, the physical properties show that the metallic valence remains at the maximum of 6 for manganese, iron, cobalt, and nickel, and then begins to decrease at copper. The maximum value of 6 corresponds to the number of good bond orbitals that can be formed by hybridization of the s, p, and d orbitals. The decrease in metallic valence beginning at copper is caused by the limited number of orbitals, as shown by the example of tin.

Tin, element 50, has 14 electrons outside of the krypton shell, and nine stable orbitals ($4d$, $5s$, $5p$). The five $4d$ orbitals, which are more stable than the $5s$ and $5p$ orbitals, are occupied by five unshared electron pairs. The remaining four electrons may separately occupy the four tetrahedral $5s5p^3$ orbitals, and be used in forming four bonds, tetrahedrally directed. In fact, gray tin, one of the two allotropic forms of the element, has the diamond structure. The tin atoms in gray tin are quadrivalent, as are the carbon atoms in diamond. They have no metallic orbital, and gray tin is not a metal, but is a metalloid.

If the tin atom were to retain one of its orbitals for use as a metallic orbital, it would be bivalent rather than quadrivalent:

Gray tin:	$5s$	$5p$	$5p$	$5p$	No metallic orbital
	↑	↑	↑	↑	Quadrivalent
White tin:	$5s$	$5p$	$5p$	$5p$	One metallic orbital
	↓↑	↑	↑	Metallic orbital	Bivalent

The ordinary allotropic form of tin, white tin, has metallic properties. The observed bond lengths indicate that the valence of tin in this form is about 2.5.

The value 2.5 can be accounted for in the following way. The magnetic properties of the iron-group elements and their alloys indicate that the number of metallic orbitals per atom in a metal is 0.72, rather than 1 (see the discussion of magnetic properties in the next section). This fractional value can be explained by the reasonable deduction (see following section) that the metal contains 28% M^+, 44% M, and 28% M^-. The ions M^- do not need a metallic orbital, because they cannot accept another electron (M^{--} would be unstable, according to the electroneutrality principle, Section 6-13). The structure of white tin can accordingly be represented in the following way:

	5*s*	5*p*	5*p*	5*p*	*Contribution to valence*
28% Sn^+	↑	↑	↑	Metallic	$3 \times 0.28 = 0.84$
44% Sn	↑↓	↑	↑	Metallic	$2 \times 0.44 = 0.88$
28% Sn^-	↑↓	↑	↑	↑	$3 \times 0.28 = 0.84$
				Metallic valence of tin	2.56

The same argument leads to the following values for the metallic valence of the elements from copper to germanium:

Cu	Zn	Ga	Ge
5.56	4.56	3.56	2.56

Copper, zinc, and gallium are metals, with properties compatible with these values of the valence. Germanium under ordinary pressure is a metalloid, with the diamond structure and valence 4. At high pressure it is converted into another form, with greatly increased electric conductivity and density corresponding to the white tin structure and valence 2.56.

Ferromagnetism and Metallic Valence

Iron, cobalt, and nickel are ferromagnetic metals. The ferromagnetism of iron corresponds to 2.2 electrons with unpaired spins per atom. The alloys of iron with a small amount of cobalt are more strongly ferromagnetic than pure iron. The ferromagnetism increases to a maximum value at about 28% cobalt, and then decreases, reaching the value corresponding to 1.7 unpaired electrons per atom for pure cobalt.

The maximum ferromagnetism for the alloy of 72% iron and 28% cobalt can be interpreted in the following way. The atoms in this alloy have the average atomic number 26.28, and hence have 8.28 electrons outside of the argon shell. These electrons may occupy nine orbitals: the five 3*d* orbitals, the 4*s* orbital, and the three 3*p* orbitals. But if all nine orbitals were available for occupancy by the electrons (6 for bond formation and the others contributing to the ferromagnetism) the number of unpaired electrons would be expected to continue to increase beyond 28% cobalt and to reach its maximum at pure cobalt, which has nine electrons outside the argon shell. The fact that the maximum ferromagnetism is reached at 28% cobalt (8.28 electrons outside of the argon shell) indicates that only 8.28 of the nine orbitals are available for occupancy. The remaining 0.72 orbital per atom is interpreted as the metallic orbital of 72% of the atoms, as discussed above.

Example 17-1. The ferromagnetism of alloys of nickel and copper decreases from the value corresponding to 0.6 unpaired electron per atom for pure nickel to 0 for the alloy with 56% copper. How is this fact interpreted?

Solution. In these alloys 8.28 orbitals are available for occupancy by the electrons outside of the argon shell. The alloy of 44% nickel and 56% copper has an average of 10.56 such electrons per atom. Of these, 6 are bonding electrons, which occupy 6 of the 8.28 orbitals. The remaining 4.56 electrons occupy the remaining 2.28 orbitals; since the electron/orbital ratio is 2, these electrons are all paired. Hence there are no unpaired electrons in this alloy, and it is not ferromagnetic.

Example 17-2. What is the metallic valence of zinc?

Solution. Zinc has 12 electrons outside the argon shell, and 8.28 orbitals for them to occupy. We place 8.28 electrons with positive spin in these orbitals, and the remaining $12 - 8.28 = 3.72$ electrons with negative spin in 3.72 of the orbitals. Hence 3.72 orbitals per atom are occupied by electron pairs, and the remaining $8.28 - 3.72 = 4.56$ orbitals per atom are occupied by single electrons. These 4.56 electrons can be used in forming bonds. Hence the metallic valence of zinc is 4.56, as stated above.

17-4. The Nature of Alloys

An *alloy* is a metallic material containing two or more elements. It may be homogeneous, consisting of a single phase, or heterogeneous, consisting of a mixture of phases. An example of a homogeneous alloy is coinage gold. An ordinary sample of coinage gold consists of small crystal grains, each of which is a solid solution of copper and gold, with structure of the sort represented in Figure 17-4. An example of a different kind of homogeneous alloy is the very hard metallic substance tantalum carbide, TaC. It is a compound, with the same structure as sodium chloride (Figure 6-19). Each tantalum atom has twelve tantalum atoms as neighbors. In addition, carbon atoms, which are relatively small, are present in the interstices between the tantalum atoms and serve to bind them together. Each

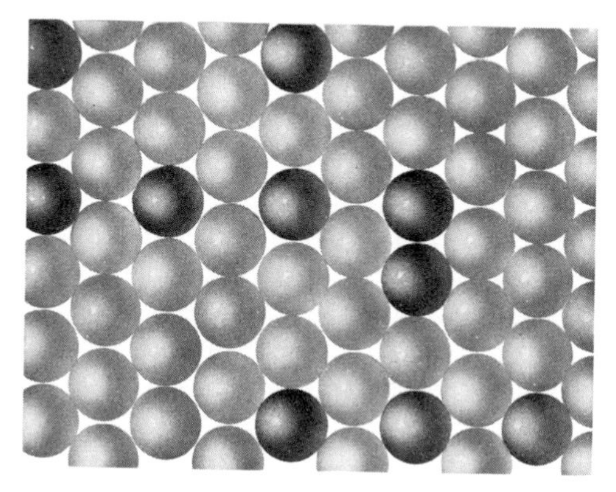

FIGURE 17-4
An alloy of gold and copper. The alloy consists of small crystals, each crystal being made of gold atoms and copper atoms in an orderly array, but with the atoms of the two different kinds distributed essentially at random among the atomic positions.

carbon atom is bonded to the six tantalum atoms that surround it. The bonds are $\frac{2}{3}$ bonds—the four covalent bonds resonate among the six positions about the carbon atom. Each tantalum atom is bonded not only to the adjacent carbon atoms but also to the twelve tantalum atoms surrounding it. The large number of bonds (nine valence electrons per TaC, as compared with five per Ta, occupying, in metallic tantalum, nearly the same volume) explains the greater hardness of the compound than of tantalum itself.

The Binary System Arsenic-Lead

The phase diagram for the binary system arsenic-lead is shown as Figure 17-5. In this diagram the vertical coordinate is the temperature, in degrees centigrade. The diagram corresponds to the pressure 1 atm. The horizontal coordinate is the composition of the alloy, represented along the bottom of the diagram in atomic percentage of lead, and along the top in mass percentage of lead. The diagram shows the temperature and composition corresponding to the presence in the alloy of different phases.

The range of temperatures and compositions represented by the region above the lines *AB* and *BC* is a region in which a single phase is present, the liquid phase, consisting of the molten alloy. The region included in the triangle *ADB* represents two phases, a liquid phase and a solid phase consisting of crystals of arsenic. The triangle *BEC* similarly represents a

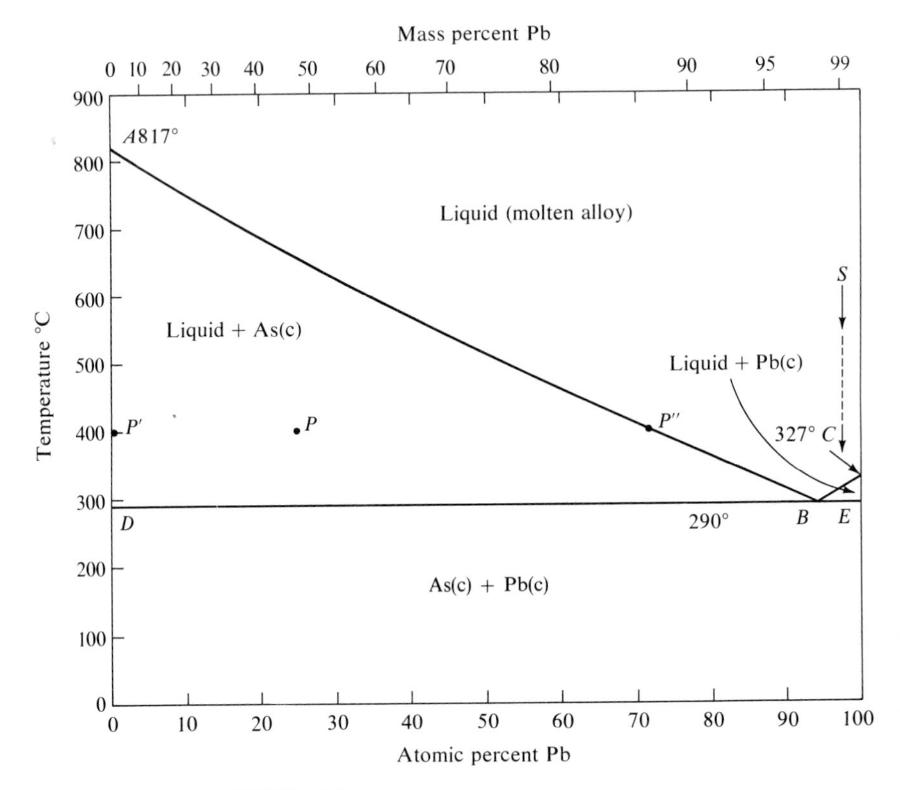

FIGURE 17-5
Phase diagram for the binary system arsenic-lead.

two-phase region, the two phases being the liquid and crystalline lead. The range below the horizontal line *DBE* consists of the two phases crystalline arsenic and crystalline lead, the alloy being a mixture of small grains of the two elements.

Let us apply the phase rule to an alloy in the one-phase region above the line *ABC*. Here we have a system of two components, and, in this region, one phase; the phase rule states that the variance should be three. The three quantities describing the system that may be varied in this region are the pressure (taken arbitrarily in this diagram as 1 atm, but capable of variation), the temperature, which may be varied through the range permitted by the boundaries of the region, and the composition of the molten alloy, which may similarly be varied through the range of compositions permitted by the boundaries of the region.

An alloy in the region *ADB*, such as that represented by the point *P*, at 35 atomic percent lead and 400°C, lies in a two-phase region, and the variance is accordingly stated by the phase rule to be two. The pressure and the temperature are the two variables; the phase rule hence states that it is not possible to vary the composition of the phases present in the alloy. The phases are crystalline arsenic, represented by the point P' directly to the left of P, and the molten alloy, with the composition P'' directly to the right of P. The composition of the molten alloy in equilibrium with crystalline arsenic at 400°C and 1 atm pressure is definitely fixed at P''; it cannot be varied.

The only conditions under which three phases can be in equilibrium with one another at the arbitrary pressure 1 atm are represented by the point B. With three phases in equilibrium with one another for this two-component system, the phase rule requires that there be only one arbitrary variable, which we have used in fixing the pressure arbitrarily at 1 atm. Correspondingly we see that the composition of the liquid is fixed at that represented by the point B, 93 atomic percent lead, and the composition of the two solid phases is fixed, these phases being pure arsenic and pure lead. The temperature is also fixed, at the value 290°C, corresponding to the point B. This point is called the *eutectic point,* and the corresponding alloy is called the *eutectic alloy,* or simply the *eutectic*. The word eutectic means melting easily; the eutectic has a sharp melting point. When a liquid alloy with the eutectic composition is cooled, it crystallizes completely on reaching the temperature 290°C, forming a mixture of very small grains of pure arsenic and pure lead, with a fine texture. When this alloy is slowly heated, it melts sharply at the temperature 290°C.

The lines in the phase diagram are the boundaries separating a region in which one group of phases are present from a region in which another group of phases are present. A line such as AB is called the *freezing-point curve, liquidus curve,* or *liquidus,* and a line such as DB is called the *melting-point curve, solidus curve,* or *solidus*. These boundary lines can be located by various experimental methods, including the method of thermal analysis, discussed in Section 17-5.

If a molten alloy of arsenic and lead with the eutectic composition is cooled, the temperature drops at a regular rate until the eutectic temperature, 290°C, is reached; the liquid then crystallizes into the solid eutectic alloy, the temperature remaining constant until crystallization is complete. The eutectic has a constant melting point, just as has either one of the pure elementary substances.

The effect of the phenomenon of depression of the freezing point (Section 9-13) in causing the eutectic melting point to be lower than the melting point of the pure metals can be intensified by the use of additional components. Thus an alloy with eutectic melting point 70°C can be made by melting together 50 mass percent bismuth (m.p. 271°C), 27% lead (m.p. 327.5°C), 13% tin (m.p. 232°C), and 10% cadmium (m.p. 321°C), and the melting point can be reduced still further, to 47°C, by the incorporation in this alloy of 18% of its mass of indium (m.p. 155°C).

It is possible, in terms of this phase diagram, to discuss the following phenomenon. A small amount, about $\frac{1}{2}\%$ by weight, of arsenic is added to lead used to make lead shot, in order to increase the hardness of the shot and also to improve the properties of the molten material. Lead shot is made by dripping the molten alloy through a sieve. The fine droplets freeze during their passage through the air, and are caught in a tank of water after they have solidified. If pure lead were used, the falling drops would solidify rather suddenly on reaching the temperature 327°C. A falling drop tends not be perfectly spherical, but to oscillate between prolate and oblate ellipsoidal shapes, as you may have noticed by observing drops of water dripping from a faucet; and hence the shot made of pure lead might be expected not to be perfectly spherical in shape. But the alloy containing $\frac{1}{2}\%$ arsenic by weight, represented by the arrow *S* (Figure 17-5), would begin to freeze on reaching the temperature 320°C, and would continue to freeze, forming small crystals of pure lead, until the eutectic temperature 290°C is reached. During this stage of its history the drop would consist of a sludge of lead crystals in the molten alloy, and this sluggish sludge would be expected to be drawn into good spherical shape by the action of the surface-tension forces of the liquid.

The Binary System Lead-Tin

The phase diagram for the lead-tin system of alloys is shown as Figure 17-6. This system rather closely resembles the system arsenic-lead, except that there is an appreciable solubility of tin in crystalline lead and a small solubility of lead in crystalline tin. The phase designated α is a solid solution of tin in lead, the solubility being 19.5 mass percent at the eutectic temperature and dropping to 2% at room temperature. The phase β is a solid solution of lead in tin, the solubility being about 2% at the eutectic temperature and extremely small at room temperature. The eutectic composition is about 62 mass percent tin, 38 mass percent lead.

The composition of *solder* is indicated by the two arrows, corresponding to ordinary plumbers' solder and to 60:40 solder. The properties of solder are explained by the phase diagram. The useful property of plumbers' solder is that it permits a wiped-joint to be made. As the solder cools it forms a sludge of crystals of the α phase in the liquid alloy, and the mechanical properties of this sludge are such as to permit it to be handled by the plumber in an effective way. The sludge corresponds to transition through the region of the phase diagram in which liquid and the α phase are present together. For plumbers' solder the temperature range involved is about 70°, from 250°C to 183°C, the eutectic temperature. The 60:40 tin-lead solder is preferred for electrical work because it has the lowest melting point, and is the least likely to lead to overheating the transistors.

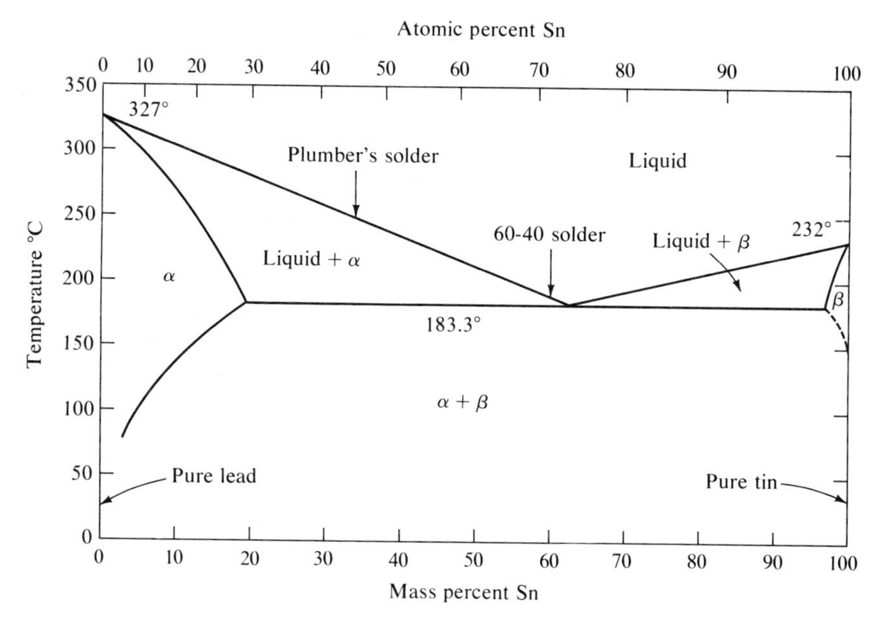

FIGURE 17-6
Phase diagram for the binary system lead-tin.

The Binary System Silver-Gold

The metals silver and gold are completely miscible with one another not only in the liquid state but also in the crystalline state. A solid alloy of silver and gold consists of a single phase, homogeneous crystals with the cubic closest-packed structure, described for copper in Chapter 2, with gold and silver atoms occupying the positions in this lattice essentially at random (Figure 17-4). The phase diagram shown as Figure 17-7 represents this situation. It is seen that the addition of a small amount of gold to pure silver does not depress the freezing point, in the normal way, but instead causes an increase in the temperature of crystallization.

The alloys of silver and gold, usually containing some copper, are used in jewelry, in dentistry, and as a gold solder.

The Binary System Silver-Strontium

A somewhat more complicated binary system—that formed by silver and strontium—is represented in Figure 17-8. It is seen that four intermetallic compounds are formed, their formulas being Ag_5Sr, Ag_5Sr_3, AgSr, and

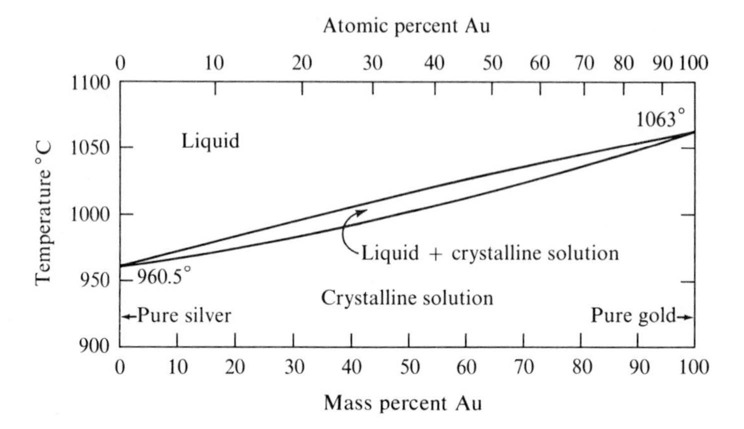

FIGURE 17-7
Phase diagram for the binary system silver-gold, showing the formation of a complete series of crystalline solutions.

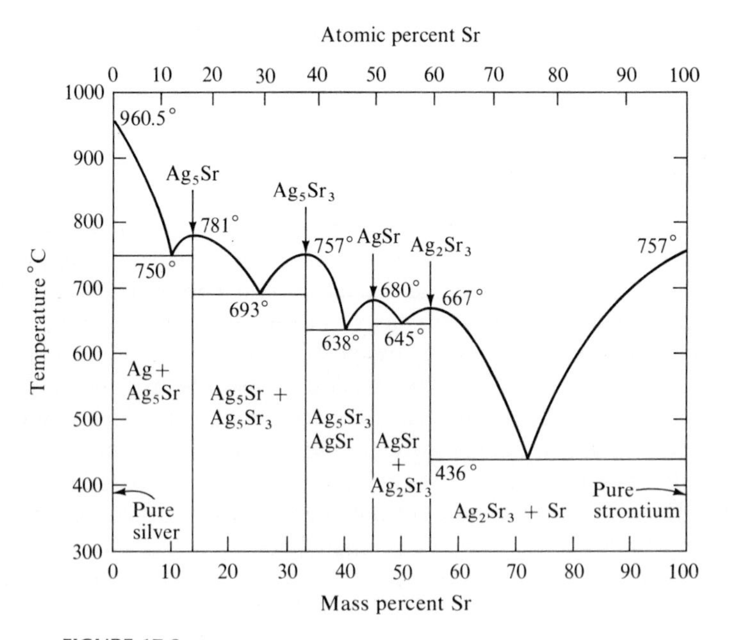

FIGURE 17-8
Phase diagram for the binary system silver-strontium, showing the formation of four intermetallic compounds.

Ag_2Sr_3. These compounds and the pure elements form a series of eutectics; for example, the alloy containing 25 weight percent strontium is the eutectic mixture of Ag_5Sr and Ag_5Sr_3.

Some other binary systems are far more complicated than this one. As many as a dozen different phases may be present, and these phases may involve variation in composition, resulting from the formation of solid solutions. Ternary alloys (formed from three components) and alloys involving four or more components are of course still more complex.

It is seen that the formulas of intermetallic compounds, such as Ag_5Sr, do not correspond in any simple way to the usually accepted valences of the element. Compounds such as Ag_5Sr can be described by saying that the strontium atom uses its two valence electrons in forming bonds with the silver atoms that surround it, and that the silver atoms then use their remaining electrons in forming bonds with other silver atoms. Some progress has been made in developing a valence theory of the structure and properties of intermetallic compounds and of alloys in general, but this field of chemistry is still in a primitive state.

17-5. Experimental Methods of Studying Alloys

About 100 years ago a metallurgical technique called *metallography* was developed as a way of investigating the phases present in alloys. This technique consists of grinding and polishing the surface of a metallic specimen, sometimes etching it with reagents (such as nitric acid or picric acid) to emphasize grain boundaries and to help to distinguish between different phases, and then examining the surface with use of an optical microscope, with a method of illumination from above. In this way the sizes and shapes of crystal grains can be studied, and the presence of grains of two or more phases can be determined in alloys that to the unaided eye appear to be homogeneous. The polished and etched surface of a piece of copper is shown in Figure 2-2, and other photomicrographs of alloys, showing different phases, can be found in Chapter 19.

During recent years much use has been made of the electron microscope in the study of metals and alloys. Very thin foils of the metal are made, sometimes by dissolving the surface of the specimen with acid until holes develop; the region adjacent to a hole may be thin enough to permit penetration by the electron beam. The structure of individual crystal grains can also be determined by observing the electron diffraction pattern from a beam of electrons transmitted through a single grain. Changes in structure that occur with time, perhaps at elevated temperature, may be followed in this way.

These techniques for studying phase transitions, although powerful, are time-consuming and difficult. A simple and easily applied technique, called *thermal analysis,* has been used for more than a century. Phase transitions are characterized by the absorption or emission of a heat of transition. The way in which the heat of transition is involved in the technique of thermal analysis can be illustrated by discussion of a simple experiment.

Samples of arsenic-lead alloys (Figure 17-5) may be made, corresponding to different compositions, from pure arsenic to pure lead. A sample of pure arsenic is placed in a crucible in a furnace and heated to above the melting point. One of the junctions of a thermocouple is inserted in the sample, to permit measurement of the temperature. When the furnace is turned off, the temperature of the sample begins to decrease, because of conduction and radiation from the sample to the furnace and from the furnace to the surrounding environment. The temperature decreases with time, as indicated by the first curve in Figure 17-9. (These curves are called *cooling curves.*) At the freezing point of arsenic, 817°C, there occurs a discontinuity in the slope of the temperature-time curve, which begins to follow a horizontal course, with slope zero. During a period of time the heat of crystallization serves to balance the heat loss to the surrounding furnace. The temperature of the horizontal section of the cooling curve is the melting point of the substance. After completion of crystallization the temperature begins to drop.

The second cooling curve in Figure 17-9, corresponding to 40 atomic percent lead, shows a decrease in temperature and then a change in slope, which represents the beginning of crystallization of arsenic from the molten alloy. The temperature continues to drop, even though arsenic is crystallizing, because the composition of the liquid alloy changes as arsenic is removed from it. The curve continues with changed slope until

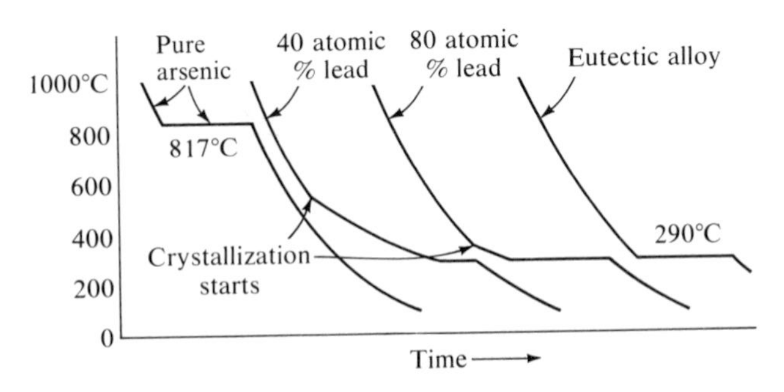

FIGURE 17-9
Cooling curves for samples of arsenic-lead alloys.

the temperature 290°C is reached, and then its slope becomes zero. The horizontal section represents the simultaneous crystallization of arsenic and lead, as separate phases, at the eutectic temperature.

The next curve, for the alloy with 80 atomic percent lead, shows a lower temperature at which crystallization of arsenic begins (that is, a lower liquidus temperature) and then a longer horizontal section, representing crystallization of a larger amount of the eutectic mixture of the two crystalline phases.

The next curve corresponds to the eutectic composition, with 93 atomic percent lead (point B on Figure 17-5). This curve is qualitatively similar to the curve for a pure substance.

The alloys in the region between the eutectic composition and pure lead give cooling curves that are like those for the alloys in the other half of the phase diagram.

The method of thermal analysis can be improved and refined by the use of two samples of metal or alloy and measurement of the temperature difference between the two samples; this is called *differential thermal analysis*. For example, suppose that an alloy of iron and cobalt has been made and that it is thought that the transition* between the α phase (ferromagnetic) and the β phase (paramagnetic, with the same body-centered crystal structure as the α phase) probably lies between 700°C and 900°C. A sample of the alloy is prepared and also a sample of a similar metal, such as copper, that does not have a phase transition in this temperature range. The samples are adjusted in weight so as to have approximately equal average values of their total heat capacity. A thermocouple is placed in each sample, and they are connected with a recording apparatus in such a way that the difference in temperature between the two samples can be continuously recorded as a function of the temperature of the first sample. The two samples are placed in a furnace and heated to 1000°C, the furnace is then turned off, and the temperature recordings are made as the samples cool. A change in slope of the differential temperature indicates the transition temperature, which can in this way be determined with considerable accuracy (0.1° to 1°).

Another important and useful method of investigating alloys is by preparing samples of different compositions and making x-ray photographs of them (especially powder photographs, which are the diffraction patterns given by a large number of small crystals in random orientation). By the analysis of the diffraction patterns the number of phases present can be determined. For example, the samples of silver-strontium alloys, with

*This transition, which does not involve a change in crystal structure but only a change in relative orientations of the magnetic moments of the atoms, occurs over a range of temperature, roughly 1°. It is called a *second-order transition*.

phase diagram represented in Figure 17-8, are found to give characteristic diffraction patterns at six compositions: pure silver, pure strontium, and the four compositions indicated by the arrows in Figure 17-8. For an alloy with intermediate composition the diffraction pattern shows the lines characteristic of two phases, with relative intensities proportional to the relative amounts of the two phases. Moreover, it is often possible by the analysis of the diffraction pattern to determine the structure of the crystal, and thus to verify the composition. It is in this way that the compound Ag_5Sr was identified.

17-6. Physical Metallurgy

In recent years emphasis has been placed on the study of a branch of metallurgy called *physical metallurgy*. In this branch of metallurgy an attempt is made to explain the physical properties, such as tensile strength, hardness, ductility, electrical and thermal conductivity, and heat capacity, of pure metals and alloys in terms of their atomic and electronic structure. One of the ultimate aims of the physical metallurgist is to be able to design alloys with any desired set of properties.

Mechanical Properties of Metals

Most metals are malleable and ductile. Instead of being smashed into splinters when struck by a hammer, a piece of metal is flattened into a sheet of foil. A crystal of a metal must hence be able to deform itself without breaking.

If a crystal of sodium chloride is deformed in such a way that the ions are moved about one ionic diameter relative to one another, then sodium ions become adjacent to sodium ions and chloride ions to chloride ions, and the repulsion of the ions of like sign causes the crystal to break into pieces. In a metal, however, the atoms are all of the same kind, and any atom can form bonds with any other atom. Moreover, the valence bonds, which resonate easily from one position to another in the crystal, can still form between neighboring atoms even if the crystal is deformed, and accordingly a crystal of a metal remains strong during deformation.

The way in which a crystal of a metal changes its shape is by *slip along glide planes*. For example, the metal zinc has the hexagonal closest-packed structure indicated in Figure 17-1. The distance between the hexagonal layers of atoms is somewhat larger than for ideal closest packing—the distance between neighboring zinc atoms in the same hexagonal layer is 266 pm, whereas that between atoms in adjacent layers is 291 pm. Accordingly, we might expect it to be easy for a hexagonal

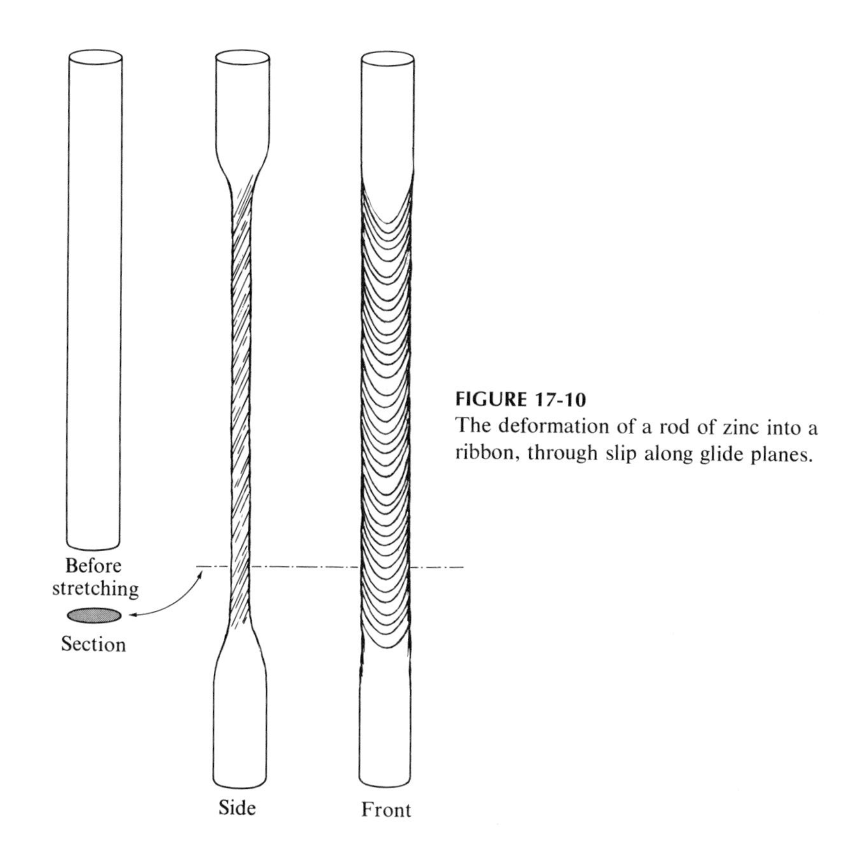

FIGURE 17-10
The deformation of a rod of zinc into a ribbon, through slip along glide planes.

layer to slip over another hexagonal layer. If a single crystal of zinc is made in the form of a round wire, with the hexagonal layers at an angle, and the ends of the wire are pulled, the wire stretches out into a ribbon, through slip along the hexagonal planes, as illustrated in Figure 17-10. Photomicrographs of a metal that has been subjected to strain often show traces of these glide planes.

The slip along a glide plane does not occur by the simultaneous motion of a whole layer of atoms relative to an adjacent layer. Instead, the atoms move one at a time. There is a flaw in the structure, where an atom is missing. The atom to one side of this flaw (which is called a *dislocation*) moves to occupy the space, and leaves a space where it was; that is, the dislocation moves in the opposite direction to the atom. When the dislocation has moved all the way across the crystal grain, the whole row of atoms has moved, and the lower part of the lower part of the crystal has slipped one atomic diameter in the direction of the strain. A description of some kinds of dislocations is given below.

Lattice Vacancies

One type of imperfection found in crystals is the *lattice vacancy* or *point imperfection:* an atom is missing at the place in the crystal lattice that is normally occupied by an atom, and the surrounding atoms have moved slightly toward this position. Lattice vacancies are formed by thermal agitation, with the number of vacancies per unit volume in the metal about equal to the number of atoms per unit volume in the vapor in equilibrium with the metal. They may also be produced in larger numbers by bombardment of the metal with high-energy particles or x-rays.

Interstitial Atoms

Another type of point defect consists of an extra atom of a metal occupying a position that in the perfect crystal would be vacant. This extra atom may be a foreign atom, usually smaller than the atoms of the metal itself, such as hydrogen, carbon, nitrogen, or oxygen in iron. Larger foreign atoms may substitute for the atoms of the metal itself. It is found by experiment that a small amount of impurity in a metal may make it brittle. For example, copper containing sulfur or arsenic is brittle, rather than malleable and ductile. One way in which the foreign atoms may produce brittleness is by interfering with the motion of dislocations through the crystal; when the dislocation reaches a sulfur atom or other foreign atom in the copper crystal, it may be stopped, and the slip may thus be prevented from continuing.

Dislocations

The most important imperfections, so far as the mechanical properties of crystals are concerned, are the various imperfections called dislocations. The ease with which dislocations move through a crystal determine to a large extent its ranges of elastic and plastic deformation under an applied stress and its ultimate yield point—that is, the stress under which the crystal fractures. One kind of dislocation, called an *edge dislocation,* is shown in Figure 17-11. An edge dislocation can be described as involving removal of one-half of a plane of atoms from the crystal.

The *screw dislocation* has an axis that is either right-handed or left-handed. A crystal containing one screw dislocation is not made up of layers of atoms parallel to one another; instead it consists of a single layer of atoms, distorted about the screw axis into a helicoid or spiral ramp.

A dislocation can move through a crystal by a succession of processes, each of which involves the motion of a single atom from one position in

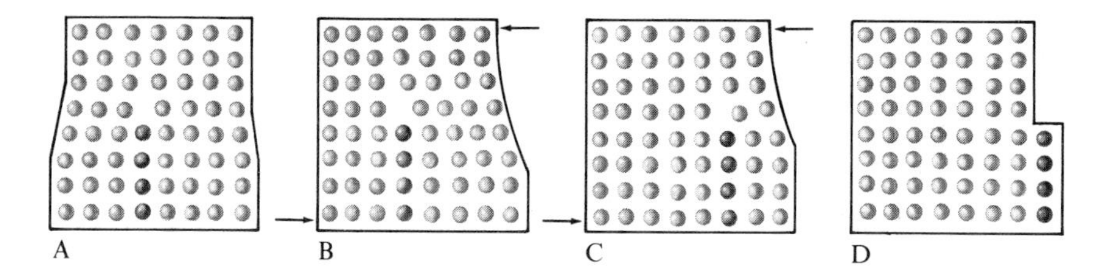

FIGURE 17-11
(A) An edge dislocation in a metal. The shaded atoms indicate the extra part-layer of atoms. (B) The crystal grain under stress. (C) Motion of the dislocation to the right; different atoms (shaded) now constitute the dislocation. (D) The dislocation has reached the edge of the grain, forming a step; by motion of the dislocation the upper part of the crystal has moved to the left relative to the lower part.

the crystal to an adjacent position. The activation energy for this process may be small enough to permit it to occur at a rapid rate, and in consequence plastic deformation of metals under stress can take place.

Many of the mechanical properties can be understood in terms of the motion of dislocations. If a stress is applied in the right way to a metal, the metal bends. When the stress is removed, the metal either returns to its original shape or is permanently deformed. In the first case the dislocations have not moved or have moved in a reversible way; that is, the applied stress has not carried them over any foreign inclusion in the lattice or caused too many of them to collide. In the second case some dislocations have moved irreversibly, so that they do not return to their original positions upon removal of the applied stress. If the applied stress is very large, dislocations will move until many of them become piled up against some barrier, such as a foreign inclusion or the boundary between adjacent crystalline grains. In the region of a dislocation pile-up the applied stress strains the bonds in such a way as to initiate failure of the material.

A boundary between crystal grains can serve as a barrier to the motion of dislocations, and in this way can decrease the plasticity and increase the hardness of a metal. If a piece of copper is hammered until the large crystal grains are broken up into small crystal grains, the crystal boundaries may interfere with slip by stopping the motion of the dislocations. This is the mechanism of the hardening of copper and other metals by *cold work* (by hammering them or otherwise working them in the cold). Heating the work-hardened metal to the temperature at which recrystallization occurs (growth of the small strained crystals to large unstrained crystals) restores plasticity; this process is called *annealing*. The recrystallization temperature is usually about one-third to one-half of the melting point of the metal (both on the absolute temperature scale).

Pure aluminum is a soft, malleable, and ductile metal. For some purposes an alloy of aluminum that is stronger, tougher, and less ductile is needed. Aluminum alloys of this sort can be made by incorporating small amounts of other metals, such as copper and magnesium. An alloy containing about 4% copper and 0.5% magnesium may strengthen the aluminum through the formation of hard, brittle crystals of the intermetallic compound $MgCu_2$. These minute crystals, interspersed through the crystals of aluminum, can serve to key the glide planes of aluminum so effectively as to improve the mechanical properties of the alloy significantly over those of the pure metal.

EXERCISES

17-1. Aluminum crystallizes in cubic closest packing. How many nearest neighbors does each atom have? Predict its metallic valence from its position in the periodic table. Would you predict it to have greater or less tensile strength than magnesium? Why?

17-2. Discuss the metallic valence of the elements rubidium, strontium, and yttrium. What would you predict about change in hardness, density, strength, and melting point in this series of elementary metals?

17-3. Compare the metallic valences of sodium, magnesium, and aluminum with their oxidation numbers in their principal compounds.

17-4. Describe the structure of tantalum carbide, TaC. Can you explain why it has much greater strength and hardness than tantalum itself?

17-5. Define alloy, intermetallic compound, phase, variance, eutectic, triple point.

17-6. State the phase rule, and given an application of it (see Section 10-10).

17-7. Cadmium (m.p. 321°C) and bismuth (m.p. 271°C) do not form solid solutions nor compounds with one another. Their eutectic point lies at 61 mass percent bismuth and 146°C. Sketch their phase diagram, and label each region to show what phases are present.

17-8. Describe the alloy that would be obtained by cooling a melt of silver containing 8 atomic percent strontium. (See Figure 17-8.)

17-9. Describe the alloy that would be obtained by cooling a melt of silver and strontium containing 50 atomic percent Ag. Would it be homogeneous or heterogeneous? Would it melt sharply, at one temperature, or over a range of temperatures?

17-10. What is the lowest temperature at which an alloy of silver and strontium can remain liquid? What is the composition of this alloy? Is the solid alloy homogeneous or heterogeneous? Does it have a sharp melting point?

17-11. Why does the Ag-Sr alloy with 75 atomic percent Sr have a lower melting point than pure strontium?

17-12. From Figure 17-8 it is seen that the silver-strontium alloy containing 1 atomic percent Sr begins to freeze at a temperature 11° less than the freezing point of pure silver. What is the mass-molar freezing-point constant of silver? (See Section 9-13.) Silver and silicon have a phase diagram resembling that shown in Figure 17-5; neither element is soluble in the other in the crystalline state. At what temperature would the Ag-Si alloy containing 1 atomic percent Si begin to freeze? (Answer: 11° below the freezing point of silver.)

17-13. The enthalpy of formation of the intermetallic compound $Mg_2Sn(c)$ (which has the fluorite structure, Figure 18-3) from Mg(c) and Sn(gray) is -74 kJ mole^{-1}, and that for $Mg_2Si(c)$ is -80 kJ mole^{-1}. Note that these substances have the composition of normal-valence compounds. To what values of the electronegativity difference do these enthalpy values correspond? (Answer: about 0.45.)

17-14. The intermetallic compounds in $Ba_2Sn(c)$ and $BaSn_3(c)$ have enthalpy of formation -379 and -189 kJ mole^{-1}, respectively, from Ba(c) and Sn(gray). Discuss these values in relation to the number of bonds between unlike atoms and the electronegativity difference.

18

Lithium, Beryllium Boron, and Silicon and their Congeners

In this chapter we shall discuss the metals and metalloids of groups I, II, III, and IV of the periodic table, and their compounds.*

The alkali metals, group I, are the most strongly electropositive elements – the most strikingly metallic. Many of their compounds have been mentioned in earlier chapters. The alkaline-earth metals are also strongly electropositive.

Boron, silicon, and germanium are metalloids, with properties intermediate between those of metals and those of nonmetals. The electric conductivity† of boron, for example, is 1×10^{-6} mho cm^{-1}; this value is intermediate between the values for metals ($4 + 10^5$ mho cm^{-1} for aluminum, for example), and those for nonmetals (2×10^{-13} for diamond, for example). They have a corresponding tendency to form oxygen acids, rather than to serve as cations in salts.

Silicon (from Latin *silex,* flint) is the second element in group IV, and is hence a congener of carbon. Silicon plays an important part in the inorganic world, similar to that played by carbon in the organic world. Most

---

*Compounds of carbon have been discussed in Chapters 13, 14, and 15.

†The electric conductivity, in mho cm^{-1}, is the current in amperes flowing through a rod with cross-section 1 cm^2 when there is an electric potential difference between the ends of the rod of 1 volt per cm length of the rod.

of the rocks that constitute the earth's crust are composed of the silicate minerals, of which silicon is the most important elementary constituent.

The importance of carbon in organic chemistry results from its ability to form carbon-carbon bonds, permitting complex molecules, with the most varied properties, to exist. The importance of silicon in the inorganic world results from a different property of the element—a few compounds are known in which silicon atoms are connected to one another by covalent bonds, but these compounds are relatively unimportant. The characteristic feature of the silicate minerals is the existence of chains and more complex structures (layers, three-dimensional frameworks) in which the silicon atoms are not bonded directly to one another but are connected by oxygen atoms. The nature of these structures is described briefly in later sections of this chapter.

18-1. The Electronic Structures of Lithium, Beryllium, Boron, and Silicon and Their Congeners

The electronic structures of the first elements of groups I, II, III, and IV are given in Table 5-4. The distribution of the electrons among the orbitals is the same in this table as in the energy-level chart, Figure 5-6.

The elements of group I have one more electron than the preceding argonon, those of group II have two more, and those of group III have three more. The outermost shell of each of these argonon atoms is an octet of electrons, two electrons in the *s* orbital and six in the three *p* orbitals of the shell. The one, two, or three outermost electrons of the metallic elements are easily removed with formation of the cations Li^+, Na^+, K^+, Rb^+, Cs^+, Be^{++}, Mg^{++}, Ca^{++}, Sr^{++}, Ba^{++}, Al^{+++}, Sc^{+++}, Y^{+++}, and La^{+++}. Each of these elements forms only one principal series of compounds, in which it has oxidation number +1 for group I, +2 for group II, or +3 for group III. The metalloid boron also forms compounds in which its oxidation number is +3, but the cation B^{+++} is not stable.

Whereas carbon is adjacent to boron in the sequence of the elements, and also silicon to aluminum, the succeeding elements of group IV of the periodic table, germanium, tin, and lead, are widely separated from the corresponding elements of group III, scandium, yttrium, and lanthanum. Germanium is separated from scandium by the ten elements of the iron transition series, tin from yttrium by the ten elements of the palladium transition series, and lead from lanthanum by the ten elements of the platinum transition series, and also the fourteen lanthanons.*

*There is some disagreement among chemists about nomenclature of the groups of the periodic system. We have described the transition elements as coming between groups III and IV in the long periods of the periodic table. An alternative that has found about as wide acceptance is to place them between groups II and III (Section 16-2).

Each of the elements of group IV has four valence electrons, which occupy *s* and *p* orbitals of the outermost shell. The maximum oxidation number of these elements is +4. All of the compounds of silicon correspond to this oxidation number. Germanium, tin, and lead form two series of compounds, representing oxidation number +4 and oxidation number +2, the latter being more important than the former for lead.

18-2. Radius Ratio, Ligancy, and the Properties of Substances

Some of the properties of substances can be discussed in a useful way in terms of the sizes of ions or atoms. Many of the substances mentioned in the later sections of this chapter and in the following chapter are compounds of metals, with small electronegativity, and nonmetals, with large electronegativity. The bonds between these atoms may have a sufficiently large amount of ionic character to justify the discussion of the substance as composed of cations and anions. Such a discussion may be helpful even for substances in which the bonds have a large amount of covalent character.

For example, let us consider the fluorides of the elements of the second short period of the periodic table. Their formulas, melting points, boiling points, heats of fusion, and heats of vaporization (or sublimation) are the following:

	NaF	MgF_2	AlF_3	SiF_4	PF_5	SF_6
M.p.	995°	1263°	>1257°	−90°	−94°	−51°C
B.p.	1704°	2227°	1257°*	−95°*	−85°	−64°*
Heat of fusion	33	58	–	7	12.8	5 kJ mole^{-1}
Heat of vaporization	209	272	322†	19‡	7.1	17.1 kJ mole^{-1}

*Temperature of sublimation of crystal at 1 atm pressure.
†Heat of sublimation of crystal.
‡Heat of vaporization at 1.74 atm.

The first three substances are crystalline solids at room temperature, with high melting and boiling points and large heats of fusion and vaporization, and the other three are gases at room temperature, with low melting and boiling points and small heats of fusion and vaporization. The pronounced change in properties between AlF_3 and SiF_4 cannot be attributed in an obvious way to the change in oxidation number, composition (number of fluorine atoms per second-row atom), or electronegativity of the second-row atom. It can, however, be accounted for by consideration of the relative sizes of the atoms.

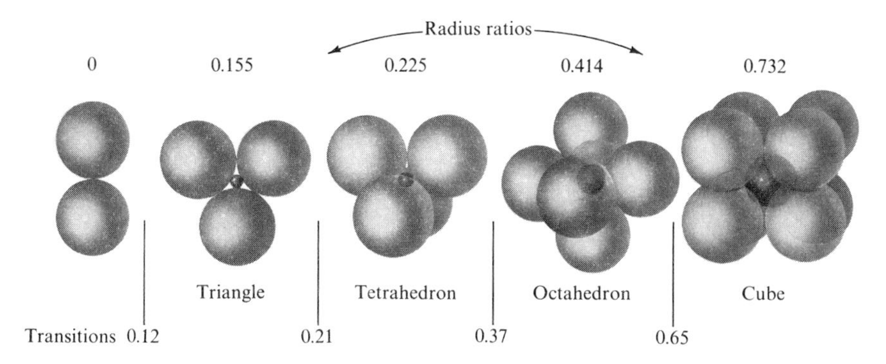

FIGURE 18-1
Linear, triangular, tetrahedral, octahedral, and cubic arrangements of anions around a central cation of increasing radius.

In Figure 18-1 there are shown several ways of arranging two or more large anions, such as the fluoride ion, around a cation. For each arrangement there is given the ratio of radius of cation to radius of anion (the *radius ratio*) corresponding to closest packing; that is, contact of anions with one another as well as with the cation, both anions and cation being considered to be spheres.

Thus for the planar triangular structure MX_3 the distances $r_M + r_X$ and $2r_X$ have the relative values $1:\sqrt{3}$, from which we calculate $r_M/r_X = 2/\sqrt{3} - 1 = 0.155$. In a similar way the values 0.225 for ligancy 4 (a tetrahedron of anions about the cation), 0.414 for ligancy 6 (octahedron), and 0.732 for ligancy 8 (cube) are obtained.

Of substances MX_2, silicon dioxide (radius ratio 0.29) forms crystals with tetrahedral coordination of four oxygen ions about each silicon ion, magnesium fluoride (radius ratio 0.48) and stannic oxide (radius ratio 0.51) form crystals with octahedral coordination of six anions around each cation (the rutile structure, Figure 18-2), and calcium fluoride (radius ratio 0.73) forms crystals with cubic coordination of eight anions around each cation (the fluorite structure, Figure 18-3). The ligancy (coordination number) increases with increase in the radius ratio, as indicated in Figure 18-1.

The increase in stability (decrease in energy) with increase in ligancy is easy to understand. Let us consider two ionic molecules, M^+X^-, with $M^+—X^-$ distance r. The electrostatic interaction energy of the electric charge $+e$ and that $-e$ in one molecule is $-e^2/r$, and for two molecules it is $-2e^2/r$. Now if the ligancy changes from 1 to 2, through the formation of a square,

$$\begin{array}{ccc} M^+ & — & X^- \\ | & & | \\ X^- & — & M^+ \end{array}$$

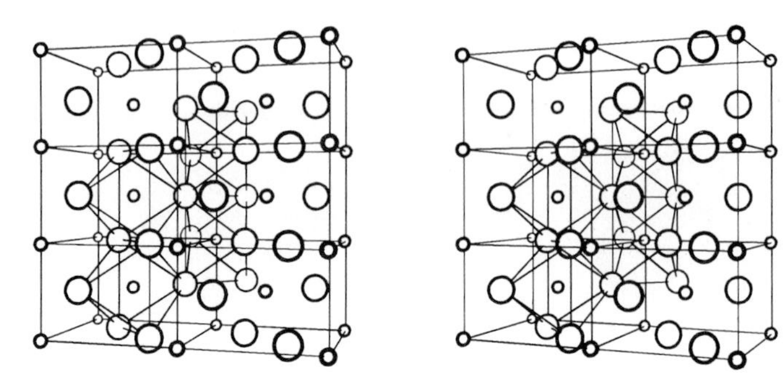

FIGURE 18-2
The structure of magnesium fluoride (stereo); this substance has high melting point and boiling point. (This structure is usually called the rutile structure; it is the structure of the mineral rutile, TiO_2.) The large spheres are fluoride ions and the small spheres magnesium ions.

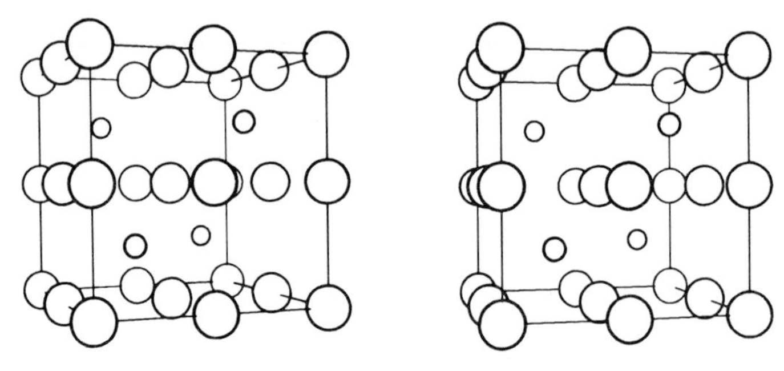

FIGURE 18-3
The structure of the fluorite crystal, CaF_2 (stereo).

and if the M^+—X^- distance retains the value r, each of the four M^+—X^- interactions contributes $-e^2/r$ and each of the two repulsions across the diagonals contributes $e^2/(\sqrt{2}\ r)$. The total electrostatic energy for the square is then $(-4 + \sqrt{2})e^2/r = -2.59e^2/r$. The square arrangement is thus 29% more stable than two separate molecules, with respect to the electrostatic interactions.

Similar calculations show that, for constant cation-anion distance, the rutile structure (ligancy 6) is 8% more stable than the quartz structure (ligancy 4), and the fluorite structure (ligancy 8) is 5% more stable than the rutile structure.

If, however, the radius ratio is less than the value given in Figure 18-1, the anions come into contact with one another, the cation-anion distance becomes larger than the contact distance, and the structure becomes unstable relative to the structure with smaller ligancy. The approximate values of the radius ratio at which the transitions occur are shown in Figure 18-1.

Germanium dioxide is an interesting example. Its radius ratio (Table 6-2) is 53 pm/140 pm = 0.38. This value is very near the transition value, 0.37, from tetrahedral to octahedral coordination, and GeO_2 is in fact dimorphous, with one crystalline form having the quartz structure (ligancy 4) and the other having the rutile structure (ligancy 6).

We can now discuss the melting points and boiling points of the second-row fluorides. The ionic radii of the cations (the radius of F^- is 136 pm) and the radius ratios are the following:

	NaF	MgF_2	AlF_3	SiF_4	PF_5	SF_6
Radius of cation	95 pm	65 pm	50 pm	41 pm	34 pm	29 pm
Radius ratio	0.70	0.48	0.37	0.30	0.25	0.21
Expected ligancy of cation	6 or 8	6	4 or 6	4	4	4 or 3

We see that for silicon tetrafluoride the expected ligancy of silicon, 4, corresponds exactly to the formula to the molecule, SiF_4. Hence we conclude that in the crystal and liquid as well as the gas the substance consists of SiF_4 molecules. The structure of the crystal as determined by x-ray diffraction is shown in Figure 18-4. The crystal is an arrangement of tetrahedral SiF_4 molecules held together only by the van der Waals forces discussed in Section 9-5. The heat of fusion and heat of vaporization are correspondingly small, and the substance accordingly melts and

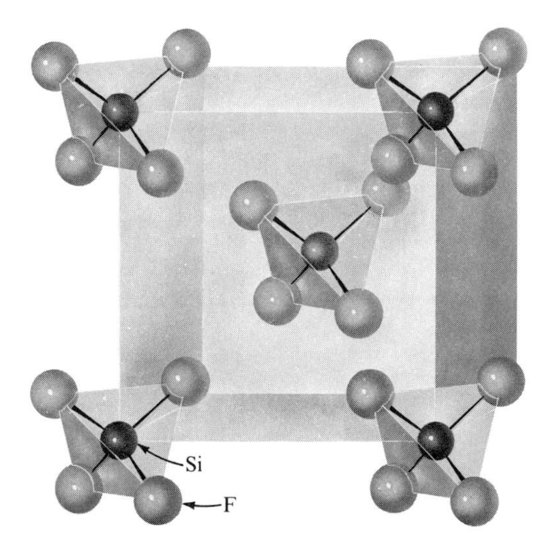

FIGURE 18-4
The structure of a molecular crystal, silicon tetrafluoride. The tetrahedral SiF_4 molecules are arranged in a body-centered cubic arrangement.

boils (in fact, sublimes at 1 atm pressure) at a low temperature. For PF_5 and SF_6 the expected ligancy is less than the number of fluorine atoms. There is accordingly some strain in the molecules—the fluorine atoms in contact with one another are under compression and the P—F and S—F bonds are stretched; but the crystals, like those of SiF_4, consist of molecules held together only by van der Waals forces, and the melting and boiling points and heats of fusion and vaporization are close to those of silicon tetrafluoride.

In AlF_3, on the other hand, the expected ligancy of aluminum is 4 or 6, and the x-ray diffraction study of the crystals has shown the ligancy to be 6. Each aluminum atom is surrounded octahedrally by six fluorine atoms, and each fluorine atom is ligated to two aluminum atoms.* The crystal MgF_2 has the rutile structure (Figure 18-2) corresponding to the expected ligancy 6 for magnesium and with each fluorine atom ligated to three magnesium atoms, and NaF has the sodium chloride structure, in which both sodium and fluorine have ligancy 6.

Fusion and vaporization of these three substances involves not just overcoming the van der Waals attractive forces, as for SiF_4, but rather the breaking of some Al—F, Mg—F, or Na—F bonds. For this reason the heats of fusion and vaporization are large and the melting points and boiling points are high.

The foregoing discussion has been based on the relative sizes of cations and anions. A closely parallel discussion could be presented based on covalent bond radii and van der Waals radii of atoms (Section 6-15). The van der Waals radius of the fluorine atom, 135 pm (Table 6-6), is nearly equal to the ionic radius of the fluoride ion, 136 pm, and the sums of covalent radii are approximately equal to the corresponding sums of ionic radii.

The Electrostatic Valence Rule

The description of a crystal or molecule in terms of cations and anions permits us to describe the bonds in a simple and useful way. The strength of each of the bonds formed by a cation can be defined as the electric charge of the cation divided by its ligancy. Thus in the crystal AlF_3, in which Al^{3+} has ligancy 6 (has six F^- ions coordinated about it), each of the six bonds formed by the aluminum ion has strength $\frac{3}{6} = \frac{1}{2}$. In the molecule SiF_4 each of the four bonds formed by SiF_4 has strength $\frac{4}{4} = 1$.

*The ligancy 6 rather than 4 for aluminum in the crystal is seen to be reasonable from the following argument. With ligancy 4 for aluminum, the composition AlF_3 would require an average $\frac{4}{3}$ aluminum atoms about each fluorine atom; that is, some fluorine atoms ligated to one aluminum atom and some to two. Such a structure is less stable than the one with ligancy 6 for aluminum and ligancy 2 for fluorine.

The electrostatic valence rule states that *the most stable structures of crystals and molecules are those in which the sum of the strengths of the bonds reaching each anion is just equal to its negative charge.*

For example, in the AlF_3 crystal each F^- is held to Al^{3+} by two bonds with strength $\frac{1}{2}$, and in the SiF_4 molecule it is held to Si^{4+} by one bond with strength 1; in each case the sum of the bond strengths equals the negative charge of the fluoride ion.*

Many of the properties of substances can be explained by the consideration of the relative sizes of ions or atoms, as illustrated above. Structural inorganic chemistry is, however, a new subject, and as yet far from precise. You may find it worth while to attempt to explain in terms of structure some of the properties of substances mentioned in later sections of this chapter and in the following chapter, but you must not become discouraged if you are unsuccessful. The fault may lie not with you but with the chemists of the present generation and earlier generations, who have not yet succeeded in the task of developing a really powerful theory of structural inorganic chemistry. If you become a chemist, you yourself may make a major contribution to the solution of this problem.

18-3. The Alkali Metals and Their Compounds

The elements of the first group—lithium, sodium, potassium, rubidium, and cesium†—are soft, silvery-white metals with great chemical reactivity. These metals are excellent conductors of electricity. Some of their physical properties are given in Table 18-1. It can be seen from the table that they melt at low temperatures—four of the five metals melt below the boiling point of water. Lithium, sodium, and potassium are lighter than water. The vapors of the alkali metals are mainly monatomic, with a small concentration of diatomic molecules, such as Li_2, in which the two atoms are held together by a covalent bond.

The alkali metals are made by electrolysis of the molten hydroxides or chlorides (Chapter 11). Because of their reactivity, the metals must be kept in an inert atmosphere or under oil. The metals are useful chemical reagents in the laboratory, and they find industrial use (especially sodium) in the manufacture of organic chemicals, dyestuffs, and lead tetraethyl (a constituent of "ethyl gasoline"). Sodium is used in sodium-vapor lamps, and, because of its large heat conductivity, in the stems of valves of

*These arguments can also be presented in terms of covalent bonds with partial ionic character.

†The sixth alkali metal, francium (Fr), element 87, has been obtained only in minute quantities, and no information has been published about its properties.

TABLE 18-1
Some Properties of the Alkali Metals

	Atomic Number	Melting Point	Boiling Point	Density ($g\ cm^{-3}$)	Metallic Radius*	Ionic Radius†	Bond Energy‡
Li	3	186°C	1336°C	0.530	155 pm	60 pm	115
Na	11	97.5°	880°	.963	190	95	75
K	19	62.3°	760°	.857	235	133	51
Rb	37	38.5°	700°	1.594	248	148	48
Cs	55	28.5°	670°	1.992	267	169	45

*For ligancy 12.
†For singly charged cation (Na^+, for example), with ligancy 6, as in the sodium chloride crystal.
‡Bond energy of $M_2(g)$, in kJ $mole^{-1}$.

reciprocating airplane engines, to conduct heat away from the valve heads. A sodium-potassium alloy is used as a cooling liquid in nuclear reactors. Cesium is used in vacuum tubes, to increase electron emission from cathodes.

Compounds of sodium are readily identified by the yellow color that they give to a flame. Lithium causes a carmine coloration of the flame, and potassium, rubidium, and cesium cause a violet coloration. These elements may be tested for in the presence of sodium by use of a blue filter of cobalt glass.

Compounds of Lithium

Lithium occurs in the mineral *spodumene*, $LiAlSi_2O_6$, *amblygonite*, $LiAlPO_4F$, and *lepidolite*, $K_2Li_3Al_5Si_6O_{20}F_4$. Lithium chloride, LiCl, is made by fusing (melting) a mineral containing lithium with barium chloride, $BaCl_2$, and extracting the fusion with water. It is used in the preparation of other compounds of lithium.

Compounds of lithium have found use in the manufacture of glass and of glazes for dishes and porcelain objects.

Compounds of Sodium

The most important compound of sodium is sodium chloride (common salt), NaCl. It crystallizes as colorless cubes, with melting point 801°C, and it has a characteristic salty taste. It occurs in seawater to the extent of 3%, and in solid deposits and concentrated brines (salt solutions) that are pumped from wells. Many million tons of the substance are obtained from these sources every year. It is used mainly for the preparation of other compounds of sodium and of chlorine, as well as of sodium metal and chlorine gas. Blood plasma and other body fluids contain about 0.9 g of sodium chloride per 100 ml.

Sodium hydroxide (caustic soda), NaOH, is a white hygroscopic (water-attracting) solid, which dissolves readily in water. Its solutions have a smooth, soapy feeling, and are very corrosive to the skin (this is the meaning of "caustic" in the name caustic soda). Sodium hydroxide is made either by the electrolysis of sodium chloride solution or by the action of calcium hydroxide, $Ca(OH)_2$, on sodium carbonate, Na_2CO_3:

$$Na_2CO_3 + Ca(OH)_2 \rightarrow CaCO_3 + 2NaOH$$

Calcium carbonate is insoluble, and precipitates out during this reaction, leaving the sodium hydroxide in solution. Sodium hydroxide is a useful laboratory reagent and a very important industrial chemical. It is used in industry in the manufacture of soap, the refining of petroleum, and the manufacture of paper, textiles, rayon and cellulose film, and many other products.

Compounds of Potassium

Potassium chloride, KCl, forms colorless cubic crystals, resembling those of sodium chloride. There are very large deposits of potassium chloride, together with other salts, at Stassfurt, Germany, and near Carlsbad, New Mexico. Potassium chloride is also obtained from Searles Lake in the Mojave Desert in California.

Potassium hydroxide, KOH, is a strongly alkaline substance, with properties similar to those of sodium hydroxide. Other important salts of potassium, which resemble the corresponding salts of sodium, are potassium sulfate, K_2SO_4, potassium carbonate, K_2CO_3, and potassium hydrogen carbonate, $KHCO_3$.

Potassium hydrogen tartrate (*cream of tartar*), $KHC_4H_4O_6$, is a constituent of grape juice; sometimes crystals of the substance form in grape jelly. It is used in making baking powder.

The principal use of potassium compounds is in *fertilizers*. Plant fluids contain large amounts of potassium ion, concentrated from the soil, and potassium salts must be present in the soil in order for plants to grow. A fertilizer containing potassium sulfate or some other salt of potassium must be used if the soil becomes depleted in this element.

The compounds of rubidium and cesium resemble those of potassium closely. They do not have any important industrial uses.

18-4. The Alkaline-earth Metals and Their Compounds

The metals of group II of the periodic table—beryllium, magnesium, calcium, strontium, barium, and radium—are called the alkaline-earth metals. Some of their properties are listed in Table 18-2. These metals are much harder and less reactive than the alkali metals because there are twice as many valence electrons. The compounds of all the alkaline-earth metals are similar in composition; they all form oxides MO, hydroxides $M(OH)_2$, carbonates MCO_3, sulfates MSO_4, and other compounds (M = Be, Mg, Ca, Sr, Ba, or Ra).

The early chemists gave the name "earth" to many nonmetallic substances. Magnesium oxide and calcium oxide were found to have an alkaline reaction, and hence were called the *alkaline earths.* The metals themselves (magnesium, calcium, strontium, and barium) were isolated in 1808 by Humphry Davy. Beryllium was discovered in the mineral beryl ($Be_3Al_2Si_6O_{18}$) in 1798 and was isolated in 1828.

TABLE 18-2
Some Properties of the Alkaline-earth Metals

Symbol	Atomic number	Atomic mass	Melting point*	Density (g cm^{-3})	Metallic radius	Ionic radius†
Be	4	9.0122	1350°C	1.86	112 pm	31 pm
Mg	12	24.312	651°	1.75	160	65
Ca	20	40.08	810°	1.55	197	99
Sr	38	87.62	800°	2.60	215	113
Ba	56	137.34	850°	3.61	222	135
Ra	88	226.04	960°	(4.45)‡	(246)‡	

*The boiling points of these metals are uncertain; they are about 600° higher than the melting points.
†For doubly charged cation with ligancy 6.
‡Estimated.

Beryllium

Beryllium is a light, silvery white metal, which can be made by electrolysis of a fused mixture of beryllium chloride, $BeCl_2$, and sodium chloride. The metal is used for making windows for x-ray tubes (x-rays readily penetrate elements with low atomic number, and beryllium metal has the best mechanical properties of the very light elements). It is also used as a constituent of special alloys. About 2% of beryllium in copper produces a hard alloy especially suited for use in springs.

The principal ore of beryllium is *beryl,* $Be_3Al_2Si_6O_{18}$. *Emeralds* are beryl crystals containing traces of chromium, which give them a green color. *Aquamarine* is a bluish-green variety of beryl.

The compounds of beryllium have little special value, except that beryllium oxide, BeO, is used in the uranium reactors in which plutonium is made from uranium (Chapter 20).

Compounds of beryllium are very poisonous. Even the dust of the powdered metal or its oxide may cause very serious illness.

Magnesium

Magnesium metal is made by electrolysis of fused magnesium chloride, and also by the reduction of magnesium oxide by carbon or by ferrosilicon (an alloy of iron and silicon). Except for calcium and the alkali metals, magnesium is the lightest metal known; and it finds use in lightweight alloys, such as *magnalium* (10% magnesium, 90% aluminum).

Magnesium reacts with boiling water, to form magnesium hydroxide, $Mg(OH)_2$, an alkaline substance:

$$Mg + 2H_2O \longrightarrow Mg(OH)_2 + H_2$$

The metal burns in air with a bright white light, to form magnesium oxide, MgO, the old name of which is *magnesia:*

$$2Mg + O_2 \longrightarrow 2MgO$$

Magnesium oxide suspended in water is used in medicine (as "milk of magnesia"), for neutralizing excess acid in the stomach and as a laxative. Magnesium sulfate, "Epsom salt," $MgSO_4{\cdot}7H_2O$, is used as a cathartic.

Magnesium carbonate, $MgCO_3$, occurs in nature as the mineral *magnesite.* It is used as a basic lining for copper converters and open-hearth steel furnaces (Chapter 19).

Calcium

Metallic calcium is made by the electrolysis of fused calcium chloride, $CaCl_2$. The metal is silvery white in color, and is somewhat harder than lead. It reacts with water, and burns in air when ignited, forming a mixture of calcium oxide, CaO, and calcium nitride, Ca_3N_2.

Calcium has a number of practical uses—as a deoxidizer (substance removing oxygen) for iron and steel and for copper and copper alloys, as a constituent of lead alloys (metal for bearings, or the sheath for electric cables) and of aluminum alloys, and as a reducing agent for making other metals from their oxides.

Calcium reacts with cold water to form calcium hydroxide, $Ca(OH)_2$, and burns readily in air, when ignited, to produce calcium oxide, CaO.

Calcium sulfate occurs in nature as the mineral *gypsum*, $CaSO_4 \cdot 2H_2O$. Gypsum is a white substance, which is used commercially for fabrication into wallboard, and conversion into *plaster of Paris*. When gypsum is heated a little above 100°C it loses three-quarters of its water of crystallization, forming the powdered substance $CaSO_4 \cdot \frac{1}{2}H_2O$, which is called plaster of Paris. (Heating to a higher temperature produces anhydrous $CaSO_4$, which reacts more slowly with water.) When mixed with water the small crystals of plaster of Paris dissolve and then crystallize as long needles of $CaSO_4 \cdot 2H_2O$. These needles grow together, and form a solid mass, with the shape into which the wet powder was molded.

Strontium

The principal minerals of strontium are strontium sulfate, *celestite*, $SrSO_4$, and strontium carbonate, *strontianite*, $SrCO_3$.

Strontium nitrate, $Sr(NO_3)_2$, is made by dissolving strontium carbonate in nitric acid. It is mixed with carbon and sulfur to make red fire for use in fireworks, signal shells, and railroad flares. Strontium chlorate, $Sr(ClO_3)_2$, is used for the same purpose. The other compounds of strontium are similar to the corresponding compounds of calcium. Strontium metal has no practical uses.

Barium

The metal barium has no significant use. Its principal compounds are barium sulfate, $BaSO_4$, which is only very slightly soluble in water and dilute acids, and barium chloride, $BaCl_2 \cdot 2H_2O$, which is soluble in water. Barium sulfate occurs in nature as the mineral *barite*.

Barium, like all other elements with large atomic number, absorbs x-rays strongly, and a thin paste of barium sulfate and water is swallowed as a "barium meal" to obtain contrasting x-ray photographs and fluoroscopic views of the alimentary tract. The solubility of the substance is so small that the poisonous action of most barium compounds is avoided.

Barium nitrate, $Ba(NO_3)_2$, and barium chlorate, $Ba(ClO_3)_2$, are used for producing green fire in fireworks.

Radium

Compounds of radium are closely similar to those of barium. The only important property of radium and its compounds is its radioactivity, which will be discussed further in Chapter 20.

18-5. Boron

Boron can be made by heating potassium tetrafluoroborate, KBF_4, with sodium in a crucible lined with magnesium oxide:

$$KBF_4 + 3Na \rightarrow KF + 3NaF + B$$

The element can also be made by heating boric oxide, B_2O_3, with powdered magnesium:

$$B_2O_3 + 3Mg \rightarrow 3MgO + 2B$$

Boron forms brilliant transparent crystals, nearly as hard as diamond. Some of its properties are given in Table 18-3.

Because of its light weight and great strength, a composite of boron fibers made by depositing boron on very fine tungsten wires and embedding them in a matrix of aluminum or epoxy resin is being used as a structural material.

Boron forms a compound with carbon, B_4C. This substance, *boron carbide,* is one of the hardest substances known and it has found extensive use as an abrasive and for the manufacture of small mortars and pestles for grinding very hard substances. The cubic form of boron nitride, BN, with a tetrahedral structure like that of diamond, has about the same hardness.

TABLE 18-3
Some Physical Properties of Elements of Groups III and IV

	Atomic Number	Atomic Mass	Density (g cm^{-3})	Melting Point	Atomic Radius*	Ionic Radius†
B	5	10.811	2.54	2300°C	80 pm	20 pm
Al	13	26.9815	2.71	660°	143	50
Sc	21	44.956	3.18	1200°	162	81
Y	39	88.905	4.51	1490°	180	93
La	57	138.91	6.17	826°	187	115
C‡	6	12.01115	3.52	3500°	77	–
Si	14	28.086	2.36	1440°	117	41
Ge	32	72.59	5.35	959°	122	53
Sn	50	118.69	7.30	232°	162	71
Pb	82	207.19	11.40	327°	175	84

*Single-bond covalent radius for B, C, Si, and Ge; metallic radius (ligancy 12) for the others.
†Section 6-10.
‡Diamond.

Boric acid, H_3BO_3, occurs in the volcanic steam jets of central Italy. The substance is a white crystalline solid, which is sufficiently volatile to be carried along with a stream of steam. Boric acid can be made by treating borax with an acid.

The principal source of compounds of boron is the complex borate minerals, including *borax*, sodium tetraborate decahydrate, $Na_2B_4O_7 \cdot 10H_2O$; *kernite*, sodium tetraborate tetrahydrate, $Na_2B_4O_7 \cdot 4H_2O$ (which gives borax when water is added); and *colemanite*, calcium hexaborate pentahydrate, $Ca_2B_6O_{11} \cdot 5H_2O$. The main deposits of these minerals are in California.

Borax is used in making certain types of enamels and glass (such as Pyrex glass, which contains about 12% of B_2O_3), for softening water, as a household cleanser, and as a flux in welding metals. The last of these uses depends upon the power of molten borax to dissolve metallic oxides, forming borates.

18-6. The Boranes. Electron-deficient Substances

The reaction of magnesium boride, Mg_3B_2, with water would be expected from simple valence theory to result in the production of molecules of boron trihydride, BH_3. Instead, the substance diborane, B_2H_6, is produced:

$$Mg_3B_2 + 6H_2O \longrightarrow 3Mg(OH)_2 + B_2H_6(g)$$

Diborane is a gas under ordinary conditions (m.p. −165.5°C, b.p. −92.5°C).

Many other boranes are known; those that have been the most thoroughly investigated have formulas B_4H_{10}, B_5H_9, B_5H_{10}, and $B_{10}H_{14}$. They have found some use as rocket fuels.

The B_2H_6 molecule has the following structure:

```
H    H    H
 \  / \  /
  B----B
 /  \ /  \
H    H    H
```

Each boron atom has ligancy 5, and two of the hydrogen atoms have ligancy 2. The bond lengths B—B = 177 pm, B—H = 133 pm (for bridging hydrogen atoms) and 119 pm (for outer hydrogen atoms) indicate that the bonds are fractional bonds, each involving less than one electron pair. There are six pairs of valence electrons in the molecule, and nine bonds; hence on the average each bond is two-thirds of a single bond. The bond lengths indicate that the six electron pairs resonate among the nine positions in such a way that the five central bonds involve about five valence electrons and the four outer bonds involve about seven valence electrons.

Some borane ions are also known, such as $B_4H_{10}^{--}$ in $Na_2B_4H_{10}$ and $B_{12}H_{12}^{--}$ in $K_2B_{12}H_{12}$. The $B_{12}H_{12}^{--}$ ion has an interesting structure: the twelve boron atoms lie at the corners of a regular icosahedron, as shown in Figure 18-5, and each boron atom forms six bonds, five to the adjacent boron atoms in the icosahedron and one, directed outward radially, to a hydrogen atom. The bond lengths indicate bond numbers about 0.5 for the thirty B—B bonds and 0.83 for the twelve B—H bonds.

$B_{10}H_{14}$ and some other borane molecules have also been found to have structures based on the icosahedron, with some of the corners not occupied by boron atoms, and with some bridging hydrogen atoms, as in diborane. The B_{12} icosahedron is also present in elementary boron and in the hard substance B_4C.

These substances can be described as *electron-deficient substances.* Electron-deficient substances are substances in which some or all of the atoms have more stable orbitals than electrons in the valence shell. The boron atom has four orbitals in its valence shell, and three valence electrons.

A characteristic feature of the structure of most electron-deficient substances is that the atoms have ligancy that is not only greater than the number of valence electrons but is even greater than the number of

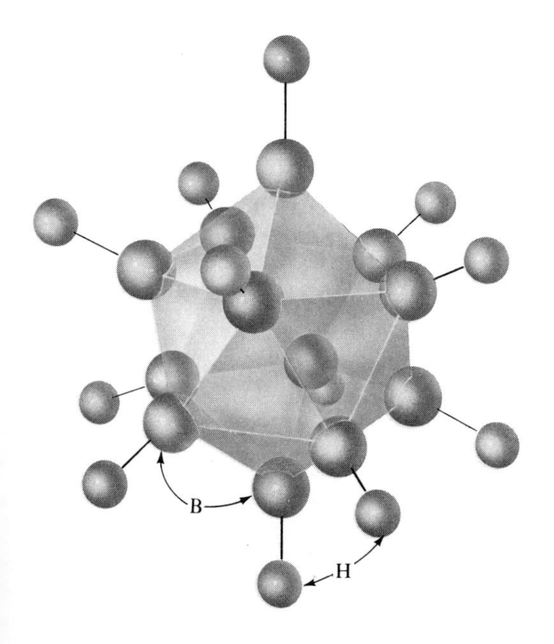

FIGURE 18-5
The structure of the icosahedral dodecaborane ion, $B_{12}H_{12}^{--}$.

stable orbitals. Thus most of the boron atoms in the tetragonal form of crystalline boron have ligancy 6. Also, lithium and beryllium, with four stable orbitals and only one and two valence electrons, respectively, have structures in which the atoms have ligancy 8 or 12. All metals can be considered to be electron-deficient substances.

Another generalization that may deserve to be called a structural principle is that an electron-deficient atom causes adjacent atoms to increase their ligancy to a value greater than the orbital number. For example, in the boranes some of the hydrogen atoms, adjacent to the electron-deficient atoms of boron, have ligancy 2.

The boron-boron bond lengths, 180 pm, are 18 pm greater than the single-bond value, 162 pm; we may say that in tetragonal boron each atom with ligancy 6 uses its three electrons to form six half-bonds, rather than three electron-pair bonds.

18-7. Aluminum and Its Congeners

Some of the physical properties of aluminum and its congeners are given in Table 18-3. Aluminum is only about one-third as dense as iron, and some of its alloys, such as duralumin (described below), are as strong as mild steel; it is this combination of lightness and strength, together with low cost, that has led to the extensive use of aluminum alloys. Aluminum is also used, in place of copper, as a conductor of electricity; its electric conductivity is about 80% that of copper.* Its metallurgy has been discussed in Chapter 11.

The metal is reactive (note its position in the electromotive-force series), and when strongly heated it burns rapidly in air or oxygen. Aluminum dust forms an explosive mixture with air. Under ordinary conditions, however, aluminum rapidly becomes coated with a thin, tough layer of aluminum oxide, which protects it against further corrosion.

Some of the alloys of aluminum are very useful. *Duralumin* or *dural* is an alloy (containing about 94.3% aluminum, 4% copper, 0.5% manganese, 0.5% magnesium, and 0.7% silicon), which is stronger and tougher than pure aluminum. It is less resistant to corrosion, however, and often is protected by a coating of pure aluminum. Plate made by rolling a billet of dural sandwiched between and welded to two pieces of pure aluminum is called alclad plate.

*The conductivity refers to the conductance of electricity by a wire of unit cross-sectional area. The density of aluminum is only 30% of that of copper; accordingly, an aluminum wire with the same weight as a copper wire with the same length conducts 2.7 times as much electricity as the copper wire with the same transmission loss.

Aluminum oxide (*alumina*), Al_2O_3, occurs in nature as the mineral *corundum*. Corundum and impure corundum (*emery*) are used as abrasives. Pure corundum is colorless. The precious stones *ruby* (red) and *sapphire* (blue or other colors) are transparent crystalline corundum containing small amounts of other metallic oxides (chromic oxide, titanium oxide). Artificial rubies and sapphires can be made by melting aluminum oxide (m.p. 2050°C) with small admixtures of other oxides, and cooling the melt in such a way as to produce large crystals. These stones are indistinguishable from natural stones, except for the presence of characteristic rounded microscopic air bubbles. They are used as gems, as bearings ("jewels") in watches and other instruments, and as dies through which wires are drawn.

Aluminum sulfate, $Al_2(SO_4)_3 \cdot 18H_2O$, may be made by dissolving aluminum hydroxide in sulfuric acid:

$$2Al(OH)_3 + 3H_2SO_4 + 12H_2O \longrightarrow Al_2(SO_4)_3 \cdot 18H_2O$$

It is used in water purification and as a mordant in dyeing and printing cloth (a *mordant* is a substance that fixes the dye to the cloth, rendering it insoluble). Both of these uses depend upon its property of producing a gelatinous precipitate of aluminum hydroxide, $Al(OH)_3$, when it is dissolved in a large amount of neutral or slightly alkaline water. The reaction that occurs is the reverse of the above reaction. In dyeing and printing cloth the gelatinous precipitate aids in holding the dye onto the cloth. In water purification it adsorbs dissolved and suspended impurities, which are removed as it settles to the bottom of the reservoir.

A solution containing aluminum sulfate and potassium sulfate, K_2SO_4, forms, on evaporation, beautiful colorless cubic (octahedral) crystals of *alum,* $KAl(SO_4)_2 \cdot 12H_2O$. Similar crystals of ammonium alum, $NH_4Al(SO_4)_2 \cdot 12H_2O$, are formed with ammonium sulfate. The alums also are used as mordants in dyeing cloth, in water purification, and in weighting and sizing paper (by precipitating aluminum hydroxide in the meshes of the cellulose fibers).

Aluminum chloride, $AlCl_3$, is made by passing dry chlorine or hydrogen chloride over heated aluminum:

$$2Al + 3Cl_2 \longrightarrow 2AlCl_3$$

$$2Al + 6HCl \longrightarrow 2AlCl_3 + 3H_2$$

The anhydrous salt is used in many chemical processes, including a cracking process for making gasoline.

Scandium, Yttrium, Lanthanum, and the Lanthanons

Scandium, yttrium, and lanthanum, the congeners of boron and aluminum, form colorless compounds similar to those of aluminum, their oxides having the formulas Sc_2O_3, Y_2O_3, and La_2O_3. These elements and their compounds have not yet found any important use.

Scandium, yttrium, and lanthanum usually occur in nature with the fourteen lanthanons, cerium (atomic number 58) to lutetium (atomic number 71).* All of these elements except promethium (which is made artificially) occur in nature in very small quantities, the principal source being the mineral *monazite*, a mixture of phosphates containing also some thorium phosphate.

The metals themselves are very electropositive, and are accordingly difficult to prepare. Electrolytic reduction of a fused oxide-fluoride mixture may be used. An alloy containing about 70% cerium and smaller amounts of other lanthanons and iron gives sparks when scratched. This alloy is widely used for cigarette lighters and gas lighters.

The sulfides cerium monosulfide, CeS, and thorium monosulfide, ThS, and related sulfides have been found valuable as refractory substances. The melting point of cerium monosulfide is 2450°C.

18-8. Silicon and Its Simpler Compounds

Elementary Silicon and Silicon Alloys

Silicon is a brittle steel-gray metalloid. Some of its physical properties are given in Table 18-3. It can be made by reduction of silicon tetrachloride by sodium:

$$SiCl_4 + 4Na \longrightarrow Si + 4NaCl$$

The element has the same crystal structure as diamond, each silicon atom forming single covalent bonds with four adjacent silicon atoms, which surround it tetrahedrally. It is used in transistors, especially for service at elevated temperatures (Section 18-13).

Silicon contaminated with carbon can be obtained by reduction of silica, SiO_2, with carbon in an electric furnace. An alloy of iron and silicon,

*Lanthanum is often considered as one of the rare-earth elements (lanthanons). For convenience, the convention is adopted here of including lanthanum as a member of group III, leaving fourteen elements in the lanthanon group.

called *ferrosilicon,* is obtained by reducing a mixture of iron oxide and silica with carbon.

Ferrosilicon, which has composition approximately FeSi, is used in the manufacture of acid-resisting alloys, such as *duriron,* which contains about 15% silicon. Duriron is used in chemical laboratories and manufacturing plants. A mild steel containing a few percent of silicon may be made which has a high magnetic permeability, and is used for the cores of electric tranformers.

Silicon Carbide

Silicon carbide, SiC, is made by heating a mixture of carbon and sand in a special electric furnace:

$$SiO_2 + 3C \rightarrow SiC + 2CO$$

The structure of this substance is similar to that of diamond, with carbon and silicon atoms alternating; each carbon atom is surrounded by a tetrahedron of silicon atoms, and each silicon atom by a tetrahedron of carbon atoms. The covalent bonds connecting all of the atoms in this structure make silicon carbide very hard. The substance is used as an abrasive.

Silicon Dioxide

Silicon dioxide (*silica*), SiO_2, occurs in nature in three different crystal forms: as the minerals *quartz* (hexagonal), *cristobalite* (cubic), and *tridymite* (hexagonal). Quartz is the most widespread of these minerals; it occurs in many deposits as well-formed crystals, and also as a crystalline constituent of many rocks, such as granite. It is a hard, colorless substance. Its crystals may be identified as right-handed or left-handed by their face development and also by the direction in which they rotate the plane of polarization of polarized light.

The structure of quartz is closely related to that of silicic acid, H_4SiO_4. In this acid silicon has ligancy 4, the silicon atom being surrounded by a tetrahedron of four oxygen atoms, with one hydrogen atom attached to each oxygen atom. Silicic acid, which is a very weak acid, has the property of undergoing condensation very readily, with elimination of water. If each of the four hydroxyl groups of a silicic acid molecule condenses with a similar hydroxyl group of an adjacent molecule, eliminating water, a structure is obtained in which the silicon atom is bonded to four surrounding silicon atoms by silicon-oxygen-silicon bonds. This process leads to a condensation product with formula SiO_2, since each silicon atom is surrounded by four oxygen atoms, and each oxygen atom serves as a neighbor to two silicon atoms. The structure of quartz and of the other forms of silica may be described as consisting of SiO_4 tetrahedra, with each oxygen atom serving as the corner of two of these tetrahedra. In order to break a crystal of quartz it is necessary to break some silicon-oxygen bonds. In this way the structure of quartz accounts for the hardness of the mineral.

Cristobalite and tridymite are similarly made from SiO_4 tetrahedra fused together by sharing oxygen atoms, with, however, different arrangements of the tetrahedra in space from that of quartz. Tridymite resembles ordinary ice (Figure 9-8) in structure, with silicon atoms in the oxygen-atom positions; cristobalite similarly resembles cubic ice. Three other crystalline modifications of silica have been discovered since 1956—keatite, coesite, and stishovite. Keatite and coesite contain SiO_4 tetrahedra somewhat distorted from the regular configuration (bond angles different from 110°28′). Coesite was discovered in the laboratory by subjecting silica to high pressure (about 30,000 atm), and was later found at Meteor Crater, Arizona, and other places where large meteorites have struck the earth. Stishovite was first made in 1961 by use of pressures of about 120,000 atm. It has a structure (like that of rutile, TiO_2; Figure 18-2) in which each silicon atom is octahedrally surrounded by six oxygen atoms. The presence of coesite and stishovite in rocks near a crater is evidence that the crater was formed by impact of a meteorite.

Silica Glass

If any form of silica is melted (m.p. about 1600°C) and the molten material is then cooled, it usually does not crystallize at the original melting point, but the liquid becomes more viscous as the temperature is lowered, until, at about 1500°C, it is so stiff that it cannot flow. The material obtained in this way is not crystalline, but is a supercooled liquid, or glass. It is called *silica glass* (or sometimes *quartz glass* or *fused quartz*). Silica glass does not have the properties of a crystal—it does not cleave, nor form crystal faces, nor show other differences in properties in different directions. The reason for this is that the atoms that constitute it are not arranged in a completely regular manner in space, but show a randomness in arrangement similar to that of the liquid.

The structure of silica glass is very similar in its general nature to that of quartz and the other crystalline forms of silica. Nearly every silicon atom is surrounded by a tetrahedron of four oxygen atoms, and nearly every oxygen atom serves as the common corner of two of these tetrahedra. The arrangement of the framework of tetrahedra in the glass is not regular, however, as it is in the crystalline forms of silica, but is irregular, so that a very small region may resemble quartz, and an adjacent region may resemble cristobalite or tridymite, in the same way that liquid silica, above the melting point of the crystalline forms, would show some resemblance to the structures of the crystals.

Silica glass is used for making chemical apparatus and scientific instruments. The coefficient of thermal expansion of silica glass is very small, so that vessels made of the material do not break readily on sudden heating or cooling. Silica is transparent to ultraviolet light, and because of this property it is used in making mercury-vapor ultraviolet lamps and optical instruments for use with ultraviolet light.

Ordinary glass is discussed in Section 18-10.

Cryptocrystalline Silica

A cryptocrystalline mineral is one in which the crystal grains are so small that they cannot be seen, even with a microscope. Silica, sometimes partially hydrated, occurs in nature in many cryptocrystalline varieties, distinguished from one another mainly by their color (usually arising from impurities). Among these are *chalcedony* (waxy luster, transparent or translucent, white, grayish, blue, brown, black), *carnelian* (a clear red or red-brown chalcedony), *chrysoprase* (an apple-green chalcedony, containing bipositive nickel), *agate* (a variegated chalcedony, either banded or cloudy), *onyx* (agate with layers in planes), *sardonyx* (onyx with some layers of carnelian, which is also called sard), *flint* (resembling chalcedony, but opaque and dull in color, usually gray, smoky-brown, or brownish-black), and *jasper* (more dull and opaque than flint; often red in color from iron(III) oxide, or yellow, gray-blue, brown-black).

The foregoing minerals are varieties of quartz. *Opal* is a cryptocrystalline variety of cristobalite, somewhat hydrated. The striking colors of opal are caused by Bragg diffraction (Appendix IV) of visible light by spherulites of cristobalite about 300 nm in diameter that have settled into a close-packed array and have been cemented together by a silicious cement with index of refraction different from that of the spherulites.

Sodium Silicate and Other Silicates

Silicic acid (orthosilicic acid), H_4SiO_4, cannot be made by the hydration of silica. The sodium and potassium salts of silicic acid are soluble in water, however, and can be made by boiling silica with a solution of sodium hydroxide or potassium hydroxide, in which it slowly dissolves. A concentrated solution of sodium silicate, called water glass, is available commercially and is used for fireproofing wood and cloth, as an adhesive, and for preserving eggs. This solution is not sodium orthosilicate, Na_4SiO_4, but is a mixture of the sodium salts of various condensed silicic acids, such as $H_6Si_2O_7$, $H_4Si_3O_8$, and $(H_2SiO_3)_n$.

A gelatinous precipitate of condensed silicic acids ($SiO_2 \cdot xH_2O$) is obtained when an ordinary acid, such as hydrochloric acid, is added to a solution of sodium silicate. When this precipitate is partially dehydrated it forms a porous product called silica gel. This material has great powers of adsorption for water and other molecules and is used as a drying agent and decolorizing agent.

Except for the alkali silicates, most silicates are insoluble in water. Many occur in nature, as ores and minerals.

18-9. The Silicate Minerals

Most of the minerals that constitute rocks and soil are silicates, which usually also contain aluminum. Many of these minerals have complex formulas, corresponding to the complex condensed silicic acids from which they are derived. These minerals can be divided into three principal classes: the *framework minerals,* with three-dimensional covalent bonding (hard minerals similar in their properties to quartz), the *layer minerals,* with two-dimensional bonding (such as mica), and the *fibrous minerals,* with one-dimensional bonding (such as asbestos).

The Framework Minerals

Many silicate minerals have tetrahedral framework structures in which some of the tetrahedra are AlO_4 tetrahedra instead of SiO_4 tetrahedra. These minerals have structures somewhat resembling that of quartz, with additional ions, usually alkali or alkaline-earth ions, introduced in the larger openings in the framework structure. Ordinary *feldspar* (*orthoclase*), $KAlSi_3O_8$, is an example of a tetrahedral aluminosilicate mineral. The aluminosilicate tetrahedral framework, $(AlSi_3O_8^-)_\infty$, extends throughout the entire crystal, giving it hardness nearly as great as that of quartz.

A characteristic feature of these tetrahedral framework minerals is that the number of oxygen atoms is just twice the sum of the number of aluminum and silicon atoms. In some of these minerals the framework is an open one, through which corridors run that are sufficiently large to permit ions to move in and out. The *zeolite minerals,* used for softening water, are of this nature. As the hard water, containing Ca^{++} and Fe^{+++} ions, passes around the grains of the mineral, these cations enter the mineral, replacing an equivalent number of sodium ions.

Some of the important minerals in soil are aluminosilicate minerals that have the property of base exchange, and that, because of this property, serve a useful function in the nutrition of the plant.

Minerals with Layer Structures

By a condensation reaction involving three of the four hydroxyl groups of each silicic acid molecule, a condensed silicic acid can be made, with composition $(H_2Si_2O_5)_\infty$, which has the form of an infinite layer, as shown in Figure 18-6. The mineral *hydrargillite*, $Al(OH)_3$, has a similar layer structure, which involves AlO_6 octahedra (Figure 18-7). More complex layers, involving both tetrahedra and octahedra, are present in other layer minerals, such as *talc*, *kaolinite* (clay), and *mica*.

In talc and kaolinite, with formulas $Mg_3Si_4O_{10}(OH)_2$ and $Al_2Si_2O_5(OH)_4$, respectively, the layers are electrically neutral, and they are loosely superimposed on one another to form the crystalline material. These layers slide over one another very readily, which gives to these minerals their characteristic properties (softness, easy cleavage, soapy feel). In mica, $KAl_3Si_3O_{10}(OH)_2$, the aluminosilicate layers are negatively charged, and positive ions, usually potassium ions, must be present between the layers in order to give the mineral electric neutrality. The electrostatic forces between these positive ions and the negatively charged layers make mica considerably harder than kaolinite and talc,

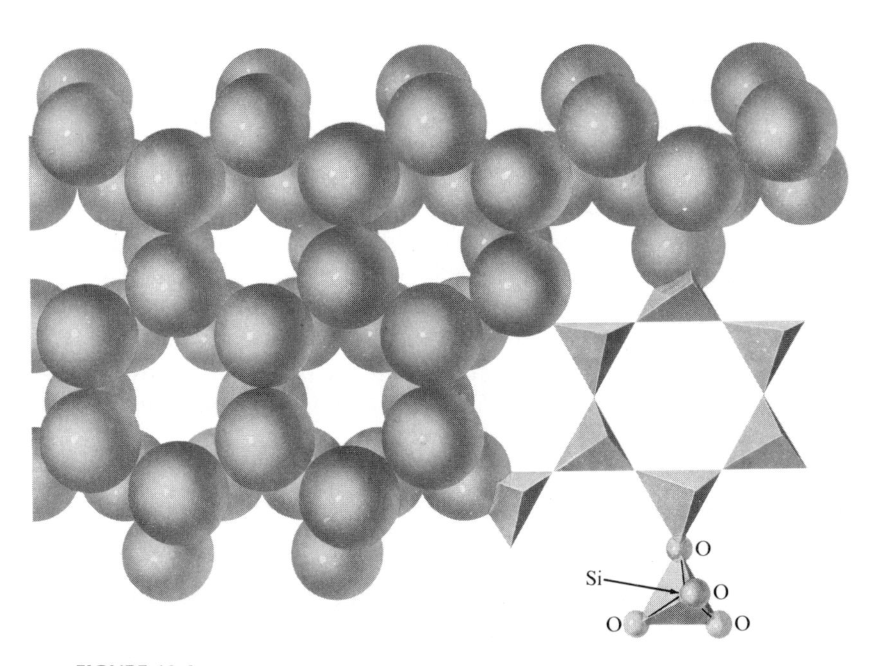

FIGURE 18-6
A portion of an infinite layer of silicate tetrahedra, as present in talc and other minerals with layer structures.

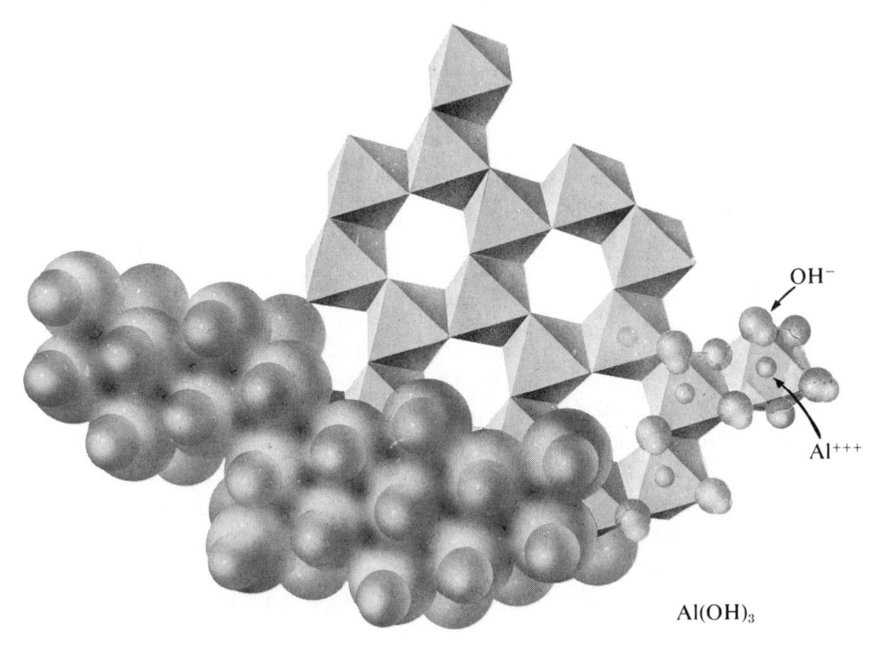

FIGURE 18-7
The crystal structure of aluminum hydroxide, $Al(OH)_3$. This substance crystallizes in layers, consisting of octahedra of oxygen atoms (hydroxide ions) about the aluminum atoms. Each oxygen atom serves as a corner for two aluminum octahedra.

but its layer structure is still evident in its perfect basic cleavage, which permits the mineral to be split into very thin sheets. These sheets of mica are used for windows in stoves and furnaces, and for electric insulation in machines and instruments.

Other layer minerals, such as *montmorillonite,* with formula approximately $AlSi_2O_5(OH)\cdot xH_2O$, are important constituents of soils, and have also found industrial uses, as catalysts in the conversion of long-chain hydrocarbons into branched-chain hydrocarbons (to make high-octane gasoline), and for other special purposes.

The Fibrous Minerals

The fibrous minerals contain very long silicate ions in the form of tetrahedra condensed into a chain. These crystals can be cleaved readily in directions parallel to the silicate chains, but not in the directions that cut the chains. Accordingly crystals of these minerals show the extraordinary property of being easily unraveled into fibers. The principal minerals of this sort, *tremolite,* $Ca_2Mg_5Si_8O_{22}(OH)_2$, and *chrysotile,* $Mg_6Si_4O_{11}(OH)_6\cdot H_2O$, are called *asbestos*. Deposits of these minerals are found, especially in South Africa, in layers several inches thick. These minerals are shredded into fibers, which are then spun or felted into asbestos yarn, fabric, and board for use for thermal insulation and as a heat-resistant structural material.

18-10. Glass

Silicate materials with important uses include glass, porcelain, glazes and enamels, and cement. Ordinary glass is a mixture of silicates in the form of a supercooled liquid. It is made by melting a mixture of sodium carbonate (or sodium sulfate), limestone, and sand, usually with some scrap glass of the same grade to serve as a flux. After the bubbles of gas have been expelled, the clear melt is poured into molds or stamped with dies, to produce pressed glassware, or a lump of the semifluid material on the end of a hollow tube is blown, sometimes in a mold, to produce hollow ware, such as bottles and flasks. *Plate glass* is made by pouring liquid glass onto a flat table and rolling it into a sheet. The sheet is then ground flat and polished on both sides. *Safety glass* consists of a sheet of tough plastic sandwiched between two sheets of glass.

Ordinary glass (soda-lime glass, soft glass) contains about 10% sodium, 5% calcium, and 1% aluminum, the remainder being silicon and oxygen. It consists of an aluminosilicate tetrahedral framework, within which are embedded sodium ions and calcium ions and some smaller complex anions. Soda-lime glass softens over a range of temperatures beginning at a dull-red heat, and can be conveniently worked in this temperature range.

Boric acid easily forms highly condensed acids, similar to those of silicic acid, and borate glasses are similar to silicate glasses in their properties. *Pyrex glass,* used for chemical glassware and baking dishes, is a boroaluminosilicate glass containing only about 4% of alkali and alkaline-earth metal ions. This glass is not as soluble in water as is soft glass, and it also has a smaller coefficient of thermal expansion than soft glass, so that it does not break readily when it is suddenly heated or cooled.

Glazes on chinaware and pottery and *enamels* on iron kitchen utensils and bathtubs consist of easily fusible glass containing pigments or white fillers such as titanium dioxide and tin dioxide.

18-11. Cement

Portland cement is an aluminosilicate powder that sets to a solid mass on treatment with water. It is usually manufactured by grinding limestone and clay to a fine powder, mixing with water to form a slurry, and burning the mixture, with a flame of gas, oil, or coal dust, in a long rotary kiln. At the hot end of the kiln, where the temperature is about 1500°C, the aluminosilicate mixture is sintered together into small round marbles, called "clinker." The clinker is ground to a fine powder in a ball mill (a rotating cylindrical mill filled with steel balls) to produce the final product.

Portland cement before treatment with water consists of a mixture of calcium silicates, mainly Ca_2SiO_4 and Ca_3SiO_5, and calcium aluminate, $Ca_3Al_2O_6$. When treated with water the calcium aluminate hydrolyzes, forming calcium hydroxide and aluminum hydroxide, and these substances react further with the calcium silicates to produce calcium aluminosilicates, in the form of intermeshed crystals.

Ordinary *mortar* for laying bricks is made by mixing sand with Portland cement. The amount of cement needed for a construction job is greatly reduced by mixing sand and crushed stone or gravel with the cement, forming the material called *concrete*. Concrete is a very valuable building material. It does not require carbon dioxide from the air in order to harden, and it will set under water and in very large masses.

18-12. Silicones

When we consider the variety of structures represented by the silicate minerals, and their resultant characteristic and useful properties, we might well expect chemists to synthesize many new and valuable silicon compounds. In recent years this has been done; many silicon compounds, especially those of the class called *silicones*, have been found to have valuable properties.

The simplest silicones are the methyl silicones. These substances exist as oils, resins, and elastomers (rubberlike substances). Methyl silicone oil consists of long molecules, each of which is a silicon-oxygen chain with methyl groups attached to the silicon atoms. A short silicone molecule would have the following structure:

H_3C O O O CH_3

Si Si Si Si

H_3C CH_3 H_3C CH_3 H_3C CH_3 H_3C CH_3

A *silicone oil* for use as a lubricating oil or in hydraulic systems contains molecules with an average of about 10 silicon atoms per molecule.

The valuable properties of the silicone oils are their very low coefficient of viscosity with temperature, ability to withstand high temperature without decomposition, and chemical inertness to metals and most reagents. A typical silicone oil increases only about sevenfold in viscosity on cooling from 100°F to −35°F, whereas a hydrocarbon oil with the same viscosity at 100°F increases in viscosity about 1800-fold at −35°F.

Resinous silicones can be made by polymerizing silicones into cross-linked molecules. These resinous materials are used for electric insulation. They have excellent dielectric properties and are stable at operating temperatures at which the usual organic insulating materials decompose rapidly. The use of these materials permits electric machines to be operated with increased loads.

Silicones may be polymerized to molecules containing 2000 or more $(CH_3)_2SiO$ units, and then milled with inorganic fillers (such as zinc oxide or carbon black, used also for ordinary rubber), and vulcanized, by heating to cause cross-links to form between the molecules, bonding them into an insoluble, infusible three-dimensional framework.

Similar silicones with ethyl groups or other organic groups in place of the methyl groups are also used.

The coating of materials with a water-repellent film has been achieved by use of the *methylchlorosilanes.* A piece of cotton cloth exposed for a second or two to the vapor of trimethylchlorosilane, $(CH_3)_3SiCl$, becomes coated with a layer of trimethylsilicyl groups, through reaction with hydroxyl groups of the cellulose:

$$(CH_3)_3SiCl + HOR \longrightarrow (CH_3)_3SiOR + HCl$$

The exposed methyl groups repel water in the way that a hydrocarbon film such as lubricating oil would. Paper, wool, silk, glass, porcelain, and other materials can be treated in this way. The treatment has been found especially useful for ceramic insulators.

18-13. Germanium

The chemistry of germanium, a moderately rare element, is similar to that of silicon. Most of the compounds of germanium correspond to oxidation number +4; examples are germanium tetrachloride, $GeCl_4$, a colorless liquid with boiling point 83°C, and germanium dioxide, GeO_2, a colorless crystalline substance melting at 1086°C.

The compounds of germanium have found little use. The element itself, a gray metalloid, has been extensively used in electronic devices.

The electrical properties of a single crystal of germanium (or silicon) can be drastically changed by alloying the element with very small amounts of other elements. As explained in the following paragraphs, these effects are the basis of the operation of the semiconductor junction rectifier, the transistor, and integrated circuits.

The electrical conductivity of pure germanium at very low temperatures is close to zero. The crystal, like that of diamond, contains atoms with ligancy 4 and a pair of electrons for every bond. We may use a two-dimensional representation:

```
  |      |      |      |
—Ge————Ge————Ge————Ge—
  |      |      |      |
  |      |      |      |
—Ge————Ge————Ge————Ge—
  |      |      |      |
  |      |      |      |
—Ge————Ge————Ge————Ge—
  |      |      |      |
```

The electrons are restricted to the bond regions, and are not free to move when an electric field is applied. At higher temperatures an electron may be promoted to an excited orbit (5*s*, for example), leaving one electron in a bond where there should be two:

```
  |       |       |       |
—Ge—————Ge—————Ge—————Ge—
  |       |       |       |
  |       |       |       |
—Ge+ ·−  Ge+————Ge·−————Ge—
  |       |       |       |
  |       |       |       |
—Ge—————Ge—————Ge—————Ge—
```

The electron left alone in the bond between two germanium atoms each with a positive charge, $Ge^{+}\cdot^{-}Ge^{+}$, is called a *hole*. The hole can contribute to the electric conductance by the motion to it of an electron from an adjacent bond, as in the following sequence:

```
—Ge+ ·− Ge+ —Ge———Ge———Ge———
—Ge———Ge+ ·− Ge+ —Ge———Ge———
—Ge———Ge———Ge+ ·− Ge+ —Ge———
—Ge———Ge———Ge———Ge+ ·− Ge+ —
```

The promoted electron also contributes to the conductance; it moves in the opposite direction to the hole.

A crystal of germanium containing some atoms of arsenic (germanium doped with arsenic) has extra electrons in the excited orbitals, because each arsenic atom contributes not only the four electrons needed for the tetrahedral bonds but also a fifth electron. Such a crystal has greater conductivity than pure germanium, and the conductivity is of the *n* type (carried by the negative electrons).

A crystal containing some atoms of aluminum, each contributing only three valence electrons, has for every aluminum atom a hole in the set of bonding electron pairs, and has conductivity of the *p* type (carried by the positive holes, which, of course, move in one direction as electrons jump in the opposite direction into them from adjacent bonding pairs).

A *p-n* junction rectifier is made by placing a *p* crystal and an *n* crystal in contact with one another, as shown in Figure 18-8. Each crystal is attached at the other end to a metal plate carrying the terminals. Both holes and electrons transfer readily to the metal plates and across the junction, and a steady current is carried when a potential is applied in such a direction as to cause both holes and electrons to move toward the junction. When the potential is reversed, however, the holes and electrons move away from the junction (bottom of Figure 18-8). There is no mechanism for rapidly producing new holes and promoted electrons at the junction—this process requires the energy to raise an electron from a bond orbital (for germanium the $4s4p^3$ tetrahedral hybrid orbitals) to an excited orbital, $5s$, and its rate is determined by the temperature (the Arrhenius exponential rate factor, Section 10-4). In consequence, the region near the junction becomes depleted of carriers and the current ceases to flow.

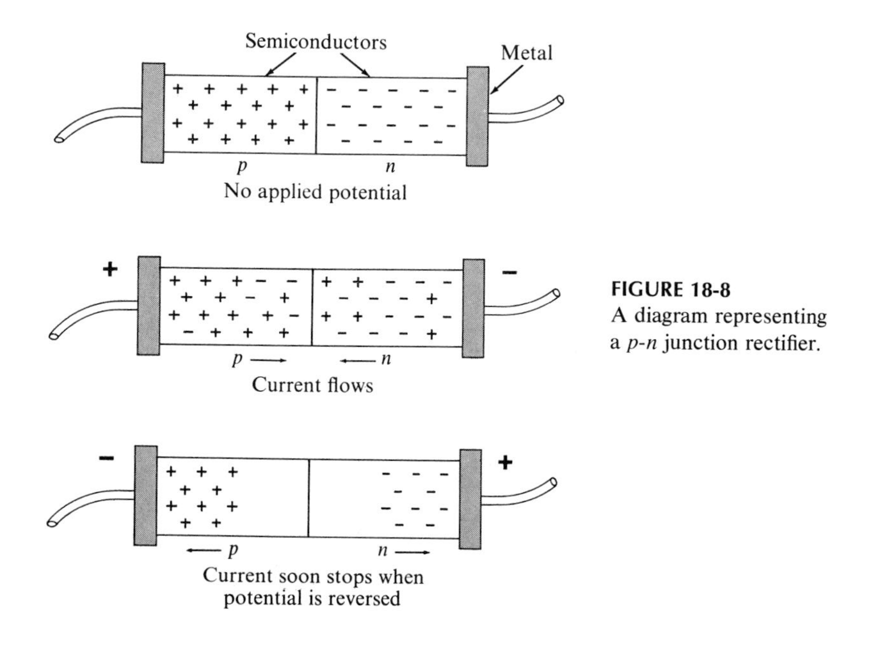

FIGURE 18-8
A diagram representing a *p-n* junction rectifier.

An *n-p-n* transistor can be made by sandwiching a *p* crystal between two *n* crystals and attaching terminals in such a way that one applied potential depletes or augments the supply of carriers for another. In this way a current in one circuit can be caused to produce a proportional current in another circuit at a higher power level.

18-14. Tin

Tin is a silvery white metal, with great malleability, permitting it to be hammered into thin sheets, called tin foil. Ordinary *white tin,* which has metallic properties, slowly changes at temperatures below 18°C to a nonmetallic allotropic modification, *gray tin,* which has the diamond structure. (The physical properties given in Table 18-3 pertain to white tin.) At very low temperatures, around −40°C, the speed of this conversion is sufficiently great that metallic tin objects sometimes fall into a powder of gray tin. This phenomenon has been called the "tin pest."

Tin finds extensive use as a protective layer for mild steel. Tin plating is done by dipping clean sheets of mild steel into molten tin, or by electrolytic deposition. Copper and other metals are sometimes also coated with tin.

The principal alloys of tin are *bronze* (tin and copper), *soft solder* (tin and lead), *pewter* (75% tin and 25% lead), and *britannia metal* (tin with small amounts of antimony and copper).

Bearing metals, used as the bearing surfaces of sliding-contact bearings, are usually alloys of tin, lead, antimony, and copper. They contain small, hard crystals of a compound such as SnSb embedded in a soft matrix of tin or lead. The good bearing properties result from orientation of the hard crystals to present flat faces at the bearing surface.

Tin is reactive enough to displace hydrogen from dilute acids, but it does not tarnish in moist air. It reacts with warm hydrochloric acid to produce stannous chloride, $SnCl_2$, and hydrogen, and with hot concentrated sulfuric acid to produce stannous sulfate, $SnSO_4$, and sulfur dioxide, the equations for these reactions being

$$Sn + 2HCl \rightarrow SnCl_2 + H_2$$

and

$$Sn + 2H_2SO_4 \rightarrow SnSO_4 + SO_2 + 2H_2O$$

With cold dilute nitric acid it forms stannous nitrate, and with concentrated nitric acid it is oxidized to a hydrated stannic acid, H_2SnO_3.

Compounds of Tin

Stannous chloride, made by solution of tin in hydrochloric acid, forms colorless crystals, $SnCl_2 \cdot H_2O$, on evaporation of the solution. In neutral solution the substance hydrolyzes, forming a precipitate of stannous hydroxychloride, $Sn(OH)Cl$. The hydrolysis in solution may be prevented by the presence of an excess of acid. Stannous chloride solution is used as a mordant in dyeing cloth.

The stannous ion is an active reducing agent, which is easily oxidized to stannic chloride, $SnCl_4$, or, in the presence of excess chloride ion, to the complex chlorostannate ion, $SnCl_6^{--}$.

Stannic chloride, $SnCl_4$, is a colorless liquid (boiling point 114°C), which fumes very strongly in moist air, producing hydrochloric acid and stannic acid, $H_2Sn(OH)_6$. Sodium stannate, $Na_2Sn(OH)_6$, contains the octahedral hexahydroxystannate ion (stannate ion). This complex ion is similar in structure to the chlorostannate ion. Sodium stannate is used as a mordant and in preparing fireproof cotton cloth and weighting silk. The cloth is soaked in the sodium stannate solution, dried, and treated with ammonium sulfate solution. This treatment causes hydrated stannic oxide to be deposited in the fibers.

18-15. Lead

Lead is a soft, heavy, dull gray metal with low tensile strength. It is used in making type, for covering electric cables, and in many alloys. The organic lead compound lead tetraethyl, $Pb(C_2H_5)_4$, is added to gasoline to prevent knock in automobile engines.

Lead forms a thin surface layer of oxide in air. This oxide slowly changes to a basic carbonate. Hard water forms a similar coating on lead, which protects the water from contamination with soluble lead compounds. Soft water dissolves appreciable amounts of lead, which is poisonous; for this reason lead pipes should not be used to carry drinking water.

There are several oxides of lead, of which the most important are lead monoxide (*litharge*), PbO, minium or red lead, Pb_3O_4, and lead dioxide, PbO_2.

Litharge is made by heating lead in air. It is a yellow powder or yellowish-red crystalline material, used in making lead glass and for preparing compounds of lead. Red lead, Pb_3O_4, can be made by heating lead in oxygen. It is used in glass making, and for making a red paint for protecting iron and steel structures. Lead dioxide, PbO_2, is a brown substance made by oxidizing a solution of sodium plumbite, $Na_2Pb(OH)_4$, with hypochlorite ion, or by anodic oxidation of lead sulfate. It is soluble in sodium hydroxide and potassium hydroxide, forming the hexahydroxyplumbate ion, $Pb(OH)_6^{--}$. The principal use of lead dioxide is in the lead storage battery.

Lead nitrate, $Pb(NO_3)_2$, is a white crystalline substance made by dissolving lead, lead monoxide, or lead carbonate in nitric acid. Lead carbonate, $PbCO_3$, occurs in nature as the mineral *cerussite.* It appears as a precipitate when a solution containing the hydrogen carbonate ion, HCO_3^-, is added to lead nitrate solution. With a more basic carbonate solution a basic carbonate of lead, $Pb_3(OH)_2(CO_3)_2$, is deposited. This basic salt, called *white lead,* is used as a white pigment in paint. For this use it is manufactured by methods involving the oxidation of lead by air, the formation of a basic acetate by interaction with vinegar or acetic acid, and the decomposition of this salt by carbon dioxide. Lead chromate, $PbCrO_4$, is also used as a pigment, under the name *chrome yellow.*

Lead sulfate, $PbSO_4$, is a white, nearly insoluble substance. Its precipitation is used as a test for either lead ion or sulfate ion in analytical chemistry.

EXERCISES

18-1. Compare the properties of elements of groups I, II, III, and IV with their electronegativities (Table 6-3). What electronegativity value separates the metals from the metalloids?

18-2. Beryllium hydroxide is essentially insoluble in water, but is soluble both in acids and in alkalis. What do you think the products of its reaction with sodium hydroxide solution are? Discuss these properties of the substance in relation to the position of beryllium in the periodic table and in the electronegativity scale.

18-3. Discuss the electronic structure of potassium fluoroborate, KBF_4. Its solution in water contains the ion BF_4^-.

18-4. What is the electronic structure of the aluminum atom? How does it explain the fact that almost all the compounds of aluminum correspond to oxidation number +3?

18-5. One of the crystalline forms of silicon carbide is cubic, with $a = 436$ pm, 4C at 0 0 0, 0 $\frac{1}{2}$ $\frac{1}{2}$, $\frac{1}{2}$ 0 $\frac{1}{2}$, $\frac{1}{2}$ $\frac{1}{2}$ 0; 4Si at $\frac{1}{4}$ $\frac{1}{4}$ $\frac{1}{4}$, $\frac{1}{4}$ $\frac{3}{4}$ $\frac{3}{4}$, $\frac{3}{4}$ $\frac{1}{4}$ $\frac{3}{4}$, $\frac{3}{4}$ $\frac{3}{4}$ $\frac{1}{4}$. What nearest neighbors does a carbon atom have? A silicon atom? At what distances? What are the values of the bond angles? Can you suggest an explanation of the great hardness of the substance?

18-6. From the electronegativities of the elements calculate a value for the standard enthalpy of formation of SiC(c). The experimental value is -111 kJ mole^{-1}. (Answer: -96 kJ mole^{-1}.)

18-7. The compound AlP has a tetrahedral structure resembling that of SiC. Would you expect it to be a possible substitute for germanium in a p-n junction rectifier? How could AlP be doped to give a p-crystal and to give an n-crystal?

18-8. The trichloride of boron has m.p. -107°C and boiling point 12.5°C, whereas that of its congener lanthanum has m.p. 870°C and very high b.p. What is the explanation of these greatly different physical properties?

18-9. Of the two crystal structures that might be expected for BaF_2, the MgF_2 structure and the CaF_2 structure, which would you assign to it on the basis of values of the ionic radii?

18-10. When liquid NaH is electrolyzed, hydrogen is evolved at the anode. Explain in terms of the electronic structure of the substance. What volume of H_2 (standard conditions) would be produced per faraday?

18-11. Do you think that lanthanum metal could be made from lanthanum oxide by reaction with aluminum powder (see Section 6-12)?

18-12. Why is the presence of coesite or stishovite near a crater taken as evidence that the crater was formed by impact of a meteorite? In what way is the principle of LeChatelier involved in your answer?

18-13. Using the Mulliken relation (Section 6-12), evaluate the electronegativities of the five alkali metals. Values of the enthalpy of ionization are given in Table 6-1, and the electron affinity of the alkali-metal atoms (for which there are no experimental values) can be taken as zero. (Answer: 0.99, 0.95, 0.80, 0.77, 0.72.)

18-14. Assign electronic structures to Si_2Cl_6 and Si_2Cl_6O. Would you expect both of these molecules to have an electric dipole moment differing from zero?

19

The Transition Metals and Their Compounds

In this chapter we shall discuss the chemistry of the transition metals—the elements that occur in the central region of the periodic table. These elements and their compounds have great practical importance. Their chemical properties are complex and interesting.

We shall begin the discussion of the transition metals with iron, cobalt, nickel, and the platinum metals, which lie in the center of the transition-metal region in the periodic table. The following sections will be devoted to the elements that lie to the right of these metals (copper, zinc, and gallium and their congeners), and those that lie to the left (titanium, vanadium, chromium, and manganese and other elements of groups IVa, Va, VIa, and VIIa of the periodic table).

19-1. The Electronic Structures and Oxidation States of Iron, Cobalt, Nickel, and the Platinum Metals

The electronic structures of iron, cobalt, nickel, and the platinum metals are given in Table 19-1, as represented in the energy-level diagram of Figure 5-6. It is seen that each of the atoms has two outermost electrons, in the 4*s* orbital for iron, cobalt, and nickel, the 5*s* orbital for ruthenium,

TABLE 19-1
The Electronic Structures of Iron, Cobalt, Nickel, and the Platinum Metals

Atomic Number	Element	*K*	*L*		*M*			*N*				*O*			*P*
		1*s*	2*s*	2*p*	3*s*	3*p*	3*d*	4*s*	4*p*	4*d*	4*f*	5*s*	5*p*	5*d*	6*s*
26	Fe	2	2	6	2	6	6	2							
27	Co	2	2	6	2	6	7	2							
28	Ni	2	2	6	2	6	8	2							
44	Ru	2	2	6	2	6	10	2	6	6		2			
45	Rh	2	2	6	2	6	10	2	6	7		2			
46	Pd	2	2	6	2	6	10	2	6	8		2			
76	Os	2	2	6	2	6	10	2	6	10	14	2	6	6	2
77	Ir	2	2	6	2	6	10	2	6	10	14	2	6	7	2
78	Pt	2	2	6	2	6	10	2	6	10	14	2	6	8	2

rhodium, and palladium, and the 6*s* orbital for osmium, iridium, and platinum. The next inner shell is incomplete, the 3*d* orbital (or 4*d*, or 5*d*) contains only six, seven, or eight electrons, instead of the full complement of ten.

It might be expected that the two outermost electrons would be easily removed, to form a bipositive ion. In fact, iron, cobalt, and nickel all form important series of compounds in which the metal is bipositive. These metals also have one or more higher oxidation states. The platinum metals form covalent compounds representing various oxidation states between +2 and +8.

Iron can assume the oxidation states +2, +3, and +6, the last being rare, and represented by only a few compounds, such as potassium ferrate, K_2FeO_4. The oxidation states +2 and +3 correspond to the ferrous ion, Fe^{++}, and ferric ion, Fe^{+++}, respectively. The ferrous ion has six electrons in the incomplete 3*d* subshell, and the ferric ion has five electrons in this subshell. The magnetic properties of the compounds of iron and other transition elements are due to the presence of a smaller number of electrons in the 3*d* subshell than required to fill this subshell. For example, ferric ion can have all five of its 3*d* electrons with spins oriented in the same direction, because there are five 3*d* orbitals in the 3*d* subshell, and the Pauli principle permits parallel orientation of the spins of electrons so long as there is only one electron per orbital. The ferrous ion is easily oxidized to ferric ion by air or other oxidizing agents. Both bipositive and terpositive iron form complexes, such as the ferrocyanide ion, $Fe(CN)_6^{----}$, and the ferricyanide ion, $Fe(CN)_6^{---}$, but they do not form complexes with ammonia.

Cobalt(II) and cobalt(III) compounds are known; the cobalt(II) ion, Co^{++}, is more stable than the cobalt(III) ion, Co^{+++}, which is a sufficiently powerful oxidizing agent to oxidize water, liberating oxygen. But the covalent cobalt(III) complexes, such as the cobalticyanide ion, $Co(CN)_6^{---}$, are very stable, and the cobalt(II) complexes, such as the cobaltocyanide ion, $Co(CN)_6^{----}$, are unstable, being strong reducing agents.

Nickel forms only one series of salts, containing the nickel ion, Ni^{++}. A few compounds of nickel with higher oxidation number are known; of these the nickel(IV) oxide, NiO_2, is important.

As was mentioned in Chapter 17, iron, cobalt, and nickel are sexivalent in the metals and their alloys. This high metallic valence causes the bonds to be especially strong, and confers valuable properties of strength and hardness on the alloys.

19-2. Iron

Pure iron is a bright silvery-white metal, which tarnishes (rusts rapidly) in moist air or in water containing dissolved oxygen. It is soft, malleable, and ductile, and is strongly magnetic (ferromagnetic). Its melting point is 1535°C, and its boiling point 3000°C (Table 19-2). Ordinary iron (alpha-iron) has the atomic arrangement shown in Figure 2-9 (the body-centered arrangement—each atom is in the center of a cube formed by the eight surrounding atoms). At 912°C alpha-iron undergoes a transition to another allotropic form, gamma-iron, which has the face-centered arrangement described for copper in Chapter 2. At 1400°C another transition occurs, to delta-iron, which has the same body-centered structure as alpha-iron.

Pure iron, containing only about 0.01% of impurities, can be made by electrolytic reduction of iron salts. It has little use; a small amount is used in analytical chemistry, and a small amount in the treatment of anemia.*

Metallic iron is greatly strengthened by the presence of a small amount of carbon, and its mechanical and chemical properties are also improved by moderate amounts of other elements, especially other transition metals. Wrought iron, cast iron, and steel are described in the following sections.

The Ores of Iron

The chief ores of iron are its oxides *hematite,* Fe_2O_3, and *magnetite,* Fe_3O_4, and its carbonate *siderite,* $FeCO_3$. The hydrated ferric oxides such as *limonite* are also important. The sulfide *pyrite*, FeS_2, is used as a source

*See hemoglobin, Section 15-4.

TABLE 19-2
Some Physical Properties of Iron, Cobalt, and Nickel

	Atomic Number	Atomic Mass	Density (g cm^{-3})	Melting Point	Boiling Point	Metallic Radius*	Enthalpy of Sublimation at 25°C
Iron	26	55.847	7.86	1,535°C	3000°C	126 pm	405 kJ $mole^{-1}$
Cobalt	27	58.9332	8.93	1,480	2900	125	439
Nickel	28	58.71	8.89	1,452	2900	124	425

*For ligancy 12.

of sulfur dioxide, but the impure iron oxide left from its roasting is not satisfactory for smelting iron, because the remaining sulfur is a troublesome impurity.

The Metallurgy of Iron

The ores of iron are usually first roasted (heated in air), in order to remove water, to decompose carbonates, and to oxidize sulfides. They are then reduced with coke, in a structure called a *blast furnace* (Figure 19-1). Ores containing limestone or magnesium carbonate are mixed with an acidic flux (containing an excess of silica), such as sand or clay, in order to make a liquid *slag*. Limestone is used as flux for ores containing an excess of silica. The mixture of ore, flux, and coke is introduced at the top of the blast furnace, and preheated air is blown in the bottom through holes called tuyeres. As the solid materials slowly descend they are converted completely into gases, which escape at the top, and two liquids, molten iron and slag, which are tapped off at the bottom. The parts of the blast furnace where the temperature is highest are water-cooled, to keep the lining from melting.

The important reactions that occur in the blast furnace are the combustion of coke to carbon monoxide, the reduction of iron oxide by the carbon monoxide, and the combination of acidic and basic oxides (the impurities of the ore and the added flux) to form slag:

$$2C + O_2 \rightarrow 2CO$$

$$3CO + Fe_2O_3 \rightarrow 2Fe + 3CO_2$$

$$CaCO_3 \rightarrow CaO + CO_2$$

$$CaO + SiO_2 \rightarrow CaSiO_3$$

The slag is a glassy silicate mixture of complex composition, idealized as calcium metasilicate, $CaSiO_3$, in the above equation.

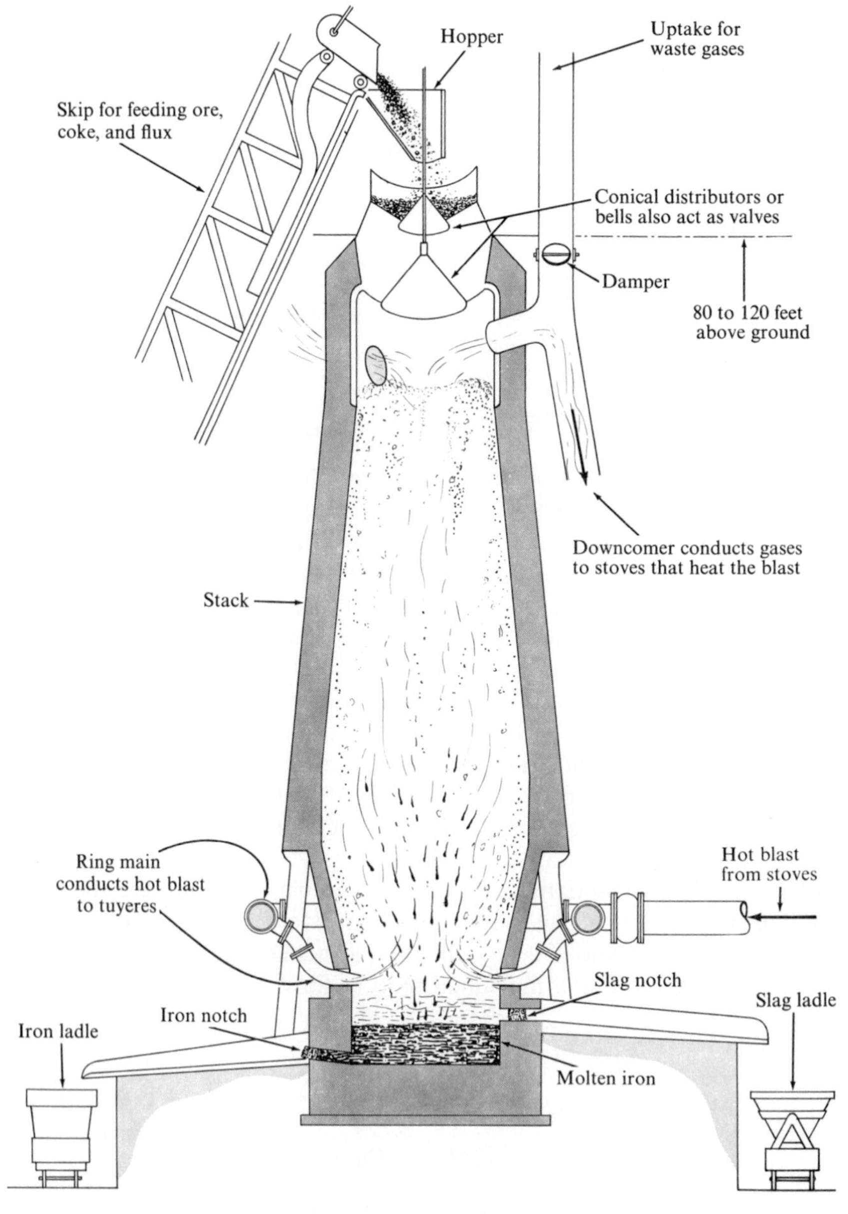

FIGURE 19-1
A blast furnace for smelting iron ore.

The hot exhaust gases, which contain some unoxidized carbon monoxide, are cleaned of dust and then are mixed with air and burned in large steel structures filled with fire brick. When one of these structures, which are called *stoves*, has thus been heated to a high temperature the burning exhaust gas is shifted to another stove and the heated stove is used to preheat the air for the blast furnace.

Cast Iron

The molten iron from the blast furnace, having been in contact with coke in the lower part of the furnace, contains several percent of dissolved carbon (usually about 3 or 4%), together with silicon, manganese, phosphorus, and sulfur in smaller amounts. These impurities lower its melting point from 1535°C, that of pure iron, to about 1200°C. This iron is often cast into bars called *pigs;* the cast iron itself is called *pig iron.*

When cast iron is made by sudden cooling from the liquid state it is white in color, and is called **white cast iron.** It consists largely of the compound *cementite,* Fe_3C, a hard, brittle substance.

Gray cast iron, made by slow cooling, consists of crystalline grains of pure iron (called *ferrite*) and flakes of graphite (Figure 19-2). Both white cast iron and gray cast iron are brittle, the former because its principal constituent, cementite, is brittle, and the latter because the tougher ferrite in it is weakened by the soft flakes of graphite distributed through it.

Malleable cast iron, which is tougher and less brittle than either white or ordinary gray cast iron, is made by heat treatment of gray cast iron of suitable composition. Under this treatment the flakes of graphite coalesce into globular particles, which, because of their small cross-sectional area, weaken the ferrite less than do the flakes (Figure 19-3).

Cast iron is the cheapest form of iron, but its usefulness is limited by its low strength. A great amount is converted into steel, and a smaller amount into wrought iron.

Wrought Iron

Wrought iron is nearly pure iron, with only 0.1% or 0.2% carbon and less than 0.5% of all impurities. It is made by melting cast iron on a bed of iron oxide in a reverberatory furnace (Figure 19-4). As the molten cast iron is stirred, the iron oxide oxidizes the dissolved carbon to carbon monoxide, and the sulfur, phosphorus, and silicon are also oxidized and pass into the slag. As the impurities are removed, the melting point of the iron rises, and the mass becomes pasty. It is then taken out of the furnace and beaten under steam hammers to force out the slag.

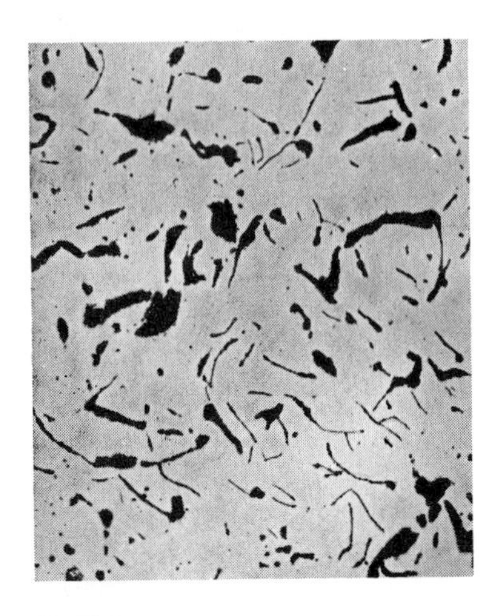

FIGURE 19-2
A photomicrograph of gray cast iron, unetched. The white background is ferrite, and the black particles are flakes of graphite. Magnification 100×. [From Malleable Founders' Society.]

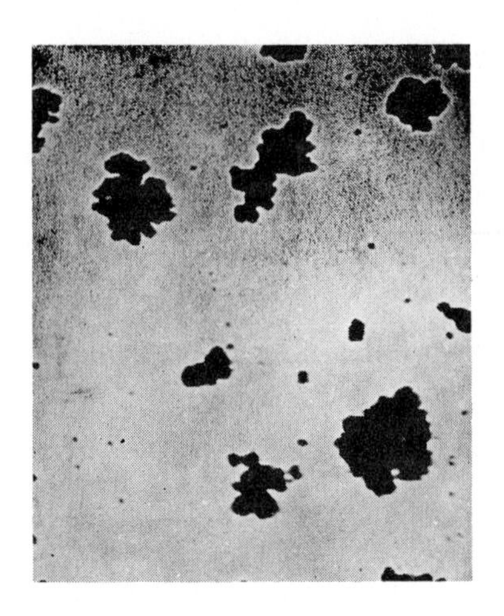

FIGURE 19-3
A photomicrograph of malleable cast iron, showing ferrite (background) and globular particles of graphite. Unetched. Magnification 100×. [From Malleable Founders' Society.]

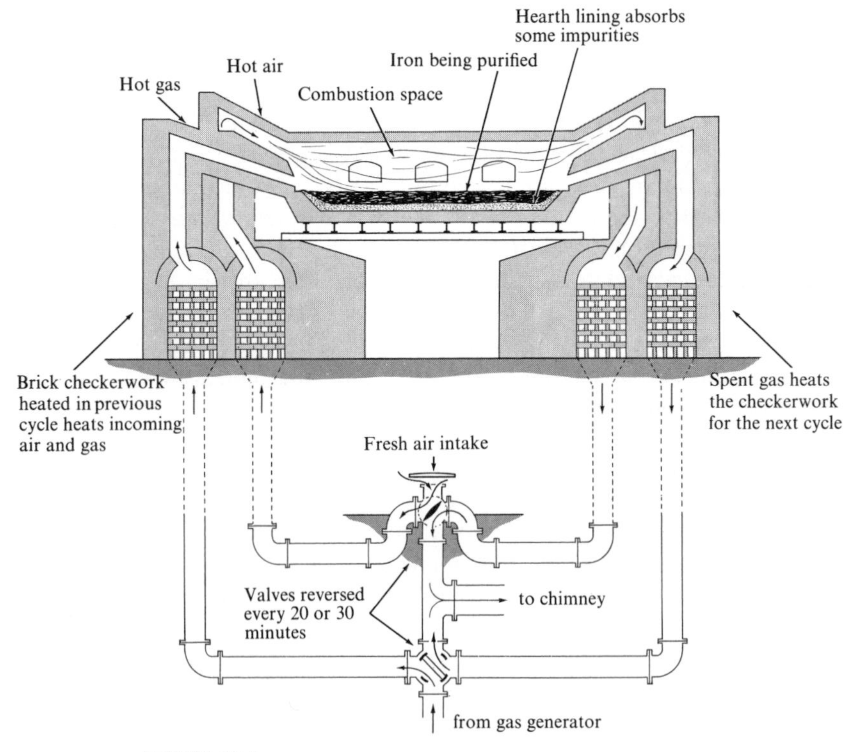

FIGURE 19-4
Reverberatory furnace, used for making wrought iron and steel.

Wrought iron is a strong, tough metal which can be readily welded and forged. In past years it was extensively used for making chains, wire, and similar objects. It has now been largely displaced by mild steel.

19-3. Steel

Steel is a purified alloy of iron, carbon, and other elements that is manufactured in the liquid state. Most steels are almost free from phosphorus, sulfur, and silicon, and contain between 0.1 and 1.5% of carbon. *Mild steels* are low-carbon steels (less than 0.2%). They are malleable and ductile, and are used in place of wrought iron. They are not hardened by being quenched (suddenly cooled) from a red heat. *Medium steels,* containing from 0.2 to 0.6% carbon, are used for making rails and structural elements (beams, girders, etc.). Mild steels and medium steels can be forged and welded. *High-carbon steels* (0.75 to 1.50% carbon) are used for making razors, surgical instruments, drills, and other tools. Medium steels and high-carbon steels can be hardened and tempered (see section on properties of steel).

Steel is made from pig iron chiefly by the *open-hearth process* (by which over 80% of that produced in the United States is made), the *Bessemer process,* and the *oxygen top-blowing process.* In each process either a basic or an acidic lining may be used in the furnace or converter. A basic lining (lime, magnesia, or a mixture of the two) is used if the pig iron contains elements, such as phosphorus, that form acidic oxides, and an acidic lining (silica) if the pig iron contains base-forming elements.

The Open-hearth Process

Open-hearth steel is made in a reverberatory furnace; that is, a furnace in which the flame is reflected by the roof onto the material to be heated. Cast iron is melted with scrap steel and some hematite in a furnace heated with gas or oil fuel. The fuel and air (sometimes enriched with oxygen) are preheated by passage through a lattice of hot brick at one side of the furnace, and a similar lattice on the other side is heated by the hot outgoing gases. From time to time the direction of flow of gas is reversed. The carbon and other impurities in the molten iron are oxidized by the hematite and by excess air in the furnace gas. Analyses are made during the run, which requires about 8 hours, and when almost all the carbon is oxidized the amount desired for the steel is added as coke or as a high-carbon alloy, usually ferromanganese or spiegeleisen. The molten steel is then cast into billets. Open-hearth steel of very uniform quality can be made, because the process can be closely checked by analyses during the several hours of the run.

The Bessemer Process

The Bessemer process of making steel was invented by an American, William Kelly, in 1852 and independently by an Englishman, Henry Bessemer, in 1855. Molten pig iron is poured into an egg-shaped converter (Figure 19-5). Air is blown up through the liquid from tuyeres in the bottom, oxidizing silicon, manganese, and other impurities and finally the carbon. In about ten minutes the reaction is nearly complete, as is seen from the change in character of the flame of burning carbon monoxide from the mouth of the converter. High-carbon alloy is then added, and the steel is poured.

The Bessemer process is inexpensive, but the steel is not as good as open-hearth steel.

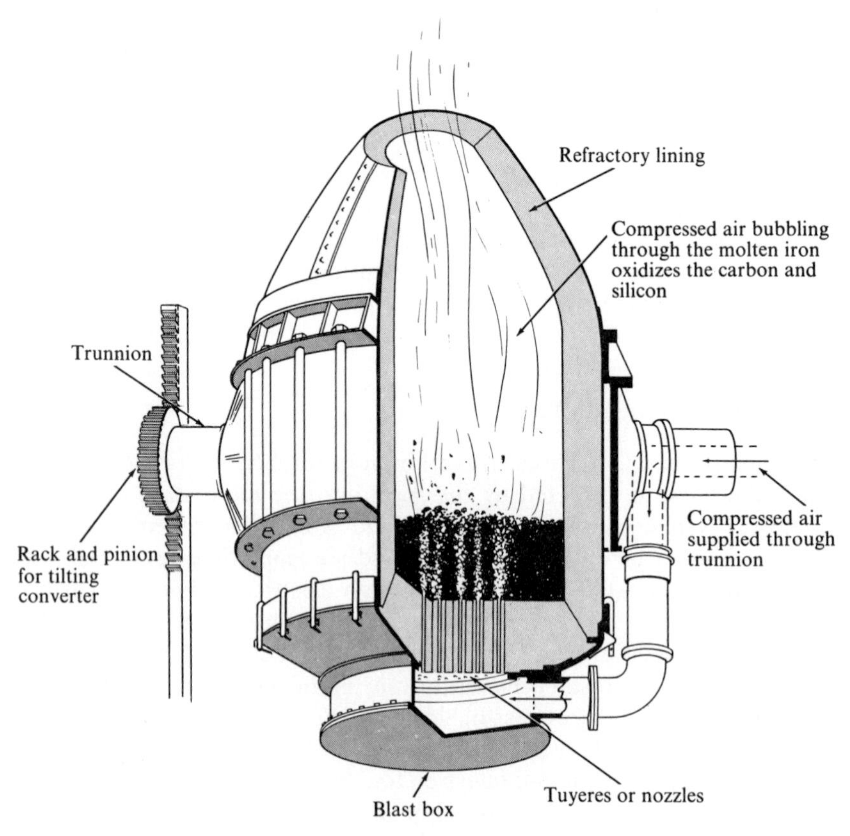

FIGURE 19-5
Bessemer converter, used for making steel from pig iron.

The Oxygen Top-blowing Process

Since 1955 an increasing fraction of the steel produced in the world has been made by a new process, the oxygen top-blowing process. Iron is placed in a converter resembling the Bessemer converter (Figure 19-5), but without the tuyeres at the base. Pure oxygen (99.5%) is then blown onto the surface of the molten metal through a long water-cooled copper lance, to oxidize carbon and phosphorus. The treatment of the charge of 50 to 250 tons is completed in 40 or 50 minutes. This process gives steel of high quality.

The Properties of Steel

When high-carbon steel is heated to bright redness and slowly cooled, it is comparatively soft. However, if it is rapidly cooled, by quenching in water, oil, or mercury, it becomes harder than glass, and brittle instead of tough. This hardened steel can be "tempered" by suitable reheating, to give a product with the desired combination of hardness and toughness. Often the tempering is carried out in such a way as to leave a very hard cutting edge backed up by softer, tougher metal.

The amount of tempering can be estimated roughly by the interference colors of the thin film of oxide formed on a polished surface of the steel during reheating: a straw color (230°C) corresponds to a satisfactory temper for razors, yellow (250°C) for pocket knives, brown (260°C) for scissors and chisels, purple (270°C) for butcher knives, blue (290°C) for watch springs, and blue-black (320°C) for saws.

These processes of hardening and tempering can be understood by consideration of the phases that can be formed by iron and carbon. Carbon is soluble in gamma-iron, the form stable above 912°C. If the steel is quenched from above this temperature there is obtained a solid solution of carbon in gamma-iron. This material, called *martensite,* is very hard and brittle. It confers hardness and brittleness upon hardened high-carbon steel. Martensite is not stable at room temperature, but its rate of conversion to more stable phases is so small at room temperature as to be negligible, and hardened steel containing martensite remains hard as long as it is not reheated.

When hardened steel is tempered by mild reheating, the martensite undergoes transformation to more stable phases. The changes that it undergoes are complex, but result ultimately in a mixture of grains of alpha-iron (ferrite) and the hard carbide Fe_3C, cementite. Steel containing 0.9 % carbon (*eutectoid steel*) changes on tempering into *pearlite,* which is composed of extremely thin alternating layers of ferrite and cementite. Pearlite is strong and tough. Steel containing less than 0.9%

carbon (*hypo-eutectoid steel*) changes on tempering into a microcrystalline metal consisting of grains of ferrite and grains of pearlite, whereas that containing more than 0.9% carbon (*hyper-eutectoid steel*) on tempering yields grains of cementite and grains of pearlite.

Alloy Steels

Many alloy steels, steel containing considerable amounts of metals other than iron, have valuable properties and extensive industrial uses. Manganese steel (12 to 14% Mn) is extraordinarily hard, and crushing and grinding machines and safes are made of it. Nickel steels have many special uses. Chromium-vanadium steel (5 to 10% Cr, 0.15% V) is tough and elastic, and is used for automobile axles, frames, and other parts. Stainless steels usually contain chromium; a common composition is 18% Cr, 8% Ni. Molybdenum and tungsten steels are used for high-speed cutting tools.

19-4. Compounds of Iron

Iron is an active metal, which displaces hydrogen easily from dilute acids. It burns in oxygen to produce ferrous-ferric oxide, Fe_3O_4. This oxide is also made by interaction with superheated steam. One method of preventing rusting involves the production of an adherent surface layer of this oxide on iron.

Iron becomes *passive* when it is dipped in very concentrated nitric acid. It then no longer displaces hydrogen from dilute acids. However, a sharp blow on the metal produces a change that spreads over the surface from the point struck, the metal once more becoming active. This production of passivity is due to the formation of a protective layer of oxide, and the passivity is lost when the layer is broken. Passivity is also produced by other oxidizing agents, such as chromate ion; safety razor blades kept in a solution of potassium chromate remain sharp much longer than blades kept in air.

When exposed to moist air, iron becomes oxidized, forming a loose coating of rust, which is a partially hydrated ferric oxide.

Ferrous Compounds

The ferrous compounds, containing bipositive iron, are usually green in color. Most of the ferrous salts are easily oxidized to the corresponding ferric salts through the action of atmospheric oxygen.

Ferrous sulfate, $FeSO_4 \cdot 7H_2O$, is made by dissolving iron in sulfuric acid, or by allowing pyrite to oxidize in air. The green crystals of the substance are efflorescent, and often have a brown coating of a ferric hydroxide-sulfate, produced by atmospheric oxidation. Ferrous sulfate is used in dyeing and in making ink. To make ink, a solution of tannic acid—a complex organic acid obtained by extraction of nut-galls—is mixed with ferrous sulfate, producing ferrous tannate. On oxidation by the air a fine black insoluble pigment is produced.

Ferric Compounds

The hydrated ferric ion, $Fe(H_2O)_6^{+++}$, is pale violet in color. The ion loses protons readily, however, and ferric salts in solution usually are yellow or brown, because of the formation of hydroxide complexes. **Ferric nitrate,** $Fe(NO_3)_3 \cdot 6H_2O$, exists as pale violet deliquescent crystals. Anhydrous **ferric sulfate,** $Fe_2(SO_4)_3$, is obtained as a white powder by evaporation of a ferric sulfate solution. **Iron alum,** $KFe(SO_4)_2 \cdot 12H_2O$, forms pale violet octahedral crystals.

Ferric ion in solution can be reduced to ferrous ion by treatment with metallic iron or by reduction with hydrogen sulfide or stannous ion.

Ferric hydroxide, $Fe(OH)_3$, is formed as a brown precipitate when alkali is added to a solution of ferric ion. When it is strongly heated ferric hydroxide is converted into **ferric oxide,** Fe_2O_3, which, as a fine powder, is called *rouge* and, as a pigment, *Venetian red.*

Complex Cyanides of Iron

Cyanide ion added to a solution of ferrous or ferric ion forms precipitates, which dissolve in excess cyanide to produce complex ions. Yellow crystals of **potassium ferrocyanide,** $K_4Fe(CN)_6 \cdot 3H_2O$, are made by heating organic material, such as dried blood, with iron filings and potassium carbonate. The mass produced by the heating is extracted with warm water, and the crystals are made by evaporation of the solution. **Potassium ferricyanide,** $K_3Fe(CN)_6$, is made as red crystals by oxidation of ferrocyanide.

The substances contain the coordinated complexes *ferrocyanide ion,* $Fe(CN)_6^{----}$, and *ferricyanide ion,* $Fe(CN)_6^{---}$, respectively, and the ferrocyanides and ferricyanides of other metals are easily made from them.

The pigments *Turnbull's blue* and *Prussian blue* are made by addition of ferrous ion to a ferricyanide solution or ferric ion to a ferrocyanide solution. The pigments which precipitate have the approximate composition $KFeFe(CN)_6 \cdot H_2O$. They have a brilliant blue color. Ferrous ion and ferrocyanide ion produce a white precipitate of $K_2FeFe(CN)_6$, whereas ferric ion and ferricyanide ion form only a brown solution.

19-5. Cobalt

Cobalt occurs in nature in the minerals *smaltite,* $CoAs_2$, and *cobaltite,* CoAsS, usually associated with nickel. The metal is obtained by reducing the oxide with aluminum.

Metallic cobalt is silvery-white, with a slight reddish tinge. It is less reactive than iron, and displaces hydrogen slowly from dilute acids. It is used in special alloys, including *Alnico,* a strongly ferromagnetic alloy of aluminum, nickel, cobalt, and iron, which is used for making permanent magnets.

Cobalt ion, $Co(H_2O)_6^{++}$, in solution and in hydrated salts is red or pink in color. **Cobalt chloride,** $CoCl_2 \cdot 6H_2O$, forms red crystals, which when dehydrated change into a deep blue powder. Writing made with a dilute solution of cobalt chloride is almost invisible, but becomes blue when the paper is warmed, dehydrating the salt. **Cobalt oxide,** CoO, is a black substance which dissolves in molten glass to give it a blue color (*cobalt glass*).

Terpositive cobalt ion is unstable, and an attempt to oxidize Co^{++} usually leads to the precipitation of **cobalt(III) hydroxide,** $Co(OH)_3$. The covalent compounds of cobalt(III) are very stable. The most important of these are **potassium cobaltinitrite,** $K_3Co(NO_2)_6$, and **potassium cobalticyanide,** $K_3Co(CN)_6$.

19-6. Nickel

Nickel occurs, with iron, in meteorites. Its principal ores are *nickelite,* NiAs, *millerite,* NiS, and *pentlandite,* (Ni,Fe)S. The metal is produced as an alloy containing iron and other elements, by roasting the ore and reducing with carbon. In the purification of nickel by the Mond process the compound **nickel tetracarbonyl,** $Ni(CO)_4$, is manufactured and then decomposed. The ore is reduced with hydrogen to metallic nickel under conditions such that the iron oxide is not reduced. Carbon monoxide is then passed through the reduced ore at room temperature; it combines with the nickel to form nickel carbonyl:

$$Ni + 4CO \rightarrow Ni(CO)_4$$

Nickel tetracarbonyl is a gas. It is passed into a decomposer heated to 150°C; the gas decomposes, depositing pure metallic nickel, and the liberated carbon monoxide is returned to be used again.

Nickel is a white metal, with a faint tinge of yellow. It is used in making alloys, including the copper-nickel alloy (75% Cu, 25% Ni) used in coinage. Iron objects are plated with nickel by electrolysis from an ammoniacal solution. The metal is still less reactive than cobalt, and displaces hydrogen only very slowly from acids.

The hydrated salts of nickel such as **nickel sulfate,** $NiSO_4 \cdot 6H_2O$, and **nickel chloride,** $NiCl_2 \cdot 6H_2O$, are green in color. **Nickel(II) hydroxide,** $Ni(OH)_2$, is formed as an apple-green precipitate by addition of alkali to a solution containing nickel ion. When heated it produces the insoluble green substance **nickel(II) oxide,** NiO. Nickel(II) hydroxide is soluble in ammonium hydroxide, forming ammonia complexes such as $Ni(NH_3)_4(H_2O)_2^{++}$ and $Ni(NH_3)_6^{++}$.

In alkaline solution nickel(II) hydroxide can be oxidized to a hydrated **nickel(IV) oxide,** $NiO_2 \cdot xH_2O$. This reaction is used in the *Edison storage cell.* The electrodes of this cell are plates coated with $NiO_2 \cdot xH_2O$ and metallic iron, which are converted on discharge of the cell into nickel(II) hydroxide and ferrous hydroxide, respectively. The electrolyte in this cell is a solution of sodium hydroxide.

19-7. The Platinum Metals

The congeners of iron, cobalt, and nickel are the *platinum metals*—ruthenium, rhodium, palladium, osmium, iridium, and platinum. Some properties of these elements are given in Table 19-3.

The platinum metals are noble metals, chemically unreactive, which are found in nature as native alloys, consisting mainly of platinum.

Ruthenium and **osmium** are iron-gray metals, the other four elements being whiter in color. Ruthenium can be oxidized to RuO_2, and even to the octavalent compound RuO_4. Osmium unites with oxygen to form osmium tetroxide ("osmic acid"), OsO_4, a white crystalline substance melting at 40°C and boiling at about 100°C. Osmium tetroxide has an

TABLE 19-3
Some Physical Properties of the Platinum Metals

	Atomic Number	Atomic Mass	Density (g cm^{-3})	Melting Point	Enthalpy of Sublimation at 25°C
Ru	44	101.07	12.36	2,450°C	670 kJ mole^{-1}
Rh	45	102.905	12.48	1,985°	577
Pd	46	106.4	12.09	1,555°	389
Os	76	190.2	22.69	2,700°	732
Ir	77	192.2	22.82	2,440°	690
Pt	78	195.09	21.60	1,755°	509

irritating odor similar to that of chlorine. It is a very poisonous substance. Its aqueous solution is used in histology (the study of the tissues of plants and animals); it stains tissues through its reduction by organic matter to metallic osmium, and also hardens the material without distorting it.

Ruthenium and osmium form compounds corresponding to various states of oxidation, such as the following: $RuCl_3$, K_2RuO_4, Os_2O_3, $OsCl_4$, K_2OsO_4.

Rhodium and **iridium** are very unreactive metals, not being attacked by aqua regia (a mixture of nitric acid and hydrochloric acid). Iridium is alloyed with platinum to produce a very hard alloy, which is used for the tips of gold pens, surgical tools, and scientific apparatus. Representative compounds are Rh_2O_3, K_3RhCl_6, Ir_2O_3, K_3IrCl_6, and K_2IrCl_6.

Palladium is the only one of the platinum metals that is attacked by nitric acid. Metallic palladium has an unusual ability to absorb hydrogen. At 1,000°C it absorbs enough hydrogen to correspond to the formula $PdH_{0.6}$.

The principal compounds of palladium are the salts of chloropalladous acid, H_2PdCl_4, and chloropalladic acid, H_2PdCl_6. The chloropalladite ion $PdCl_4^{--}$, is a planar ion, consisting of the palladium atom with four coplanar chlorine atoms arranged about it at the corners of a square. The chloropalladate ion, $PdCl_6^{--}$, is an octahedral covalent complex ion.

Platinum is the most important of the palladium and platinum metals. It is grayish-white in color, and is very ductile. It can be welded at a red heat, and melted in an oxyhydrogen flame. Because of its very small chemical activity it is used in electrical apparatus and in making crucibles and other apparatus for use in the laboratory. Platinum is attacked by chlorine and dissolves in a mixture of nitric and hydrochloric acids. It also interacts with fused alkalis, such as potassium hydroxide, but not with alkali carbonates.

The principal compounds of platinum are the salts of chloroplatinous acid, H_2PtCl_4, and chloroplatinic acid, H_2PtCl_6. These salts are similar in structure to the corresponding palladium salts. Both palladium and platinum form many other covalent complexes, such as the platinum(II) ammonia complex ion, $Pt(NH_3)_4^{++}$.

A finely divided form of metallic platinum, called *platinum sponge,* is made by strongly heating ammonium chloroplatinate, $(NH_4)_2PtCl_6$. *Platinum black* is a fine powder of metallic platinum made by adding zinc to chloroplatinic acid. These substances have very strong catalytic activity, and are used as catalysts in commercial processes, such as the oxidation of sulfur dioxide to sulfur trioxide. Platinum black causes the ignition of a mixture of illuminating gas and air or hydrogen and air as a result of the heat developed by the rapid chemical combination of the gases in contact with the surface of the metal.

19-8. The Electronic Structures and Oxidation States of Copper, Silver, and Gold

The three metals copper, silver, and gold compose group Ib of the periodic table. These metals all form compounds representing oxidation state +1, as do the alkali metals, but aside from this they show very little similarity in properties to the alkali metals. The alkali metals are very soft and light, and very reactive chemically, whereas the metals of the copper group are much harder and heavier and are rather inert, sufficiently so to occur in the free state in nature and to be easily obtainable by reducing their compounds, sometimes simply by heating.

The electronic structures of copper, silver, and gold, as well as those of zinc and gallium and their congeners, are given in Table 19-4.

It is seen that copper has one outer electron, in the 4*s* orbital of the *N* shell, zinc has two outer electrons, in the 4*s* orbital, and gallium has three outer electrons, two in the 4*s* orbital and one in the 4*p* orbital. The congeners of these elements also have one, two, or three electrons in the outermost shell. The shell next to the outermost shell in each case contains 18 electrons; this is the *M* shell for copper, zinc, and gallium, the *N* shell for silver, cadmium, and indium, and the *O* shell for gold, mercury, and thallium. This shell is called an *eighteen-electron shell.*

The electrons in the outermost shell are held loosely, and can be easily removed. The resulting ions, Cu^+, Zn^{++}, Ga^{+++}, etc., have an outer shell of eighteen electrons, and are called *eighteen-shell ions.* If these elements either lose their outermost electrons, forming eighteen-shell ions, or share the outermost electrons with other atoms, the resulting oxidation state is +1 for copper, silver, and gold, +2 for zinc, cadmium, and mercury, and +3 for gallium, indium, and thallium.

TABLE 19-4
Electronic Structures of Copper, Zinc, and Gallium and Their Congeners

Atomic Number	Element	*K*	*L*		*M*			*N*				*O*			*P*	
		1*s*	2*s*	2*p*	3*s*	3*p*	3*d*	4*s*	4*p*	4*d*	4*f*	5*s*	5*p*	5*d*	6*s*	6*p*
29	Cu	2	2	6	2	6	10	1								
30	Zn	2	2	6	2	6	10	2								
31	Ga	2	2	6	2	6	10	2	1							
47	Ag	2	2	6	2	6	10	2	6	10		1				
48	Cd	2	2	6	2	6	10	2	6	10		2				
49	In	2	2	6	2	6	10	2	6	10		2	1			
79	Au	2	2	6	2	6	10	2	6	10	14	2	6	10	1	
80	Hg	2	2	6	2	6	10	2	6	10	14	2	6	10	2	
81	Tl	2	2	6	2	6	10	2	6	10	14	2	6	10	2	1

These are important oxidation states for all of these elements; there are, however, also some other important oxidation states. The cuprous ion, Cu^+, is unstable, and the cuprous compounds, except the very insoluble ones, are easily oxidized. The cupric ion, Cu^{++} (hydrated to $Cu(H_2O)_4^{++}$), occurs in many copper salts, and the cupric compounds are the principal compounds of copper. In the cupric ion the copper atom has lost two electrons, leaving it with only seventeen electrons in the *M* shell. In fact, the $3d$ electrons and the $4s$ electrons in copper are held by the atom with about the same energy—you may have noticed that the electronic structure given in Table 19-4 for copper differs from that given in the energy-level diagram, Figure 5-6, in that in the diagram copper is represented as having two $4s$ electrons and only nine $3d$ electrons.

The unipositive silver ion, Ag^+, is stable, and forms many salts. A very few compounds have also been made containing bipositive and terpositive silver. These compounds are very strong oxidizing agents. The stable oxidation state +1 shown by silver corresponds to the electronic structure of the element as given in Table 19-4. The Ag^+ ion is an eighteen-shell ion.

The gold(I) ion, Au^+, and the gold(III) ion, Au^{+++}, are unstable in aqueous solution. The stable gold(I) compounds and gold(III) compounds contain covalent bonds, as in the complex ions $AuCl_2^-$ and $AuCl_4^-$.

The chemistry of zinc and cadmium is especially simple, in that these elements form compounds representing only the oxidation state +2. This oxidation state is closely correlated with the electronic structures shown in Table 19-4; it represents the loss or the sharing of the two outermost electrons. The ions Zn^{++} and Cd^{++} are eighteen-shell ions.

Mercury also forms compounds (the mercuric compounds) representing the oxidation state +2. The mercuric ion, Hg^{++}, is an eighteen-shell ion. In addition, mercury forms a series of compounds, the mercurous compounds, in which it has oxidation number +1. The electronic structure of the mercurous compounds is discussed in Section 19-18.

19-9. The Properties of Copper, Silver, and Gold

The metallurgy of copper has been discussed in Chapter 11. Copper is a red, tough metal with a moderately high melting point (Table 19-5). It is an excellent conductor of heat and of electricity when pure, and it finds extensive use as an electric conductor. Pure copper that has been heated is soft, and can be drawn into wire or shaped by hammering. This "cold work" (of drawing or hammering) causes the metal to become hard, because the crystal grains are broken into much smaller grains, with grain

TABLE 19-5
Some Physical Properties of Copper, Silver, and Gold

	Atomic Number	Atomic Mass	Density (g cm^{-3})	Melting Point	Boiling Point	Metallic Radius	Color
Copper	29	63.54	8.97	1083°C	2,310°C	128 pm	Red
Silver	47	107.870	10.54	960.5°	1,950°	144	White
Gold	79	196.967	19.42	1063°	2,600°	144	Yellow

boundaries that interfere with the process of deformation and thus strengthen the metal. The hardened metal can be made soft by heating ("annealing"), which permits the grains to coalesce into large grains.

Silver is a soft, white metal, somewhat denser than copper, and with a lower melting point. It is used in coinage, jewelry, and tableware, and as a filling for teeth.

Gold is a soft, very dense metal, which is used for jewelry, coinage, dental work, and scientific and technical apparatus. Gold is bright yellow by reflected light; very thin sheets are blue or green. Its beautiful color and fine luster, which, because of its inertness, are not affected by exposure to the atmosphere, are responsible for its use for ornamental purposes. Gold is the most malleable and most ductile of all metals; it can be hammered into sheets only 100 nm thick, and drawn into wires 2 μm in diameter.

Alloys of Copper, Silver, and Gold

The transition metals find their greatest use in alloys. Alloys are often far stronger, harder, and tougher than their constituent elementary metals. The alloys of copper and zinc are called *brass,* those of copper and tin are called *bronze,* and those of copper and aluminum are called *aluminum bronze.* Many of these alloys have valuable properties. Copper is a constituent also of other useful alloys, such as beryllium copper, coinage silver, and coinage gold.

Coinage silver in the United States contains 90% silver and 10% copper. This composition also constitutes *sterling silver* in the United States. British sterling silver is 92.5% silver and 7.5% copper.

Gold is often alloyed with copper, silver, palladium, or other metals. The amount of gold in these alloys is usually described in *carats,* the number of parts of gold in 24 parts of alloy—pure gold is 24 carat. American coinage gold is 21.6 carat and British coinage gold is 22 carat. *White gold,* used in jewelry, is usually a white alloy of gold and nickel.

19-10. The Compounds of Copper

With the more electronegative nonmetals copper tends to form compounds of copper(II) (cupric compounds). For example, the heat of reaction of cuprous chloride with chlorine to form cupric chloride is positive:

$$CuCl(c) + \tfrac{1}{2}Cl_2(g) \longrightarrow CuCl_2(c) + 71 \text{ kJ mole}^{-1}$$

With sulfur and iodine, in which the bonds have little ionic character (electronegativity of copper, 1.9; of sulfur and iodine, 2.5), the cuprous compounds are the more stable.

Cupric Compounds

The hydrated **cupric ion,** $Cu(H_2O)_4^{++}$, is an ion with light blue color that occurs in aqueous solutions of cupric salts and in some of the hydrated crystals. The most important cupric salt is **copper sulfate,** which forms blue crystals, $CuSO_4 \cdot 5H_2O$. The metal copper is not sufficiently reactive to displace hydrogen ion from dilute acids (it is below hydrogen in the electromotive-force series, Chapter 11), and copper does not dissolve in acids unless an oxidizing agent is present. However, hot concentrated sulfuric acid is itself an oxidizing agent, and can dissolve the metal, and dilute sulfuric acid also slowly dissolves it in the presence of air:

$$Cu + 2H_2SO_4 + 3H_2O \longrightarrow CuSO_4 \cdot 5H_2O + SO_2$$

or

$$2Cu + 2H_2SO_4 + O_2 + 8H_2O \longrightarrow 2CuSO_4 \cdot 5H_2O$$

Copper sulfate, which has the common names *blue vitriol* and *bluestone,* is used in copper plating, in printing calico, in electric cells, and in the manufacture of other compounds of copper.

Cupric chloride, $CuCl_2$, can be made as yellow crystals by direct union of the elements. The hydrated salt, $CuCl_2 \cdot 2H_2O$, is blue-green in color, and its solution in hydrochloric acid is green. The blue-green color of the salt is due to its existence as a complex,

$$\begin{array}{c} OH_2 \\ | \\ Cl—Cu—Cl \\ | \\ OH_2 \end{array}$$

in which the chlorine atoms are bonded directly to the copper atom. The green solution contains ions $CuCl_3(H_2O)^-$ and $CuCl_4^{--}$. All of these ions are planar, the copper atom being at the center of a square formed by the four attached groups. The planar configuration is shown also by other complexes of copper, including the deep-blue ammonia complex, $Cu(NH_3)_4^{++}$.

Cuprous Compounds

Cuprous ion, Cu^+, is so unstable in aqueous solution that it undergoes auto-oxidation-reduction into copper and cupric ion:

$$2Cu^+ \rightarrow Cu + Cu^{++}$$

Very few cuprous salts of oxygen acids exist. The stable cuprous compounds are either insoluble crystals containing covalent bonds or covalent complexes.

When copper is added to a solution of cupric chloride in stong hydrochloric acid a reaction occurs that results in the formation of a colorless solution containing cuprous chloride complex ions such as $CuCl_2^-$:

$$CuCl_4^{--} + Cu \rightarrow 2CuCl_2^-$$

This complex ion involves two covalent bonds, its electronic structure being

$$\left[:\ddot{\underset{..}{Cl}}—Cu—\ddot{\underset{..}{Cl}}:\right]^-$$

Other cuprous complexes, $CuCl_3^{--}$ and $CuCl_4^{---}$, also exist.

If the solution is diluted with water a colorless precipitate of **cuprous chloride,** CuCl, forms. This precipitate also contains covalent bonds, each copper atom being bonded to four neighboring chlorine atoms and each chlorine atom to four neighboring copper atoms, with use of the outer electrons of the chloride ion. The structure is closely related to that of diamond, with alternating carbon atoms replaced by copper and chlorine (Figure 7-1).

19-11. The Compounds of Silver

Silver oxide, Ag_2O, is obtained as a dark-brown precipitate on the addition of sodium hydroxide to a solution of silver nitrate. It is slightly soluble, producing a weakly alkaline solution of silver hydroxide:

$$Ag_2O + H_2O \rightarrow 2Ag^+ + 2OH^-$$

Silver oxide is used in inorganic chemistry to convert a soluble chloride, bromide, or iodide into the hydroxide. For example, cesium chloride solution can be converted into cesium hydroxide solution in this way:

$$2Cs^+ + 2Cl^- + Ag_2O + H_2O \rightarrow 2AgCl + 2Cs^+ + 2OH^-$$

This reaction proceeds to the right because silver chloride is much less soluble than silver oxide.

The **silver halides**—AgF, AgCl, AgBr, and AgI—can be made by adding silver oxide to solutions of the corresponding halogen acids. Silver fluoride is very soluble in water, and the other halogenides are nearly insoluble. Silver chloride, bromide, and iodide form as curdy precipitates when the ions are mixed. They are respectively white, pale yellow, and yellow in color, and on exposure to light they slowly turn black, through photochemical decomposition. Silver chloride and bromide dissolve in ammonium hydroxide solution, forming the **silver ammonia complex** $Ag(NH_3)_2^+$ (Chapter 16); silver iodide does not dissolve in ammonium hydroxide. These reactions are used as qualitative tests for silver ion and the halide ions.

Other complex ions formed by silver, such as the silver cyanide complex $Ag(CN)_2^-$ and the silver thiosulfate complex $Ag(S_2O_3)_2^{---}$, have been mentioned in Chapter 16.

Silver nitrate, $AgNO_3$, is a colorless, soluble salt made by dissolving silver in nitric acid. It is used to cauterize sores. Silver nitrate is easily reduced to metallic silver by organic matter, such as skin or cloth, and is for this reason used in making indelible ink.

Silver ion is an excellent antiseptic, and several of the compounds of silver are used in medicine because of their germicidal power.

19-12. Ice Nucleation and Rain Making

Some liquids, such as water and glycerol, can remain as liquids when they are cooled below the normal freezing point, the temperature at which the liquid and the crystalline substance are in equilibrium. Crystallization begins with the formation of a very small crystal, a nucleus, which then grows. Very small crystals, however, have a higher free energy than large crystals, because of the instability of their surfaces; this surface energy causes the vapor pressure of small crystals to be greater than that of large crystals, and hence causes the melting point to be lower. Freezing occurs rapidly when small crystals of the substance are added as "seeds" to the supercooled liquid.

Solutions, such as of sodium acetate in water or of water vapor in air, can also be supercooled. Rain or snow sometimes do not fall from air that is supersaturated with water vapor because of the absence of small droplets of water or small ice crystals that can serve as seeds.

Rain or snow might be made to form by seeding with crushed ice. It is hard, however, to crush ice to a powder, and hence no large number of seeds can be distributed. It was discovered by the American scientist Irving Langmuir that minute crystals of silver iodide, formed by condensation of silver iodide vapor, can serve as seeds for ice crystals. This discovery is the basis of the silver iodide method of making rain or snow by seeding supersaturated parts of the atmosphere.

A silver iodide crystal has a surface pattern that closely resembles that of an ice crystal; it was this resemblance that led Langmuir to try silver iodide as a seeding material. The silver atoms and iodine atoms occupy the positions of alternate oxygen atoms in the ice structure (Figure 9-8), and the silver-iodine distance, 280 pm, is only 1.5 percent greater than the oxygen-oxygen distance in ice.

The covalent radius of copper is 16 pm less than that of silver (see the metallic radii, Table 19-5). Accordingly we would expect that a crystalline solution of cuprous iodide in silver iodide with copper and silver atoms in ratio 1 to 3 would have the same average bond length as the oxygen-oxygen distance in ice, and would be the most effective seeding material. Experiment has in fact shown that the maximum seeding effect occurs at this composition, verifying the hypothesis that the crystals serve as nuclei about which the ice crystals grow.

19-13. Photochemistry and Photography

Many chemical reactions are caused to proceed by the effect of light. For example, a dyed cloth may fade when exposed to sunlight because of the destruction of molecules of the dye under the influence of the sunlight. Reactions of this sort are called *photochemical reactions.* A very important photochemical reaction is the conversion of carbon dioxide and water into carbohydrate and oxygen in the leaves of plants, where the green substance chlorophyll serves as a catalyst.

One law of photochemistry, discovered by Grotthus in 1818, is that *only light that is absorbed is photochemically effective.* Hence a colored substance must be present in a system that shows photochemical reactivity with visible light. In the process of natural photosynthesis this substance is chlorophyll.

The second law of photochemistry, formulated in 1912 by Einstein, is that *one molecule of reacting substance may be activated and caused to react by the absorption of one photon.* In some systems, such as material containing rather stable dyes, many photons are absorbed by the molecules for each molecule that is decomposed; the fading of the dye by light is a slow and inefficient process in these materials. In some simple systems the absorption of one photon results in the reaction or decomposition of one molecule.

There are also chemical systems in which a *chain of reactions* may be set off by one light quantum. An example is the photochemical reaction of hydrogen and chlorine. A mixture of hydrogen and chlorine kept in the dark does not react at room temperature. When, however, it is illuminated with blue light, reaction immediately begins. Hydrogen is transparent to all visible light; chlorine, which owes its yellow-green color to its strong absorption of blue light, is the photochemically active constituent in the mixture. The absorption of a photon of blue light by a chlorine molecule splits the molecule into two chlorine atoms:

$$Cl_2 + h\nu \longrightarrow 2Cl$$

These chlorine atoms initiate a chain of reactions, as described in Section 10-6:

$$Cl + H_2 \longrightarrow HCl + H$$

$$H + Cl_2 \longrightarrow HCl + Cl$$

It may be observed that the mixture of hydrogen and chlorine explodes when exposed to blue light. The chain of reactions may be broken through the recombination of chlorine atoms to form chlorine molecules; this reaction occurs on the collision of two chlorine atoms with the wall of the vessel containing the gas or with another atom or molecule in the gas.

A photochemical reaction of much geophysical and biological importance is the formation of ozone from oxygen. Oxygen is practically transparent to visible light and to light in the near ultraviolet region, but it strongly absorbs light in the far ultraviolet region—in the region from 160 nm to 240 nm. Each photon that is absorbed dissociates an oxygen molecule into two oxygen atoms:

$$O_2 + h\nu \longrightarrow 2O$$

A reaction that does not require absorption of a photon then follows:

$$O + O_2 \longrightarrow O_3$$

Accordingly there are produced two molecules of ozone, O_3, for each photon absorbed. In addition, however, the ozone molecules can be destroyed by combining with oxygen atoms, or by a photochemical reaction. The reaction of combining with an oxygen atom is

$$O + O_3 \longrightarrow 2O_2$$

The reactions of photochemical production of ozone and destruction of ozone lead to a photochemical equilibrium, which maintains a small concentration of ozone in the oxygen being irradiated. The layer of the atmosphere in which the major part of the ozone is present is about 15 miles above the earth's surface; it is called the *ozone layer.*

The geophysical and biological importance of the ozone layer results from the absorption of light in the near ultraviolet region, from 240 nm to 360 nm, by the ozone. The photochemical reaction is

$$O_3 + h\nu \longrightarrow O + O_2$$

This reaction permits ozone to absorb ultraviolet light so strongly as to remove practically all of the ultraviolet light from the sunlight before it reaches the earth's surface. The ultraviolet light that it absorbs is photochemically destructive toward many of the organic molecules necessary in life processes, and if the ultraviolet light of sunlight were not prevented by the ozone layer from reaching the surface of the earth life in its present form could not exist.

Photography

A photographic film is a sheet of cellulose acetate coated with a thin layer of gelatin in which very fine grains of silver bromide are suspended. This layer of gelatin and silver bromide is called the *photographic emulsion.* The silver halides are sensitive to light, and undergo photochemical decomposition. The gelatin increases this sensitivity, apparently because of the sulfur which it contains.

When the film is briefly exposed to light some of the grains of silver bromide undergo a small amount of decomposition, perhaps forming a small particle of silver sulfide on the surface of the grain. The film can then be *developed* by treatment with an alkaline solution of an organic reducing agent, such as Metol or hydroquinone, the *developer.* This causes the silver bromide grains that have been sensitized to be reduced to metallic silver, whereas the unsensitized silver bromide grains remain unchanged. By this process the developed film reproduces the pattern of the light that exposed it. This film is called the *negative,* because it is darkest (with the greatest amount of silver) in the places that were exposed to the most light.

The undeveloped grains of silver halogenide are next removed, by treatment with a fixing bath, which contains thiosulfate ion, $S_2O_3^{--}$ (from sodium thiosulfate, "hypo," $Na_2S_2O_3 \cdot 5H_2O$). The soluble silver thiosulfate complex is formed:

$$AgBr + 2S_2O_3^{--} \longrightarrow Ag(S_2O_3)_2^{---} + Br^-$$

The fixed negative is then washed. Care must be taken not to transfer the negative from a used fixing bath, containing a considerable concentration of silver complex, directly to the wash water, as insoluble silver thiosulfate might precipitate in the emulsion:

$$2Ag(S_2O_3)_2^{---} \rightarrow Ag_2S_2O_3(c) + 3S_2O_3^{--}$$

Since there are three ions on the right, and only two on the left, dilution causes the equilibrium to shift toward the right.

A positive print can be made by exposing print paper, coated with a silver halide emulsion, to light that passes through the superimposed negative, and then developing and fixing the exposed paper.

The Chemistry of Color Photography

The electromagnetic waves of light of different colors have different wavelengths. In the visible spectrum these wavelengths extend from a little below 400 nm (violet in color) to nearly 800 nm (red in color). The sequence of colors in the visible region is shown in the diagram next to the top of Figure 19-6.

The visible spectrum is only a very small part of the complete spectrum of the electromagnetic waves. At the top of Figure 19-6 other parts are indicated. Ordinary x-rays have wavelengths approximately 100 pm. Even shorter wavelengths are possessed by the gamma rays that are produced in radioactive decompositions and through the action of cosmic rays. The ultraviolet region, not visible to the eye, consists of light somewhat shorter in wavelength than violet light, and the infrared consists of wavelengths somewhat longer than red. Then there come the microwave regions, approximately 1 cm, and the longer radiowaves.

When gases are heated or are excited by the passage of an electric spark, the atoms and molecules in the gases emit light of definite wavelengths. The light that is emitted by an atom or molecule under these conditions is said to constitute its *emission spectrum*. The emission spectra of the alkali metals, mercury, and neon are shown in Figure 19-6. The emission spectra of elements, especially of the metals, can be used for identifying them; *spectroscopic chemical analysis* is an important technique of analytical chemistry.

When white light (light containing all wavelengths in the visible region) is passed through a substance, light of certain wavelengths may be absorbed by the substance. The solar spectrum is shown in Figure 19-6. It consists of a background of white light, produced by the very hot gases in the sun, on which there are superimposed some dark lines, resulting from absorption of certain wavelengths by atoms in the cooler surface

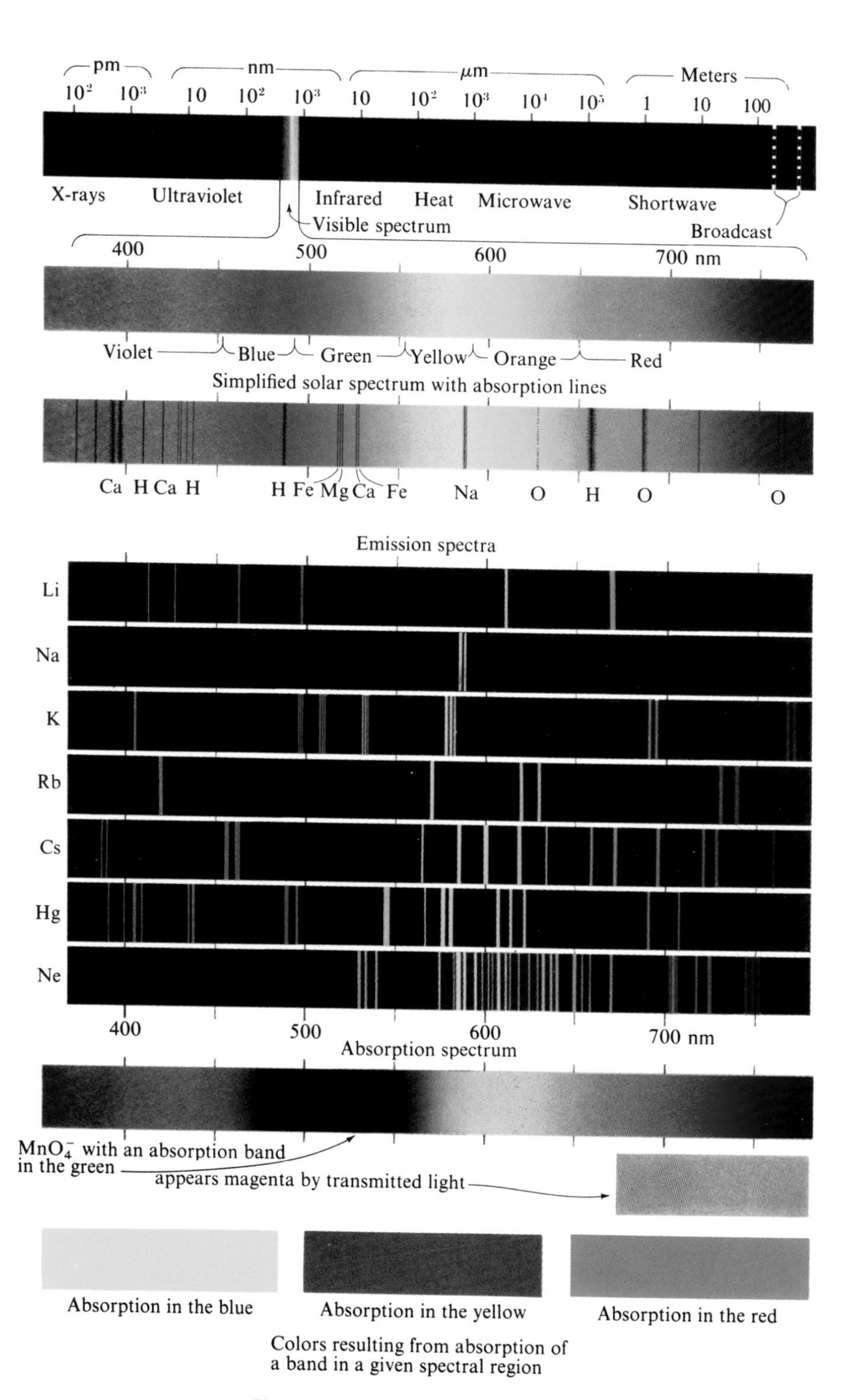

FIGURE 19-6
Emission spectra and absorption spectra.

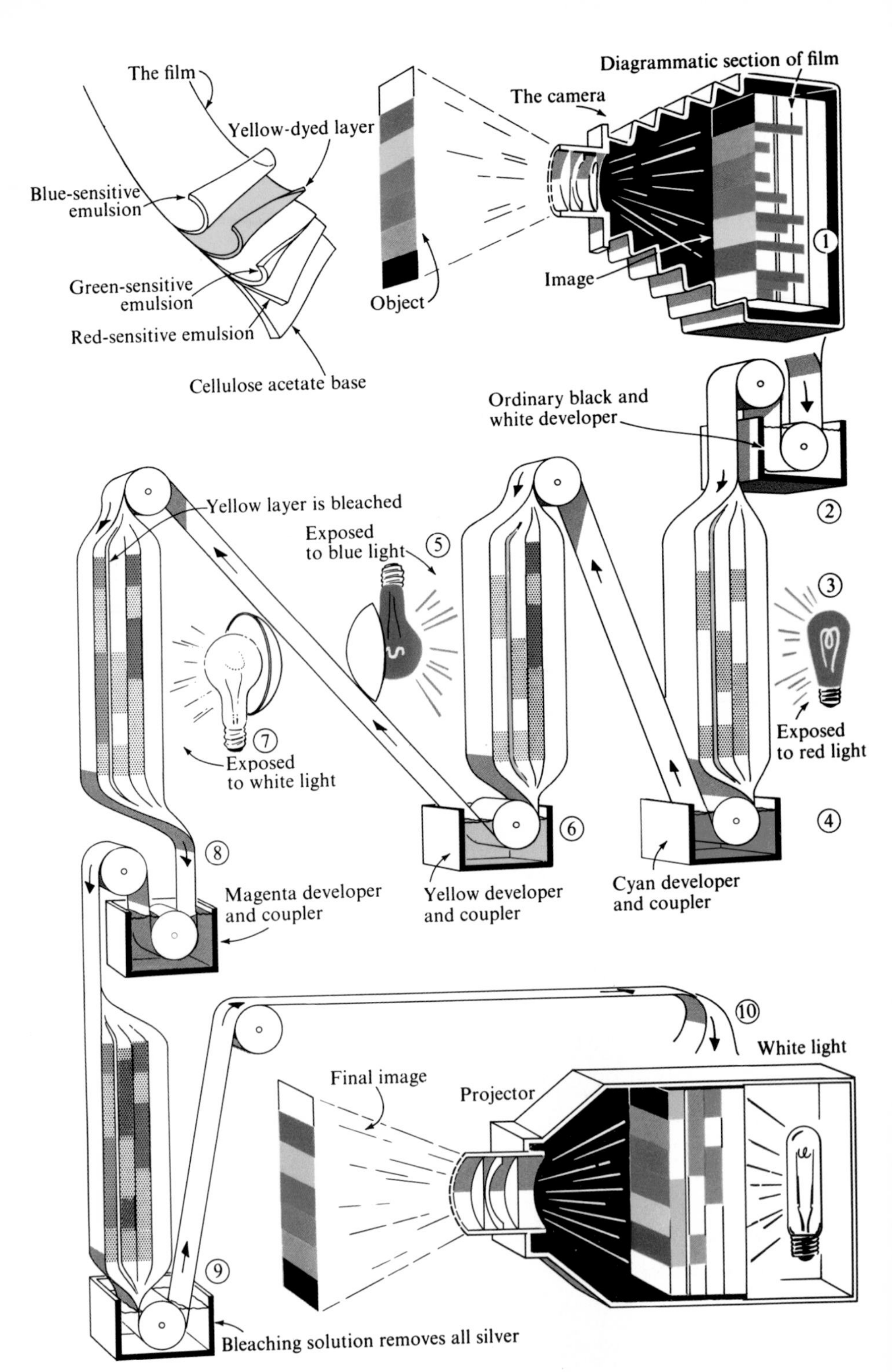

FIGURE 19-7
The Kodachrome process of color photography.

layers of the sun. It is seen that the yellow sodium lines, which occur as bright lines in the emission spectrum of sodium atoms, are shown as dark lines in the solar spectrum.

Molecules and complex ions in solution and in solid substances sometimes show sharp line spectra, but usually show rather broad absorption bands, as is indicated for the permanganate ion near the bottom of Figure 19-6. The permanganate ion has the power of absorbing light in the green region of the spectrum, permitting the blue-violet light and red light to pass through. The combination of blue-violet and red light appears magenta in color. We accordingly say that permanganate ion has a magenta color.

The human eye does not have the power of completely differentiating between light of one wavelength and that of another wavelength in the visible spectrum. Instead, it responds to three different wavelength regions in different ways. All of the colors that can be recognized by the eye can be composed from three fundamental colors. These may be taken as red-green (seen by the eye as yellow), which is complementary to blue-violet; blue-red, or magenta, which is complementary to green; and blue-green, or cyan, which is complementary to red. Three *primary colors,* such as these, need to be used in the development of any method of color photography.

An important modern method of color photography is the *Kodachrome method,* developed by the Kodak Research Laboratories. This method is illustrated in Figure 19-7. The film consists of several layers of emulsion, superimposed on a cellulose acetate base. The uppermost layer of photographic emulsion is the ordinary photographic emulsion, which is sensitive to blue and violet light. The second layer of photographic emulsion is a green-sensitive emulsion. It consists of a photographic emulsion that has been treated with a magenta-colored dye, which absorbs green light and sensitizes the silver bromide grains, thus making the emulsion sensitive to green light as well as to blue and violet light. The third photographic emulsion, red-sensitive emulsion, has been treated with a blue dye, which absorbs red light, making the emulsion sensitive to red light as well as to blue and violet (but not to green). Between the first layer and the middle layer there is a layer of yellow filter, containing a yellow dye, which during exposure prevents blue and violet light from penetrating to the lower layers. Accordingly when such a film is exposed to light the blue-sensitive emulsion is exposed by blue light, the middle emulsion is exposed by green light, and the bottom emulsion is exposed by red light.

The exposure of the different layers of photographic emulsion in the film is illustrated diagrammatically as Process 1 in Figure 19-7.

The development of Kodachrome film involves several steps, which are represented as Processes 2 to 9. First (Process 2) the Kodachrome film after exposure is developed with an ordinary black and white developer, which develops the silver negative in all three emulsions. Then, after simple washing in water (not shown in the figure) the film is exposed through the back to red light, which makes the previously unexposed silver bromide in the red-sensitive emulsion capable of development (Process 3). The film then passes into a special developer, called cyan developer and coupler (Process 4). This mixture of chemical substances has the power of interacting with the exposed silver bromide grains in such a way as to deposit a cyan dye in the bottom layer, at the same time that the silver bromide grains are reduced to metallic silver. The cyan dye is deposited only in the regions occupied by the sensitized silver bromide grains. The next process (Process 5) consists in exposure to blue light from the front of the negative. The blue light is absorbed by the yellow dye, and so affects only the previously unexposed grains in the first emulsion, the blue-sensitive emulsion. This emulsion is then developed in a special developer (Process 6), a yellow developer and coupler, which deposits a yellow dye in the neighborhood of these recently exposed grains. The film is then exposed to white light, to sensitize the undeveloped silver bromide grains in the middle emulsion, the yellow layer is bleached, the middle emulsion is developed with a magenta developer and coupler (Process 8), and the deposited metallic silver in all three solutions is removed by a bleaching solution (Process 9), leaving only a film containing deposited cyan, yellow, and magenta dyes in the three emulsion layers, in such a way that by transmitted light the originally incident colors are reproduced (Process 10).

19-14. The Compounds of Gold

$KAu(CN)_2$, the potassium salt of the complex **gold(I) cyanide ion** $Au(CN)_2^-$, with electronic structure

$$[:N\equiv C-Au-C\equiv N:]^-$$

is an example of a gold(I) compound.* The **gold(I) chloride** complex $AuCl_2^-$ has a similar structure, and the **halides,** AuCl, AuBr, and AuI, resemble the corresponding halides of silver.

Gold dissolves in a mixture of concentrated nitric and hydrochloric

*The gold(I) and gold(III) compounds are often called *aurous* and *auric* compounds, respectively.

acids to form **hydrogen aurichloride,** $HAuCl_4$. This acid contains the aurichloride ion, $AuCl_4^-$, a square planar complex ion:

$$\left[\begin{array}{ccc} & :\ddot{\text{Cl}}: & \\ & | & \\ :\ddot{\text{Cl}}- & \text{Au} & -\ddot{\text{Cl}}: \\ & | & \\ & :\ddot{\text{Cl}}: & \end{array}\right]^-$$

Hydrogen aurichloride can be obtained as a yellow crystalline substance, which forms salts with bases. When heated it forms **gold(III) chloride,** $AuCl_3$, and then gold(I) gold(III) chloride, Au_2Cl_4, and then gold(I) chloride, AuCl. On further heating all the chlorine is lost, and pure gold remains.

19-15. Color and Mixed Oxidation States

The gold halides provide examples of an interesting phenomenon—the *deep, intense color often observed for a substance that contains an element in two different oxidation states.* Gold(I) gold(III) chloride, Au_2Cl_4, is intensely black, although both gold(I) chloride and gold(III) chloride are yellow. Cesium gold(I) gold(III) bromide, $Cs_2^+[AuBr_2]^-[AuBr_4]^-$, is deep black in color and both $CsAuBr_2$ and $CsAuBr_4$ are much lighter. Black mica (biotite) and black tourmaline contain both ferrous and ferric iron. Prussian blue is ferrous ferricyanide; ferrous ferrocyanide is white, and ferric ferricyanide is light yellow. When copper is added to a light green solution of cupric chloride, a deep brownish-black solution is formed, before complete conversion to the colorless cuprous chloride complex.

The theory of this phenomenon is not understood. The very strong absorption of light is presumably connected with the transfer of an electron from one atom to another of the element present in two valence states.

19-16. The Properties and Uses of Zinc, Cadmium, and Mercury

Zinc is a bluish-white, moderately hard metal. It is brittle at room temperature, but is malleable and ductile between 100° and 150°C, and becomes brittle again above 150°C. It is an active metal, above hydrogen in the electromotive-force series, and it displaces hydrogen even from dilute acids. In moist air zinc is oxidized and becomes coated with a tough film of basic zinc carbonate, $Zn_2CO_3(OH)_2$, which protects it from further corrosion. This behavior is responsible for its principal use, in protecting iron from rusting. Iron wire or sheet iron is *galvanized* by cleaning with sulfuric acid or a sandblast, and then dipping in molten zinc; a thin layer of zinc adheres to the iron. Galvanized iron in some shapes is made by electroplating zinc onto the iron pieces.

Zinc is also used in making alloys, the most important of which is *brass* (the alloy with copper), and as a reacting electrode in dry cells and wet cells.

Cadmium is a bluish-white metal of pleasing appearance. It has found increasing use as a protective coating for iron and steel. The cadmium plate is deposited electrolytically from a bath containing the cadmium cyanide complex ion, $Cd(CN)_4^{--}$. Cadmium is also used in some alloys, such as the low-melting alloys needed for automatic fire extinguishers. *Wood's metal,* which melts at 65.5°C, contains 50% Bi, 25% Pb, 12.5% Sn, and 12.5% Cd. Because of the toxicity of compounds of elements of this group, care must be taken not to use cadmium-plated vessels for cooking, and not to inhale fumes of zinc, cadmium, or mercury.

Mercury is the only metal that is liquid at room temperature (cesium melts at 28.5°C, and gallium at 29.8°C). It is unreactive, being below hydrogen in the electromotive-force series. Because of its unreactivity, fluidity, high density, and high electric conductivity it finds extensive use in thermometers, barometers, and many special kinds of scientific apparatus.

The alloys of mercury are called *amalgams*. Amalgams of silver, gold, and tin are used in dentistry. Mercury does not wet iron, and it is usually shipped and stored in iron bottles, called flasks, which hold 76 lbs of the metal.

The low melting points and small values of the heats of sublimation of zinc and its congeners (Table 19-6) are attributed to the fact that the gas

TABLE 19-6
Some Physical Properties of Zinc, Cadmium, and Mercury

	Atomic Number	Atomic Mass	Density ($g\ cm^{-3}$)	Melting Point	Boiling Point	Metallic Radius	Color	Enthalpy of Sublimation at 25°C
Zinc	30	65.37	7.14	419.4°C	907°C	138 pm	Bluish-white	131 kJ mole^{-1}
Cadmium	48	112.40	8.64	320.9°	767°	154	Bluish-white	113
Mercury	80	200.59	13.55	−38.89°	356.9°	157	Silvery-white	61

atoms in the normal state contain only completed subshells of electrons and hence have no unpaired electrons that can be used to form chemical bonds. The first excited state of the zinc atom is less stable than the normal state by 385 kJ $mole^{-1}$. The zinc atom in this excited state has two unpaired electrons ($4s4p$), corresponding to bivalence.

19-17. Compounds of Zinc and Cadmium

The *zinc ion,* $Zn(H_2O)_4^{++}$, is a colorless ion formed by solution of zinc in acid. It is poisonous to bacteria, and is used as a disinfectant. It is one of the metal ions required in small amounts by human beings (as well as animals) for good health. The normal intake by an adult, obtained from meat and, in smaller amounts, from fruits and vegetables, is 10 to 15 mg per day. The zinc ion is a coenzyme for a number of enzymes, including human carbonic anhydrase (in erythrocytes) and human alcohol dehydrogenase (in the liver).

Zinc ion forms tetraligated complexes readily, such as $Zn(NH_3)_4^{++}$, $Zn(CN)_4^{--}$, and $Zn(OH)_4^{--}$. The white precipitate of **zinc hydroxide,** $Zn(OH)_2$, that forms when ammonium hydroxide is added to a solution containing zinc ion dissolves in excess ammonium hydroxide, forming the zinc ammonia complex. The zinc hydroxide complex, $Zn(OH)_4^{--}$, which is called **zincate ion,** is similarly formed on solution of zinc hydroxide in an excess of strong base; zinc hydroxide is amphiprotic.

Zinc sulfate, $ZnSO_4 \cdot 7H_2O$, is used as a disinfectant and in dyeing calico, and in making *lithopone,* which is a mixture of barium sulfate and zinc sulfide used as a white pigment in paints:

$$Ba^{++}S^{--} + Zn^{++}SO_4^{--} \longrightarrow BaSO_4 + ZnS$$

Zinc oxide, ZnO, is a white powder (yellow when hot) made by burning zinc vapor or by roasting zinc ores. It is used as a pigment (zinc white), as a filler in automobile tires, adhesive tape, and other articles, and as an antiseptic (zinc oxide ointment).

Zinc sulfide, ZnS, is the only white sulfide among the sulfides of the common transition metals.

The compounds of cadmium are closely similar to those of zinc. **Cadmium ion,** Cd^{++}, is a colorless ion, which forms complexes ($Cd(NH_3)_4^{++}$, $Cd(CN)_4^{--}$) similar to those of zinc. The cadmium hydroxide ion, $Cd(OH)_4^{--}$, is not stable, and **cadmium hydroxide,** $Cd(OH)_2$, is formed as a white precipitate by addition even of concentrated sodium hydroxide to a solution containing cadmium ion. The precipitate is soluble in ammonium hydroxide or in a solution containing cyanide ion. **Cadmium**

oxide, CdO, is a brown powder obtained by heating the hydroxide or burning the metal. **Cadmium sulfide,** CdS, is a bright yellow precipitate obtained by passing hydrogen sulfide through a solution containing cadmium ion; it is used as a pigment (*cadmium yellow*).

19-18. Compounds of Mercury

The mercuric compounds, in which mercury is bipositive, differ somewhat in their properties from the corresponding compounds of zinc and cadmium. The differences are due in part to the very strong tendency of the mercuric ion Hg^{++}, to form covalent bonds. Thus the covalent crystal **mercuric sulfide,** HgS, is far less soluble than cadmium sulfide or zinc sulfide.

Mercuric nitrate, $Hg(NO_3)_2$ or $Hg(NO_3)_2 \cdot \frac{1}{2}H_2O$, is made by dissolving mercury in hot concentrated nitric acid:

$$Hg + 4HNO_3 \rightarrow Hg(NO_3)_2 + 2NO_2 + 2H_2O$$

It hydrolyzes on dilution, unless a sufficient excess of acid is present, to form basic mercuric nitrates, such as $HgNO_3OH$, as a white precipitate.

Mercuric chloride, $HgCl_2$, is a white crystalline substance usually made by dissolving mercury in hot concentrated sulfuric acid, and then heating the dry mercuric sulfate with sodium chloride, subliming the volatile mercuric chloride:

$$Hg + 2H_2SO_4 \rightarrow HgSO_4 + SO_2 + 2H_2O$$

$$HgSO_4 + 2NaCl \rightarrow Na_2SO_4 + HgCl_2$$

A dilute solution of mercuric chloride (about 0.1%) is used as a disinfectant. Any somewhat soluble mercuric salt would serve equally well, except for the tendency of mercuric ion to hydrolyze and to precipitate basic salts. Mercuric chloride has only a small tendency to hydrolyze because its solution contains only a small concentration of mercuric ion, the mercury being present mainly as un-ionized covalent molecules:

$$:\ddot{\underset{..}{Cl}}—Hg—\ddot{\underset{..}{Cl}}:$$

The electronic structure of these molecules, which have a linear configuration (Figure 19-8), is analogous to that of the gold(I) chloride complex, $AuCl_2^-$. The ease of sublimation of mercuric chloride (melting point 275°C, boiling point 301°C) results from the stability of these molecules.

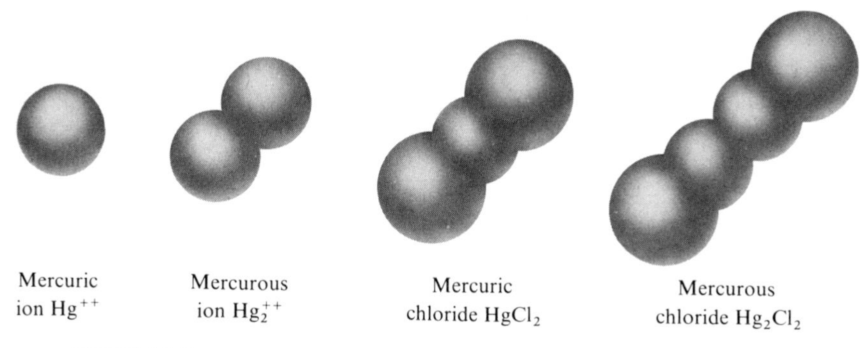

FIGURE 19-8
The structure of the mercuric ion, mercurous ion, mercuric chloride molecule, and mercurous chloride molecule. In the mercurous ion and the two molecules the atoms are held together by covalent bonds.

Mercuric chloride, like other soluble salts of mercury, is very poisonous when taken internally. The mercuric ion combines strongly with proteins; in the human body it acts especially on the tissues of the kidney, destroying the ability of this organ to remove waste products from the blood. Egg white and milk are swallowed as antidotes; their proteins precipitate the mercury in the stomach.

With ammonium hydroxide, mercuric chloride forms a white precipitate, $HgNH_2Cl$:

$$HgCl_2 + 2NH_3 \longrightarrow HgNH_2Cl(c) + NH_4^+ + Cl^-$$

Mercuric sulfide, HgS, is formed as a black precipitate when hydrogen sulfide is passed through a solution of a mercuric salt. It can also be made by rubbing mercury and sulfur together in a mortar. The black sulfide (which also occurs in nature as the mineral *metacinnabarite*) is converted by heat into the red form (cinnabar). Mercuric sulfide is the most insoluble of metallic sulfides. It is not dissolved even by boiling concentrated nitric acid, but it does dissolve in aqua regia, under the combined action of the nitric acid, which oxidizes the sulfide to free sulfur, and hydrochloric acid, which provides chloride ion to form the stable complex $HgCl_4^{--}$:

$$3HgS + 12HCl + 2HNO_3 \longrightarrow 3HgCl_4^{--} + 6H^+ + \tfrac{3}{8}S_8 + 2NO + 4H_2O$$

Mercuric oxide, HgO, is formed as a yellow precipitate by adding a base to a solution of mercuric nitrate or as a red powder by heating dry mercuric nitrate or, slowly, by heating mercury in air. The yellow and red forms seem to differ only in grain size; it is a common phenomenon that red crystals (such as potassium dichromate or potassium ferricyanide) form a yellow powder when they are ground up. Mercuric oxide liberates oxygen when it is strongly heated.

Mercuric fulminate, $Hg(CNO)_2$, is made by dissolving mercury in nitric acid and adding ethyl alcohol, C_2H_5OH. It is a very unstable substance, which detonates when it is struck or heated, and it is used for making detonators and percussion caps.

Mercurous nitrate, $Hg_2(NO_3)_2$, is formed by reduction of a mercuric nitrate solution with mercury:

$$Hg^{++} + Hg \rightarrow Hg_2^{++}$$

The solution contains the **mercurous ion,** Hg_2^{++}, a colorless ion which has a unique structure; it consists of two mercuric ions plus two electrons, which form a covalent bond between them (Figure 19-8):

$$2Hg^{++} + 2e^- \rightarrow [Hg:Hg]^{++} \quad \text{or} \quad [Hg—Hg]^{++}$$

Mercurous chloride, Hg_2Cl_2, is an insoluble white crystalline substance obtained by adding a solution containing chloride ion to a mercurous nitrate solution:

$$Hg_2^{++} + 2Cl^- \rightarrow Hg_2Cl_2(c)$$

It is used in medicine under the name *calomel.* The mercurous chloride molecule (Figure 19-8) has the linear covalent structure

$$:\ddot{\underset{..}{Cl}}—Hg—Hg—\ddot{\underset{..}{Cl}}:$$

The precipitation of mercurous chloride and its change in color from white to black on addition of ammonium hydroxide are used as the test for mercurous mercury in qualitative analysis. The effect of ammonium hydroxide is due to the formation of finely divided mercury (black) and mercuric aminochloride (white) by an auto-oxidation-reduction reaction:

$$Hg_2Cl_2 + 2NH_3 \rightarrow Hg + HgNH_2Cl + NH_4^+ + Cl^-$$

Mercurous sulfide, Hg_2S, is unstable, and when formed as a brownish-black precipitate by action of sulfide ion on mercurous ion it immediately decomposes into mercury and mercuric sulfide:

$$Hg_2^{++} + S^{--} \rightarrow Hg_2S \rightarrow Hg + HgS$$

19-19. Gallium, Indium, and Thallium

The elements of group IIIb—gallium, indium, and thallium—are rare and have little practical importance. Their principal compounds represent oxidation state +3; thallium also forms compounds in which it has oxidation number +1. Gallium is liquid from 29°C, its melting point, to 1700°C, its boiling point. It has found use as the liquid in quartz-tube thermometers, which can be used to above 1200°C.

19-20. The Electronic Structures of Titanium, Vanadium, Chromium, and Manganese and Their Congeners

The electronic structures of the elements of groups IIIa, IVa, Va, and VIa, as represented in the energy-level diagram (Figure 5-6), are given in Table 19-7. Each of the elements has either one electron or two electrons in the *s* orbital of the outermost shell. In addition, there are two, three, four, or five electrons in the *d* orbitals of the next inner shell. Reference to Figure 5-6 shows that the heaviest elements of these groups—thorium, protactinium, uranium, and neptunium—are thought to have the additional two to five electrons, respectively, in the 5*f* subshell, rather than the 6*d* subshell.

The oxidation state +2, corresponding to the loss of two electrons, is an important one for all of these elements. In particular, the elements in

TABLE 19-7
Electronic Structures of Titanium, Vanadium, Chromium, and Manganese and Their Congeners

Atomic Number	Element	*K*	*L*		*M*			*N*				*O*			*P*
		1*s*	2*s*	2*p*	3*s*	3*p*	3*d*	4*s*	4*p*	4*d*	4*f*	5*s*	5*p*	5*d*	6*s*
22	Ti	2	2	6	2	6	2	2							
23	V	2	2	6	2	6	3	2							
24	Cr	2	2	6	2	6	5	1							
25	Mn	2	2	6	2	6	5	2							
40	Zr	2	2	6	2	6	10	2	6	2		2			
41	Nb	2	2	6	2	6	10	2	6	4		1			
42	Mo	2	2	6	2	6	10	2	6	5		1			
43	Tc	2	2	6	2	6	10	2	6	5		2			
72	Hf	2	2	6	2	6	10	2	6	10	14	2	6	2	2
73	Ta	2	2	6	2	6	10	2	6	10	14	2	6	3	2
74	W	2	2	6	2	6	10	2	6	10	14	2	6	4	2
75	Re	2	2	6	2	6	10	2	6	10	14	2	6	5	2

the first long period form the ions Ti^{++}, V^{++}, Cr^{++}, and Mn^{++}. Several other oxidation states, involving the loss or sharing of additional electrons, are also represented by compounds of these elements. The maximum oxidation state is that corresponding to the loss or sharing of all of the electrons in the *d* orbitals of the next inner shell, as well as the two electrons in the outermost shell. Accordingly, the maximum oxidation numbers of titanium, vanadium, chromium, and manganese are +4, +5, +6, and +7, respectively.

19-21. Titanium, Zirconium, Hafnium, and Thorium

The elements of group IVa of the periodic system are titanium, zirconium, hafnium, and thorium. Some of the properties of the elementary substances are given in Table 19-8.

Titanium occurs in the minerals *rutile,* TiO_2, and *ilmenite,* $FeTiO_3$. It forms compounds representing oxidation states +2, +3, and +4. Pure **titanium dioxide,** TiO_2, is a white substance. As a powder it has great power of scattering light, which makes it an important pigment. It is used in special paints and in face powders. Crystals of titanium dioxide (rutile) colored with small amounts of other metal oxides have been made recently for use as gems. **Titanium tetrachloride,** $TiCl_4$, is a molecular liquid

TABLE 19-8
Some Properties of Titanium, Vanadium, Chromium, and Manganese and Their Congeners

	Atomic Number	Atomic Mass	Density (g cm^{-3})	Melting Point	Boiling Point	Metallic Radius*
Titanium	22	47.90	4.44	1,800°C	3,000°C	147 pm
Vanadium	23	50.942	6.06	1,700°	3,000°	134
Chromium	24	51.996	7.22	1,920°	2,330°	127
Manganese	25	54.9380	7.26	1,260°	2,150°	126
Zirconium	40	91.22	6.53	1,860°		160
Niobium	41	92.906	8.21	2,500°		146
Molybdenum	42	95.94	10.27	2,620°	4,700°	139
Hafnium	72	178.49	13.17	2,200°		136
Tantalum	73	180.948	16.76	2,850°		146
Tungsten	74	183.85	19.36	3,382°	6,000°	139
Rhenium	75	186.2	21.10	3,167°		137
Thorium	90	232.038	11.75	1,850°	3,500°	180
Uranium	92	238.03	18.97	1,690°		152

*For ligancy 12.

at room temperature. On being sprayed into air it hydrolyzes, forming hydrogen chloride and fine particles of titanium dioxide; for this reason it is sometimes used in making smoke screens:

$$TiCl_4 + 2H_2O \longrightarrow TiO_2 + 4HCl$$

Titanium metal is very strong, light (density 4.44 g cm^{-3}), refractory (melting point 1,800°C), and resistant to corrosion. Since 1950 it has been produced in quantity, and has found many uses for which a light, strong metal with high melting point is needed; for example, it is used in airplane wings where the metal is in contact with exhaust flame.

Zirconium occurs in nature principally as the mineral *zircon*, $ZrSiO_4$. Zircon crystals are found in a variety of colors—white, blue, green, and red—and because of its beauty and hardness the mineral is used as a semiprecious stone. The principal oxidation state of zirconium is +4; the states +2 and +3 are represented by only a few compounds.

Hafnium is closely similar to zirconium, and natural zirconium minerals usually contain a few percent of hafnium. The element was not discovered until 1923, and it has found little use.

Thorium is found in nature as the mineral *thorite*, ThO_2, and in *monazite sand*, which consists of thorium phosphate mixed with the phosphates of the lanthanons (Section 18-7). The principal use of thorium is in the manufacture of gas mantles, which are made by saturating cloth fabric with thorium nitrate, $Th(NO_3)_4$, and cerium nitrate, $Ce(NO_3)_4$. When the treated cloth is burned, there remains a residue of thorium dioxide and cerium dioxide, ThO_2 and CeO_2, which has the property of exhibiting a brilliant white luminescence when it is heated to a high temperature. Thorium dioxide is also used in the manufacture of laboratory crucibles, for use at temperatures as high as 2300°C.

19-22. Vanadium, Niobium, Tantalum, and Protactinium

Vanadium is the most important element of group Va. It finds extensive use in the manufacture of special steels. Vanadium steel is tough and strong, and is used in automobile crank shafts and for similar purposes. The principal ores of vanadium are *vanadinite*, $Pb_5(VO_4)_3Cl$, and *carnotite*, $K(UO_2)VO_4 \cdot \frac{3}{2}H_2O$. The latter mineral is also important as an ore of uranium.

The chemistry of vanadium is very complex. The element forms compounds representing the oxidation states +2, +3, +4, and +5. The hydroxides of bipositive and terpositive vanadium are basic, and those of the higher oxidation states are amphiprotic. The compounds of vanadium are striking for their varied colors. The bipositive ion, V^{++}, has a deep violet color; the terpositive compounds, such as **potassium vanadium alum,** $KV(SO_4)_2 \cdot 12H_2O$, are green; the dark-green substance **vanadium dioxide,** VO_2, dissolves in acid to form the blue *vanadyl ion,* VO^{++}. **Vanadium(V) oxide,** V_2O_5, an orange substance, is used as a catalyst in the contact process for making sulfuric acid. **Ammonium metavanadate,** NH_4VO_3, which forms yellow crystals from solution, is used for making preparations of vanadium(V) oxide for the contact process.

Niobium (columbium) and **tantalum** usually occur together, as the minerals *columbite,* $FeNb_2O_6$, and *tantalite,* $FeTa_2O_6$. Niobium finds use as a constituent of alloy steels and as a superconductor. **Tantalum carbide,** TaC, a very hard substance, is used in making high-speed cutting tools.

Protactinium is a radioactive element that occurs in minute amounts in all uranium ores.

19-23. Chromium and its Congeners

The principal oxidation states of chromium are +3 and +6. The maximum oxidation number, +6, corresponds to the position of the element in the periodic table.

The most important ore of chromium is *chromite,* $FeCr_2O_4$. The element was not known to the ancients, but was discovered in 1798 in lead chromate, $PbCrO_4$, which occurs in nature as the mineral *crocoite.*

The metal can be prepared by reducing chromic oxide with metallic aluminum (Section 11-9). Metallic chromium is also made by electrolytic reduction of compounds, usually chromic acid in aqueous solution.

Chromium is a silvery-white metal, with a bluish tinge. It is a very strong metal, with a high melting point, 1830°C. Because of its high melting point it resists erosion by the hot powder gases in big guns, the linings of which are accordingly sometimes plated with chromium.

Although the metal is more electropositive than iron, it easily assumes a passive (unreactive) state, by becoming coated with a thin layer of oxide, which protects it against further chemical attack. This property and its pleasing color are the reasons for its use for plating iron and brass objects, such as plumbing fixtures.

Ferrochrome, a high-chromium alloy with iron, is made by reducing chromite with carbon in the electric furnace. It is used for making alloy steels. The alloys of chromium are very important, especially the *alloy steels.* The chromium steels are very hard, tough, and strong. Their properties can be attributed to the high metallic valence (6) of chromium and to an interaction between unlike atoms that in general makes alloys harder and tougher than elementary metals. They are used for armor plate, projectiles, safes, etc. Ordinary *stainless steel* contains 14 to 18% chromium, and usually 8% nickel.

Chromium in its highest oxidation state (+6) does not form a hydroxide. The corresponding oxide, CrO_3, a red substance called **chromium(VI) oxide,** has acid properties. It dissolves in water to form a red solution of **dichromic acid,** $H_2Cr_2O_7$:

$$2CrO_3 + H_2O \rightarrow H_2Cr_2O_7 \rightleftarrows 2H^+ + Cr_2O_7^{--}$$

The salts of dichromic acid are called **dichromates;** they contain the dichromate ion, $Cr_2O_7^{--}$. Sexivalent chromium also forms another important series of salts, the **chromates,** which contain the ion CrO_4^{--}.

The chromates and dichromates are made by a method that has general usefulness for preparing salts of an acidic oxide—the method of *fusion with an alkali hydroxide or carbonate.* The carbonate functions as a basic oxide by losing carbon dioxide when heated strongly. Potassium carbonate is preferred to sodium carbonate because potassium chromate and potassium dichromate crystallize well from aqueous solution, and can be easily purified by recrystallization, whereas the corresponding sodium salts are deliquescent and are difficult to purify.

A mixture of powdered chromite ore and potassium carbonate slowly forms **potassium chromate,** K_2CrO_4, when strongly heated in air. The oxygen of the air oxidizes chromium to the sexipositive state, and also oxidizes the iron to ferric oxide:

$$4FeCr_2O_4 + 8K_2CO_3 + 7O_2 \rightarrow 2Fe_2O_3 + 8K_2CrO_4 + 8CO_2$$

On addition of an acid, such as sulfuric acid, to a solution containing chromate ion, CrO_4^{--}, the solution changes from yellow to orange-red in color, because of the formation of dichromate ion, $Cr_2O_7^{--}$:

$$\underset{\text{Yellow}}{2CrO_4^{--}} + 2H^+ \rightleftarrows \underset{\text{Orange-red}}{Cr_2O_7^{--}} + H_2O$$

The reaction can be reversed by the addition of a base:

$$\underset{\text{Orange-red}}{Cr_2O_7^{--}} + 2OH^- \rightleftarrows \underset{\text{Yellow}}{2CrO_4^{--}} + H_2O$$

At an intermediate stage* both chromate ion and dichromate ion are present in the solution, in chemical equilibrium.

*There is also present in the solution some hydrogen chromate ion, $HCrO_4^-$

$$H^+ + CrO_4^{--} \rightleftarrows HCrO_4^-$$

The chromate ion has a tetrahedral structure. The formation of dichromate ion involves the removal of one oxygen ion O^{--} (as water), by combination with two hydrogen ions, and its replacement by an oxygen atom of another chromate ion.

Both chromates and dichromates are strong oxidizing agents, the chromium being easily reduced from +6 to +3 in acid solution. **Potassium dichromate,** $K_2Cr_2O_7$, is a beautifully crystallizable bright-red substance used considerably in chemistry and industry. A solution of this substance or of chromium(VI) oxide, CrO_3, in concentrated sulfuric acid is a very strong oxidizing agent, which serves as a cleaning solution for laboratory glassware.

Large amounts of **sodium dichromate,** $Na_2Cr_2O_7 \cdot 2H_2O$, are used in the tanning of hides, to produce "chrome-tanned" leather. The chromium forms an insoluble compound with the leather protein.

Lead chromate, $PbCrO_4$, is a bright yellow, practically insoluble substance that is used as a pigment (*chrome yellow*).

When ammonium dichromate, $(NH_4)_2Cr_2O_7$, a red salt resembling potassium dichromate, is ignited, it decomposes to form a green powder, **chromium(III) oxide,** Cr_2O_3:

$$(NH_4)_2Cr_2O_7 \longrightarrow N_2 + 4H_2O + Cr_2O_3$$

This reaction involves the reduction of the dichromate ion by ammonium ion. Chromium(III) oxide is also made by heating sodium dichromate with sulfur, and leaching out the sodium sulfate with water:

$$Na_2Cr_2O_7 + S \longrightarrow Na_2SO_4 + Cr_2O_3$$

It is a very stable substance, which is resistant to acids and has a very high melting point. It is used as a pigment (*chrome green,* used in the green ink for paper money).

Reduction of a dichromate in aqueous solution produces **chromium(III) ion,** Cr^{+++} (really the hexahydrated ion, $[Cr(H_2O)_6]^{+++}$), which has a violet color. The salts of this ion are similar in formula to those of aluminum. *Chrome alum,* $KCr(SO_4)_2 \cdot 12H_2O$, forms large violet octahedral crystals.

Molybdenum

The principal ore of molybdenum is *molybdenite,* MoS_2, which occurs especially in a great deposit near Climax, Colorado. This mineral forms shiny black plates, closely similar in appearance to graphite.

Molybdenum metal is used to make filament supports in radio tubes and for other special uses. It is an important constituent of alloy steels.

The chemistry of molybdenum is complicated. It forms compounds corresponding to oxidation numbers +6, +5, +4, +3, and +2.

Molybdenum(VI) oxide, MoO_3, is a yellow-white substance made by roasting molybdenite. It dissolves in alkalis to produce molybdates, such as **ammonium molybdate,** $(NH_4)_6Mo_7O_{24}\cdot 4H_2O$. This reagent is used to precipitate orthophosphates, as the substance $(NH_4)_3PMo_{12}O_{40}\cdot 18H_2O$.

Tungsten

Tungsten (also called *wolfram*) is a strong, heavy metal, with very high melting point (3370°C.) It has important uses, as filaments in electric light bulbs, for electric contact points in spark plugs, as electron targets in x-ray tubes, and, in tungsten steel (which retains its hardness even when very hot), for making cutting tools for high-speed machining.

The principal ores of tungsten are *scheelite,* $CaWO_4$, and *wolframite,* $(Fe,Mn)WO_4$.*

Tungsten forms compounds in which it has oxidation number +6 (tungstates, including the minerals mentioned above), +5, +4, +3, and +2. **Tungsten carbide,** WC, is a very hard compound that is used for the cutting edge of high-speed tools.

Uranium

Uranium is the rarest metal of the chromium group. Its principal ores are *pitchblende,* U_3O_8, and *carnotite,* $K_2U_2V_2O_{12}\cdot 3H_2O$. Its most important oxidation state is +6 (**sodium diuranate,** $Na_2U_2O(OH)_{12}$; **uranyl nitrate,** $UO_2(NO_3)_2\cdot 6H_2O$; etc.).

Before 1942 uranium was said to have no important uses—it was used mainly to give a greenish-yellow color to glass and glazes. In 1942, however, exactly one hundred years after the metal was first isolated, uranium became one of the most important of all elements. It was discovered in that year that uranium could be made a source of nuclear energy, liberated in tremendous quantity at the will of man.

Nuclear Fission

Ordinary uranium contains two isotopes,† U^{238} (99.3%) and U^{235} (0.7%). When a neutron collides with a U^{235} nucleus it combines with it, forming a U^{236} nucleus. This nucleus is unstable, and it immediately decomposes spontaneously by splitting into two large fragments, plus several neutrons.

---

*The formula $(Fe,Mn)WO_4$ means a solid solution of $FeWO_4$ and $MnWO_4$, in indefinite ratio.

†A minute amount, 0.006%, of a third isotope, U^{234}, is also present.

Each of the two fragments is itself an atomic nucleus, the sum of their atomic numbers being 92, the atomic number of uranium.

This nuclear fission is accompanied by the emission of a very large amount of energy—about 2×10^{13} J* per gram-atom of uranium decomposed (235 g of uranium). This is about 2,500,000 times the amount of heat evolved by burning the same weight of coal, and about 12,000,000 times that evolved by exploding the same weight of nitroglycerine. The large numbers indicate the very great importance of uranium as a source of energy; one ton of uranium (prewar price about $5000) can produce the same amount of energy as 2,500,000 tons of coal.

The heavier uranium isotope, U^{238}, also can be made to undergo fission, but by an indirect route—through the transuranium elements. These elements are discussed in Chapter 20.

19-24. Manganese and Its Congeners

The principal oxidation states of manganese are +2, +3, +4, +6, and +7. The maximum oxidation number, +7, corresponds to the position of the element in the periodic table (group VIIa).

The principal ore of manganese is *pyrolusite,* MnO_2. Pyrolusite occurs as a black massive mineral and also as a very fine black powder. Less important ores are *braunite,* Mn_2O_3 (containing some silicate); *manganite,* $MnO(OH)$; and *rhodochrosite,* $MnCO_3$.

Impure manganese can be made by reducing manganese dioxide with carbon:

$$MnO_2 + 2C \rightarrow Mn + 2CO$$

Manganese is also made by the aluminothermic process:

$$3MnO_2 + 4Al \rightarrow 2Al_2O_3 + 3Mn$$

Manganese alloy steels are usually made from special high-manganese alloys prepared by reducing mixed oxides of iron and manganese with coke in a blast furnace (see Section 19-2). The high-manganese alloys (70 to 80% Mn, 20 to 30% Fe) are called *ferromanganese,* and the low-manganese alloys (10 to 30% Mn) are called *spiegeleisen.*

Manganese is a silvery-gray metal, with a pinkish tinge. It is reactive, and displaces hydrogen even from cold water. Its principal use is in the manufacture of alloy steel.

---

*This amount of energy weighs about 0.25 g, by the Einstein equation $E = mc^2$ (E = energy, m = mass, c = velocity of light). The material products of the fission are 0.25 g lighter than the gram-atom of U^{235}.

Manganese dioxide (pyrolusite) is the only important compound of quadripositive manganese. This substance has many uses, most of which depend upon its action as an oxiding agent (with change from Mn^{+4} to Mn^{+2}) or as a reducing agent (with change from Mn^{+4} to Mn^{+6} or Mn^{+7}).

Manganese dioxide oxidizes hydrochloric acid to free chlorine, and is used for this purpose:

$$MnO_2 + 2Cl^- + 4H^+ \longrightarrow Cl_2 + Mn^{++} + 2H_2O$$

Its oxidizing power also underlies its use in the ordinary dry cell (Chapter 11).

When manganese dioxide is heated with potassium hydroxide in the presence of air it is oxidized to **potassium manganate,** K_2MnO_4:

$$2MnO_2 + 4KOH + O_2 \longrightarrow 2K_2MnO_4 + 2H_2O$$

Potassium manganate is a green salt that can be dissolved in a small amount of water to give a green solution, containing potassium ion and the *manganate ion,* MnO_4^{--}. The manganates are the only compounds of Mn^{+6}. They are powerful oxidizing agents, and are used to a small extent as disinfectants.

The manganate ion can be oxidized to *permanganate ion,* MnO_4^-, which contains Mn^{+7}. The electron reaction for this process is

$$MnO_4^{--} \longrightarrow MnO_4^- + e^-$$

In practice this oxidation is carried out electrolytically (by anodic oxidation) or by use of chlorine:

$$2MnO_4^{--} + Cl_2 \longrightarrow 2MnO_4^- + 2Cl^-$$

The process of auto-oxidation-reduction is also used; manganate ion is stable in alkaline solution, but not in neutral or acidic solution. The addition of any acid, even carbon dioxide (carbonic acid), to a manganate solution causes the production of permanganate ion and the precipitation of manganese dioxide:

$$\underset{\text{Green}}{3MnO_4^{--}} + 4H^+ \longrightarrow \underset{\text{Magenta}}{2MnO_4^-} + MnO_2 + 2H_2O$$

When hydroxide is added to the mixture of the purple solution and the brown or black precipitate, a clear green solution is again formed, showing that the reaction is reversible.

This reaction serves as another example of Le Chatelier's principle: the addition of hydrogen ion, which occurs on the left side of the equation, causes the reaction to shift to the right.

Potassium permanganate, $KMnO_4$, is the most important chemical compound of manganese. It forms deep purple-red prisms, which dissolve readily in water to give a solution intensely colored with the magenta color characteristic of permanganate ion. The substance is a powerful oxidizing agent, which is used as a disinfectant. It is an important chemical reagent, especially in analytical chemistry.

On reduction in acidic solution the permanganate ion accepts five electrons, to form the manganese(II) ion:

$$MnO_4^- + 8H^+ + 5e^- \longrightarrow Mn^{++} + 4H_2O$$

In neutral or basic solution it accepts three electrons, to form a precipitate of manganese dioxide:

$$MnO_4^- + 2H_2O + 3e^- \longrightarrow MnO_2 + 4OH^-$$

A one-electron reduction to manganate ion can be made to take place in strongly basic solution:

$$MnO_4^- + e^- \longrightarrow MnO_4^{--}$$

Permanganic acid, $HMnO_4$, is a strong acid that is very unstable. Its anhydride, **manganese(VII) oxide,** can be made by the reaction of potassium permanganate and concentrated sulfuric acid:

$$2KMnO_4 + H_2SO_4 \longrightarrow K_2SO_4 + Mn_2O_7 + H_2O$$

It is an unstable, dark-brown oily liquid.

The manganese(III) ion, Mn^{+++}, is a strong oxidizing agent, and its salts are unimportant. The insoluble oxide, Mn_2O_3, and its hydrate, $MnO(OH)$, are stable. When manganese(II) ion is precipitated as hydroxide, $Mn(OH)_2$, in the presence of air, the white precipitate is rapidly oxidized to the brown compound $MnO(OH)$:

$$Mn^{++} + 2OH^- \longrightarrow \underset{\text{White}}{Mn(OH)_2}$$

$$4Mn(OH)_2 + O_2 \longrightarrow \underset{\text{Brown}}{4MnO(OH)} + 2H_2O$$

Manganese(II) ion, Mn^{++} or $[Mn(H_2O)_6]^{++}$, is the stable cationic form of manganese. The hydrated ion is pale rose-pink in color. Representative salts are $Mn(NO_3)_2 \cdot 6H_2O$, $MnSO_4 \cdot 7H_2O$, and $MnCl_2 \cdot 4H_2O$. These salts and the mineral *rhodochrosite,* $MnCO_3$, are all rose-pink or rose-red. Crystals of rhodochrosite are isomorphous with calcite.

With hydrogen sulfide manganese(II) ion forms a light pink precipitate of **manganese sulfide,** MnS:

$$Mn^{++} + H_2S \longrightarrow MnS + 2H^+$$

Technetium

No stable isotopes of this element exist. One short-lived radioactive isotope, which emits gamma rays, is used for brain scans.

Rhenium

The element rhenium, atomic number 75, was discovered by the German chemists Walter Noddack and Ida Tacke in 1925. The principal compound of rhenium is potassium perrhenate, $KReO_4$, a colorless substance. In other compounds all oxidation numbers from +7 to −1 are represented: examples are Re_2O_7, ReO_3, $ReCl_5$, ReO_2, Re_2O_3, $Re(OH)_2$.

Neptunium

Neptunium, element 93, was first made in 1940, by E. M. McMillan and P. H. Abelson, at the University of California, by the reaction of a neutron with U^{238}, to form U^{239}, and the subsequent emission of an electron from this nucleus, increasing the atomic number by 1:

$$_{92}U^{238} + {_0}n^1 \longrightarrow {_{92}}U^{239}$$

$$_{92}U^{239} \longrightarrow e^- + {_{93}}Np^{239}$$

Neptunium is important as an intermediate in the manufacture of plutonium (Chapter 20).

EXERCISES

19-1. What is the characteristic feature of the electronic structure of the transition metals?

19-2. What are the most common valence states of the transition metals shown in their compounds? Give the electronic structures of iron, cobalt, and nickel in these valence states.

19-3. In terms of constituents, describe the microscopic (on an atomic scale) structures of tempered eutectoid, hypo-eutectoid, and hyper-eutectoid steel.

19-4. What are the differences between white cast iron, gray cast iron, malleable cast iron, and wrought iron?

19-5. What are the constituent differences between cast iron and carbon steel?

19-6. Describe martensite, ferrite, cementite, and pearlite.

19-7. What are the usual oxidation states of iron? Cobalt? Nickel? Give examples of each.

19-8. What are the uses of the platinum metals?

19-9. What are the stable oxidation states of copper, zinc, and gallium and their congeners?

19-10. What is the bonding electron structure of $CuCl_2 \cdot 2H_2O$? What orbitals of the copper atom are used in bonds?

19-11. Silver nitrate is used as a disinfectant. It turns black on contact with organic materials. What is the reaction?

19-12. Verify that a solid solution containing one part cuprous iodide and three parts silver iodide has an average metal-iodine bond length equal to the oxygen-oxygen bond length of ice.

19-13. Give the reactions for the formation and decomposition of ozone in the upper atmosphere. Why are these reactions important for life?

19-14. A dilute solution of acetic acid is often used as a stop bath between the developing and fixing of photographic emulsions. Why? What is the reaction?

19-15. What are the primary colors of Kodachrome and their complementary colors?

19-16. Describe the reactions and oxidation states that take place on heating $HAuCl_4$.

19-17. What is the likely explanation of the deep, intense color shown by substances containing the same element in two oxidation states?

19-18. What are the chief uses of metallic zinc, cadmium, and mercury?

19-19. Can you advance a likely explanation of the low melting points and boiling points of zinc, cadmium, and mercury, based on their electronic structure?

19-20. What is the normal oxidation state of zinc? It is usually tetrahedrally coordinated. From the electronic structure of the ion, why are zinc compounds usually white whereas transition metal compounds in general are colored?

19-21. Describe the electronic structure of the mercurous ion, the mercuric ion, the mercurous chloride molecule, and the mercuric chloride molecule. Compare the total number of electrons surrounding each mercury atom with the number in the nearest argonon. What hybrid orbitals are used in bond formation?

19-22. What is an important oxidation state of titanium, vanadium, chromium, and manganese? What are the electronic structures of these elements in this oxidation state? What are the maximum oxidation states shown by these elements? What are the electronic structures corresponding to these states?

19-23. Explain why $TiCl_4$ is more effective in making smoke screens over the ocean than over dry land.

19-24. Make a diagram listing compounds representative of the various important oxidation levels of chromium and manganese.

19-25. What reduction product is formed when dichromate ion is reduced in acidic solution? When permanganate ion is reduced in acidic solution? When permanganate ion is reduced in basic solution? Write the electron reactions for these three cases.

19-26. Write equations for the reduction of dichromate ion by (*a*) sulfur dioxide; (*b*) ethyl alcohol, C_2H_5OH, which is oxidized to acetaldehyde, H_3CCHO; (*c*) iodide ion, which is oxidized to iodine.

19-27. Write an equation for the chemical reaction that occurs on fusion of a mixture of chromite ($FeCr_2O_4$), potassium carbonate, and potassium chlorate (which forms potassium chloride).

19-28. Write the chemical equations for the preparation of potassium manganate and potassium permanganate from manganese dioxide, using potassium hydroxide, air, and carbon dioxide.

19-29. What property of tungsten makes it suitable for use as the filament material in electric light bulbs?

19-30. What chemical reactions are taking place when a violet solution of chrome alum on treatment with hydrochloric acid turns green in color?

20

The Chemistry of Fundamental Particles and Nuclei

During recent years there has been a great increase in our knowledge of the world. Atoms have been found to consist of electrons and nuclei, and the atomic nuclei have been found to consist of protons and neutrons. Moreover, in addition to the electron, the proton, and the neutron, many other particles have been discovered.

The field of science dealing with the nature and the reactions of these particles is developing very rapidly at the present time. Work in this field of science has been carried out largely by physicists, but the reactions by means of which the particles are created, converted into others, and destroyed are in a general way similar to chemical reactions, and we may be justified in considering the study of these reactions and the properties of the particles themselves as constituting the field of the chemistry of particles.

At the present time about 200 particles are known, of which 34 may be classed as fundamental. This number includes 6 (the photon, the graviton, two neutrinos, and two antineutrinos) that move only with the speed of light, and 28 that move only at speeds less than the speed of light. In accordance with the theory of relativity, the particles that move only at the speed of light have zero rest-mass, whereas the others have finite rest-mass.

Much of the knowledge about the fundamental particles has been obtained during the last decade. The scientists who have been working in this field have made many completely unexpected discoveries, which are changing our ways of thinking about the world. Just as the discoveries in the field of atomic and molecular science, discussed in earlier chapters, and the field of nuclear science, to be discussed later in this chapter, have had profound effects upon our daily lives, changing the nature of our civilization and especially the methods of waging war, so may we expect that the new knowledge about fundamental particles will in the course of time have equally profound effects upon our lives. If Benjamin Franklin were alive today, he might well say "It is impossible to imagine the height to which may be carried during the next *twenty* years the power of man over matter."

20-1. The Classification of the Fundamental Particles

At the present time it is convenient to classify the thirty-four fundamental particles in the following way:

8 baryons (the proton, the neutron, and six heavier particles)
8 antibaryons
8 mesons and antimesons
8 leptons and antileptons
The photon
The graviton

Most of the fundamental particles can be described as constituting either *matter* or *antimatter*. The existence of these two kinds of matter was predicted, on the basis of relativistic quantum mechanics, by P. A. M. Dirac (born 1902), the English theoretical physicist who first developed a theory of quantum mechanics compatible with the theory of relativity. His prediction has been thoroughly confirmed by experiment. Every electrically charged particle has a counterpart that is identical with it in some properties and opposite to it in others: the masses and spins are identical, but the electric charges are opposite. For example, the electron, which constitutes a part of ordinary matter, and the positron, which is the antielectron, have opposite electric charges, $-e$ and $+e$, respectively; their masses are the same; and each has a spin represented by the spin quantum number $\frac{1}{2}$, which permits two ways of orienting the spinning particle in a magnetic field. Some neutral particles have antiparticles and some are their own antiparticles. Whenever a particle and the corresponding antiparticle come together they annihilate each other. Their masses are totally converted into high-energy light waves or, in some cases, into lighter particles moving with great speeds. The Einstein equation $E=mc^2$ gives the amount of energy that is released when a particle and its antiparticle annihilate one another with formation of radiant energy. The neutral particles that are their own antiparticles decay very rapidly.

Antimatter does not exist, except fleetingly, on earth. Particles of antimatter are created by collisions, as described in the following section, and the antiparticles are then rapidly destroyed as they react with particles of ordinary matter with which they collide.

There is the possibility that some regions of the universe, perhaps some nebulae, are composed of antimatter. The hydrogen atom in such a region consists of a positron moving about an antiproton. The collision between an antimatter nebula and a nebula composed of ordinary matter would result in the liberation of a tremendous amount of radiant energy, and might be recognized by astronomers.

Fermions and Bosons

The elementary particles may be divided into two classes on the basis of the magnitude of their spin. The electron can be described as having spin $\frac{1}{2}$. It has an angular momentum determined by the spin quantum number $\frac{1}{2}$, and in a magnetic field it can orient its angular momentum with component either $+\frac{1}{2}$ or $-\frac{1}{2}$ in the direction of the field (the unit of angular momentum is the Bohr unit $h/2\pi$). It was mentioned in Chapter 5 that two electrons cannot occupy the same orbital in an atom unless they have opposite orientations of their spin; that is, they cannot be in exactly the same quantum state, as they would be if they occupied the same orbital and both had positive orientation of the spin. This is the expression of the Pauli exclusion principle.

Particles that have spin $\frac{1}{2}$ are called *fermions,* named after the physicist Enrico Fermi. In accordance with the Pauli exclusion principle, no two identical fermions can be in exactly the same quantum state.

The baryons, antibaryons, leptons, and antileptons are all fermions. Particles with spins $\frac{3}{2}$, $\frac{5}{2}$, . . . are also fermions.

Particles with integral spin (0, 1, 2, . . .) are called *bosons,* named after the Indian physicist S. N. Bose. They interact with one another in a way that permits two or more particles to be in exactly the same quantum state. The photon, the graviton, and the mesons are bosons. The mesons all have spin 0. The photon has spin 1. The graviton, which is the quantum of the gravitational field, is expected to have spin 2.

The **photon** or light quantum is now accepted as one of the fundamental particles. Newton discussed both a corpuscular theory and a wave theory of light. During the nineteenth century a great emphasis was given to the wave theory of light in connection with experiments on the diffraction of light. Then in 1905 Einstein pointed out that a number of puzzling experimental results could be interpreted in a simple way if it were assumed that light (visible light, ultraviolet light, radio waves, gamma rays, etc.) has some of the properties of particles (Section 3-10). He called these "particles" of light "light quanta," and the name photon has since come into use. The amount of energy constituting a light quantum is determined by the frequency of the light; it is $E = h\nu$.

The properties of light cannot be described completely by analogy with either ordinary waves or ordinary particles. In the discussion of some phenomena the description of light as wave motion is found to be the more useful, and in the discussion of other phenomena the description of light in terms of photons is preferred (Sections 3-11, 3-12). This wave-particle duality applies also to matter. Electrons, protons, neutrons, and other material particles have been found to have some properties that we usually correlate with wave motion. For example, a beam of electrons or a beam of neutrons can be diffracted in the same way as a beam of x-rays. Electron diffraction and neutron diffraction have turned out to be valuable techniques for investigating the structure of crystals and gas molecules. The wavelength associated with an electron, a neutron, or other particle depends on its rest-mass and the speed with which it is traveling. It is given by the de Broglie equation, $\lambda = h/mv$, in which λ is the wavelength of the particle, h is Planck's constant, m is the mass, and v is the speed (Section 3-11).

The main distinction between photons and material particles with finite rest-mass is that in a vacuum photons travel always at constant speed, the speed of light, whereas particles with finite rest-mass are able to travel at various speeds relative to the observer, up to a maximum of the speed of light, which for these particles would correspond to infinite energy.

The symbol used for the photon is γ (the Greek letter gamma). This symbol was originally used for γ-rays, which are photons of high energy liberated in the course of the radioactive decomposition of nuclei.

The value 1 for the spin of the photon is connected with the polarization of light, discussed in Chapter 6. Right-handed circularly polarized light corresponds to a component $+1$ of the spin in the direction of motion of light, and left-handed circularly polarized light to the component -1.

Photons may be emitted or absorbed by an oscillating electric dipole, such as a negatively charged electron rotating around a positively charged proton. It might be thought that a system of two masses, such as the earth and the moon, rotating about their common center would emit gravitational quanta. These gravitational quanta are called gravitons. Some studies of the properties of gravitational waves are now being made, but the existence of the graviton, a quantized gravitational wave, has not yet been verified by experiment.

20-2. The Discovery of the Fundamental Particles

The **electron** has been discussed throughout this book. It was the first of the fundamental particles to be recognized, having been discovered by J. J. Thomson in 1897. It is present in ordinary matter, and is easily separated from the atomic nuclei to which it is ordinarily attached.

The **proton** was observed as positively charged rays in a discharge tube in 1886 by the Germany physicist E. Goldstein. The nature of the rays was not at first understood. In 1898 the German physicist W. Wien made a rough determination of their ratio of charge to mass, and accurate measurements of this sort, which verified the existence of protons as independent particles in a discharge tube containing ionized hydrogen at low pressure, were made by J. J. Thomson in 1906.

The next particle to be discovered (aside from the photon) was the **positron** (the antielectron), found in 1932 by the American physicist Carl D. Anderson (born 1905). The positrons were found among the particles produced by the interaction of cosmic rays with matter. They are identical with electrons except that their electric charge is $+e$ instead of $-e$.

The mass of the electron corresponds, according to the Einstein equation $E = mc^2$, to the energy 510,976 electron volts (0.510976 MeV; 1 MeV is 1 million electron volts). Hence the annihilation of an electron and a positron liberates 1.022 MeV of energy, which might be in the form of two photons, each with the energy 0.511 MeV and corresponding wavelength 2.426 pm.

A rapidly moving electron that strikes the anode in an x-ray tube is suddenly slowed down, and much of its energy is converted into a photon of x-radiation. If its kinetic energy is greater than 1.022 MeV, this amount of energy may be converted into an electron-positron pair. Electron-positron pair production can be carried out in this way in the laboratory with use of particles that have been given large amounts of kinetic energy in a particle accelerator, as described later in this section. The positrons that were first observed by Anderson were produced, together with electrons, by the impingement of cosmic-ray particles against particles of ordinary matter. Cosmic rays are described later in this section.

The **neutron** was discovered in 1932 by the English physicist James Chadwick (1891–1974). It had been observed in 1930 by two German investigators, Bothe and Becker, that a very penetrating radiation is produced when beryllium metal is bombarded with alpha particles from radium. Bothe and Becker considered the radiation to consist of γ-rays. Frédéric Joliot and his wife Irène Joliot-Curie then discovered that this radiation from beryllium, when passed through a block of paraffin or other substance containing hydrogen, produces large number of protons. Because of the difficulty of understanding how protons could be produced by γ-rays, Chadwick carried out a series of experiments that led to the discovery that the rays from beryllium are in fact composed of particles with no electric charge and with mass approximately equal to that of the proton. Because they have no electric charge, neutrons interact with other forms of matter very weakly, except at very small distances, less than 10^{-14} m.

The existence of the **antiproton** was verified in 1955 by Segré, Chamberlain, Wiegand, and Ypsilantis, by use of a particle accelerator (the Berkeley synchrotron) that could generate particles with energy 6 GeV (the GeV, giga-electron volt, is 1,000 MeV). The mass of the proton-antiproton pair is 1836 times that of the electron-positron pair, and accordingly 1836×1.022 MeV = 1,876 MeV of energy is needed to produce this pair of heavier particles.* The antiproton has negative electric charge, mass equal to the charge of the proton, and spin $\frac{1}{2}$.

The discovery of some of the other fundamental particles will be described in later sections.

Cosmic Rays

Cosmic rays are particles of very high energy that reach the earth from interstellar space or other parts of the cosmos or that are produced in the earth's atmosphere by the rays from outer space. The discovery that ionizing radiation on the earth's surface comes from outer space was made by the Austrian physicist Victor Hess, who made measurements of the amount of ionization in the earth's atmosphere during balloon ascents to a height of 15,000 feet in 1911 and 1912. Many of the fundamental particles in addition to the positron were discovered in the course of studies of cosmic rays.

Cosmic rays that impinge on the outer part of the atmosphere consist of protons and the nuclei of heavier atoms moving with great speed. The cosmic rays that reach the earth's surface consist in large part of mesons, positrons, electrons, and protons produced by reaction of the fast protons and other atomic nuclei with atomic nuclei in the atmosphere.

Some of the phenomena produced by cosmic rays can be explained only if it is assumed that particles are present with energy in the range

*This is the amount of energy needed for proton-antiproton pair production by collision of two similar particles moving with equal speeds in opposite directions in the coordinate system of the laboratory. A much larger amount of energy—nearly 6 GeV—must be imparted to a particle in order that a pair may be produced when it collides with a stationary particle. Some experiments are now being made in which two beams of particles are directed against one another.

from 10^{15} to 10^{20} eV. The great accelerators that have been or are being built (following section) produce or will produce particles with energies in the range 10^{6} to 10^{12} eV. There is no way known at present to accelerate particles to energies as great as those of the fastest particles in cosmic rays, and accordingly the study of cosmic rays will probably continue to yield information about the universe that cannot be obtained in any other way.

Particle Accelerators

In recent years great progress has been made in the laboratory production of high-speed particles. The first efforts to accomplish this involved the use of transformers. Different investigators built transformers and vacuum tubes operating to voltages as high as three million volts, in which protons, deuterons, and helium nuclei could be accelerated. In 1931 an electrostatic generator was developed by R. J. Van de Graaff, an American physicist, involving the carrying of electric charges to the high-potential electrode on a moving insulated belt. Van de Graaff generators have been built and operated to produce potential differences up to fifteen million volts.

The **cyclotron** was invented by the American physicist Ernest Orlando Lawrence (1901–1958) in 1929. In the cyclotron positive ions (protons, deuterons, or other light nuclei) are given successive accelerations by repeatedly falling through a potential difference of a few thousand volts. The charged particles are caused to move in circular paths by a magnetic field, produced by a large magnet between whose pole pieces the apparatus is placed (Figure 20-1). Cyclotrons can be used to accelerate particles to about 100 MeV, but the relativistic change in mass of the particle then causes it to get out of phase with the alternating electric field, so that higher energies cannot be obtained.

A new accelerator, the **synchrotron,** in which a number of the particles are injected and the frequency of the alternating field is adjusted to compensate for the relativistic change in mass, was proposed by the Russian physicist V. Veksler and independently by the American physicist E. M. McMillan in 1945. By use of the synchrotron principle particles have now been accelerated to about 300 GeV, and plans are at present being made for construction of giant accelerators to produce particles in the range 500 GeV to 1000 GeV.

The reactions of particles can be observed by the study of the tracks of the particles in a cloud chamber or a bubble chamber. The **cloud chamber,** which was invented by the English physicist C. T. R. Wilson (1869–1959) in 1911, is a chamber containing air saturated with water vapor. When the air is suddenly expanded by increasing the volume of the chamber by moving a piston, the air is cooled and becomes supersaturated, so

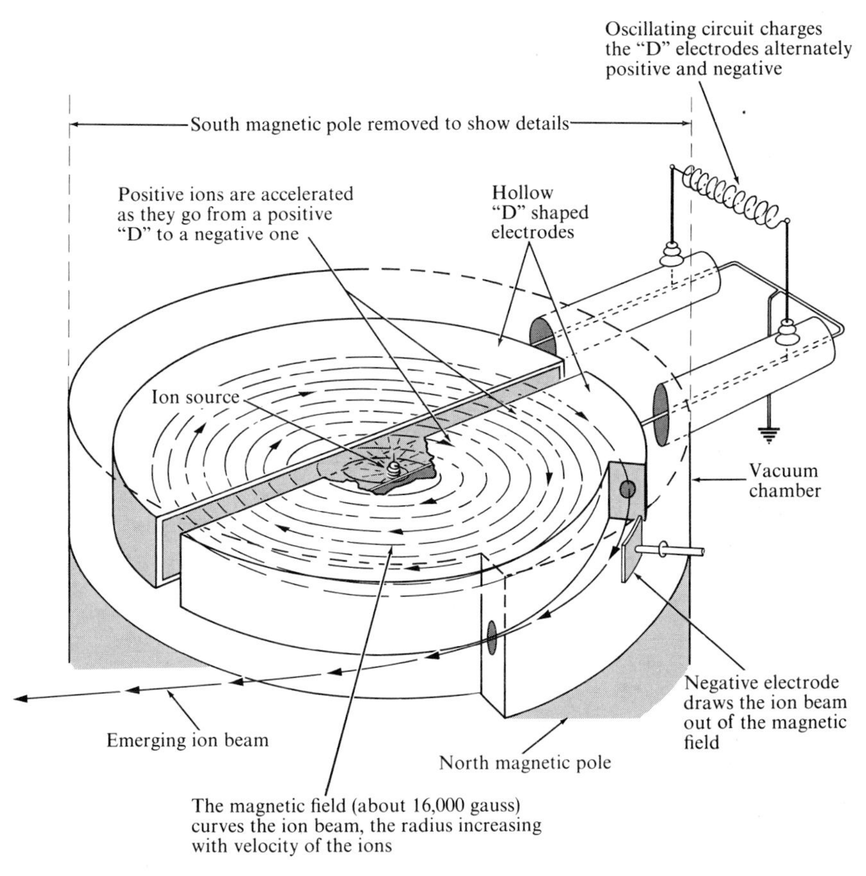

FIGURE 20-1
Diagram showing how the cyclotron works.

that droplets of water form. These droplets tend to form around the ions that are produced as high-energy electrically charged particles traverse the gas, and thus the droplets define the paths of the particles. Neutral particles do not form paths, but their presence can sometimes be detected by the presence of paths radiating from a point where the neutral particle underwent a reaction that produced high-energy charged particles. The **bubble chamber,** invented by the American physicist D. A. Glaser (born 1926) in 1952, has found extensive use in recent years. It is a chamber containing a liquid held at a temperature slightly above its boiling point. The ions formed by high-energy particles traversing the liquid serve as centers of formation of small vapor bubbles, which define the tracks of the particles.

FIGURE 20-2
A direct view and a mirror view of two electron-positron pairs produced by a cosmic-ray photon near the nucleus of a lead atom in a lead plate 1 cm thick in a cloud chamber. This photograph was made about 1934 by Carl D. Anderson, the discoverer of the positron. There is a magnetic field present, which causes the paths of the electron and the positron to curve in opposite directions.

A cloud-chamber photograph is shown as Figure 20-2 and a bubble-chamber photograph as Figure 20-3.

Another instrument for defining the tracks of high-energy particles is the **spark chamber.** This instrument contains a gas, and a series of metal plates that can be electrically charged to such a potential between alternate plates as nearly to cause a spark to pass from one plate to an adjacent one. If a track of ions is formed by a high-energy particle the spark follows the track, and can be photographed. In the 1962 neutrino experiment (Section 20-5) a spark chamber 10 feet by 6 feet by 4 feet was used, containing 90 aluminum plates 4 feet square and 1 inch thick, $\frac{1}{2}$ inch apart, with neon as the gas.

20-3. The Forces between Nucleons. Strong Interactions

In 1932, when the neutron was discovered, it was recognized that the heavier atomic nuclei can be described as being built of protons and neutrons, with the electric charge equal to the number of protons and the mass number equal to the sum of the number of protons and the number of neutrons; that is, equal to the number of nucleons, with a nucleon either a proton or a neutron. The question immediately arose as to the nature of the forces holding the neutrons and protons together. If electrostatic forces were the only forces operating between nucleons the heavier nuclei would break up, because of the electrostatic repulsion between the protons.

It was evident that the force of attraction between nucleons must be a strong force at small distances, stronger than the repulsion due to the positive electric charges on the protons, and a weak force at large distances, weaker than the electrostatic repulsion. Careful studies of the

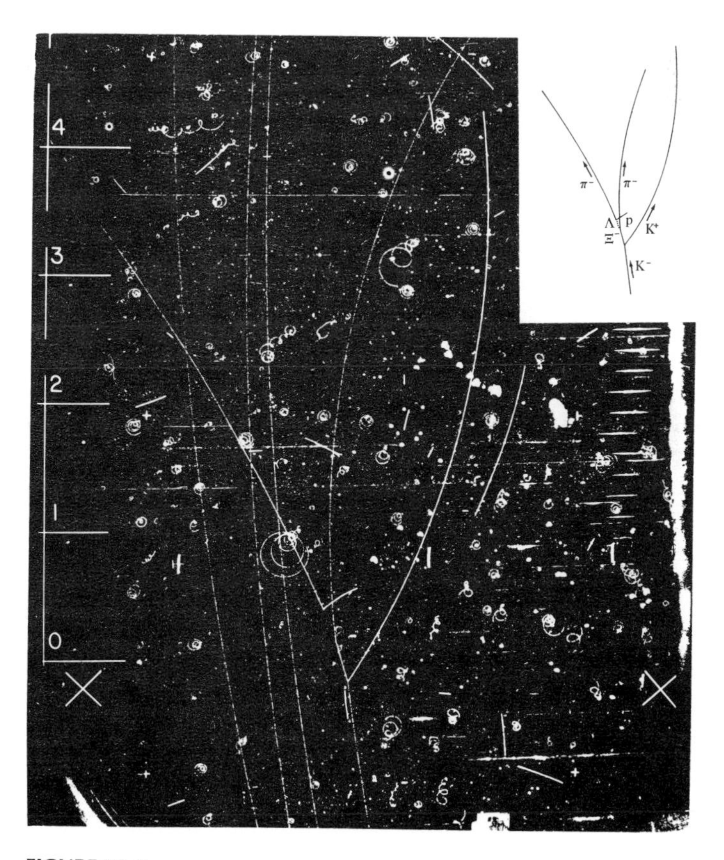

FIGURE 20-3
An event recorded in the 72-inch liquid-hydrogen bubble chamber of the University of California (L. W. Alvarez and coworkers). The incident particle is a negative kaon, in a beam of these particles. By collision with a proton it forms a positive kaon and a negative xion. The negative xion then decomposes to form a lambda particle and a negative pion. The lambda particle, which is neutral, produces no track. It is shown as decomposing to form a proton and a negative pion.

size of the heavier nuclei and the scattering of nucleons from one another led to the discovery that two nucleons attract one another with approximately a constant force when they are less than 1.4 fm apart, and that the internucleonic force, other than electrostatic repulsion of protons, drops rapidly to zero at distances greater than 1.4 fm.

The idea of action at a distance is not a satisfying one; instead, physicists have developed a **quantum theory of force fields,** in which the field at some distance from its source is thought of as carried to that point by a messenger or quantum of the field. For electrostatic attraction and repulsion these messengers are the photons, and for gravitational attraction they are described as being gravitons. In 1935 the Japanese physicist Hideki Yukawa (born 1907) proposed an answer to the question of the mechanism of the force of attraction between nucleons. He pointed out that, whereas messengers that have zero rest-mass, such as photons and gravitons, can extend their influence to infinity, a messenger with finite rest-mass could reach only a limited distance from the particle. He suggested that messengers of this sort are involved in the interaction of nucleons, and from the known range of internucleonic force, 1.4 fm, he calculated that the rest-mass of these particles should be about 274 times the electronic mass. These particles, which are intermediate in mass between electrons and nucleons, are called **mesons** (Greek *mesos,* middle).

Let us consider two nucleons a small distance apart, less than 1.4 fm. as shown in A in Figure 20-4. One of the nucleons produces or emits a messenger particle, a meson, which travels with a speed close to that of light and is destroyed in the neighborhood of the second particle. This process of production and destruction of the messenger particle gives rise to the force of attraction.

If, however, the two particles are farther apart than 1.4 fm, as in B, the emitted messenger particle is not able to traverse the distance between the particles, but instead turns back and disappears. There is accordingly no interaction between the nucleons at the larger distance. The reason that the range of the messenger particles is restricted can be understood by consideration of the uncertainty principle.

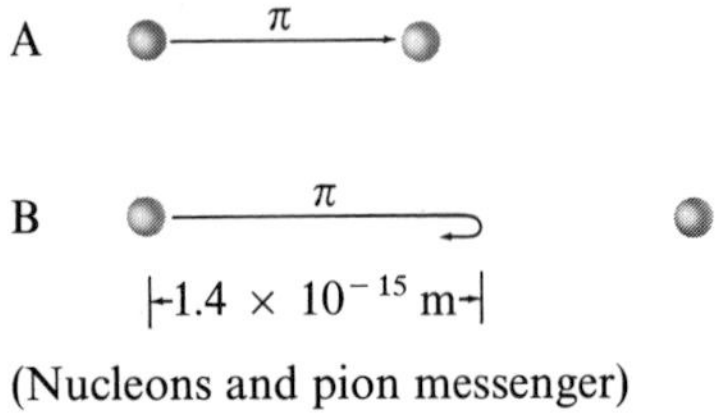

A π

B π

$\vdash 1.4 \times 10^{-15}\,\mathrm{m} \dashv$

(Nucleons and pion messenger)

FIGURE 20-4
A diagram illustrating the range of internucleonic forces.

Let us consider the reaction

$$p^+ \rightleftarrows p^+ + \pi^0$$

Here we use π^0 to represent a messenger particle; the mesons responsible in the main for internuclear forces are called **pions.** This reaction, the reaction of a proton to form a proton plus a pion, violates the principle of conservation of mass-energy. Until the uncertainty principle was discovered no reaction of this sort would ever have been considered.

However, because of the uncertainty relation between energy and time we may consider a reaction such as this, which violates the principle of conservation of mass-energy, provided that the length of time during which we consider the reaction to be taking place is less than the time Δt given by the uncertainty principle. We make use of the equation $\Delta E \cdot \Delta t = h/2\pi$ given in Section 3-13. The length of time in which we are interested is the time required for a particle moving with a speed close to that of light to move the distance 1.4×10^{-15} m. The corresponding value of ΔE, the uncertainty for mass-energy, is $h/2\pi$ divided by this time Δt, $1.4 \times 10^{-15}/3 \times 10^8$ (the speed of light), which is $\Delta t = 0.47 \times 10^{-23}$ s. The value of ΔE is accordingly $1.05 \times 10^{-34}/0.47 \times 10^{-23} = 2.24 \times 10^{-11}$ J, and the corresponding value in mass units, obtained by dividing by c^2 (since $E = mc^2$), is 2.49×10^{-25} g, which is 274 times the mass of the electron. Yukawa accordingly stated that the short range of internucleonic forces could be explained by assuming that the interactions are carried out by particles with mass about 274 times the mass of the electron. No such particles were known at that time.

In 1936 particles with mass 207 times the mass of the electron and with either a positive or a negative charge were discovered by Anderson and Neddermeyer and independently by Street and Stevenson in the course of cosmic-ray experiments. These particles, which are now called **muons,** were at first thought to be the Yukawa particles. However, if they were responsible for the internucleonic forces they would interact strongly with nucleons. This strong interaction should cause them to react with a period of time about 10^{-23} sec when in the neighborhood of a nucleon. Muons were found to decompose in free space, their half-life being about 10^{-6} sec, and their rate of decomposition was found not to be greatly changed when a beam of muons was passed through solid substances and the muons were thus subjected to the influence of nucleons; hence they could not be the Yukawa particles.

When this last experiment was carried out, in 1945, the physicists were again at a loss to account for internucleonic forces, but not for long, because the strongly interacting mesons, which were named pions, were soon discovered. Cosmic-ray experiments with use of stacks of photographic emulsions to detect the tracks of the charged particles, carried out in 1947 by the British physicist C. F. Powell (1903–1969) and his coworkers, led to the discovery of three particles, the positive pion, neutral pion, and negative pion, with masses 273.3 for π^+ and π^- and 264.3 for π^0 and with the properties of strong interaction with nucleons that had been predicted by Yukawa. There is now no doubt that the internucleonic forces that operate in atomic nuclei involve pions. It has been shown by experiment that charged pions as well as neutral pions are involved in the internucleonic forces. The equations for the charged-pion forces are

$$p^+ \rightleftarrows n + \pi^+$$

$$n \rightleftarrows p^+ + \pi^-$$

Other particles, especially the rho and omega particles (Section 20-10), are probably also involved in the internucleonic forces.

20-4. The Structure of Nucleons

The proton and the neutron are closely similar in properties except that the proton has positive charge and the neutron is electrically neutral. The mass of the neutron is only about 0.1% greater than that of the proton. Both particles have spin $\frac{1}{2}$. The proton-proton, proton-neutron, and neutron-neutron internucleonic forces at small distances are essentially the same. Because of these facts, the idea arose some years ago that the proton and the neutron are simply two states of one particle, the nucleon.

We have pointed out in Section 20-1 that an electron in an atom may have two orientations of its spin relative to the direction of a magnetic field or of the angular momentum vector produced by its orbital motion. These two directions, represented by $+\frac{1}{2}$ and $-\frac{1}{2}$, respectively, are said to give rise to a doublet. The doublet is associated with the spin quantum number $\frac{1}{2}$. This suggested that the proton and the neutron may constitute an *electric-charge doublet.* It has been suggested that the nucleon has an intrinsic electric charge with magnitude $+\frac{1}{2}$ (in units e) and an electric-charge vector with magnitude $\frac{1}{2}$ which can assume two orientations (not in ordinary, three-dimensional space, but in some undefined space) such as to contribute either $+\frac{1}{2}$ to the resultant charge, to produce the proton, or to contribute $-\frac{1}{2}$, to produce the neutron. The proton and the neutron, according to this picture, constitute the two states of the electric-charge doublet of a nucleon with intrinsic charge $+\frac{1}{2}$ and electric-charge vector $\frac{1}{2}$. Similarly, the antiproton and the antineutron constitute the two states of the corresponding type of antimatter, the antinucleon, with intrinsic charge $-\frac{1}{2}$ and electric-charge vector $\frac{1}{2}$.

In 1961, some experimental results providing support for this picture of the proton and the neutron were reported by Robert Hofstadter and his coworkers at Stanford University and by a group of investigators at Cornell University. These physicists studied the scattering of high-speed electrons by protons and neutrons, and were able to interpret their experiments to determine the distribution of electric charge with the proton and the neutron.

They reported that both the proton and the neutron can be described as involving a central ball of positive charge, somewhat less than $0.5e$, extending to the radius about 0.3 fm. Surrounding the ball is a shell, extending to about 1 fm, and with positive charge $+\frac{1}{2}e$ for the proton and negative charge $-\frac{1}{2}e$ for the neutron. In addition, there is a fringe of positive electricity in both the proton and the neutron, amounting to about $0.15e$ and extending to about 1.5 fm.

It is possible that the fringe represents ephemeral mesons that constitute the mechanism of production of the strong internucleonic interactions. Except for the cloud of mesons surrounding it, the nucleon can be described, in its two states, the proton and the neutron, as consisting of a central ball of positive charge, $+\frac{1}{2}e$, which may be identified with the the intrinsic charge of the nucleon, and a shell, $+\frac{1}{2}e$ for the proton and $-\frac{1}{2}e$ for the neutron, representing the component of the electric-charge vector.

These results about the structure of the nucleon give exciting promise of great future developments in the understanding of the fundamental nature of the universe.

Several other charge doublets, corresponding to the two aspects of an electric-charge vector $\frac{1}{2}$ (also called *isotopic spin*), are known. In addition, as will be seen in the tables given in the following sections, there are several charge triplets that are known, groups of three particles with closely similar properties except for their electric charge, $+1$, 0, and -1. These charge triplets can be described as the three states of a single particle with electric-charge vector 1, which can have the component $+1$, 0, or -1. The three pions, π^+, π^0, π^-, constitute such a charge triplet (Section 20-9).

20-5. Leptons and Antileptons

We begin the tabulation of the fundamental particles by discussing the leptons and antileptons. There are eight of these particles known. Some of their properties are given in Table 20-1. Except for the muon and antimuon, they are stable particles. The word lepton is from the Greek *leptos*, small.

TABLE 20-1
Leptons and Antileptons*

Name	Electric Charge +1	0	−1	Mass	Xenicity (strangeness)	Spin
Electron			e^-	0.511 MeV	0	$\frac{1}{2}$
Muon			μ^-	105.66	0	$\frac{1}{2}$
Electron neutrino		ν		0	0	$\frac{1}{2}$R†
Muon neutrino		ν'		0	0	$\frac{1}{2}$R†
Positron	$\bar{e}^+$			0.511	0	$\frac{1}{2}$
Antimuon	$\bar{\mu}^+$			105.66	0	$\frac{1}{2}$
Electron antineutrino		$\bar{\nu}$		0	0	$\frac{1}{2}$L†
Muon antineutrino		$\bar{\nu}'$		0	0	$\frac{1}{2}$L†

*The electron, muon, and neutrino have lepton number +1; the positron, antimuon, and antineutrino have lepton number −1; all other particles have lepton number 0.
†The spin of the neutrinos corresponds to a right-handed screw, that of the antineutrinos to a left-handed screw.

The muon, μ^-, was the first particle with mass intermediate between the electron and the proton to be discovered. It is present in cosmic rays. It is made by the following reaction:

$$\bar{\pi}^- \longrightarrow \mu^- + \bar{\nu}'$$

The positive muon, the antimuon ($\bar{\mu}^+$), is made by a similar reaction from the positive pion. Both the positive pion and the negative pion are present in cosmic rays. They decompose rapidly, with half-life about 2.56 $\times 10^{-8}$ sec, to form muons. The muon and the antimuon themselves decompose, to form an electron (or positron), a neutrino, and an antineutrino:

$$\mu^- \longrightarrow e^- + \nu' + \bar{\nu}$$

$$\bar{\mu}^+ \longrightarrow \bar{e}^+ + \nu + \bar{\nu}'$$

The muon and the antimuon have no significance with respect to internucleonic forces. Their nature is uncertain. It is possible that they represent an excited state of the electron and positron.

Neutrinos and Antineutrinos. Weak Interactions

The neutrino is a particle with zero rest-mass and spin $\frac{1}{2}$; it differs from the photon primarily in the value of the spin (the photon has spin 1). The existence of the neutrino was proposed in 1927 by W. Pauli, in order to account for the apparent lack of conservation of energy in the process of emission of a β particle (an electron) by a radioactive nucleus, as discussed in Section 20-13. It had been observed that all radioactive nuclei of the same kind that emitted an α particle, such as radium 226 (Figure 20-6), shoot out their particles with the same energy, as expected from the law of conservation of mass-energy, but that, on the other hand, radioactive atoms that emit β particles, such as ^{214}Pb, emit the β particles with varying energies. Pauli, and later Fermi, suggested that another particle, with small or zero rest-mass, is also emitted when the nucleus undergoes radioactive decay with emission of a β particle, and that the energy of the reaction is divided between the β particle and the other particle, which Fermi named the neutrino.

In 1934 Fermi developed his theory of β decay, in order to explain the puzzling observation that some radioactive nucleides shoot out an electron in the course of radioactive decomposition, although they were supposed to be composed only of protons and neutrons. He pointed out that atoms emit photons when they change from one quantum state to another, although it is not believed that the atoms contain the photons; instead, it is accepted that the photon is created at the time when it is emitted. Fermi suggested that the electrons, the β particles, are created when the radioactive nucleus undergoes decomposition, and that at the same time one of the neutrons inside the nucleus becomes a proton, and a neutrino (or, rather, an antineutrino) is emitted.

The fundamental reaction of the Fermi theory is

$$n \longrightarrow p^+ + e^- + \bar{\nu}$$

This is the reaction of decomposition of the free neutron (Table 20-4). The free neutron decomposes with a half-life of 1040 s. In many nuclei the neutron is made stable by interaction with other nucleons, but in some nuclei it remains unstable, and this reaction takes place.

Neutrinos interact only very weakly with other particles, and the existence of the neutrino was not verified by experiment until 1956. In that year the American physicists Reines and Cowan showed that neutrinos from a nuclear reactor passing through a liquid-hydrogen bubble chamber cause a reaction to take place that is approximately the reverse of the decay of a neutron:

$$\bar{\nu} + p^+ \longrightarrow n + \bar{e}^+$$

The decay of a neutron into a proton, an electron, and a neutrino cannot be explained by strong interactions (Section 20-3) or by electromagnetic forces. Fermi assumed that another kind of interaction, called weak interaction, occurs among some particles. It is about 10^{-15} times as strong as the strong interactions that occur between nucleons and similar particles, and it leads to reaction times of the order of 10^{-8} s, instead of the time 10^{-23} s that applies to strong interactions.

Neutrinos and antineutrinos have spin $\frac{1}{2}$, but they have an extraordinary property that was discovered in 1957 as a result of the work of the Chinese physicists Tsung-Dao Lee (born 1926) and Chen Ning Yang (born 1922), working in the United States. These theoretical physicists and the experimental physicists whom they inspired found that the neutrino, which has spin $\frac{1}{2}$, always orients its spin in the direction of its motion, so that it moves through space with the speed of light as though it were a right-handed propeller. The antineutrino always orients its spin in the opposite direction, and moves as though it were a left-handed propeller.

In 1960 it was proposed by several physicists, in order to explain a number of experimental observations in a simple way, that there are two neutrinos and two antineutrinos, with somewhat different properties. It was postulated that one neutrino (ν) and one antineutrino ($\bar{\nu}$) have a close relation of some sort to the electron and positron, and the other neutrino (ν') and antineutrino ($\bar{\nu}'$) have a similar relation to the muon and antimuon. Experimental verification of this hypothesis was obtained in 1962 by a difficult experiment carried out by a group of Columbia University and Brookhaven National Laboratory scientists. As mentioned above, Reines and Cowan had shown that a neutrino produced by a reaction involving electrons reacts with a proton to produce a neutron and an electron. In the 1962 experiment it was shown that neutrinos produced by the decomposition of muons react with protons to produce only muons, and not electrons:

$$\bar{\nu}' + p^+ \longrightarrow n + \bar{\mu}^+$$

We shall call the two neutrinos the *electron neutrino*, ν, and the *muon neutrino*, ν'. At the present time nothing can be said about their nature, to explain the difference in their properties in terms of a difference in structure.

20-6. Mesons and Antimesons

The known mesons and antimesons, eight in number, are listed in Table 20-2. The kaons are the antiparticles of the antikaons, and the two charged pions are antiparticles of one another. The neutral pion is its own antiparticle, and the eta particle is its own antiparticle. All of the mesons are unstable; their decay reactions will be discussed in Section 20-8.

TABLE 20-2
Mesons and Antimesons*

Name	Electric Charge +1	0	−1	Mass	Intrinsic Charge	Charge Spin	Xenicity (strangeness)	Spin
Eta		η^0		550 MeV	0	0	0	0
Kaons		K^0	K^-	497.8, 494	$-\frac{1}{2}$	$\frac{1}{2}$	−1	0
Antikaons	$\bar{K}^+$	$\bar{K}^0$		494, 497.8	$+\frac{1}{2}$	$\frac{1}{2}$	+1	0
Pions	π^+	π^0	π^-	139.6, 135, 139.6	0	1	0	0

*The muon was originally named the meson, and then the μ meson, but it is now placed in the lepton class. Mesons and antimesons have baryon number 0 and lepton number 0. All the particles listed in this table have spin 0 (zero angular momentum). The positive pion and the negative pion are the antiparticles of one another. The neutral pion is its own antiparticle, and the eta is its own antiparticle. The inclusion of eta in this set of particles is somewhat arbitrary; see Section 20-10.

The pions and kaons were discovered in experiments with cosmic rays, and their properties have been determined by use both of cosmic rays and of high-energy particles produced by particle accelerators. The pions were discovered by Powell and his collaborators, as mentioned in Section 20-3. The kaons were discovered about 1950 by many investigators.

20-7. Baryons and Antibaryons

The baryons include the nucleons and heavier particles. Eight baryons and eight antibaryons are listed in Table 20-3. The word baryon is from the Greek *barys,* heavy. The word hyperon (Greek *hyper,* beyond) is also used; it refers to the baryons other than the proton and the neutron.

The baryons other than the proton and the neutron were discovered in the period between 1950 and 1960 by use of cosmic rays and particle accelerators. Their masses range from 1115 to 1318 MeV. All baryons are fermions, obeying the Pauli exclusion principle. Many heavier baryons, with spin $\frac{3}{2}$, $\frac{5}{2}$, . . . , have also been observed. They represent excited states (rotational states) of the fundamental baryons.

20-8. The Decay Reactions of the Fundamental Particles

Most of the fundamental particles decompose spontaneously. The exceptions, the stable particles, comprise the proton, and antiproton, the electron, the positron, and the particles that move with the speed of light.

TABLE 20-3
Baryons and Antibaryons*

Name	Electric Charge			Mass	Intrinsic Charge	Charge Vector	Xenicity (strangeness)	Spin
	+1	0	−1					
Xi particles		Ξ^0	Ξ^-	1311, 1318.4 MeV	$-\frac{1}{2}$	$\frac{1}{2}$	−2	$\frac{1}{2}$
Sigma particles	Σ^+	Σ^0	Σ^-	1189.4, 1191.5, 1196	0	1	−1	$\frac{1}{2}$
Lambda particle		Λ		1115.4	0	0	−1	$\frac{1}{2}$
Nucleons (proton, neutron)	p^+	n		938.2, 939.5	$+\frac{1}{2}$	$\frac{1}{2}$	0	$\frac{1}{2}$
Xi antiparticles	$\bar{\Xi}^+$	$\bar{\Xi}^0$		1318.4, 1311	$+\frac{1}{2}$	$\frac{1}{2}$	+2	$\frac{1}{2}$
Sigma antiparticles	$\bar{\Sigma}^+$	$\bar{\Sigma}^0$	$\bar{\Sigma}^-$	1196, 1191.5, 1189.4	0	1	+1	$\frac{1}{2}$
Lambda antiparticle		$\bar{\Lambda}$		1115.4	0	0	+1	$\frac{1}{2}$
Antineutron, antiproton		$\bar{n}$	p^-	939.5, 938.2	$-\frac{1}{2}$	$\frac{1}{2}$	0	$\frac{1}{2}$

*Baryons have baryon number +1. Antibaryons have baryon number −1. Both have lepton number 0.

Even though many of the fundamental particles were discovered only a few years ago, a tremendous amount of information has been gained about their properties and the reactions by which they are produced, changed into other forms of matter, and destroyed. The reactions by which the unstable particles decay are listed in Table 20-4, which also gives the values of the half-life. All of these decay reactions are unimolecular reactions, the nature of which has been discussed in Chapter 10.

TABLE 20-4
Reactions of Decay of Particles

	Reaction	Ratio (%)	Half-life (seconds)
Baryons:	$\Xi^0 \rightarrow \Lambda + \pi^0$		$\sim 2 \times 10^{-10}$
	$\Xi^- \rightarrow \Lambda + \pi^-$		2×10^{-10}
	$\Sigma^+ \rightarrow p^+ + \pi^0$	46 ± 6	0.8×10^{-10}
	$n + \pi^+$	54 ± 6	
	$\Sigma^0 \rightarrow \Lambda + \gamma$		$\sim 10^{-20}$
	$\Sigma^- \rightarrow n + \pi^-$		1.6×10^{-10}
	$\Lambda \rightarrow p^+ + \pi^-$	63 ± 3	2.4×10^{-10}
	$n + \pi^0$	37 ± 3	
	$n \rightarrow p^+ + e^- + \bar{\nu}$		1040
Mesons:	$\eta^0 \rightarrow \pi^+ + \pi^0 + \pi^-$		$\sim 10^{-23}$
	$K_1{}^0 \rightarrow \pi^+ + \pi^-$	78 ± 6	1.0×10^{-10}*
	$\pi^0 + \pi^0$	21 ± 6	
	$K_2{}^0 \rightarrow \pi^+ + \pi^-$	78 ± 6	6×10^{-8}
	$\pi^0 + \pi^0$	22 ± 6	
	$K^- \rightarrow \mu^- + \bar{\nu}'$	59 ± 2	1.22×10^{-8}
	$\pi^0 + \pi^-$	26 ± 2	
	$\pi^+ + \pi^- + \pi^-$	5.7 ± 0.3	
	$\pi^0 + \pi^0 + \pi^-$	1.7 ± 0.3	
	$e^- + \bar{\nu} + \pi^0$	4.2 ± 0.4	
	$\mu^- + \bar{\nu}' + \pi^0$	4.0 ± 0.8	
	$\pi^+ \rightarrow \bar{\mu}^+ + \nu'$	100	2.56×10^{-8}
	$\bar{e}^+ + \nu$	0.013	
	$\pi^0 \rightarrow \gamma + \gamma$		2×10^{-15}
Leptons:	$\mu^- \rightarrow e^- + \nu' + \bar{\nu}$		10^{-6}

*In a beam of neutral kaons K^0 and antikaons $\bar{K}^0$ the particles decompose at two rates, to give the same products. This behavior is explained by saying that the beam contains particles $K_2{}^0$ that are in the quantum state corresponding to symmetric resonance of K^0 and $\bar{K}^0$ and also particles $K_1{}^0$ that are in the quantum state corresponding to antisymmetric resonance of K^0 and $\bar{K}^0$. The kaon-antikaon pair is the only pair known to have this property.

Conservation Principles

By analyzing the tracks produced by individual particles in cloud chambers, stacks of photographic emulsions, and bubble chambers, and by other methods of detecting particles, the decay of individual particles has been studied, and it has been found that in every case there is conservation of mass-energy and conservation of momentum. Other conservation principles have also been found to be adhered to rigorously, as follows:

Conservation of angular momentum
Conservation of electric charge
Conservation of baryon number
Conservation of lepton number

The principle of conservation of electric charge is illustrated by the decay reactions given in Table 20-4. For example, the lambda particle, which is a hyperon, with mass somewhat greater than that of a nucleon, can decompose either to form a proton and a negative pion or to form a neutron and a neutral pion. In the first case the lambda particle, which is neutral, forms a positively charged particle and a negatively charged particle; in the second case it forms two neutral particles.

A more complicated example, also given in Table 20-4, is the decomposition of the negative kaon. This particle has been observed to decompose in six different ways. Five of the reactions of decomposition lead to the formation of a negatively charged particle and one or two neutral particles. The sixth reaction leads to the formation of a positively charged particle, a positive pion, and two negatively charged particles, negative pions. Hence in each of the six reactions there is conservation of electric charge.

There is also conservation of the baryon number in every reaction. The baryons have baryon number +1 and the antibaryons have baryon number −1; all other particles have baryon number 0. In the various processes of formation of baryons and antibaryons they are always formed in pairs, one baryon and one antibaryon. Similarly, the decomposition of a baryon always leads to the formation of another baryon, plus other particles with 0 baryon number. Thus the negative xi particle is observed to decompose to form a lambda particle, which has baryon number +1, and a negative pion, which has baryon number 0.

Leptons, which include the electron, the neutrino, and the muon, have lepton number +1, and antileptons have lepton number −1; all other particles have lepton number 0. There is rigorous conservation of the lepton number in all reactions.

There are also some conservation principles that are observed to hold for strong interactions but not for weak interactions. This matter is discussed in the following section.

20-9. Strangeness (Xenicity)

A great contribution to the understanding of the nature of the fundamental particles was made in the period between 1953 and 1956 by the American physicist Murray Gell-Mann and the Japanese physicist K. Nishijima, working independently. The classification of the fundamental particles given in Tables 20-1, 20-2, and 20-3 is in considerable part due to their efforts. This classification is based upon the concept of charge multiplets and the concept of strangeness. Neither of these concepts can be said to be thoroughly understood at the present time, and it is likely that some additional great contributions will be made in the near future.

In Section 20-4 it was pointed out that the close similarity in properties of the neutron and the proton, except for electric charge, suggests that these two particles represent two aspects of the same particle, the nucleon. The nucleon may be said to have intrinsic electric charge $+\frac{1}{2}$ and electric-charge vector $\frac{1}{2}$, which can have the component $+\frac{1}{2}$ or $-\frac{1}{2}$ in ordinary space, leading to the resultant electric charge +1 for the proton and 0 for the neutron. The proton and neutron can then be described as a charge doublet.

The diagram in Figure 20-5 shows that the 24 particles represented in the diagram constitute three charge singlets, six doublets, and three triplets. The charge singlets have electric-charge vector equal to 0, and intrinsic charge 0. The doublets all have electric-charge vector equal to $\frac{1}{2}$; the nucleon, antinucleon, kaons, and xions have intrinsic charge $+\frac{1}{2}$ or $-\frac{1}{2}$. A charge doublet can thus have electric charges either 0 and +1 or 0 and −1. The triplets, with electric-charge vector 1 and intrinsic charge 0, have electric charges +1, 0, and −1, corresponding to the three orientations of the charge vector.

The idea of strangeness was introduced by Gell-Mann and Nishijima to explain in a rough way the rates of decay reactions. Some of the unstable particles are expected to decay by virtue of the strong interactions (Section 20-3), and this decomposition should be very rapid, with half-lives of the order of 10^{-23} s. An example is the decay of the η^0 particle, to form three pions; its half-life is about 10^{-23} s.

Many other particles, however, are observed to have much longer half-lives, of the order of 10^{-9} s. These particles accordingly live 10^{14} times as long as predicted for them on the basis of the theory of strong interactions.

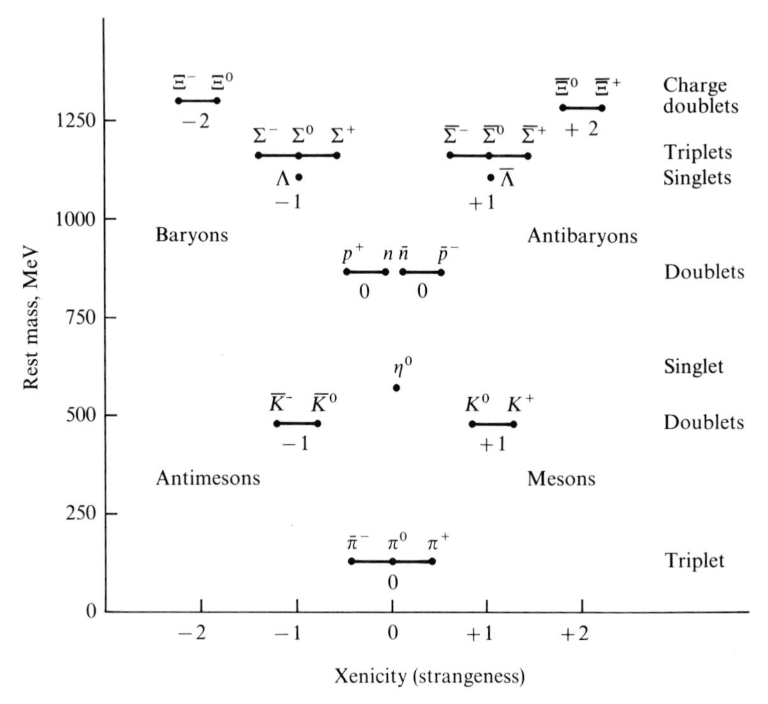

FIGURE 20-5
A diagram representing the masses and xenicities of some of the elementary particles.

Gell-Mann and Nishijima suggested that a characteristic property, which is called strangeness, should be assigned to the particles, such that there is conservation of strangeness for reactions involving strong interactions but violation of conservation of strangeness for weak interactions. *Xenicity** may be a better name for this property than strangeness.

The values of the xenicity are shown in Figure 20-5. Pions, the eta particle, nucleons, and antinucleons have xenicity 0. The kaons, antilambda particle, and antisigma particles have xenicity +1, and the antikaons, lambda particle, and sigma particles have xenicity −1. The antixions have xenicity +2, and the xions have xenicity −2.

The conservation principle is that for strong interactions there must be conservation of xenicity; the sum of the xenicities for the reactants equals the sum of the xenicities for the products. Reactions in which the sum of the xenicities changes by one unit can occur as a result of weak interactions, but these reactions are slow. Reactions in which the sum changes by two units are very slow.

---

*From Greek *xenos,* stranger.

The eta particle has xenicity 0, and thc pions have xenicity 0. There is, accordingly, no change in xenicity accompanying the decay of the eta particle, and the reaction is very fast.

Table 20-4 contains many examples of reactions in which there is a change in xenicity. The negative antikaon, $\bar{K}^-$, has xenicity −1. It can decay in six ways, to form pions and leptons (the muon, the electron, an antineutrino), all of which have xenicity 0. The total half-life for these various reactions is 1.22×10^{-8} s, far longer than the half-life for the eta decomposition, and this long half-life is attributed to the change in xenicity.

20-10. Resonance Particles and Complexes

In 1952 it was found by Enrico Fermi and his coworkers, who were studying the scattering of a beam of pions by protons, that the scattering is much larger when the pions have about 200 MeV of kinetic energy than for smaller or greater amounts of kinetic energy. This observation was interpreted as showing that there is a strong interaction between the pion and the proton, which can be described as corresponding to the formation of a short-lived particle or complex, to which the symbol N^* has been assigned:

$$\pi + p \rightleftarrows N^*$$

The mass of N^* is about 1237 MeV. Its half-life is about 10^{-23} s; and, in accordance with the uncertainty relation between energy and time (Section 3-13), the mass (energy) of the particle is not well defined, but has an uncertainty of about ±60 MeV. (The half-life 10^{-23} s is in fact calculated from the observed distribution function for the mass of the N^* complex.)

Since 1952 about 100 of these short-lived particles or complexes have been discovered. They are called *resonance particles* or *resonance complexes*. One of them, η^0, has been included in our listing of the mesons (Table 20-2). It is produced by reaction of a pion and a neutron (within a deuteron):

$$\pi^+ + d^+ \longrightarrow \eta^0 + p^+ + p^+$$

It is the lightest of the resonance particles, with mass 550 MeV. It decomposes, with half-life 10^{-23} s, by the reaction

$$\eta^0 \longrightarrow \pi^+ + \pi^- + \pi^0$$

The next lightest resonance particles are the ρ particles, ρ^+, ρ^0, and ρ^-, with mass 760 MeV. They constitute a charge triplet (inherent charge 0, charge vector 1, angular momentum spin 1). They are formed by the following reactions:

$$\pi^+ + p^+ \longrightarrow \rho^+ + p^+$$

$$p^+ + \bar{p}^- \longrightarrow \rho^0 + \pi^+ + \pi^-$$

$$\pi^- + p^+ \longrightarrow \rho^- + p^+$$

They decompose, with half-life about 10^{-23} s, as follows:

$$\rho^+ \longrightarrow \pi^+ + \pi^0$$

$$\rho^0 \longrightarrow \pi^+ + \pi^- \quad \text{or} \quad \rho^0 \longrightarrow \pi^0 + \pi^0$$

$$\rho^- \longrightarrow \pi^- + \pi^0$$

The only other known resonance particle that resembles the eta particle and the rho particles in having lepton number 0, baryon number 0, and strangeness 0 is the ω^0 particle, which has mass 790 MeV, inherent charge 0, charge vector 0, and angular momentum spin 1. It has been observed to be formed in the following ways:

$$p^+ + \bar{p}^- \longrightarrow \omega^0 + \pi^+ + \pi^-$$

$$\pi^+ + d^+ \longrightarrow \omega^0 + p^+ + p^+$$

It has half-life 4×10^{-23} s, corresponding to two ways of decomposing:

$$\omega^0 \longrightarrow \pi^+ + \pi^- + \pi^0$$

or

$$\omega^0 \longrightarrow \pi^0 + \gamma$$

Other resonance particles or complexes that decompose into pions and kaons or into pions or kaons and one of the baryons are also known. Their masses lie in the range from 880 MeV to 2000 MeV.

At the present time there is no satisfactory theory of these particles. It seems likely, however, that some of them may be classed with the fundamental particles (see the following section), and that others may be described as complexes of two or more fundamental particles, possibly with resonance among several structures, roughly analogous to the resonance of molecules among several valence-bond structures.

In November 1974 it was reported by investigators in Stanford University and the University of California, Berkeley, that a new particle, formed by collision of a fast-moving electron and a fast-moving positron, had been observed. The electron and positron, moving in opposite directions, had been accelerated to the energy just over 1550 MeV, and the mass of the new particle was found to be 3105 MeV. The independent discovery of the same particle by another method was announced at the same time from Massachusetts Institute of Technology and Brookhaven National Laboratory. A second particle, with similar properties and also obtained by collision of an electron and a positron, was discovered later in November 1974 by the Stanford-Berkeley group. Its mass is 3695 MeV. The existence of these particles had not been predicted, and their nature is not clear at the present time. It is likely that studies of these new particles will lead to greater understanding of the fundamental particles and the forces between them.

20-11. Quarks

Theoretical physicists are now attempting to develop a structural theory of the fundamental particles. At the present time the most promising idea is that mesons and baryons are made of quarks, each meson being a diquark (a compound of a quark and an antiquark), and each baryon being a triquark. The idea was developed independently in 1964 by Murray Gell-Mann and George Zweig.

The three quarks are the positive quark, p, the negative quark, n, and the strange quark, λ. (Note that we use p and n for the quarks, *p* and *n* for the nucleons.) All three are fermions, with spin $\frac{1}{2}$; p has electric charge $+\frac{2}{3}$, and n and λ have charge $-\frac{1}{3}$. The antiquarks $\bar{\text{p}}$, $\bar{\text{n}}$, and $\bar{\lambda}$ have charges $-\frac{2}{3}$, $+\frac{1}{3}$, and $+\frac{1}{3}$, respectively. Each quark has baryon number $+\frac{1}{3}$ (each antiquark $-\frac{1}{3}$); λ has xenicity 1 and $\bar{\lambda}$ has xenicity −1.

The failure so far to observe any single quarks among the products of high-energy reactions indicates that they have very large mass, several thousand MeV. The effective mass of λ in diquarks is about 145 MeV greater than that of n and p; that of n is about 4 MeV greater than that of p.

Let us consider the diquarks with baryon number 0—that is, the compounds of a quark and an antiquark. The most stable diquarks are expected to be those in which both particles are in a 1*s* orbital, as they move about their common center of mass. The quarks and antiquarks are different particles; hence the Pauli exclusion principle does not forbid parallel spins for a quark and its antiquark, and a $1s^2$ diquark can have resultant spin 0 or resultant spin 1. The mesons π^+, π^0, and π^- are p$\bar{\text{n}}$, p$\bar{\text{p}}$, (or n$\bar{\text{n}}$), and $\bar{\text{p}}$n, respectively, and various other mesons are similarly represented. The proton is represented by p^2n and the neutron by pn^2. The strange quark λ (and its antiparticle $\bar{\lambda}$) are found in the mesons and baryons with xenicity different from zero.

The nature of the strange quark λ is not as yet understood. One possibility is that λ bears the same structural relation to n that the muon bears to the electron; this relation, however, is also not yet clear.

20-12. Positronium, Muonium, Mesonic Atoms

In 1953 it was observed that a positron and an electron combine to form a pseudo-atom, somewhat similar to the hydrogen atom. In the hydrogen atom the electron can be described as moving around an essentially stationary nucleus, the proton. In the pseudo-atom formed by a positron and an electron, which has been given the name *positronium,* the two particles have the same mass, so that they carry out similar motions about their center of mass, the point midway between the two.

It was found by the American physicist Martin Deutsch that there are two kinds of positronium. The kind in which the spin of the positron is antiparallel to that of the electron is called parapositronium, and that in which the two spins are parallel is called orthopositronium. Parapositronium decomposes with destruction of the positron and the electron and production of two photons, its half-life being 0.9×10^{-10} s. Orthopositronium decomposes with production of three photons, and half-life 1.0×10^{-7} s. The existence of positronium was detected by the observation of a delay between its production (by decomposition of sodium 22, which emits positrons) and its annihilation. The time of delay was found to correspond to the sum of two first-order reactions, with the values of the half-life given above.

Muonium, a pseudo-atom involving a negative muon moving about a proton, has also been observed. Other mesonic atoms, having structures similar to ordinary atoms but with a muon or other meson replacing one of the electrons, have also been observed. For example, muonic neon is a neon atom with a negative muon in place of an electron.

A muonic molecule ion, $[H^+\mu^-D^+]^+$, in which a proton and a deuteron are held together by a negative muon, has also been made. The proton and the deuteron are sufficiently close together, about 0.3 pm apart, to permit reaction between them, liberating the muon and producing a helium-3 nucleus plus an additional muon, with release of 5.4 MeV of energy. The use of a mesonic molecule of this sort might possibly permit the controlled release of energy through nuclear fusion.

20-13. Natural Radioactivity

The field of nuclear chemistry deals with the reactions that involve changes in atomic nuclei. This field began with the discovery of radioactivity and the work of Pierre and Marie Curie on the chemical nature of the radioactive substances. After some decades, during which natural radioactivity was rather thoroughly investigated, a great increase in knowledge resulted through the discovery of artificial radioactivity.

Nuclear chemistry has now become a large and important branch of science. About 1000 radioactive nucleides (isotopes) have been made in the laboratory, whereas only about 272 stable nucleides and 55 unstable (radioactive) nucleides have been detected in nature. The use of radioactive isotopes as "tracers" has become a valuable technique in scientific and medical research. The controlled release of nuclear energy provides an important new supply of energy.

After their discovery of polonium and radium in 1898 (Chapter 3), the Curies found that radium chloride could be separated from barium chloride by fractional precipitation of the aqueous solution by addition of alcohol, and by 1902 Madame Curie had prepared 0.1 g of nearly pure radium chloride, with radioactivity about 3,000,000 times that of uranium. Within a few years it had been found that natural radioactive materials emit three kinds of rays capable of sensitizing the photographic plate (Chapter 3). These rays—alpha rays, beta rays, and gamma rays—are affected differently by a magnetic field (Figure 3-10). Alpha rays are the nuclei of helium atoms, moving at high speeds; beta rays are electrons, also moving at high speeds; and gamma rays are photons, with very short wavelengths.

It was soon discovered that the rays from radium and other radioactive elements cause regression of cancerous growths. These rays also affect normal cells, "radium burns" being caused by overexposure; but often the cancerous cells are more sensitive to radiation than normal cells, and can be killed by suitable treatment without serious injury to normal tissues. The medical use in the treatment of cancer is the main use for radium. Since about 1950, considerable use has also been made of the artificial radioactive isotope cobalt 60 as a substitute for radium (Section 20-15).

Through the efforts of many investigators the chemistry of the radioactive elements of the uranium series and the thorium series was unraveled during the first two decades of the twentieth century, and that of the neptunium series during a few years from 1939 on.

The Uranium Series of Radioactive Disintegrations

When an alpha particle (He^{++}) is emitted by an atomic nucleus the nuclear charge decreases by two units; the element hence is transmuted into the element two columns to the left in the periodic table. Its mass number (atomic mass) decreases by 4, the mass of the alpha particle. When a beta particle (an electron) is emitted by a nucleus the nuclear charge is increased by one unit, with no change in mass number (only a very small decrease in atomic mass); the element is transmuted into the element one column to its right. No change in atomic number or mass number is caused by emission of a gamma ray.

The nuclear reactions in the *uranium-radium series* are shown in Figure 20-6. The principal isotope of uranium, ^{238}U, constitutes 99.28% of

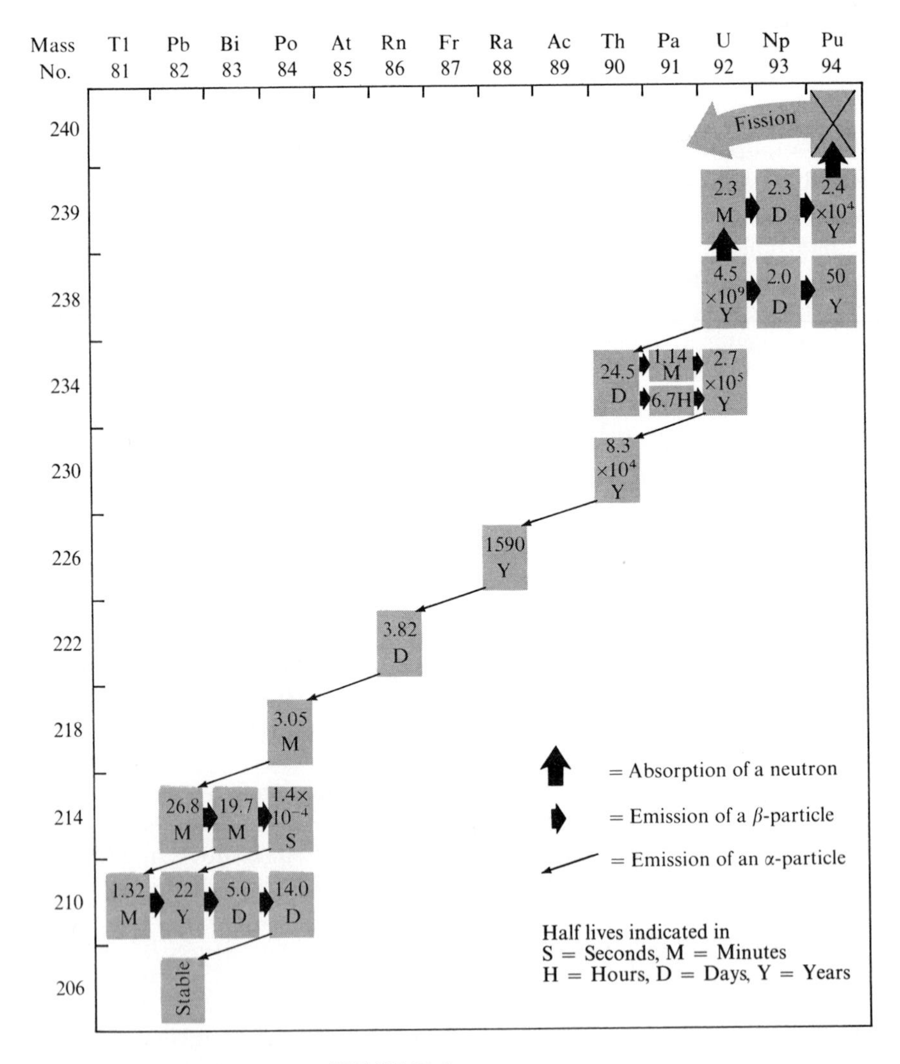

FIGURE 20-6
The uranium-radium series.

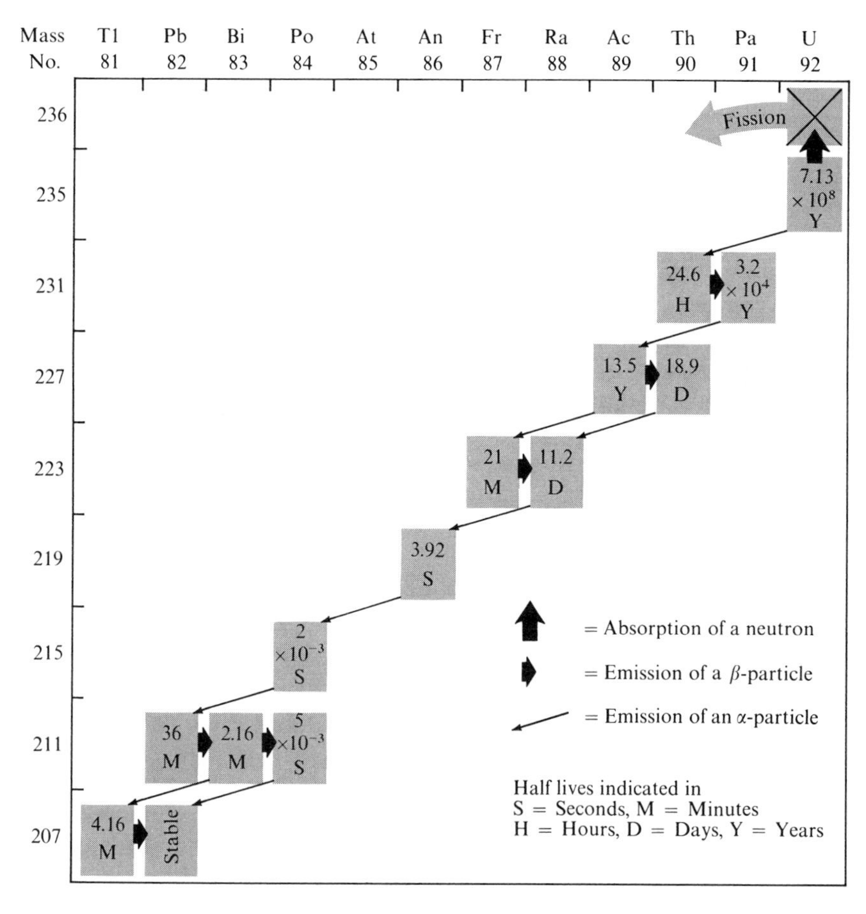

FIGURE 20-7
The uranium-actinium series.

the natural element. This isotope has a half-life of 4,500,000,000 years. It decomposes by emitting an alpha particle and forming ^{234}Th. This isotope of thorium undergoes decomposition with β-emission, forming ^{234}Pa, which in turn forms ^{234}U. Five successive α-emissions then occur, giving ^{214}Pb, which ultimately changes to ^{206}Pb, a stable isotope of lead.

The *uranium-actinium series,* shown in Figure 20-7, is a similar series beginning with ^{235}U, which occurs to the extent of 0.71% in natural uranium. It leads, through the emission of seven alpha particles and four beta particles, to the stable isotope ^{207}Pb.

The Thorium Series

The third natural radioactive series begins with the long-lived, naturally occurring isotope of thorium, ^{232}Th, which has half-life 1.39×10^{10} years (Figure 20-8). It leads to another stable isotope of lead, ^{208}Pb.

The Neptunium Series

The fourth radioactive series (Figure 20-9) is named after its longest-lived member, which is ^{237}Np.

The nature of radioactive disintegration within each of the four series—the emission of β-particles, with mass nearly zero, or of α-particles, with

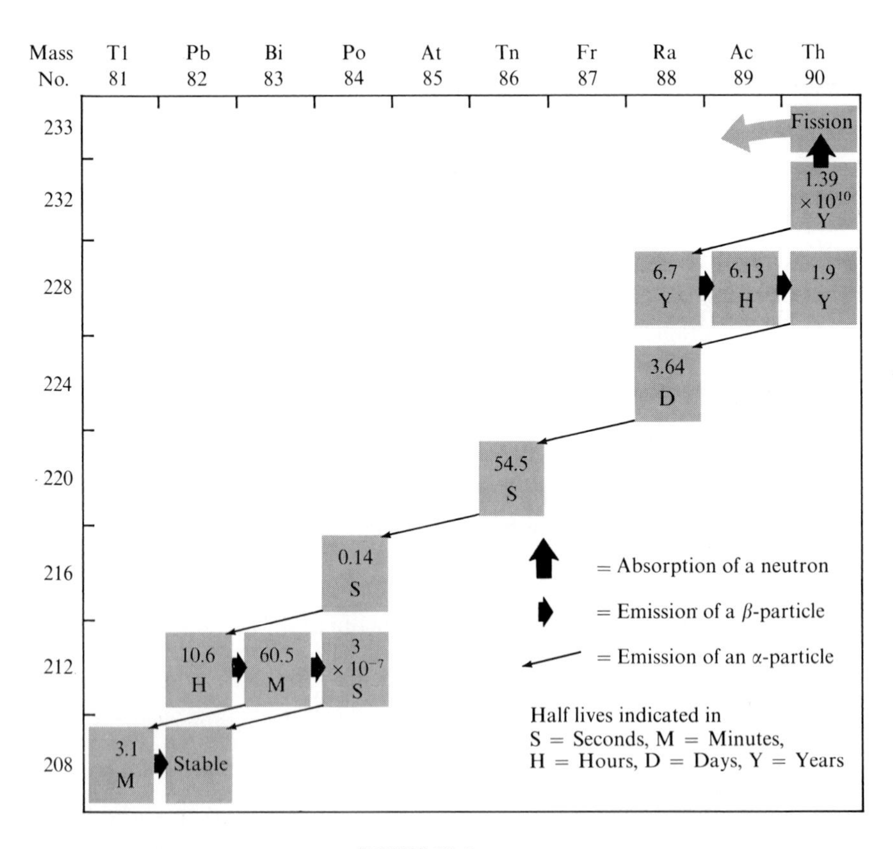

FIGURE 20-8
The thorium series.

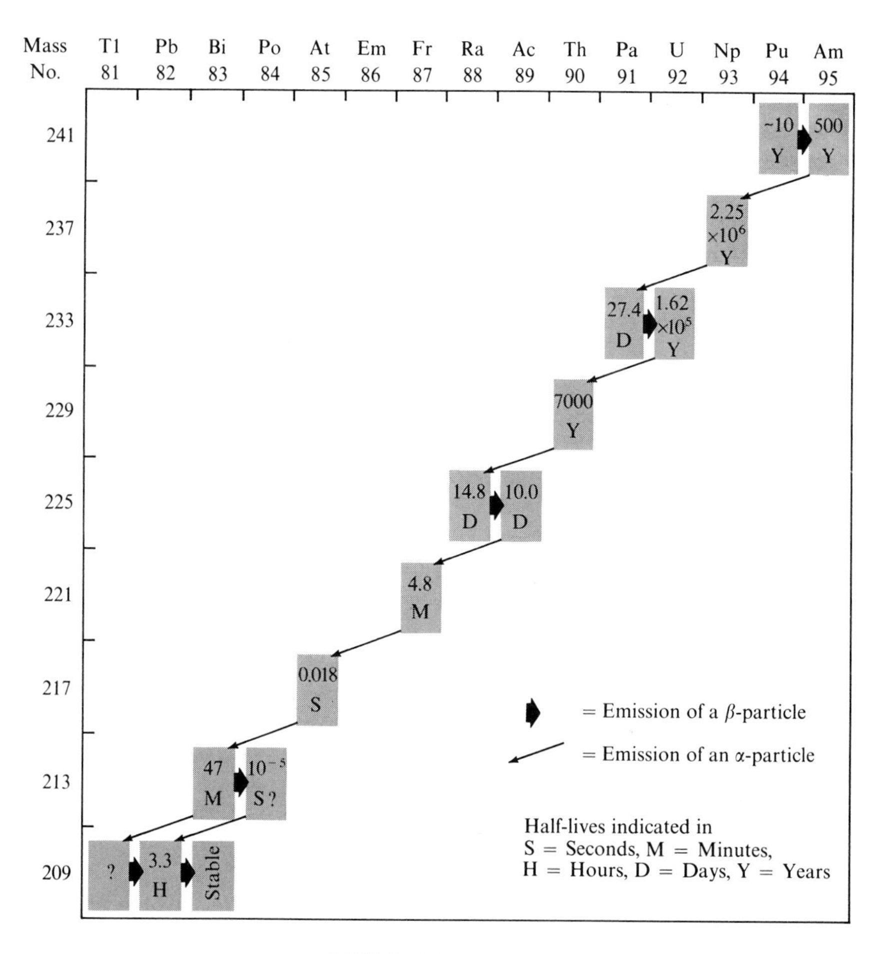

FIGURE 20-9
The neptunium series.

mass 4—is such that all the members of a series have mass numbers differing by a multiple of 4. The four series can hence be classified as follows (n being integral):

The $4n$ series = the thorium series

The $4n + 1$ series = the neptunium series

The $4n + 2$ series = the uranium-radium series

The $4n + 3$ series = the uranium-actinium series

20-14. The Age of the Earth

Measurements made on rocks containing radioactive elements can be interpreted to provide values of the age of the rocks, and hence of the age of the earth; that is, the time that has elapsed since the oldest rocks were laid down. For example, 1 g of ^{238}U would in its half-life of 4.5×10^9 years decompose to leave 0.5000 g of ^{238}U and to produce 0.0674 g of helium and 0.4326 g of ^{206}Pb. (Each atom of ^{238}U that decomposes forms eight atoms of helium, with total mass 32, leaving one atom of ^{206}Pb.) If analyses showed that the nucleides were present in a rock in the ratios of these numbers, the rock would be assumed to be 4.5×10^9 years old. The $^{235}U/^{207}Pb$ ratio, the $^{232}Th/^{208}Pb$ ratio, the $^{40}K/^{40}Ar$ ratio, and the $^{87}Rb/^{87}Sr$ ratio are also being used for determining the ages of rocks. Ages as great as 4.6×10^9 years have been determined for rocks found in Finland, Canada, and Africa, the moon, and meteorites. The present estimate of the age of the crust of the earth and other parts of the solar system is 4.6×10^9 years. There is also evidence that the cosmic process of building heavy nuclei from light ones was still going on as recently as 4.9×10^9 years ago.

20-15. Artificial Radioactivity

Stable atoms can be converted into radioactive atoms by bombardment with particles traveling at high speeds. In the early experimental work the highspeed particles used were alpha particles from ^{214}Bi (called radium C). The first nuclear reaction produced in the laboratory was that between alpha particles and nitrogen, carried out by Rutherford and his collaborators in the Cavendish Laboratory at Cambridge in 1919. The nuclear reaction that occurs when nitrogen is bombarded with alpha particles is the following:

$$^{14}_{7}N + ^{4}_{2}He \rightarrow ^{17}_{8}O + ^{1}_{1}H$$

In this reaction a nitrogen nucleus reacts with a helium nucleus, which strikes it with considerable energy, to form two new nuclei, a ^{17}O nucleus and a proton.

The ^{17}O nucleus is stable, so that this nuclear reaction does not lead to the production of artificial radioactivity. Many other elements, however, undergo similar reactions with the production of unstable nuclei, which then undergo radioactive decomposition.

Many nuclear reactions result from the interaction of nuclei and neutrons. The early experiments with neutrons were carried out by use of a mixture of radon and beryllium metal. The alpha particles from radon react with the beryllium isotope ^{9}Be to produce neutrons in the following ways:

$$^{9}_{4}Be + ^{4}_{2}He \longrightarrow ^{12}_{6}C + ^{1}_{0}n$$

$$^{9}_{4}Be + ^{4}_{2}He \longrightarrow 3\ ^{4}_{2}He + ^{1}_{0}n$$

Neutrons are also prepared by reactions in the cyclotron and in uranium reactors.

Manufacture of the Transuranium Elements

The first transuranium element to be made was a neptunium isotope, $^{239}_{93}Np$. This nucleide was made by E. M. McMillan and P. H. Abelson in 1940 by bombarding uranium with high-speed deuterons:

$$^{238}_{92}U + ^{2}_{1}H \longrightarrow ^{239}_{92}U + ^{1}_{1}H$$

$$^{239}_{92}U \longrightarrow ^{239}_{93}Np + e^-$$

The first isotope of plutonium to be made was ^{238}Pu, by the reactions

$$^{238}_{92}U + ^{2}_{1}H \longrightarrow ^{238}_{93}Np + 2\ ^{1}_{0}n$$

$$^{238}_{93}Np \longrightarrow ^{238}_{94}Pu + e^-$$

The ^{238}Np decomposes spontaneously, emitting electrons. Its half-life is 2.0 days.

During and since World War II some quantity, of the order of one million kilograms, of the nucleide ^{239}Pu has been manufactured. This nucleide is relatively stable; it has a half-life of about 24,000 years. It slowly decomposes with the emission of alpha particles. It is made by the reaction of the principal isotope of uranium, ^{238}U, with a neutron, to form ^{239}U, which then undergoes spontaneous radioactive decomposition with emission of an electron to form ^{239}Np, which in turn emits an electron spontaneously, forming ^{239}Pu:

$$^{238}_{92}U + ^{1}_{0}n \longrightarrow ^{239}_{92}U$$

$$^{239}_{92}U \longrightarrow ^{239}_{93}Np + e^-$$

$$^{239}_{93}Np \longrightarrow ^{239}_{94}Pu + e^-$$

Plutonium and the next four transuranium elements—americium, curium, berkelium, and californium—were discovered by G. T. Seaborg and his collaborators at the University of California in Berkeley. Americium has been made as ^{241}Am by the following reactions:

$$^{238}_{92}U + ^{4}_{2}He \longrightarrow ^{241}_{94}Pu + ^{1}_{0}n$$

$$^{241}_{94}Pu \longrightarrow ^{241}_{95}Am + e^{-}$$

This nucleide slowly undergoes radioactive decomposition, with emission of alpha particles. Its half-life is 500 years. Curium is made from plutonium 239 by bombardment with helium ions accelerated in the cyclotron:

$$^{239}_{94}Pu + ^{4}_{2}He \longrightarrow ^{242}_{96}Cm + ^{1}_{0}n$$

The nucleide ^{242}Cm is an alpha-particle emitter, with half-life about 5 months. Other isotopes of curium have also been made. One is ^{240}Cm, made by bombarding plutonium, ^{239}Pu, with high-speed helium ions:

$$^{239}_{94}Pu + ^{4}_{2}He \longrightarrow ^{240}_{96}Cm + 3\ ^{1}_{0}n$$

Using only very small quantities of the substances, Seaborg and his collaborators succeeded in obtaining a considerable amount of information about the chemical properties of the transuranium elements. They have found that, whereas uranium is similar to tungsten in its properties, in that it has a pronounced tendency to assume oxidation state +6, the succeeding elements are not similar to rhenium, osmium, iridium, and platinum, but show an increasing tendency to form ionic compounds in which their oxidation number is +3. This behavior is similar to that of the rare-earth metals.

20-16. The Kinds of Nuclear Reactions

Many different kinds of nuclear reactions have now been studied. Spontaneous radioactivity is a nuclear reaction in which the reactant is a single nucleus. Other known nuclear reactions involve a proton, a deuteron, an alpha particle, a neutron, or a photon (usually a gamma ray) interacting with the nucleus of an atom. The products of a nuclear reaction may be a heavy nucleus and a proton, an electron, a deuteron, an alpha particle, a neutron, two or more neutrons, or a gamma ray. In addition, there occurs a very important type of nuclear reaction in which a very heavy nucleus, made unstable by the addition of a neutron, breaks up into two parts of comparable size, plus several neutrons. This process of fission has been mentioned in Chapter 19 and is described in a later section of the present chapter.

The following radioactive decompositions are examples of the different ways in which an unstable nucleus can decompose:

	Half-life
${}^{3}_{1}H_2 \longrightarrow e^- + {}^{3}_{2}He_1 + \bar{\nu}$	12.26 y
${}^{148}_{64}Gd_{84} \longrightarrow \alpha^{++} + {}^{144}_{62}Sm_{82}$	130 y
${}^{15}_{8}O_7 \longrightarrow \bar{e}^+ + {}^{15}_{7}N_8 + \nu$	124 s
${}^{37}_{18}Ar_{19} + e^- \longrightarrow {}^{37}_{17}Cl_{20} + \nu$	35.0 d
${}^{5}_{3}Li_2 \longrightarrow p^+ + {}^{4}_{2}He_2$	10^{-21} s
${}^{5}_{2}He_3 \longrightarrow n + {}^{4}_{2}He_2$	2×10^{-21} s

The first two reactions are found for many of the heavy radioactive nucleides. Alpha emission has been observed also for a number of the neutron-rich nucleides in the rare-earth region. The third reaction, positron emission, occurs for most neutron-rich nucleides, many of which also decompose by electron capture (the fourth reaction). (Electron capture is classed as a spontaneous decomposition because the electrons are always available in the atom for capture; it is the *s* electrons, principally 1*s*, that are captured; they are the only electrons with finite probability at the nucleus.) The last two reactions, proton and neutron emission, occur only rarely.

Most β decompositions (either e^+ or e^-) are accompanied (immediately followed) by the emission of γ rays. The β decomposition may take place to one or more of the excited states of the product nucleus, which then drops to the normal state by γ emission. A simple example is shown in Figure 20-10. A great amount of information about energy levels of nuclei has been obtained by measuring the wavelengths of the photons (the γ rays) and the maximum kinetic energy of the β rays (the maximum corresponds to zero energy for the neutrino).

Nuclear reactions are caused to take place by bombarding nuclei with photon, neutrons, protons, deuterons, tritons (${}^3H^+$), trelions (${}^3He^{++}$), helions (alpha particles), or heavier nuclei. An example is the production of ${}^{32}P$ by bombarding ordinary phosphorus, ${}^{31}P$, with 10-MeV deuterons:

$$ {}^{31}_{15}P_{16} + {}^{2}_{1}H_1 \longrightarrow {}^{32}_{15}P_{17} + {}^{1}_{1}H_0 $$

Nuclear reactions are usually written in a shorter way:

$$ {}^{31}P(d, p){}^{32}P $$

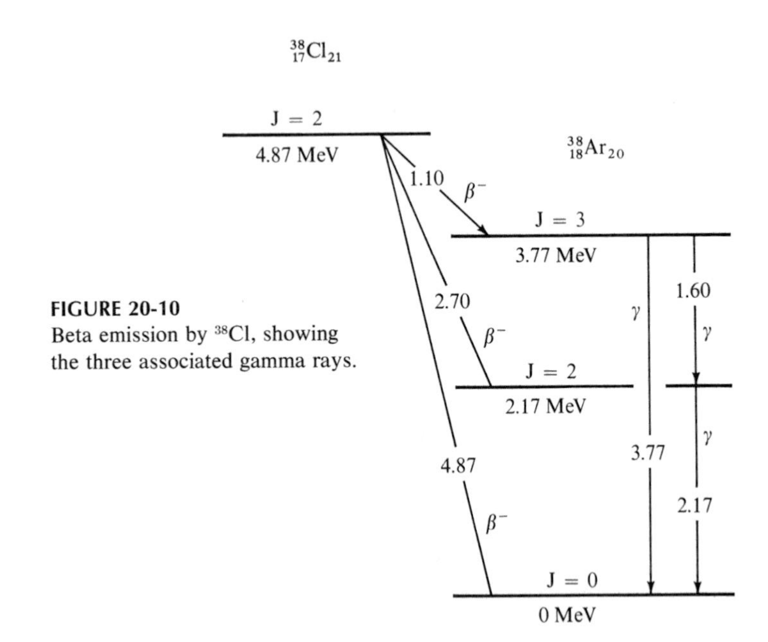

FIGURE 20-10
Beta emission by ^{38}Cl, showing the three associated gamma rays.

The symbol (*d*, *p*) means that the bombarding (reacting) particle is a deuteron and the emitted particle is a proton. The following are examples of other nuclear reactions of this general type:

$^{6}Li(n, \alpha)^{3}H$	$^{11}Be(d, p)^{12}Be$
$^{7}Li(n, \gamma)^{8}Li$	$^{11}Be(t, p)^{13}Be$
$^{9}Be(n, p)^{9}Li$	$^{10}B(p, 2n)^{9}C$
$^{9}Be(d, 2p)^{9}Li$	$^{10}B(p, \gamma)^{11}C$
$^{6}Li(d, n)^{7}Be$	$^{10}B(\alpha, n)^{13}N$
$^{10}B(p, \alpha)^{7}Be$	$^{23}Na(p, 3n)^{21}Mg$
$^{6}Li(^{3}He, n)^{8}B$	$^{141}Pm(^{12}C, 4n)^{149}Tb$

20-17. The Use of Radioactive Elements as Tracers

A valuable technique for research is the use of both radioactive and nonradioactive isotopes as tracers.* By the use of these isotopes an element can be observed in the presence of large quantities of the same element.

*Analysis for nonradioactive isotopes used as tracers, such as ^{15}N, is carried out by use of a mass spectrograph.

For example, one of the earliest uses of tracers was the experimental determination of the rate at which lead atoms move around through a crystalline sample of the metal lead. This phenomenon is called *self-diffusion.* If some radioactive lead is placed as a surface layer on a sheet of lead, and the sample is allowed to stand for a while, it can then be cut up into thin sections parallel to the original surface layer, and the radioactivity present in each section can be measured. The presence of radioactivity in layers other than the original surface layer shows that lead atoms from the surface layer have diffused through the metal.

Perhaps the greatest use for radioactive nucleides as tracers will continue to be in the field of biology and medicine. The human body contains such large amounts of the elements carbon, hydrogen, nitrogen, oxygen, sulfur, and others, that it is difficult to determine the state of organic material in the body. An organic compound containing a radioactive nucleide, however, can be traced through the body. An especially useful radioactive nucleide for these purposes is carbon 14. This isotope of carbon has a half-life of about 5000 years. It undergoes slow decomposition with emission of beta rays, and the amount of the isotope present in a sample can be followed by measuring the beta activity. Large quantities of ^{14}C can be readily made in a nuclear reactor, by the action of slow neutrons on nitrogen.

$$^{14}_{7}N + ^{1}_{0}n \longrightarrow ^{14}_{6}C + ^{1}_{1}H$$

The process can be carried out by running a solution of ammonium nitrate into the nuclear reactor, where it is exposed to neutrons. The carbon that is made in this way is in the form of the hydrogen carbonate ion, HCO_3^-, and it can be precipitated as barium carbonate by adding barium hydroxide solution. The samples of radioactive carbon are very strongly radioactive, containing as much as 5% of the radioactive isotope.

The Unit of Radioactivity, the Curie

It has been found convenient to introduce a special unit in which to measure amounts of radioactive material. The unit of radioactivity is called the *curie.* One curie of any radioactive substance is an amount of the substance such that 3.70×10^{10} atoms of the substance undergo radioactive disintegration per second.

The curie is a rather large unit. One curie of radium is approximately one gram of the element. (The curie was orginally defined in such a way as to make a curie of radium equal to one gram, but because of improvement in technique it has been found convenient to define it instead in the way given above.)

It is interesting to point out that in a disintegration chain of radioactive elements in a steady state all of the radioactive elements are present in the same radioactive amounts. For example, let us consider one gram of the element radium, in a steady state with the first product of its decomposition, radon (^{222}Rn), and the successive products of disintegration (see Figure 20-6). The rate at which radon is being produced is proportional to the amount of radium present, one atom of radon being produced for each atom of radium that undergoes decomposition. The number of atoms of radium that undergo decomposition in unit time is proportional to the number of atoms of radium present; the decomposition of radium is a unimolecular reaction. When the system has reached a steady state the number of atoms of radon present remains unchanged, so that the rate at which radon is itself undergoing radioactive decomposition must be equal to the rate at which it is being formed from radium. Hence the radon present in a steady state with one gram of radium itself amounts to one curie.

The amount of radon present in a steady state with one gram of radium can be calculated by consideration of the first-order reaction-rate equations discussed in Chapter 10. The reaction-rate constant for the decomposition of radium is inversely proportional to its half-life. Hence when a steady state exists, and the number of radium atoms undergoing decomposition is equal to the number of radon atoms undergoing decomposition, the ratio of the number of radon atoms and radium atoms present must be equal to the ratio of their half-lives.

20-18. Dating Objects by Use of Carbon 14

One of the most interesting recent applications of radioactivity is the determination of the age of carbonaceous materials by measurement of their carbon-14 radioactivity. This technique of radiocarbon dating, which was developed by an American physical chemist, Willard F. Libby, permits the dating of samples containing carbon with an accuracy of around 200 years. At the present time the method can be applied to materials that are not over about 50,000 years old.

Carbon 14 is being made at a steady rate in the upper atmosphere. Cosmic-ray neutrons transmute nitrogen into carbon 14, by the reaction given in the preceding section. The radiocarbon is oxidized to carbon dioxide, which is thoroughly mixed with the nonradioactive carbon dioxide in the atmosphere, through the action of winds. The steady-state concentration of carbon 14 built up in the atmosphere by cosmic rays is about one atom of radioactive carbon to 10^{12} atoms of ordinary carbon. The carbon dioxide, radioactive and nonradioactive alike, is absorbed by plants, which fix the carbon in their tissues. Animals that eat the plants also similarly fix the carbon, containing 1×10^{-12} part radiocarbon, in their tissues. When a plant or animal dies, the amount of radioactivity of the carbon in its tissues is determined by the amount of radiocarbon present, which is the amount corresponding to the steady state in the atmosphere. After 5,760 years (the half-life of carbon 14), however, half of the carbon 14 has undergone decomposition, and the radioactivity of the material is only half as great. After 11,520 years only one-quarter of the original radioactivity is left, and so on. Accordingly, by determining the radioactivity of a sample of carbon from wood, flesh, charcoal, skin, horn, or other plant or animal remains, the number of years that have gone by since the carbon was originally extracted from the atmosphere can be determined.

In applying the method of radiocarbon dating, a sample of material containing about 30 g of carbon is burned to carbon dioxide, which is then reduced to elementary carbon in the form of lamp black. The beta-ray activity of the elementary carbon is then determined, with the use of Geiger counters, and compared with the beta-ray activity of recent carbon, which is 15.3 ± 0.1 decompositions per minute per gram of carbon. The age of the sample is then calculated by the use of the equation for a first-order reaction (Chapter 10). The method was checked by measurement of carbon from the heartwood of a giant Sequoia tree, for which the number of tree rings showed that 2928 ± 50 years had passed since the wood was laid down. This check was satisfactory, as were also similar checks with other carbonaceous materials, such as wood in 1st Dynasty Egyptian tombs 4900 years old, whose dating was considered to be reliable.

The method of radiocarbon dating has now been applied to several thousand samples. One of the interesting conclusions that has been reached is that the last glaciation of the northern hemisphere occurred about 11,400 years ago. Specimens of wood from a buried forest in Wisconsin, in which all of the tree trunks are lying in the same direction as though pushed over by a glacier, were found to have an age of $11{,}400 \pm 700$ years. The age of specimens of organic materials laid down during the last period of glaciation in Europe was found to be $10{,}800 \pm 1200$ years. Many samples of organic matter, charcoal, and other carbonaceous material from human camp sites in the western hemisphere have been dated as extending to 11,400 years ago; a very few older ones (30,000 years) have been reported.

The eruption of Mt. Mazama in southern Oregon, which formed the crater now called Crater Lake, was determined to have occurred 6453 ± 250 years ago, by the dating of charcoal from a tree killed by the eruption. Three hundred pairs of woven rope sandals found in Fort Rock Cave, Oregon, which had been covered by an earlier eruption, were found to be 9053 ± 350 years old. The Lescaux Cave near Montignac, France, contains some remarkable paintings made by prehistoric man; charcoal from camp fires in this cave was found to have the age 15,516 ± 900 years. Linen wrappings from the Dead Sea scrolls of the Book of Isaiah, recently found in a cave in Palestine and thought to be from about the first or second century B.C., were dated 1917 ± 200 years old.

20-19. The Properties of Nucleides

The nucleides of the various elements show many interesting properties. Most of the known nucleides corresponding to the first ten elements are listed in Table 20-5.

For most elements other than those that form part of the natural radioactive series the distribution of nucleides for an element has been found to be the same for all natural occurrences. The average natural distribution is shown in the fifth column of the table.

Some striking regularities are evident, especially for the heavier elements. The elements of odd atomic number have only one or two natural nucleides, whereas those of even atomic number are much richer in nucleides, many having eight or more. It is also found that the odd elements are much rarer in nature than the even elements. The elements with no stable isotopes (technetium, atomic number 43; astatine, atomic number 85, promethium, atomic number 61) have odd atomic numbers.

Binding Energy

Consideration of the masses of the nucleides shows that they are not additive. Thus the mass of the ordinary hydrogen atom is 1.007825 d, and that of the neutron is 1.008665 d. If the helium atom were made from two hydrogen atoms and two neutrons without change in mass, its mass would be 4.032980 d, but it is in fact less, only 4.002604 d. The masses of the heavier atoms are also less than they would be if they were composed of hydrogen atoms and neutrons without change in mass.

The loss in mass accompanying the formation of a heavier atom from hydrogen atoms and neutrons shows that these reactions are strongly exothermic. A very large amount of energy is evolved in the formation of the heavier atoms from hydrogen atoms and neutrons, an amount given by the Einstein equation $E = mc^2$. The more stable the heavy nucleus, the larger is the decrease in mass from that of the neutrons and protons from which the nucleus may be considered to be made.

The decrease in mass accompanying the reaction of two hydrogen atoms and two neutrons to form a ^{4}He atom is 0.030376 *d*, which is equal

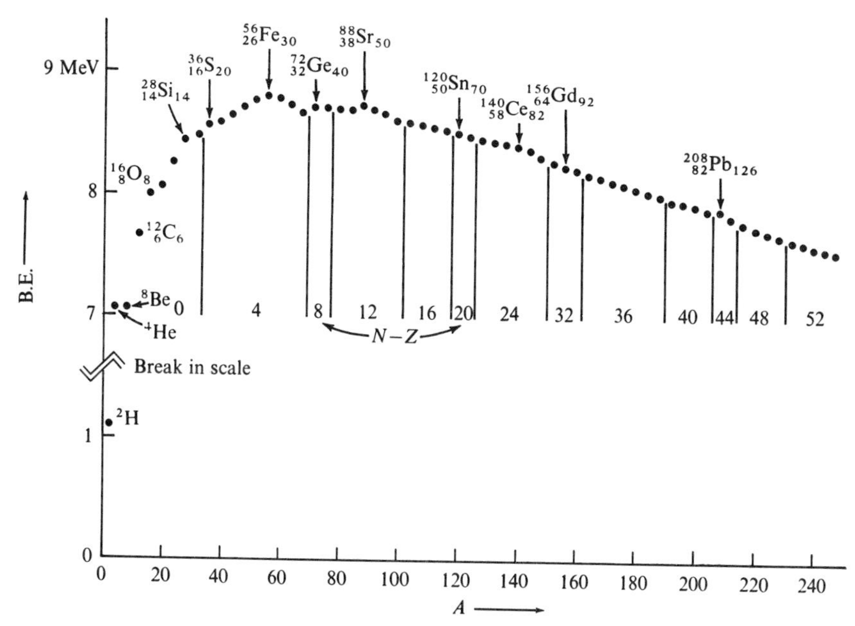

FIGURE 20-11
The binding energy per nucleon in stable even-even nuclei with A a multiple of 4.

to 28.294 MeV. It is customary to express the binding energy in MeV per nucleon (7.073 MeV per nucleon for ^{4}He). The binding energy per nucleon as a function of the number of nucleons (the mass number) is shown for some stable nucleides in Figure 20-11. It is seen that the elements of the first long period of the periodic table, between chromium and zinc, lie at the maximum of the curve, and can accordingly be considered to be the most stable of all the elements. If one of these elements were to be converted into other elements, the total mass of the other elements would be somewhat greater than that of the reactants, and accordingly energy would have to be added in order to cause the reaction to occur. On the other hand, either the heavier or the lighter elements could undergo a nuclear reaction to form the elements with mass numbers in the neighborhood of 60, and these nuclear reactions would be accompanied by the evolution of a large amount of energy.

Scientists have attempted to develop a theory of the origin of nuclear species on the basis of the extensive information now available about nuclear reactions. One idea is that the elements have been produced by synthesis from hydrogen by a succession of neutron captures interspersed where necessary by decrease in atomic number through β decay. There is convincing astronomical evidence that the universe is expanding. The

TABLE 20-5
Isotopes of the Lighter Elements

Atomic Number	Name	Mass Number	Mass*	Percent Abundance	Half-Life†	Radiation
0	Electron	0	0.0005486			
0	Neutron	1	1.008665		12 m	e^-
1	Proton	1	1.007276			
1	Hydrogen	1	1.007825	99.985		
		2	2.014102	0.015		
		3	3.014949		12.26 y	
2	Alpha	4	4.001507			e^-
2	Helium	3	3.016030	0.00013		
		4	4.002604	~100		
		5	5.012296		2×10^{-21} s	n
		6	6.018900		0.79 s	e^-
		7			60×10^{-6} s	e^-
3	Lithium	5	5.012541		$\sim 10^{-21}$ s	
		6	6.015126	7.42		
		7	7.016005	92.58		
		8	8.022488		0.85 s	
		9	9.027300		0.17 s	e^-
4	Beryllium	6	6.019780		$\geq 4 \times 10^{-21}$ s	
		7	7.016931		53 d	γ
		8	8.005308		$\sim 3 \times 10^{-16}$ s	
		9	9.012186	100		
		10	10.013535		2.7×10^{6} y	e^-
		11	11.021660		13.6 s	e^-, γ
5	Boron	8	8.024612		0.78 s	e^+
		9	9.013335		$\geq 3 \times 10^{-19}$ s	
		10	10.012939	19.6		
		11	11.009305	80.4		
		12	12.014353		0.020 s	e^-, γ
		13	13.017779		0.035 s	e^-
6	Carbon	10	10.016830		19 s	e^+, γ
		11	11.011433		20.5 m	e^+
		12	12.000000	98.89		
		13	13.003354	1.11		
		14	14.003242		5760 y	e^-
		15	15.010600		2.25 s	e^-, γ
		16	16.014702		0.74 s	e^-

TABLE 20-5 *(continued)*

Atomic Number	Name	Mass Number	Mass*	Percent Abundance	Half-Life†	Radiation
7	Nitrogen	12	12.018709		0.011 s	e^+
		13	13.005739		10.0 m	e^+
		14	14.003074	99.63		
		15	15.000108	0.37		
		16	16.006089		7.35 s	e^-, γ
		17	17.008449		4.14 s	e^-
8	Oxygen	14	14.008597		71 s	e^+, γ
		15	15.003072		124 s	e^+
		16	15.994915	99.759		
		17	16.999133	0.037		
		18	17.999160	0.204		
		19	19.003577		29 s	e^-, γ
		20	20.004071		14 s	e^-, γ
9	Fluorine	16	16.011707		$\sim 10^{-19}$ s	
		17	17.002098		66 s	e^+
		18	18.000950		111 m	e^+
		19	18.998405	100		
		20	19.999986		11 s	e^-, γ
		21	20.999972		5 s	e^-
10	Neon	18	18.005715		1.46 s	e^+, γ
		19	19.001892		18 s	e^+
		20	19.992440	90.92		
		21	20.993849	0.257		
		22	21.991384	8.82		
		23	22.994475		38 s	e^-, γ
		24	23.993597		3.38 m	e^-, γ

*Carbon-12 scale.
†s = second, m = minute, y = year.

light from distant galaxies contains spectral lines that can be identified, but their frequencies are not those observed in the laboratory; instead, there is a shift in wavelength to the red (the red shift). The same fractional shift in wavelength is observed for all of the spectral lines and for the continuum in the optical spectrum and also for all radio waves emitted by a particular galaxy. This fact is the basis for the belief that the red shift is due to the Doppler effect (the dependence of observed frequency on the relative velocities of emitter and observer) and that the distant galaxies are receding from us. It was discovered by the American astronomers Hubble and Humason about 40 years ago that the red shift is greatest for the most distant galaxies. The magnitude of the velocity of the galaxies as deduced from the red shift and distances of the galaxies fixes the time of creation of the universe at about 15×10^9 years ago.

The American scientist George Gamow postulated that at that time, the beginning, the universe consisted of a huge ball of neutrons bathed in radiation, which immediately began to expand because of its great internal energy. Some of the neutrons then began to decay to form protons, electrons, and neutrinos, liberating 0.78 MeV of energy per neutron. The protons could then capture neutrons to form deuterons, and, in the neutron-capture theory of the origin of nucleides, the process of neutron capture would continue, and would build up the distribution of nucleides that is observed.

There are, however, some difficulties about this theory. One is that there are no stable nucleides with mass 5 or mass 8, and thus no synthesis of the elements beyond these masses through neutron capture alone is possible; the synthesis stops when all of the hydrogen has been turned into helium 4.

An alternative theory of nucleide synthesis is that this synthesis has taken place and is still taking place in the center of stars. This theory has been supported principally by the British astrophysicist Fred Hoyle. The problem of the instability of the nucleides with mass 5 and mass 8 is overcome by way of reactions such as the following:

$$3\ {}^4\text{He} \rightleftarrows {}^{12}_{6}\text{C}^* \rightarrow {}^{12}_{6}\text{C} + \gamma$$

At a temperature of 100 million degrees and density of 10,000 g cm^{-3} in the center of a star, there is an equilibrium involving three alpha particles and an excited state of the carbon-12 nucleus, with energy 7.653 MeV greater than the normal state of the nucleus. The excited ^{12}C nucleus can change to the normal state by emission of a photon. Various other known nuclear reactions can then lead to the synthesis of all of the heavier nucleides.

Magic Numbers

In 1934 W. M. Elsasser, then working in France, discussed evidence showing that there is a special stability associated with certain numbers of protons and certain numbers of neutrons. These numbers, which are known as *magic numbers,* are 2, 8, 20, 28, 50, 82, and 126. The numbers can be correlated with the subshell numbers that have been developed in the discussion of the electrons in atoms. The magic number 2 is, of course, the number of fermions that can occupy a $1s$ orbital. The magic number 8 may be described as the number of fermions occupying a $1s$ orbital and the three $2p$ orbitals; in nuclei, it is to be expected that a particle in a $2p$ orbital would be more stable than in a $2s$ orbital. The number 20 similarly corresponds to pairs of fermions occupying the $1s$ orbital, the three $2p$ orbitals, the $2s$ orbital, and the five $3d$ orbitals. The larger magic numbers presented a puzzle that was solved fourteen years later (Sections 20-20 and 20-21). The first evidence for the magic numbers was provided by the values of the binding energy and properties closely related to it. An example, for magic number 50, is shown in Figure 20-12, which represents the binding energy for pairs of neutrons in the neighborhood of $N = 50$. It is seen that the binding energy drops sharply at $N = 50$; the value for $50 \rightarrow 52$ is much less than for $48 \rightarrow 50$.

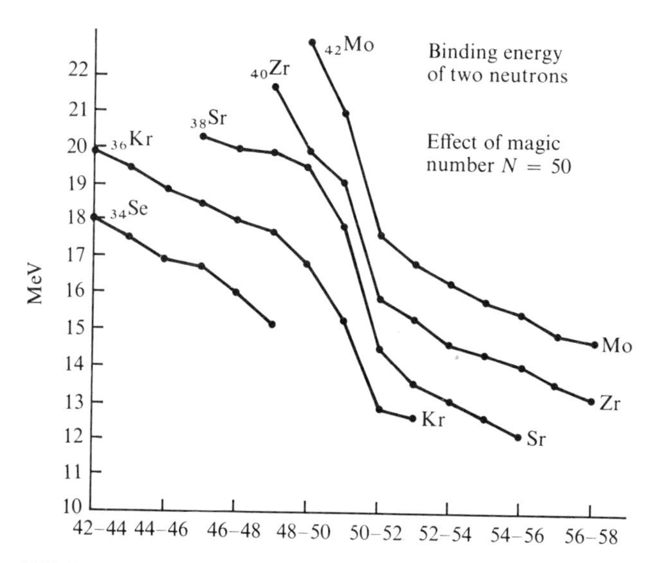

FIGURE 20-12
The energy of binding of two neutrons for nuclei in the region 42 to 58 for N, showing the decreased binding energy for $N > 50$.

Radii of Nuclei

The results of the experiments on the scattering of helions (alpha particles) by gold foil were interpreted by Rutherford and his coworkers (Section 3-4) as showing that the interaction of the helion and the heavier nucleus shows no deviation from Coulomb repulsion at distances larger

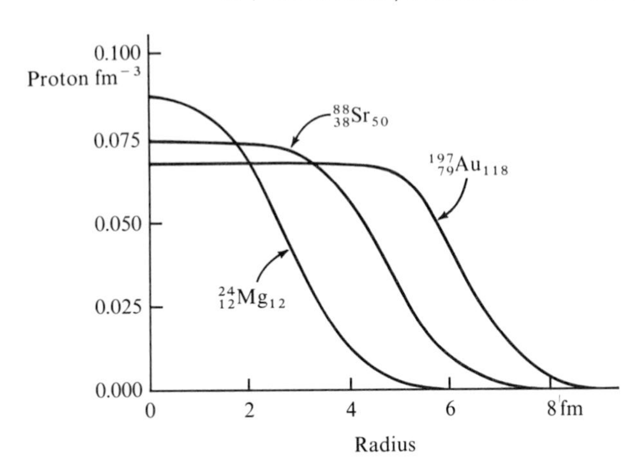

FIGURE 20-13
Proton density distribution for three nuclei, determined from the observed scattering of high-energy electrons (100 to 250 MeV). The neutron density function is nearly the same (with the factor N/Z).

than about 10 fm. Other experiments have led to rather accurate values of the sizes of nuclei and to knowledge of the probability distribution function of nucleons within the nuclei. The scattering of high-energy electrons, studied especially by the American physicist Robert Hofstadter (born 1915) and his associates, has led to results such as those shown in Figure 20-13. The nucleonic density is found to be constant, with value about 0.17 nucleons per fm^3, throughout the central region of every nucleus (except the lightest ones); it then drops to zero over the radial increment 2 fm (from density 90% to 10% of the maximum). The radius (measured to density 50% of maximum) is proportional to the cube root of the nucleon number:

$$R = 1.07\ A^{1/3}\ \text{fm} \qquad (20\text{-}1)$$

The observed constant nucleonic density suggests that the nucleons are packed together in a nucleus in a way resembling the packing of molecules in a liquid, rather than in a gas (see Section 20-21).

20-20. The Shell Model of Nuclear Structure

In 1933 the American physicist J. H. Bartlett, Jr. (born 1904) suggested that the protons and the neutrons in a nucleus could be assigned to orbitals (about the center of mass) resembling the electron orbitals in an atom. This orbital model of the nucleus accounts reasonably well for the smaller magic numbers 2, 8, and 20: ^4He is assigned the configuration $1s^2$ for both neutrons and protons, ^{16}O the configuration $1s^2 1p^6$, and ^{40}Ca the configuration $1s^2 1p^6 1d^{10} 2s^2$. (The principal quantum number is conventionally taken as $n = 1, 2, 3, \ldots$ for each value of l, for nucleonic orbitals, rather than $n = l + 1, l + 2, \ldots$, the convention for electronic orbitals.) The other magic numbers, 28, 50, 82, and 126, were finally interpreted by a refinement of the orbital model, called the *shell model*, that was developed by the American physicist Maria Goeppert Mayer (1906–1972) and the German physicist J. Hans D. Jensen (born 1907) and his collaborators in 1948.

The characteristic feature of the shell model is the assumption that each nucleon couples its orbital angular momentum vector (with quantum number l) and its spin vector (with quantum number $s = \frac{1}{2}$) to form a resultant spin-orbit angular momentum vector, with quantum number j equal to $l + \frac{1}{2}$ or $l - \frac{1}{2}$. The total angular momentum of the nucleus, quantum number I, is the resultant of the j-vectors for all the nucleons. This sort of coupling of angular momenta is called jj coupling. The nucleonic subsubshell with $j = l + \frac{1}{2}$ lies below that for $j = l - \frac{1}{2}$, as shown in Figure 20-14. It is seen that each of the larger magic numbers corresponds to completed shells plus a subsubshell, $(1f\frac{7}{2})^8$, for example, for 28. There are, of course, $2j + 1$ protons or neutrons in a completed j subsubshell, corresponding to the $2j + 1$ values of m_j (from $-j$ to $+j$).

The shell model leads rather directly to the observed values of the spin and parity of most nuclei. The nuclei $^{17}_{8}O_9$ and $^{17}_{9}F_8$, for example, have an odd (unpaired) nucleon outside the completed-shell structure of $^{16}_{8}O_8$. From Figure 20-14 we assign this nucleon to $1d\frac{5}{2}$. Both ^{17}O and ^{17}F have $I^P = \frac{5}{2}^+$ for the normal state, and $\frac{1}{2}^+$ (corresponding to $2s$) as the first excited state. (Note that $s, d, g, \ldots$ orbitals have even parity and $p, f, \ldots$ have odd parity.) The nuclei $^{41}_{20}Ca_{21}$ and $^{41}_{21}Sc_{20}$, with an odd nucleon in $1f\frac{7}{2}$, have $I^P = \frac{7}{2}^-$ in the normal state.

20-21. The Helion-Triton Model

Another useful model of nuclear structure, called the alpha-particle model or the helion-triton model, is based on the assumption that the nucleons in a nucleus can be considered to be grouped together into helions or tritons, occupying localized $1s$ orbitals. For ^{16}O, for example, the 8 protons and the 8 neutrons could be described as forming four helions, which are arranged at the corners of a tetrahedron. In the shell model the protons and the neutrons would be described as occupying the $1s$ orbital and the three $1p$ orbitals. These four orbitals can be hybridized, as described in Chapter 5, to form four localized tetrahedral orbitals, each of which is concentrated about one of the four corners of a tetrahedron.

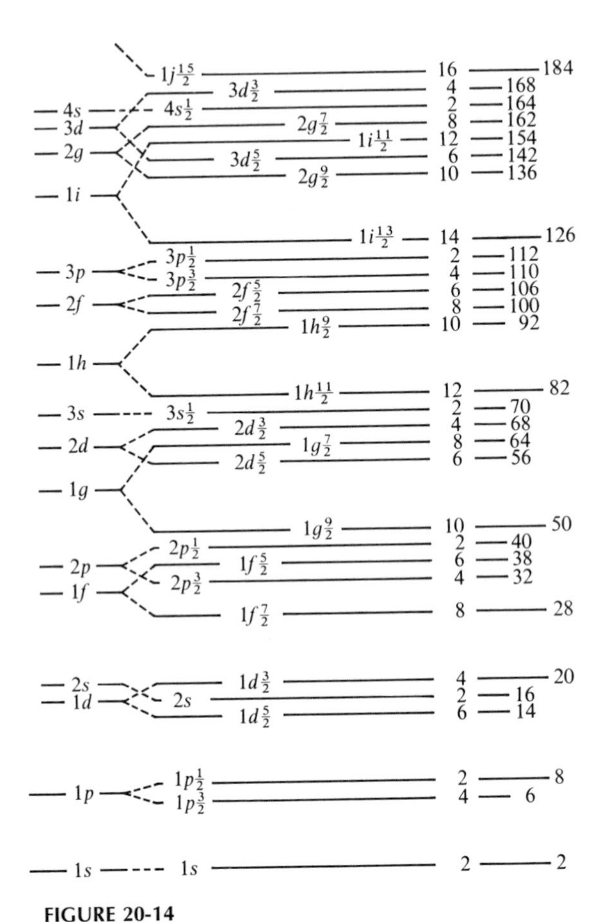

FIGURE 20-14
Sequence of energy levels for spin-orbit coupling. $j = l + \frac{1}{2}$ or $j = l - \frac{1}{2}$, for protons and neutrons, as given by the Mayer-Jensen shell model of nuclear structure.

Many properties of nuclei can be simply explained on this basis. An example is provided by ^{20}Ne. The properties of this nucleus and related nuclei (such as ^{21}Ne and ^{21}Na), indicate that the ^{20}Ne nucleus has a prolate deformation from spherical shape; that is, it is elongated. The structure expected for an aggregate of five helions is the trigonal bipyramid, five helions in positions corresponding to those of the chlorine atoms in phosphorus pentachloride (Figure 7-6), in agreement with the observed prolate deformation.

Nuclei with a larger number of helions and tritons might be expected to have a structure in which one helion or triton is at the center of the nucleus, and the others are arranged in a layer about it. This structure would, of course, be expected for 13 spheres—a close-packing of 12 spheres around a central sphere. A layer structure can be assigned to nuclei, with the helions and tritons placed in an inner core, an outer core layer, and an outermost layer, called the mantle, by assigning the nucleonic orbitals to the successive layers in the way indicated in Figure 20-15. This assignment is based on the shell-model sequence of subsubshells, with the assumption that a subsubshell that occurs with only one value of the principal quantum number contributes to the outer layer of helions and tritons, one with two values of the principal quantum number (such as 1*s* and 2*s*) contributes to the mantle and the next inner layer, and one with three values of the principal quantum number contributes to the mantle, the outer core, and the inner core.

The shell-model diagrams for the magic numbers, with shells and subsubshells assigned to the different layers of the nucleus, are shown in Figure 20-16. The numbers of helions and tritons in these successive layers approximate the numbers for close-packed aggregates of spheres.

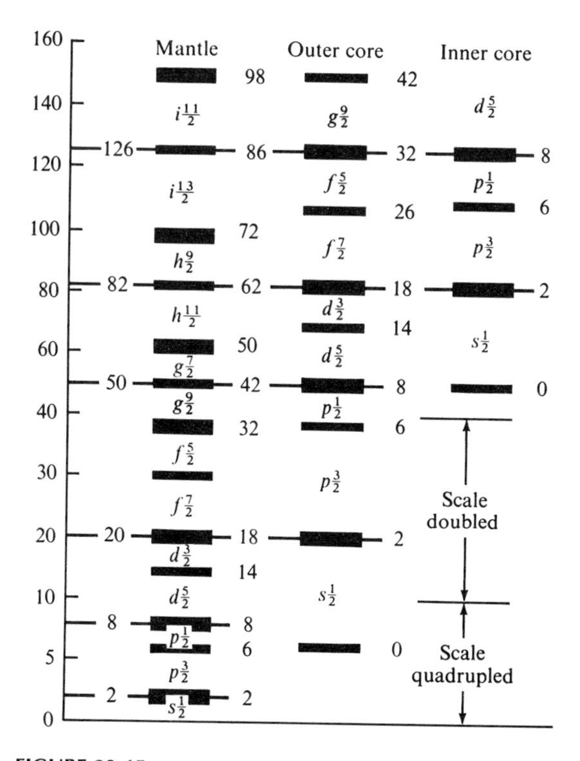

FIGURE 20-15
The sequence of nucleon energy levels (overlapping ranges), with assignment to successive layers (inner core, outer core, mantle) on the basis of the principal quantum number.

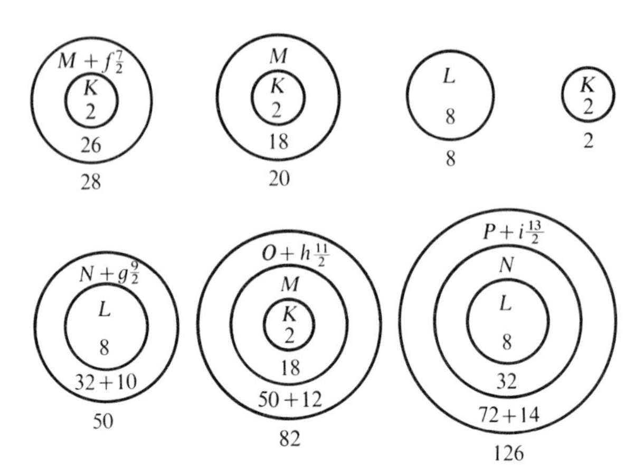

FIGURE 20-16
The magic-number structures of nuclei.

Prolate Deformation of Heavy Nuclei

The observed properties of many of the heavier nuclei have been interpreted as showing that the nuclei are not spherical but are permanently deformed. The principal ranges of deformation are from neutron number 90 to 116 and from 140 on. Most of the deformed nuclei are described as prolate ellipsoids of revolution, with major radii 20% to 40% larger than the minor radii.

The helion-triton model provides a simple explanation of the onset of prolate deformation at neutron number 90. This value corresponds to 45 helions and tritons. The magic number $N = 82$ corresponds to 41, which, from Figures 20-15 and 20-16, are assigned to the successive layers in the following way: 1 in the inner core, 9 in the outer core, and 31 in the mantle. The closest packing of spheres of approximately equal size is 1 sphere in the inner core, 12 in the outer core, and 32 in the mantle, a total of 45, as indicated in Figure 20-17A. An additional sphere would have to be placed on the surface of this nearly spherical aggregate, or be introduced into the inner core, as indicated in Figure 20-17B. We conclude that at neutron number 90 an extra pair of neutrons is introduced in the core (occupying a $p\frac{3}{2}$ orbital), causing the core to contain two tritons or helions, and thus leading to prolate deformation of the nucleus as a whole.

20-22. Nuclear Fission and Nuclear Fusion

The instability of the heavy elements relative to those of mass number around 60, as shown by the binding-energy curve, suggests the possibility

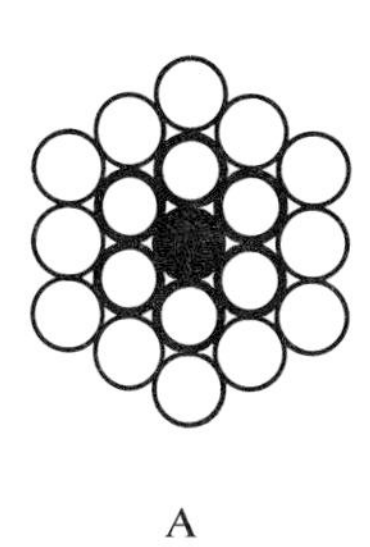

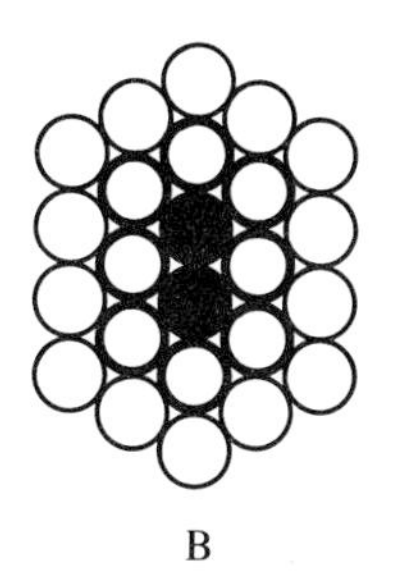

FIGURE 20-17
Two-dimensional representation of packing of spheres around (A) one central sphere and (B) two central spheres. In three dimensions icosahedral packing of spheres of nearly the same size (within 10%) places 12 in the second layer about 1 central sphere and 32 in the outer layer.

of spontaneous decomposition of the heavy elements into fragments of approximately half-size. This fission has been accomplished.

It was reported on January 6, 1939, by the German physicists O. Hahn and F. Strassmann that barium, lanthanum, cerium, and krypton seemed to be present in substances containing uranium that had been exposed to neutrons. Within two months more than forty papers were then published on the fission of uranium. It was verified by direct calorimetric measurement that a very large amount of energy is liberated by fission—over 20×10^{12} J per mole. Since a kilogram of uranium contains 4.26 gram-atoms, the complete fission of 1 kg of this element, or a similar heavy element, produces about 0.8×10^{14} J. This may be compared with the heat of combustion of 1 kg of coal, which is approximately 4×10^7 J. Thus uranium as a source of energy is 2 million times more valuable than coal.

Uranium 235 and plutonium 239, which can be made from ^{238}U, are capable of undergoing fission when exposed to slow neutrons. It was also shown by the Japanese physicist Y. Nishina in 1939 that the thorium isotope ^{232}Th undergoes fission under the influence of fast neutrons.

It has been customary to use the *megaton* as a unit of energy released in nuclear fission (or fusion, discussed below). One megaton is equal to 8×10^{15} J, the energy of explosion of one million tons of the ordinary explosive TNT. The energy of fission of 100 kg of uranium or plutonium is one megaton.

Uranium and thorium have become important sources of heat and energy. There are large amounts of these elements available—the amount of uranium in the earth's crust has been estimated as 4 parts per million and the amount of thorium as 12 parts per million. The deposits are distributed all over the world.

The fission reactions can be chain reactions. These reactions are initiated by neutrons. A nucleus ^{235}U, for example, may combine with a neutron to form ^{236}U. This isotope is unstable, and undergoes spontaneous fission, into two smaller particles; the protons in the ^{236}U nucleus are divided between the two daughter nuclei (Figure 20-18). These daughter nuclei also contain most of the neutrons originally present in the ^{236}U nucleus. Since, however, the ratio of neutrons to protons is greater in the heavier nuclei than in those of intermediate mass, the fission is also accompanied by the liberation of a few free neutrons. The neutrons that are thus liberated may then combine with other ^{235}U nuclei, forming additional ^{236}U nuclei, which themselves undergo fission. A reaction of this sort, the products of which cause the reaction to continue, is a chain reaction or an autocatalytic reaction (Section 10-6).

The fission of some nuclei, such as those in the neighborhood of ^{208}Pb (gold, thallium, lead, bismuth), is symmetric. When one of these elements is bombarded with 20-MeV deuterons it undergoes fission to give two daughter nuclei with nearly the same values of Z and N, the distribution function having a half-width at half maximum of about 15 mass-number units. The heavier nuclei show unsymmetric fission, as illustrated in Figure 20-18 for ^{236}U.

If a mass of ^{235}U or ^{239}Pu weighing a few kilograms and in a suitable form, such as a hollow sphere, is suddenly compressed into a small volume, the autocatalytic fission of the nuclei occurs nearly completely, and the amount of energy 0.01 megaton per kilogram that undergoes fission is released. An ordinary *atomic bomb* (nuclear bomb) consists of a few kilograms of ^{235}U or ^{239}Pu and a mechanism for suddenly compressing the metal. Each of the atomic bombs exploded over Hiroshima and Nagasaki in August 1945 had explosive energy of about 0.020 megatons

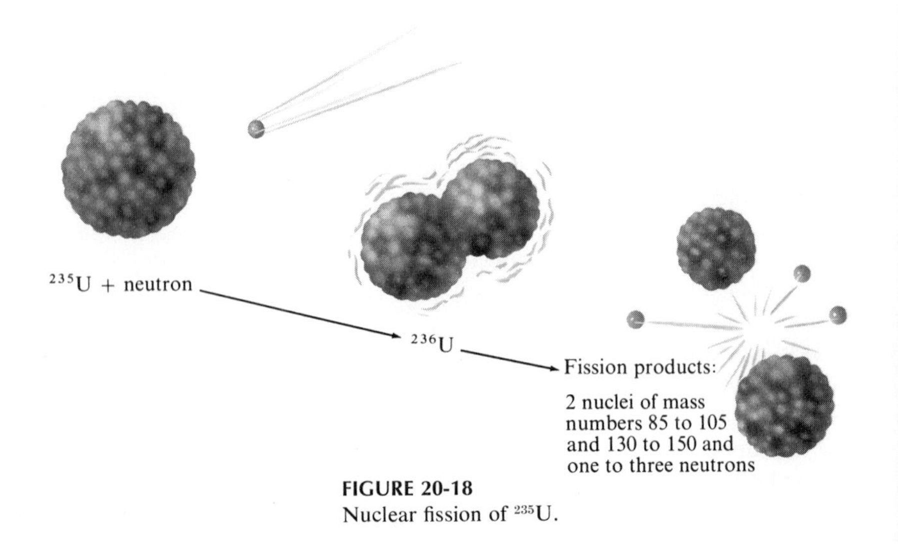

FIGURE 20-18
Nuclear fission of ^{235}U.

(20 kilotons). Modern nuclear weapons involving fission alone have explosive energy 0.001 megaton to 0.1 megaton.

The process of *nuclear fusion* also may liberate energy. From the binding-energy diagram we see that the fission of a very heavy nucleus converts about 0.1% of its mass into energy. Still larger fractions of the mass of very light nuclei are converted into energy by their fusion into heavier nuclei. The process 4H → He, which is the principal source of the energy of the sun, involves the conversion of 0.7% of the mass into energy. The similar reaction of a deuteron and a triton to form a helium nucleus and a neutron is accompanied by the conversion of 0.4% of the mass into energy:

$$^{2}_{1}H + ^{3}_{1}H \rightarrow ^{4}_{2}H + ^{1}_{0}n$$

It was found by experiment (in 1952) that these materials surrounding an ordinary atomic bomb undergo reaction at the temperature of many millions of degrees produced by the reaction. Tritium is, however, inconvenient and expensive to use because it is radioactive and unstable (half-life 12 years). In 1953 it was shown that a fission-fusion bomb could be made by placing some of the stable solid substance *lithium deuteride*, LiD, about an ordinary fission bomb. Some of the reactions that occur are the following:

$$^{2}D + ^{2}D \rightarrow ^{3}He + n$$

$$^{2}D + ^{2}D \rightarrow ^{3}H + p$$

$$n + ^{6}Li \rightarrow ^{4}He + ^{3}H$$

$$^{3}H + D \rightarrow ^{4}He + n$$

$$p + ^{7}Li \rightarrow ^{4}He + ^{4}He$$

The amount of energy released in the fusion of lithium deuteride is about 60 megatons per ton of the material undergoing fusion, as compared with 10 megatons per ton of uranium undergoing fission. The largest nuclear bomb so far exploded, the Soviet bomb of November 1961, was a fission-fusion bomb with explosive energy of about 60 megatons, about 10 times the total for all bombs used in the Second World War.

The standard present-day nuclear weapons are three-stage fission-fusion-fission bombs (superbombs). An ordinary 20-megaton superbomb (50:50 fission-fusion) has as its first stage material (the detonator) a few kilograms of plutonium, with some ordinary explosive to compress it suddenly. The second-stage material is about 150 kg of lithium deuteride, which is surrounded by a shell of ordinary uranium metal (^{238}U, the third-stage material) weighing somewhat more than 1000 pounds. The ^{238}U undergoes fission through reaction with fast neutrons. The total weight of a 20-megaton bomb is about 1.5 tons.

The *manufacture of plutonium* is carried out by a controlled chain reaction. A piece of ordinary uranium contains 0.71% of ^{235}U. An occasional neutron strikes one of these atoms, causing it to undergo fission and release a number of neutrons. The autocatalytic reaction does not build up, however, if the piece of uranium is small, because the neutrons escape, and some of them may be absorbed by impurities, such as cadmium, the nuclei of which combine very readily with neutrons.

However, if a large enough sample of uranium is taken, nearly all of the neutrons that are formed by the fission remain within the sample of uranium, and either cause other ^{235}U nuclei to undergo fission, or are absorbed by ^{238}U, converting it into ^{239}U, which then undergoes spontaneous change to ^{239}Pu. This is the process used in practice for the manufacture of plutonium. A large number of lumps of uranium are piled together, alternately with bricks of graphite, in a structure called a reactor. The first uranium reactor ever constructed, built at the University of Chicago and put into operation on December 2, 1942, contained 5600 kg of uranium metal. Cadmium rods were held in readiness to be introduced into cavities in the reactor, and to serve to arrest the reaction by absorbing neutrons, whenever there was danger of its getting out of hand.

The large reactors that were put into operation in September, 1944, at Hanford, Washington, were of such size as to permit the fission reaction to proceed at the rate corresponding to an output of energy of 1,500,000 kilowatts.

The significance of the uranium reactors as a source of radioactive material can be made clear by a comparison with the supply of radium. About 1000 curies (1000 grams) of radium has been separated from its ores and put into use, mainly for medical treatment. The rate of operation mentioned above for the reactors at Hanford represent the fission of about 5×10^{20} nuclei per second, forming about 10×10^{20} radioactive atoms. The concentration of these radioactive atoms will build up until they are undergoing decomposition at the rate at which they are being formed. Since 1 curie corresponds to 3.70×10^{10} disintegrating atoms per second, these reactors develop a radioactivity of approximately 3×10^{10} curies—that is, about thirty million times the radioactivity of all the radium that has been so far isolated from its ores.

The foregoing calculation illustrates the great significance of the fissionable elements as a source of radioactive material. Their significance as a source of energy has also been pointed out, by the statement that 1 kg of uranium or thorium is equivalent to 2 million kg of coal.

EXERCISES

20-1. In what decade was the electron discovered? The photon? The proton? The positron? The neutron? The muon? The pion?

20-2. What experiments (Hofstadter) show that the proton and the neutron are not point particles (as the electron seems to be) but have an extension in space and seem to be made of several other particles?

20-3. Discuss conservation of baryon number in relation to the experiment by Segré and others in 1955 in which the antiproton was first observed.

20-4. The mass of the antiproton corresponds by the Einstein relation $E = mc^2$ to the energy 938 MeV. Why was bombardment of nuclei with particles with more than twice this kinetic energy needed in order to make an antiproton?

20-5. How are alpha rays, beta rays, and gamma rays distinguished from one another by their behavior on passing through a magnetic field?

20-6. By how much is position in the periodic table changed when a radioactive atom emits an alpha particle? A beta particle? A gamma particle?

20-7. In the series of decompositions from uranium ($Z = 92$) to lead ($Z = 82$) a nucleus may undergo seven alpha emissions. How many transformations with emission of a beta particle must it undergo?

20-8. In 1914 the American chemist T. W. Richards reported the chemical atomic mass of lead from an ore of uranium to be about 2 d less than that of lead from an ore of thorium. Can you explain this difference?

20-9. By counting the number of alpha particles emitted in a measured time by a minute sample of radium, it was found to correspond to 3.70×10^{10} disintegrations per second for 1 g of the metal. To what value of the half-life does this lead?

20-10. The nucleus $^{40}_{20}Ca_{20}$ is described as doubly magic. What shells of orbitals are occupied by pairs of protons (with opposed spins) and pairs of neutrons in this nucleus?

20-11. The American chemist W. D. Harkins in 1920 used the chemical atomic masses of light elements of even atomic number (up to $Z = 16$, sulfur) to help him to the conclusion that their nuclei are composed of alpha particles. Can you reproduce his reasoning? Note that very few isotopes had been identified by that time.

20-12. Harkins also suggested that the neutron, a particle with mass 1 d and zero electric charge, existed as a constituent of nuclei. What light element led him to this conclusion, through the value of its chemical atomic mass?

20-13. Similarly, Harkins suggested that ${}^{2}_{1}H_1$ (a proton plus a neutron) and ${}^{3}_{1}H_2$ (a proton plus two neutrons) are present in some nuclei. What light elements led him to each of these conclusions?

20-14. By reference to Figure 20-11, obtain a rough value for the change in binding energy per nucleon accompanying the fission of a nucleus with *A* approximately 240, and check the statement in Section 20-22 that nuclear fission releases about 20×10^{12} J per mole.

APPENDIX I

Units of Measurement

BASIC IS UNITS

Physical quantity	Name of unit	Symbol for unit
length	meter	m
mass	kilogram	kg
time	second	s
electric current	ampere	A
thermodynamic temperature	degree Kelvin	K or °K
luminous intensity	candela	cd
SUPPLEMENTARY UNITS*		
plane angle	radian	rad
solid angle	steradian	sr

*These units are dimensionless.

FRACTIONS AND MULTIPLES*

Fraction	Prefix	Symbol	Multiple	Prefix	Symbol
10^{-1}	deci	d†	10	deka	da†
10^{-2}	centi	c†	10^{2}	hecto	h†
10^{-3}	milli	m	10^{3}	kilo	k
10^{-6}	micro	μ	10^{6}	mega	M
10^{-9}	nano	n	10^{9}	giga	G
10^{-12}	pico	p	10^{12}	tera	T
10^{-15}	femto	f			
10^{-18}	atto	a			

*Compound prefixes should not be used. Thus 10^{-9} meter is represented by 1 nm, not 1 mμm. The attaching of a prefix to a unit in effect constitutes a new unit, so that $1\ km^2 = 1\ (km)^2 = 10^6\ m^2$, not $1\ k(m^2) = 10^3\ m^2$.

†To be restricted as much as possible.

DERIVED IS UNITS WITH SPECIAL NAMES

Physical quantity	Name of unit	Symbol for unit	Definition of unit
energy	joule	J	$kg\ m^2\ s^{-2} = N\ m$
force	newton	N	$kg\ m\ s^{-2} = J\ m^{-1}$
power	watt	W	$kg\ m^2\ s^{-3} = J\ s^{-1}$
electric charge	coulomb	C	$A\ s$
electric potential difference	volt	V	$kg\ m^2\ s^{-3}\ A^{-1} = J\ A^{-1}\ s^{-1}$
electric resistance	ohm	Ω	$kg\ m^2\ s^{-3}\ A^{-2} = V\ A^{-1}$
electric capacitance	farad	F	$A^2\ s^4\ kg^{-1}\ m^{-2} = A\ s\ V^{-1}$
magnetic flux	weber	Wb	$kg\ m^2\ s^{-2}\ A^{-1} = V\ s$
inductance	henry	H	$kg\ m^2\ s^{-2}\ A^{-2} = V\ s\ A^{-1}$
magnetic flux density	tesla	T	$kg\ s^{-2}\ A^{-1} = V\ s\ m^{-2}$
luminous flux	lumen	lm	cd sr
illumination	lux	lx	$cd\ sr\ m^{-2}$
frequency	hertz	Hz	cycle per second
customary temperature, t	degree Celsius	°C	t °C = T °K − 273.15°

EXAMPLES OF OTHER DERIVED IS UNITS

Physical quantity	IS unit	Symbol for unit
area	square meter	m^2
volume	cubic meter	m^3
density	kilogram per cubic meter	$kg\ m^{-3}$
velocity	meter per second	$m\ s^{-1}$
angular velocity	radian per second	$rad\ s^{-1}$
acceleration	meter per second squared	$m\ s^{-2}$
pressure	newton per square meter	$N\ m^{-2}$
kinematic viscosity, diffusion coefficient	square meter per second	$m^2\ s^{-1}$
dynamic viscosity	newton second per square meter	$N\ s\ m^{-2}$
electric field strength	volt per meter	$V\ m^{-1}$
magnetic field strength	ampere per meter	$A\ m^{-1}$
luminance	candela per square meter	$cd\ m^{-2}$

UNITS TO BE ALLOWED IN CONJUNCTION WITH IS

Physical quantity	Name of unit	Symbol for unit	Definition of unit
length	parsec	pc	30.87×10^{15} m
area	barn	b	10^{-28} m^2
	hectare	ha	10^4 m^2
volume	liter	l	10^{-3} $m^3 = dm^3$
pressure	bar	bar	10^5 N m^{-2}
mass	tonne (metric ton)	t	10^3 kg = Mg
magnetic flux density (magnetic induction)	gauss	G	10^{-4} T
radioactivity	curie	Ci	37×10^9 s^{-1}
energy	electronvolt	eV	1.6021×10^{-19} J

The common units of time, such as hour or year, will persist, and also, in appropriate contexts, the angular degree.

EXAMPLES OF UNITS CONTRARY TO IS, WITH THEIR EQUIVALENTS

Physical quantity	Name of unit	Equivalent
length	ångström	10^{-10} m
	inch	0.0254 m
	foot	0.3048 m
	mile	1.60934 km
area	square inch	645.16 mm^2
	square foot	0.092903 m^2
	acre	4046.9 m^2
volume	cubic inch	1.63871×10^{-5} m^3
	cubic foot	0.028317 m^3
mass	pound	0.4535924 kg
	ounce	28.3495 g
	grain	64.799 mg
density	pound/cubic inch	2.76799×10^4 kg m^{-3}
force	dyne	10^{-5} N
	poundal	0.138255 N
pressure	atmosphere	101.325 kN m^{-2}
	torr	133.322 N m^{-2}
energy	erg	10^{-7} J
	calorie (thermochemical)	4.184 J
	British thermal unit	1055.06 J
	kilowatt hour, kW h	3.6 MJ
power	horse power	745.700 W

APPENDIX **II**

Values of Some Physical and Chemical Constants

Based on the ^{12}C Scale

Avogadro's number	$N = 0.60229 \times 10^{24}$ mole^{-1}
Velocity of light	$c = 2.997925 \times 10^{8}$ m s^{-1}
Mass of electron	$m = 0.91083 \times 10^{-30}$ kg
Electronic charge	$e = 0.160206 \times 10^{-18}$ C
Faraday	$F = Ne = 96490$ C mole^{-1}
Dalton	$d = 1.66033 \times 10^{-27}$ kg
Planck's constant	$h = 0.66252 \times 10^{-33}$ J s
Angular-momentum quantum	$\hbar = h/2\pi = 0.105443 \times 10^{-33}$ J s
Mass of proton	$m_p = 1.67239 \times 10^{-27}$ kg
Mass of neutron	$m_n = 1.67470 \times 10^{-27}$ kg
Boltzmann constant	$k = 13.805 \times 10^{-24}$ J K^{-1}
Gas constant	$R = Nk = 8.3146$ J K^{-1} mole^{-1}
Gas constant	$R = 0.08206$ l atm K^{-1} mole^{-1}
Standard molar gas volume	273.15° $R = 22.415$ l
Celsius temperature	t° C = T°K − 273.15°
Atmospheric pressure	1 atm = 101.325 kN m^{-2}
Electron volt	1 eV = 96.4905 kJ mole^{-1}

RELATIONS AMONG ENERGY QUANTITIES

1 eV = 0.160206×10^{-18} J
1 eV = 96.4905 kJ mole^{-1}
1 eV = 23.0618 kcal mole^{-1}
1 erg = 1×10^{-7} J
1 liter atm = 9.869×10^{-3} J
1 cal = 4.184 J

The energy of a photon with wavelength 100 pm is 12398 eV = 1.1963×10^{8} J mole^{-1}.

The energy of a photon with wave-number 1 cm^{-1} (reciprocal of wavelength in cm) is 1.2398×10^{-4} eV = 11.963 J mole^{-1}.

A photon with wavelength 123.98 pm, wave-number (λ^{-1}) 8066 cm^{-1}, frequency 2.418×10^{18} Hz has energy 1 eV.

APPENDIX III

The Kinetic Theory of Gases

During the nineteenth century the concepts that atoms and molecules are in continual motion and that the temperature of a body is a measure of the intensity of this motion were developed. The idea that the behavior of gases could be accounted for by considering the motion of the gas molecules had occurred to several people (Daniel Bernoulli in 1738, J. P. Joule in 1851, A. Kronig in 1856), and in the years following 1858 this idea was developed into a detailed kinetic theory of gases by Clausius, Maxwell, Boltzmann, and many later investigators. The subject is discussed in courses in physics and physical chemistry, and it forms an important part of the branch of theoretical science called statistical mechanics.

In a gas at temperature T the molecules are moving about, different molecules having at a given time different speeds v and different kinetic energies of translational motion $\frac{1}{2}mv^2$ (m being the mass of a molecule). It has been found that the average kinetic energy per molecule, $\frac{1}{2}m[v^2]_{\text{average}}$, is the same for all gases at the same temperature, and that its value increases with the temperature, being directly proportional to T.

The average (root-mean-square*) velocity of hydrogen molecules at 0°C is 1.84×10^3 m s^{-1}. At higher temperatures the average velocity is greater; it reaches twice as great a value, 3.68×10^3 m s^{-1}, for hydrogen molecules at 820°C, corresponding to an increase by 4 in the absolute temperature.

Since the average kinetic energy is equal for different molecules, the average value of the square of the velocity is seen to be inversely proportional to the mass of the molecule, and hence the average velocity (root-mean-square average) is inversely proportional to the square root of the molecular mass. The molecular mass of oxygen is just 16 times that of hydrogen; accordingly, molecules of oxygen move with a speed

*The root-mean-square average of a quantity x is the square root of the average value of the square of the quantity, $\left\{\left(\sum_{i=1}^{n} x_i^2\right)\Big/ n\right\}^{1/2}$.

just one quarter as great as molecules of hydrogen at the same temperature. The average speed of oxygen molecules at 0°C is 0.46×10^3 m s^{-1}.

The explanation of Boyle's law given by the kinetic theory is simple. A molecule on striking the wall of the container of the gas rebounds, and contributes momentum to the wall; in this way the collisions of the molecules of the gas with the wall produce the gas pressure, which balances the external pressure applied to the gas. If the volume is decreased by 50%, each of the molecules strikes the wall twice as often, and hence the pressure is doubled. The explanation of the law of Charles and Gay-Lussac is equally simple. If the absolute temperature is doubled, the speed of the molecules is increased by the factor $\sqrt{2}$. This causes the molecules to make $\sqrt{2}$ times as many collisions as before, and each collision is increased in force by $\sqrt{2}$, so that the pressure itself is doubled ($\sqrt{2} \times \sqrt{2} = 2$) by doubling the absolute temperature. Avogadro's law is also explained by the fact that the average kinetic energy is the same at a given temperature for all gases.

The Effusion and Diffusion of Gases; the Mean Free Paths of Molecules

There is an interesting dependence of the rate of effusion of a gas through a small hole on the molecular mass of the gas. The speeds of motion of different molecules are inversely proportional to the square roots of their molecular masses. If a small hole is made in the wall of a gas container, the gas molecules will pass through the hole into an evacuated region outside at a rate determined by the speed at which they are moving (these speeds determine the probability that a molecule will strike the hole). Accordingly, the kinetic theory requires that the rate of *effusion* of a gas through a small hole be inversely proportional to the square root of its molecular mass. This law was discovered experimentally before the development of the kinetic theory—it was observed that hydrogen effuses through a porous plate four times as rapidly as oxygen.

In the foregoing discussions we have ignored the appreciable sizes of gas molecules, which cause the molecules to collide often with one another. In an ordinary gas, such as air at standard conditions, a molecule moves only about 50 nm, on average, between collisions; that is, its *mean free path* under these conditions is only about two hundred times its own diameter.

The value of the mean free path is significant for phenomena that depend on molecular collisions, such as the viscosity and the thermal conductivity of gases. Another such phenomenon is the *diffusion* of one gas through another or through itself (such as of radioactive molecules of a gas through the nonradioactive gas). In the early days of kinetic theory it was pointed out by skeptics that it takes minutes or hours for a gas to diffuse from one side of a quiet room to the other, even though the molecules are attributed velocities of about a kilometer per second. The explanation of the slow diffusion rate is that a molecule diffusing through a gas is not able to move directly from one point to another a long distance away, but instead is forced by collisions with other molecules to follow a tortuous path, making only slow progress in its resultant motion. Only when diffusing into a high vacuum can the gas diffuse with the speed of molecular motion.

The Distribution Law for Molecular Velocities

In 1860 the English physicist James Clerk Maxwell (1831–1879) derived an equation that correctly gives the fraction of gas molecules with velocities in the range v to $v + dv$. This equation is called the *Maxwell distribution law* (or *Maxwell-Boltzmann distribution law*) for molecular velocities. In a perfect gas at temperature T, containing N molecules, each with mass m, we ask how many molecules dN have velocities lying between v and $v + dv$. The velocity v may be described as a vector with components v_x, v_y, and v_z in velocity space. The volume of the spherical shell bounded by the surfaces v and $v + dv$ is $4\pi v^2 dv$. It was found by Maxwell through analysis of the transfer of momentum from one molecule to another during a molecular collision that this volume element must be multiplied by the exponential factor $\exp(-\frac{1}{2}mv^2/kT)$. The normalizing factor $(m/2\pi kT)^{3/2}$ is also needed in order that the integral of dN over all velocities ($v = 0$ to $v = \infty$) should be equal to N. The distribution law for molecular velocities is

$$dN = 4\pi N \left(\frac{m}{2\pi kT}\right)^{3/2} \exp\left(-\frac{\frac{1}{2}mv^2}{kT}\right) v^2\, dv \qquad \text{(III-1)}$$

The distribution function calculated for helium atoms at 100 K and also for helium atoms at 400 K is shown as Figure III-1. We see that in Equation III-1 the mass and the absolute temperature occur only in the ratio m/T. Accordingly, the two curves that are shown apply also to methane, CH_4, with molecular weight four times that of helium, at temperatures four times as great, 400 K and 1600 K, respectively.

The maximum for the distribution function occurs at the value of v called the most probable velocity, v_{mp}; it is equal to $(2kT/m)^{1/2}$, which has the value $128.95(T/M)^{1/2}$ m s^{-1} (here M is the molecular mass). This value is represented by a vertical line on the curve for helium at 400 K.

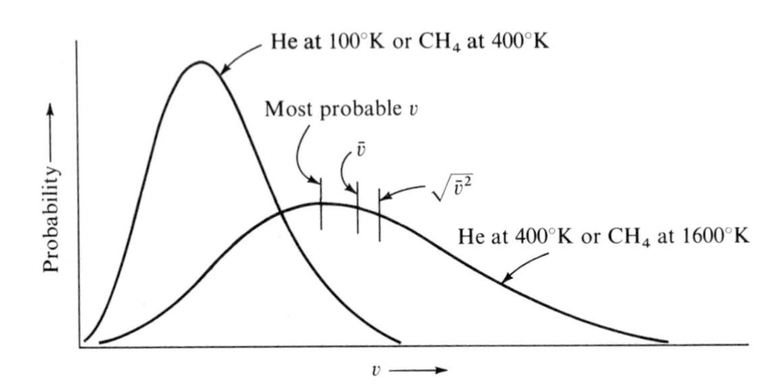

FIGURE III-1
The velocity distribution function for helium atoms at 100 K and for helium atoms at 400 K. These two curves also apply to methane at temperatures 400 K and 1600 K, respectively.

The average value of the velocity is $(8kT/\pi m)^{1/2}$, which is equal to $145.51(T/M)^{1/2}$ m s^{-1}. The root-mean-square value of the velocity, which is the square root of the average value of v^2, is equal to $(3kT/m)^{1/2}$, with value $157.94(T/M)^{1/2}$ m s^{-1}.

We see that *the average kinetic energy per molecule,* $\frac{1}{2}m(v^2)_{\text{average}}$, *is equal to* $\frac{3}{2}kT$. It accordingly has the same value for all gases at the same temperature, as stated in the second paragraph of this section.

This result, called the *equipartition of energy,* is one of the most important consequences of the kinetic theory. Some of the ways in which it can be used in the discussion of the properties of gases have been mentioned above.

APPENDIX **IV**

X-Rays and Crystal Structure

One of the most important ways of studying the structures of substances was developed in 1912 and 1913 by W. Lawrence Bragg (1890–1971) and his father, William H. Bragg (1862–1942), in England. W. L. Bragg, who was a student in Cambridge University, developed the theory of x-ray diffraction (the Bragg equation, described below) and used it in November 1912 to determine the structure of sphalerite, the cubic form of zinc sulfide, by analyzing the x-ray diffraction photographs of sphalerite that had been published by the German physicist Max von Laue (1879–1960). His father then devised the x-ray spectrometer (Figure IV-1), and within a year W. L. and W. H. Bragg had determined the exact atomic arrangements for many crystals and also the wavelengths of the characteristic x-ray lines emitted by several elements serving as targets in the x-ray tubes. In the Bragg technique a beam impinges on the face of a crystal, such as the cleavage face of a salt crystal. An instrument for detecting x-rays (in their original experiments the Braggs used an ionization chamber, but in modern work a Geiger counter or scintillation counter may be used) was then placed as shown in the figure.

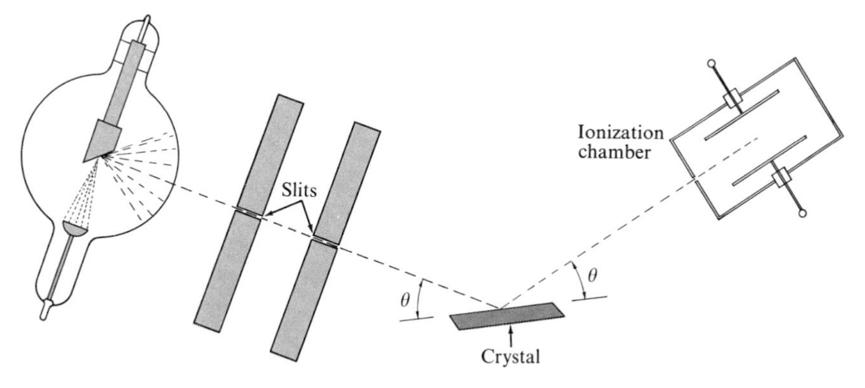

FIGURE IV-1
The Bragg ionization-chamber technique of investigating the diffraction of x-rays by crystals.

The simple theory of diffraction of x-rays by crystals developed by Lawrence Bragg is illustrated in Figures IV-2 and IV-3. He pointed out that if the beam of rays incident on a plane of atoms and the scattered beam are in the same vertical plane and at the same angle with the plane, as shown in Figure IV-2, the conditions for reinforcement are satisfied. This sort of scattering is called specular reflection—it is similar to reflection from a mirror. He then formulated the conditions for reinforcement of the beams specularly reflected from one plane of atoms and the beams specularly reflected from another plane of atoms separated from

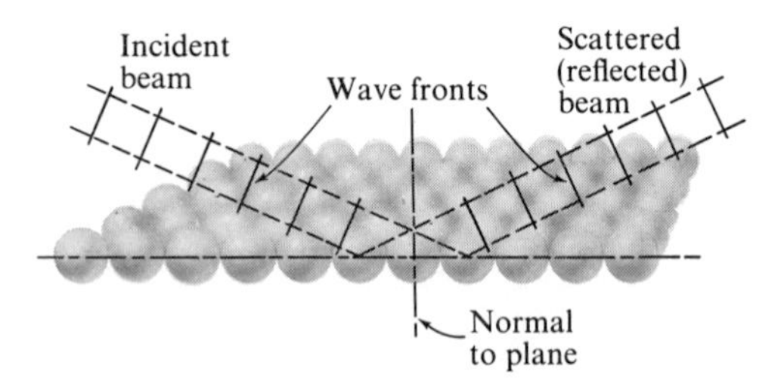

FIGURE IV-2
Diagram showing the equality of path lengths when the conditions for specular reflection from a layer are satisfied.

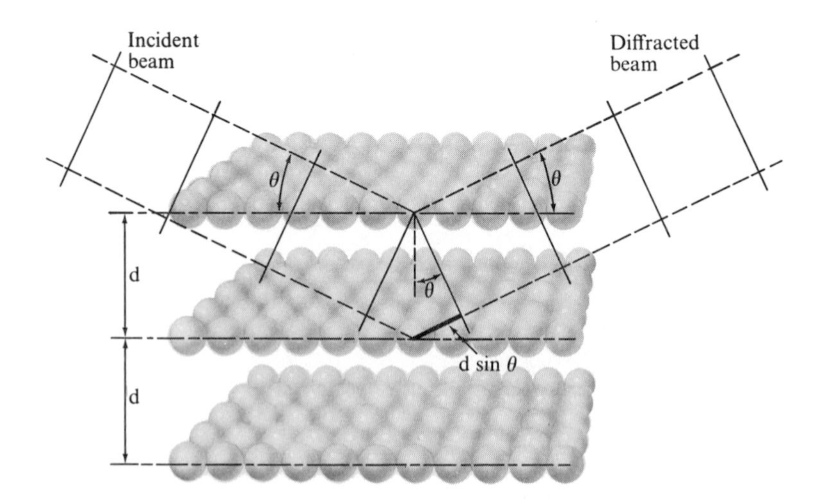

FIGURE IV-3
Diagram illustrating the derivation of the Bragg equation for the diffraction of x-rays by crystals.

it by the interplanar distance d. This situation is illustrated in Figure IV-3. We see that the difference in path is equal to $2d \sin \theta$, in which θ is the Bragg angle (the angle between the incident beam and the plane of atoms). In order to have reinforcement, this difference in path $2d \sin \theta$ must be equal to the wavelength λ or an integral multiple of this wavelength—that is, to $n\lambda$, in which n is an integer. We thus obtain the *Bragg equation* for the diffraction of x-rays:

$$n\lambda = 2d \sin \theta \qquad \text{(IV-1)}$$

The way in which the structure of crystals was then determined is illustrated in Figure IV-4. Here we show a simple cubic arrangement of atoms, as seen along one of the cube faces. It is evident that there are layers of atoms, shown by their traces in the plane of the paper, with spacings d_1, d_2, d_3, . . . , which are in the ratios $1:2^{-\frac{1}{2}}:5^{-\frac{1}{2}}$, ... Since the relative values of the spacings d_1, d_2, d_3, . . . could be determined without knowledge of the wavelengths of x-rays, but simply as inversely proportional to their values of $\sin \theta$, the nature of the atomic arrangement could be discovered by the Bragg experiments.

A reproduction of some of the first experimental measurements made by the Braggs is shown as Figure IV-5. It is seen that there occurs a pattern of reflections that is repeated at values of $\sin \theta$ representing the values 1, 2, and 3 for the integer n, which is called the order of the reflection. The pattern shows that there were present in the beam of x-rays produced by the x-ray tube a shorter wavelength and a somewhat longer wavelength, with greater intensity of the x-rays of longer wavelength than of the shorter wavelength.

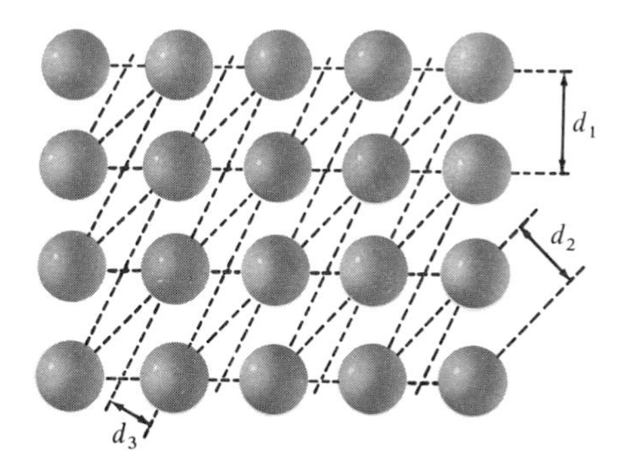

FIGURE IV-4
Spacings between different rows of atoms in a two-dimensional crystal.

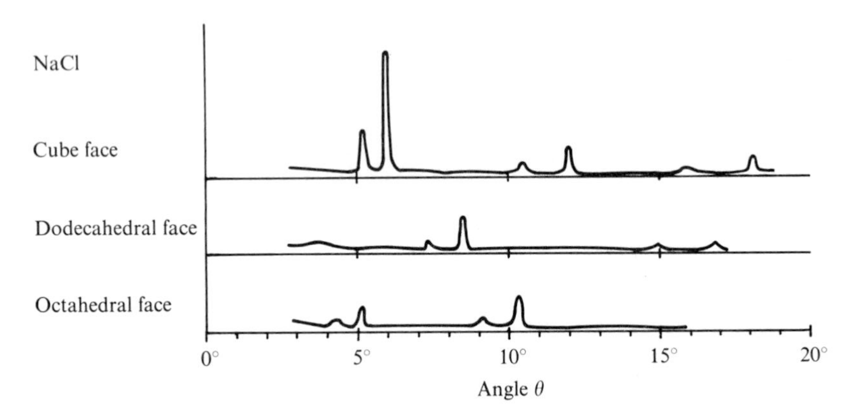

FIGURE IV-5
Experimental data obtained by the Braggs for the diffraction of x-rays by the sodium chloride crystal.

The x-ray diffraction method has provided knowledge of the crystal structures of many crystals, including molecular crystals. The values of the interatomic distances provide information about the nature of the bonds between the adjacent atoms. The technique has become a very powerful one, and it is now sometimes used for determining the complete molecular structures of substances, in place of the older chemical methods of decomposing substances into simpler substances.

APPENDIX V

Bond Energy and Bond-dissociation Energy

In Section 6-12 it was mentioned that a set of bond-energy values can be found such that their sum over all the bonds of a molecule that can be satisfactorily represented by a single valence-bond structure is equal to the heat of formation of the gaseous substance from atoms of the elements. Such a set of bond-energy values is given in Table V-1. Most of the values are reliable to about 3 kJ $mole^{-1}$, and calculations made with their use can be trusted to within about 3 kJ $mole^{-1}$ per bond (see the example at the end of this Appendix).

Values for single bonds not given in the table can be estimated roughly from the relation to the electronegativity difference:

$$E(\mathrm{A{-}B}) = \tfrac{1}{2}\{E(\mathrm{A{-}A}) + E(\mathrm{B{-}B}) \\ + 100(x_\mathrm{A} - x_\mathrm{B})^2 - 6.5(x_\mathrm{A} - x_\mathrm{B})^4 \text{ kJ mole}^{-1} \qquad \text{(V-1)}$$

For example, as yet no determination has been made of the heat of formation of Br_2O, which is an unstable substance for which the structure

```
 ..
:Br:
  |   ..
:O—Br:
 ..   ..
```

may be written with confidence. From Equation V-1 and the electronegativity values (Table 6-3) we obtain the value 215 kJ $mole^{-1}$ for the Br—O bond energy:

$$E(\mathrm{Br{-}O}) = \tfrac{1}{2}(193 + 143) + 100(3.5 - 2.8)^2 - 6.5(3.5 - 2.8)^4 \\ = 168 + 49 - 2 = 215 \text{ kJ mole}^{-1}$$

Hence we obtain 430 kJ $mole^{-1}$ for the heat of formation of $Br_2O(g)$ from atoms. By use of the values of the enthalpy of the atoms relative

TABLE V-1
Bond-energy Values for Single Bonds

H—H	436 kJ mole^{-1}	C—H	415 kJ mole^{-1}	Si—Cl	396 kJ mole^{-1}
Li—Li	111	Si—H	295	Si—Br	289
Na—Na	75	N—H	391	Si—I	213
K—K	55	P—H	322	Ge—Cl	408
Rb—Rb	52	As—H	245	N—O	175
Cs—Cs	45	O—H	463	N—F	270
B—B	225	S—H	368	N—Cl	200
C—C	344	Se—H	277	P—O	360
Si—Si	187	Te—H	241	P—F	486
Ge—Ge	157	H—F	563	P—Cl	317
Sn—Sn	143	H—Cl	432	P—Br	266
N—N	159	H—Br	366	P—I	218
P—P	217	H—I	299	As—O	311
As—As	134	B—C	312	As—F	466
Sb—Sb	126	B—O	460	As—Cl	288
Bi—Bi	105	B—S	276	As—Br	236
O—O	143	B—F	582	As—I	174
S—S	266	B—Cl	388	O—F	212
Se—Se	184	B—Br	310	O—Cl	210
Te—Te	168	C—Si	290	O—Br	217
F—F	158	C—N	292	O—I	241
Cl—Cl	243	C—O	350	S—Cl	277
Br—Br	193	C—S	259	S—Br	239
I—I	151	C—F	443	Se—Cl	243
Li—H	245	C—Cl	328	Cl—F	251
Na—H	202	C—Br	276	Br—F	249
K—H	182	C—I	240	Br—Cl	218
Rb—H	167	Si—O	432	I—F	281
Cs—H	175	Si—S	227	I—Cl	210
B—H	331	Si—F	590	I—Br	178

to the elements in their standard states (Table V-3) we can then obtain a predicted value of the standard heat of formation:

$$2Br(g) + O(g) \rightarrow Br_2O(g) + 430 \text{ kJ mole}^{-1}$$

$$Br_2(l) \rightarrow 2Br(g) - 224 \text{ kJ mole}^{-1}$$

$$\tfrac{1}{2}O_2(g) \rightarrow O(g) - 249 \text{ kJ mole}^{-1}$$

$$Br_2(l) + \tfrac{1}{2}O_2(g) \rightarrow Br_2O(g) - 43 \text{ kJ mole}^{-1}$$

The heat of formation is thus predicted to have a small negative value. The negative value may explain the ease with which the substance decomposes.

Bond energy values for multiple bonds are given in Table V-2.

The values in Table V-3, which are the values of ΔH for the reaction X(standard state) $\rightarrow$ X(g), are useful in converting values of the heat of formation from atoms.

TABLE V-2
Bond-energy Values for Multiple Bonds

C=C	615 kJ mole^{-1}	(−73)*	C≡C	812 kJ mole^{-1}	(−220)
N=N	418	(+100)	N≡N	946	(+469)
O=O†	402	(+116)	C≡N	890	(+14)
C=N	615	(+31)	P≡P	490	(−161)
C=O	725	(+25)			
C=S	477	(−41)			

*Values in parentheses are A=B − 2(A—B) and (A≡B) − 3(A—B). A positive value shows that a structure with double or triple bonds is more stable than one with single bonds.

†This value is for the first excited state of O_2, a singlet state that may be considered to involve a double bond.

TABLE V-3
Enthalpy (in kJ mole^{-1}) of Monatomic Gases of Elements Relative to the Elements in Their Standard States

H	218.0								
Li	160.7	C	715.0	N	472.7	O	249.2	F	78.9
Na	107.8	Si	443.5	P	333.9	S	279	Cl	121.0
K	89.2	Ge	328	As	254	Se	202	Br	111.9
Rb	86	Sn	301	Sb	254	Te	199	I	106.9
Cs	79	Pb	194	Bi	208				

Bond-dissociation Energy

The bond-dissociation energy of a bond in a molecule is the energy required to break that bond alone—that is, to split the molecule into the two parts that were previously connected by that bond. For example, the bond-dissociation energy of the C—C bond in ethane, $H_3C—CH_3$, is the enthalpy of dissociation of ethane into two methyl radicals, CH_3.

For diatomic molecules the bond energy and the bond-dissociation energy are the same. For polyatomic molecules they are in general different. The sum of successive bond-dissociation energies is, of course, equal to the sum of the bond energies.

APPENDIX **VI**

Values of the Standard Enthalpy of Formation of Some Substances

Values of the standard enthalpy of formation at 25°C of some substances from the elements in their standard states are given in Tables VI-1 to VI-10. These values for the most part have been taken from the reference book *Selected Values of Chemical Thermodynamic Properties,* circular of the U.S. Bureau of Standards No. 500 (1952). Values are also given in *Handbook of Chemistry and Physics* and other reference books. A selection of values for compounds of the transition metals, which are not included in the following tables, can be found in *General Chemistry,* L. Pauling, W. H. Freeman and Company, San Francisco, 3rd edition, 1970.

Enthalpies of reaction can be determined by use of instruments such as the bomb calorimeter shown in Figure VI-1. A weighed sample of combustible material is placed in the bomb, and oxygen is introduced under pressure. The temperature of the surrounding water is recorded, and the sample is ignited by passing an electric current through a wire embedded in it. The heat liberated by the reaction causes the entire system inside the insulating material to increase in temperature. After enough time has elapsed to permit the temperature of this material to become uniform, the temperature is again recorded. From the rise in temperature and the total water equivalent of the calorimeter (that is, the weight of water that would require the same amount of heat to cause the temperature to rise one degree as is required to cause a rise in temperature of one degree of the total material of the calorimeter inside the insulation), the amount of heat liberated in the reaction can be calculated. Correction must, of course, be made for the amount of heat introduced by the electric current that produced the ignition and for the fact that the reaction has not taken place at constant pressure.

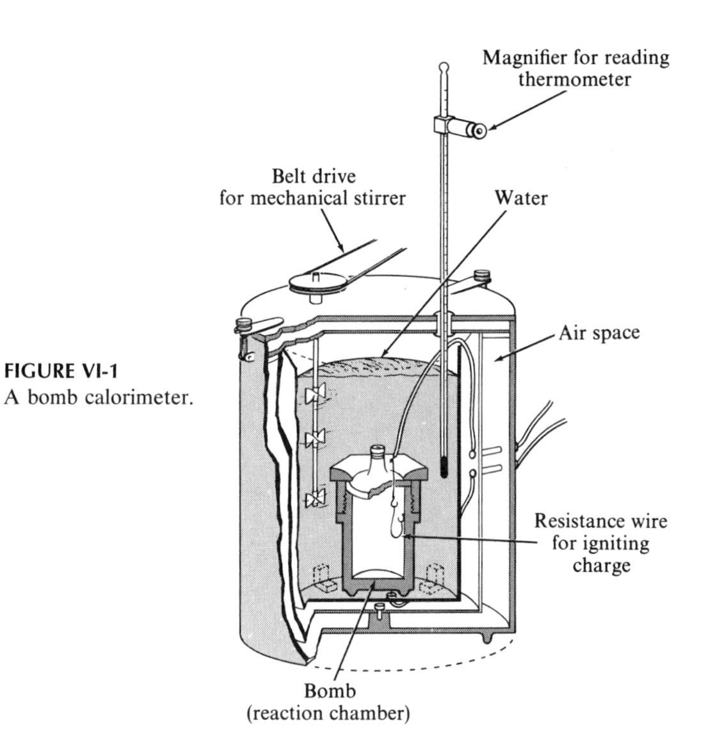

FIGURE VI-1
A bomb calorimeter.

TABLE VI-1
Standard Enthalpy of Compounds of Hydrogen and Oxygen at 25°C (kJ $mole^{-1}$)

$e^-(g)$	0*	$O^+(g)$	1567	$OH^-(g)$	−133
$H_2(g)$	0	$O^-(g)$	110	$OH^-(aq)$	−230
$O_2(g)$	0	$O_2^+(g)$	1184	$H_2O(g)$	−242
$H^+(aq)$	0	$O_2(aq)$	−16	$H_2O(l)$	−286
$H(g)$	218	$O_3(g)$	142	$H_2O_2(g)$	−133
$O(g)$	249	$O_4(g)$	−0.7	$H_2O_2(l)$	−188
$H^+(g)$	1536	$OH(g)$	42	$H_2O_2(aq)$	−191
$H^-(g)$	149				

*Values of the enthalpy of formation of elements in a standard state, of electron gas, and of hydrogen ion in aqueous solution are arbitrarily taken to be zero.

TABLE VI-2
Standard Enthalpy of Formation of Compounds of Alkali Metals at 25°C (kJ $mole^{-1}$)

	M = Li	Na	K	Rb	Cs
M(g)	155	109	90	86	79
M^+(g)	681	611	515	495	461
M^+(aq)*	−278	−240	−251	−246	−248
M_2(g)	199	142	129	124	113
M_2O(c)	−596	−416	−361	−330	−318
MH(g)	128	125	126	138	121
MH(c)	−90	−57	−57	−59	−84
MF(c)	−612	−569	−563	−549	−531
MCl(c)	−409	−411	−436	−431	−433
MBr(c)	−350	−360	−392	−389	−395
MI(c)	−271	−288	−328	−328	−337
M_2S(c)		−373	−418	−348	−339
M_2Se(c)	−381	−264	−332		

*Relative to assumed value 0 for H^+(aq).

TABLE VI-3
Standard Enthalpy of Formation of Compounds of Alkaline-earth Metals at 25°C (kJ $mole^{-1}$)

	M = Be	Mg	Ca	Sr	Ba
M(g)	321	150	193	164	176
M^+(g)	1226	894	789	719	684
M^{++}(g)	2989	2351	1940	1790	1656
M^{++}(aq)*	−389	−462	−543	−546	−538
MO(c)	−611	−602	−636	−590	−558
MF_2(c)		−1102	−1215	−1215	−1200
MCl_2(c)	−512	−642	−795	−828	−860
MBr_2(c)	−370	−518	−675	−716	−755
MI_2(c)	−212	−360	−535	−567	−602
MS(c)	−234	−347	−482	−473	−485
MSe(c)			−313	−329	−310
MTe(c)		−209			
MH_2(c)			−195	−177	−172

*Relative to assumed value 0 for H^+(aq).

TABLE VI-4
Standard Enthalpy of Formation of Compounds of Boron, Aluminum, and Their Congeners at 25°C (kJ $mole^{-1}$)

	M = B	Al	Sc	Y	La
M(g)	407	314	389	431	368
M^{+}(g)	1213	897	1028	1067	916
M^{++}(g)	3646	2720	2278	2269	2025
M^{+++}(g)	7312	5468	4674	4252	3881
M^{+++}(aq)*		−525	−623	−703	−737
M_2O_3(c)	−1264	−1670			−1916
MF_3	−1110(g)	−1301(c)			
MCl_3	−418(l)	−695(c)	−924(c)	−982(c)	−1103(c)
MBr_3	−221(l)	−526(c)	−751(c)		
MI_3(c)		−315		−599	−700
M_2S_3(c)	−238	−509			−1284

*Relative to assumed value 0 for H^{+}(aq).

TABLE VI-5
Standard Enthalpy of Formation of Compounds of Silicon, Germanium, Tin, and Lead at 25°C (kJ $mole^{-1}$)

	M = Si	Ge	Sn*	Pb
M(g)	368	328	301	194
MO	−113(g)	−95(g)	−286(c)	−219(c)
MO_2(c)	−859	−537	−581	−277
MH_4(g)	−62			
MF_2(c)				−663
MF_4	−1548(g)			−930(c)
MCl_2(c)			−350	−359
MCl_4(l)	−640	−544	−545	
MBr_2(c)			−266	−277
MBr_4	−398(l)		−406(c)	
MI_2(c)			−144	−175
MI_4(c)	−132			
MS(c)			−78	−94

*White tin is the standard state; the value for gray tin is 2.5 kJ $mole^{-1}$.

TABLE VI-6
Standard Enthalpy of Carbon Compounds at 25°C (kJ mole^{-1})

C(graphite)	0	C_2H_2(g)	227	CH_2O(g)	formaldehyde	−116
C(diamond)	1.9	C_2H_4(g)	52	CH_3CHO(g)	acetaldehyde	−166
C(g)	718	C_2H_6(g)	−85	$(CH_3)_2CO$(g)	acetone	−216
C^+(g)	1806	CF_4(g)	−912	HCOOH(g)	formic acid	−363
C_2(g)	982	CCl_4(g)	−107	CH_3COOH(g)	acetic acid	−435
CO(g)	−111	$CHCl_3$(g)	−100	CH_3OH(g)	methanol	−201
CO^+(g)	1224	CH_2Cl_2(g)	−88	C_2H_5OH(g)	ethanol	−237
CO_2(g)	−394	CH_3Cl(g)	−82	$(CH_3)_2O$(g)	dimethyl ether	−185
CH(g)	595	CBr_4(g)	50	C_3H_6(g)	cyclopropane	38
CH_2(g)	397	CS_2(g)	115	C_6H_{12}(g)	cyclohexane	−126
CH_3(g)	134	COS(g)	−137	C_6H_{10}(g)	cyclohexene	−6
CH_4(g)	−75	$(CH_3)_2S$(g)	−38	C_6H_6(g)	benzene	83

TABLE VI-7
Standard Enthalpy of Some Nitrogen Compounds at 25°C (kJ mole^{-1})

N_2(g)	0	NH(g)	331	NO_2^-(aq)	−106
N(g)	473	NH_3(g)	−46	NO_3^-(aq)	−207
N^+(g)	1883	NH_3(aq)	−81	NH_2OH(c)	−107
NO(g)	90	NH_4^+(g)	628	NH_4OH(aq)	−367
NO_2(g)	34	NH_4^+(aq)	−133	$H_2N_2O_2$(aq)	−57
NO_3(g)	54	N_2H_4(l)	50	NH_4NO_3(c)	−365
N_2O(g)	82	HN_3(g)	294	NF_3(g)	−114
N_2O_3(g)	84	N_3^-(aq)	245	NCl_3(in CCl_4)	229
N_2O_4(g)	44	HNO_2(aq)	−119	NH_4F(c)	−467
N_2O_5(g)	13	HNO_3(l)	−173	NH_4Cl(c)	−315

TABLE VI-8
Standard Enthalpy of Some Compounds of Phosphorus, Arsenic, Antimony, and Bismuth at 25°C (kJ $mole^{-1}$)

	X = P	As	Sb	Bi
X(c)*	0	0	0	0
X(g)	315	254	254	208
X^+(g)	1380	1273	1094	917
X_2(g)	142	124	218	249
X_4(g)	55	149	204	
XO(g)	−41	20	188	67
X_4O_6(c)	−1640	−1314	−1409	−1154
X_4O_{10}(c)	−2984	−1829	−1961	
XH_3(g)	9	171		
HXO_3(c)	−955			
H_3XO_2(aq)	−609			
H_3XO_3(aq)	−972	−742		
H_3XO_4(aq)	−1289	−899	−902	
XO_4^{---}(aq)	−1279	−870		
$X_2O_7^{----}$(aq)	−2276			
XCl_3(g)	−255	−299	−315	−271
XCl_5(g)	−343		−393	
XCl_3O(g)	−592			
XBr_3	−150(g)	−195(c)	−260(c)	
XBr_5(c)	−276			
XBr_3O(c)	−479			
XI_3(c)	−46	−57	−96	
XN(g)	−85	29	311	

*The standard states are white phosphorous (cubic, P_4) and hexagonal arsenic, antimony, and bismuth. The enthalpy of black phosphorus is −43 kJ $mole^{-1}$.

TABLE VI-9
Standard Enthalpy of Compounds of Sulfur, Selenium, and Tellurium at 25°C (kJ mole^{-1})

	X = S	Se	Te
X(c)*	0	0	0
X(g)	279	202	199
X^+(g)	1284	1149	1074
X^{--}(g)	524		
X^{--}(aq)	42	132	
X_2(g)	125	139	172
X_6(g)	106		
X_8(g)	101		
XO(g)	6	40	180
XO_2	−297(g)	−230(c)	−325(c)
XO_3(g)	−395		
H_2X(g)	−20	86	154
H_2XO_3(aq)	−633	−512	−605
H_2XO_4	−811(l)	−538(c)	
H_2XO_4(aq)	−908	−608	−697
XCl_2(g)		−41	
XCl_2(l)	−60		−84
XF_6(g)	−1209	−1029	−1318

*The standard states are orthorhombic sulfur (S_8 molecules) and hexagonal selenium and tellurium (long chains of atoms).

TABLE VI-10
Standard Enthalpy of Halogen Compounds at 25°C (kJ mole^{-1})

	X = F	Cl	Br	I
X_2(g)	0	0	31	62
X_2			0(l)	0(c)
X_2(aq)		−25	−5	21
X(g)	77	121	112	107
X^+(g)	1764	1378	1261	1120
X^-(g)	−256	−229	−218	−193
X^-(aq)	−329	−167	−121	−56
HX(g)	−269	−92	−36	26
KX(c)	−563	−436	−392	−328
X_2O(g)	23	76		
HXO(aq)		−118		−159
HXO_2(aq)		−52		
HXO_3(aq)		−98	−40	−230
HXO_4(aq)		−131		
H_5XO_6(aq)				−766

APPENDIX **VII**

Selected Readings

FROM *Scientific American*

The *Scientific American* Offprints listed below by number are available from bookstores or from W. H. Freeman and Company, 660 Market Street, San Francisco, California 94104, and 58 Kings Road, Reading, England RG1 3AA.

19 *The Contraction of Muscle,* H. E. Huxley, November 1958.
31 *The Structure of Protein Molecules,* L. Pauling, R. B. Corey, and R. Hayward, July 1954.
54 *Nucleic Acids,* F. H. C. Crick, September 1957.
80 *The Chemical Structure of Proteins,* W. H. Stein and S. Moore, February 1961.
81 *Chromatography,* W. H. Stein and S. Moore, March 1951.
100 *Tracers,* M. D. Kamen, February 1949.
123 *The Genetic Code,* F. H. C. Crick, October 1962.
153 *The Genetic Code: II,* M. W. Nirenberg, March 1963.
171 *Polyribosomes,* A. Rich, December 1963.
201 *What Holds the Nucleus Together?,* H. A. Bethe, September 1953.
204 *Dislocations in Metals,* F. B. Cuff, Jr. and L. McD. Schetky, July 1955.
210 *The Origin of the Elements,* W. A. Fowler, September 1956.
213 *Elementary Particles,* M. Gell-Mann and E. P. Rosenbaum, July 1957.
217 *The Atomic Nucleus,* R. Hofstadter, July 1956.
228 *The Structure of the Nucleus,* M. G. Mayer, March 1951.
242 *The Synthetic Elements I,* G. T. Seaborg and I. Perlman, April 1950.
243 *The Synthetic Elements II,* G. T. Seaborg and A. Ghiorso, December 1956.
244 *The Antiproton,* E. Segrè and C. G. Wiegand, June 1956.
253 *The Age of the Elements in the Solar System,* J. H. Reynolds, November 1960.
263 *Frozen Free Radicals,* C. M. Herzfeld and A. M. Bass, March 1957.
293 *The Synthetic Elements: III,* G. T. Seaborg and A. R. Fritsch, April 1963.
325 *X-Ray Crystallography,* Sir Lawrence Bragg, July 1968.
1012 *The Evolution of Hemoglobin,* E. Zuckerkandl, May 1965.
1026 *The Mechanism of Muscular Contraction,* H. E. Huxley, December 1965.
1052 *The Genetic Code: III,* F. H. C. Crick, October 1966.
1055 *The Three-dimensional Structure of an Enzyme Molecule,* D. C. Phillips, November 1966.
1061 *The Repair of DNA,* P. C. Hanawalt and R. H. Haynes, February 1967.
1074 *Gene Structure and Protein Structure,* C. Yanofsky, May 1967.
1075 *Molecular Isomers in Vision,* R. Hubbard and A. Kropf, June 1967.
1124 *The Synthesis of DNA,* A. Kornberg, October 1968.
1185 *The Structure and Function of Antibodies,* G. M. Edelman, August 1970.
1193 *The Carbon Cycle,* B. Bolin, September 1970.
1225 *Elastic Fibers in the Body,* R. Ross and P. Bornstein, June 1971.
1235 *Prostaglandins,* J. E. Pike, November 1971.
1256 *Cyclic AMP,* I. Pastan, August 1972.
1259 *Lactose and Lactase,* N. Kretchmer, October 1972.

Index